STATISTICS

MAKING SENSE OF DATA

Chester L. Olson
CAMROSE LUTHERAN COLLEGE

wcb
Wm. C. Brown Publishers
Dubuque, Iowa

Library of Congress Catalog Card Number: 86–26628

ISBN 0–697–06895–1

Printed in the United States of America by Wm. C. Brown Publishers
2460 Kerper Boulevard, Dubuque, IA 52001

10 9 8 7 6 5 4 3 2

CONTENTS

* This symbol identifies material that is not prerequisite for later topics in this book.

GLOSSARY OF SYMBOLS

CHAPTER 1

x = a variable
y = a variable
x_i = ith value of x
π = 3.14159...
e = 2.71828...
N = number of observations in population
n = number of observations in sample
Σ = "the sum of"
k = a constant

CHAPTER 2

f = frequency
k = number of categories

CHAPTER 3

$x_{\max}$ = maximum value of x
$x_{\min}$ = minimum value of x
$\bar{x}$ = sample mean of x
n = number of observations in sample
$\bar{y}$ = sample mean of y
μ = population mean
N = number of observations in population
k = number of categories or groups
w = weight
$\tilde{x}$ = sample median
$\tilde{\mu}$ = population median
$\dot{x}$ = sample mode
$\dot{\mu}$ = population mode
$Q_{j/m}$ = quantile
Q_j = jth quartile
D_j = jth decile
C_j = jth centile

CHAPTER 4

σ^2 = population variance
s^2 = sample variance
σ = population standard deviation
s = sample standard deviation
z = standard score

CHAPTER 5

$f(x)$ = function of x
y = dependent variable
x = independent variable
b = slope, sample regression coefficient
a = y-intercept, sample regression constant
$\hat{y}$ = fitted value of y
SS = sum of squares
r^2 = coefficient of determination
r = sample correlation coefficient

CHAPTER 6

$A, B, \ldots$ = events
$P(A)$ = probability of A
m = number of outcomes
n = number of trials or observations
p = probability
q = $1 - p$
$P(B|A)$ = probability of B given A
S = sample space
A' = complement of event A

CHAPTER 7

$P(x)$ = probability function
$f(x)$ = probability density function
$F(x)$ = distribution function
$E(x)$ = expected value of x
λ = mean of Poisson variable
e = 2.71828...
z = standard normal deviate
A = upper-tail area of a distribution

CHAPTER 8

$\hat{\mu}$ = mean of empirical sampling distribution
$\hat{\sigma}$ = standard deviation of empirical sampling distribution

CHAPTER 10

α = significance level
$e_{\alpha/2}$ = margin of error at $1 - \alpha$ confidence level
t = the variable in the t distribution
df = number of degrees of freedom

CHAPTER 11

H_0 = null hypothesis
H_A = alternative hypothesis
P = significance test P-value
d = difference in paired samples
β = Type II error probability

CHAPTER 12

χ^2 = the variable in the chi-square distribution
F = the variable in the F distribution

CHAPTER 13

ρ = population correlation coefficient
$z(r)$ = Fisher's transformation of r
ln = natural logarithm
SP = sum of products
$\mathrm{cov}(x,y)$ = sample covariance of x and y
$\mathrm{var}(x)$ = sample variance of x
$\mathrm{Var}(x)$ = population variance of x
β = population regression coefficient
MS = mean square
s_e = standard error of estimation
A = population regression constant
s_f = standard error of a forecast mean
m = number of observations

CHAPTER 14

p = population proportion
$\hat{p}$ = sample proportion
$q = 1 - p$
$\hat{q} = 1 - \hat{p}$
o = observed frequency
e = expected frequency
X^2 = Pearson's test statistic
sbr = number of sample-based restrictions
$A,B,\ldots$ = factors or variables
$k_A,k_B,\ldots$ = number of categories of variable $A,B,\ldots$ or number of levels of Factor $A,B,\ldots$
$a,b,\ldots$ = subscripts running from 1 to $k_A,k_B,\ldots$

CHAPTER 15

T = total
MS = mean square
n_R = number of replications of each treatment combination

CHAPTER 16

r_S = Spearman's rank correlation coefficient
d = difference in ranks
ρ_S = population rank correlation coefficient
D = difference between numbers of positives and negatives in sign test
R_j = sum of ranks in jth group
t = number of tied observations in a set
H = Kruskal-Wallis test statistic
W = sum of signed ranks
X^2 = Friedman's rank test statistic

CHAPTER 17

EMV = expected monetary value
EU = expected utility
EVPI = expected value of perfect information
$\mathrm{Cov}(x,y)$ = population covariance of x and y

CHAPTER 18

HM = harmonic mean
log = common logarithm

CHAPTER 20

$\bar{d}$ = comparison
c_j = coefficient of a comparison
δ = population mean difference
g = degree of polynomial
I = independent variable
Q = studentized range statistic

CHAPTER 21

b_0 = sample regression constant
$b_1,b_2,\ldots,b_g$ = sample regression coefficients
g = number of predictors
D = denominator determinant in Cramer's rule
R^2 = coefficient of multiple determination
R = multiple correlation coefficient
$\beta_1,\beta_2,\ldots,\beta_g$ = population regression coefficients
c = number of variables held constant
antilog = common antilogarithm

PREFACE

Intended Audience

This book was written to serve as the textbook for courses in applied statistics at the elementary level. Requiring only high school algebra, it is intended for students in agriculture, biology, business, economics, education, environmental studies, health and medicine, psychology, and the social sciences. In addition, it is suitable for other students in the arts and sciences, since principles of making sense of data form an important component of a liberal arts education today.

Objectives

The book's major objectives are to help the members of its intended audience to become more intelligent consumers of statistical information, and to provide them with the basic statistical tools they will need for undergraduate research projects and courses in their own area of specialization. Although the book is not specifically directed toward the statistics major, I hope too that some students will find the excitement of the field contagious and be inspired to take further courses in statistics.

Level of Presentation

Today's students require—and are able to master—a solid and thorough, yet readable, introduction to the principles and methods of applied statistics. In fact, the mathematical demands of this book are modest. Proofs are omitted in favor of verbal explanations, and worked examples are numerous. At the same time, I have sought to avoid conceptual oversimplifications, on the grounds that truth makes more sense in the long run than fiction.

Approach

To keep the presentation readable without sacrificing content, I have been guided by several principles of cognitive psychology that are known to facilitate learning. In general,

- we learn things better if underlying principles and relationships are explained so as to foster understanding rather than rote memory only;
- we learn things better in the context of concrete examples within the realm of our experience; and
- we bring a lifetime of experience to any new learning situation, and we interpret the new situation in light of our past experience. (For example, students' intuitive interpretations of technical terms and concepts must sometimes be unlearned before new meanings are secure.)

In many ways, studying statistics is like studying a language, and students need opportunities to practice the vocabulary. If, for example, they confuse variables and values of variables, or statistics and parameters, explanations of subsequent concepts based on these terms will be confusing. Therefore, exercises have been included pertaining to terminology, choice of methods, and interpretation of results, in addition to the usual numerical exercises. Students will find the list of terms and concepts at the end of each chapter useful for review. Also, there is an extensive glossary of statistical terms at the end of the book and a brief glossary of symbols inside the front and back covers.

This book's general approach, involving thorough verbal explanations, numerous worked examples relevant to the students' experience, and attention to the language of statistics, is especially important in the present computer age. Readily available computer software and preprogrammed calculators make it relatively easy to perform a wide range of statistical calculations. What is required now is to ensure that the consumers of the statistical output understand and correctly interpret the results. Although computer output is illustrated for some topics, this book does not assume student access to computers, nor does it dictate which, if any, specific computer software should be used for instruction. Rather, it explains and demonstrates basic principles and methods of applied statistics so that the student, whether working by hand or by computer, will be able to make proper sense of the data.

Special Features

Every chapter of the text includes section exercises and chapter problems. The section exercises are gathered together at the end of each chapter for convenience, but they are grouped and labeled according to section. They provide numerical practice so that students can become comfortable with the mechanics of a method before attempting to apply it in a real-data problem. The chapter problems, which follow the section exercises at the end of each chapter, provide practice in applying the concepts and in identifying which of the chapter's methods is appropriate for which kinds of questions.

I consider it important to use real data, as much as possible, for examples and problems. If the study of statistics is to help us make sense of empirical data, we ought to have some empirical data to make sense of. Use of artificial data tends to lead to the mistaken view that observations begin with neat columns of numbers waiting to be popped into a formula and that statistics consists in popping them in and cranking out a numerical answer. Even real data can leave the same impression if the student does not understand how they were collected and so has to take them as unquestionable givens. Consequently, most of the examples and problems in this book involve real data collected by students working on projects in their first statistics course, projects designed by the students themselves to answer questions of interest to them. Such data are apt to be readily understood by students, so that it is easier to show how statistics is relevant to every step of an empirical investigation from the formulation of the question to the presentation of the results. As an added benefit, the data sets usually consist of numbers that are easy to work with and that involve manageable calculations. Larger sets of real data have been drawn from various sources to provide more challenging problems where appropriate.

Most chapters begin with an example describing a published application of the chapter's methods and/or principles in an important, real-world problem. These examples demonstrate the relevance of the course material to significant and interesting questions in science, business, and everyday life.

The applications in the examples and problems are not limited to any one discipline. I believe that students gain a better understanding of statistics if they see that the same basic principles and methods find application in many areas.

With the increasing prevalence of electronic calculators and personal computers, there seems to be a corresponding increase in confusion as to how many digits to report in a numerical answer. The topic of how many digits to report is not developed rigorously, but some basic principles are introduced in the text, and guidelines that are consistent with usual statistical practice are presented in Appendix 1. These principles and guidelines should discourage both premature rounding and the mindless reporting of as many digits as the calculator provides.

I have found that students are often interested in information about the lives of famous statisticians. Accordingly, I have included brief biographies of ten major contributors to the development of the science of statistics: Bernoulli, Fisher, Galton, Gauss, Gosset, Laplace, Legendre, de Moivre, Pearson, and Tukey.

Organization

Examination of the table of contents reveals a fairly conventional list of topics for first-year courses. The presentations of analysis of variance and regression analysis are somewhat more extensive than usual, in view of the fact that these analyses are the ones that students are most likely to encounter later in the form of computer software in connection with undergraduate research projects in their area of specialization. The text has been written to accommodate three kinds of courses in applied statistics:

a. the student's first one-semester course: Chapters 1–16 deal with topics commonly taught in the one-semester service course in applied statistics. Some chapters or sections may be omitted and perhaps others added, depending on the needs of the students and the interests of the instructor. The remainder of the text serves as a useful reference for students' later undergraduate applications of statistics in their own area of specialization.
b. the student's first full-year course: There is ample material here for a full-year course, with sufficient flexibility for giving more or less emphasis to any of several topics, especially analysis of variance and regression analysis.
c. the student's second one-semester course: A variety of packages of topics is possible, depending on the student's first course.

It takes a certain amount of time for students to assimilate statistical concepts. I have found that this assimilation is facilitated in full-year courses and two-semester sequences if the material is arranged so that each major topic is met more than once, rather than in one large dose. Thus, the notion of distributions is introduced in Chapter 2 and explored further in Chapter 7; measures of location and dispersion are introduced in Chapters 3 and 4 and taken up again in Chapter 17; correlation and regression are the topics of Chapters 5, 13, and 21; the logic of classical inference is introduced in Chapter 9 and taken up again in Chapters 10 and 11 after some examples; the analysis of variance is introduced in Chapter 15 and explored in Chapters 18–20. This pedagogically useful organization has been accomplished within a rather standard overall structure:

Chapter 1. Introduction.

Chapters 2–5. Descriptive Statistics.

Chapters 6–8. Probability Theory.

Chapters 9–16. Inferential Statistics.

Chapters 17–21. Further Topics in Applied Statistics.

Various alternative orders of the chapters are possible. For example,

Chapter 5 may be delayed to accompany Chapter 13;
Chapter 13 may be delayed to accompany Chapter 21;
Chapter 14 or 15 may follow immediately after Chapter 12;
portions of Chapter 16 may follow Chapter 5, 11, 12, or 14;
Chapter 17 may be introduced in or after Chapter 7;
Chapters 18–20 may follow immediately after Chapter 15;
Chapter 21 may follow Chapter 13.

I hesitate to label any sections as optional, since different courses have very different needs. However, sections or subsections labeled with an asterisk (*) can be identified as ones whose omission will not unduly jeopardize subsequent material.

Acknowledgments

Two essential ingredients in writing a statistics textbook are, first, sufficiently little foresight to realize how much work it will be and, second, an exceptionally understanding spouse. I have both.

Many other people have also contributed to the successful completion of this project. My students over the years have provided the data for most of the examples and problems, as well as valuable feedback concerning earlier drafts of the manuscript. The numerous reviewers engaged by Allyn and Bacon, Inc., gave many helpful criticisms and suggestions. Rhonda Amsel of McGill University commented on much of the manuscript and provided the computer output. Joseph Zaremba of the State University of New York College at Geneseo reviewed parts of the manuscript and provided additional examples and problems. Joel Katz and Naomi Kestenbaum of McGill University checked the answers to all the exercises and problems as they produced the Solutions Manuals. The responsibility for any errors or ambiguities that remain is, of course, my own. Readers who wish to make suggestions for the next edition are invited to send them to me at Camrose Lutheran College, Camrose, Alberta, Canada T4V 2R3.

I am particularly indebted to the editorial staff and the professional associates of Allyn and Bacon, Inc., for their skill at turning a manuscript into a book. Carol Nolan-Fish, Mary Beth Finch, and Sally Lifland all worked with patience, perseverance, and good humor in overseeing the many details throughout the course of this project.

A number of copyright holders have kindly granted permission to reproduce tables or examples in this book. Specific acknowledgments appear with the reproduced material. I am grateful to the Literary Executor of the late Sir Ronald A. Fisher, F.R.S., to Dr. Frank Yates, F.R.S., and to Longman Group Ltd., London, for permission to reprint Table VII and portions of Table XXIII from their book *Statistical Tables for Biological, Agricultural and Medical Research* (6th edition, 1974). Output from SAS procedures is printed with permission of SAS Institute Inc., Cary, NC 27511-8000, copyright © 1985.

TO THE STUDENT

Success in your first statistics course is virtually guaranteed if you master the vocabulary and keep up with the exercises and problems. To assist you with the vocabulary, a list of terms and concepts is provided for review after you have read each chapter. In addition, an extensive glossary at the end of the book gives concise definitions of the statistical terms introduced in the text.

At the end of each chapter, there are exercises for numerical practice and problems for applying the concepts. Answers to odd-numbered exercises and problems appear at the back of the book. The worked examples in the text are intended to help you with the exercises and problems and will provide maximum benefit if you work through the calculations in the worked examples to check your understanding. To save time, a hand-held calculator is a good investment. It does not have to be an expensive one: Besides addition, subtraction, multiplication, and division, the most needed functions are square roots, squares, and logarithms. The model that calculator manufacturers often label as their "scientific" or "statistical" calculator has other functions that may prove useful to you if you take time to learn how to use them, but a less expensive calculator is all that is assumed in this book.

Statistics exercises are not like calisthenics and jogging, which are good for you no matter how your mind wanders as long as you go through the motions. To benefit from the exercises and problems, it is essential that you *think* about the problem, with special emphasis on determining why the method that led to the correct answer was the appropriate method for the problem and what the result tells you about the original question that the data were collected to answer. A few minutes of careful thought as you complete each exercise or problem can make a tremendous difference in the benefits derived from your exercise time.

Sometimes students get discouraged if they do not see the answer to a problem in statistics immediately, thinking there is something wrong with them. Such students should be reminded that most of the interesting questions in life do not have mechanical answers, and genuine statistical questions are usually interesting. The main business of the statistician is thinking, not mechanically plugging in mathematical formulas, and even expert statisticians have to take time to think when confronted with a new problem. You will find that statistical thinking is facilitated by learning some terms that summarize statistical concepts, and that it will become easier with practice. The exercises and problems are intended to help you become more fluent in the language of statistics and to give you some experience in making sense of data.

CHAPTER 1

INTRODUCTION

EXAMPLE 1.1 The concepts and methods of statistics are applied to solve many interesting and important problems. Consider the following topics found in a collection of forty-six essays illustrating the impact of statistics on our lives:

the safety of anesthetics used in surgery,
the effects of smoking on health,
estimating sizes of whale populations for conservation purposes,
the effect of speed limits on traffic-accident rates,
projecting winners in live television coverage of election results,
deciding who wrote famous works sometimes ascribed to different authors,
estimating demand for a new product,
interpreting college entrance examination results,
the value of bunts and intentional walks in baseball,
obtaining accurate census information,
weather forecasting,
designing and using computers.

The diversity of these topics should dispel any notion that statistics is the mere accumulation of facts and figures for purely academic or archival purposes. Statistics is a fascinating discipline with significant impact on our everyday lives.

(Source: Tanur *et al.*, 1978.)

When the word *statistics* was first introduced about 200 years ago, it referred to a branch of political science involving the collection and study of political facts and figures, especially numerical data describing the "state" or nation, such as its number of inhabitants or the total value of its production of goods and services. That definition is no longer adequate. Over the years, the discipline of statistics has developed and changed to such an extent that today it makes substantial contributions to the quality of our lives in almost every area of life, even though many of us are not aware of the role played by statistics in these improvements.

1.1 WHAT IS STATISTICS?

Perhaps you think the heading should ask, "What *are* statistics?" Most adults in our society have a pretty good idea of what statistics are. We are confronted with statistics almost every day: results of public opinion polls about the current popularity of various politicians and parties, national and regional unemployment rates, cost-of-living indexes, batting records of baseball players and scoring races among hockey players, and reports of profits or losses by oil companies or by automobile manufacturers. Such numbers make the news regularly. Common knowledge has it that statistics are numbers, and that they are to be greeted with anything from enthusiasm and interest, to awe and acceptance, to boredom and suspicion. But fewer people know what statistics *is*.

A Definition of Statistics

What then is statistics?

> **Statistics** is the science of answering questions on the basis of empirical data. **(1.1)**

The term **empirical data** refers to data originating in observation or experience. Thus Expression (1.1) identifies statistics as the science that deals with methods for answering questions through the proper collection and interpretation of data originating in observation or experience. The observations and experiences that constitute the data may be obtained in the course of scientific research, business activity, or everyday life, with the proper interpretation depending on the method of collection. In any case, statistics is the science of making sense of data.

It is worth pointing out that the science of statistics is not defined in terms of numerical tabulations or computations, although these are sometimes involved. It is defined in terms of its objective of making sense of—drawing correct conclusions from—the data of experience, and so it includes both the collection and the interpretation of empirical data.

a. Under the broad label of **collection of empirical data** are included
the design of empirical investigations,
the planning of what observations to make, and
the adequacy of the observing and recording of the data.

b. Under the label of **interpretation of empirical data** are included
the analysis ("number-crunching") and summarizing of the data,
the drawing of conclusions, and
the reporting and presentation of the findings.

In other words, statistics is relevant to every step of an empirical investigation, from the formulation of the question to the presentation of the results.

The Statistician's Approach to Data

Since the purpose of the science of statistics is to extract meaningful information from empirical data, statistics is concerned with methods for the *proper* collection and interpretation of data. Statisticians are sometimes maligned with remarks to the effect that "you can prove anything with statistics," and former British Prime Minister Benjamin Disraeli (1804–81) is still frequently quoted as having said that "there are three kinds of lies: lies, damned lies, and statistics." Of course, the possibility that something can be misused does not preclude the possibility of benefit from proper use. Statistics can indeed be misused: Liars can lie with numbers just as they can lie with words. However, statisticians are concerned with finding out what the numbers can and cannot legitimately tell us. Thus, they are keenly interested in how the data are collected, since that will limit the use to which the numbers may be put and the interpretations that are acceptable.

EXAMPLE 1.2 "George A. Birmingham, the novelist and essayist, . . . was a clergyman in Ireland and, as such, he was pestered by bishops and other authorities to fill in recurring questionnaires. He took particular umbrage against the annual demand from the education office to report the dimensions of his village schoolroom. In the first and second years, he duly filled in the required figures. On the third year, he replied that the schoolroom was still the same size as before. Schoolrooms are not trees, he observed, they do not grow. The education office badgered him with reminders until Birmingham finally filled in the figures.

"But he didn't put in the same figures as before. He doubled the dimensions of his schoolroom. Nobody queried it. So he went on doubling the measurements until 'in the course of five or six years that schoolroom became a great deal larger than St. Paul's Cathedral.' But nobody at the education office was at all concerned. So, the next year, the Canon suddenly reduced the dimensions of his colossal classroom 'to the size of an American tourist trunk. . . . It would have been impossible to get three children, without a teacher, in that schoolroom.' And nobody took the slightest notice, for nobody needed the information. But the system did, and the system had to be satisfied."

(Source: Ryan, 1977.)

With any data, one should always ask, How were these data collected? Are they based on information that the observers would know correctly, or would take the trouble to find out? And would the observers see the collection of the information as being sufficiently important that they would be sure to report accurately?

Even when there is no deliberate misrepresentation, the statistician recognizes that some observations are more trustworthy than others.

EXAMPLE 1.3 A psychologist interviewed parents in 47 families who were taking part in a long-term study of child development. Each family had a three-year-old child at the time of the interview and had been taking part in the study since the birth of that child. Observations made during the course of the three-year study had included such milestones as bladder training and end of bottle-feeding. When their child was approximately three years old, the parents were individually asked to recall as best they could some of these same facts.

There were substantial errors in the parents' recall, even though the events to be recalled were significant ones in parents' lives and even though the events were relatively recent. For example, mothers recalled bladder training to have begun an average of 5 months later than it actually had (which was at about 16 months on the average), while fathers' recall was off an average of 6 months. Similarly, mothers and fathers recalled the end of bottle-feeding as having occurred 3 months and 4 months, respectively, earlier than it actually had occurred, which was typically around 24 months.

(Source: Robbins, 1963.)

Example 1.3 illustrates that memory can be a fickle friend. Consequently, **retrospective reports**—reports from memory—include an additional source of possible error beyond that found in immediate reports of current events.

Sometimes our knowledge of current facts is not as accurate as we think either.

EXAMPLE 1.4 In a student project for an introductory course in statistics, eight male college students were asked to report their height to the nearest tenth of an inch for subsequent verification by actual measurement. Only one of the eight men was within .1 inch (.3 cm) of being correct. The other seven were off by anywhere from .5 to 1.9 inches (1.3 to 4.7 cm), all but two of them thinking they were taller than they actually were.

Readers may find it instructive to check their knowledge of their own height. To do so, make a note of how tall you think you are and then, standing shoeless with your back against a wall, have a friend place a book vertically on your head against the wall (to make a right angle with the wall) and make a light pencil mark on the wall just below the book. Measure carefully the exact distance from the floor to the mark. While you're at it, check whether your friend is any more accurate about his/her own height than you were about yours.

(Source: Student project by R. Schulz.)

Example 1.4 illustrates the general principle that **verbal report** is not equivalent to **direct measurement**. Verbal reports indicate what people *say* is the case, but not necessarily what actually *is* the case.

Whether the data are collected by verbal report or by direct measurement, we cannot necessarily assume that they shed any light on the question at hand. The statistician's goal is to distinguish sense from nonsense in empirical data and thereby to contribute to wiser decisions in our everyday lives and to a better understanding of the natural and social worlds in which we live.

The Work of Statisticians

Designing, planning, analyzing, interpreting—these activities by which we defined statistics may sound rather abstract, but it must be remembered that statistics involves the application of these activities to answer questions on the basis of empirical data. As illustrated in Example 1.1, statistics is applied in a wide range of subject areas. A brochure entitled *Careers in Statistics,* published

"Perfect! Six feet two inches and 195 lbs."

Figure 1.1 HERMAN, by Jim Unger, 7 September 1981. Copyright 1981, Universal Press Syndicate. Reprinted with permission. All rights reserved.

by the American Statistical Association, points out that careers in statistics may be found in such fields as actuarial science, agriculture and fisheries, biology, business, economics, education, engineering, health and medicine, psychology, quality control, and social sciences, and that new opportunities are developing in such fields as anthropology, archaeology, computer science, history, library science, law, and public policy. Some expertise in one or more of these substantive areas is a definite asset—some would say an absolute necessity!—for successful statistical work.

Certain basic statistical principles and methods are common to virtually all fields to which statistics is applied, so that while labels such as *business statistics* and *psychological statistics* are sometimes used to indicate the area of application, these are simply different facets of **applied statistics**. The alternative to applied statistics is called **theoretical statistics** or **mathematical statistics**, which is concerned with the development of mathematical theory relevant to statistical problems. In practice, it may be difficult to distinguish between these two orientations to statistics. Whether their orientation is applied or theoretical, statisticians work with other scientists in diverse fields to determine how best to collect and interpret empirical data to answer particular questions.

Statistics as a whole, whether theoretical or applied, is most appropriately viewed as an applied science, in much the same way that engineering is. Unlike mathematics, statistics cannot exist apart from real-world questions in need of answers, as the definition in Expression (1.1) should make clear. True, there are statisticians who specialize in the mathematics of statistics—mathematical statisticians—but theirs is theoretical work to improve the collection and interpretation of empirical data for the purpose of answering questions. Statistics begins with empirical questions to be answered, rather than with axioms or postulates, and in that sense it is always an applied science.

1.2 POPULATIONS AND SAMPLES

An important distinction is made in statistics according to whether one is dealing with all possible data relevant to the question under investigation, or whether one is limited to a subset of the hypothetically possible data relevant to the question.

A **population** is defined as the set of *all* observations (or other things) relevant to the question being asked. **(1.2)**

A **sample** is a set of *some,* but not all, of the observations (or other things) relevant to the question being asked. **(1.3)**

Thus, a sample is part of a population, and there are many different samples that could possibly be drawn from any population having more than a few elements.

Notice that the distinction between populations and samples is based on the question being asked. The same observations may constitute either a population or a sample, depending on the question being asked of the data.

EXAMPLE 1.5 The following data were collected from the 11 students who wrote the first class quiz in a certain sociology course (Sociology 322) in September 1979.

	Reported number of hours spent studying for the quiz	Quiz mark (%)
Student 1	3	78
Student 2	2.5	100
Student 3	2.5	98
Student 4	2	98
Student 5	2	95
Student 6	2	93
Student 7	2	78
Student 8	2	70
Student 9	1.5	70
Student 10	1.5	52
Student 11	.5	90

If our ultimate interest in these data is to determine such things as (a) the average reported study time of those 11 students for that particular quiz, (b) the average mark on the first quiz in that particular class, and (c) the relationship between reported study time and first quiz mark for the students in that class, then the data constitute a population. As long as the questions do not go beyond reported study times and marks obtained by the 11 students in that particular sociology quiz in that particular year, we have all the relevant observations; they are the population for such questions.

On the other hand, if we wanted to know such things as (d) the average study time that students would report for their first class quiz in any senior course, or (e) the average first-quiz mark in Sociology 322 in recent years, or (f) the general relationship between reported study time and marks in class quizzes, then the given data would constitute a sample from a larger population. For such questions, we do not have all the relevant observations; those questions pertain to a more general situation than the specific cases observed. Just which other observations are relevant is different for the different questions, and it depends too on whether we are talking about one particular college only or all the colleges in a given region. In any case, other observations in the population would at least include, for question (d), reported study times in other courses; for question (e), first-quiz marks from other years; and for question (f), data from other students.

(Source: Student project by S. Broen.)

In order to determine whether a given set of data is a population or a sample—and, if it is a sample, what the relevant population is—we need to know precisely what question is being asked of the data.

In science as well as in everyday life, sample data are more the rule than the exception. Scientific experiments rarely are run to find out what will happen on one specific occasion, but rather are run to determine how a certain type of material or organism, of which the actual material or organism used in the experiment is intended to be representative, can be expected to respond under similar circumstances in the future. Reports of unemployment rates, consumer price indexes, and public opinion polls are likewise based on samples, because it is either impractical or impossible to collect all of the observations that constitute the relevant populations. The extent to which the sample adequately represents the population is obviously an important consideration in all such situations.

1.3 THE SCIENCE OF STATISTICS

We have seen that, as a science, statistics is the body of knowledge concerning methods for answering questions through the proper collection and interpretation of empirical data. Here we included methods pertaining to all the stages of empirical investigations: designing, planning, observing, recording, analyzing, summarizing, drawing conclusions, reporting, and presenting. Statistics in this sense lies at the intersection of mathematics and scientific methodology. It is sometimes thought of as a branch of mathematics and at other times as the scientific-methodology subdiscipline of each of the disciplines in which it is applied, but it is probably more accurate to view it as a body of knowledge in its own right, combining observation and mathematics to answer particular questions.

There are two main aspects to the science of statistics—descriptive statistics and inferential statistics—distinguished by the nature of the question that is asked of the data.

Descriptive Statistics

Descriptive statistics deals with methods of characterizing or summarizing a given set of data, without attempting to draw conclusions that go beyond those data.

EXAMPLE 1.6 Calculation of the average reported number of hours of studying by the 11 sociology students in Example 1.5 would fall into the realm of descriptive statistics. The average, 2.0 hours, merely summarizes or describes one aspect of the given data and involves no uncertainty. The average is simply a description of one feature of the data.

The resulting description is called either a *parameter* or a *statistic*, depending on whether the observations are regarded as a population or as a sample, respectively.

A **parameter** is defined as a numerical description that summarizes information about a *population*. **(1.4)**

A **statistic** is a numerical description that summarizes information about a *sample*. **(1.5)**

The goal of descriptive statistics is to answer empirical questions by summarizing and presenting data in ways that display the relevant information clearly and accurately.

Inferential Statistics

Complementing descriptive statistics, the other main aspect of the science of statistics is **inferential statistics**, which deals with methods for making appropriate inferences or generalizations about populations on the basis of samples. In this connection, a statistic is used as an **estimate** of a parameter.

EXAMPLE 1.7 In Examples 1.5 and 1.6, suppose that the instructor of Sociology 322 wanted to know to what extent any students who might take Sociology 322 from her would be motivated to study for the first quiz. She might ask, "How long do students claim to study for the first quiz in Sociology 322, when I teach it the way I did in the class of Example 1.5?" The question deals with the average study time that would be reported by the hypothetical population of all potential students for the course. The 11 students actually in the course would be considered a sample from that hypothetical population of potential students. If we use the sample average, 2.0 hours, as an estimate of the average reported study time of the hypothetical population, we are in the realm of inferential statistics. There is some uncertainty associated with the value 2.0 hours when we use it as an estimate of a parameter, because it is based on only a part of the population.

Whereas descriptive statistics seeks to summarize and display the relevant information in a given set of data, the problem in inferential statistics is to determine what can be said about all the relevant data (the population) on the basis of only some of the relevant data (a sample).

Whenever a question is to be answered on the basis of sample data, part of the answer must address itself to the issue of how well the sample represents the population. If not all the population data are available, there is bound to be some uncertainty in the answer, and yet, as noted at the end of Section 1.2, it is frequently necessary to rely on samples. Attempting to minimize and to measure the inevitable uncertainty in samples constitutes a major concern of inferential statistics.

1.4 STATISTICS AS NUMERICAL DATA

In this book, we will be using the word *statistics* in two senses:

a. the science of answering empirical questions, as defined in Expression (1.1), and

b. numerical descriptions of samples, according to the definition in Expression (1.5).

As noted in Section 1.1, the everyday understanding of the term *statistics* does not correspond to its usage in this book but rather tends to ignore (a) and to be more general than (b).

In everyday usage, the term *statistics* is used to refer to virtually any numerical data, including not only numerical descriptions of samples—which we also call statistics, according to Expression (1.5)—but also

c. numerical descriptions of populations—which we will call parameters, according to Expression (1.4)—as well as

d. numerical descriptions of individual items in a sample or population—which we will refer to as **observations**, or simply as the **raw data**.

We will avoid the use of the term *statistics* in its everyday sense in order to emphasize the technical meaning of statistic, referred to in (b) above, and to try to minimize ambiguity. When we refer to numerical data as statistics in this book, we are referring to numerical descriptions of samples.

1.5 THE ART OF STATISTICS

Sections 1.1 and 1.2 introduced the important principle that the appropriate statistical methods to use and the appropriate conclusions to draw from a given set of data are critically dependent upon how the data were collected and what questions are being asked of the data. Practicing statistics is not simply a matter of plugging numbers into formulas and cranking out numerical answers. Rather it involves determining what observations should be made to answer particular questions, what calculations should be performed, and what limitations are imposed on the results by virtue of the way the observations were made. These are questions for which there are no mechanical formulas but rather whose solution depends on understanding certain statistical concepts or principles. Clear and careful thinking, not formula memorization or computational prowess, is the coin required in the realm of statistics. The art or skill in the practice of statistics involves recognizing what information can and cannot be legitimately extracted from particular sets of empirical observations.

The development of good data-side manner requires experience with real-world data, preferably in areas of application with which one has enough familiarity to understand how the data were collected, since judging the adequacy of data is an important component of statistical skill. It is important to realize that statistics is not chiefly about numbers, whose origin is either mysterious or irrelevant. Rather it is chiefly about empirical questions whose answers, sought through observation of the world of nature and human affairs, depend on the adequacy and relevance of the data. Accordingly, many of the examples and exercises in this book involve real data collected by students working on projects in their first statistics course—projects designed by the students themselves to answer questions of interest to them. These examples are intended not only to provide some experience with real data, but also to help the reader to see the operation of statistical principles in everyday life.

Virtually any set of data collected to answer an empirical question has limitations—sometimes only minor limitations, sometimes major ones. However, the impossibility of perfect data does not prevent the skilled statistician from obtaining useful and correct information from empirical data, by taking into account those limitations. When presented with empirical data (in or out of class!), the student of statistics is well advised to get into the habit of thoughtfully assessing what qualifications must be placed on the conclusions because of the way in which the data were collected.

1.6 NOTATION

Statistical concepts, someone has said, are really not much more than common sense translated into mathematical language. Some people think they have difficulty with statistics because they lack common sense, or at least mathematical sense, but more often it is simply the mathematical notation that throws some people off. It is not that the notation is difficult—in fact, it is not difficult—but you must take time to become familiar with it. Thus, it is bound to take you longer per page to read your statistics text than to read, say, a history text. You are already familiar with the language in which your history text is written, but you can expect it to take some time to become familiar with the notational language and technical terminology used in your statistics text.

For the most part, notation will be introduced in the text where we first have need of it. In this section, a few points are noted that are relevant throughout the text.

Symbols

Most good literature involves symbolism, and statistics is no exception. It is customary to represent numerical observations and other quantities by letters such as x, y, and z.

EXAMPLE 1.8 We could represent the numbers of books on three particular shelves in the college library as $x = 23$, $y = 16$, and $z = 27$. Then the total number of books on those shelves is $x + y + z = 23 + 16 + 27 = 66$.

Variables

It would be very cumbersome if we had to use a different letter for each observation, as in Example 1.8, and we would run out of letters if we ever had more than 26 observations in one investigation. Thus, we usually use the same letter for all measurements of one type or, as we shall say, for all the values of one *variable,* and we distinguish among the individual observations by subscripts attached to the letter. A variable is something that varies from one individual (or case) to another:

> A **variable** is a characteristic that may distinguish one item from another or, more precisely, a dimension along which items may differ from each other. **(1.6)**

An individual observation is called a **value of the variable**. Letters such as x and y are used to represent different variables, and subscripted letters such as x_1, x_2, x_3, y_1, y_2, and y_3 are used to represent specific values of those variables. (The subscripted letters are pronounced x-sub-one, x-sub-two, and so on.)

EXAMPLE 1.9 As part of a student project, the number of children in each of four families was recorded as follows.

	Number of children	
	Girls	**Boys**
Family 1	1	2
Family 2	3	0
Family 3	6	5
Family 4	2	3

Let x represent the number of girls in a family, and y the number of boys. The number of girls in a family is one variable, and the number of boys is another variable. We can write the values of the variables as

$$\begin{array}{cccc} x_1 = 1 & x_2 = 3 & x_3 = 6 & x_4 = 2 \\ y_1 = 2 & y_2 = 0 & y_3 = 5 & y_4 = 3 \end{array} \quad (1.7)$$

Notice that the subscript merely identifies *which* value of the variable; the subscript is not the variable's value but simply tells us which observation (in this case, which family) is being referred to. The symbol x_3, for example, designates the number of girls (variable x) in the third family (subscript 3), and the value of the variable for that family is $x_3 = 6$. The symbol y_4 designates the number of boys (variable y) in the fourth family (subscript 4), and the value of the variable for that family is $y_4 = 3$. We could use words or other labels as identifying subscripts (for example, x_{Smith} instead of x_3, or $y_{\text{fourth-family}}$ instead of y_4), but it is usually more convenient simply to use the successive integers 1, 2, 3, and so on, to label the individual observations.

For the x-variable (number of girls), the values of the variable are represented by the general term x_i for $i = 1, 2, 3, 4$. Similarly, y_i is the number of boys for the ith family, that is, the number of boys in some particular family, although we do not know which family until the value of i is specified.

(Source: Student project by T. Webb.)

The general term referring to the individual items that might be observed in an empirical investigation is **observational unit**. (In Example 1.9, the observational units are the families.) When the observational unit is a living organism, such as a person or an animal, it may be referred to as a **subject** in the investigation.

An **observational unit** is an item or element—such as an individual, or a specimen, or an institution—that, when observed in an investigation, yields one observation. **(1.8)**

If just one variable is observed for each observational unit in the investigation, the observations are said to be **univariate**. If two variables are observed for each observational unit, we speak of **bivariate** observations. (In Example 1.9, the observational units are the families, and the observations are bivariate, because two variables were observed for each unit. The example involves four bivariate observations.) In general, if two or more variables are observed for each observational unit, the observations are said to be **multivariate**. Bivariate observations are multivariate observations in which just two variables are observed for each observational unit.

Constants

A **constant** is a quantity that does not vary in a particular problem. The constants $\pi \simeq 3.14159$ and $e \simeq 2.71828$ (the base of natural logarithms) are fairly familiar. The number of observations (4 in Example 1.9) is frequently a constant for a particular investigation, and it is typically denoted by N if the entire population is included or by n if the observations constitute a sample. The number of variables observed, the number of conditions or experimental groups in an investigation, and the parameters of the relevant population are other examples of constants in many statistical problems.

Summation

The use of subscripted variables allows the introduction of **summation notation** to express in a concise form the sum of the values of a variable. The Greek capital letter sigma (Σ) is used to mean "the sum of." It is used as follows:

$$x_1 + x_2 + x_3 + x_4 = \sum_{i=1}^{4} x_i \tag{1.9}$$

The right-hand side of Expression (1.9) is read, "The sum of the values x-sub-i, with i running from 1 to 4." The sigma tells us that we are to add the x's. The values above and below the sigma, called the **limits of summation**, tell us which values x_i to add. The **lower limit of summation** in Expression (1.9) indicates that we start with x_1, the x value where $i = 1$; then we add to that each successive x_i (that is, x_2, x_3, and so on) until the subscript i reaches the **upper limit of summation**, which is 4 in Expression (1.9). In general,

$$\sum_{i=1}^{n} x_i = x_1 + x_2 + \cdots + x_n \tag{1.10}$$

EXAMPLE 1.10 For the observations in Expression (1.7),

$$\sum_{i=1}^{3} x_i = x_1 + x_2 + x_3 = 1 + 3 + 6 = 10$$

This is the total number of girls in the first three families. The total number of boys in the last two families is

$$\sum_{i=3}^{4} y_i = y_3 + y_4 = 5 + 3 = 8$$

Often the context makes it clear which observations are to be added, and the sigma sign is written with no limits of summation above or below it. Then it is understood that all relevant values of the variable (usually for subscripts running from 1 to n or N) are to be added. In other words, if the context makes clear that we are referring to n observations, Expression (1.10) may be written

$$\sum x_i = x_1 + x_2 + \cdots + x_n \tag{1.11}$$

In other cases, only the upper limit of summation is shown, $\sum^n x_i$, if it is clear that the lower limit of summation is $i = 1$. Furthermore, if the context makes clear that x is a variable (rather than a constant), the subscript i may be dropped, so that Expression (1.11) becomes

$$\sum x = \sum^{n} x = x_1 + x_2 + \cdots + x_n \tag{1.12}$$

EXAMPLE 1.11 For the observations in Expression (1.7), the expression $\sum x$ would be evaluated as

$$\sum x = x_1 + x_2 + x_3 + x_4 = 1 + 3 + 6 + 2 = 12$$

This is the total number of girls in the four families. The total number of boys in the four families is

$$\sum y = 2 + 0 + 5 + 3 = 10$$

The following paragraphs give a few useful characteristics of summation as denoted by sigma. In these paragraphs, k is a constant, and the variables x and y take on values x_i and y_i for $i = 1, 2, \ldots, n$.

Characteristic 1. The summation of a sum equals the sum of the summations; that is,

$$\begin{aligned}\sum_{i=1}^{n} (x_i + y_i) &= (x_1 + y_1) + (x_2 + y_2) + \cdots + (x_n + y_n) \\ &= (x_1 + x_2 + \cdots + x_n) + (y_1 + y_2 + \cdots + y_n) \\ &= \sum_{i=1}^{n} x_i + \sum_{i=1}^{n} y_i\end{aligned} \tag{1.13}$$

EXAMPLE 1.12 For the observations in Expression (1.7), Expression (1.13) says that whether we add the numbers of children in the four families or the total number of girls and the total number of boys, we will obtain the same total number of children in the four families. Adding the numbers of children

$$\sum (x_i + y_i) = 3 + 3 + 11 + 5 = 22$$

gives the same result as adding the total number of girls to the total number of boys:

$$\sum x + \sum y = 12 + 10 = 22$$

Characteristic 2. The summation of a difference equals the difference of the summations; that is,

$$\sum_{i=1}^{n} (x_i - y_i) = \sum_{i=1}^{n} x_i - \sum_{i=1}^{n} y_i \tag{1.14}$$

The demonstration of this equality parallels that for sums.

EXAMPLE 1.13 For the observations in Expression (1.7), Expression (1.14) says that the sum of the excess of girls over boys in each family

$$\sum (x_i - y_i) = (1 - 2) + (3 - 0) + (6 - 5) + (2 - 3) = 2$$

is equal to the difference between the total number of girls and the total number of boys:

$$\sum x_i - \sum y_i = 12 - 10 = 2$$

In other words, there are 2 more girls than boys in these four families combined.

Characteristic 3. The summation of a constant times a variable equals the constant times the summation of the variable; that is,

$$\sum_{i=1}^{n} kx_i = kx_1 + kx_2 + \cdots + kx_n$$
$$= k(x_1 + x_2 + \cdots + x_n) \qquad \textbf{(1.15)}$$
$$= k \sum x_i$$

EXAMPLE 1.14 In Example 1.4, eight male college students made the following errors (shown here in centimeters, ignoring signs) in verbal reports of their height:

$x_1 = 1.7$	$x_2 = .2$	$x_3 = 1.3$	$x_4 = 4.7$
$x_5 = 1.8$	$x_6 = 2.6$	$x_7 = 1.3$	$x_8 = 1.9$

What is the total of these errors in millimeters?

Expression (1.15) simply says that we can calculate this total in either of two ways. We can convert each error to millimeters and add the converted errors:

$$\sum kx_i = (10)(1.7) + (10)(.2) + \cdots + (10)(1.9) = 155$$

Or we can add the original errors and convert the total to millimeters:

$$k \sum x_i = 10(1.7 + .2 + \cdots + 1.9) = 155$$

Either way, the total of the eight errors (ignoring signs) is 155 mm.

(Source: Student project by R. Schulz.)

Characteristic 4. The summation of a constant over n terms is simply n times the constant:

$$\sum_{i=1}^{n} k = k + k + \cdots + k \quad \text{(that is, the sum of } n\ k\text{'s)}$$
$$= nk \qquad \textbf{(1.16)}$$

The distinction between Expressions (1.12) and (1.16) is that in the former case x is a variable, which takes on various values x_i, whereas in the latter case k is a constant, which does not vary for different values of i.

EXAMPLE 1.15 The principle in Expression (1.16) is rather like that in the story of the farmhand who explained his amazing ability to count a flock of sheep so quickly by saying that he just counted the legs and divided by four. Similarly, whether one counts n sheep and multiplies by $k = 4$ legs per sheep, or whether one simply counts all the legs $\sum_{i=1}^{n} k$, one arrives at the same total.

Characteristic 5. Adding a constant to a summation is not equivalent to adding a constant to each value of the variable; that is,

$$\sum_{i=1}^{n} x_i + k \neq \sum_{i=1}^{n} (x_i + k) \qquad \textbf{(1.17)}$$

The left-hand side tells us to add k to the sum of the x's, whereas the right-hand side tells us to add k to each x_i and then sum:

$$\sum_{i=1}^{n} (x_i + k) = \sum_{i=1}^{n} x_i + nk \tag{1.18}$$

EXAMPLE 1.16 For the observations in Expression (1.7),

$$\sum x_i + 5 = 1 + 3 + 6 + 2 + 5 = 12 + 5 = 17$$

whereas

$$\sum (x_i + 5) = (1 + 5) + (3 + 5) + (6 + 5) + (2 + 5) = 12 + (4)(5) = 32$$

Characteristic 6. Adding the products of corresponding values of x_i and y_i is not equivalent to multiplying the sums of x and y; that is,

$$\sum x_i y_i \neq (\sum x_i)(\sum y_i) \tag{1.19}$$

The left-hand side tells us to multiply the ith value of x times the ith value of y and then sum those products, whereas the right-hand side tells us to sum the x's, sum the y's, and then multiply those totals.

EXAMPLE 1.17 For the observations in Expression (1.7),

$$\sum xy = (1)(2) + (3)(0) + (6)(5) + (2)(3) = 2 + 0 + 30 + 6 = 38$$

whereas

$$\sum x \sum y = (1 + 3 + 6 + 2)(2 + 0 + 5 + 3) = (12)(10) = 120$$

Characteristic 7. The sum of the squares of the values of the variable is not equal to the square of the sum; that is,

$$\sum x_i^2 \neq (\sum x_i)^2 \tag{1.20}$$

The left-hand side tells us to square each value of x and then sum those squares, whereas the right-hand side tells us to square the sum of the x's.

EXAMPLE 1.18 For the observations in Expression (1.7),

$$\sum x^2 = 1^2 + 3^2 + 6^2 + 2^2 = 1 + 9 + 36 + 4 = 50$$

whereas

$$(\sum x)^2 = (1 + 3 + 6 + 2)^2 = 12^2 = 144$$

OMIT

* 1.7 A CASE IN POINT: THE CONSUMER PRICE INDEX

Let us consider the familiar Consumer Price Index as an everyday example that will allow us to illustrate many of the terms and concepts introduced in this chapter. Changes in the Consumer Price Index are commonly reported in the news media. We may hear, for example, that the annual inflation rate is 4%, based on the Consumer Price Index. The Consumer Price Index is a statistic, that is, a numerical description of a sample. The most important point to be illustrated in this section is that making proper sense of a numerical description requires some knowledge of how the data on which it is based were collected. Many people think that the Consumer Price Index tells them much more than it does!

Many aspects of business and economic conditions are conveniently summarized as changes over time. These include changes in consumer prices, wholesale prices, wages, productivity, employment, and export and import quantities, both for the country as a whole as well as for particular regions and particular industries or sectors of the economy. Information about such changes is used to guide and to evaluate government policies, to facilitate planning by business and industry, to assist in union-management negotiations, and to keep the private citizen informed about economic conditions. The Consumer Price Index is just one of a large family of economic indicators called **index numbers**.

Index Numbers

An **index number** is a percentage that indicates the value of a variable in one time period relative to its value in another time period. The reference point in time is called the **base period**, and the time being described by the index number is called the **given period**. Specifically, an index number is the value in the given period expressed as a percentage of the value in the base period, but with the percent sign dropped:

$$\text{Index} = \frac{\text{given-period value}}{\text{base-period value}} \times 100 \tag{1.21}$$

EXAMPLE 1.19 A consumer used 7660 kWh of electric power in 1978 and 8090 kWh in 1982. With 1978 as the base year, the consumer's electricity consumption index for 1982 was

$$\frac{8090}{7660} \times 100 = 105.6$$

If the base year is 1978, the electricity consumption index for 1978 itself is necessarily 100:

$$\frac{7660}{7660} \times 100 = 100.0$$

* This symbol identifies material that is not prerequisite for later topics in this book.

Thus, the base year for a given-year index is often indicated by a parenthetical notation such as in the following statement: The 1982 electricity consumption index was 105.6 (1978 = 100). This statement tells us that consumption in 1982 was 5.6% higher than in 1978. (Guidelines for how and when to round and how many digits to report are given in Appendix 1.)

If the price of electricity was 3.15 cents/kWh in 1978 and 4.00 cents/kWh in 1982, the electricity price index for 1982 was

$$\frac{4.00}{3.15} \times 100 = 127.0 \qquad (1978 = 100)$$

Thus the price of electricity in 1982 was 27% higher than in 1978. This particularly simple form of price index, based on just one commodity, is sometimes called a **price relative**, to distinguish it from more general price indexes that combine information from a variety of commodities.

Not all index numbers are as simple as those in Example 1.19. Indexes such as the Consumer Price Index are intended to reflect the average change in values for a large set of items, such as prices of consumer goods and services.

EXAMPLE 1.20 The following table shows further data for the consumer in Example 1.19.

	Price		Quantity	
	1978	**1982**	**1978**	**1982**
Electric power	3.15¢/kWh	4.00¢/kWh	7660 kWh	8090 kWh
Home heating fuel (coal)	$12.40/ton	$21.10/ton	6.5 tons	7.0 tons
Local telephone service	$6.75/month	$7.25/month	12 months	12 months

How could we compute a single price index for these three commodities? Expression (1.21) gives us a general formula for an index number, but in this case the given-period and base-period values must each be based on three different prices. We need to find a single value to summarize the price information for the three commodities each year. It would not make sense to take a simple average of the prices, because the prices could have been given in different units that would have yielded different averages. (For example, we could have listed prices for a pound, rather than a ton, of coal: .62¢/lb and 1.055¢/lb.) In addition, price alone does not indicate whether the commodity represents a major or a modest expense, because it ignores the quantities consumed. We can solve both of these problems by basing the index on the total amount spent, computed as the price times the quantity for each item, summed over all items.

The solution proposed in the last paragraph raises another question: Which quantities should we use? If we take the total amount spent in 1982 and compare it with the total amount spent in 1978, we will not be able to tell whether the change is due to changes in price, or to changes in quantities purchased, or to both. If our index is to be a price index, reflecting changes in price, we should compare the total value of a fixed "market basket" of goods and services at 1978 prices with the total value of the *same* market basket at 1982 prices.

To compute a single price index for a set of commodities, the most common approach, due to German economist Étienne Laspeyres in 1871, is to compare the total cost of the base-period quantities at base-period prices with the total cost of the same quantities at given-period prices. Substituting these total costs for the values in Expression (1.21) yields the **Laspeyres price index** for the given period:

$$\text{Laspeyres price index} = \frac{\sum (\text{given-year prices} \times \text{base-year quantities})}{\sum (\text{base-year prices} \times \text{base-year quantities})} \times 100 \qquad (1.22)$$

where the summations are taken over the entire set of commodities. The Laspeyres price index expresses the total value of the base-period market basket at given-period prices as a percentage of its value at base-period prices.

EXAMPLE 1.21 Consider again the data in Example 1.20. Taking 1978 as the base year, we compute the Laspeyres price index for these three commodities in 1982 via Expression (1.22) using the 1978 quantities:

$$\text{1982 Laspeyres price index} = \frac{(100)[(.0400)(7660) + (21.10)(6.5) + (7.25)(12)]}{(.0315)(7660) + (12.40)(6.5) + (6.75)(12)}$$

$$= \frac{(100)(\$530.55)}{\$402.89} = 131.7$$

(Notice that the price of electricity was first converted from cents to dollars so that the total costs would all be in the same units, namely, dollars.) The 1982 Laspeyres price index of 131.7 (1978 = 100) indicates that the same basket of goods and services that cost \$100 in 1978 would have cost \$131.70 in 1982.

The Consumer Price Indexes in both the United States and Canada are modified Laspeyres indexes (1967 = 100 in the United States and 1981 = 100 in Canada). In each case, the index includes literally hundreds of commodities representing typical consumer expenditures in various categories, such as food, clothing, housing, and transportation. It is computed both nationally and locally for specific urban areas in every region of the country.

Adjustment by an Index Number

One common and important use of index numbers is in the adjustment of values recorded in one time period to make them comparable with values recorded in another time period. This procedure is known as **adjustment by an index number**. In general, the adjusted value is equal to the current value divided by the appropriate index number converted from percentage to decimal form (that is, the index number over 100):

$$\text{Adjusted value} = \frac{\text{current value}}{\text{appropriate index number}/100} \qquad (1.23)$$

When the values being adjusted are incomes, the adjusted values are sometimes called **real incomes**, or **purchasing power** in constant dollars.

EXAMPLE 1.22 Consider an individual who earned \$22,370 in 1978 and \$30,140 in 1982. If an appropriate index of consumer prices for this individual was 128.6 in 1978 and 175.0 in 1982 (1975 = 100), what change in purchasing power has she experienced from 1978 to 1982?

The individual's purchasing power in 1975 dollars may be determined via Expression (1.23):

$$\text{1978 real income} = \frac{\$22{,}370}{128.6/100} = \frac{\$22{,}370}{1.286} = \$17{,}395 \qquad \text{in 1975 dollars}$$

$$\text{1982 real income} = \frac{\$30{,}140}{175.0/100} = \frac{\$30{,}140}{1.750} = \$17{,}223 \qquad \text{in 1975 dollars}$$

Thus, her purchasing power was decreased by

$$\$17{,}395 - \$17{,}223 = \$172 \qquad \text{in 1975 dollars}$$

[or (\$172)(175.0/100) = \$301 in 1982 dollars]. This is a loss of \$172/\$17,395 = 1.0% in purchasing power from 1978 to 1982.

Limitations of Price Indexes

For many aspects of business and economic conditions, index numbers are convenient summaries of changes over time. However, to avoid incorrect applications of index numbers, it is important to remember how they are obtained and what changes they do and do not reflect.

Let us list some of the **limitations of price indexes**, with specific reference to the Consumer Price Index. Analogous considerations apply to other index numbers as well.

1. *Choice of items.* The items included in the Consumer Price Index constitute a relatively small sample from the population of possible items that might be included. Not only are there many different products from which to choose, but many of them come in a variety of brands and sizes. The contents of the Consumer Price Index basket of goods and services were determined, and are occasionally revised, by careful and extensive surveys of purchasing patterns. The result is the market basket of an "average" consumer. The index applies only to that market basket as a whole and does not necessarily apply to individual items whether in or out of the basket.

2. *Choice of quantities of the items.* How many restaurant meals per month should be included in the Consumer Price Index? How many dozen eggs? How much gasoline? The Consumer Price Index answered these questions, like those about the choice of items, by taking a large sample from the population of consuming families and determining average quantities purchased of the various items. The index is thus based on one particular set of quantities of particular items. To the extent that one buys different quantities, the Consumer Price Index will not apply.

3. *Choice of price information.* Prices vary from store to store and even from day to day within the same store for some commodities. The Consumer Price Index attempts to observe appropriate samples of retail outlets and times of purchase, but the appropriateness of the index depends on how closely one's purchasing patterns approximate the sampled pattern of times and places of purchase.

4. *Changes in purchasing patterns.* To the extent that consumers change their purchasing patterns in response to price changes, the Consumer Price Index will not apply to them. When the price of one commodity goes up, consumers may substitute a less expensive item or defer their purchase until the price comes back down. Similarly, consumers may adjust the quantities purchased as prices go up or down, and they may decide when and where to purchase on the basis of announced sale prices. Such changes in purchasing patterns are not reflected in the Consumer Price Index.

5. *Changes in quality.* The Consumer Price Index reflects changes in the price of a fixed basket of goods and services, but there is no very satisfactory way of accounting for changes in the quality of the items in the basket over the years. Perhaps some observed increases in prices are due to product improvement through technological advances. In other cases, observed price increases may underestimate the inflationary effect because of inferior quality associated with cost-cutting measures in production. Either way, the change in quality is not reflected in the Consumer Price Index.

6. *Inapplicability to individuals.* The Consumer Price Index reflects the price changes experienced by some hypothetical family with "average" buying patterns. This hypothetical family is partly renting an apartment and partly paying off a mortgage on its own home; it heats its home partly with natural gas, partly with oil, and partly with electric heat; each month, it travels certain fixed proportions of the time by air, train, bus, private car, and taxi. In short, the Consumer Price Index does not necessarily reflect the price changes experienced by any particular individual, family, or group.

Price changes and inflation rates are not unitary concepts that can be definitively measured by one all-purpose number.

In spite of their limitations, index numbers, when properly understood, provide useful summaries of large and otherwise unwieldy masses of information about changing economic conditions. The example of the Consumer Price Index illustrates the way in which the correct interpretation and use of statistical information depends critically on knowledge of the source of the data. Statistics is concerned with both how to determine what questions can be answered by existing data and how to collect appropriate data to answer additional questions of interest.

STUDY MATERIAL FOR CHAPTER 1

Terms and Concepts

Section 1.1

statistics
empirical data
collection of empirical data
interpretation of empirical data
retrospective report
verbal report
direct measurement
applied statistics
theoretical statistics
mathematical statistics

Section 1.2

population
sample

Section 1.3

descriptive statistics
parameter
statistic
inferential statistics
estimate

Section 1.4

observation
raw data

Section 1.6

variable
value of a variable
observational unit
subject
univariate observation
bivariate observation
multivariate observation
constant
summation notation

limits of summation
lower limit of summation
upper limit of summation
$\sum x_i + k$
$\sum (x_i + k)$
$\sum xy$
$(\sum x)(\sum y)$
$\sum x^2$
$(\sum x)^2$

* Section 1.7
index number
base period
given period
price relative
Laspeyres price index
adjustment by an index number
real income
purchasing power
limitations of price indexes

Key Formulas

Section 1.6

(1.10) $$\sum_{i=1}^{n} x_i = x_1 + x_2 + \cdots + x_n$$

(1.13) $$\sum (x + y) = \sum x + \sum y$$

(1.14) $$\sum (x - y) = \sum x - \sum y$$

(1.15) $$\sum kx = k \sum x$$

(1.16) $$\sum^{n} k = nk$$

* Section 1.7

(1.21) $$\text{Index} = \frac{\text{given-period value}}{\text{base-period value}} \times 100$$

(1.22) $$\text{Laspeyres price index} = \frac{\sum (\text{given-year prices} \times \text{base-year quantities})}{\sum (\text{base-year prices} \times \text{base-year quantities})} \times 100$$

(1.23) $$\text{Adjusted value} = \frac{\text{current value}}{\text{appropriate index number}/100}$$

Exercises for Numerical Practice

Section 1.6

1/1. The following are values of three variables:

$x_1 = 7$	$y_1 = 2$	$z_1 = 4$
$x_2 = 2$	$y_2 = 8$	$z_2 = 1$
$x_3 = 4$	$y_3 = 0$	$z_3 = 0$
$x_4 = 3$	$y_4 = 4$	$z_4 = 5$
$x_5 = 8$	$y_5 = 1$	$z_5 = 3$

Calculate each of the following:

a. $\sum_{i=1}^{5} x_i$ **b.** $\sum_{i=2}^{4} y_i$ **c.** $\sum z$

d. $\sum (x_i + y_i)$ **e.** $\sum x_i + \sum y_i$ **f.** $\sum 2z_i$

g. $\sum x_i - 1$ **h.** $\sum (x_i - 1)$ **i.** $\sum x_i z_i$

j. $(\sum x)(\sum z)$ **k.** $\sum y^2$ **l.** $(\sum y)^2$

1/2. Given the values of the variables in Exercise 1/1, calculate each of the following:

a. $\sum_{i=1}^{3} x_i$ **b.** $\sum y$ **c.** $\sum_{i=4}^{5} z_i$

d. $\sum (y_i - z_i)$ **e.** $\sum y_i - \sum z_i$ **f.** $\sum 5y_i$

g. $\sum z_i + 10$ **h.** $\sum (z_i + 10)$ **i.** $\sum y_i z_i$

j. $(\sum y)(\sum z)$ **k.** $\sum z^2$ **l.** $(\sum z)^2$

1/3. The following are values of three variables:

$x_1 = 4$	$y_1 = 7$	$z_1 = 4$
$x_2 = 0$	$y_2 = 4$	$z_2 = 9$
$x_3 = 2$	$y_3 = 0$	$z_3 = 2$
$x_4 = 3$	$y_4 = 1$	$z_4 = 7$
$x_5 = 2$	$y_5 = 0$	$z_5 = 5$
$x_6 = 4$	$y_6 = 8$	$z_6 = 0$

Using these values, calculate each of the items (a) through (l) from Exercise 1/2.

1/4. Given the values of the variables in Exercise 1/3, calculate each of the items (a) through (l) from Exercise 1/1.

1/5. The following are values of three variables.

x	y	z
9	8	3
7	0	5
4	0	3
5	9	7
3	8	5

Using these values, calculate each of the items (a) through (l) from Exercise 1/1.

1/6. Given the values of the variables in Exercise 1/5, calculate each of the items (a) through (l) from Exercise 1/2.

* Section 1.7

1/7. The following table shows one month's expenditures on certain food products by a consumer in each of two years.

	Price		Quantity	
	1979	**1983**	**1979**	**1983**
Milk	$ .60/liter	$ .70/liter	24 liters	29 liters
Cheese	$4.50/kg	$6.00/kg	2 kg	1.5 kg
Eggs	$1.10/dozen	$1.20/dozen	3 dozen	2 dozen

Calculate the Laspeyres price index for 1983 with 1979 as the base year.

1/8. The following table shows the annual expenditures by a small business for four commodities in each of two years.

	Price per unit		Number of units purchased	
Commodity	**1975**	**1985**	**1975**	**1985**
A	$110	$150	400	500
B	$ 8	$ 10	1000	1300
C	$230	$190	20	60
D	$ 40	$ 80	200	120

Calculate the Laspeyres price index for 1985 with 1975 as the base year.

1/9. A business executive received an annual salary of $35,000 in 1980 and $36,000 in 1981. If an appropriate index of consumer prices for this individual was 246.8 in 1980 and 272.4 in 1981, with 1967 = 100:

a. What was the percentage change from 1980 to 1981 in the purchasing power of the executive's annual salary?

b. What was the executive's 1981 salary in 1980 dollars?

1/10. A worker received $10 per hour in 1979 and $13 per hour in 1983. If an appropriate index of consumer prices for this individual was 152.1 in 1979 and

188.7 in 1983 (1973 = 100):

a. What was the percentage change from 1979 to 1983 in the purchasing power of the worker's hourly wage?

b. What was the worker's 1983 hourly wage in 1979 dollars?

Problems for Applying the Concepts

1/11. In Example 1.3:

a. What variable(s) did the researcher observe?

b. What were the observational units?

c. Identify the sample and the population.

d. Identify (in words) an example of a statistic and an example of a parameter.

1/12. In Example 1.4:

a. What variable(s) did the researcher observe?

b. What were the observational units?

c. Identify the sample and the population.

d. Identify (in words) an example of a statistic and an example of a parameter.

1/13. In Example 1.2:

a. What variable(s) was Birmingham asked to observe?

b. What observational units was the education office using?

c. Which of the following was Birmingham asked to record on the questionnaire: retrospective report, verbal report, direct measurement, population, sample, parameter, statistic, estimate, observation, raw data, variable, value of a variable, observational unit, subject?

1/14. In Example 1.5:

a. What variable(s) did the researcher observe?

b. What were the observational units?

c. Which of the following are illustrated by the numbers in the table of Example 1.5: retrospective report, verbal report, direct measurement, parameter, statistic, estimate, observation, raw data?

d. In what ways does the sample not represent well the population for each of the questions labeled (d), (e), and (f) in Example 1.5?

1/15. The prices paid (in dollars) for three items at a fast-food outlet are as follows: $x_1 = 1.30$ for a hamburger, $x_2 = .75$ for an order of French fries, and $x_3 = .65$ for a soft drink.

a. What variable is represented by x?

b. What are the observational units?

c. State in words what $\sum x$ means, and calculate its value.

d. State in words what $6 \sum x$ means, and calculate its value.

1/16. A home gardener froze fresh peas in three batches weighing (in grams) $y_1 = 500$, $y_2 = 560$, and $y_3 = 600$.

a. What variable is represented by y?

b. What are the observational units?

c. State in words what $\sum y$ means, and calculate its value.

d. State in words what $.001 \sum y$ means, and calculate its value.

1/17. The dimensions (in centimeters) of three rectangular rugs are as follows.

	Length (x)	Width (w)
Rug 1	90	60
Rug 2	80	45
Rug 3	120	75

State in words what $\sum wx$ means, and calculate its value.

1/18. The following are the prices and quantities sold of three food items during a fixed time period at a fast-food outlet.

	Price (p)	Quantity (q)
Hamburger	$1.30	78
Order of French fries	.75	62
Soft drink	.65	103

State in words what $\sum pq$ means, and calculate its value.

* **1/19.** The following table gives the average weekly earnings for production and nonsupervisory workers in selected nonagricultural industries in the United States from 1963 to 1983, along with the Consumer Price Index (1967 = 100).

Year	Weekly earnings	Consumer Price Index
1963	$ 88.46	91.7
1967	101.84	100.0
1971	127.31	121.3
1975	163.53	161.2
1979	219.91	217.4
1983	280.35	298.4

a. Adjust the weekly earnings for each year by the Consumer Price Index to obtain the real earnings in constant 1967 dollars.

b. What can you conclude about average weekly earnings from 1963 to 1983?

(Source: U.S. President, *Economic Report of the President,* February 1984, pp. 265, 282.)

* **1/20.** Average family income is shown in the following table for U.S. families classified by race and Spanish origin.

Year	White	Black	Spanish origin
1975	$14,268	$ 8,779	$ 9,551
1983	25,757	14,506	16,956

a. Adjust the average family income for each year by the Consumer Price Index given in Problem 1/19 to obtain the real family income in constant 1967 dollars.

b. Calculate the percentage change in real income for each of the three groups.

(Source: U.S. Department of Commerce, Bureau of the Census, *Statistical Abstract of the United States 1985,* p. 446.)

References

Robbins, L. C. "The Accuracy of Parental Recall of Aspects of Child Development and of Child Rearing Practices." *Journal of Abnormal and Social Psychology* 66 (1963): 261–70.

Ryan, P. "Get Rid of the People, and the System Runs Fine." *Smithsonian,* September 1977, p. 140.

Tanur, J. M.; Mosteller, F.; Kruskal, W. H.; Link, R. F.; Pieters, R. S.; Rising, G. R.; and Lehmann, E. L., eds. *Statistics: A Guide to the Unknown.* 2nd ed. San Francisco: Holden-Day, 1978.

Further Reading

Committee of Presidents of Statistical Societies. *Careers in Statistics.* 3rd ed. Washington, DC: American Statistical Association, 1980. Single copies of this booklet are available free upon request from the American Statistical Association, 806 Fifteenth Street, N.W., Washington, DC 20005.

Hooke, R. *How to Tell the Liars from the Statisticians.* New York: Marcel Dekker, 1983. In a series of light, nonmathematical essays of 1 to 3 pages each, Hooke explains a wide variety of statistical concepts with the aid of illustrations from everyday life. This book is easy reading and a good way to learn something about the field of statistics. As the title might suggest, the book is especially recommended for those who experience a negative emotional response whenever they hear words such as *statistics* or *statistician*.

Huff, D. *How to Lie with Statistics.* New York: Norton, 1954. Chapter 1 of this entertaining exposé of statistical misdemeanors deals with the importance of knowing how the data were collected.

Tanur, J. M., *et al.,* eds. *Statistics: A Guide to the Unknown.* 2nd ed. San Francisco: Holden-Day, 1978. As noted in Example 1.1, this book contains 46 interesting essays describing applications of statistics in a wide variety of areas. The essays are brief and highly readable, assuming no expertise in statistics. Most of the essays are also available in the following three books, each of which consists of a subset of the 46 essays:

Tanur, J. M., *et al.,* eds. *Statistics: A Guide to Business and Economics.* San Francisco: Holden-Day, 1976.

Tanur, J. M., *et al.*, eds. *Statistics: A Guide to Political and Social Issues*. San Francisco: Holden-Day, 1977.

Tanur, J. M., *et al.*, eds. *Statistics: A Guide to the Biological and Health Sciences*. San Francisco: Holden-Day, 1977.

PART I

DESCRIPTIVE STATISTICS

e have seen that statistics is the science of answering empirical questions and that it deals with methods for the proper collection and interpretation of empirical data. The field as a whole can be subdivided into descriptive statistics and inferential statistics, depending on whether one's purpose is to describe the observational units that were actually observed or, on the other hand, to generalize by drawing inferences about observational units that were not observed.

The object of descriptive statistics, which we will be discussing in the next four chapters, is to characterize or summarize sets of data in some more useful and manageable form than a simple list of every observation. The way in which we organize or summarize the data depends on what questions we are asking of the data.

EXAMPLE 2.0 The following list shows the years in which the 61 monarchs of England and Great Britain assumed the throne, along with their lengths of reign to the nearest year.

829	Egbert	10	1066	William I	21	1547	Edward VI	6
839	Ethelwulf	19	1087	William II	13	1553	Mary I	5
858	Ethelbald	2	1100	Henry I	35	1558	Elizabeth I	44
860	Ethelbert	6	1135	Stephen	19	1603	James I	22
866	Ethelred I	5	1154	Henry II	35	1625	Charles I	24
871	Alfred	28	1189	Richard I	10	1660	Charles II	25
899	Edward	25	1199	John	17	1685	James II	3
924	Athelstan	16	1216	Henry III	56	1689	William III	13
940	Edmund I	6	1272	Edward I	35	1689	Mary II	6
946	Edred	9	1307	Edward II	20	1702	Anne	12
955	Edwy	3	1327	Edward III	50	1714	George I	13
959	Edgar	17	1377	Richard II	22	1727	George II	33
975	Edward	4	1399	Henry IV	13	1760	George III	59
978	Ethelred II	37	1413	Henry V	9	1820	George IV	10
1016	Edmund II	0	1422	Henry VI	39	1830	William IV	7
1016	Canute	19	1461	Edward IV	22	1837	Victoria	63
1035	Harold I	5	1483	Edward V	0	1901	Edward VII	9
1040	Hardecanute	2	1483	Richard III	2	1910	George V	25
1042	Edward	24	1485	Henry VII	24	1936	Edward VIII	1
1066	Harold II	0	1509	Henry VIII	38	1936	George VI	15
						1952	Elizabeth II	?

If we want to know who followed whom on the English throne, or who reigned at a particular time, the preceding list is a good description of the English monarchy. However, other questions call for other descriptions, and in general the order in which the observations were made is not the most informative organization of empirical data.

For example, questions about how long English monarchs hold office are not so readily answered from the chronological list. What is the typical length of reign? Are most reigns close to the typical length, or are there many exceptions?

Is 20 years unusually long, or 10 years unusually short? Have there been many reigns longer than that of the current queen, Elizabeth II? We can improve upon the chronological list to answer questions about length of reign.

Similarly, the chronological list does not make it easy to answer questions about the frequency of occurrence of the various first names of English monarchs. How many different names have been used? What is the most frequently occurring name? The answers to such questions are contained in the chronological list, but they are largely hidden. Other ways of presenting the information will make the answers more apparent.

(Source: *World Almanac 1983*, pp. 402–403.)

There is no single correct way to present or summarize a set of data. Different descriptions of the data are appropriate for answering different questions, and often more than one description is desirable. The methods of descriptive statistics help us to extract from a set of data the information relevant to particular questions and to present that information in ways that make the data easier to understand and less subject to misunderstanding.

CHAPTER 2

FREQUENCY DISTRIBUTIONS

EXAMPLE 2.1 Each year thousands of North Americans are admitted to mental institutions for the first time in their lives. What disorders are most common? The following table shows the diagnoses of all individuals who were first admitted to a psychiatric inpatient facility in Canada in 1978.

Number of individuals admitted in 1978 to a Canadian psychiatric inpatient facility for the first time in their lives. Classification is by the diagnosed disorder.

Diagnosis	Number	Percentage
Nonpsychotic mental disorders		
Depressive neurosis	14,014	23%
Alcoholism and drug dependence	10,852	18%
Other nonpsychotic mental disorders	13,714	22%
Psychoses		
Affective psychoses	6,152	10%
Schizophrenia	6,132	10%
Other psychoses	6,704	11%
Mental retardation	1,418	2%
Other	2,075	3%
Total	61,061	100%

We can readily see that depressive neurosis is the most frequent single diagnosis upon first admission and that alcoholism and drug dependence are also frequent. By comparison, there are relatively few first admissions for mental retardation.

(Source: Statistics Canada, 1981, pp. 42–45.)

Numbers are an inescapable part of our lives. When a child is born, its proud parents announce the relevant numbers: 7 pounds, 15 ounces, or 3600 grams. As children grow, one of the first questions they learn to answer is "How old are you?" and their oft-rehearsed answer is a number. Similarly, we keep track of our progress in school with numbers (grade 1, grade 2, and so on), we count our marbles and later our dollars, and we keep track of our height and especially our weight. Numbers help us to keep track of many of the things that we feel it is important to keep track of in our everyday lives.

Governments, businesses, and sciences likewise have things to keep track of that are important in their everyday corporate lives, and numbers are again helpful. When decisions must be made based on these things that have been recorded, the problem often is not one of too few numbers but rather of how to expand our understanding as rapidly as we expand our data base. When we have hundreds or even thousands of observations, how can we organize them so we can understand the information they contain that is relevant for our present concerns? There are some simple ways.

2.1 COUNTING OBSERVATIONS IN CATEGORIES

One way of summarizing and presenting even very large collections of observations, without losing too much of the original detail, is in the form of a frequency distribution, such as in the table of Example 2.1.

> A **frequency distribution of a variable** is a description of the frequencies with which various mutually exclusive and exhaustive categories of observation occur. **(2.1)**

The words **mutually exclusive** mean that no observation can be in more than one category; being in one category "excludes" the possibility of being in any other category. The word **exhaustive** means that the categories include all the observations; no observation is left out. The categories in a frequency distribution have often been called **classes**, but because that word has so many other meanings, we will continue to use the less ambiguous term **categories**. Thus, a frequency distribution shows how many observations fall in each category.

Quantitative Frequency Distributions

You have undoubtedly seen many frequency distributions portrayed in one form or another, although the term itself may be new to you.

EXAMPLE 2.2 The following table portrays a frequency distribution of the length-of-reign variable for the English monarchs listed in Example 2.0. (The present queen has been omitted.) The left-hand column is headed by the name of the variable and consists of the mutually exclusive and exhaustive values of the variable—mutually exclusive and exhaustive so that each monarch is counted in exactly one category. The right-hand column gives the frequency or count for each category.

How long do English monarchs hold office? The tabulation of the frequency distribution makes it easier to see that about half have reigned less than 15 years, although 15 of the 60 have reigned at least 25 years and one (Victoria) for 60–64 years. Elizabeth II, queen since 1952, has so far ruled longer than about 80% of the previous monarchs. These are some of the features of the data that were not apparent in the raw data as given in Example 2.0.

Monarchs of England and Great Britain from 829–1952, classified according to their length of reign.

Length of reign in years	Number of monarchs
0– 4	10
5– 9	11
10–14	8
15–19	7
20–24	8
25–29	4
30–34	1
35–39	6
40–44	1
45–49	0
50–54	1
55–59	2
60–64	1
Total	60

If the categories of a frequency distribution are defined by numerical values of a variable (as in Example 2.2, in which the categories are 0–4, 5–9, and so on), then it is called a **quantitative frequency distribution** or **numerical frequency distribution.** The variable in such a distribution is referred to as a **quantitative variable** or **numerical variable.**

The categories in a quantitative frequency distribution are labeled by their **category limits.** For example, the first category in the table of Example 2.2 has 0 as its **lower category limit** and 4 as its **upper category limit**; the second category has 5 and 9, respectively; and so on. The term **category boundary** refers to the hypothetical dividing point halfway between adjacent categories. Specification of the category boundaries normally requires the use of one more digit than for the category limits. The category boundaries in Example 2.2 are −.5, 4.5, 9.5, 14.5, 19.5, . . . , 59.5, and 64.5.

The difference between successive category boundaries is called the **category interval** of the corresponding category. If the category intervals are equal for all categories, the common category interval is called the **category interval of the distribution**. In Example 2.2, every category has a category interval of 5 (for example, $9.5 - 4.5 = 5$), and we say that the category interval of the distribution is 5.

If we need a single number to represent a category, the category's midpoint, which is called the **category mark** of that category, is often used. It may be calculated in any of several ways, but perhaps the simplest is to add the lower and upper limits of the category and divide by 2. Performing this operation $\left(\text{for example, } \frac{0+4}{2} = 2,\ \frac{5+9}{2} = 7, \ldots\right)$ for each category in Example 2.2, we obtain 2, 7, 12, 17, . . . , 57, and 62 as the category marks.

Qualitative Frequency Distributions

The alternative to a quantitative frequency distribution is a **nominal frequency distribution** or **qualitative frequency distribution**. In a qualitative frequency distribution, the categories are defined by verbal labels or words, rather than by numerical values. The variable in this case is called a **nominal variable** or **qualitative variable**.

EXAMPLE 2.3 The following table portrays the frequency distribution of English monarchs' names, compiled from the data of Example 2.0. The left-hand column is headed by the name of the variable and consists of mutually exclusive and exhaustive categories. Again, the "mutually exclusive and exhaustive" property of the categories is required to ensure that each observation is counted in one and only one category. Since the categories are defined verbally (Edward, Henry, and so on), rather than numerically, this is a qualitative frequency distribution. The right-hand column gives the frequency or count for each category.

Frequency distribution of first names of English monarchs from 829 to the present.

name of variable

Monarch's name	Number of monarchs
Edward	11
Henry	8
George	6
William	4
Richard	3
Charles	2
James	2
Elizabeth	2
Mary	2
Harold	2
Edmund	2
Ethelred	2
Names occurring once	15
Total	61

→ FREQ

The tabulation of the frequency distribution makes it easier to see certain features in the data. For example, the most frequently occurring name is Edward, and a total of 27 different names have been used, of which 15 have occurred only once.

In this chapter, we will consider a variety of ways of tabulating, presenting, and interpreting frequency distributions. We begin with some suggestions about counting.

* 2.2 TALLYING

It is often useful to have a quick way of counting items in several categories simultaneously. Such a procedure allows us to go through a list, such as the

* This symbol identifies material that is not prerequisite for later topics in this book.

lengths of reign in Example 2.0, just one time and, at the end, to have a count of the number of observations in each category. The procedure for such **tallying** involves making a mark, called a **tally mark**, next to the appropriate category label for each observation.

Table 2.1 illustrates the tally marks corresponding to counts from 1 through 20 for three different methods of tallying. The tallying method most widely used in North America consists of four vertical lines crossed by one horizontal or diagonal line. Unfortunately, this may also be the most error-prone of the three methods in Table 2.1, because of the danger of mistaking ||| or ||||| for ||||, especially as one's vigilance diminishes during work on a large set of data. Once ||| or ||||| is erroneously crossed to become ~~|||~~ or ~~|||||~~, it is difficult to detect the error. With practice, the other two methods in Table 2.1 seem less liable to such errors.

The alternative method of tallying by fives involves a vertical mark and a

TABLE 2.1 Three methods of tallying.

Count	Common method of tallying by fives	Alternative method of tallying by fives	Tallying by tens
1	|	|	.
2	||	+	. .
3	|||	⁎	⁖
4	||||	∗	∷
5	𝍸	⊛	| ∷
6	𝍸 |	⊛ |	└˙
7	𝍸 ||	⊛ +	⊔
8	𝍸 |||	⊛ ⁎	□
9	𝍸 ||||	⊛ ∗	⧄
10	𝍸 𝍸	⊛ ⊛	⊠
11	𝍸 𝍸 |	⊛ ⊛ |	⊠ .
12	𝍸 𝍸 ||	⊛ ⊛ +	⊠ . .
13	𝍸 𝍸 |||	⊛ ⊛ ⁎	⊠ ⁖
14	𝍸 𝍸 ||||	⊛ ⊛ ∗	⊠ ∷
15	𝍸 𝍸 𝍸	⊛ ⊛ ⊛	⊠ | ∷
16	𝍸 𝍸 𝍸 |	⊛ ⊛ ⊛ |	⊠ └˙
17	𝍸 𝍸 𝍸 ||	⊛ ⊛ ⊛ +	⊠ ⊔
18	𝍸 𝍸 𝍸 |||	⊛ ⊛ ⊛ ⁎	⊠ □
19	𝍸 𝍸 𝍸 ||||	⊛ ⊛ ⊛ ∗	⊠ ⧄
20	𝍸 𝍸 𝍸 𝍸	⊛ ⊛ ⊛ ⊛	⊠ ⊠

horizontal mark, followed by two diagonal marks to form a star, which is circled to complete the tally of five.

Tallying by tens is especially useful for large sets of data. (This method is described in the book by J. W. Tukey cited in Further Reading at the end of this chapter.) The first four tally marks are distinct dots (or tiny circles, if you are an incurably faint dotter!) marking the corners of a square. The order of marking the corners makes no difference. The next four tally marks are the sides of the square, again in any order, and the final two tally marks are diagonals inside the square. With this method, one is unlikely to incorporate an extra mark in a tally of ten, and failure to complete a set of ten, such as by forgetting the diagonals, is generally rather easy to spot. It does not take long to learn the sequence—four dots, four lines, two diagonals—sufficiently well to read and write the tallies fluently.

Tallying is a preliminary or scratch-paper procedure; it does not appear in the final report of a statistical analysis, to be read by others. (For example, tallying was used to obtain the nominal frequency distribution reported in Example 2.3.) You should, therefore, use whatever method of tallying works best for you. If you find yourself making errors with one method, it may be that one of the others would suit you better.

* 2.3 STEM-AND-LEAF DISPLAYS

Tallying is especially useful for tabulating nominal frequency distributions, but another descriptive tool is often preferable for numerical frequency distributions. This tool, called the **stem-and-leaf display**, is one of many simple yet highly effective methods devised by John W. Tukey (see Biography 2.1) for exploring data to discover what they can reveal about the observed variable(s). Like tallying, the stem-and-leaf display is unlikely to be found in the final report about a set of data, but with quantitative variables it should usually be among the first things to appear on a scratch pad after the data have been collected.

BIOGRAPHY 2.1 John Wilder Tukey (1915–)

J. W. Tukey was born on 16 June 1915 in New Bedford, Massachusetts. He attended Brown University in Providence, Rhode Island, where he earned a bachelor's degree (Sc.B.) in 1936 and a master's degree (Sc.M.) in 1937. He immediately went on to Princeton University in Princeton, New Jersey, for another master's degree (M.A.) in 1938 and a doctorate (Ph.D.) in mathematics in 1939. Beginning in 1939, he taught at Princeton University, first in the Department of Mathematics and then in the Department of Statistics, rising to the chair of Donner Professor of Science in 1976. Concurrently, beginning in 1945, he was a member of the technical staff at Bell Telephone Laboratories in Murray Hill, New Jersey, reaching the position of Associate Executive Director of Research in 1961. He retired from both positions in 1985.

Tukey's publications have included books on topology (1940), statistical problems in the Kinsey report on human sexual behavior (1954), communications engineering (1958), and exploratory data analysis (1977). In his book *Exploratory Data Analysis,* he likens the exploratory statistician's work to that of a detective looking for clues. He has invented and promoted the use of numerous relatively simple but effective techniques for exploring data. His techniques are known not only for their effectiveness and simplicity of application, but also for their colorful names. Besides stem-and-leaf displays, there are box-and-whisker plots, c'ranks, flogs (folded logarithms), double flogs, froots (folded square roots), fences, hinges, roughs, smooths, and the vacuum cleaner (for cleaning up messy data).

(Sources: *American Men and Women of Science,* 1979; *Who's Who in America,* 1980.)

Simple Stem-and-Leaf

If we were going to use one of the tallying methods to count the number of observations in each of several categories of a quantitative variable, we would first decide on the categories and write their labels (for example, 0–4, 5–9, . . .) one under the other, each on a separate line. Then we would make a tally mark on the appropriate line for each observation. We would end up with a tally in which each line consists of a category label and a set of tally marks.

In a stem-and-leaf display, each line is called a **stem**. In place of the category labels used in tallying, we write what is called the **starting part** of the observation, and in place of the tally marks, we write the **leaves** from the observations. The **starting part** of an observation is its left-most digit or digits, and the **leaf** consists of one or more subsequent digits.

In **simple stem-and-leaf**, a different starting part appears on each line (stem), and a leaf consists of a single digit.

EXAMPLE 2.4 The Test of Standard Written English (TSWE) is published by Educational Testing Service as a measure of basic skills in English composition. The following observations are the raw scores on the TSWE obtained by 54 students at the beginning of their first-year university English course in September 1981:

31.25	16.25	23.00	19.00	38.75
46.25	41.25	33.75	27.50	33.75
32.75	36.25	40.00	27.50	24.50
13.25	40.00	28.75	19.00	34.75
38.75	28.75	31.25	37.50	21.00
17.50	35.00	28.75	36.25	42.50
29.25	28.00	20.00	40.00	36.25
47.50	23.75	26.25	8.25	19.50
27.50	23.75	21.75	40.00	43.75
43.75	42.50	31.25	25.50	28.75
40.00	16.25	23.75	35.00	

Looking over the list, we see that the scores range from about 8 to 47. Let us construct a stem-and-leaf display in which the tens digit (the second one to the left of the decimal point) of each observation is the starting part, and the units digit

(immediately to the left of the decimal) is the leaf. We begin by writing the starting parts in a vertical column, followed by a vertical line to separate the starting parts from the leaves:

0	
1	
2	
3	
4	

Now, for each observation, we record the units digit to the right of the appropriate tens digit. For example, the first observation is 31.25, so we record a 1 to the right of the starting part 3: 3 | 1. The second observation is 46.25, so we attach a leaf of 6 to the stem labeled 4: 4 | 6. The third observation is 32.75, so we add a second leaf to the 3-stem, to the right of the previous leaf: 3 | 12. Notice that we do not bother to round 32.75 to 33: A leaf of 2 means at least 2 but less than 3. The fourth observation is 13.25, giving us 1 | 3, and the fifth observation, 38.75, puts a third leaf on the 3-stem, so we have 3 | 128. Continue through all the observations in this manner, recording a leaf for each observation. The result is a **simple stem-and-leaf display** of the TSWE scores with a category interval of 10:

0	8
1	3766999
2	9788333880613775418
3	128653117658346
4	673010200023

It is always a good idea to check one's work in statistics, although there is no great virtue in excessive compulsiveness. One simple check in a stem-and-leaf display is to count the number of leaves on each stem, record that number (called the **check count**) in parentheses to the right of the stem, and verify that the check counts add up to the total number of observations:

Stem	Leaves	Check count
0	8	(1)
1	3766999	(7)
2	9788333880613775418	(19)
3	128653117658346	(15)
4	673010200023	(12)
		(54)

The sum 54 equals the total number of observations, so it seems we have not missed any nor recorded any twice. If we wanted to do further checking, we could go back through the original observations again, putting a dot over each leaf as we verify our work.

(Source: Unpublished TSWE data courtesy of H. Prest.)

The stem-and-leaf is a form of tabulation of a frequency distribution. In Example 2.4, we categorized 54 observations into five categories with a category interval of 10. Observations less than 10 were recorded on the first line, those at least 10 but less than 20 on the second line, and so on. The check counts gave us the frequency in each category of observation. The beauty of a stem-and-leaf display is that even though very little of the original information is lost, it is much easier to extract information. Not only are all of the individual values of the variable still available, but one can readily determine such values as the highest, the lowest, the most common sorts of values, and the number of observations that are above or below any given value.

The simple stem-and-leaf is not limited to situations where the starting part is a single digit. If a set of observations ranged from, say, 983 to 1022, the starting parts could be 98|, 99|, 100|, 101|, and 102|. Furthermore, the leaf need not be the units digit. We could use the starting parts given in the sentence before last for observations ranging from 9837 to 10226 by dropping the units digit and using the tens digit for the leaf. Similarly, observations between 9.83 and 10.22 might use starting parts 9.8|, 9.9|, 10.0|, 10.1|, and 10.2|.

EXAMPLE 2.5 For some purposes, it may be desirable to retain the decimal places in the TSWE scores from Example 2.4 and to use a finer grouping (narrower categories) so we do not have so many observations in a single category. By moving the units digit over to the starting part and using the tenths digit for the leaf, we obtain a **simple stem-and-leaf display** with a category interval of 1 for the TSWE scores:

8	2	(1)
9		
10		
11		
12		
13	2	(1)
14		
15		
16	22	(2)
17	5	(1)
18		
19	005	(3)
20	0	(1)
21	70	(2)
22		
23	7707	(4)
24	5	(1)
25	5	(1)
26	2	(1)

27	555	(3)
28	70777	(5)
29	2	(1)
30		
31	222	(3)
32	7	(1)
33	77	(2)
34	7	(1)
35	00	(2)
36	222	(3)
37	5	(1)
38	77	(2)
39		
40	00000	(5)
41	2	(1)
42	55	(2)
43	77	(2)
44		
45		
46	2	(1)
47	5	(1)
		(54)

Again, we did not bother rounding .75 to .8. Extra digits are normally truncated rather than rounded in a stem-and-leaf display (but not elsewhere).

Stretched Stem-and-Leaf

The simple stem-and-leaf is limited to situations where the categories are defined by the right-most digit of the starting part. There is just one line for each starting part, so the categories are hundreds, or tens, or ones, or tenths, and so on. Thus, the simple stem-and-leaf is limited to category intervals such as 1, 10, 100, or .1, .01, and the like, as in Examples 2.4 and 2.5.

The **stretched stem-and-leaf** uses two stems for each starting part, with leaves 0 through 4 recorded on the starting part's upper stem and leaves 5 through 9 on the lower stem. The stretched stem-and-leaf therefore produces category intervals such as 5, 50, 500, or .5, .05, and the like.

EXAMPLE 2.6 In Example 2.4, we categorized the 54 TSWE scores into five categories with a category interval of 10 by using the tens digit of each observation as the starting part. A rather coarse grouping resulted, especially since most of the observations were in the upper three categories. Perhaps we have lost too much detail in this summary. On the other hand, the category interval of 1 that we used in Example 2.5 seems narrower than necessary, and there is perhaps too much detail. The stretched stem-and-leaf provides a compromise with a category interval of 5.

We begin by writing each starting part twice in the column of starting parts. Then we proceed as before, except that we record leaves 0–4 and 5–9 on the upper and lower stems, respectively, of the appropriate starting part. The following is a **stretched stem-and-leaf display** for the 54 TSWE scores of Example 2.4:

Stem	Leaves	Count
0		
0	8	(1)
1	3	(1)
1	766999	(6)
2	33301341	(8)
2	97888867758	(11)
3	1231134	(7)
3	86576586	(8)
4	3010200023	(10)
4	67	(2)
		(54)

It is instructive to compare the stretched stem-and-leaf above with the simple stem-and-leaf of Example 2.4. First, notice why the term *stretched* is used: We have stretched out the observations to take a closer look at them in a display with more detail. Second, notice the general appearance of the data in the two displays. The simple stem-and-leaf of Example 2.4 shows that the most frequently observed category is 20.00–29.75 (this seems to be the most typical performance on the test) and that the frequencies decrease as one moves in either direction from the most frequent category. The stretched stem-and-leaf, on the other hand, shows that there are in fact two main clusters of students: One group, with scores in the neighborhood of 25–29, may perhaps be characterized as "average"; the other group, with scores around 40–44, may include those who have pretty well mastered the basic composition skills tested by the TSWE and so score near the maximum possible mark of 50.

Squeezed Stem-and-Leaf

The simple stem-and-leaf produces category intervals such as .1, 1, 10, and 100 by using one stem for each starting part. The stretched stem-and-leaf produces category intervals such as .5, 5, 50, and 500 by using two stems for each starting part.

The **squeezed stem-and-leaf** uses five stems for each starting part and produces category intervals such as .2, 2, 20, and 200. The top stem for each starting part is labeled with a lower-case o as a reminder that leaves of 0 and 1 are recorded on that stem (o stands for zero and one; use an asterisk * if your o looks like a 0). The second stem is labeled with a lower-case t as a reminder that leaves of 2 and 3 belong there (t for two and three). The third stem is labeled f and carries leaves of 4 and 5. The fourth stem, labeled s, is for leaves of 6 and 7. The bottom stem for each starting part is labeled e and carries leaves of 8 and 9. J. W. Tukey, the inventor of stem-and-leaf displays, has remarked on how well suited the English digit names are for the squeezed stem-and-leaf.

EXAMPLE 2.7 In Example 2.5, we categorized the 54 TSWE scores into 40 categories with a category interval of 1, using a simple stem-and-leaf display. We do not obtain much of a summary if we use 40 categories for just 54 observations. However, we may want a closer look at the data than that afforded by the 9 categories of the stretched stem-and-leaf display of Example 2.6, with its category interval of 5. The squeezed stem-and-leaf provides another option, with a category interval of 2.

We begin by writing the starting parts, each followed by the letters o, t, f, s, and e, in a column, in order to give us five lines (stems) for each starting part. Then we proceed through the observations of Example 2.4, recording each leaf (the units digit) on the appropriate stem, to construct the **squeezed stem-and-leaf display** of the 54 TSWE scores:

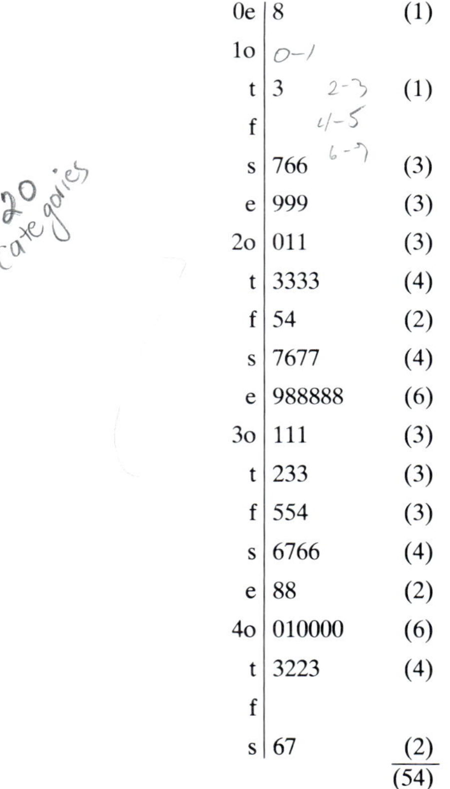

0e	8	(1)
1o		
t	3	(1)
f		
s	766	(3)
e	999	(3)
2o	011	(3)
t	3333	(4)
f	54	(2)
s	7677	(4)
e	988888	(6)
3o	111	(3)
t	233	(3)
f	554	(3)
s	6766	(4)
e	88	(2)
4o	010000	(6)
t	3223	(4)
f		
s	67	(2)
		(54)

This display has 20 categories with a category interval of 2. Whether it is better for these data than the stretched stem-and-leaf of Example 2.6 depends largely on the use to which the data are to be put and on the specific questions that are being asked of the data. The two main clusters of students in the vicinities of 28 and 40 continue to show up in the squeezed stem-and-leaf, although perhaps not quite as clearly as in the stretched stem-and-leaf. On the other hand, the extremely high and extremely low scores stand out somewhat more clearly.

The use of the term *squeezed* is explained by a comparison of the squeezed stem-and-leaf, based on five categories for each tens digit, with the simple stem-and-leaf of Example 2.5, based on one category for each units digit (and hence ten categories for each tens digit). The squeezed stem-and-leaf squeezes two units digits into one stem, rather than using a separate stem for each units digit.

Choosing a Stem-and-Leaf

Do not bother looking for any important conceptual difference between "stretching" and "squeezing" a stem-and-leaf. There is none. Squeezing two units digits into one stem is equivalent to stretching the tens-digit starting part into five stems, and stretching the tens-digit starting part into two stems is equivalent to squeezing five units digits into one stem. The terms *simple, stretched,* and *squeezed* are simply colorful names applied to stem-and-leaf displays that use 1, 2, or 5 stems, respectively, for each starting part.

The stem-and-leaf display is an easy, yet highly versatile and informative, technique for quickly organizing and displaying a collection of numerical data. Other variations of the technique are possible. For example, if it is necessary to keep track of all the digits in the observations, we may use **two-digit leaves** or **three-digit leaves**: The observations 46.25 and 47.50 would appear as 4 | 62, 75 with two-digit leaves or as 4 | 625, 750 with three-digit leaves, successive leaves being separated by commas. After you have had additional experience with the stem-and-leaf, further variations may occur to you in response to the needs of particular data.

There is no mechanical way of determining which is the "best" form of stem-and-leaf to use for a particular set of data. The stem-and-leaf allows you to explore the data in a variety of ways in the privacy of your own scratch pad to see which way provides the most information about the question at hand. One will often try more than one type of stem-and-leaf with a given set of data before deciding on the best summaries to report. The guiding criterion is that the data should be presented in ways that make the relevant information easier to understand and less susceptible to misinterpretation.

In any event, because the stem-and-leaf display is not self-explanatory, it is used in the rough work only and not in the final report or presentation of the data. The following sections deal with methods appropriate in the latter situation.

2.4 PRESENTATION OF FREQUENCY DISTRIBUTIONS

Frequency distributions may be presented in either tabular or graphical form. In general, tabular representations are better for reporting details about a set of data, while graphical representations are better for communicating an overall impression of the data.

Frequency Tables

We have already discussed tabular representations of frequency distributions in connection with Examples 2.2 and 2.3, which you may wish to review at this time. Such representations are called **frequency tables**.

One variation replaces the column of frequencies by a column containing the corresponding percentages of the total number of observations. This **percentage frequency table** is the tabular representation of a **percentage distribution**. The percentage distribution may be easier to grasp, especially if the actual frequencies are large, but it loses information about the actual frequencies of the categories. If the actual frequencies and percentage information are both important, both columns may be included in the same table, as illustrated in the table of Example 2.1.

EXAMPLE 2.8 As a further illustration, the following table shows the frequency and percentage distributions of the 54 TSWE raw scores originally given in Example 2.4. We obtain it by counting the number of scores in each category. The methods of Section 2.2 or, especially, 2.3 can make the counting easier. (The table here is actually just a more self-explanatory form of the stretched stem-and-leaf display of the same data in Example 2.6.)

Frequency and percentage distributions of raw scores on the Test of Standard Written English by 54 students beginning their first-year university English course.

TSWE raw score	Number of students	Percentage of students
5.0– 9.75	1	2%
10.0–14.75	1	2%
15.0–19.75	6	11%
20.0–24.75	8	15%
25.0–29.75	11	20%
30.0–34.75	7	13%
35.0–39.75	8	15%
40.0–44.75	10	19%
45.0–49.75	2	4%
Total	54	100%

Compared to the raw data of Example 2.4 and the stem-and-leaf display of Example 2.6, the frequency table here suffers some loss of information. For instance, from the table we cannot know whether the highest score obtained was 45.0 or 49.75 or exactly what it was, but only that it was in the category 45.0–49.75. Nonetheless, the table provides an easily understood summary that is useful for many purposes.

The percentage column is not particularly helpful in this case, primarily because the actual numbers of students are small enough to be readily understood. However, it would be helpful if we wanted to compare this distribution with that from a different group of students, where the groups were not of equal size. For another example where the percentage information facilitates comprehension, see Example 2.1.

Notice that the total of the percentage column of the above table is recorded as 100%, obtained as 54/54, and not as the sum of the rounded percentages, which would introduce rounding error. Guidelines for how and when to round and how many digits to report are given in Appendix 1.

In general, the left-hand column in a frequency table lists the mutually exclusive and exhaustive categories of the relevant variable or characteristic of the observational units, and the frequency column shows the frequency with which each category was observed. Typically, the variable is represented by the symbol x, and the frequency by f. (The observed variable x in Example 2.8 is "TSWE raw score," and the frequency f is "number of students.") The term *observation* has a very specific meaning in statistics (see Section 1.4). In the present context, the observations are the values of the x-variable (the students' TSWE raw scores, in the example). The frequencies are not the observations; rather, the frequencies indicate the number of observations in each category. The total number of observations is given at the bottom of the table as the sum of the frequencies:

$$\sum_{j=1}^{k} f_j = n \tag{2.2}$$

where k is the number of categories, f_j is the frequency of the jth category, and n is the number of observations (54 in the example).

Histograms with Equal Category Intervals

A commonly used form of graphical representation of a quantitative frequency distribution is the **histogram**. When the category intervals are all equal, a histogram consists of a horizontal axis labeled with x-variable values, a vertical axis labeled *frequency*, and a rectangle of the appropriate height over each category. The base of each rectangle extends from one category boundary to the next. Sometimes the actual frequency is written above each rectangle to avoid any possible difficulty in reading exact values from the graph.

EXAMPLE 2.9 Figure 2.1 shows the histogram corresponding to the frequency table of TSWE raw scores as given in Example 2.8. The histogram shows at a glance the general nature of the distribution. For example, the tendency of the scores to "bunch up" in two regions along the variable shows up clearly.

In spite of its apparent simplicity, the histogram can be misleading if used improperly. The difficulty arises when the category intervals are not all equal, a problem to which we will return in Section 2.5.

Frequency Polygons with Equal Category Intervals

The **frequency polygon** provides an alternative graphical representation of a quantitative frequency distribution. It differs from a histogram in that frequency is indicated by a dot of appropriate height at the midpoint (category mark) of each category, rather than by a rectangle whose base is the interval between successive category boundaries, as in the histogram. Adjacent dots are then joined by straight lines, and a category with a frequency of zero is usually appended at each end to complete the picture.

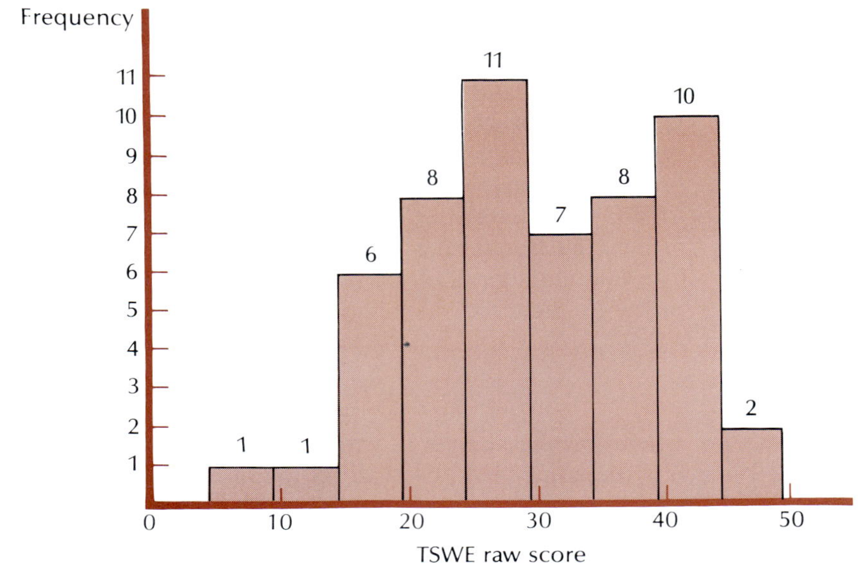

Figure 2.1
Histogram of the distribution of raw scores on the Test of Standard Written English by 54 students beginning their first-year university English course.

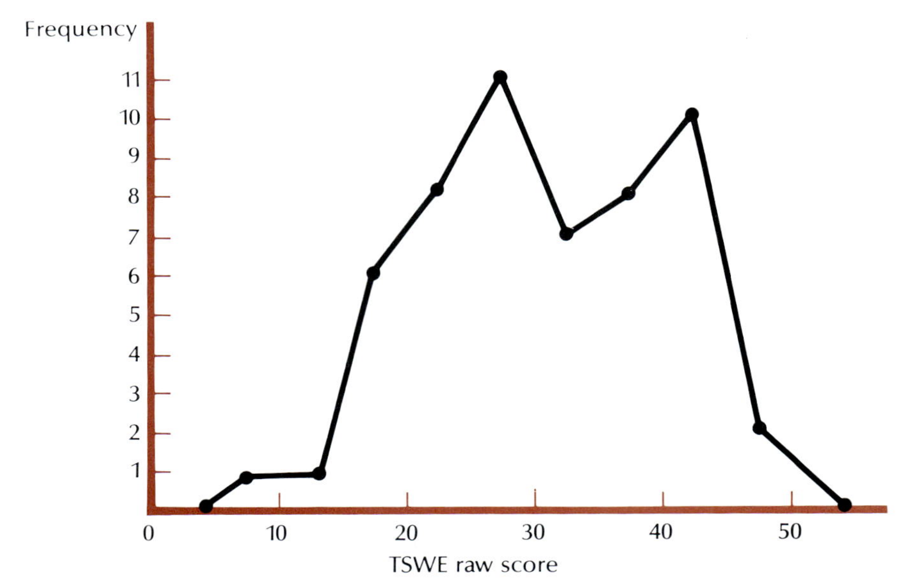

Figure 2.2
Frequency polygon of the distribution of raw scores on the Test of Standard Written English by 54 students beginning their first-year university English course.

EXAMPLE 2.10 Figure 2.2 shows the frequency polygon corresponding both to the histogram in Figure 2.1 and to the frequency table of TSWE scores in Example 2.8. Notice that the frequencies are plotted at the category marks: 2.375, 7.375, . . . , 52.375.

As in the case of the histogram, certain adjustments (to be discussed in Section 2.5) must be made in the frequency polygon if the category intervals are not all equal. The choice between the histogram and the frequency polygon is largely a matter of taste.

Bar Charts

The **bar chart** or **bar graph** is a graphical representation that is appropriate for qualitative frequency distributions. It differs from a histogram in that a slight space is usually left between adjacent rectangles to indicate the lack of continuity from one category of a nominal variable to the next.

EXAMPLE 2.11 Figure 2.3 is a bar chart of the distribution of diagnostic category of first admissions to Canadian mental institutions in 1978. It is derived from the frequency distribution as given in the table of Example 2.1.

Pie Charts

The **pie chart** is an alternative form of graphical representation of a qualitative frequency distribution. The total frequency is represented by a circle with radii drawn to divide the "pie" in the same proportions as the categories divide the total frequency. Each sector is labeled with the category name, accompanied by the corresponding frequency and/or percentage. For optimal readability, the sectors should be portrayed in order from largest to smallest, clockwise from the 12 o'clock position.

Since there are 360° in a complete circle, each category is allocated a sector bounded by radii whose angle at the center of the circle is the appropriate percentage of 360°. Each percentage point of increase in frequency increases the angle by 3.6° (that is, 1% of 360°).

EXAMPLE 2.12 Figure 2.4 is a pie chart of the distribution of diagnostic category of first admissions to Canadian mental institutions in 1978. It corresponds both to the bar chart in Figure 2.3 and to the frequency distribution in the table of Example 2.1. The angles between adjacent radii are determined by the proportions of observations in the various categories; for example, the central angle for the depressive-neurosis sector is $360° \times 14{,}014/61{,}061 = 83°$. Once the angles have been calculated, the radii are drawn with the aid of a protractor.

The choice between the bar chart and the pie chart for a graphical representation of a qualitative frequency distribution is largely a matter of taste. However, a pie chart tends to emphasize the relative frequencies (percentages) somewhat more than does the bar chart.

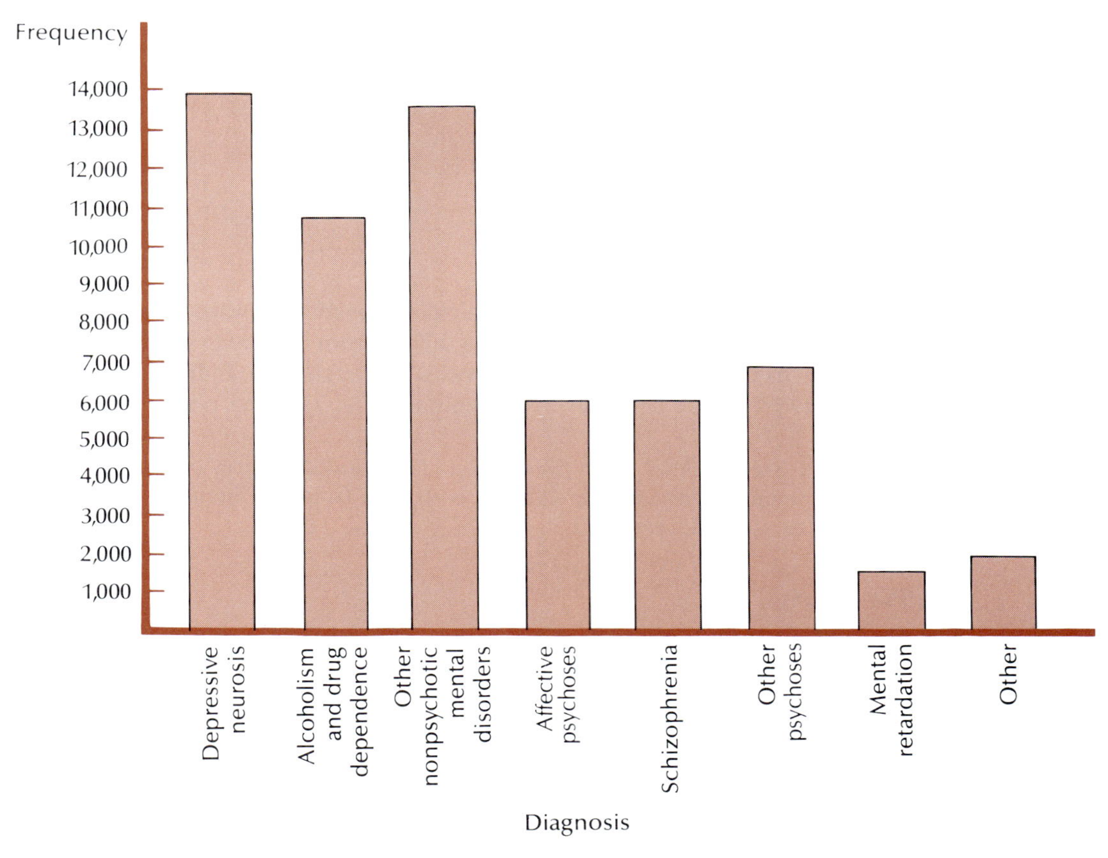

Figure 2.3
Bar chart of the distribution of diagnostic category of individuals admitted in 1978 for the first time in their lives to a Canadian psychiatric inpatient facility.

Guidelines for Presenting Frequency Distributions

There are no absolute rules to dictate how to represent a frequency distribution. Keep in mind that *the object is to make the data easier to comprehend and less liable to misunderstanding.* This principle may be translated into general guidelines for setting up the categories for presenting a frequency distribution, but these should not be taken as dogma. In each case, the statistician must spend some time trying possibilities and, above all, *thinking* about how to present the data so that the frequency distribution is a useful and honest summary. The following guidelines provide a good starting point.

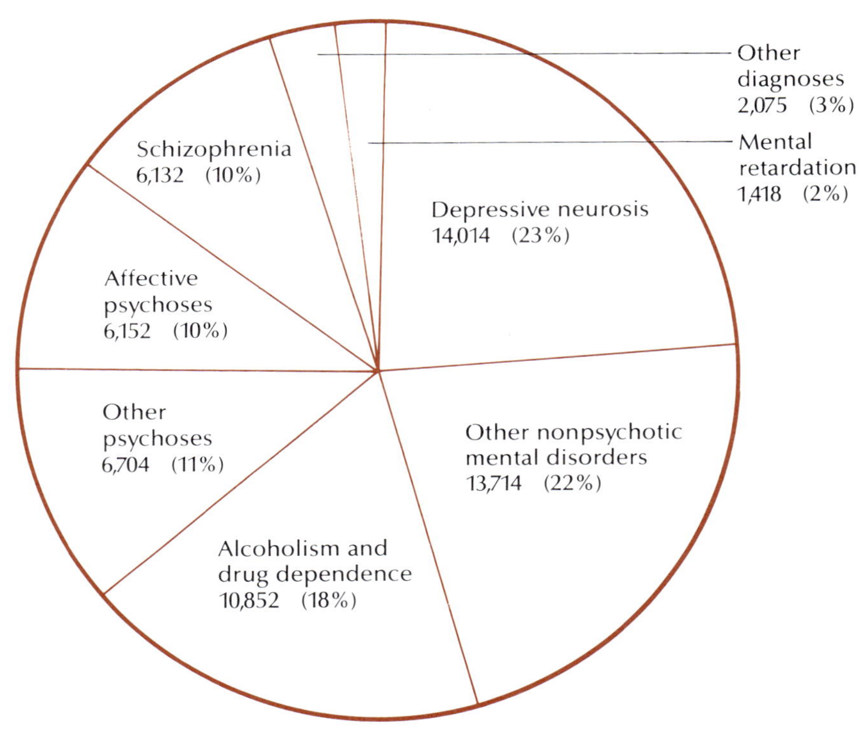

Figure 2.4
Pie chart of the distribution of diagnostic category of individuals admitted in 1978 for the first time in their lives to a Canadian psychiatric inpatient facility.

Guideline 1. Select an appropriate number of categories—not too many or too few. If you have too many, your reader will not be able easily to comprehend the relevant information—you will not have summarized sufficiently. (For example, the general nature of the distribution of the 54 TSWE scores in Example 2.4 does not seem as easy to grasp from the 40-category stem-and-leaf display in Example 2.5 as from the 9-category display in Example 2.6.) If you have too few categories, you throw away too much information—you will have missed some things that the data were trying to tell you. (For example, the 9-category display of TSWE scores in Example 2.6 is more revealing than the 5-category display in Example 2.4.) A greater number of observations may warrant a greater number of categories. Some authors suggest that you should have between 10 and 20 categories; others say between 5 and 15. In any case, you should choose on the basis of what seems reasonable for your particular data, in light of the questions that the data are intended to answer. The next guideline may also have some bearing on the decision as to what seems a reasonable number of categories for a particular set of data.

Guideline 2. For quantitative variables, make the category intervals of all categories equal, choosing a category interval that is easy to work with, such as 1, 2, 5, 10, and the like. Of course, there may be exceptions to this guideline. However, if the questions being asked of the data can be answered through the use of equal category intervals, the resulting presentation is often easier to read and less apt to be misleading than one based on unequal category intervals. We will consider some of the exceptions and what to do about them in Section 2.5.

Guideline 3. For quantitative variables, make the *lower* category limit of each category a multiple of the category interval. Experience suggests that frequency tables and histograms are easier to read when this guideline is followed. (For example, the category interval in the table of Example 2.8 and in Figure 2.1 is 5, and the lower category limits are 5, 10, 15, . . . , 45, as recommended by Guideline 3.)

The following are some practical suggestions for constructing a representation of a quantitative frequency distribution:

- Begin by scanning the data to get a rough idea of the range of the observations.
- Then jot down one or more stem-and-leaf displays to identify the largest and smallest observations and to give an overview of the data that helps you decide what convenient category interval would yield a reasonable number of categories.
- Finally, identify the categories by category limits having as many decimal places as in the original data, so that no observation can fall between categories.

Section 2.4 has introduced several ways of presenting frequency distributions. In fact, the number of ways of representing distributions using tables or graphs is limited only by the ingenuity and honesty of the presenters. With so many possibilities available, you may be wondering how to decide which is the best representation to use in a particular case. There is no black-and-white answer to the question. We must be guided by our objective of making the data easier to comprehend and less liable to misunderstanding. How to do that in any given case will depend on the nature of the data and the problem to be solved by the data.

2.5 SPECIAL CONSIDERATIONS WITH UNEQUAL CATEGORY INTERVALS

The use of unequal category intervals can cause the representation of a frequency distribution to be misleading. Consider the following example.

EXAMPLE 2.13 The distribution of total income on all Canadian taxable individual income tax returns for 1979 is represented below in tabular form and in graphical form in Figure 2.5.

Frequency and percentage distributions of total income on Canadian taxable individual income tax returns for 1979.

Total income	Number of taxable returns	Percentage
$ 0 but less than $ 5,000	147,503	1.6%
$ 5,000 but less than $ 10,000	2,272,996	24 %
$ 10,000 but less than $ 15,000	2,552,118	27 %
$ 15,000 but less than $ 25,000	3,101,545	33 %
$ 25,000 but less than $ 50,000	1,232,668	13 %
$ 50,000 but less than $100,000	131,353	1.4%
$100,000 or more	27,603	.3%
Total	9,465,786	100 %

If we look first at the table, we see that the greatest frequency (and percentage) is for the fourth category, $15,000 to $25,000. This suggests that the most common taxable income for 1979 was in the interval from $15,000 to $25,000. However, if we look more closely, we see that the first three categories have a category interval of $5,000, whereas the category interval for the fourth category is $10,000. The wider interval may account for the category's greater frequency and, in any event, makes it difficult to compare with other categories. The category intervals continue to change as

Figure 2.5
Incorrect histogram of the distribution of total income on Canadian taxable individual income tax returns for 1979.

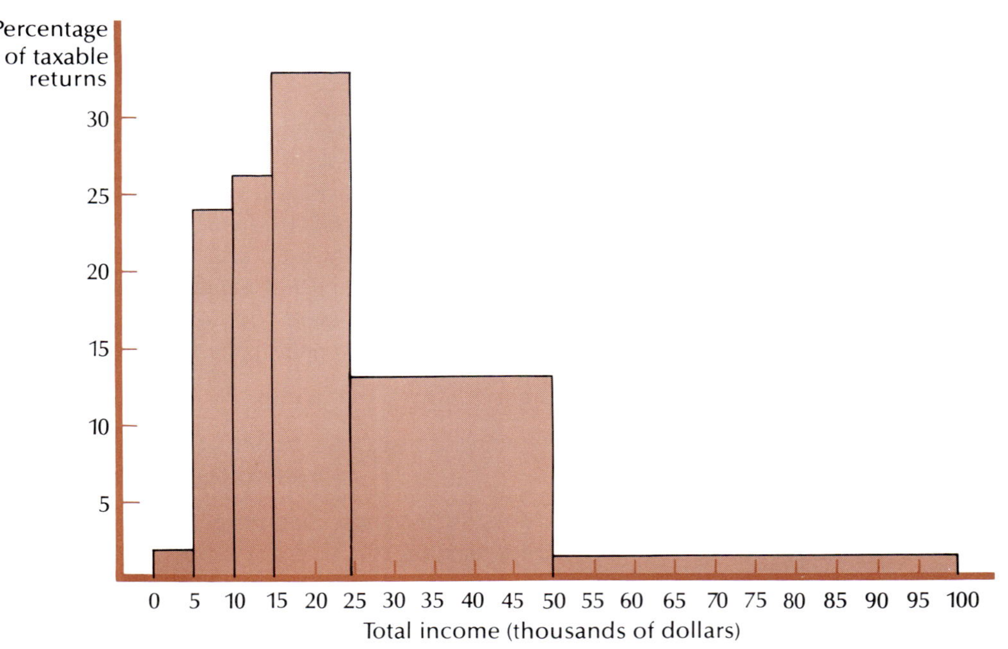

we continue reading down the list of categories, so that the frequencies (and hence the percentages) are not directly comparable with each other.

The same problem appears in a somewhat different guise in the histogram in Figure 2.5. The rectangles in Figure 2.5 are drawn over the appropriate, but changing, category intervals. However, in a two-dimensional drawing we intuitively tend to compare areas rather than heights. Thus, Figure 2.5 gives the erroneous impression that many more people had total incomes above \$15,000 than below \$15,000, when in fact there were slightly more below \$15,000 than above. The fact that 53% of the distribution is to the left of \$15,000 and 47% to the right is at odds with the impression given by Figure 2.5. Figure 2.5 is an incorrect drawing of the histogram for this distribution.

(Source: Revenue Canada, 1981.)

To represent a frequency distribution with unequal category intervals in such a way that it is not misleading, we must calculate each category's frequency relative to an unchanging reference interval. The interval chosen is arbitrary, but it must be the same for every category so as to give values that are comparable. This adjusted frequency is called the **frequency density**, because it shows how densely the actual frequency is concentrated over a given region of the x-variable. (Physical density indicates how much mass is concentrated in a given volume; hence the terminology.) For each category, we define

$$\text{Frequency density} = \frac{\text{category frequency}}{\text{number of reference intervals in the category interval}} \tag{2.3}$$

In a percentage distribution, the percentages must likewise be adjusted to yield **percentage frequency densities**.

How are the frequency densities used? In frequency tables, the safest procedure is to append an additional column containing the densities. In histograms and frequency polygons, the vertical axis is relabeled as "frequency density," and density (rather than actual frequency) is plotted for each category. If desired, the actual frequency may still be recorded above the appropriate rectangle in the histogram to permit exact readings. The effect of this adjustment is to *use area to represent frequency,* as shown in the correspondence between the following two analogous relationships:

$$\text{Area} = \text{height} \times \text{width}$$

$$\text{Category frequency} = \text{frequency density} \times \text{number of reference intervals in the category interval} \tag{2.4}$$

Expression (2.4) follows directly from the definition of frequency density in Expression (2.3), with terms rearranged.

One special case of unequal category intervals should be singled out for a separate comment. Sometimes the first category in a distribution is given as "less than so much" with no lower limit specified, and sometimes the last category in a distribution is given as "more than so much" with no upper limit specified. Such categories, called **open categories**, should be avoided if at all possible, because they make it very difficult to calculate any further descriptions of the data. For example, the frequency density of an open category cannot be calculated according to Expression (2.3), because the number of reference intervals in an open-category interval is undefined. If there are a few extreme observations at one or both ends of the distribution, it generally would be better to have a category interval that is different from the others but defined by a given limit than to have one that is both unequal and undefined. Alternatively, a small number of extreme observations may be listed individually and categories chosen to suit the rest.

EXAMPLE 2.14 The following table shows how the table of Example 2.13 may be amended to avoid creating an erroneous impression due to the unequal category intervals. Here $5,000 has been used as the reference interval. Each frequency density in the third column is calculated by dividing the frequency in the second column by the number of reference intervals in the corresponding category. The percentage frequency density in the fifth column is calculated not from the rounded percentages in the fourth column, but rather from the exact percentages as determined from the second column.

Frequency and percentage distributions of total income on Canadian taxable individual income tax returns for 1979, with adjustment for unequal category intervals.

	Frequency distribution		Percentage distribution	
Total income	**Number of taxable returns**	**Frequency per $5,000**	**Percentage of taxable returns**	**Percentage points per $5,000**
$ 0 but less than $ 5,000	147,503	147,503	1.6%	1.6
$ 5,000 but less than $ 10,000	2,272,996	2,272,996	24 %	24
$ 10,000 but less than $ 15,000	2,552,118	2,552,118	27 %	27
$ 15,000 but less than $ 25,000	3,101,545	1,550,772	33 %	16
$ 25,000 but less than $ 50,000	1,232,668	246,534	13 %	2.6
$ 50,000 but less than $100,000	131,353	13,135	1.4%	.1
$100,000 or more	27,603	?	.3%	?
Total	9,465,786		100 %	

The frequency densities recorded in the third column (labeled "frequency per $5,000") of the table can now be compared with one another to obtain a correct impression of the shape of the distribution. (The density for the open category cannot be calculated.) Similarly, the percentage frequency densities in the last column of the table may be compared with one another. Notice that no total is recorded beneath either of the density columns; the total would have no meaning. It should be noted, too, that it is not always necessary to present all five columns corresponding to those shown in the table: A frequency distribution may be presented without the percentage distribution, and vice versa.

Figure 2.6 is a corrected version of the incorrect histogram shown in Figure 2.5, and Figure 2.7 shows the corresponding frequency polygon. These figures give a correct impression of the shape of the distribution. For example, it now appears that about half of the returns were for incomes less than $15,000 and half were above $15,000, as was indeed the case. In both diagrams, frequency is represented by area rather than directly

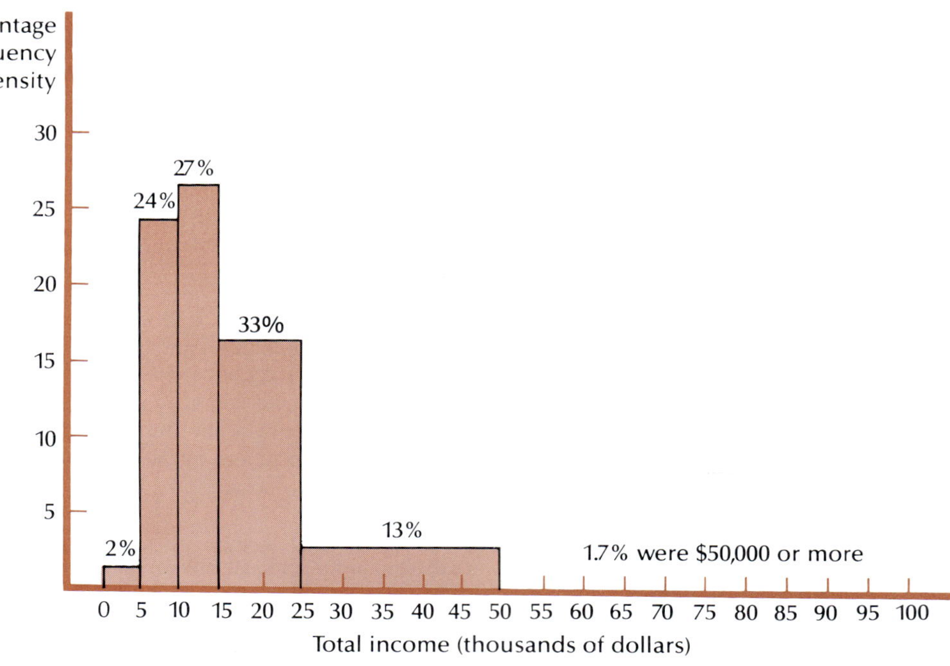

Figure 2.6
Histogram of the distribution of total income on Canadian taxable individual income tax returns for 1979, with correct handling of unequal category intervals.

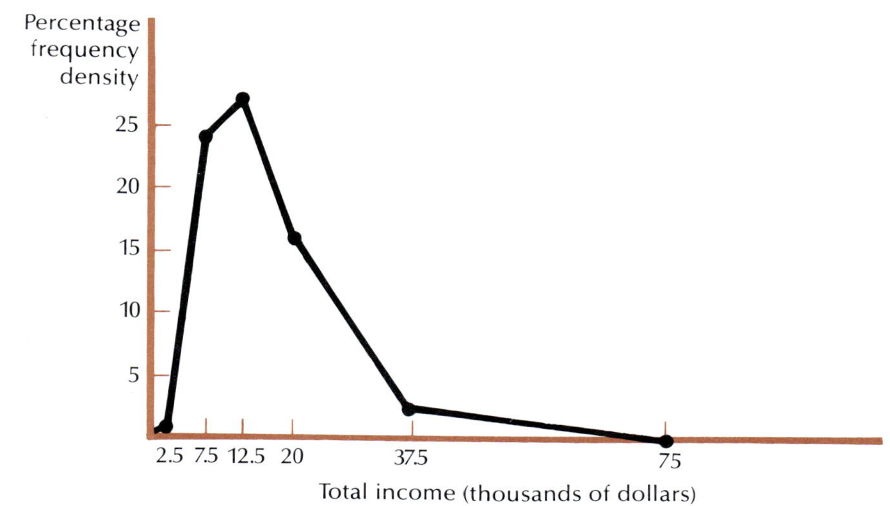

Figure 2.7
Frequency polygon of the distribution of total income on Canadian taxable individual income tax returns for 1979, with correct handling of unequal category intervals.

by height. The open category, $100,000 or more, has necessarily been omitted from the histogram and frequency polygon: Open categories cannot be diagramed. The omission seems justified in this case because the open category contains such a small proportion of the observations. The histogram in Figure 2.6 includes a note stating the contents of the open category.

2.6 CUMULATIVE DISTRIBUTIONS

In a **cumulative frequency distribution**, the category frequencies of a quantitative frequency distribution are cumulated, either from low values of the x-variable to high values or from high to low. If cumulated from low to high in a **"less than" distribution**, each cumulative frequency includes all categories of observation less than a given x-value. If cumulated from high to low in an **"or more" distribution**, each cumulative frequency includes all categories of observation greater than or equal to a given x-value. A tabular representation of a cumulative distribution is called a **cumulative frequency table**.

EXAMPLE 2.15 The following tables are cumulative **"less than"** and **"or more" frequency tables** for the 54 TSWE raw scores of Example 2.4. We derive these tables from the table of Example 2.8 by cumulating the frequencies given there.

Cumulative "less than" frequency distribution of raw scores on the Test of Standard Written English by 54 students beginning their first-year university English course.

TSWE raw score	Cumulative number of students
Less than 5.0	0
Less than 10.0	1
Less than 15.0	2
Less than 20.0	8
Less than 25.0	16
Less than 30.0	27
Less than 35.0	34
Less than 40.0	42
Less than 45.0	52
Less than 50.0	54

Cumulative "or more" frequency distribution of raw scores on the Test of Standard Written English by 54 students beginning their first-year university English course.

TSWE raw score	Cumulative number of students
5.0 or more	54
10.0 or more	53
15.0 or more	52
20.0 or more	46
25.0 or more	38
30.0 or more	27
35.0 or more	20
40.0 or more	12
45.0 or more	2
50.0 or more	0

We obtain a **cumulative percentage distribution** by converting each cumulative frequency in a cumulative frequency distribution (such as those in Example 2.15) to a percentage of the total number of observations.

The graphical representation of a cumulative distribution is called an **ogive**. An ogive is analogous to a frequency polygon, differing only in that (a) the data points are portrayed at the lower category limits (rather than category marks) along the horizontal axis, and (b) the vertical axis is labeled "cumulative frequency." The cumulative character of the ogive gives it a typical gentle-S shape.

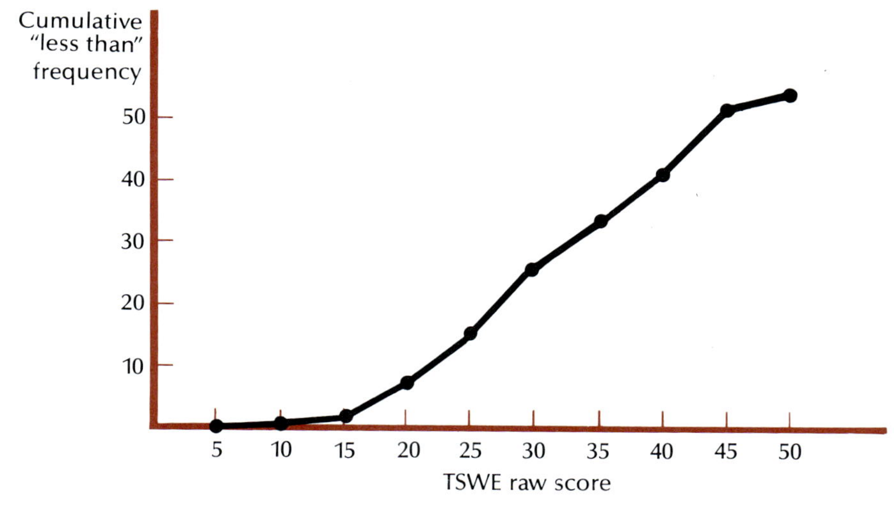

Figure 2.8 Ogive of the "less than" distribution of raw scores on the Test of Standard Written English by 54 students beginning their first-year university English course.

EXAMPLE 2.16 Figures 2.8 and 2.9 are the ogives corresponding to the cumulative distributions portrayed in Example 2.15. Notice that the data points are portrayed at the lower category limits: 5.0, 10.0, . . . , 50.0.

The curve in Figure 2.8 shows, on the vertical axis, how many students obtained a TSWE score less than any given value on the horizontal axis. The ogive in Figure 2.9

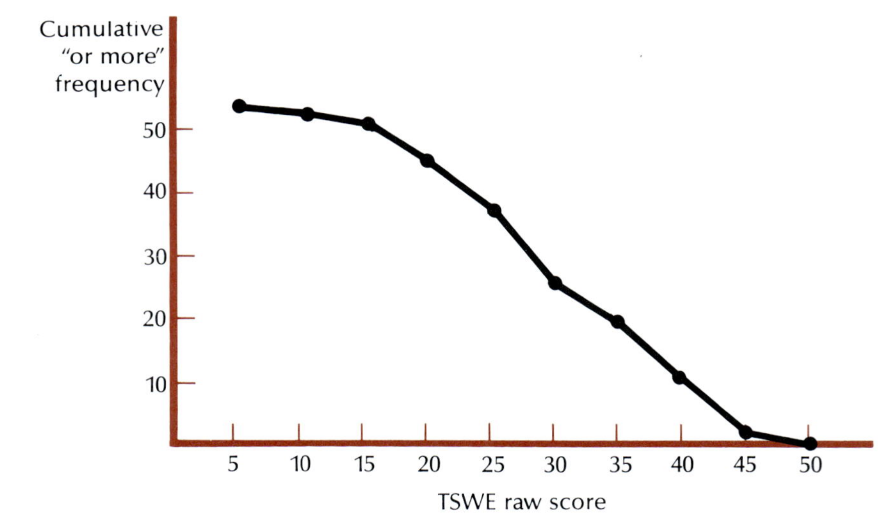

Figure 2.9 Ogive of the "or more" distribution of raw scores on the Test of Standard Written English by 54 students beginning their first-year university English course.

shows, on the vertical axis, how many students obtained a TSWE score of any given value or more on the horizontal axis.

No adjustment is required in the ogive in order to handle distributions with unequal category intervals.

2.7 CHARACTERISTICS OF FREQUENCY DISTRIBUTIONS

We will meet distributions of many different variables in this book, and there will be times when we will need to refer to certain features of these

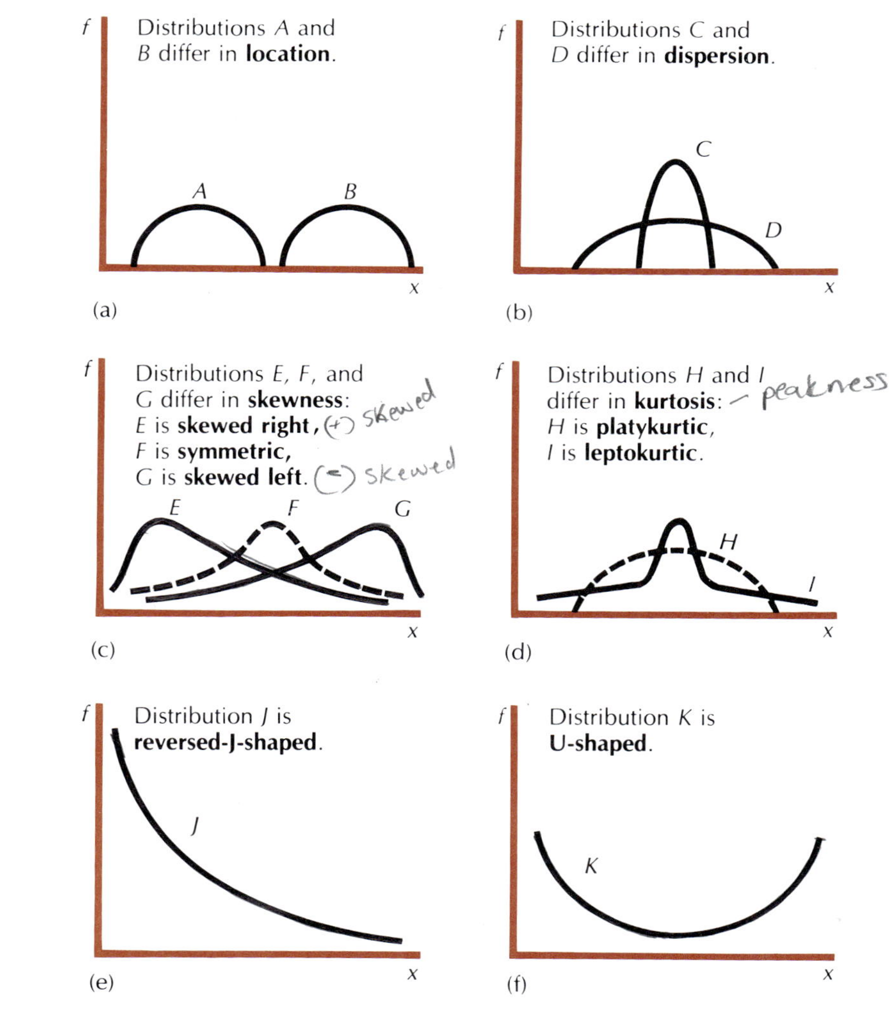

Figure 2.10 Stylized drawings to illustrate some terminology used in describing frequency distributions.

distributions. In this section, we introduce some of the terminology for describing distributions, such as those pictured in Figures 2.2 and 2.7. The terms are illustrated in Figure 2.10 by stylized drawings of distributions.

Two distributions of a variable are said to differ in **location** (on the real-number line) if the values of the variable in one distribution tend to be larger than the values of the variable in the other distribution. See Figure 2.10(a).

Two distributions of a variable are said to differ in **dispersion** if the values of the variable in one distribution tend to be more spread out along the real-number line than the values of the variable in the other distribution. See Figure 2.10(b).

A distribution is said to be **symmetric** if it forms a left–right mirror image of itself when folded at its middle; otherwise it is said to be **skewed**. If it has a short tail on the left and a long tail pointing to the right, it is said to be **skewed right** or **positively skewed**. If the long tail points to the left, the distribution is said to be **skewed left** or **negatively skewed**. See Figure 2.10(c).

The peakedness of a distribution relative to an arbitrary standard is referred to as **kurtosis**. A more or less symmetric distribution with long tails, weak shoulders, and a peaked appearance near the middle is said to be **leptokurtic**, or to have **positive kurtosis**. A more or less symmetric distribution with short tails, heavy shoulders, and a flat appearance near the middle is said to be **platykurtic**, or to have **negative kurtosis**. See Figure 2.10(d). The familiar bell-shaped distribution, which will be discussed in detail in Section 7.4, has **zero kurtosis**.

As illustrated in Figure 2.10(e), a positively skewed distribution resembling (vaguely) a backward J is called a **reversed-J-shaped** distribution. As shown in Figure 2.10(f), a platykurtic distribution whose central peak is so flattened as to be a valley is called a **U-shaped** distribution.

STUDY MATERIAL FOR CHAPTER 2

Terms and Concepts

Section 2.1

frequency distribution
mutually exclusive
exhaustive
class
category
quantitative frequency distribution
numerical frequency distribution
quantitative variable

numerical variable
category limits
lower category limit
upper category limit
category boundary
category interval
category interval of the distribution
category mark
nominal frequency distribution
qualitative frequency distribution
nominal variable
qualitative variable

* Section 2.2

tallying
tally mark

* Section 2.3

stem-and-leaf display
stem
starting part
leaf
simple stem-and-leaf
check count
stretched stem-and-leaf
squeezed stem-and-leaf
two-digit leaves, three-digit leaves

Section 2.4

frequency table
percentage frequency table
percentage distribution
histogram
frequency polygon
bar chart
bar graph
pie chart

Section 2.5

frequency density
percentage frequency density
open category

Section 2.6

cumulative frequency distribution
"less than" distribution
"or more" distribution
cumulative frequency table
cumulative percentage distribution
ogive

Section 2.7

location
dispersion
symmetric
skewed
skewed right
positively skewed
skewed left
negatively skewed
kurtosis
leptokurtic
positive kurtosis
platykurtic
negative kurtosis
zero kurtosis
reversed-J-shaped
U-shaped

Key Formulas

Section 2.4

(2.2) $$\sum_{j=1}^{k} f_j = n$$

Section 2.5

(2.3) $$\text{Frequency density} = \frac{\text{category frequency}}{\text{number of reference intervals in the category interval}}$$

Exercises for Numerical Practice

* Section 2.3

2/1. There are 40 observations in the following set of data:

16.3	13.2	0.0	1.6	1.3	18.6	14.7	1.5	23.0	15.2
3.4	12.6	20.3	.3	10.2	1.9	10.6	3.5	3.5	16.8
26.4	16.0	27.2	11.8	27.4	23.1	11.6	7.0	7.0	27.0
9.0	3.9	7.1	18.7	16.8	5.5	13.3	3.9	1.8	3.9

Make three stem-and-leaf displays of these data as follows:

a. with a category interval of 10 and single-digit leaves;
b. with a category interval of 5 and single-digit leaves;
c. with a category interval of 2 and single-digit leaves.

2/2. A sample of 50 observations was recorded:

59	394	316	398	523	874	370	531	579	344
397	610	239	362	435	620	387	548	406	254
613	374	300	355	241	291	238	340	346	329
788	445	325	301	379	323	424	573	389	544
354	468	454	388	456	416	646	361	349	376

Make three stem-and-leaf displays of these data as follows:

a. with a category interval of 100 and single-digit leaves;
b. with a category interval of 50 and single-digit leaves;
c. with a category interval of 200 and single-digit leaves.

Section 2.4

2/3. Using a category interval of 5 for the data in Exercise 2/1, construct:

a. a frequency table, including percentage frequencies;
b. a histogram;
c. a frequency polygon.

2/4. Using a category interval of 100 for the data in Exercise 2/2, construct:

a. a frequency table, including percentage frequencies;
b. a histogram;
c. a frequency polygon.

Section 2.5

2/5. The following frequency table has unequal category intervals.

Category	Frequency
0– 9	36
10–19	98
20–49	52
50–99	14
Total	200

Calculate the frequency density and percentage frequency density using a reference interval of 10, and draw the correct histogram.

2/6. The following frequency table has unequal category intervals.

Category	Frequency
1.00– 1.99	33
2.00– 2.99	21
3.00– 9.99	11
10.00–99.99	15
Total	80

Calculate the frequency density and percentage frequency density using a reference interval of 1.00, and draw the correct histogram.

Section 2.6

2/7. Convert the frequency table in Exercise 2/3 into a cumulative "less than" distribution, and draw the ogive.

2/8. Convert the frequency table in Exercise 2/4 into a cumulative "less than" distribution, and draw the ogive.

Problems for Applying the Concepts

2/9. A few years ago, a now-defunct restaurant in Winnipeg proclaimed on its placemats a record of service something like this:

Number of satisfied customers	2
Number of meals returned to the kitchen	672
Number of meals served hot	3
Number of customers giving birth while waiting for meal	16

a. In what ways does this table violate the definition of a frequency distribution?

b. Assuming the numbers were correct and were taken seriously, how could the table mislead the reader?

2/10. Answer the following question for each part: Does this set of categories satisfy the requirements of the definition of a frequency distribution, and if not, why not?

a. Interest rate paid by borrowers:

0.0–10.0%
10.0–15.0%
15.0–20.0%
20.0% or more

b. Price per gallon of gasoline:

Less than $1.00
$1.00 but less than $1.10
$1.10 but less than $1.20
$1.20 or more

c. Hourly wage:

Less than $4.00
$ 5.00– 9.00
$10.00–14.00
$15.00 or more

2/11. The following table shows a frequency distribution of pizza-bill totals, obtained from records of the pizza orders originating in one college residence during September and October 1980.

Amount of pizza bill	Number of bills
$ 0.00– 4.99	1
$ 5.00– 9.99	31
$10.00–14.99	19
$15.00–19.99	8
$20.00–24.99	2
$25.00–29.99	1
Total	62

a. What were the observational units?

b. What was the observed variable?

c. What was the largest observed value of the variable?

d. How many pizza bills were less than $10.00?

e. How many pizza bills were less than $9.00?

f. State the category limits.

g. State the category boundaries.

h. State the category intervals.

i. State the category marks.

j. These 62 bills were actually selected at random from a total of 142 bills collected in the residence in the given months. If we want to know the average of all 142 bills, the 62 bills serve as which of the following: sample, population, statistic, or parameter?

(Source: Student project by K. Ree.)

2/12. The following table shows a frequency distribution of prices of houses for sale in the city of Camrose, Alberta. Data are included for all houses advertised (with a price listed) in *The Camrose Booster* of 10 August 1982, with the exception of units in multiresidence buildings and brand new houses.

Listed price	Number of houses
$ 30,000– 39,900	3
$ 40,000– 49,900	5
$ 50,000– 59,900	9
$ 60,000– 69,900	10
$ 70,000– 79,900	23
$ 80,000– 89,900	8
$ 90,000– 99,900	5
$100,000–119,900	6
$120,000–149,900	1
$150,000–199,900	2
Total	72

a. What were the observational units?
b. What was the observed variable?
c. What was the least expensive house price advertised?
d. How many houses were advertised for at least $150,000?
e. How many houses were advertised for more than $125,000?
f. State the category limits.
g. State the category boundaries.
h. State the category intervals.
i. State the category marks.
j. If we used the *Booster* data to estimate the average asking price of all houses for sale in the city on 10 August 1982 (whether advertised in the *Booster* or not), then which of the following would we be trying to estimate: sample, population, statistic, or parameter?

(Source: *The Camrose Booster,* Camrose, Alberta, 10 August 1982.)

2/13. Suppose that a frequency distribution of the lengths of reigns of English monarchs, based on the data given in Example 2.0, has 7, 22, 37, 52, and 67 as its category marks.

a. What is the category interval of the distribution?
b. What are the category limits?
c. What are the category boundaries?

2/14. Suppose that a frequency distribution of the lengths of reigns of English monarchs, based on the data given in Example 2.0, has 5.5, 17.5, 29.5, 41.5, 53.5, and 65.5 as its category marks.

a. What is the category interval of the distribution?
b. What are the category limits?
c. What are the category boundaries?

2/15. For the data in Example 2.0 concerning the length of reigns of English monarchs:

* **a.** Make a simple stem-and-leaf display. What is the category interval?
* **b.** Make a stretched stem-and-leaf display. What is the category interval?
* **c.** Make a squeezed stem-and-leaf display with category intervals of 20 years.
* **d.** What does each of the displays in Parts (a), (b), and (c) show you about lengths of reign?

e. Construct a frequency table with a category interval of 10, draw the histogram, and describe the shape of the distribution.

f. Construct a frequency table with a category interval of 5, draw the frequency polygon, and describe the shape of the distribution.

g. Construct the cumulative "or more" frequency table with category intervals of 10, and draw the ogive.

h. Construct the cumulative "less than" percentage frequency table with category intervals of 5, and draw the ogive.

2/16. The following are pizza-bill totals (in dollars), obtained from records of pizza orders originating in one college residence during September and October 1980.

5.60	7.05	5.00	5.85	15.30
6.80	5.85	13.00	10.55	10.30
25.45	9.30	8.70	4.50	10.30
10.70	16.80	12.40	8.10	9.45
9.45	6.10	10.40	5.25	20.95
9.50	5.60	9.50	16.75	6.20
14.70	9.55	11.30	8.10	5.35
10.65	17.65	8.25	6.20	11.70
15.90	12.20	10.55	18.00	15.05
8.25	5.35	8.85	10.05	10.50
6.45	9.30	10.00	10.90	
24.50	6.80	18.60	6.70	
6.55	5.00	10.55	13.05	

* **a.** Make a stem-and-leaf display with category intervals of \$1. Is the display simple, stretched, or squeezed?

* **b.** Make a stem-and-leaf display with category intervals of \$2. Is the display simple, stretched, or squeezed?

* **c.** Make a stem-and-leaf display with category intervals of \$5. Is the display simple, stretched, or squeezed?

* **d.** What does each of the displays in Parts (a), (b), and (c) show you about the distribution of pizza-bill totals?

e. Construct a frequency table with category intervals of \$2, draw the frequency polygon, and describe the shape of the distribution.

f. Construct the cumulative "less than" percentage frequency table with category intervals of \$2, and draw the ogive.

(Source: Student project by K. Ree.)

2/17. A student bought 25 new textbooks at her college bookstore in 1979–80 for the following prices:

21.65	14.65	21.55	1.15	8.90
32.50	16.95	11.95	8.90	7.15
20.35	1.50	6.95	3.95	4.55
6.95	5.00	2.50	12.45	11.95
8.95	19.95	5.95	5.95	7.60

a. What were the observational units?

b. What was the observed variable?

* **c.** Make a stem-and-leaf display with category intervals of \$10. Is the display simple, stretched, or squeezed?

* **d.** Make a stem-and-leaf display with category intervals of \$5. Is the display simple, stretched, or squeezed?

* **e.** Make a stem-and-leaf display with category intervals of \$2. Is the display simple, stretched, or squeezed?

* **f.** Which of the displays in Parts (c), (d), and (e) seems best to you for giving an impression of the prices of this student's textbooks? Why?
g. Construct a frequency table with category intervals of $10, draw the corresponding histogram, and describe the shape of the distribution.
h. Construct a frequency table with category intervals of $5, draw the frequency polygon, and describe the shape of the distribution.
i. Construct the cumulative "less than" percentage frequency table with category intervals of $5, and draw the ogive.

(Source: Student project by L. Warman.)

2/18. Seventeen female college students provided the following information about the total amount of their summer earnings (in dollars, rounded to the nearest 100 dollars) from May through August 1980.

3,700	1,000	3,400	800
1,500	1,200	1,200	400
1,800	3,500	400	300
2,300	2,500	1,000	1,400
1,000			

a. What were the observational units?
b. What was the observed variable?
c. Construct a frequency table with category intervals of $1,000, and draw the corresponding histogram.
d. State the category boundaries for the distribution in (c).
e. State the category marks for the distribution in (c).
f. Construct the cumulative "or more" percentage frequency table with category intervals of $1,000, and draw the ogive.
g. Construct the cumulative "less than" percentage frequency table with category intervals of $1,000, and draw the ogive.

(Source: Student project by C. Hay.)

2/19. The following table shows the distribution of types of regular season tickets sold for a local concert series in 1981–82 (excluding institutional block purchases).

Type (and color) of season ticket	Number of tickets sold
Adult (red)	116
Senior (gold)	76
Family (white)	39
Student (blue)	4
Total	235

a. What is the observed variable?
b. What are the observational units?
c. What is the value of the variable where the frequency is lowest?
d. Construct a percentage frequency table.

e. Construct a pie chart and a bar chart, and compare them to see which seems more informative to you.

(Source: Camrose Overture Series, 1981–82 subscription list.)

2/20. The following table shows the formal religious affiliation of each U.S. president.

President	Affiliation	President	Affiliation
Washington	Episcopalian	Arthur	Episcopalian
J. Adams	Unitarian	Cleveland	Presbyterian
Jefferson	None	B. Harrison	Presbyterian
Madison	Episcopalian	McKinley	Methodist
Monroe	Episcopalian	T. Roosevelt	Dutch Reformed
J. Q. Adams	Unitarian	Taft	Unitarian
Jackson	Presbyterian	Wilson	Presbyterian
Van Buren	Dutch Reformed	Harding	Baptist
W. Harrison	Episcopalian	Coolidge	Congregationalist
Tyler	Episcopalian	Hoover	Quaker
Polk	Presbyterian	F. Roosevelt	Episcopalian
Taylor	Episcopalian	Truman	Baptist
Fillmore	Unitarian	Eisenhower	Presbyterian
Pierce	Episcopalian	Kennedy	Roman Catholic
Buchanan	Presbyterian	L. Johnson	Disciples of Christ
Lincoln	None	Nixon	Quaker
A. Johnson	None	Ford	Episcopalian
Grant	Methodist	Carter	Baptist
Hayes	None	Reagan	Disciples of Christ
Garfield	Disciples of Christ		

a. Tally the number of presidents of each denomination, and construct a frequency table, with accompanying percentages.

b. What is the observed variable?

c. What are the observational units?

d. What is the value of the variable where the frequency is highest?

e. Construct a bar chart and a pie chart for these data, and compare them to see which seems more informative to you.

(Source: *CBS News Almanac 1978,* p. 146.)

2/21. For the data on housing prices in Problem 2/12:

a. Calculate the frequency density for each category.

b. Draw the histogram of the distribution.

c. Construct the cumulative "less than" percentage frequency table, and draw the ogive.

2/22. The following table shows the fuel consumption ratings of 559 different 1982 car models. The ratings are in liters per 100 km, and they include vehicles ranging from the Volkswagen Rabbit Diesel, which is rated as needing only 4.3 liters of fuel to travel 100 km under ideal test conditions, through the common "Big Four" vehicles, to the Rolls-Royce Camargue and Corniche, which each require 23.2 liters of fuel to go 100 km under ideal test conditions.

Fuel consumption rating (liters per 100 km)	Number of car models
4.0– 4.9	3
5.0– 5.9	14
6.0– 6.9	69
7.0– 7.9	88
8.0– 8.9	129
9.0– 9.9	78
10.0–11.9	131
12.0–14.9	43
15.0–24.9	4
Total	559

a. What was the observed variable?

b. What were the observational units?

c. State the category boundaries.

d. State the category intervals.

e. Construct the percentage frequency table, including percentage frequency densities.

f. In what way do the percentage frequency densities correct the impression given by the percentage frequencies?

g. Draw the frequency polygon of the percentage distribution, and describe its shape.

h. Construct the cumulative "less than" percentage frequency table, and draw the ogive.

(Source: Transport Canada, 1982, pp. 38–49.)

2/23. The following table shows the distribution of adjusted gross income on all U.S. taxable individual income tax returns for 1977.

Amount of adjusted gross income	Number of taxable returns (in thousands)
Less than $1,000 (incl. losses)	7.2
$ 1,000 but less than $ 5,000	4,962.2
$ 5,000 but less than $ 10,000	15,926.5
$ 10,000 but less than $ 15,000	13,994.6
$ 15,000 but less than $ 25,000	19,025.2
$ 25,000 but less than $ 50,000	9,041.9
$ 50,000 but less than $100,000	1,118.1
$100,000 but less than $500,000	267.2
$500,000 or more	7.1
Total	64,350.0

a. Construct the percentage frequency table, including percentage frequency densities.

In what way do the percentage frequency densities correct the impression given by the percentage frequencies?

Draw the histogram of the percentage distribution. How can this histogram be drawn to avoid a lot of empty space in the diagram?

(Source: *World Almanac 1980,* p. 47.)

he following table shows the distribution of amount of total sales for the 50 rgest corporations in the world, on the basis of sales in 1984. These 50 rporations range from Exxon, which was the largest, having total sales of)0.854 billion, through Toyota Motors with sales of $24.110 billion, to Procter & amble, which was the fiftieth, having sales of $12.946 billion.

Total sales	Number of corporations
$12 billion but less than $14 billion	6
$14 billion but less than $16 billion	12
$16 billion but less than $18 billion	6
$18 billion but less than $22 billion	9
$22 billion but less than $30 billion	7
$30 billion but less than $50 billion	4
$50 billion but less than $92 billion	6
Total	50

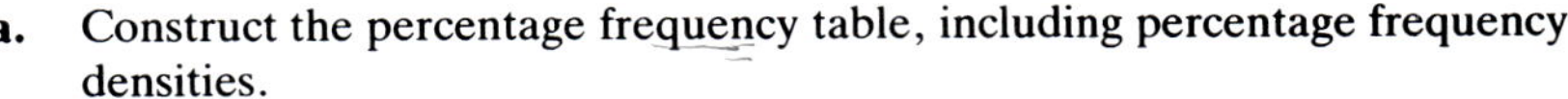

a. Construct the percentage frequency table, including percentage frequency densities.

b. In what way do the percentage frequency densities correct the impression given by the percentage frequencies?

c. Draw the percentage frequency polygon for these data.

d. How would the percentage frequency polygon in Part (c) be altered if it included the 100 largest corporations?

(Source: Loos & Moulton, 1985, p. 179.)

References

American Men and Women of Science: Physical and Biological Sciences. 14th ed. New York: Bowker, 1979.

Loos, C., and Moulton, D. J. "The World's Largest Industrial Corporations." *Fortune* 112 (19 August 1985).

Revenue Canada. *1981 Taxation Statistics Analyzing 1979 T1 Individual Tax Returns.* 1981.

Statistics Canada. *Mental Health Statistics Volume 1: Institutional Admissions and Separations, 1978.* December 1981.

Transport Canada. *Fuel Consumption Guide: 1982 Models* (Spring ed.), February 1982.

Who's Who in America. 41st ed. Chicago: Marquis Who's Who, 1980.

The World Almanac & Book of Facts 1980, 1983. New York: Newspaper Enterprise Association, 1979, 1982.

Further Reading

Ehrenberg, A. S. C. *Data Reduction: Analysing and Interpreting Statistical Data.* Rev. reprint. Chichester, England: Wiley, 1978. Chapters 1–3 present excellent advice concerning the presentation and interpretation of data in tables and graphs.

Huff, D. *How to Lie with Statistics.* New York: Norton, 1954. Chapters 5 and 6 identify and cleverly illustrate some pitfalls to avoid when portraying data graphically (and when reading such portrayals).

Reichmann, W. J. *Use and Abuse of Statistics.* London: Methuen, 1962. Chapter 3 discusses graphical representations of data.

Tukey, J. W. *Exploratory Data Analysis.* Reading, MA: Addison-Wesley, 1977. Chapter 1 is an account of stem-and-leaf displays, written by their inventor.

CHAPTER 3

MEASURES OF LOCATION

EXAMPLE 3.1 The United Nations was established in 1945 to promote international peace and cooperation. Most of the nations of the world are members, but the cost of membership is not the same for all countries. The United Nations budget, which is approved by the General Assembly, includes an assessment for each member nation according to its means. How large are these assessments? The following table shows the individual assessments to the 1981 United Nations budget from its 152 members at that time:

Assessment	Freq.	Countries
$167,347,460	1	U.S.A.
63,412,040	1	U.S.S.R.
54,728,589	1	Japan
47,473,338	1	West Germany
35,762,106	1	France
25,479,071	1	United Kingdom
19,709,147	1	Italy
18,737,973	1	Canada
10,454,418	1	Australia
9,711,754	1	Spain
9,311,858	1	Netherlands
9,254,730	1	China
8,340,682	1	Ukrainian S.S.R.
7,940,788	1	East Germany
7,483,763	1	Sweden
7,255,252	1	Brazil
7,083,868	1	Poland
6,969,612	1	Belgium
4,741,621	1	Czechoslovakia
4,455,981	1	Argentina
4,341,725	1	Mexico
4,227,469	1	Denmark
4,056,086	1	Austria
3,713,318	1	Iran
3,427,678	1	India
3,313,422	1	Saudi Arabia
2,856,398	2	Norway, Venezuela
2,742,142	1	Finland

(Continued)

TABLE (*cont'd*)

Assessment	Freq.	Countries
2,399,375	2	South Africa, Yugoslavia
2,227,991	1	Byelorussian S.S.R.
1,999,479	1	Greece
1,885,223	1	Hungary
1,722,628	1	Turkey
1,542,455	1	New Zealand
1,428,199	1	Israel
1,313,943	1	Libya
1,199,687	1	Romania
1,142,560	1	Kuwait
1,085,431	1	Portugal
914,047	4	Bulgaria, Indonesia, Ireland, Nigeria
685,535	2	Algeria, Iraq
628,408	2	Colombia, Cuba
571,279	3	Philippines, Thailand, United Arab Emirates
514,151	1	Malaysia
457,023	1	Singapore
399,896	3	Chile, Egypt, Pakistan
342,767	1	Peru
285,639	2	Luxembourg, Morocco
228,511	2	Bangladesh, Uruguay
171,384	8	Dominican Republic, Ghana, Iceland, Ivory Coast, Lebanon, Trinidad & Tobago, Tunisia, Vietnam
114,256	11	Costa Rica, Ecuador, Gabon, Guatemala, Jamaica, Panama, Qatar, Sri Lanka, Syria, Zaire, Zambia
57,128	71	Afghanistan, Albania, Angola, Bahamas, Bahrain, Barbados, Benin, Bhutan, Bolivia, Botswana, Burma, Burundi, Cambodia, Cameroon, Cape Verde, Central African Republic, Chad, Comoros, Congo, Cyprus, Djibouti, Dominica, El Salvador, Equatorial Guinea, Ethiopia, Fiji, Gambia, Grenada, Guinea, Guinea-Bissau, Guyana, Haiti, Honduras, Jordan, Kenya, Laos, Lesotho, Liberia, Madagascar, Malawi, Maldives, Mali, Malta, Mauritania, Mauritius, Mongolia, Mozambique, Nepal, Nicaragua, Niger, Oman, Papua New Guinea, Paraguay, Rwanda, Sao Tomé & Principe, Senegal, Seychelles, Sierra Leone, Solomon Islands, Somalia, St. Lucia, Sudan, Surinam, Swaziland, Tanzania, Togo, Uganda, Upper Volta, Western Samoa, Yemen Arab Republic, Yemen (People's Democratic Republic)
Total	152	

In what ways can we summarize the detailed data above to indicate clearly and simply how large the assessments were for 1981? We might note first that the smallest assessment was $57,128, and the largest was $167,347,460. In fact, half of the members were assessed $114,256 or less, and the most common single assessment was $57,128, which was assigned to 71 members. We can determine also that if the total budget of $596,000,000 had been demanded equally from all 152 members, each member would have been assessed approximately $3,900,000. This chapter deals with methods, such as those used in this paragraph, for summarizing collections of data to indicate how large the observations are.

(Source: *Information Please Almanac 1982*, p. 325.)

Things can usually be described in any of several ways, and the more complex a thing is, the more ways in which it can be correctly described. A cloud is described one way by the poet and quite another way by the meteorologist. Neither description is wrong, but the descriptions differ according to the describer's purpose. Sometimes one might need several descriptions of the same item in order to get a complete picture. In describing a doughnut, the optimist talks about the dough and the glaze, while the pessimist talks about the hole.

Collections of numerical data can likewise be described in many ways, depending on what features of the data are relevant to the problem at hand. Frequency distributions provide one way of describing data, but in many situations it may be convenient to have a more concise description of a set of observations. A frequency distribution indicates something about the whereabouts of every single observation, but there also exist a variety of useful ways of summarizing or characterizing a collection of data by a single number. This single number may indicate how large the observations are (location), how spread out they are (dispersion), or what shape the distribution is (skewness and kurtosis). In this chapter we will study measures of how large the observations are, and in Chapter 4 we will introduce measures of how spread out they are. Although numerical measures of skewness and kurtosis are available, they have relatively little application in elementary statistics, and qualitative descriptions of the shapes of distributions, such as were introduced in Section 2.7, will be sufficient for our purposes.

What qualifies as a measure of location? In fact, location is probably the aspect of data that people describe more often than any other. This class of statistical description is well within the realm of our everyday experience.

EXAMPLE 3.2 The following 12 statements illustrate the widespread use of **measures of location**:

- **a.** Today's temperature was above the normal high of 20° for this date.
- **b.** The highest point on earth is Mt. Everest, reported to be 8,848 m (29,028 ft) tall.
- **c.** Each year, thousands of college and university students measure their academic success in terms of their grade point average.
- **d.** In the 1984 U.S. presidential election, Ronald Reagan received more votes than any other candidate.
- **e.** As of 1985, the world record for the mile run was 3 minutes and 46.31 seconds, held by Steve Cram of Britain.
- **f.** Half of the buyers of new Ford Mustangs in the late 1970s were no more than 29 years old.
- **g.** Half of the buyers of new Lincolns in the late 1970s were at least 53 years old.

h. An investor received an average rate of return of 25% on his stocks in 1981.

i. The lowest official temperature ever recorded in North America was −63°C (−81°F) at Snag, Yukon, on 3 February 1947.

j. If the total 1981 United Nations budget had been demanded equally from all 152 member nations, each member would have been assessed $3.9 million.

k. The language spoken by more people in the world than any other is Mandarin (Chinese), with approximately twice as many speakers as English.

l. The major league baseball record for most home runs in one season was set at 61 by Roger Maris of the New York Yankees in 1961.

Every one of these 12 statements makes reference to a measure of location.

(Sources: Hershman & Levenson, 1979; *Information Please Almanac 1982; World Almanac 1986.*)

For quantitative variables, the term **measure of location** refers to a single number that indicates how large the observations in a population or sample are. It is defined more generally as follows:

> A **measure of location** is any single value that describes a set of data by indicating where the observations are located on the real-number line or among the possible values of the variable. **(3.1)**

In this chapter we will introduce measures of location under the general headings of (a) extremes, (b) averages, and (c) quantiles.

3.1 THE EXTREMES

Perhaps the simplest measures of location in a finite set of observations of a numerical variable x are the two extreme values, or **maximum** and **minimum**. The **maximum**, denoted by x_{max}, is the largest observation in the set, and the **minimum**, denoted by x_{min}, is the smallest. They are measures of location because each indicates where the observations are located: All the observations are less than or equal to the maximum, and all are greater than or equal to the minimum.

The maximum and minimum values of a variable are often of interest precisely because they are extreme. The well-known *Guinness Book of World Records* is full of such maxima and minima, and daily newspapers provide other examples in their sports pages. In addition to their use as records, the extremes may provide an appropriate summary of a set of data for certain other special purposes. For example, a comparison shopper uses the minimum to summarize the set of prices at which a given product is offered for sale, and a professor looks at (among other things) the maximum enrollment in recent years to help determine how many copies of the textbook to order for a course. Thus, the maximum and minimum are sometimes called for directly by the question being asked of the data.

In cases where the problem is the more general one of describing a set of data, where the extremes play no special role in the question being asked, the extremes can sometimes be of use. The maximum and minimum are easily explained and understood, and they constitute an unambiguous statement about the location of the observations. However, they can be misleading if one of the observations is very different from the majority. Indeed, it is a good practice routinely to examine the extremes of any distribution, in order to check for errors in observation or recording.

3.2 THE ARITHMETIC MEAN AS ONE KIND OF AVERAGE

The term *average* has a broader meaning in statistics than it has in everyday usage.

> An **average** of a set of observations is any single value that is most representative or typical of the whole set. **(3.2)**

Averages are sometimes called **measures of central tendency** or **measures of central location** because they identify a central or typical point in the data. Notice that the definition in Expression (3.2) does not restrict the term *average* to the everyday average, such as you would calculate, for example, to find your average mark on 3 quizzes (the sum of the 3 marks, divided by 3). That is one kind of average, but the definition in Expression (3.2) refers to *any* value that is most representative of the data. There are, in fact, many different types of average, each one being "most representative" in its own way. Six of them are introduced in this chapter: mean, median, mode, midrange, midquartile, and trimean.

The everyday average is the **arithmetic mean**, which is the sum of the observations in a set, divided by the number of observations. There are other means, such as the geometric mean and the harmonic mean, but the arithmetic mean is used so much more frequently that when one refers simply to the *mean,* it is understood to be the arithmetic mean.

Sample Means and Population Means

To express the definition of the mean in standard notation, we need to distinguish between the sample mean and the population mean. For a variable x, the **mean of a sample** is written $\bar{x}$ (read "x-bar") and is computed as

$$\bar{x} = \frac{\sum^{n} x_i}{n} \qquad \textbf{(3.3)}$$

where n is the sample size. Similarly, for a variable y the sample mean is written $\bar{y}$, and so on.

EXAMPLE 3.3 A sample of 15 college students were asked how many vehicles their immediate family owned. They reported the following numbers:

1, 8, 4, 3, 3, 4, 1, 2, 2, 2, 4, 1, 3, 2, 3

Based on $n = 15$ observations, the sample mean defined in Expression (3.3) is

$$\bar{x} = \frac{\sum x_i}{15} = \frac{1 + 8 + 4 + \cdots + 3}{15} = \frac{43}{15} = 2.867$$

or 2.9 vehicles. (Guidelines for rounding and how many digits to report are given in Appendix 1.)

What does the sample mean tell us? We can say that the families of these college students owned an average of 2.9 vehicles each, but we do not have enough information to say much more than that. The use of a formula should not stop us from asking questions about the data: What is meant by *vehicle*? (Possibilities include cars, vans, farm trucks, tractors, motorcycles, and so on.) What is meant by *immediate family*? (Does it include married siblings living away from home?) How and where was the sample obtained? (It might make a difference whether it was on an urban campus or a rural campus, in the college parking lot or college library, and so on.) The possibility of calculating a summary statistic to several decimal places should not distract us from the importance of knowing how the data were collected.

(Source: Student project by D. Swane.)

The **mean of a population** is designated by μ (the lower-case Greek letter mu) and is computed as

$$\mu = \frac{\sum^{N} x_i}{N} \tag{3.4}$$

where N is the number of observations in the population. (Throughout this book, population formulas will be given as if N were finite. We do so in order to present elementary statistical concepts without a calculus prerequisite. The underlying concepts are not limited to finite populations.) In cases where there may be ambiguity, a subscript is attached to the symbol μ to specify which variable. For example, the population means for variables x and y may be designated by μ_x and μ_y, respectively.

EXAMPLE 3.4 What has been the typical length of reign of English monarchs? There have been 60 English monarchs who have completed their reign, and the number of years that each reigned (x_i, for i running from 1 to 60) is given in Example 2.0. For the question at hand, these 60 monarchs (or, more precisely, their lengths of reign) constitute the population. If we use the arithmetic mean, defined in Expression (3.4), as a measure of the typical length of reign, we obtain an answer, based on $N = 60$ observations, of

$$\mu = \frac{\sum x_i}{60} = \frac{10 + 19 + 2 + \cdots + 15}{60} = \frac{1112}{60} = 18.5333$$

or 18.5 years.

In terms of the computations, Expressions (3.3) and (3.4) are equivalent: We add the observations in the set and divide by the number of observations. The notational distinction between sample means and population means does not indicate different computations, but it is important for the same reason that the distinction between samples and populations is important, as described in Section 1.2. Use of the symbol μ indicates that the mean has been calculated on the basis of knowledge of all the observations relevant to the question being asked; there is no uncertainty about μ, except that associated with any rounding of the numerical value. The symbol $\bar{x}$ indicates a mean based on knowledge of some, but not all, of the observations relevant to the question at hand; we recognize that the value of $\bar{x}$ may have been somewhat different for a different sample of the relevant observations. Whether we have all of the relevant observations depends on the question being asked of the data, but as noted in Section 1.2, sample data are more the rule than the exception. Even an apparently complete set of data (for example, the quiz marks of all students in a class) may usefully be regarded, for some purposes, as a sample from a hypothetical population consisting of all the observations that might have been made under different but supposedly equivalent circumstances (for example, different times of day, different days). In practice, we usually will use the sample symbol $\bar{x}$ in our calculations and the population symbol μ to refer to the unknown true mean value underlying our sample of observations.

The Meaning of the Mean

In Expression (3.2), we said that an average is a value that is most representative or typical of a whole set of data. In what sense is the arithmetic mean representative or typical of the entire set of observations from which it is calculated? It is representative in the sense that if we replace every observation with the mean, we obtain the same total. In other words, the mean gives the value that each observation would have if the total were distributed equally among all the observations. We may verify this interpretation algebraically by multiplying both sides of Expression (3.3) by n, to obtain

$$n\bar{x} = \sum^{n} x_i \tag{3.5}$$

Expression (3.5) says that the sum of the n observations is equal to n times the mean.

EXAMPLE 3.5 Example 3.3 reported a study in which the families of 15 college students were found to own a total of 43 vehicles for a mean of $\bar{x} = 2.86\dot{6}$, or approximately 2.9 vehicles per family. To apply Expression (3.5), we must recognize that the symbol $\bar{x}$ in the equation is the mathematically defined exact value, not the rounded value used for purposes of reporting results. Expression (3.5) tells us that if each of the 15 families in the sample had $2.86\dot{6}$ vehicles, the total number of vehicles would remain 43. Adding fifteen $2.86\dot{6}$'s together yields 43:

$$n\bar{x} = (15)(2.86\dot{6}) = 43 = \sum^{n} x_i$$

(If we used the rounded mean, 2.9, we would obtain $15 \times 2.9 = 43.5$, which suffers from rounding error. For purposes of calculation, we always use *unrounded* values, if they are available. Even if we write down a rounded value as the result of a particular calculation, it is the unrounded value that we enter in our calculator for further computations. Guidelines for how and when to round and how many digits to report are given in Appendix 1.)

Expression (3.5) also indicates that knowledge of the mean implies certain information about the observations. For example, if the mean number of vehicles in 15 families is 2.9, we know that at most 4 families could have 10 vehicles, because the total number of vehicles in the sample is limited, by the mean, to $15 \times 2.9 \simeq 43$.

By analogy, we may think of the sum of the observations $\sum x_i$ as the total "weight" of the observations, in which case Expression (3.5) tells us that putting n observations at the mean would give the same total weight. The mean is thus the center of gravity of the observations. It is positioned such that any observations on one side of the mean are balanced by one or more observations on the other side of the mean. This is not to say that there need to be equal *numbers* of observations on both sides of the mean—in fact, there often are not. Rather, if we determine how far each observation is from the mean (that is, calculate $x_i - \bar{x}$, where observations x_i less than the mean $\bar{x}$ yield negative values of $x_i - \bar{x}$, and observations x_i greater than the mean $\bar{x}$ yield positive values), then the sum of the positive signed differences will equal the sum of the negative signed differences. In other words, the sum of the distances above the mean will "balance" the sum of the distances below the mean, and the sum of all those differences must be zero, the positives balancing the negatives. This is demonstrated algebraically for samples as follows:

$$\begin{aligned}
\sum^n (x_i - \bar{x}) &= \sum^n x_i - \sum^n \bar{x} && \text{by Characteristic 2 of Section 1.6} \\
&= \sum^n x_i - n\bar{x} && \text{by Characteristic 4 of Section 1.6} \qquad (3.6) \\
&= n\bar{x} - n\bar{x} && \text{by Expression (3.5)} \\
&= 0
\end{aligned}$$

One consequence of the center-of-gravity property of the arithmetic mean is that a small number of extreme observations in a data set may pull the mean so much toward the extreme—shift the balance so much—that it no longer seems representative of the majority of the observations.

EXAMPLE 3.6 What is the typical distance between the college campus and the homes of students living in college residences? Thirteen students living in college residences reported the following numbers when asked how many miles their home was from the college campus:

35, 107, 120, 72, 60, 7000, 700, 160, 16, 360, 111, 63, 361

The sample mean is

$$\bar{x} = \frac{35 + 107 + \cdots + 361}{13} = \frac{9165}{13} = 705$$

or approximately 700 miles.

Closer scrutiny of the data reveals that only 1 of the 13 distances (namely, 7000 miles, reported by an overseas student) is greater than the sample mean; the other 12 are less than the mean. The mean is representative of the data in the center-of-gravity sense: If all 13 students lived 705 miles from the campus, the total distance ($13 \times 705 = 9165$) would equal the total of the 13 actual distances ($\sum x_i = 9165$). However, the fact that the sample mean value of 705 miles does not seem typical of the students in the sample suggests that we want an average that is representative of the data in some other sense. The one extremely large observation pulls the mean up so high that it does not seem an appropriate answer to the question, How far are residential college students' homes from campus?

(Source: Student project by C. Kooyman.)

It must be stressed that in spite of any extreme observations that may be present in a set of data, the mean remains, by definition, representative in the center-of-gravity sense. The question is whether, for such data, we want an average that is representative in that sense. In Section 3.3 we will consider two averages that are representative of data in other senses.

Weighted Means

The center-of-gravity property of the mean is summarized in Expression (3.5), which says that $n\bar{x} = \sum x_i$. This same expression also enables us to calculate a mean in some situations where the raw data are not available.

EXAMPLE 3.7 The milk from each of 10 Holstein dairy cows was analyzed for butterfat content as part of a herd improvement program. The results are shown separately for cows of different ages.

Age of cow	Number of cows	Mean butterfat content (%)
Four years	5	3.95
Five years	3	3.80
Ten years	2	3.70

What is the mean butterfat content for the entire sample of 10 cows? To calculate the mean of 10 observations, we must divide the total of the 10 observations by 10. How can we find the total of the 10 observations without having access to the original observations?

Expression (3.5), which says that n times the mean yields the total of the n observations, enables us to calculate the total of the 5 observations for the four-year-old cows as $5 \times 3.95 = 19.75$. Similarly, the total of the 3 observations for the five-year-old cows is $3 \times 3.80 = 11.40$, and the total of the 2 observations for the ten-year-old cows is $2 \times 3.70 = 7.40$. The total of the 10 observations is therefore $19.75 + 11.40 +$

7.40 = 38.55, and the mean butterfat content for the entire sample of 10 cows is 38.55/10, or approximately 3.86%.

The steps used in calculating the overall sample mean $\bar{x}$ in the preceding paragraph may be summarized as follows:

$$\bar{x} = \frac{(5)(3.95) + (3)(3.80) + (2)(3.70)}{5 + 3 + 2} = \frac{38.55}{10} = 3.855 \quad (3.7)$$

Notice that this is *not* equal to the mean of the three subsample means, which is

$$\frac{3.95 + 3.80 + 3.70}{3} = 3.8167$$

or approximately 3.82%. The correct mean in Expression (3.7) is larger than 3.82, because it takes into account the fact that there are 5 cows represented by the relatively high submean of 3.95 (so 3.95 is counted 5 times) and only 2 cows represented by the low submean of 3.70 (so 3.70 is counted just 2 times). It gives more weight to 3.95 than to 3.70, in order to reflect the actual composition of the entire sample.

(Source: Student project by S. Bailer.)

In general, the mean $\bar{x}$ of an entire sample may be calculated from the means $\bar{x}_j$ of k subsamples (where j runs from 1 to k) as follows:

$$\bar{x} = \frac{\sum n_j \bar{x}_j}{\sum n_j} \quad (3.8)$$

where n_j is the number of observations in the jth subsample. Expression (3.8) should not be frightening; we have already demonstrated its use in Expression (3.7). The numerator in Expression (3.8) is the sum of all the observations in the entire sample, obtained by adding the sums from the separate subsamples. The denominator is the number of observations in the entire sample, obtained by adding the number of observations from each subsample. Expression (3.8) is thus an application of the principle that a mean is equal to the sum of the observations, divided by the number of observations.

From another point of view, Expression (3.8) may be considered to be a special case of a **weighted mean**. From this point of view, it is a weighted mean of the three subsample means: Each of the subsample means is weighted by the number of observations in the subsample, and the sum of these weighted values is divided by the sum of the weights. In general, the **weighted mean** $\bar{x}_{\text{weighted}}$ of a set of values x_i is defined as

$$\bar{x}_{\text{weighted}} = \frac{\sum w_i x_i}{\sum w_i} \quad (3.9)$$

where w_i is the weight applied to the ith value x_i. Expression (3.8) is the special case of Expression (3.9) when the x's are subsample means and the weights are subsample sizes, but the weighted mean is useful in other applications as well.

EXAMPLE 3.8 At many colleges and universities, students' grade point averages are weighted means. Course grades may be assigned on a three-point system, a four-point system, or some other system, depending on the institution, and the grade point average is then a weighted mean of the course grades, where each grade is weighted by the credit rating or credit value of the course.

Consider, for example, the following performance of a student at a university where grades are assigned on a four-point system (A = 4, B = 3, C = 2, D = 1, F = 0).

Course	Credits	Grade
Art history	6	3
Chemistry	4	2
English	3	4
Geography	3	1
History	3	0
Philosophy	6	4
Psychology	6	3
Credits attempted	31	

The student's grade point average may be calculated by means of Expression (3.9) as the weighted mean of the grades:

$$\bar{x}_{\text{weighted}} = \frac{(6)(3) + (4)(2) + (3)(4) + (3)(1) + (3)(0) + (6)(4) + (6)(3)}{6 + 4 + 3 + 3 + 3 + 6 + 6}$$

$$= \frac{83}{31} = 2.7$$

Grades obtained in courses with higher credit ratings are weighted more heavily than those obtained in courses with low credit values.

(Source: McGill University, *1977–78 Arts and Science Calendar,* p. 12.)

* Approximating the Mean of Grouped Data

We have seen how to calculate the arithmetic mean for situations where the raw data are available. Sometimes, however, the data available to us have already been tabulated into a frequency distribution, as in the following example.

EXAMPLE 3.9 A class of 41 students wrote a statistics exam and obtained the following scores, out of a possible score of 60.

Test score	Number of students
20–24	2
25–29	2
30–34	3
35–39	13
40–44	10
45–49	7
50–54	2
55–59	2
Total	41

* This symbol identifies material that is not prerequisite for later topics in this book.

Data such as those in Example 3.9 are called **grouped data** because the individual observations have been grouped into the categories of a frequency distribution. The mean generally cannot be calculated exactly from grouped data, but a good approximation is usually possible.

The difficulty in calculating summary statistics from grouped data is that we do not know exactly where the various observations are located within each category. To calculate the mean of a set of grouped data, we need to identify a specific value to use for each observation. The usual procedure is as follows:

> To approximate the mean of grouped data, assume that every observation is equal to its category mark. **(3.10)**

Substituting the resulting n values into the usual formula for the mean yields an acceptable approximation in many applications.

EXAMPLE 3.10 Let us calculate the sample mean of the distribution of statistics test scores presented in Example 3.9. We treat the 2 observations in the category 20–24 as if they both fell at the category mark 22, the 2 observations in the category 25–29 as if they both fell at the category mark 27, and so on. The category marks are shown in the third column of the following table.

Test score x	Frequency	Category mark	Category mark times frequency
20–24	2	22	44
25–29	2	27	54
30–34	3	32	96
35–39	13	37	481
40–44	10	42	420
45–49	7	47	329
50–54	2	52	104
55–59	2	57	114
	$n = 41$		$\sum x \simeq 1642$

Using the corresponding category mark in place of each observation, we can approximate the sum of the observations by

$$\sum x \simeq (22 + 22) + (27 + 27) + (32 + 32 + 32) + \cdots + (57 + 57)$$

However, this procedure is the same as multiplying each category mark by its frequency and adding those products:

$$\begin{aligned}\sum x &\simeq \sum(\text{category mark times frequency})\\ &= (22)(2) + (27)(2) + (32)(3) + (37)(13) + (42)(10) + (47)(7) + (52)(2) + (57)(2)\\ &= 1642\end{aligned}$$

We may then approximate the sample mean using this sum; that is,

$$\bar{x} = \frac{\sum x}{n} \simeq \frac{1642}{41} = 40.0$$

The mean score on the test was 40.0.

It is worth noting that the approximation for the mean of grouped data may be viewed as a weighted mean of the category marks, weighted by the frequencies; see Expression (3.9). However, the most important point about the approximation for grouped data is that it is based on the usual formula for individual data. We simply use the category mark as our best guess of the value of all the observations in each category.

3.3 OTHER AVERAGES

The arithmetic mean is representative of a set of data in the sense that it is the value that every observation would have to have if the sum $\sum x_i$ were distributed equally among all the observations. If there is one unusually large observation in the data, the sum of the observations is correspondingly enlarged, which in turn raises the mean. If there is one unusually small observation in the data, the sum is correspondingly reduced, which in turn reduces the mean. It is because the mean's representativeness is defined in terms of the sum of the observations that it is susceptible to disturbance due to even one extreme observation. Other averages are representative of the data in other ways.

The Median

The **median**, which we will designate by the symbol $\tilde{x}$ (read "x-tilde") for samples and by $\tilde{\mu}$ (read "mu-tilde") for populations, is an average that is less influenced than the mean by a few extreme observations.

> The **median** is the value such that half of the observations are greater than or equal to the median and half of the observations are less than or equal to the median. **(3.11)**

Unlike the mean, which takes into account exactly how large each observation is, the median counts each observation equally, regardless of its magnitude.

To determine the value of the median, first arrange the observations in either ascending or descending order; then find the value of the middle observation, if there is one, or the mean of the middle two observations as follows. In a sample of n observations, calculate $n/2$, which gives the number of observations on each side of the median, and count off that number of observations from each end of the ordered observations. If n is even, $n/2$ will be an integer, and the median is taken to be halfway between the middle two observations. If n is odd, $n/2$ will be an integer-plus-a-half, and the median is the value of the middle observation, where the integer part of $n/2$ is the number of observations on each side. Notice that $n/2$ is not itself a formula for the median, but rather it tells us how many observations are on each side of the median after we have arranged them in ascending or descending order.

EXAMPLE 3.11 Example 3.6 gave the following reported distances in miles from a college campus to the homes of 13 students living in college residences:

35, 107, 120, 72, 60, 7000, 700, 160, 16, 360, 111, 63, 361

To find the median, we first arrange the observations in ascending or descending order. Jotting down a stem-and-leaf display (Section 2.3) on a scratch pad is often helpful, especially as the number of observations increases. In this case, the observations in ascending order are as follows:

16, 35, 60, 63, 72, 107, 111, 120, 160, 360, 361, 700, 7000

Since the sample size is $n = 13$, the number of observations on each side of the median is given by $n/2 = 13/2 = 6.5$. Thus, there are 6 observations on each side of the middle observation. Counting off 6 observations from each end as shown by the circled numbers, we find that the median of the 13 observations is 111; in symbols, $\tilde{x} = 111$.

① ② ③ ④ ⑤ ⑥ ⑥ ⑤ ④ ③ ② ①

16, 35, 60, 63, 72, 107, 111, 120, 160, 360, 361, 700, 7000

Half of the observations — Median $\tilde{x} = 111$ — Half of the observations

As calculated in Example 3.6, the mean of these 13 numbers is 705. The mean is substantially larger than the median in this case because of the one very large observation, 7000. The median would remain the same whether the largest observation were 7000 or 701, whereas the mean would be altered drastically by such a change.

EXAMPLE 3.12 Ten unmarried male college students aged 17–19 years were asked their opinion as to what, in general, is an appropriate age for a person to get married. Their answers were as follows:

26, 22, 23, 25, 27, 28, 25, 23, 24, 23

What was the median response?

Arranging the observations in ascending order, we have

22, 23, 23, 23, 24, 25, 25, 26, 27, 28

Since there are $n = 10$ observations, the number of observations on each side of the median is $n/2 = 10/2 = 5$. Counting off 5 observations from each end as shown by the circled numbers, we find that the middle two observations are 24 and 25. The median is the mean of these middle two observations, $\tilde{x} = (24 + 25)/2 = 24.5$.

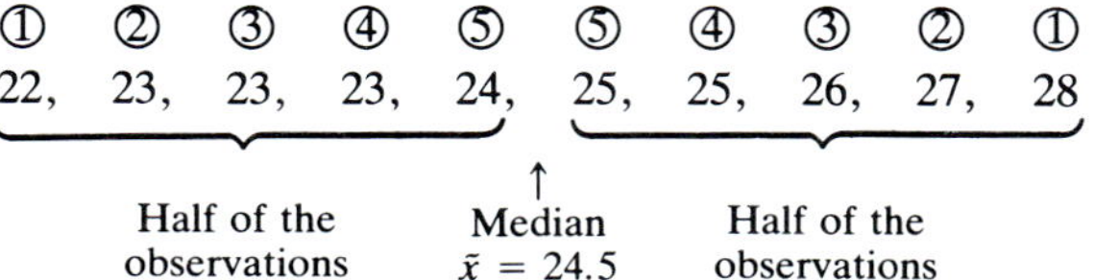

Half of the students suggested an age less than 24.5 years, and half suggested an age greater than 24.5 years.

Notice that age is an unusual variable in that a person is said to be, say, 24 years old from his 24th birthday until the day before his 25th birthday, that is, from age 24.0 to age 24.997 years. We might therefore expect the "average 24-year-old" to have lived 24.5 years. If the median of 24.5 means half a year older than the "average 24-year-old," then the median should be interpreted as meaning "about the time of the 25th birthday." Special care in interpretation is needed whenever the observed variable is age in years. Reported ages have a built-in bias of half a year, on the average.

(Source: Student project by H. Asfeldt.)

The **median** is representative of the entire set of observations in the sense that there are equal numbers of observations above it and below it. The median is neither in the upper half nor in the lower half of the observations; it separates the halves.

In terms of a histogram, the median divides the total area covered by the rectangles into two equal parts such that half the total area (representing half the frequency) is to the left of the median and half the total area (representing the other half of the observations) is to the right of the median.

EXAMPLE 3.13 Figure 3.1 is a histogram of the data from Example 3.12 concerning opinions about an appropriate age for marriage. Notice that the median, 24.5, divides the total area of the histogram into two halves of equal areas.

Each observation contributes the same amount to the total area of the rectangles in a histogram. It follows that when the median separates the responses into equal halves, it also partitions the total area in the histogram into equal halves. Similarly, the median divides the total area under a frequency polygon into two equal halves. Since frequency (that is, number of observations) is represented by area in histograms and frequency polygons, the median, which divides the total frequency in half, divides the area in half.

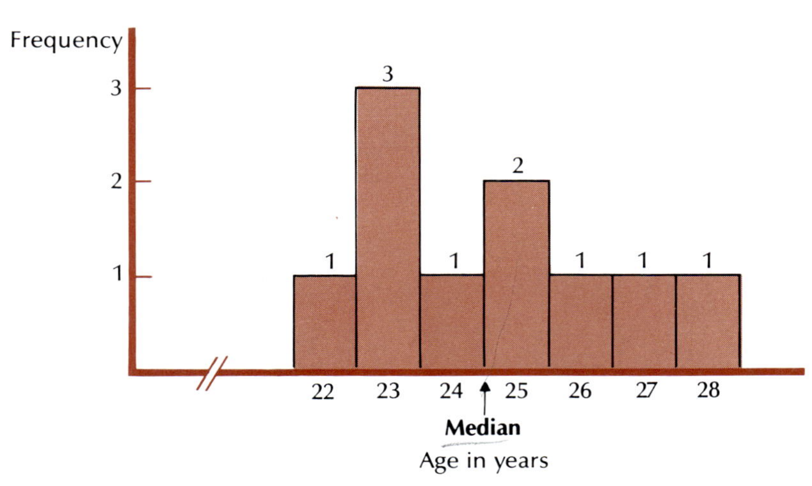

Figure 3.1 Histogram of the responses of unmarried male college students concerning their opinion of an appropriate age for marriage.

The Mode

An average is a single number that is most representative or typical of a set of data. The median is typical of a set of observations in the sense that there are equal numbers of observations above it and below it. The **mode**, on the other hand, is typical in that it is the most common single value in the set. We will designate the mode by the symbol $\dot{x}$ (read "x-dot") for samples and by $\dot{\mu}$ (read "mu-dot") for populations.

> The **mode** is the observation value that occurs more frequently than any other value in a set of observations. **(3.12)**

The mode may be thought of as the most "popular" observation.

EXAMPLE 3.14 As reported in Example 3.12, the responses of 10 unmarried male college students concerning their opinion of an appropriate age for marriage were as follows:

$$26,\quad 22,\quad 23,\quad 25,\quad 27,\quad 28,\quad 25,\quad 23,\quad 24,\quad 23$$

The mode of these numbers, or the modal response, is 23 because the observation 23 occurs more frequently than any other value of the variable. You can easily see this fact in the histogram of these data in Figure 3.1. The observation value with the highest frequency is the mode, 23.

It follows directly from the definition of the mode that in a histogram or frequency polygon, the mode is the x-value corresponding to the highest point or peak of the graph. A distribution with a single peak is a **unimodal** distribution. (Figures 2.6 and 2.7 portray a unimodal distribution.)

There may be more than one mode in a set of data. If two values occur with equal frequency such that each is more frequent than any other value,

Figure 3.2
Stylized drawings to illustrate modality in frequency distributions.

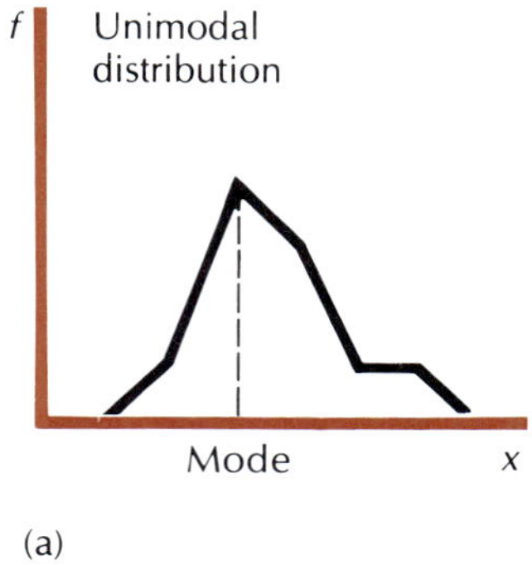

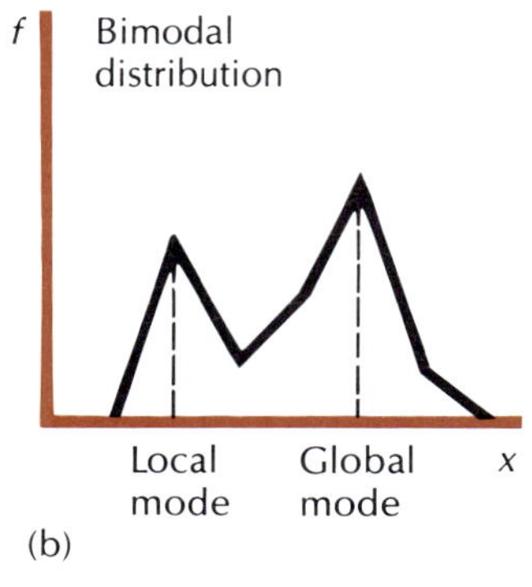

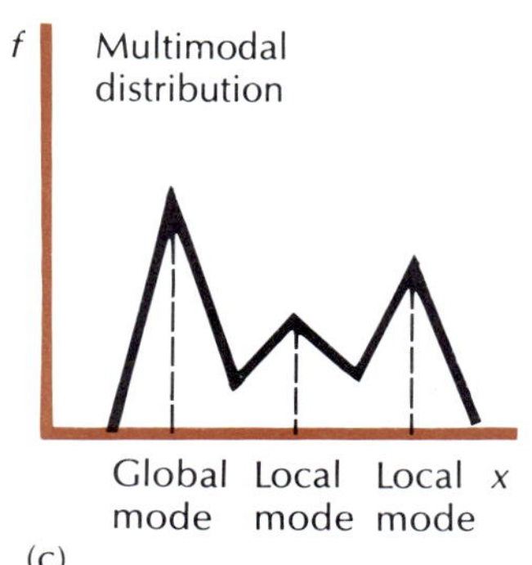

both values are modes and the distribution is said to be **bimodal**. In general, distributions with more than one mode are **multimodal**. A histogram or frequency polygon of a multimodal distribution has more than one peak—specifically, one peak for each mode. A distribution is referred to as bimodal or multimodal even if the peaks are not all of equal height, especially if they are reasonably prominent. The highest peak may then be thought of as the overall or **global mode**, while any lesser peaks are **local modes**. (Figures 2.1 and 2.2 portray a bimodal distribution.) Figure 3.2 illustrates distributions that differ in modality.

The presence of more than one mode in a distribution may be a sign that the observations have come from different populations or that there are different types of observational units represented in the data. One illustration of this principle was discussed at the end of Example 2.6.

EXAMPLE 3.15 As another illustration, grade distributions in statistics courses are often bimodal, suggesting that the course is rather easy for some students and more difficult for others, with relatively few students in between. Perhaps once a student understands certain concepts, any application is fairly easy, but without enough work to attain that insight, all problems are difficult.

One inconvenient property of the mode is that in some sets of data it does not exist: If all the values in a distribution occur an equal number of times, there is no mode. In particular, if no value of the variable occurs more than once, the mode does not exist for the distribution.

EXAMPLE 3.16 In an area where both mule deer and white-tailed deer are hunted, the following dressed weights (in pounds) were reported for a sample of 5 three-point bucks of each species:

Mule deer:	140,	155,	120,	160,	130
White-tailed deer:	135,	155,	120,	140,	110

The average weights of the two species cannot be compared in terms of the mode, because there is no mode in either group of 5 observations. (They could be compared in terms of the mean or the median.)

What is the mode of the entire sample of 10 observations? There are, in fact, three modes because three observation values occur twice and the other values only once each. The modes are 120, 140, and 155. The mode is not a very useful summary in this case. The mean or median would be more informative.

(Source: Student project by B. Reil.)

On the other hand, the mode can be determined in some distributions where other averages are not defined. Unlike the mean and the median, the mode can be used for qualitative as well as for quantitative variables. Thus, we may speak of the modal eye color (to indicate the most frequently occurring eye color in a group), or the modal choice of detergent brands (the most popular brand), or the modal ethnic origin (the largest ethnic group), even though the mean and median are not defined for qualitative variables such as these. In each case, the mode is representative of the entire set of observations in the sense of being the most frequently observed value of the variable.

EXAMPLE 3.17 An observer recorded the color of hair of each student who entered the college library during a four-hour period. The results were reported as 15 black, 27 blond, 44 brown, and 13 red. The observed variable was hair color, and the mode, or the modal hair color, was brown, because that value of the variable was recorded more often than any other.

(Source: Student project by D. Mostowich.)

The Relationship of Mean, Median, and Mode

Which is the best average to use? There is no simple answer. To a large extent, it depends on the particular questions that we are asking of the data. Each average is representative of the set of observations in a different sense, and so each answers a different question.

a. The **mean** is the value that every observation would have if the sum $\sum x_i$ were divided equally among all the observations.

b. The **median** is the value in the middle of the observations, separating the upper 50% of the observations from the lower 50% of the observations.

c. The **mode** is the most frequently observed value of the variable.

We try to choose the average that is representative of the data in the sense implied by the question at hand.

How should we visualize the relationship of mean, median, and mode? In a unimodal distribution that is symmetric, the mean, median, and mode are all equal, as illustrated in Figure 3.3(b). The mode is the x-value under the highest point on the graph. If the distribution is symmetric, then the area to the left of the mode must equal the area to the right of the mode, and hence the median equals the mode. Similarly, if the distribution is symmetric, every observation at a given distance above the mode must be balanced by an observation at the same distance below the mode, and hence the mean equals the mode.

A negatively skewed distribution may be thought of as the result of adding a few unusually small observations to a symmetric distribution, as illustrated in Figure 3.3(a). This addition has the effect of reducing the mean, and to a lesser extent the median, without changing the mode, yielding the typical order of mean, median, mode, in a negatively skewed unimodal distribution. (This ordering may be disrupted in multimodal or irregularly shaped distributions.) Similarly, a positively skewed distribution may be thought of as the result of adding a few abnormally large observations to a symmetric distribution, as illustrated in Figure 3.3(c). This addition has the effect of increasing the mean, and to a lesser extent the median, without changing the mode, giving the typical order of mode, median, mean, in a positively skewed unimodal distribution. In moderately skewed unimodal distributions, it often turns out that the distance between the mode and the median is about twice the distance

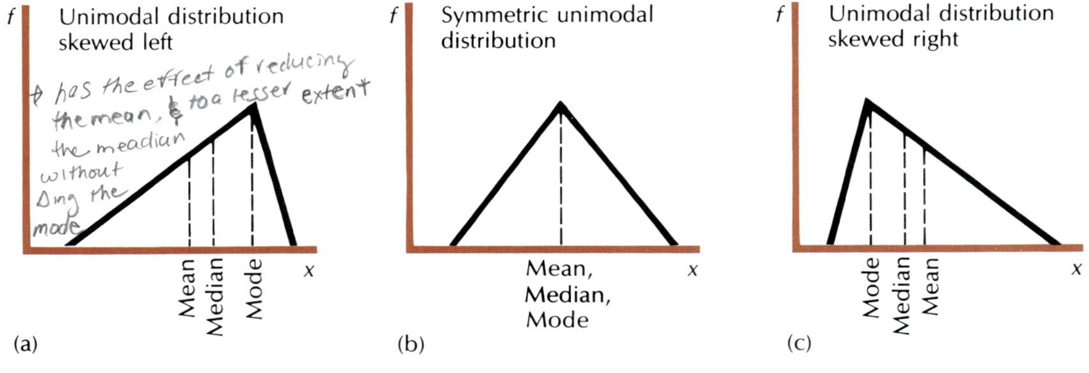

Figure 3.3
Stylized drawings showing the relationship of mean, median, and mode in unimodal distributions.

between the median and the mean, as pictured in Figure 3.3, but this relationship breaks down with extreme skewness and with irregularly shaped distributions. In any case, the **mode** is the x-value under the highest point of the graph; the **median** is pulled toward the long tail, to divide the total area into two equal halves; and the mean is displaced even farther toward the long tail, to divide the sum of the x's equally among the observations.

EXAMPLE 3.18 Students living in college residences often individualize their rooms by putting posters (and other things) on their walls. Ten residence rooms of male students were selected at random on a college campus, and the number of posters displayed in each room was counted and recorded as follows:

$$8, \quad 1, \quad 7, \quad 5, \quad 1, \quad 8, \quad 2, \quad 1, \quad 2, \quad 18$$

What was the average number of posters in this sample of rooms?

The arithmetic mean is

$$\bar{x} = \frac{8 + 1 + \cdots + 18}{10} = \frac{53}{10} = 5.3 \text{ posters per room}$$

This value is the average in the sense that there were enough posters altogether for each room to have 5.

Arranging the observations in ascending order, we count off $n/2 = 10/2 = 5$ observations from each end to find the middle two observations, 2 and 5. The median is halfway between the middle two observations; that is,

$$\tilde{x} = \frac{2 + 5}{2} = 3.5 \text{ posters per room}$$

This value is the average in the sense that half the rooms had 4 or more posters, and the other half had 3 or fewer posters.

The most frequently observed number of posters in a room was

$$\dot{x} = 1$$

which is the mode. This value is the average in the sense that it was the individual x-value that was most likely to be observed in any given room.

It is not always easy to know what is meant by a question such as, "What is the typical number of posters in male college students' rooms?" In this particular case, we might be inclined to answer on the basis of the median: about 3 or 4. The answer involves a judgment as to what was intended by the question.

(Source: Student project by J. Kruschel.)

* 3.4 QUANTILES

A simple way of describing the location of a set of observations is to indicate what proportion of the observations in the set is less than some specified value, as the following example shows.

EXAMPLE 3.19 Many colleges, universities, and graduate and professional schools require prospective students to take certain standard tests, such as the Scholastic Aptitude Test (SAT), the Graduate Record Examination (GRE), or the Medical College Admission Test (MCAT), in order to be considered for admission.

Consider, for example, a student applying for admission to graduate study in psychology. If she were told that she scored at the 87th percentile of the GRE Advanced Test in psychology, she would know that she did better than 87% (and worse than 13%) of the students in a reference group who took that test.

In this section we introduce a family of measures of location of which "87th percentile" in Example 3.19 is an example. The general term for this kind of measure of location is **quantile** or, in some references, **fractile** or **measure of relative location.** In any case, it is a value that cuts off a certain lower fraction, say, j/m, of a distribution:

The **quantile** $Q_{j/m}$ is the value of the variable such that j $m$ths of the observations are less than $Q_{j/m}$. **(3.13)**

To determine the value of the quantile $Q_{j/m}$, first arrange the observations in either ascending or descending order; then find the smallest j $m$ths of the observations as follows. In a sample of n observations, calculate $n \times j/m$, which gives the number of observations that are less than $Q_{j/m}$. Then n minus that number is the number of observations that are greater than $Q_{j/m}$. To find $Q_{j/m}$, count off the integer part of these numbers of observations from the low and high ends, respectively, of the ordered observations. If $n \times j/m$ is an integer, the quantile $Q_{j/m}$ is taken to be halfway between the two observations found by counting from both ends as indicated. If $n \times j/m$ is not an integer, the quantile $Q_{j/m}$ is taken to be the value of the one observation remaining after counting in from both ends as indicated. [There is a certain arbitrariness in the determination of quantiles in small sets of data. The method described above is recommended as a straightforward application of the definition in Expression (3.13).]

The median is probably the best-known member of the quantile family, and indeed we used the method in the preceding paragraph to calculate medians in Section 3.3. The median, $\tilde{x} = Q_{1/2}$, is the value of the variable such that half (that is, $j/m = 1/2$) of the observations are less than the median. See Expression (3.11).

The extremes may also be considered to be special cases of quantiles. The minimum of a set of observations, $x_{\text{min}} = Q_{0/n}$, is the value of the variable such that none (that is, $j/m = 0/n$) of the n observations is less than the minimum. Similarly, the maximum of a set of n observations, $x_{\text{max}} = Q_{n/n}$, is the value of the variable such that all of the other observations are less than the maximum.

Quartiles, Deciles, and Centiles

We have seen that the quantile that divides a distribution into two halves containing equal numbers of observations is the median, $\tilde{x} = Q_{1/2}$.

The three quantiles that divide a distribution into four quarters containing equal numbers of observations are the **quartiles**. The first quartile, designated by $Q_1 = Q_{1/4}$, cuts off a lower-tail area of 25% of the distribution. The second quartile, designated by $Q_2 = Q_{2/4}$, cuts off the lower 50% of the distribution and hence equals the median. The third quartile, $Q_3 = Q_{3/4}$, separates the lower 75% of the distribution from the upper 25%. (Notice that the symbol Q stands for *quartile* if its subscript is a single digit 1, 2, or 3, and for *quantile* if its subscript is a fraction j/m.)

The nine quantiles that divide a distribution into ten ordered segments containing equal numbers of observations are the **deciles**, denoted by $D_1 = Q_{1/10}$, $D_2 = Q_{2/10}, \ldots, D_9 = Q_{9/10}$.

The 99 quantiles that divide a distribution into 100 ordered segments containing equal numbers of observations are called the **centiles** or, more

Figure 3.4
Schematic drawings of distributions, showing proportions of areas corresponding to (a) quartiles, (b) deciles, and (c) centiles. In Part (c), the vertical line at C_j moves across the distribution from left to right as j increases from 1 to 99.

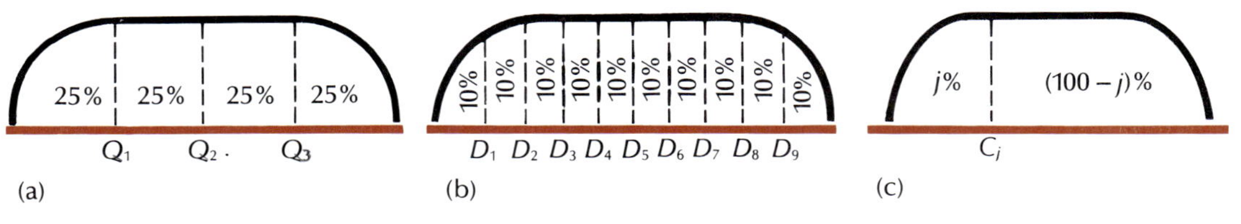

awkwardly, the **percentiles**. From smallest to largest, these 99 centiles are $C_1 = Q_{1/100}$, $C_2 = Q_{2/100}, \ldots, C_{99} = Q_{99/100}$. Notice that the fiftieth centile, the fifth decile, the second quartile, and the median all refer to the same value:

$$Q_{50/100} = C_{50} = D_5 = Q_2 = \tilde{x}$$

The quartiles, deciles, and centiles are illustrated in Figure 3.4.

EXAMPLE 3.20 The following table shows the estimated 1984 sizes, in millions of persons, of the 13 members of the Organization of Petroleum Exporting Countries (OPEC).

OPEC member	Size (millions of persons)
Algeria	21.351
Ecuador	9.091
Gabon	.958
Indonesia	169.442
Iran	43.280
Iraq	15.000
Kuwait	1.758
Libya	3.684
Nigeria	88.148
Qatar	.276
Saudi Arabia	10.794
United Arab Emirates	1.523
Venezuela	18.552

To find any of the quantiles, we first arrange the observations in ascending or descending order. A stem-and-leaf display (Section 2.3) is often helpful, especially if there are very many observations. Here are the OPEC countries from largest to smallest:

Indonesia	169.442
Nigeria	88.148
Iran	43.280
Algeria	21.351
Venezuela	18.552
Iraq	15.000
Saudi Arabia	10.794
Ecuador	9.091
Libya	3.684
Kuwait	1.758
United Arab Emirates	1.523
Gabon	.958
Qatar	.276

Let us use Expression (3.13) and the method described thereafter to find the median size of the 13 OPEC countries. The median cuts off a lower fraction of $j/m = 1/2$ of the observations. Hence, the number of observations below $Q_{1/2}$ is $n \times j/m = 13 \times 1/2 = 6.5$, leaving the remaining $13 - 6.5 = 6.5$ observations above $Q_{1/2}$. Counting 6 (the integer part of 6.5) from each end of the ordered observations above, we come down from the top to the value 15.000 and up from the bottom to

9.091. The one observation between these two is 10.794, so the median size of the 13 OPEC countries in 1984 was 10.794 million persons, which was the size of Saudi Arabia. Half of the other 12 OPEC members were smaller than Saudi Arabia and half were larger.

According to Expression (3.13), the first quartile $Q_1 = Q_{1/4}$ of OPEC member sizes separates the smallest $n \times j/m = 13 \times 1/4 = 3.25$ observations from the top $13 - 3.25 = 9.75$ observations. Counting off 3 (the integer part of 3.25) observations from the bottom and 9 (the integer part of 9.75) observations from the top, we come up from the bottom to the value 1.523 and down from the top to 3.684. The one observation between these two is 1.758, so the first quartile Q_1 is 1.758 million persons, the size of Kuwait. To say that Kuwait was at the first quartile of OPEC member sizes indicates that one-quarter of the other 12 OPEC countries were smaller than Kuwait and three-quarters of them were larger.

Similarly, the third quartile $Q_3 = Q_{3/4}$ of OPEC country sizes separates the smallest $13 \times 3/4 = 9.75$ observations from the top $13 - 9.75 = 3.25$ observations. Thus, the observation between the 3 largest ones and the 9 smallest ones is the third quartile, $Q_3 = 21.351$ million persons. Algeria was at the third quartile, with three-quarters of the other 12 OPEC countries being smaller and one-quarter larger.

Deciles and centiles are determined in the same fashion. For example, to find the ninth decile $D_9 = Q_{9/10}$, we first calculate $n \times j/m = 13 \times 9/10 = 11.7$ and $13 - 11.7 = 1.3$. Then we find the observation between the smallest 11 and the largest 1, namely, 88.148: Nigeria was at the ninth decile. Likewise, to find the 45th centile $C_{45} = Q_{45/100}$, we first calculate $n \times j/m = 13 \times 45/100 = 5.85$ and $13 - 5.85 = 7.15$. Then we find the observation between the smallest 5 and the largest 7 observations, namely, 9.091: Ecuador was at the 45th centile.

To illustrate the case where $n \times j/m$ is an integer, let us calculate the quantile $Q_{8/13}$. This quantile separates the smallest $13 \times 8/13 = 8$ observations from the largest $13 - 8 = 5$ observations. Counting off 8 observations from the bottom and 5 from the top, we arrive at the values 15.000 and 18.552, respectively. The quantile $Q_{8/13}$ is halfway between these two observations:

$$Q_{8/13} = \frac{15.000 + 18.552}{2} = 16.776 \text{ million persons}$$

To determine at which centile any given observation is located, we calculate what percentage of the other $n - 1$ observations is less than the given observation. For example, Saudi Arabia was at the 50th centile of OPEC country sizes, because 6 of the other 12 observations were less than Saudi Arabia's value of 10.794 million persons. Iran was at the 83rd centile, because 10 of the other 12 observations (10/12 = 83%) were less than Iran's value of 43.280 million persons.

(Source: *World Almanac 1986,* pp. 535–626.)

Quantile Measures of Central Location

A measure of central location, or average, is a single value that, in some sense, is most representative or typical of an entire set of observations. In this section, we note four measures of central location based on quantiles.

The first and most widely used of the four is the **median**, which we have discussed previously. Although all of the quantiles are measures of location, the median is the only one of them that, by itself, is a measure of *central* location.

The **midrange** is defined as the arithmetic mean of the extremes:

$$\text{Midrange} = \frac{x_{\text{min}} + x_{\text{max}}}{2} \tag{3.14}$$

Although it has some value as a quick approximation, the midrange is rather unstable because of its reliance on just two of the observations in the set.

The **midquartile** is not as susceptible to distortion by one or two aberrant observations as is the midrange. It is defined as the arithmetic mean of the first and third quartiles:

$$\text{Midquartile} = \frac{Q_1 + Q_3}{2} \tag{3.15}$$

The **trimean** is coming to be used in place of the median in a number of applications, because it takes into account both the median and the behavior of the observations above and below the median. It is defined as the arithmetic mean of the median and the midquartile:

$$\begin{aligned}\text{Trimean} &= \frac{\text{median} + \text{midquartile}}{2} \\ &= \frac{Q_1 + 2Q_2 + Q_3}{4}\end{aligned} \tag{3.16}$$

Recall that the median equals the second quartile Q_2. The trimean provides a relatively reliable measure of central location.

EXAMPLE 3.21 Example 3.20 presented data concerning the sizes, in millions of persons, of the 13 members of OPEC. The median size was shown to be 10.794 million persons. In comparison, the mean, which is most representative in a different sense from that of the median, was 29.527 million persons.

Let us consider the midrange as an alternative measure of central location for these data. The midrange is (.276 + 169.442)/2 = 84.859 million persons, a value that hardly seems most typical of the observations. The single very large value 169.442 for Indonesia has undue influence in the calculation of the midrange, yielding a potentially misleading measure of central location. In general, the midrange is apt to be of dubious value in skewed data (such as these) and hence should be used with caution.

The midquartile provides a more stable measure of central location for these data than does the midrange. The first and third quartiles are $Q_1 = 1.758$ and $Q_3 = 21.351$, and hence the midquartile is (1.758 + 21.351)/2 = 11.554 million persons.

The trimean is an alternative measure of central location that is halfway between the median and the midquartile; in this case, trimean = (10.794 + 11.554)/2 = 11.174 million persons. As indicated in Expression (3.16), it may also be calculated as

$$\text{Trimean} = \frac{Q_1 + 2Q_2 + Q_3}{4} = \frac{1.758 + (2 \times 10.794) + 21.351}{4} = 11.174$$

Approximating the Quantiles of Grouped Data

The quantiles generally cannot be calculated exactly from grouped data, but a good approximation is usually possible. As in the case of approximating the mean of grouped data (Section 3.2, page 83), we need to make some assumption as to how the observations are distributed within each category. Here we make a different assumption than for the mean. The usual procedure is as follows:

> To **approximate the quantiles of grouped data,** assume that the observations are uniformly distributed within each category. **(3.17)**

With this convention, any quantile $Q_{j/m}$ can be determined unambiguously as the value of the variable such that j/m of the total frequency is less than that value. The general procedure is first to determine in which category the required quantile falls, and then to determine what proportion of the way through that category we must go to reach the given quantile. Graphically, the approximation consists of finding the value such that the given fraction of the area of the histogram (j/m of n) is to the left of that value.

EXAMPLE 3.22 Figure 3.5 is a histogram of the distribution of statistics test scores presented in Example 3.9. The total frequency of $n = 41$ is represented by the total area of the bars of the histogram.

Let us calculate the median of the distribution of these test scores. The median is the value such that $j/m = 1/2$ of the total frequency is less than the median, and hence half of the total frequency is greater than the median. In terms of the histogram, the median is the value along the x-axis such that half of the total area of the histogram is to the left of the median and half of the total area is to the right of the median.

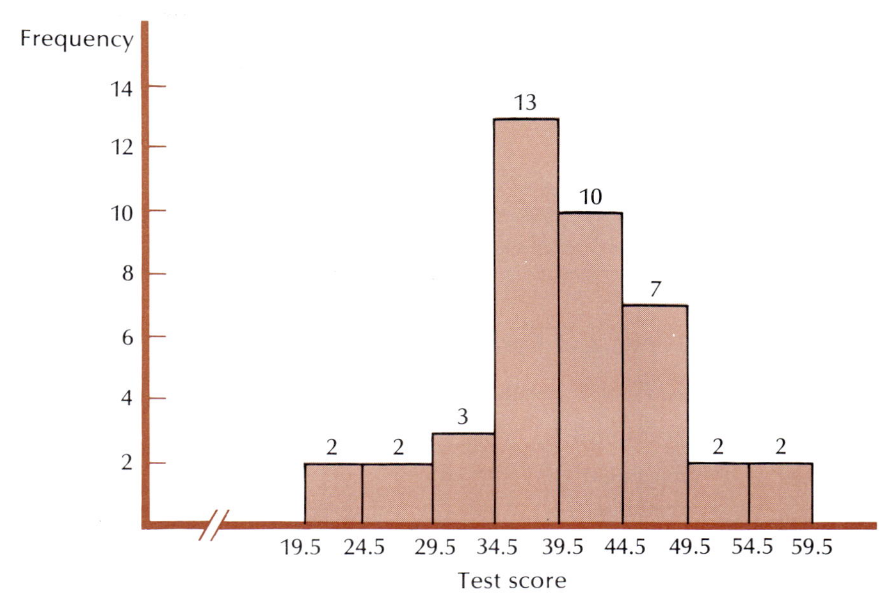

Figure 3.5 Histogram of the distribution of test scores (out of a possible score of 60) by 41 statistics students.

The total frequency (area) of the histogram is 41, and the median divides it into a left area of 20.5 and a right area of 20.5. Adding frequencies in the left tail of the histogram in Figure 3.5, we find a total of 20 in the first four categories and another 10 observations in the fifth category (40–44). Thus, the median must fall in the category 40–44, or between 39.5 and 44.5. To divide the total frequency into two halves of 20.5, we must partition the 10 observations in the category 40–44 so that .5 is to the left of the median and 9.5 are to the right. Thus, the median is .5/10 of the way from 39.5 to 44.5, under the assumption of Expression (3.17) that the observations are distributed uniformly within the category. In other words, the median is approximated by

$$\tilde{x} \approx 39.5 + \left(\frac{.5}{10}\right)(44.5 - 39.5) = 39.75$$

Let us find the first quartile Q_1 of the distribution. The first quartile separates the total frequency of 41 into a left area of $(\frac{1}{4})(41) = 10.25$ and a right area of $(\frac{3}{4})(41) = 30.75$. Adding frequencies in the left tail of the histogram, we find a total of $2 + 2 + 3 = 7$ in the first three categories. To divide the total frequency into 10.25 on the left and 30.75 on the right, we must partition the 13 observations in the category 35–39 so that 3.25 are to the left of Q_1 and 9.75 are to the right. In other words, we take Q_1 to be 3.25/13 of the way from 34.5 to 39.5; the first quartile is approximated as

$$Q_1 \approx 34.5 + \left(\frac{3.25}{13}\right)(39.5 - 34.5) = 35.75$$

As another example, the 98th centile C_{98} divides the total frequency of 41 into a left area of $(.98)(41) = 40.18$ and a right area of $(.02)(41) = .82$. Thus, the 98th centile is in the category 55–59, since 39 of the observations are in the lower categories. According to the uniformity assumption in Expression (3.17), the 2 observations in the category 55–59 should be partitioned so that 1.18 are to the left of C_{98} and .82 is to the right. Thus, C_{98} is taken to be 1.18/2 of the way from 54.5 to 59.5, yielding

$$C_{98} \approx 54.5 + \left(\frac{1.18}{2}\right)(59.5 - 54.5) = 57.45$$

STUDY MATERIAL FOR CHAPTER 3

Terms and Concepts

measure of location

Section 3.1

maximum
minimum

Section 3.2

average
measure of central tendency

measure of central location
arithmetic mean
mean
weighted mean
* grouped data
* approximating the mean of grouped data

Section 3.3

median
mode
unimodal
bimodal
multimodal
global mode
local mode
relationship of mean, median, and mode

* Section 3.4

quantile
fractile
measure of relative location
quartile
decile
centile
percentile
midrange
midquartile
trimean
approximating the quantiles of grouped data

Key Formulas

Section 3.2

(3.3) $$\bar{x} = \frac{\sum^{n} x_i}{n}$$

(3.4) $$\mu = \frac{\sum^{N} x_i}{N}$$

(3.5) $$n\bar{x} = \sum^{n} x_i$$

(3.6) $$\sum (x_i - \bar{x}) = 0$$

(3.8) $$\bar{x} = \frac{\sum n_j \bar{x}_j}{\sum n_j}$$

(3.9) $$\bar{x}_{\text{weighted}} = \frac{\sum w_i x_i}{\sum w_i}$$

* Section 3.4

(3.14) $$\text{Midrange} = \frac{x_{\min} + x_{\max}}{2}$$

(3.15) $$\text{Midquartile} = \frac{Q_1 + Q_3}{2}$$

(3.16) $$\text{Trimean} = \frac{\text{median} + \text{midquartile}}{2} = \frac{Q_1 + 2Q_2 + Q_3}{4}$$

Exercises for Numerical Practice

Section 3.2

3/1. Calculate the mean of each of the following samples:

a. 7, 4, 9, 9, 2, 9, 0, 8, 7, 1, 3.
b. 12, 7, 13, 5, 8, 11, 13, 12, 14, 7, 12, 7.

3/2. Calculate the mean of each of the following samples:

a. 5, 5, 10, 12, 13, 8, 5, 11, 12, 7, 13, 7, 5.
b. 1, 7, 5, 1, 2, 0, 4, 4, 0, 0, 2, 2, 2, 0.

3/3. Calculate the sum of the observations in each case:

a. 20 observations with a mean of 36.8.
b. 100 observations with a mean of 1.79.

3/4. Calculate the sum of the observations in each case:

a. 40 observations with a mean of .861.
b. 50 observations with a mean of 201.3.

3/5. Calculate the weighted mean of the x's.

x	Weight
1	60
2	30
3	10

3/6. Calculate the weighted mean of the x's.

x	Weight
4	1
5	1
6	2
7	6

* **3/7.** Calculate the mean value of x for a sample summarized as follows.

x	Frequency
0– 2	16
3– 5	21
6– 8	10
9–11	4
12–14	1

* **3/8.** Calculate the mean value of x for a sample summarized as follows.

x	Frequency
2.5– 4.9	2
5.0– 7.4	7
7.5– 9.9	20
10.0–12.4	11

Section 3.3

3/9. Calculate the median and the mode for each of the samples in Exercise 3/1.

3/10. Calculate the median and the mode for each of the samples in Exercise 3/2.

* Section 3.4

3/11. Calculate the statistics specified below for this sample:

81, 39, 77, 67, 10, 51, 92, 78, 87, 56, 59, 82

a. first, second, and third quartiles.
b. 10th and 80th centiles.
c. midrange.
d. midquartile.
e. trimean.

3/12. Calculate the statistics specified below for this sample:

68, 29, 17, 7, 81, 33, 19, 11, 1, 46, 21, 30, 6, 5, 3, 15

a. first, second, and third quartiles.
b. 20th and 70th centiles.
c. midrange.
d. midquartile.
e. trimean.

3/13. Calculate the following statistics for the sample in Exercise 3/7:

a. first, second, and third quartiles.
b. 40th and 95th centiles.

3/14. Calculate the following statistics for the sample in Exercise 3/8:

a. first, second, and third quartiles.
b. 33rd and 83rd centiles.

Problems for Applying the Concepts

3/15. Identify which measure of location is referred to in each of the statements (a) through (f) in Example 3.2.

3/16. Identify which measure of location is referred to in each of the statements (g) through (l) in Example 3.2.

3/17. Statement (f) in Example 3.2 ("Half of the buyers of new Ford Mustangs in the late 1970s were no more than 29 years old") was based on observation of only a small proportion of all the buyers of new Ford Mustangs. Identify each of the following items in the context of the statement:

a. observational units.
b. observed variable.
c. population.
d. sample.
e. statistic.
f. parameter.

3/18. Example 3.1 presents the assessments to the 1981 United Nations budget from its 152 members at that time. Identify each of the following items in the example:

a. observational units.
b. observed variable.
c. population or sample (whichever is in the example).
d. statistic or parameter (whichever is in the example).

3/19. Example 3.7 dealt with the butterfat content of the milk of 10 dairy cows. What situation can you suggest where the mean of 10 such observations would be considered to be:

a. the sample mean $\bar{x}$?
b. the population mean μ?

3/20. Example 3.6 dealt with the distance home for 13 college students. What situation can you suggest where the mean of 13 such observations would be considered to be:

a. the sample mean $\bar{x}$?
b. the population mean μ?

3/21. An English professor assigned a 308-page novel to be read in three weeks. The following are the numbers of pages read by the assigned date by a sample of 14 students, as reported to another student in the same English course:

308, 308, 42, 76, 308, 212, 185, 72, 239, 12, 308, 36, 308, 48

Identify:

a. the observational units.
b. the observed variable.

Find:

c. the mean.
d. the median.
e. the mode.

(Source: Student project by R. Slowski.)

3/22. Fourteen female students who had been living in a college residence for one month reported the number of letters they had received through the postal system during that month:

5, 0, 1, 1, 7, 10, 1, 1, 0, 6, 0, 0, 0, 0

Identify:

a. the observational units.

b. the observed variable.

Find:

c. the mean.

d. the median.

e. the mode.

(Source: Student project by S. Schaefer.)

3/23. A waitress recorded the amount of the tip (in cents) that she received for each of 15 tables served during the supper hour at a department store restaurant:

80, 100, 100, 50, 50, 50, 50, 20, 50, 200, 100, 100, 125, 0, 0

Identify:

a. the observational units.

b. the observed variable.

Find:

c. the maximum.

d. the minimum.

e. the mean.

f. the median.

g. the mode.

(Source: Student project by J. Leibel.)

3/24. The total amounts (in dollars) purchased during evening shopping by 15 customers at a central cash register in a department store were as follows:

8.99, 34.88, 18.47, 48.09, 17.97, 52.10, 15.81, 36.99,
6.94, 69.99, 6.79, 9.10, 19.99, 1.10, 6.96

Identify:

a. the observational units.

b. the observed variable.

Find:

c. the maximum.

d. the minimum.

e. the mean.

f. the median.

g. the mode.

(Source: Student project by D. Schultz.)

3/25. In November 1985, a news magazine reported that there had been eight periods of business recovery in the U.S. economy since World War II, in addition to the one then in progress. The durations (in months) of these business recoveries were as follows.

Period	Number of months
October 1945–November 1948	37
October 1949–July 1953	45
May 1954–August 1957	39
April 1958–April 1960	24
February 1961–December 1969	106
November 1970–November 1973	36
March 1975–January 1980	58
July 1980–July 1981	12
Since November 1982	Not yet ended at the time of the report

Based on the eight recoveries that had ended by the time of the report, specify:

a. the observational units.
b. the observed variable.
c. the maximum.
d. the minimum.
e. the mean.
f. the median.
g. the mode.

(Source: Karmin & Elbo, 1985.)

3/26. The numbers of refrigerators sold by 11 appliance stores in the same city during the month of July were as follows:

4, 2, 6, 10, 3, 2, 3, 7, 6, 3, 15

Identify:

a. the observational units.
b. the observed variable.

Find:

c. the mean.
d. the median.
e. the mode.

3/27. An individual sold four shares of a certain stock over a period of time for a mean price of $21 per share. She has one share remaining to sell, but she would like to wait until the price of this stock rises high enough so that she can realize a mean price for all five shares of $23 per share. How high must the price of the stock rise in order for her to obtain this average?

3/28. The mean mark of a class of 6 students on a midterm physics exam was 63%.

a. What is the maximum number of students who could have received a mark of 100% on the exam?

b. What is the maximum number who could have received a mark of 50% or less on the exam?

3/29. A student's 11 hardcover textbooks had a mean price of \$15.45, and her 14 paperback textbooks had a mean price of \$7.14. What was the mean price of all her textbooks?
(Source: Student project by L. Warman.)

3/30. A 29-acre hay field yielded 175 bales per acre in the first cutting of 1979, while a 25-acre field yielded 108 bales per acre. What was the mean yield per acre?
(Source: Student project by S. Bailer.)

3/31. During a period of high interest rates, an individual had \$200 in a savings account earning 13% interest, \$300 in a term deposit at 14%, and \$1000 in government savings bonds at 19% interest. What was the average rate of return on these savings?

3/32. *Fortune* magazine published a recommended portfolio of six stocks, called the Shearson Portfolio, as follows.

Stock	Number of shares	Price per share
IBM	200	\$126.625
Boeing	400	46.625
Frank B. Hall	500	27.625
Intel	500	26.00
Telecredit	500	24.50
Upjohn	100	111.375

What is the average price per share of the stocks in this portfolio?
(Source: *Fortune*: *1986 Investor's Guide*, p. 109.)

3/33. Consider the distribution of pizza-bill totals (in dollars) obtained from records of pizza orders originating in a college residence. Why might we expect such a distribution to be multimodal?

3/34. Consider the distribution of price of different textbooks for sale in a college bookstore. Why might we expect such a distribution to be bimodal?

* **3/35.** The following table shows the number of home runs in the 1982 season by each baseball team in the American League.

Team	Home runs
Baltimore	179
Boston	136
California	186
Chicago	136
Cleveland	109
Detroit	177
Kansas City	132
Milwaukee	216
Minnesota	148
New York	161
Oakland	149
Seattle	130
Texas	115
Toronto	106

Calculate:

a. the first, second, and third quartiles.

b. the fourth, fifth, and sixth deciles.

c. the 33rd, 50th, and 77th centiles.

d. California's centile, Milwaukee's centile, and Texas' centile.

e. the median.

f. the midrange.

g. the midquartile.

h. the trimean.

(Source: *World Almanac 1983,* p. 908.)

* **3/36.** The following table shows the number of home runs in the 1982 season by each baseball team in the National League.

Team	Home runs
Atlanta	146
Chicago	102
Cincinnati	82
Houston	74
Los Angeles	138
Montreal	133
New York	97
Philadelphia	112
Pittsburgh	134
St. Louis	67
San Diego	81
San Francisco	133

Calculate:

a. the first, second, and third quartiles.

b. the first, fifth, and sixth deciles.

c. the 9th, 50th, and 67th centiles.

d. Atlanta's centile, Chicago's centile, and Los Angeles' centile.

e. the median.

f. the midrange.

g. the midquartile.

h. the trimean.

(Source: *World Almanac 1983,* p. 904.)

* **3/37.** The following table gives the mid-1985 salaries of the 50 state governors in the United States, recorded to the nearest $1000, exclusive of any expense allowances.

Salary	Number of states	States
$100,000	1	New York
91,000	2	North Carolina, Texas
85,000	4	California, Minnesota, New Jersey, Pennsylvania
82,000	2	Alaska, Michigan
79,000	1	Georgia
75,000	5	Maryland, Massachusetts, Missouri, Virginia, Wisconsin
73,000	1	Louisiana
72,000	1	West Virginia
70,000	4	Delaware, Florida, Oklahoma, Wyoming
69,000	1	Alabama
68,000	1	Tennessee
66,000	1	Indiana
65,000	5	Connecticut, Kansas, Nevada, North Dakota, Ohio
64,000	1	Iowa
63,000	2	Mississippi, Washington
62,000	1	Arizona
60,000	6	Colorado, Kentucky, New Hampshire, New Mexico, South Carolina, Utah
59,000	1	Hawaii
58,000	1	Illinois
55,000	1	Oregon
53,000	1	South Dakota
50,000	3	Idaho, Rhode Island, Vermont
49,000	1	Montana
40,000	1	Nebraska
35,000	2	Arkansas, Maine

Calculate:

a. the first, second, and third quartiles.
b. the 6th and 95th centiles.
c. Illinois's centile and California's centile.
d. the median.
e. the midrange.
f. the midquartile.
g. the trimean.

(Source: *World Almanac 1986,* p. 252.)

* **3/38.** For the data on United Nations assessments in Example 3.1, calculate:

a. the first, second, and third quartiles.
b. the 60th and 90th centiles.
c. Canada's centile and Mexico's centile.
d. the median.
e. the midrange.

f. the midquartile.
g. the trimean.

* **3/39.** For the data on residential house prices in Problem 2/12 (Chapter 2), calculate:

a. the mean.
b. the first, second, and third quartiles.
c. the 2nd and 98th centiles.
d. the median.
e. the midquartile.
f. the trimean.

* **3/40.** For the data on pizza bills in Problem 2/11 (Chapter 2), calculate:

a. the mean.
b. the first, second, and third quartiles.
c. the 5th and 95th centiles.
d. the median.
e. the midquartile.
f. the trimean.

References

Hershman, A., and Levenson, M. "The Big Clout of Two Incomes." *Dun's Review,* April 1979.

Information Please Almanac 1982. New York: Simon and Schuster, 1981.

Karmin, M. W., and Elbo, R. G. "Recovery's 3rd Birthday—And Still Plugging Along." *U.S. News & World Report,* 4 November 1985, pp. 51, 54.

The World Almanac and Book of Facts 1983, 1986. New York: Newspaper Enterprise Association, 1982, 1985.

Further Reading

Campbell, S. K. *Flaws and Fallacies in Statistical Thinking.* Englewood Cliffs, NJ: Prentice-Hall, 1974. Chapter 5 discusses averages.

Huff, D. *How to Lie with Statistics.* New York: Norton, 1954. Chapter 2 discusses averages.

CHAPTER 4

MEASURES OF DISPERSION

EXAMPLE 4.1 One of the most widely prescribed drugs in North America and Europe is the tranquilizer diazepam, better known by its trade name Valium. Because of its widespread use, it is important to identify any unwanted side effects so that appropriate precautions may be taken.

Researchers in the Netherlands conducted a carefully controlled investigation of diazepam's effects on a person's highway driving performance. In particular, the driving performance of nine police driving instructors was measured during actual highway driving in a car that was specially equipped for accurate recording of speed and left–right position on the highway. The results showed that the *average* left–right position of the vehicle was well within the lane boundaries whether the driver was under the influence of diazepam or not. However, for eight of the nine drivers, the *variability* in the vehicle's left–right position was greater under the influence of diazepam than in the control conditions; that is, the drivers tended to weave from side to side more than usual, sometimes even infringing on the adjacent lane of traffic. Obviously, it is not sufficient to be in one's own lane "on the average," if the average is achieved by drifting across the lane boundaries to the left and to the right about equally often. The researchers concluded that "potentially dangerous impairment was inferred from the reactions of some subjects."

(Source: O'Hanlon *et al.*, 1982.)

An average is not always the relevant summary to answer the questions being asked of a set of data. As the preceding example illustrates, variability or dispersion is sometimes of interest in its own right, quite apart from any interest that may attend the average. In the illustration, even if the average of the measurements of the car's distance from the lane boundary looked reasonable, there would still be considerable interest in how spread out the measurements were, or in the fluctuation of the distances.

The term **measure of dispersion** is used to refer to a number that summarizes data in terms of their variability or amount of fluctuation around the average—how dispersed or scattered the observations are.

> A **measure of dispersion** is any single number that describes a set of data by indicating how spread out the observations are along the real-number line. **(4.1)**

In this chapter we will introduce measures of dispersion under the general headings of (a) the range, (b) the mean deviation, (c) the standard deviation, and (d) quantile measures of dispersion. The emphasis will be on the third of these, the standard deviation.

4.1 THE RANGE

An obvious measure of the variability or dispersion of a finite set of data is the difference between the largest value and the smallest value in the set. This difference, called the **range**, is computed as

$$\text{Range} = x_{\max} - x_{\min} \tag{4.2}$$

EXAMPLE 4.2 Ten residence rooms of male students were selected at random on a college campus, and the number of posters displayed in each room was counted and recorded as follows:

8, 1, 7, 5, 1, 8, 2, 1, 2, 18

In Example 3.18, we saw that the median number of posters in these rooms was 3.5. The median is an average, and it tells us that a typical number of posters in a room was about 3 or 4, but it does not tell us anything about how many rooms were close to that average.

As a measure of dispersion, the range indicates whether all the rooms had about the same number of posters, or whether there was considerable variability among the rooms. Calculated as the difference between the extremes, the range of these data is

$$\text{Range} = x_{\max} - x_{\min} = 18 - 1 = 17$$

The range tells us that there were considerable differences among the rooms in terms of the number of posters displayed.

(Source: Student project by J. Kruschel.)

The **range** can be a useful measure of dispersion, especially in small sets of data. Its main advantage is its simplicity, which facilitates both calculation and comprehension. It is easily explained to a nonstatistical audience.

One drawback of the range relates to the fact that it is calculated from the extremes. We saw in Section 3.1 that the maximum and minimum sometimes give a misleading impression of a set of data if there are one or two unusually large or unusually small observations in the set. As a result, the range is rather unstable in the sense that a single wild observation—one aberrant value—can change the value of the range drastically. (In Example 4.2, if the investigation had stopped after 9 rather than 10 rooms, the range would have been $8 - 1 = 7$, rather than 17, conveying a substantially different impression of the variability in the data.)

The fact that the range is based on information from only the two extreme observations implies that it tells us nothing about the dispersion of the observations between the extremes.

EXAMPLE 4.3 Each of the following two artificial data sets contains nine observations, arranged in ascending order:

Data set A:	1, 4, 5, 5, 5, 5, 5, 6, 9
Data set B:	1, 1, 1, 2, 5, 8, 9, 9, 9

In both cases, the mean and median are 5 and the range is $9 - 1 = 8$. However, the dispersion of the observations is obviously different in the two cases: In data set A, seven of the nine observations are within 1 unit of the mean and median; in data set B, only one of the observations is.

A measure of dispersion based on information from all the observations not only would detect differences in dispersion between the extremes but also would tend to be a more stable measure.

Besides the practical difficulties that result from reliance on only two observations, the range also suffers a theoretical disadvantage in that it lacks certain mathematical properties that are convenient for some of the uses to which we will want to put a measure of dispersion. Nonetheless, used with caution, the range should not be overlooked as a quick and easy measure of dispersion.

4.2 MEAN DEVIATION

One approach to including every observation in the measurement of dispersion is to consider how far each observation x_i is from some reference point such as the mean $\bar{x}$. The differences $x_i - \bar{x}$, one of which can be calculated for each observation, are referred to as **deviations from the mean**, or sometimes simply as **deviations**. The greater the dispersion or variability in a set of data, the larger in absolute value these deviations will be on the average.

It is easy to demonstrate that the mean of the actual deviations is not a measure of dispersion, since it always equals zero. Adding the deviations in a sample of n observations and dividing by the number n of deviations, we obtain the mean of the actual deviations as

$$\frac{\sum (x_i - \bar{x})}{n} = \frac{0}{n} = 0 \qquad \text{by Expression (3.6)} \tag{4.3}$$

This equation follows directly from the center-of-gravity property of the mean: The sum of the positive signed deviations is, by definition of the mean, exactly balanced by the sum of the negative signed deviations.

For a measure of dispersion, the relevant feature of deviations from the mean is their absolute value, ignoring the sign. The sign indicates the *direction* of the deviation—above or below the mean—but only the *size* of the deviation is relevant in a measure of dispersion. The arithmetic mean of the absolute values of the deviations from the mean is a measure of dispersion called the **mean deviation**:

$$\text{Mean deviation} = \frac{\sum^{n} |x_i - \bar{x}|}{n} \tag{4.4}$$

for samples of n observations x_i with a mean of $\bar{x}$, or

$$\text{Mean deviation} = \frac{\sum^{N} |x_i - \mu|}{N} \tag{4.5}$$

for populations of N observations x_i with a mean of μ. The term **mean deviation** is actually short for *mean absolute deviation.* It is an average amount of dispersion from the mean of a set of data.

EXAMPLE 4.4 In Example 3.18, we found that the mean of the number of posters in 10 residence rooms was 5.3. The calculation of the mean deviation may now be summarized as follows for the same data.

Number of posters x_i	Deviation $x_i - \bar{x}$	Absolute deviation $\|x_i - \bar{x}\|$
8	2.7	2.7
1	−4.3	4.3
7	1.7	1.7
5	−.3	.3
1	−4.3	4.3
8	2.7	2.7
2	−3.3	3.3
1	−4.3	4.3
2	−3.3	3.3
18	12.7	12.7
$\sum x_i = 53$	$\sum (x_i - \bar{x}) = 0.0$	$\sum \|x_i - \bar{x}\| = 39.6$
$\bar{x} = \frac{53}{10} = 5.3$		Mean deviation $= \frac{39.6}{10} = 3.96$

The mean deviation is approximately 4.0. It tells us that although some rooms differed from the mean by more than 4 posters and some by less than 4 posters, a typical difference between an observation and the mean was about 4 posters.

In Example 4.2, we found that the range of these data was 17; but if we had omitted the last room, the range would have dropped to 7. The use, in the mean deviation, of information from all the observations lessens the effect of the occasional unusual value. In this case, by omitting the last room and recalculating the mean, we find that the mean deviation is reduced only from 4.0 to 2.8.

Although the mean deviation provides a reasonable measure of dispersion, it is used rather rarely. The involvement of absolute values creates complications in more advanced theoretical work related to topics in inferential statistics. Fortunately, there is another simple way—a way that has happier theoretical properties than taking absolute values—to dispose of the signs of the deviations.

4.3 VARIANCE AND STANDARD DEVIATION

The most commonly used measures of dispersion in statistics are the **standard deviation** and its close relative, the **variance**. Like the mean deviation, these measures are based on the values of the deviations from the mean, which indicate how far each observation x_i is away from the mean. However, the standard deviation and variance are distinguished from the mean deviation by their use of squared deviations rather than absolute values to dispose of the signs of the deviations. Let us consider the variance first.

Variance

It is irrelevant to a measure of dispersion whether an observation is above or below the mean; what matters is how far it is from the mean, not which direction. Thus, we can circumvent the insistence of the deviations from the mean to add up to zero [see Expression (3.6)] by squaring each of the deviations. This operation retains the critical information contained in the deviations in that squared deviations continue to reflect distance from the mean: The greater the distance between an observation and the mean, the greater the value of the squared deviation will be. Moreover, observations equidistant from the mean are converted to equal squared deviations; for example, deviations of −3 and of +3 each take on the value 9 as a squared deviation. Thus, squared deviations can form the basis for a measure of dispersion.

The **variance of a population** is usually defined as the mean of the squared deviations from the mean. It is denoted by the symbol σ^2 (read "sigma-squared," because σ is the lower-case Greek letter sigma), and the definition may therefore be written symbolically as follows for a population of N observations x_i with mean μ:

$$\sigma^2 = \frac{\sum^{N} (x_i - \mu)^2}{N} \tag{4.6}$$

If we compare Expression (4.6) with (4.5), we find that the only change is the use of squaring instead of absolute values. The **variance** is simply an average of the squared deviations.

In practice, we have sample data more often than population data, and sample data call for a slightly different formula. The **variance of a sample** of n observations is denoted by the symbol s^2 and is defined as

$$s^2 = \frac{\sum^{n} (x_i - \bar{x})^2}{n - 1} \tag{4.7}$$

Notice that the divisor in the formula for the sample variance is $n - 1$ rather than n. The reason for this adjustment is related to the use of the sample variance s^2 in inferential statistics. Division by n in the formula for the sample variance would tend on the average to underestimate the population variance σ^2 by a factor of $(n - 1)/n$, giving a **biased estimate** of σ^2. Thus, the sample variance is defined with a divisor of $n - 1$ so that s^2 will be an **unbiased estimate** of σ^2. (An unbiased estimate of a parameter is one which neither underestimates nor overestimates the parameter on the average, although it is not necessarily exactly correct on any one particular occasion.) Nonetheless, we can still think of the variance as a kind of average of squared deviations.

EXAMPLE 4.5 Continuing with the data from Example 4.4, let us calculate the sample variance of the number of posters displayed in 10 residence rooms, to illustrate the use of Expression (4.7). The calculations may be summarized as follows:

Number of posters x_i	Deviation $x_i - \bar{x}$	Squared deviation $(x_i - \bar{x})^2$
8	2.7	7.29
1	−4.3	18.49
7	1.7	2.89
5	−.3	.09
1	−4.3	18.49
8	2.7	7.29
2	−3.3	10.89
1	−4.3	18.49
2	−3.3	10.89
18	12.7	161.29
$\sum x_i = 53$	$\sum (x_i - \bar{x}) = 0.0$	$\sum (x_i - \bar{x})^2 = 256.10$
$\bar{x} = \frac{53}{10} = 5.3$		$s^2 = \frac{256.10}{10 - 1} = 28.46$

As a partial check on the computations, we verify that the sum of the deviations in the second column equals zero as prescribed by Expression (3.6).

The sample variance, $s^2 = 28.46$, is not convenient for interpretation because it is in squared units. The variance tells us that the average of the squared deviations is approximately 28 for these data, but we need to express this value in a different form to understand better what it tells us about the data.

As we will see in later chapters, the variance has many important applications in inferential statistics. However, its usefulness as a descriptive statistic is curtailed by the fact that its value is not in the same units as the observations themselves but rather in squared units. A more easily interpreted measure of dispersion is the standard deviation, to which we now turn.

Standard Deviation

We have seen that the variance is an average of the squared deviations of the observations from the mean. The **standard deviation** is the positive square root of the variance, where the operation of taking the square root converts the variance value back into the original units of measurement of the observations.

In symbols, the **standard deviation of a population** of N observations x_i with mean μ is given by

$$\sigma = \sqrt{\frac{\Sigma (x_i - \mu)^2}{N}} \tag{4.8}$$

The **standard deviation of a sample** of n observations is

$$s = \sqrt{\frac{\Sigma (x_i - \bar{x})^2}{n - 1}} \tag{4.9}$$

These formulas are simply the positive square roots of Expressions (4.6) and (4.7), respectively. (It should be noted that the fact that s^2 is an unbiased estimate of σ^2 does not imply that s is an unbiased estimate of σ; in fact, it is not. But since s^2 is an unbiased estimate of σ^2, s is a widely used and, as it turns out, not unreasonable estimate of σ.)

Sometimes a subscript is attached to the standard deviation symbol to identify the variable under consideration; for example, σ_x and σ_y indicate the population standard deviations of the variables x and y, respectively, and s_x and s_y indicate the sample standard deviations of variables x and y.

EXAMPLE 4.6 In Example 4.5, we found that the variance of the number of displayed posters per room was $s^2 = 28.46$. Hence the standard deviation is $s = \sqrt{28.46} = 5.33$, or approximately 5.3 posters. (It is just coincidence that the standard deviation happens to have approximately the same value as the mean in these data.)

Although the standard deviation is generally larger than the mean deviation (which we found to be 4.0 for these data in Example 4.4), the two measures are interpreted in much the same way. The standard deviation is an alternative kind of "average" deviation, indicating how much the observations tend to deviate from the mean.

EXAMPLE 4.7 Example 4.3 introduced the following two artificial data sets:

Data set A:	1,	4,	5,	5,	5,	5,	5,	6,	9
Data set B:	1,	1,	1,	2,	5,	8,	9,	9,	9

Both have a mean of 5 and a range of 8, but their internal dispersions are very different. Their sample standard deviations are 2.1 and 3.8, respectively, reflecting the fact that most of the observations in data set A are closer to the mean than are the corresponding observations in data set B. Hence, the "standard" size of deviation is smaller in data set A than in B.

As suggested in Examples 4.6 and 4.7, the standard deviation may be interpreted as a kind of average deviation if we remember that the term *average* has a broader meaning in statistics than in common parlance (see Section 3.2). Generally, some deviations are larger than the standard deviation and some smaller; the standard deviation gives the amount of deviation that is in some sense an "average" deviation or a "typical" deviation, and hence the "standard" deviation. When interpreting the standard deviation as an average amount of deviation, however, keep in mind that the particular kind of average represented by the standard deviation is one that gives a value larger than the mean deviation (except in certain limiting cases, where they are equal). For approximately symmetric unimodal distributions, the mean deviation often turns out to be about 80% of the value of the standard deviation. This figure should not be taken too literally but is cited to help clarify what kind of "average" deviation the standard deviation is.

Computational Formulas

The formulas for the variance and standard deviation given in Expressions (4.6), (4.7), (4.8), and (4.9) are called **definitional formulas**, because they provide rather transparent symbolic statements of the corresponding definitions. They indicate that variance is the mean of squared deviations from the mean and that the standard deviation involves taking the square root of the variance in order to express, in the original units of measurement, a standard amount of deviation from the mean. The definitional formulas are helpful in understanding the concepts being defined.

The definitional formulas are not always the simplest ones to work with from a computational point of view. Straightforward algebraic manipulation allows derivation of **computational formulas** that are algebraically equivalent to the corresponding definitional formulas (that is, they are not just approximations) but that simplify hand calculations in many cases. For example, the computational formula for the sample variance may be derived from the definitional formula in Expression (4.7) as follows:

$$s^2 = \frac{\sum^n (x_i - \bar{x})^2}{n - 1} \qquad \text{from Expression (4.7)}$$

$$= \frac{\sum^n (x_i^2 - 2x_i\bar{x} + \bar{x}^2)}{n - 1} \qquad \text{by expanding the square}$$

$$= \frac{\sum^n x_i^2 - 2\bar{x}\sum^n x_i + \sum^n \bar{x}^2}{n - 1} \qquad \text{by Characteristics 1, 2, and 3 of Section 1.6}$$

$$= \frac{\sum^n x_i^2 - 2\bar{x}\sum^n x_i + n\bar{x}^2}{n - 1} \qquad \text{by Characteristic 4 of Section 1.6}$$

$$= \frac{\sum x_i^2 - 2\bar{x}(n\bar{x}) + n\bar{x}^2}{n - 1} \qquad \text{by Expression (3.5)}$$

$$= \frac{\sum x_i^2 - 2n\bar{x}^2 + n\bar{x}^2}{n - 1}$$

$$= \frac{\sum x_i^2 - n\bar{x}^2}{n - 1}$$

$$= \frac{\sum x_i^2 - n\left(\frac{\sum x_i}{n}\right)^2}{n - 1} \qquad \text{by Expression (3.3)}$$

$$= \frac{\sum x_i^2 - \frac{(\sum x_i)^2}{n}}{n - 1}$$

Such algebraic manipulation yields the following **computational formulas**.

Population variance:

$$\sigma^2 = \frac{\sum x^2 - \frac{(\sum x)^2}{N}}{N} \tag{4.10}$$

Sample variance:

$$s^2 = \frac{\sum x^2 - \frac{(\sum x)^2}{n}}{n - 1} \tag{4.11}$$

Population standard deviation:

$$\sigma = \sqrt{\frac{\sum x^2 - \frac{(\sum x)^2}{N}}{N}} \tag{4.12}$$

Sample standard deviation:

$$s = \sqrt{\frac{\sum x^2 - \frac{(\sum x)^2}{n}}{n - 1}} \tag{4.13}$$

A major advantage of these computational formulas is that they bypass the step of subtracting the mean from every observation before squaring. This advantage is most significant if the mean does not work out evenly or if there are very many observations.

EXAMPLE 4.8 To illustrate the use of the computational formulas, let us calculate the sample standard deviation of the residence-room poster data from Example 4.5. Notice that the only quantities we require in Expression (4.13) are the sum of the observations, the sum of the squared observations, and the sample size. The calculations may be organized in the following steps, as illustrated in the table: First, list the observations in a column (or store them in a memory bank in your calculator) and find their sum. Second, list the squared observations in a column (or store them in a memory bank in your calculator) and find their sum. Substitute the appropriate values into Expression (4.13).

Number of posters x	Squared observation x^2
8	64
1	1
7	49
5	25
1	1
8	64
2	4
1	1
2	4
18	324
$\sum x = 53$	$\sum x^2 = 537$

$$s = \sqrt{\frac{537 - \frac{(53)^2}{10}}{10 - 1}}$$

$$= \sqrt{\frac{256.10}{9}}$$

$$= \sqrt{28.46}$$

$$= 5.33$$

The standard deviation obtained from the computational formula is exactly equal to that obtained in Example 4.6 from the definitional formula, as it must be.

In addition to their advantages in terms of time and effort, the computational formulas are also less likely to lead to rounding errors than are the definitional formulas. The mean that appears in the definitional formulas refers to the mathematically defined exact value, not the rounded value that would be used for reporting results. This particular temptation to premature rounding does not exist in the computational formulas. Although there may be exceptions in unusual cases, such as when the observations are so large that their squares exceed the capacity of the calculating equipment, in most cases the computational formulas require less work and are more likely to lead to correct answers.

Even if computational formulas are used, questions often arise concerning when to round numerical values and when not to round. Two points should be mentioned here. The first is **general advice for calculating:**

> Always use *unrounded* quantities in all stages of all calculations. **(4.14)**

The quantities in statistical formulas, whether definitional or computational, are the mathematically defined, exact values, so wherever possible we substitute unrounded quantities into formulas for calculation.

The second point is a **guideline for reporting means and standard deviations:**

> After all calculations have been completed, report the standard deviation rounded to two significant figures, and report the mean rounded to the same digit position as the accompanying standard deviation is rounded. **(4.15)**

This advice and other matters concerning how to round and how many digits to report are discussed in Appendix 1.

EXAMPLE 4.9 The profit earned by a farmer who buys, feeds, and subsequently sells cattle depends in large part on the cattle's weight gain between buying and selling. Weight gains during three months of pasture grazing were recorded for a sample of 12 Hereford steers purchased shortly after weaning. Calculation of the mean, variance, and standard deviation of the weight gains may be summarized as follows:

Weight gain (pounds) x	x^2
205	42,025
285	81,225
190	36,100
305	93,025
295	87,025
365	133,225
250	62,500
245	60,025
275	75,625
260	67,600
290	84,100
235	55,225
$\sum x = 3{,}200$	$\sum x^2 = 877{,}700$

$$\bar{x} = \frac{3{,}200}{12} = 266.6667$$

$$s^2 = \frac{877{,}700 - \dfrac{(3{,}200)^2}{12}}{12 - 1} = \frac{24{,}366.66667}{11} = 2215.151515$$

$$s = \sqrt{2215.151515} = 47.0654$$

The calculations show that the mean weight gain by the 12 steers during three months of pasture grazing was 267 pounds, and the standard deviation was 47 pounds. Roughly speaking, we can interpret the standard deviation as indicating that it was not especially unusual for weight gains in this sample to be as little as 267 − 47 = 220 pounds or as much as 267 + 47 = 314 pounds. Some weight gains were closer to the mean than those amounts, and some were farther from the mean; but 47 pounds was a "standard" amount of difference from the mean of 267 pounds.

(Source: Student project by P. Vandersluis.)

* Approximating the Standard Deviation of Grouped Data

The standard deviation generally cannot be calculated exactly from grouped data, where the individual observations are grouped into the categories of a frequency distribution. However, a good approximation is usually possible. The usual procedure is identical to that for approximating the mean [Expression (3.10)]:

> To **approximate the mean and the standard deviation** of grouped data, assume that every observation is equal to its category mark. **(4.16)**

We simply substitute the resulting n values into the usual formulas for the mean and the standard deviation.

EXAMPLE 4.10 Example 3.10 illustrated the approximation of the mean of grouped data, using the first four columns of the following table. The fourth column yielded an approximation of the

* This symbol identifies material that is not prerequisite for later topics in this book.

sum of the observations by using the relevant category mark in place of each observation.

Test score x	Frequency	Category mark	Category mark times frequency	(Category mark)^2	(Category mark)^2 times frequency
20–24	2	22	44	484	968
25–29	2	27	54	729	1,458
30–34	3	32	96	1,024	3,072
35–39	13	37	481	1,369	17,797
40–44	10	42	420	1,764	17,640
45–49	7	47	329	2,209	15,463
50–54	2	52	104	2,704	5,408
55–59	2	57	114	3,249	6,498
	$n = 41$		$\sum x \simeq 1{,}642$		$\sum x^2 \simeq 68{,}304$

Similarly, the sum of the squared observations may be approximated by

$$\sum x^2 \simeq (22^2 + 22^2) + (27^2 + 27^2) + \cdots + (57^2 + 57^2)$$

which is the same as multiplying each squared category mark by its frequency and adding those products:

$$\sum x^2 \simeq \sum [(\text{category mark})^2 \times \text{frequency}]$$

$$= (22^2)(2) + (27^2)(2) + (32^2)(3) + \cdots + (57^2)(2) = 68{,}304$$

Therefore, the approximate standard deviation is

$$s = \sqrt{\frac{\sum x^2 - \dfrac{(\sum x)^2}{n}}{n - 1}} \simeq \sqrt{\frac{68{,}304 - \dfrac{1{,}642^2}{41}}{40}} = 8.0$$

It should be emphasized that the calculations of the mean and the standard deviation from grouped data are based on the same formulas as for individual data. The only difference is that each observation (whose exact value is not available from the frequency tabulation) is replaced by its category mark as a sort of best guess as to its value. The precision of the resulting approximation depends on such factors as the number of observations, the size of the category intervals, and the shape of the distribution, but it is often found to be satisfactory in cases where the only data available are the grouped data.

4.4 INTERPRETATION OF THE STANDARD DEVIATION

Relationship of the Standard Deviation to the Range

It is important to understand what the standard deviation tells us about a set of data. In this subsection we will look at several simple sets of data to illustrate the relationship between the value of the standard deviation and the observations from which it was computed.

The standard deviation (a standard amount of deviation from the mean) must always be smaller than the range (the distance from one extreme to the other). Although there is no general formula relating the standard deviation and the range, it can be shown that the population standard deviation can never be more than one-half of the population range, and the sample standard deviation can never be more than $1/\sqrt{2} = 71\%$ of the sample range. Conversely, the range is always at least twice the population standard deviation and at least $\sqrt{2} = 1.4$ times the sample standard deviation. As the number of observations increases, the standard deviation tends to become a slightly smaller proportion of the range, but here is a general **rule of thumb relating the standard deviation and the range:**

> For small to moderate numbers of observations, we usually find that the standard deviation is roughly 1/4 of the range or, conversely, that the range is roughly 4 times the standard deviation. **(4.17)**

This rule of thumb is not intended to give exact values: "roughly 1/4" should be taken to mean "typically between 1/6 and 1/2," and "roughly 4" as "typically between 2 and 6." Taken in this vein, Expression (4.17) can be applied in cases with up to about 100 or so observations. It provides a ballpark figure for checking your calculation of a standard deviation, as well as some guidance for interpreting standard deviations.

EXAMPLE 4.11 Consider first the fictitious observations 5, 6, 7, 8, and 9. You can readily verify that the mean of these numbers is $(5 + 6 + 7 + 8 + 9)/5 = 7$. The deviations of the observations from the mean are -2, -1, 0, 1, and 2, respectively. Since the standard deviation is an average amount of deviation from the mean, we know that it cannot be larger than 2, which is the maximum amount by which any of the observations deviates from the mean. We might expect an average amount of deviation to be between 1 and 2 in this case, and indeed you can verify that the sample standard deviation is $s = 1.6$. The range, which is $9 - 5 = 4$ for these data, is $4/1.6 = 2.5$ times the sample standard deviation, in keeping with the general rule of thumb in Expression (4.17).

EXAMPLE 4.12 Now consider the fictitious observations 45, 46, 47, 48, and 49. Their mean is 47, and the deviations from the mean are -2, -1, 0, 1, and 2, respectively. Although the mean is different from that in Example 4.11, the spread or dispersion of the numbers is exactly the same, and hence the sample standard deviation is again $s = 1.6$.

EXAMPLE 4.13 Similarly, the numbers 198, 199, 200, 201, and 202 have the same dispersion as the previous two sets of numbers, so that once again the sample standard deviation turns out to be $s = 1.6$.

As the three preceding examples suggest, adding a constant to every observation or subtracting a constant from every observation in a set of data does not change the standard deviation. Such addition or subtraction changes the mean (and other measures of location) by a corresponding amount, but it has no effect on deviations from the mean or on measures of dispersion.

EXAMPLE 4.14 The numbers 50, 60, 70, 80, and 90 are each 10 times the respective numbers in Example 4.11. Their mean is 70, and the deviations from the mean are −20, −10, 0, 10, and 20, respectively. These deviations are each 10 times the corresponding deviations in Example 4.11, and the sample standard deviation is $s = 16$, which is likewise exactly 10 times that of Example 4.11.

Example 4.14 illustrates the general principle that the formula for the standard deviation gives the standard deviation in the same measurement units as the observations. If the five numbers in Example 4.11 were observations measured in centimeters, and those of Example 4.14 were the same observations measured in millimeters, then the standard deviation would be given correctly as either 1.6 cm or 16 mm.

EXAMPLE 4.15 Without performing the calculation, what would you expect to be the approximate standard deviation of the numbers 6.21, 6.29, 6.31, 6.31, 6.45? A simple approach is to use the rule of thumb relating the range to the standard deviation, as given in Expression (4.17). Observing that the range is .24, we would expect the standard deviation to be about 1/4 of that value, or about .06, perhaps between .04 and .12. As you may verify, the sample standard deviation is actually $s = .086$, consistent with the rule of thumb.

EXAMPLE 4.16 Finally, consider the 101 consecutive integers 800, 801, 802, . . . , 899, 900. The range of these numbers is 900 − 800 = 100, and Expression (4.17) suggests that the standard deviation should be in the vicinity of 25, or at least between about 17 and 50. The sample standard deviation turns out to be $s = 29$.

These examples should help clarify what information about the data is conveyed by the standard deviation. As you calculate further standard deviations in the exercises, you will find that evaluating whether your answer is a reasonable one for the data at hand can alert you to computational errors. Likewise, it is a good practice to take time to observe the relationship between the correct standard deviation and the data from which it was computed.

* The Bienaymé–Chebyshev Inequality

We have seen that the standard deviation indicates a standard or average amount by which observations deviate from the mean. In addition, its typical relationship to the range in small sets of data can assist us in understanding what the standard deviation tells us about the spread or dispersion of a set of observations. A more precise statement of the relationship between the standard deviation and the data from which it is computed can be proved as a formal theorem. The theorem was initially proved in 1853 by the French mathematician Irénée Jules Bienaymé (1796–1878) and later independently discovered and extended by the Russian mathematician Pafnuti L. Chebyshev (1821–1894). It is now known as the **Bienaymé–Chebyshev inequality**.

The Bienaymé–Chebyshev inequality may be thought of as an elaboration of the general principle that the standard deviation is a standard amount by which observations deviate from the mean. In effect, because the minimum possible distance from the mean is zero, and because any deviations greater than the standard amount of deviation must somehow be offset by other deviations less than the standard amount of deviation, knowledge of the standard deviation sets a limit on the number of observations that can be very far from the mean. The **Bienaymé–Chebyshev inequality** may be stated as follows:

> In any population or sample, no more than $1/z^2$ of the observations differ from the mean by z or more times the standard deviation (for any $z > 1$); that is, the maximum proportion of observations in the tails of a distribution is $1/z^2$. **(4.18)**

The theorem is illustrated in Figure 4.1, where the value of the standard deviation is abbreviated SD. The value of z is arbitrary and is set at whatever number of standard deviation units away from the mean we want to consider. As z becomes larger, the shaded areas in the diagram become smaller.

Table 4.1 illustrates Expression (4.18) for selected values of z. The middle column shows that the farther we move (counting in standard deviation units) from the mean of a distribution, the smaller the percentage of observations that can possibly be so far away from the mean. You should check your understanding of Expression (4.18) by using it to obtain the values in the middle column of the table. Notice in the first row of the table that for $z = 1.0$, the inequality tell us nothing; the inequality provides information only for z-values greater than 1.0.

Because the Bienaymé–Chebyshev inequality says how many observations can be *far* from the mean, it has implications for how many must be relatively *close* to the mean. The right-hand column of Table 4.1 illustrates these implications for central areas by means of percentages that are simply the

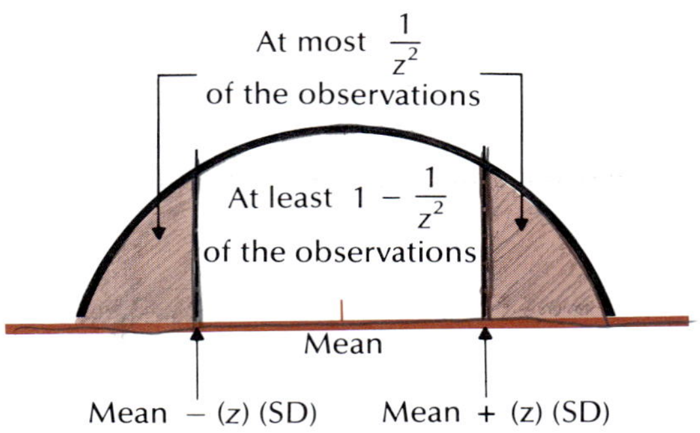

Figure 4.1 Schematic illustration of the distribution of observations according to the Bienaymé–Chebyshev inequality. SD is the standard deviation, and z is set so that z times the standard deviation is the distance from the mean that we want to consider.

TABLE 4.1 Tail areas and corresponding central areas of any distribution according to the Bienaymé–Chebyshev inequality.

z	Percentage of observations differing from the mean by z or more times the standard deviation	Percentage of observations differing from the mean by less than z times the standard deviation
1.0	No more than 1/1 = 100% in the tails	At least 0/1 = 0% in the central area
2.0	No more than 1/4 = 25% in the tails	At least 3/4 = 75% in the central area
3.0	No more than 1/9 = 11% in the tails	At least 8/9 = 89% in the central area
4.0	No more than 1/16 = 6% in the tails	At least 15/16 = 94% in the central area
5.0	No more than 1/25 = 4% in the tails	At least 24/25 = 96% in the central area

complements of those for tail areas. As we count in standard deviation units in both directions away from the mean of any distribution of observations, the Bienaymé–Chebyshev inequality guarantees us certain increasing percentages of the observations in the central area. There may actually be more observations in the central area than the percentage calculated from the inequality, but the inequality guarantees us *at least* the specified percentage.

The important point for the present discussion is that the size of the standard deviation implies certain limits on the number of observations that can be various distances away from the mean. The Bienaymé–Chebyshev theorem is called an inequality because it does not say that *exactly* $1/z^2$ of the observations differ from the mean by z or more times the standard deviation, but rather that *no more than* $1/z^2$ of the observations differ by that much. In many cases there will be considerably fewer observations that far from the mean—sometimes none—but there cannot be more. This information can be useful in interpreting a reported standard deviation, at least in a rough way.

EXAMPLE 4.17 The sample of 12 Hereford steers in Example 4.9 had a mean weight gain of 267 pounds and a standard deviation of 47 pounds.

The Bienaymé-Chebyshev inequality tells us that no more than 1/4 = 25% of the 12 steers (that is, no more than 3 steers) had weight gains that were 2 or more standard deviation units away from the mean. Since the standard deviation was calculated as 47.0654 pounds, 2 standard deviation units is 2 × 47.0654 = 94.1308 pounds. Thus, 2 standard deviation units below the mean is 266.6667 − 94.1308 = 172.5359 pounds; and 2 standard deviation units above the mean is 266.6667 + 94.1308 = 360.7975 pounds. The inequality, therefore, asserts that at least 12 − 3 = 9 steers in this sample had weight gains between 173 and 361 pounds and, conversely, no more than 3 steers were at or beyond those limits. In fact, only one steer was beyond those limits, consistent with the inequality's claim.

The preceding paragraph illustrated the Bienaymé–Chebyshev inequality for an arbitrarily chosen z-value of 2.0. We could have used any other z-value (greater than 1.0) that may be of interest in the particular problem at hand. For example, a farmer may want to know what percentage of Hereford steers might gain less than 200 pounds when fed under the conditions of this investigation. A weight gain of 200 pounds is 200 − 266.6667 = −66.6667, or 66.6667 pounds below the mean. This value is 66.6667/47.0654 = 1.42 standard deviation units below the mean. From Expression (4.18), we know that no more than $1/1.42^2$ = 50% of the observations differ from the mean by 1.42 or more times the standard deviation. Thus, the inequality tells us that no

more than 50% or 6 of the 12 steers could have gained 200 pounds or less. In fact, only one did, again consistent with the inequality's claim.

In practice, we would not apply the Bienaymé–Chebyshev inequality to data sets where we have access to all the individual observations, as we did in Example 4.17. We would just look at the actual observations to see how many fell in various intervals, perhaps displaying the information in a frequency table or histogram. Example 4.17 allowed us to verify the correctness of the inequality's claims for a particular set of data. The usual application of the inequality occurs when data have been summarized and we want to interpret a reported standard deviation.

EXAMPLE 4.18 In a test of highway driving ability, drivers were monitored for their ability to maintain a constant speed of 100 km/h when instructed to do so. A group of police driving instructors performed this task with an overall mean speed of 100.0 km/h and a standard deviation of 1.9 km/h.

How well did they do? Their mean speed was exactly correct, and the fact that the "standard" amount of deviation from the mean was only 1.9 km/h tells us that for the most part they kept quite close to the designated speed. The Bienaymé–Chebyshev inequality allows us to quantify this statement. We can consider, for example, how much of the time the speed was between 95 and 105 km/h. These values are 5 km/h from the mean, or $5/1.9 = 2.63$ standard deviations from the mean. Using Expression (4.18), we find that $1/2.63^2 = 14\%$ is the maximum tail area, so the speed was in the central area between 95 and 105 km/h at least 86% of the time. Be sure to notice the words *at least*; the speed may have been within the 95–105 interval considerably more than 86% of the time, but the Bienaymé–Chebyshev inequality guarantees that it was there at least 86% of the time.

(Source: O'Hanlon *et al.*, 1982.)

The Bienaymé–Chebyshev inequality is quite general, applying to any set of observations, regardless of the nature of the distribution—regardless, for example, of whether the distribution is symmetric or skewed, leptokurtic or platykurtic. This generality is both an advantage and a disadvantage of the inequality. It is an advantage in that it allows us to make statements that are guaranteed to be correct even if we know nothing about the shape of the distribution—and often we can only guess at the shape of a population distribution. It is a disadvantage in that it results in much weaker statements than can be made by taking into account the shape of the distribution. Later in this book we will consider some standard distributions, knowledge of which will allow us to make much stronger statements in cases where they are applicable.

EXAMPLE 4.19 In the general North American population, IQ test scores have a mean of 100 and a standard deviation of 15. On the basis of the Bienaymé–Chebyshev inequality, we can say that at least 75% of the population score between 70 and 130 (that is, within 2 standard deviations from the mean), and that no more than 25% are outside those limits. If we ask about the percentage who score 130 or more, the inequality allows us to say only that it is no more than 25%: It specifies nothing about how the 25% who are outside the 70–130 limits are divided between the two tails of the distribution.

The approximate shape of this particular distribution is in fact known. We will be studying the distribution in Chapter 7, but for now we simply note that knowledge of the distribution enables us to say that approximately 95% (rather than at least 75%) of the population score between 70 and 130. Moreover, approximately 2.5% score 130 or more, and approximately 2.5% score 70 or less.

The Bienaymé–Chebyshev inequality's statements are correct, but considerably more precise statements can be made when the shape of the relevant distribution is known.

Rule of Thumb for Symmetric Unimodal Distributions

We have said that the standard deviation is some kind of average amount by which observations deviate (or differ) from their mean. It follows that deviations greater than the *standard* deviation must in some sense be balanced by deviations less than the *standard* deviation. However, the smallest deviation possible is no deviation at all—that is, a deviation of zero for an observation equal to the mean. If there are observations very far from the mean—that is, with large deviations—then the value of the standard deviation will be large enough so that the "balance" of deviations greater and smaller than the *standard* deviation is maintained.

The preceding principles dictate a certain pattern that statisticians have observed in the relationship between many data sets and their standard deviation. The statement of this pattern represents an approximation based on experience, and it is summarized in the following **rule of thumb for**

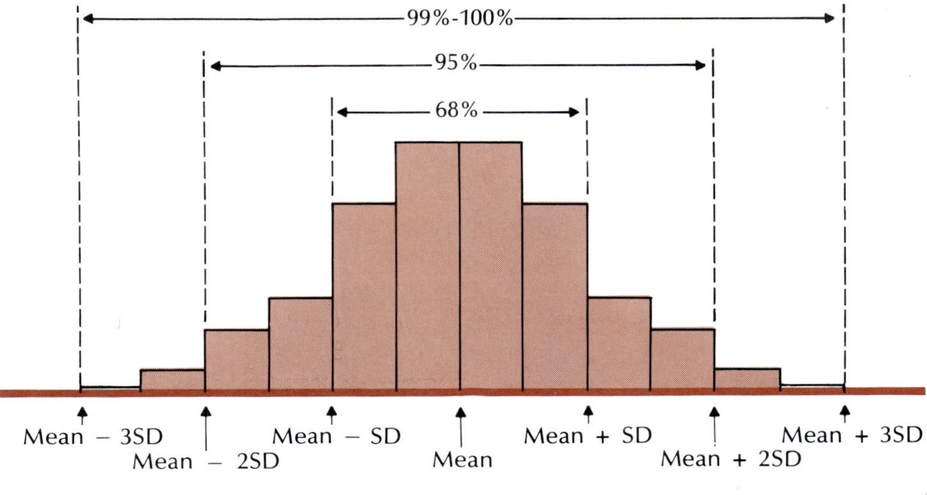

Figure 4.2 Schematic illustration of the distribution of observations according to the empirical rule of thumb, where SD is the value of the standard deviation.

interpreting the standard deviation in symmetric unimodal distributions:

> For distributions that are more or less symmetric and unimodal, approximately 68% of the observations fall within 1.0 standard deviation above and below the mean, approximately 95% fall within 2.0 standard deviations above and below the mean, and approximately 99–100% of the observations fall within 3.0 standard deviations above and below the mean. **(4.19)**

This rule of thumb, illustrated in Figure 4.2, is sometimes called the **empirical rule** because it is an approximation based on experience rather than a mathematically exact result. It provides a reasonably good approximation for many shapes of distribution, as long as there is a greater concentration of observations near the center of the distribution than in both tails.

EXAMPLE 4.20 In a test of highway driving ability, drivers were monitored for their ability to maintain a constant speed of 100 km/h when instructed to do so (see Example 4.18). A group of police driving instructors performed this task with an overall mean speed of 100.0 km/h and a standard deviation of 1.9 km/h.

The driving instructors had a mean speed that was right on target, and it is likely that slight departures from the target speed were more common than larger discrepancies. Thus, the empirical rule of thumb can be used to interpret the standard deviation of 1.9 km/h. We can say that the speed was between 98.1 and 101.9 (that is, 100.0 ± 1.9) km/h approximately 68% of the time, that it was between 96.2 and 103.8 (that is, $100.0 \pm 2 \times 1.9$) km/h about 95% of the time, and that it was between 94.3 and 105.7 (that is, $100.0 \pm 3 \times 1.9$) km/h approximately 99–100% of the time.

(Source: O'Hanlon *et al.*, 1982.)

Notice that the percentages claimed in this subsection are phrased in terms of "approximately," whereas the more conservative percentages claimed by the Bienaymé–Chebyshev inequality in the previous subsection were phrased in terms of "at least." This difference in terminology is important. The Bienaymé–Chebyshev inequality is a mathematically exact result that applies to any set of data whatsoever, regardless of the shape of the distribution. The empirical rule of thumb in this subsection is widely used in practice as a reasonable approximation for distributions that are more or less symmetric and unimodal.

Standard Scores

In working with the Bienaymé–Chebyshev inequality [Expression (4.18)] and the empirical rule of thumb [Expression (4.19)], we talked about observations as being a certain number of standard deviations away from the mean or as being closer to the mean than a distance of z times the standard deviation. In other words, we used the standard deviation as a unit of measurement, and we referred to distances such as 2 standard deviations from the mean and 2.63 standard deviations from the mean.

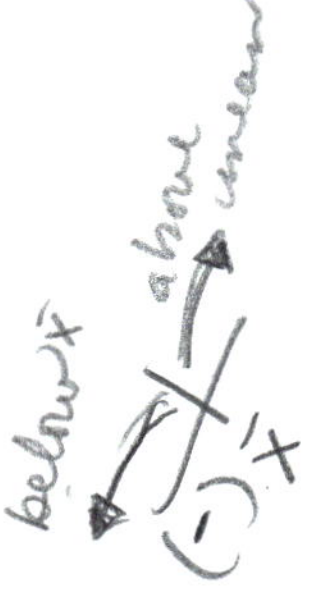

There are in fact numerous occasions in statistics where it is convenient to use the standard deviation of a set of observations as the measurement unit for those observations. Observations in which the standard deviation is the unit of measurement are called **standard scores**. The **standard score** or ***z*-score** corresponding to a particular observation in a set of data is simply the number of standard deviation units that the observation is away from the mean, with a negative value indicating that the observation is below the mean and a positive value indicating that the observation is above the mean. The standard score z_i for an observation x_i from a sample with a mean of $\bar{x}$ and a standard deviation of s may be calculated as

$$z_i = \frac{x_i - \bar{x}}{s} \tag{4.20}$$

For an observation x_i from a population with a mean of μ and a standard deviation of σ, the standard score z_i is given by

$$z_i = \frac{x_i - \mu}{\sigma} \tag{4.21}$$

The z-score simply changes the units of measurement of the observation from the original units to standard deviation units. The z-score is said to express the observation in **standard units**.

EXAMPLE 4.21 Example 4.9 reported an investigation in which 12 Hereford steers had a mean weight gain of 267 pounds with a standard deviation of 47 pounds.

What is the standard score of the steer that gained 365 pounds? For calculations, we use the unrounded values of the mean and standard deviation, for reasons outlined in Appendix 1. Performing the operations in Expression (4.20) one step at a time, we find that an observation of 365 pounds is $365 - 266.6667 = 98.3333$ pounds above the mean, and therefore $98.3333/47.0654 = 2.09$ standard units above the mean. The standard score is 2.09.

What is the weight gain in standard units of the steer that gained 190 pounds? Using Expression (4.20), we find that

$$z = \frac{190 - 266.6667}{47.0654} = \frac{-76.6667}{47.0654} = -1.63$$

The weight gain of this steer is -1.63 standard units, the negative sign indicating that this observation is below the mean.

Because the standard deviation represents a standard amount of deviation from the mean, scores in standard deviation units provide information about the relative magnitude of the observations. A z-score of, say, -1.05 is below the mean, but not by an unusual amount relative to the other observations in the data set. A z-score of 2.09 represents an observation that is above the mean by about twice the standard amount of deviation. In view of the empirical rule of thumb in Expression (4.19), observations that are more than about two standard deviations from the mean are often considered to be somewhat deviant relative to the other observations in the data set.

It is conventional to report z-scores to two decimal places. Wherever possible, calculations should be based on exact values rather than on values that have been rounded for purposes of reporting. Excessive rounding of the mean and standard deviation values used in the calculations reduces the number of correct digits in the z-score. Guidelines for how to round and how many digits to report are discussed in Appendix 1.

EXAMPLE 4.22 Male students trying out for a college volleyball team took a series of fitness tests. Subsequently, a Jump Index and a Sprint Index were calculated for each candidate to indicate his performance in two areas of importance in volleyball. The Jump Index had a mean of .68 and a standard deviation of .13 among these candidates, while the Sprint Index had a mean of .58 and a standard deviation of .09.

With no prior knowledge as to what values these indexes should have in a successful candidate, what can we say about the performance of a candidate with a Jump Index of .42 and a Sprint Index of .70? We apply Expression (4.20) to obtain standard scores:

$$z_{\text{Jump Index}} = \frac{.42 - .68}{.13} = -\frac{.26}{.13}$$
$$= -2.0$$

$$z_{\text{Sprint Index}} = \frac{.70 - .58}{.09} = \frac{.12}{.09}$$
$$= 1.3$$

(Additional decimal places in the z-scores cannot be known with any certainty in this example because of the rounding of the input information.) This candidate's jumping performance was 2.0 standard deviations below the mean, but his sprinting performance was 1.3 standard deviations above the mean. The empirical rule of thumb in Expression (4.19) suggests that he was probably in the extreme 5% (not in the central 95%) of candidates with respect to jumping (specifically, the lowest $2\frac{1}{2}$%) and in the extreme 32% (not in the central 68%) of candidates with respect to sprinting (specifically, the top 16%).

What can we say about the performance of a candidate with a Jump Index of .64 and a Sprint Index of .60? At first glance, it may appear that his jumping was slightly better than his sprinting, but the observations are easier to interpret if they are expressed in standard units:

$$z_{\text{Jump Index}} = \frac{.64 - .68}{.13} = -\frac{.04}{.13}$$
$$= -.3$$

$$z_{\text{Sprint Index}} = \frac{.60 - .58}{.09} = \frac{.02}{.09}$$
$$= .2$$

The standard scores indicate that his jumping ability was slightly worse than the average of the group, while his sprinting was slightly better. Thus, he did relatively better in sprinting than in jumping.

(Source: Student project by H. Young.)

What are the mean and the standard deviation of a set of observations that are expressed in standard units? The answer to this question follows directly from the definition of a standard score as the number of standard deviation units that a point is away from the mean: Whatever the mean of the observations in the original units (pounds, km/h, or whatever), we label the mean as 0 standard units, because it is 0 standard deviations away from the mean. And whatever the standard deviation of the observations in the original units, we define 1 standard unit to be equal to that quantity, because the standard deviation is the unit of measurement in standard scores.

A conversion from original units to standard units is analogous to a conversion from the Fahrenheit to the Celsius temperature scale, in that it changes the zero point and changes the size of the measurement unit but expresses the same reality. Specifically, standard units are defined in such a way that any set of observations expressed in standard units has a mean of 0 standard units and a standard deviation of 1 standard unit.

EXAMPLE 4.23 Example 4.20 described a driving experiment in which a group of drivers maintained a mean speed of 100.0 km/h with a standard deviation of 1.9 km/h. The following lines show part of a speedometer dial labeled in various units to show their relationship. For convenience, the horizontal line is calibrated with vertical strokes 1 standard deviation (SD) apart.

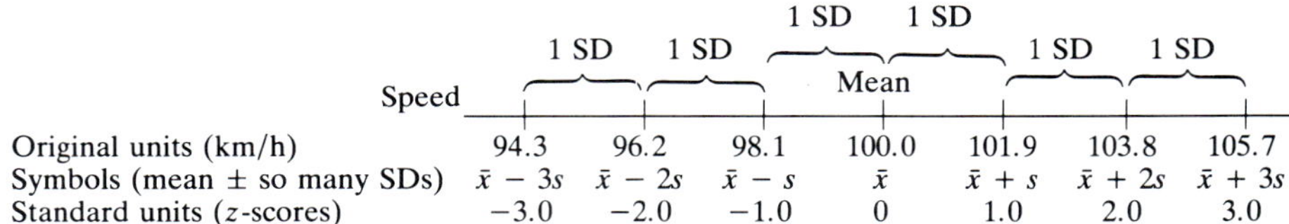

Notice that each z-score is simply the number of standard deviations that a point is away from the mean. Expressed in standard units, the mean is 0 and the standard deviation is 1.

Some readers may prefer a more algebraic demonstration of the mean and the standard deviation of a standardized variable. Consider a sample of observations x_i with mean $\bar{x}$, measuring location, and standard deviation s, measuring dispersion. Subtracting a constant from every observation changes the location of the entire distribution on the real-number line but does not change the dispersion of the distribution. Thus, subtraction of $\bar{x}$ from every observation x_i gives us a set of relocated observations $x_i - \bar{x}$ with mean $\bar{x} - \bar{x} = 0$ and standard deviation s. Next, dividing every observation by a constant changes the scale or units of measurement, not only for the observations x_i with mean $\bar{x}$, measuring location, and standard deviation s, measuring dispersion. Subtracting a constant from every observation changes set of rescaled observations $(x_i - \bar{x})/s$ with mean $0/s = 0$ and standard deviation $s/s = 1$. But this kind of relocating and rescaling is precisely what defines a standard score [see Expression (4.20)], so any standardized variable has a mean of 0 and a standard deviation of 1.

* 4.5

BOX PLOTS

A simple but highly informative graphical method has been devised by John W. Tukey (see Biography 2.1) to summarize the location and dispersion of a set of data and to identify any unusual observations that may warrant special consideration. The resulting graphical summary is called a **box plot** or, in Tukey's original terminology, a *box-and-whisker plot.* The measure of location shown on such a plot is the median, and the measure of dispersion is based on the quartiles, which were introduced in Section 3.4.

The next subsection identifies three measures of dispersion based on quantiles, and the following subsection describes how these and certain other values are portrayed in a box plot.

Quantile Measures of Dispersion

We saw in Section 3.4 that the extremes are members of the quantile family. Thus, the **range**, which was defined in Section 4.1 as the difference between the maximum and the minimum in a set of observations, may be considered to be the simplest quantile measure of dispersion. However, as measures of dispersion go, it is rather unstable due to its dependence on the values of the two extreme observations.

We can improve the stability by replacing the extremes with the first and third quartiles to obtain the **interquartile range**:

$$\text{Interquartile range} = Q_3 - Q_1 \tag{4.22}$$

The interquartile range takes in the central 50% of the observations.

We obtain an alternative measure of dispersion by taking one-half of the interquartile range. This measure, called the **semi-interquartile range** or the **quartile deviation**, is given by

$$\text{Quartile deviation} = \frac{Q_3 - Q_1}{2} \tag{4.23}$$

Half of the observations differ from the midquartile by more than the amount of the quartile deviation, and half by less. In this sense, the quartile deviation is a typical amount of deviation of the observations from the average.

EXAMPLE 4.24 The table on page 132, from Example 3.20, shows the estimated 1984 sizes, in millions of persons, of the 13 members of the Organization of Petroleum Exporting Countries (OPEC).

The mean size of these countries is 30 million persons with a standard deviation of 49 million persons. The range of country sizes is $169.442 - .276 \simeq 169$ million persons, which is approximately three to four times the standard deviation, as is typical in relatively small sets of data [Expression (4.17)].

The first and third quartiles of these data were found in Example 3.20 to be $Q_1 = 1.758$ million persons and $Q_3 = 21.351$ million persons. The interquartile range is therefore $21.351 - 1.758 = 19.593$ million persons, indicating that half of the OPEC countries had population sizes within an interval of 19.593 million persons.

OPEC member	Size (millions of persons)
Indonesia	169.442
Nigeria	88.148
Iran	43.280
Algeria	21.351
Venezuela	18.552
Iraq	15.000
Saudi Arabia	10.794
Ecuador	9.091
Libya	3.684
Kuwait	1.758
United Arab Emirates	1.523
Gabon	.958
Qatar	.276

The quartile deviation is one-half the interquartile range, or (21.351 − 1.758)/2 = 9.796 million persons. Half of the countries differed from the midquartile (11.554 million persons, from Example 3.21) by less than 9.796 million persons, and half differed by more than 9.796 million persons.

(Source: *World Almanac 1986,* pp. 535–626.)

Both the quartile deviation and the standard deviation can be characterized as representing a typical amount of deviation of the observations from the average, but they are "typical" in different ways, and they are based on different types of "average." For example, one quartile deviation above and below the midquartile (an average) takes in 50% of the observations in *any* distribution, whereas one standard deviation above and below the mean (another kind of average) takes in about 68% of the observations in distributions that are more or less symmetric and unimodal [Expression (4.19)]. In general, the quartile deviation is smaller than the standard deviation. In distributions where the empirical rule of thumb [Expression (4.19)] for interpreting the standard deviation applies, the quartile deviation is about two-thirds of the value of the standard deviation.

Adjacent and Outside Values

The following values are shown on a **box plot**:

a. the median,

b. the first and third quartiles,

c. the **adjacent values**, and

d. the **outside values**, if any.

In this subsection we define adjacent and outside values and show how to construct and read box plots.

We know that the median is the value of the middle observation (when the observations are in ascending or descending order), and that the central 50% of the observations are between the first and third quartiles, Q_1 and Q_3. The

adjacent values are defined so that most of the observations are between the upper and lower adjacent values, and any that are not can be said to be unusually far from the average:

> The **upper adjacent value** is the value of the largest observation that is less than or equal to $Q_3 + (1.5 \times \text{interquartile range})$. **(4.24)**

> the **lower adjacent value** is the value of the smallest observation that is greater than or equal to $Q_1 - (1.5 \times \text{interquartile range})$. **(4.25)**

If we think of the first and third quartiles as being one quartile deviation below and above the midquartile, then the lower and upper adjacent values are the values of the most extreme observations within four quartile deviations below and above the midquartile.

Any observations beyond the adjacent values are known as **outside values**. Outside values are unusually far from the average. Each one should be examined in an effort to account for its unusual value relative to the rest of the distribution, including the possibility of errors in observation or recording.

To construct a box plot, begin with a vertical scale calibrated appropriately for the variable under consideration. Draw a rectangular box with its ends at the first and third quartiles and with a horizontal bar at the median. The "whiskers" consist of vertical dashed lines extending from the box to horizontal bars marking the adjacent values. Plot outside values individually, and identify them by appropriate labels.

EXAMPLE 4.25 Example 4.24 gave the estimated 1984 sizes, in millions of persons, of the 13 members of the Organization of Petroleum Exporting Countries (OPEC). The quartiles and interquartile range were calculated as follows:

$$Q_1 = 1.758$$
$$Q_2 = 10.794$$
$$Q_3 = 21.351$$
$$\text{Interquartile range} = 19.593$$

To find the upper adjacent value according to Expression (4.24), we need to calculate

$$Q_3 + (1.5 \times \text{interquartile range}) = 21.351 + (1.5 \times 19.593) = 50.740$$

From the data of Example 4.24, we see that the upper adjacent value is therefore 43.280 (for Iran), because that is the value of the largest observation that is less than or equal to 50.740. The observations for Indonesia and Nigeria are outside values.

Similarly, to apply Expression (4.25), we need to calculate

$$Q_1 - (1.5 \times \text{interquartile range}) = 1.758 - (1.5 \times 19.593) = -27.632$$

The lower adjacent value is therefore .276, because that is the value of the smallest observation that is greater than or equal to −27.632. There are no outside values beyond the lower adjacent value.

The box plot for these data is shown in Figure 4.3.

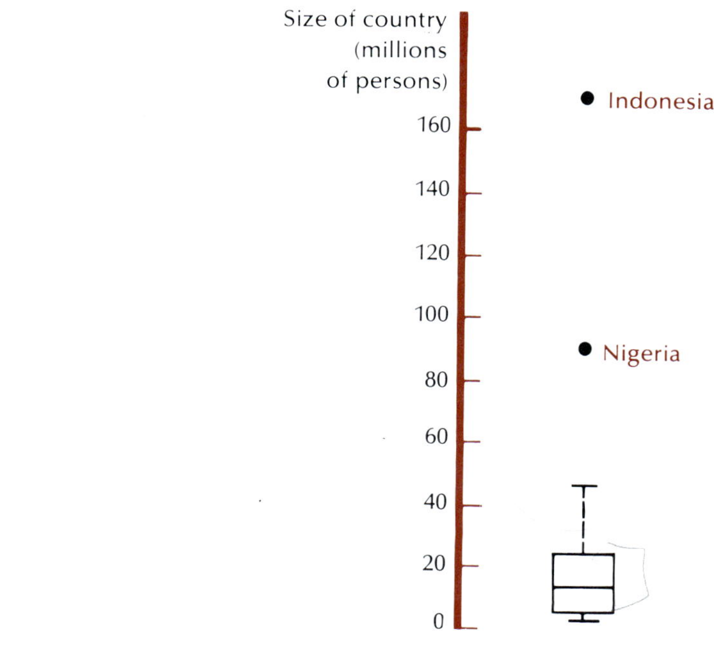

Figure 4.3 Box plot of 1984 population sizes, in millions of persons, for 13 OPEC member nations.

In summary, the following information can be read from a **box plot**:

1. measures of location
   a. The **median** is shown by a bar partitioning the box.
   b. The **first and third quartiles** are at the ends of the box.
   c. The **midquartile** is not marked but falls at the middle of the box.
2. measures of dispersion
   a. The **range** is shown as the distance between the two extreme values.
   b. The **interquartile range** is shown as the length of the box.
   c. The **quartile deviation** is not marked but is half the length of the box.
3. distributional information
   a. 50% of the observations are in the box, with 25% in each of the two compartments.
   b. The main body of observations falls between the **adjacent values**, marked by the ends of the whiskers.
   c. Individually identified observations are **outside values**, which are relatively unusual and may warrant special attention for verification or explanation.

STUDY MATERIAL FOR CHAPTER 4

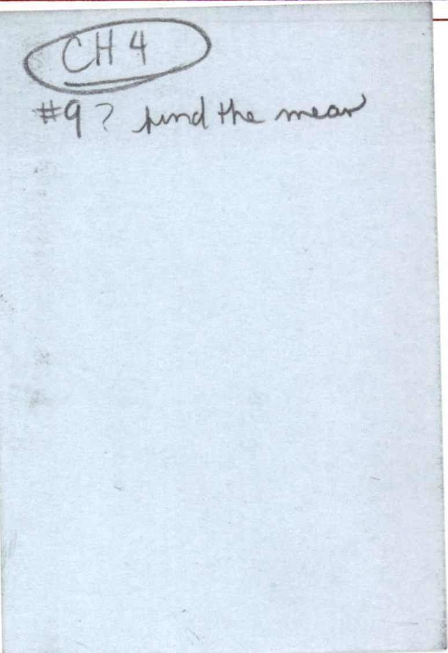

Terms and Concepts

measure of dispersion

Section 4.1

range

Section 4.2

deviation from the mean
deviation
mean deviation

Section 4.3

variance of a population
variance
variance of a sample
biased estimate
unbiased estimate
standard deviation
standard deviation of a population
standard deviation of a sample
definitional formula
computational formula
general advice for calculating
guideline for reporting means and standard deviations
* approximating the standard deviation of grouped data

Section 4.4

rule of thumb relating standard deviation and range
* Bienaymé–Chebyshev inequality
rule of thumb for interpreting the standard deviation in symmetric unimodal distributions
empirical rule
standard score
z-score
standard units

* Section 4.5

interquartile range
semi-interquartile range
quartile deviation

box plot
adjacent values
upper adjacent value and lower adjacent value
outside values

Key Formulas

Section 4.1

(4.2) Range $= x_{\max} - x_{\min}$

Section 4.2

(4.4) $$\text{Mean deviation} = \frac{\sum^{n} |x_i - \bar{x}|}{n}$$

Section 4.3

(4.6) $$\sigma^2 = \frac{\sum^{N} (x_i - \mu)^2}{N}$$

(4.7) $$s^2 = \frac{\sum^{n} (x_i - \bar{x})^2}{n - 1}$$

(4.8) $$\sigma = \sqrt{\frac{\sum (x_i - \mu)^2}{N}}$$

(4.9) $$s = \sqrt{\frac{\sum (x_i - \bar{x})^2}{n - 1}}$$

(4.11) $$s^2 = \frac{\sum x^2 - \frac{(\sum x)^2}{n}}{n - 1}$$

(4.13) $$s = \sqrt{\frac{\sum x^2 - \frac{(\sum x)^2}{n}}{n - 1}}$$

Section 4.4

* **(4.18)** $$\text{Maximum proportion of observations in tails} = \frac{1}{z^2}$$

(4.20) $$z_i = \frac{x_i - \bar{x}}{s}$$

(4.21) $$z_i = \frac{x_i - \mu}{\sigma}$$

* Section 4.5

(4.22) Interquartile range $= Q_3 - Q_1$

(4.23) $$\text{Quartile deviation} = \frac{Q_3 - Q_1}{2}$$

Exercises for Numerical Practice

Section 4.1

4/1. Calculate the range of the following sample: 7, 5, 9, 8.

4/2. Calculate the range of the following sample: 3, 30, 9, 1.

Section 4.2

4/3. Calculate the mean deviation of the data in Exercise 4/1.

4/4. Calculate the mean deviation of the data in Exercise 4/2.

Section 4.3

4/5. For the data in Exercise 4/1:

- **a.** Use Expression (4.7) to calculate the variance.
- **b.** Use Expression (4.11) to calculate the variance.
- **c.** Calculate the standard deviation as the square root of the variance.

4/6. For the data in Exercise 4/2:

- **a.** Use Expression (4.7) to calculate the variance.
- **b.** Use Expression (4.11) to calculate the variance.
- **c.** Calculate the standard deviation as the square root of the variance.

4/7. Use Expression (4.13) to calculate the standard deviation of each of the following samples:

- **a.** 75, 71, 78.
- **b.** 0, 12, 1, 0, 6, 0, 3.

4/8. Use Expression (4.13) to calculate the standard deviation of each of the following samples:

- **a.** 101, 104, 102.
- **b.** 2, 5, 0, 20, 1, 9.

* **4/9.** Calculate the mean and standard deviation of x for a sample summarized as follows.

x	**Frequency**
0– 2	6
3– 5	17
6– 8	32
9–11	3

* **4/10.** Calculate the mean and standard deviation of x for a sample summarized as follows.

x	**Frequency**
0– 9	34
10–19	18
20–29	10
30–39	5
40–49	2

Section 4.4

4/11. Given a sample of observations 502, 501, 507, 502, 509:

- **a.** Calculate the range.
- **b.** Use Expression (4.17) to approximate the standard deviation.
- **c.** Use Expression (4.13) to calculate the standard deviation, and compare the result to your approximation in Part (b).

4/12. Given a sample of observations 5, 0, 2, 30, 4, 96, 8, 1:

- **a.** Calculate the range.
- **b.** Use Expression (4.17) to approximate the standard deviation.
- **c.** Use Expression (4.13) to calculate the standard deviation, and compare the result to your approximation in Part (b).

* **4/13.** For a set of data with a mean of 50 and a standard deviation of 10, what does the Bienaymé–Chebyshev inequality say about the percentage of the observations that are:

- **a.** more than 2 standard deviations away from the mean?
- **b.** more than 4 standard deviations away from the mean?
- **c.** between 30 and 70?
- **d.** between 10 and 90?

* **4/14.** For a set of data with a mean of 400 and a standard deviation of 50, what does the Bienaymé–Chebyshev inequality say about the percentage of the observations that are:

- **a.** more than 3 standard deviations away from the mean?
- **b.** more than 5 standard deviations away from the mean?
- **c.** between 250 and 550?
- **d.** between 150 and 650?

4/15. For a more or less symmetric and unimodal set of data with a mean of 50 and a standard deviation of 10, what does the empirical rule say about the percentage of observations that are:

- **a.** between 40 and 60?
- **b.** between 30 and 70?
- **c.** between 20 and 80?

4/16. For a more or less symmetric and unimodal set of data with a mean of 400 and a standard deviation of 50, what does the empirical rule say about the percentage of observations that are:

- **a.** between 350 and 450?
- **b.** between 300 and 500?
- **c.** between 250 and 550?

4/17. A set of data has a mean of 50.0 and a standard deviation of 10.0. Express each of the following observations as a standard score:

- **a.** 71.7
- **b.** 38.9
- **c.** 50.2

d. 41.4
e. 50.0

4/18. A set of data has a mean of 2700 and a standard deviation of 100. Express each of the following observations as a standard score:

a. 2437
b. 2700
c. 2756
d. 2895
e. 2601

4/19. A set of data has a mean of 250.0 and a standard deviation of 20.0. Calculate the value of the observation corresponding to each of the following z-scores:

a. 0.00
b. −1.00
c. 2.34
d. .75
e. −.17

4/20. A set of data has a mean of 100.0 and a standard deviation of 15.0. Calculate the value of the observation corresponding to each of the following z-scores:

a. 2.00
b. −.86
c. −1.33
d. 0.00
e. .44

* Section 4.5

4/21. The following values were recorded for a sample of observational units identified by the letters A through J:

A	B	C	D	E	F	G	H	I	J
40	44	42	50	21	41	48	65	55	33

a. Calculate the first, second, and third quartiles.
b. Calculate the interquartile range and the quartile deviation.
c. Determine the adjacent values for the box plot.
d. Identify the outside values for the box plot.
e. Draw the box plot.

4/22. The following values were recorded for a sample of observational units identified by the letters A through N:

A	B	C	D	E	F	G	H	I	J	K	L	M	N
74	70	77	59	40	100	73	76	62	66	72	69	25	68

a. Calculate the first, second, and third quartiles.
b. Calculate the interquartile range and the quartile deviation.
c. Determine the adjacent values for the box plot.

d. Identify the outside values for the box plot.

e. Draw the box plot.

Problems for Applying the Concepts

4/23. A sample of 10 unmarried female college students (aged 17–19 years) were asked their opinion as to what, in general, is an appropriate age for a person to get married. Their answers were as follows:

25, 23, 19, 25, 23, 30, 23, 27, 26, 23

a. Calculate the range.

b. Based on the range, what would you guess to be the approximate value of the standard deviation?

c. Use Expression (4.11) to calculate the sample variance.

d. Use Expression (4.13) to calculate the sample standard deviation, and then compare the result to your approximation in Part (b).

e. Write a sentence to summarize the data in terms of the mean and the standard deviation.

(Source: Student project by H. Asfeldt.)

4/24. A sample of 10 residence rooms of female students were selected at random on a college campus, and the number of posters displayed in each room was counted and recorded as follows:

14, 12, 30, 12, 8, 6, 21, 6, 14, 8

a. Calculate the range.

b. Based on the range, what would you guess to be the approximate value of the standard deviation?

c. Use Expression (4.11) to calculate the sample variance.

d. Use Expression (4.13) to calculate the sample standard deviation, and then compare the result to your approximation in Part (b).

e. Write a sentence to summarize the data in terms of the mean and the standard deviation.

(Source: Student project by J. Kruschel.)

4/25. In an investigation of physical fitness of female college students, the following resting heart rates were recorded (in beats per minute) for a sample of 15 students while they were sitting watching television:

76, 80, 60, 74, 62, 72, 76, 70, 72, 68, 74, 71, 82, 63, 66

a. Calculate the range.

b. Based on the range, what would you guess to be the approximate value of the standard deviation?

c. Calculate the sample variance.

d. Calculate the sample standard deviation, and then compare the result to your approximation in Part (b).

e. Subtract 60 from every observation and calculate the standard deviation of these modified data. What is the relationship between this result and that of

Part (d), and what property of the standard deviation is responsible for this relationship?

f. Write a sentence to summarize the original data in terms of the mean and the standard deviation.

(Source: Student project by A. Graham.)

4/26. Eight students agreed to take part in an investigation of the physical fitness of college students. The time that each student took to run a distance of 100 meters was recorded to the nearest second as follows:

20, 16, 12, 14, 17, 12, 12, 15

a. Calculate the range.

b. Based on the range, what would you guess to be the approximate value of the standard deviation?

c. Calculate the sample variance.

d. Calculate the sample standard deviation, and then compare the result to your approximation in Part (b).

e. Subtract 12 from every observation, and calculate the standard deviation of these modified data. What is the relationship between this result and that of Part (d), and what property of the standard deviation is responsible for this relationship?

f. Write a sentence to summarize the original data in terms of the mean and standard deviation.

(Source: Student project by T. Mammo.)

4/27. A sample of eight male college students (aged 18–19 years) reported their hourly wage rates (in dollars) for summer employment in 1985:

4.75, 8.50, 17.23, 9.00, 7.00, 11.69, 6.25, 7.50

a. Calculate the sample standard deviation of the hourly wage rates in dollars.

b. Convert each observation to cents (for example \$4.75 is 475 cents), and calculate the sample standard deviation of the hourly wage rates in cents.

c. What is the relationship between the answers in Parts (a) and (b)? What does this relationship tell us about the standard deviation?

d. Write a sentence to summarize the data in terms of the mean and the standard deviation.

(Source: Student project by D. Sweeney.)

4/28. A sample of 9 home-grown Netted Gem potatoes were selected haphazardly from a sack and weighed to the nearest gram as follows:

232, 170, 208, 233, 304, 514, 186, 188, 126

a. Calculate the standard deviation of the weights measured in grams.

b. Convert each observation to kilograms, and calculate the standard deviation of the weights measured in kilograms.

c. What is the relationship between the answers in Parts (a) and (b)? What does this relationship tell us about the standard deviation?

d. Write a sentence to summarize the data in terms of the mean and the standard deviation.

(Source: Student project by J. Sheets.)

* **4/29.** Calculate the mean and the standard deviation of the residential house prices in Problem 2/12 (Chapter 2).

* **4/30.** Calculate the mean and the standard deviation of the pizza-bill totals in Problem 2/11 (Chapter 2).

4/31. For the hourly wage rates given in Problem 4/27:

a. Express each observation in standard units.

b. Calculate the mean and the standard deviation of the sample of z-scores in Part (a).

4/32. For the running times given in Problem 4/26:

a. Express each observation in standard units.

b. Calculate the mean and the standard deviation of the sample of z-scores in Part (a).

4/33. The mean mark on a calculus test was 77, and the standard deviation was 21. Using the information provided by the relevant z-score:

a. How should a student with a mark of 85 interpret her mark?

b. How should a student with a mark of 74 interpret his mark?

4/34. The mean mark on a chemistry exam was 60, and the standard deviation was 18. Using the information provided by the relevant z-score:

a. How should a student with a mark of 84 interpret his mark?

b. How should a student with a mark of 72 interpret her mark?

4/35. Altex is a late-maturing variety of canola (low-acid rapeseed), and Candle is an early-maturing variety. In the Lake Demay district of central Alberta, the mean yield of Altex in 1980 was found to be 37 bushels per acre with a standard deviation of 9 bushels per acre, and the mean yield of Candle was found to be 28 bushels per acre with a standard deviation of 7 bushels per acre. One farmer in the district who planted both varieties in 1980 obtained a yield of 47 bushels per acre of Altex and 35 bushels per acre of Candle. Express these yields in standard units to determine which variety performed relatively better (relative to its average yield) on his land.
(Source: Student project by V. Lindstrand.)

4/36. To facilitate planning, a certain college cafeteria maintains accurate records of the numbers of students who eat each meal in the cafeteria. The daily mean number of students eating breakfast in the cafeteria on a weekday (Monday through Friday) during the regular academic year was found to be 147.2 with a standard deviation of 14.6. During an examination period, the mean was found to be 171.6 with a standard deviation of 16.6. Which would be the relatively more unusual event: a breakfast attendance of 128 during a regular week, or an attendance of 160 during an exam week?
(Source: Student project by D. Stewart.)

* **4/37.** With respect to Problem 4/35, what does the Bienaymé–Chebyshev inequality indicate about the percentage of farmers in the investigation whose Altex yield

was

a. between 24 and 50 bushels per acre?
b. 20 bushels per acre or less?
c. 55 bushels per acre or more?

* **4/38.** With respect to Problem 4/36, what does the Bienaymé–Chebyshev inequality indicate about the percentage of regular weekdays that have breakfast attendances of

a. between 120 and 175 students?
b. 100 or fewer students?
c. at least 200 students?

4/39. With respect to Problem 4/35, what does the empirical rule indicate about the percentage of farmers in the investigation whose Altex yield was

a. between 10 and 64 bushels per acre?
b. 28 bushels per acre or less?
c. 55 bushels per acre or more?

4/40. With respect to Problem 4/36, what does the empirical rule indicate about the percentage of regular weekdays that have breakfast attendances of

a. between 118 and 176 students?
b. between 104 and 191 students?
c. at least 162 students?

* **4/41.** The mid-1985 salaries of the 50 state governors in the United States are recorded in Problem 3/37 (Chapter 3).

a. Calculate the interquartile range and the quartile deviation.
b. Determine the adjacent values for the box plot.
c. Identify the outside values for the box plot.
d. Draw the box plot.

* **4/42.** The following table lists the population densities of the 63 states, provinces, and continental territories and districts of the United States and Canada. The densities are recorded as the number of people per square mile of land area, based on 1984–85 population estimates.

a. Calculate the first, second, and third quartiles.
b. Calculate the interquartile range and the quartile deviation.
c. Determine the adjacent values for the box plot.
d. Identify the outside values for the box plot.
e. Draw the box plot.

(Source: *World Almanac 1986,* pp. 632–56, 697–701.)

Density	Place	Density	Place	Density	Place
9300	District of Columbia	159.7	Michigan	20.1	Utah
78.6	Alabama	52.5	Minnesota	57.2	Vermont
.88	Alaska	54.9	Mississippi	141.7	Virginia
26.9	Arizona	72.6	Missouri	64.6	Washington
45.2	Arkansas	5.66	Montana	81.1	West Virginia
163.9	California	21.0	Nebraska	87.5	Wisconsin
30.6	Colorado	8.3	Nevada	5.3	Wyoming
648.7	Connecticut	108.2	New Hampshire	9.4	Alberta
309.3	Delaware	999.2	New Jersey	8.0	British Columbia
202.7	Florida	11.7	New Mexico	5.0	Manitoba
100.5	Georgia	370.8	New York	25.8	New Brunswick
161.7	Hawaii	126.3	N. Carolina	4.0	Newfoundland
12.1	Idaho	9.9	North Dakota	43.1	Nova Scotia
206.5	Illinois	262.4	Ohio	26.2	Ontario
152.3	Indiana	47.9	Oklahoma	58.2	Prince Edward Island
52.0	Iowa	27.8	Oregon	12.5	Quebec
29.8	Kansas	264.7	Pennsylvania	4.6	Saskatchewan
93.9	Kentucky	917.1	Rhode Island	.12	Yukon
99.3	Louisiana	109.3	S. Carolina	.04	Northwest Territories
37.4	Maine	9.3	South Dakota		
439.7	Maryland	114.1	Tennessee		
740.9	Massachusetts	61.0	Texas		

References

O'Hanlon, J. F.; Haak, T. W.; Blaauw, G. J.; and Riemersma, J. B. J. "Diazepam Impairs Lateral Position Control in Highway Driving." *Science* 217 (1982): 79–81.

The World Almanac and Book of Facts 1986. New York: Newspaper Enterprise Association, Inc., 1985.

Further Reading

Campbell, S. K. *Flaws and Fallacies in Statistical Thinking*. Englewood Cliffs, NJ: Prentice-Hall, 1974. Chapter 7 provides a brief but convincing demonstration of the dangers of ignoring dispersion.

Tukey, J. W. *Exploratory Data Analysis*. Reading, MA: Addison-Wesley, 1977. Chapter 2 includes an account of "box-and-whisker plots," written by their inventor.

CHAPTER 5

DESCRIPTIONS OF RELATIONSHIP

EXAMPLE 5.1 Tooth decay, or dental caries, is known to be caused by bacteria, and not directly by sugar or other things we eat. In general, eating a lot of sugary foods increases tooth decay, but the precise role of diet in the incidence of tooth decay is still a topic of considerable dental research.

Data have been collected from a variety of countries to study the relationship between sugar consumption and tooth decay. In some studies, with countries as the observational units, one observed variable is the yearly per capita consumption of sugar in a country, and a second observed variable is the average number of decayed (or missing or filled) teeth of 12-year-old children in the country. Using the methods of the preceding chapters, researchers can calculate the mean and standard deviation of the consumption variable and from that determine which countries are above and below average in sugar consumption. Similarly, they can calculate the mean and the standard deviation of the tooth-decay variable and from that determine which countries are above and below average in number of decayed teeth. However, means and standard deviations do not tell us anything about the relationship between sugar consumption and dental caries. Do residents of countries with lower sugar consumption tend to have fewer decayed teeth, and do residents of countries with higher sugar consumption tend to have more tooth decay? The question is not about either variable separately, but about the relationship between the two observed variables. Using methods to be introduced in this chapter, researchers have answered in the affirmative. For example, 12-year-old children in countries with low sugar consumption, such as China and Ethiopia, have been found to average less than one decayed tooth, whereas children in countries with high sugar consumption, such as Finland and Australia, have been found to average more than 10 decayed teeth.

Other investigators have examined the relationship between sugar consumption and dental caries via other approaches. A careful review of the various approaches concluded that "frequent or high intake of sugary foods predisposes to dental decay. The relation is not always clear-cut, and most studies have important methodological problems and limitations. . . . Human diets . . . vary in food items eaten and the frequency and sequence of eating, and these factors can affect the cariogenicity [decay-promoting role] of a food. Therefore, reported correlations must be interpreted with caution."

(Source: Newbrun, 1982.)

The descriptive methods of Chapters 2, 3, and 4 are used for summarizing data that consist of the values of a single variable for each observational unit.

The examples presented in those chapters involved variables such as

- length of reign (of English monarchs),
- individual income (on taxable income tax returns),
- butterfat content (of milk from Holstein cows),
- dressed weight (of hunted deer),
- number of posters (in college residence rooms), and
- highway driving speed (of police driving instructors).

We have seen how to summarize such univariate data in terms of frequency distributions, measures of location, and measures of dispersion. Because they apply to univariate observations, or to values of one variable at a time, the methods discussed so far are sometimes called **univariate methods**.

In many situations, multivariate observations are recorded; that is, more than one variable is recorded for each observational unit. The following are examples of such multivariate observations:

a. per capita sugar consumption and incidence of tooth decay (in various countries);

b. price and length (of textbooks);

c. mortgage interest rate and number of residential construction starts (for different years);

d. total yearly income and yearly expenditures on consumer durables (for various families);

e. total floor area and fair market value (of houses);

f. university statistics grade, out-of-class study time for statistics, and high school mathematics mark (of students in introductory statistics);

g. number of years in business, number of accidents reported, and shift length (of companies in the petrochemical industry);

h. barley yield, amount of rainfall, and amount of fertilizer applied (for various farms);

i. number of minutes played, number of points scored, and height (of basketball players);

j. weight, engine size, and fuel consumption (of various car models).

When more than one variable is measured for each observational unit, we frequently want to know more than just, say, the mean and the standard deviation of each variable separately. We may—and usually should—still compute those univariate descriptions for each variable separately, but we might also want to know how one variable relates to another. Do the variables tend to vary together, or do they vary independently of each other? Are the variables related or unrelated? In this chapter we introduce certain **bivariate methods** relevant to such questions.

5.1 SIMPLE LINEAR REGRESSION

Let us begin by reviewing some elementary mathematical notation for describing the relationship between two variables x and y. Then we will define the term *regression* which appears in the heading above.

Functions

The expression

$$y = f(x) \tag{5.1}$$

indicates that the variable y is a **function** of the variable x.

> A **function** may be thought of as a rule associating each possible value of x with exactly one value of y. **(5.2)**

Once the nature of the function is specified, we can calculate the value of y for any given value of x. Because the function tells us about y in terms of x, the variable y is referred to as the **dependent variable**, and x is called the **independent variable**. We could write Expression (5.1) as follows:

$$\text{Dependent variable} = f(\text{independent variable}) \tag{5.3}$$

The value of the dependent variable y is thought of as depending on the value of the other variable, x.

EXAMPLE 5.2 Example 5.1 described an investigation of the relationship between a country's annual per capita consumption of sugar and the incidence of tooth decay in the country. Since the purpose of the investigation was to examine the effect of diet on the incidence of tooth decay, or how tooth decay depends on diet, we can say that the researchers were studying incidence of tooth decay as a function of sugar consumption. Sugar consumption is the independent variable, and number of decayed teeth is the dependent variable.

When a function takes the form

$$y = bx + a \tag{5.4}$$

we have the general form of the equation of a straight line, where

y is the dependent variable (the dimension to be described or predicted on the basis of x);

x is the independent variable (the dimension or characteristic on the basis of which y is described);

b is the **slope** of the line ("rise" divided by "run"); and

a is the **y-intercept** of the line (the value of y where the line intersects the y-axis).

Expression (5.4) is called a **linear equation in two unknowns**, the unknowns being x and y.

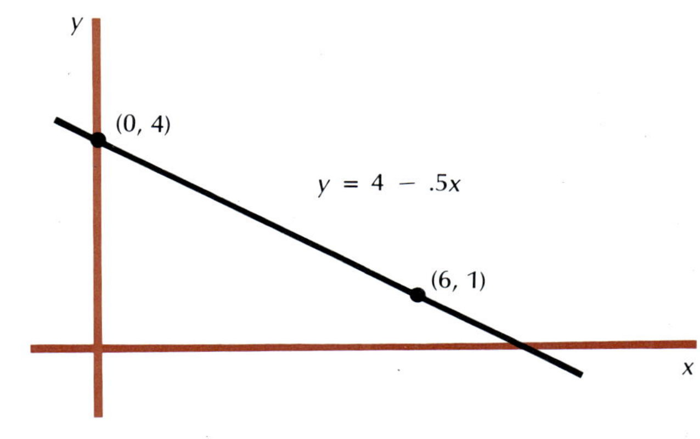

Figure 5.1 The linear equation $y = 4 - .5x$.

EXAMPLE 5.3 Consider the linear equation $y = 4 - .5x$. It is conventional to plot such an equation with the independent variable (x) on the horizontal axis and the dependent variable (y) on the vertical axis. To plot the equation of a straight line, we need find only two points on the line and draw the line through them. Any two points will do. For example, when $x = 0$, $y = 4 - (.5)(0) = 4$; and when $x = 6$, $y = 4 - (.5)(6) = 1$. The line is shown in Figure 5.1.

You may verify in the diagram that the line $y = 4 - .5x$ has a slope of $b = -.5$ (that is, it goes down .5 unit for every unit of increase in x) and a y-intercept of $a = 4$.

EXAMPLE 5.4 To aid him in planning his time, a worker in a local printing office kept detailed records of the orders he printed. Using methods to be introduced shortly, he found that the length of time required to print a job could be well described as a linear function of the number of copies to be printed. In particular, to predict how long a job would take him, he could use the equation $y = .53x + .32$, where y is the time required for the job (in hours) and x is the number of copies to be printed (in thousands). The equation is plotted in Figure 5.2.

The equation describes printing time as a function of the number of copies: Given the number of copies to be printed, the printer wanted to predict how much time would be required. Printing time is the dependent variable, and number of copies is the independent variable. The slope is $b = .53$ hour per thousand copies. In other words, each increase of 1000 copies in the size of the order adds an average of .53 hour, or about 32 minutes, to the time to complete the job. The y-intercept is $a = .32$ hour; this is the value of y when $x = 0$ copies. It may be interpreted as indicating that the average time taken for things other than the actual printing, such as setting up and cleaning up, is .32 hour, or about 19 minutes, for each order. The equation $y = .53x + .32$ thus suggests that on the average there is a constant time of .32 hour required no matter what the size of the job, plus an additional .53 hour for each thousand copies printed.

In addition to supplying the information provided directly by the slope and y-intercept values, the equation $y = .53x + .32$ enables us to predict the printing time for orders of various sizes. For an order of 2200 copies, for example, $x = 2.2$ thousand copies, and the predicted printing time would be $y = (.53)(2.2) + .32 = 1.486$, or

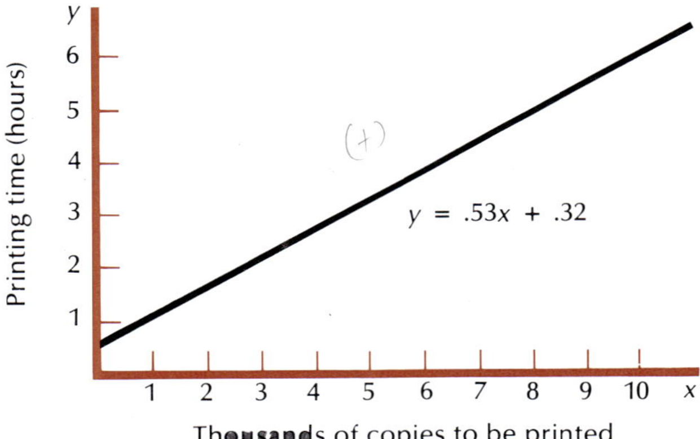

Figure 5.2 Printing time as a function of the number of copies to be printed.

about 1.5 hours. For an order of 5000 copies, $x = 5.0$ thousand copies, and the expected printing time would be $y = (.53)(5.0) + .32 = 2.97$, or about 3 hours.
(Source: Student project by D. Hutchinson.)

The equation of a straight line provides a simple way of expressing a relationship between two variables. Once we learn how to read it, it allows us to see the following at a glance:

a. whether the relationship is direct (positive slope) or inverse (negative slope);

b. the number of units change in the dependent variable y for each unit increase in the independent variable x (indicated by the value of the slope); and

c. what happens when $x = 0$ (indicated by the value of the y-intercept).

Of course, not all functional relationships in society and nature are so simple as the linear relationship; many relationships in the world around us do not translate into straight lines on graphs. Nonetheless, the simple linear equation $y = bx + a$ finds many important applications because

a. many observed relationships are essentially linear;

b. such an equation often provides a good approximation, within the range of interest, to otherwise complex relationships; and

c. many complex relationships can be described by linear equations after the variables are redefined, for example by taking logarithms of one or both variables.

Example 5.4 illustrates that the linear equation in two unknowns is not difficult to use once we know the equation. But suppose that we did not know the equation and simply had, say, 25 bivariate observations (such as printing time and number of copies for each of 25 completed orders). How could we determine the appropriate linear equation to describe one variable in terms of the other? If all 25 points fell exactly in a straight line, the task would be easy, but with real data there is almost certain to be some degree of scatter in the points. The points are apt to look more like a cloud or swarm than like a straight line, even if the underlying relationship is essentially linear. We turn now to this problem of determining the values of b and a in the linear equation representing the relationship between two variables in real data.

Regression Analysis

The study of the functional relationships among observed variables is called **regression analysis**. The term **regression** is something of an historical accident. It was introduced in 1885 in an important paper by Sir Francis Galton (see Biography 5.1). He studied heights of sons as a function of heights of fathers and observed what he called "regression toward mediocrity": Taller-than-average fathers tended, on the average, to have sons slightly shorter than themselves, and shorter-than-average fathers tended, on the average, to have sons slightly taller than themselves. It is now recognized that if one takes an observation (such as a father's height) near one extreme of a distribution, then another observation (such as the son's mature height) from the distribution is more likely to be on the less extreme side of the first observation, simply because there is more of the distribution on that side. The error of giving an empirical explanation (such as regression in heights from one generation to the next) for this statistical phenomenon is now known as the **regression fallacy**. (Think of the giants there would be after a few generations if tall men were in fact equally likely to have sons taller or shorter than themselves, and if their taller sons also were equally likely to have sons taller or shorter than themselves, and so on! Conversely, think of how little dispersion in heights there would be after a few generations if there were a genetic "regression toward mediocrity.") Nonetheless, Galton's terminology still labels the statistical method that he helped to develop, and the functional relationship between two or more variables, when determined from observed values of those variables, is called **regression**.

BIOGRAPHY 5.1 Francis Galton (1822–1911)

A cousin of Charles Darwin, Francis Galton was born on 16 February 1822 in Birmingham, England. He studied medicine at King's College in London from 1839 to 1840 and then at Trinity College in Cambridge, where he received his degree in 1844. During a long and distinguished career, he received honors and awards—including knighthood in 1909—for his scientific contributions in heredity, anthropology, geography, and meteorology.

Galton is best known for his work in the area of heredity, one of his central interests being the heritability of intellectual ability. He is regarded as the founder of the science of eugenics, which is concerned with the improvement of a race or breed by selective breeding, but his comments on human eugenics are so blatantly racist and ethnocentric that it is astonishing to realize that they were written by a learned man only about a century ago. On 10 September 1885 he introduced his theory of "regression toward mediocrity" in the anthropology section (of which he was president at the time) of the British Association for the Advancement of Science meeting in Aberdeen, Scotland. He also originated the modern technique of weather-mapping; devised a system of fingerprint identification; and invented a device, known as a Galton whistle, for the determination of the upper limit of hearing for high-pitched tones.

Galton married Louisa Jane Butler on 1 August 1853. He died on 17 January 1911 in Haslemere, Surrey, England, at the age of 89.

(Sources: Galton, 1869; Stigler, 1980; *World Who's Who in Science,* 1968.)

In this chapter, we limit our discussion of regression analysis to **simple linear regression**, which deals with relationships that can be described by the equation of a straight line, $y = bx + a$. The objective of simple linear regression analysis may be characterized as follows: Given a sample of bivariate observations, determine the "best" values of b and a to describe y as a straight-line function of x.

EXAMPLE 5.5 Can a person's weight as a young adult be predicted from his or her birth weight? This question calls for an expression of adult weight as a function of birth weight. Therefore, weight as a young adult is the dependent variable (y), and birth weight is the independent variable (x).

To discover what the relationship is between these two variables, we will first need a sample of bivariate observations consisting of birth weight and young adult weight for each of a number of individuals. In one study, sixteen 18-year-old females were asked to give their weight at birth and their present weight. Because this study deals with verbal reports rather than direct measurements, the results will have to be interpreted with caution, but they should be sufficiently accurate to indicate the general nature of the relationship, if any. The observations for the 16 individuals were as follows.

Individual	Birth weight (oz), x	Adult weight (lb), y
1	112	120
2	118	135
3	118	125
4	118	138
5	101	115
6	93	118
7	128	145
8	118	120
9	140	144
10	108	130
11	118	120
12	123	135
13	112	120
14	127	132
15	112	125
16	134	135

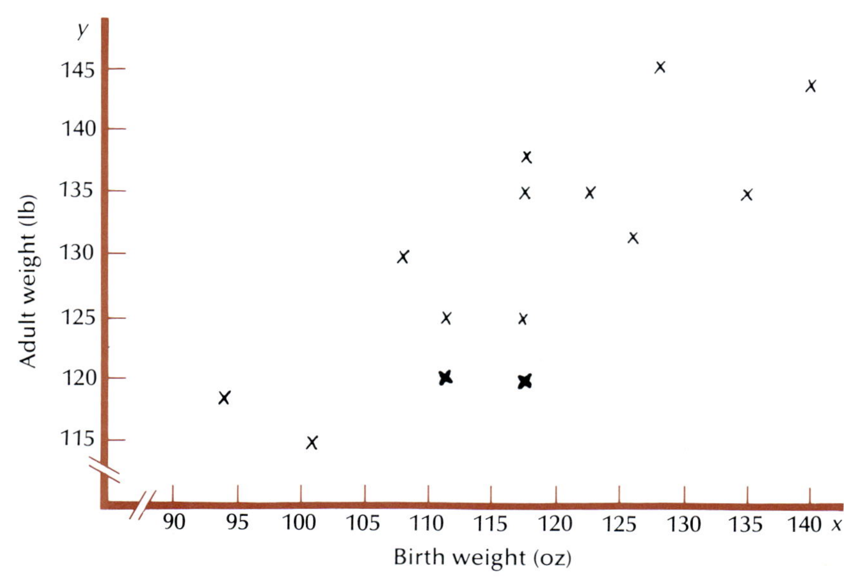

Figure 5.3 Adult weight as a function of birth weight for sixteen 18-year-old females.

The 16 bivariate observations are represented by the 16 points plotted in Figure 5.3.

It is apparent from Figure 5.3 that there is a general tendency for greater adult weights to be associated with greater birth weights, but, as is typical in real data, it is by no means a perfect relationship. The points may be described as being scattered around an imaginary straight line rising from the lower left to the upper right. This kind of straight-line description is called the **simple linear regression** of adult weight on birth weight. Simple linear regression analysis will provide us with a way of determining the values of the slope b and the y-intercept a so that we can write the equation of the "best" straight-line description of the data and draw the line on the graph.

(Source: Student project by C. Hochhausen.)

It is a good practice to begin any regression analysis by plotting the data on graph paper, each bivariate observation being represented by a point (x_i, y_i). Such a plot of the data points is called a **scatter diagram**, or a **scattergram**; Figure 5.3 is a scatter diagram. The scatter diagram allows us to see the general nature of the relationship between the variables. We use it to make an initial judgment as to whether it is reasonable to describe the relationship by a straight line, or whether a more complex kind of description would be more appropriate. In addition, the scatter diagram allows us to identify any deviant points, so we can check for errors in recording or for the possibility that such an observation does not actually belong to the population relevant to the problem at hand.

Once we have decided that a straight-line description may provide a useful summary of the relationship, our problem is to find the most appropriate values of b and a in the linear equation, $y = bx + a$. One way to find the values of b and a is to adjust a straightedge on the scatter diagram to draw a single straight line that appears to pass as close as possible to all the points, and then to read the values of b and a from the graph. **Fitting by eye**, as this procedure is called, may be adequate for some purposes, but its subjectivity prevents its general acceptance. Someone else looking at the same data may feel that a somewhat different line seems to fit the data better, and even the person who draws the line cannot define precisely what makes it the best line except that it "looks about right"—hardly a statement to convince the scientific community or investment companies.

The procedure that is usually used to choose the best-fitting straight line is based on the **method of least squares**, which was first published by the French mathematician Adrien Legendre (see Biography 5.2), although it had been discovered more than 10 years earlier by Carl Gauss (see Biography 7.1). A line resulting from the method of least squares—a procedure that is by no means limited to the straight-line case—is called a **least-squares line**.

BIOGRAPHY 5.2 Adrien Marie Legendre (1752–1833)

Adrien Marie Legendre was born in Paris, France, on 18 September 1752. After studying at the Collège Mazarin in Paris, he was appointed as a professor of mathematics at the École Militaire and later at the École Normale, as well as to various minor government positions.

Legendre's primary research area was elliptic integrals. He applied his research to topics such as the paths of projectiles, the orbits of comets, and geodesy, which deals with the exact determination of points and areas on the curved surface of the earth. In 1794 he published a widely used geometry text. His book on new methods for the determination of comet orbits, which was published in 1806, included the first published treatment of the method of least squares. He was unaware at the time that Carl Gauss (see Biography 7.1) had already invented the method in 1795 at the age of 18, but had never published it. In spite of his many and varied contributions to mathematics, Legendre was soon followed by brilliant younger mathematicians such as Abel, Jacobi, and Gauss, whose further developments in the same areas often overshadowed his own.

Legendre died in Paris on 10 January 1833 at the age of 80.

(Sources: Bell, 1937; *Encyclopaedia Britannica,* 1959, 1977.)

We refer to the y-value described by the least-squares line as the **fitted value** $\hat{y}$ (read "y-hat"), so that for the ith observational unit, the observed value of the dependent variable is y_i and the fitted value is $\hat{y}_i$. The fitted value $\hat{y}$ corresponding to a particular value of x represents the average y-value for that value of x, according to the regression equation. Any particular observation (x_i, y_i) may have an observed y-value (y_i) that is above or below the average of observations at that particular value of x. The difference between the observed value y_i and the fitted value $\hat{y}_i$ described by the line is called an **error of estimation**:

$$\text{Error of estimation} = \text{observed value} - \text{fitted value} = y_i - \hat{y}_i \tag{5.5}$$

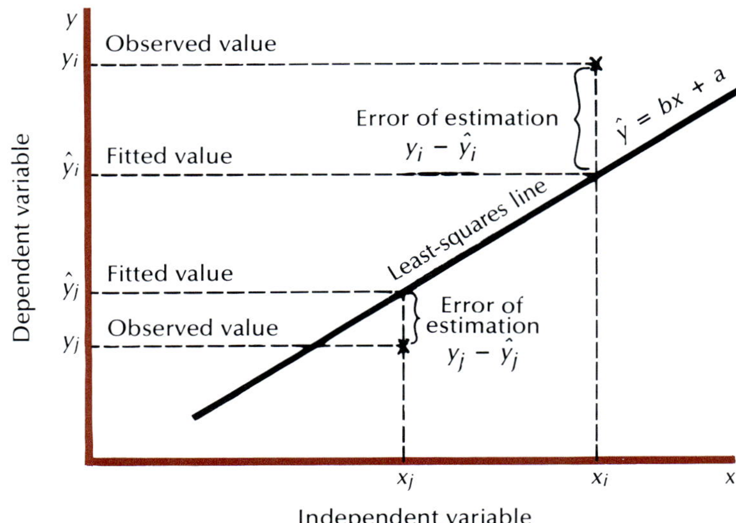

Figure 5.4 Incomplete scattergram illustrating observed values, fitted values, and errors of estimation of the dependent variable. Only two of the n data points are shown.

These quantities are illustrated in Figure 5.4 for two of the points in a scatter diagram; the remaining points have been deleted to avoid cluttering the picture. Every observed value y_i, including those not shown in Figure 5.4 (see, for example, the complete scattergram in Figure 5.3), may be thought of as consisting of a fitted value plus an error of estimation.

How do we decide where to position a straight line to describe y as a linear function of x? We would, of course, like the errors of estimation to be as small as possible for every observation, but whenever we move the line to reduce the error for one observation (for example, in Figure 5.3), we are apt to move farther from another observation. We need some objective criterion for determining just where the best line is to be drawn. The criterion in general use is the sum of the squared errors of estimation:

$$\sum^{n} (y_i - \hat{y}_i)^2 \tag{5.6}$$

which is called the **least-squares criterion**. The essence of the method of least squares is to choose the line of fitted values $\hat{y}$ so as to minimize the least-squares criterion. In the case of simple linear regression, we choose the straight line of fitted values

$$\hat{y} = bx + a \tag{5.7}$$

so as to make the quantity in Expression (5.6) as small as possible. In terms of the diagram in Figure 5.4, the method of least squares involves finding the line such that the sum of the squares of the *vertical* or y-value deviations from the points to the line is as small as possible. (Minimizing the sum of the squared *horizontal* or x-value deviations would yield a different regression line. Thus, it is important to identify the dependent variable y correctly.)

The use of squared deviations from fitted values in the least-squares criterion in Expression (5.6) corresponds to the use of squared deviations from the mean in the definitions of variance and standard deviation in Section 4.3. The operation of squaring preserves information about the amount of deviation while disposing of information about the direction of deviation. Although this criterion has many convenient properties, it is not the only possible criterion, and when we talk about the "best-fitting" line, we should always specify whether we mean "best in the least-squares sense" or best by some other criterion.

Differential calculus can be used to show that the values of b and a that minimize the least-squares criterion

$$\sum (y_i - \hat{y}_i)^2 = \sum [y_i - (bx_i + a)]^2 \tag{5.8}$$

are given by

$$b = \frac{\sum xy - \dfrac{\sum x \sum y}{n}}{\sum x^2 - \dfrac{(\sum x)^2}{n}} \tag{5.9}$$

$$a = \frac{\sum y - b \sum x}{n} = \bar{y} - b\bar{x} \tag{5.10}$$

The values of b and a calculated by Expressions (5.9) and (5.10) may be substituted into Expression (5.7) to give us an equation $\hat{y} = bx + a$ for any sample consisting of a bivariate observation (x_i, y_i) for each observational unit. (Our discussion has been in terms of sample data and sample notation, because the formulas are applied to data that are regarded as sample data. For example, regression data are often viewed as a sample from the population of possible data that hypothetically could have been generated by the underlying true relationship.)

EXAMPLE 5.6 Let us illustrate the use of Expressions (5.9) and (5.10) by calculating the simple linear regression of reported adult weight (y) on recalled birth weight (x) for the data given in Example 5.5. The calculations may be organized as shown in the table on page 156.

Applying Expressions (5.9) and (5.10), we obtain

$$b = \frac{242{,}972 - \dfrac{(1{,}880)(2{,}057)}{16}}{222{,}964 - \dfrac{(1{,}880)^2}{16}} = \frac{1{,}274.5}{2{,}064} = .61749$$

$$a = \frac{2{,}057 - (.61749)(1{,}880)}{16} = 56.007$$

Birth weight x	Adult weight y	xy	x^2
112	120	13,440	12,544
118	135	15,930	13,924
118	125	14,750	13,924
118	138	16,284	13,924
101	115	11,615	10,201
93	118	10,974	8,649
128	145	18,560	16,384
118	120	14,160	13,924
140	144	20,160	19,600
108	130	14,040	11,664
118	120	14,160	13,924
123	135	16,605	15,129
112	120	13,440	12,544
127	132	16,764	16,129
112	125	14,000	12,544
134	135	18,090	17,956
$\sum x = 1{,}880$	$\sum y = 2{,}057$	$\sum xy = 242{,}972$	$\sum x^2 = 222{,}964$

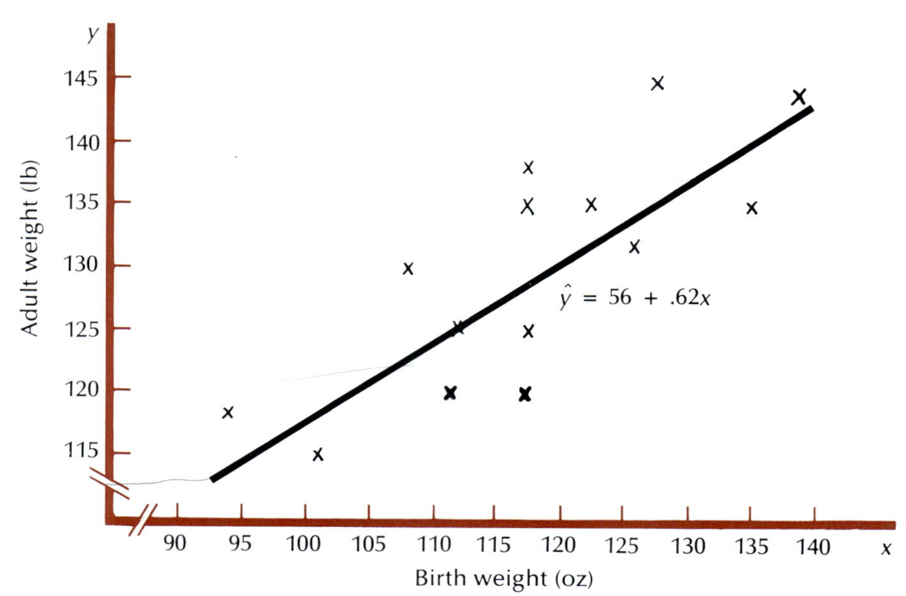

Figure 5.5 Scattergram of adult weight as a function of birth weight for sixteen 18-year-old females, with least-squares line.

For these 18-year-old females, the least-squares regression line describing reported adult weight (y) in pounds as a linear function of recalled birth weight (x) in ounces is therefore

$$\hat{y} = 56 + .62x \tag{5.11}$$

as illustrated in Figure 5.5. (Guidelines for how to round and how many digits to report are given in Appendix 1.)

Interpretation of the Simple Linear Regression Equation

What information is provided by the linear regression equation $\hat{y} = bx + a$? The following comments about interpretation should be kept in mind.

Comment 1, Concerning Extrapolation. The regression equation $\hat{y} = bx + a$ describes y as a linear function of x between the smallest and largest values of x that were observed, but it does not indicate whether it is safe to extrapolate to values of x beyond those observed. Extrapolation is always risky and must be justified (if and when it can be) by knowledge of the process being described, not on statistical grounds.

EXAMPLE 5.7 Example 5.6 gave the regression of adult weight (in pounds) on birth weight (in ounces) as $\hat{y} = 56 + .62x$ for 18-year-old females. The observed values of recalled birth weight (x) ranged from 93 to 140 ounces. For premature babies weighing, say, 64 ounces at birth, the equation predicts an average weight of 96 pounds at 18 years of age. Such a prediction is not justified by the data since it is based on extrapolation beyond the range of x-values that were observed. The fact that the relationship between these two variables is approximately linear within the range observed does not imply that the descriptive line remains straight when extended to lighter or heavier birth weights. A judgment as to whether the extrapolation is reasonable must be based on one's knowledge of birth weights and adult weights, that is, knowledge brought to the analysis from outside, rather than from the statistical analysis.

As this example demonstrates, the existence of a regression equation is not a license for its unrestricted application.

Comment 2, Concerning the Slope. The slope b, which is sometimes called the **regression coefficient** in a regression equation, indicates by its sign whether the variables x and y are directly (positive slope) or inversely (negative slope) related. In addition, the value of b is the average number of units of change in y for each unit of increase in x. This aspect of the interpretation of the regression equation was illustrated in Example 5.4.

Comment 3, Concerning the y-Intercept. The y-intercept a, which is sometimes called the **regression constant** in a regression equation, is an estimate of the average value of y when x equals 0. The warnings about extrapolation in Comment 1 are relevant here whenever $x = 0$ is beyond the extremes of the observed values of x. Knowledge of the process being described is required to judge the interpretability of the y-intercept in cases where extrapolation is involved.

EXAMPLE 5.8 Interpretation of the y-intercept was illustrated for the printing-time data in Example 5.4. The interpretation involved extrapolation, but the fact that the printing process did in fact include a constant time for setting up and cleaning up, plus an incremental time for each copy printed, made the interpretation meaningful.

EXAMPLE 5.9 In the regression of adult weight on birth weight in Example 5.6 ($\hat{y} = 56 + .62x$), the extrapolation to interpret the y-intercept is most certainly not justified. It makes no sense to predict that an infant with a birth weight of $x = 0$ ounces will weigh $\hat{y} = 56$ pounds as an 18-year-old! The y-intercept is not directly interpretable in this case.

As an alternative, a convenient value at or near the minimum observed value of x is sometimes useful in place of $x = 0$ in the interpretation of the regression equation. The minimum value of x in Example 5.6 was 93 ounces. Since a weight of 96 ounces is an even 6 pounds, we might interpret the regression equation $\hat{y} = 56 + .62x$ as follows: The predicted young-adult weight of a 6-pound female baby is $\hat{y} = 56 + (.62)(96) \simeq 115$ pounds, and this predicted weight increases by an average of .62 pound for each additional ounce of birth weight.

In some cases, the absurdity of the y-intercept is itself interpretable; for example, a negative value of the y-intercept when y is a positive-valued variable defined at $x = 0$ suggests that the relationship of x and y is not linear over the entire observed range of x.

EXAMPLE 5.10 A study was conducted to investigate the relationship between the amount of grain lost in harvesting and the speed at which the combine is run. The testing was done by operating a combine at various speeds over a pan on the ground and counting the number of kernels of grain that fell into the pan at each speed. The results are shown in the form of a scatter diagram and least-squares line in Figure 5.6.

The y-intercept is the impossible value of $a = -1.8$ kernels, an impossible value

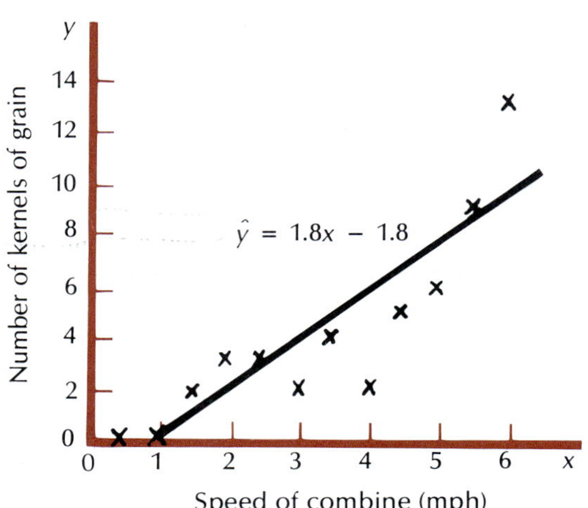

Figure 5.6 Scattergram of amount of grain lost as a function of combine speed.

because the number of kernels in the pan cannot be negative. However, the negative value should suggest to us that the relationship between the two variables is curvilinear rather than linear, and inspection of the scattergram confirms this suspicion. The straight-line description seems reasonably good within the range from about 1 to 6 mph, but a better description could probably be provided by a nonlinear function.

(Source: Student project by D. Cameron.)

Comment 4, Concerning the Noninterchangeability of x and y. Because the least-squares equation $\hat{y} = bx + a$ estimates average values of the variable y for given values of x, solving the equation for x does *not* yield least-squares estimates of x for given values of y. The least-squares procedure that has been presented above is for the **regression of *y* on *x***, and it involves minimizing the sum of squared errors of estimation of y. If one wanted to describe x as a function of y, the least-squares approach would minimize a different criterion, namely, the sum of squared errors of estimation of x, yielding a different regression line, called the **regression of *x* on *y***.

EXAMPLE 5.11 The regression of y on x in Example 5.6 was $\hat{y} = 56 + .62x$. It may be shown that the regression of x on y, with the terms rearranged to facilitate comparison with the previous equation, is $y = 4 + 1.06\hat{x}$.

In practice, we need deal only with the regression of y on x, provided that we always label the dependent variable as y. The important point here is simply that in least-squares regression the labeling of the variables as x and y is not arbitrary. The regression of y on x ($\hat{y} = bx + a$) describes y as a function of x, and it should not be turned around to estimate values of x for given values of y.

Comment 5, Concerning Mixtures of Populations. If one or more of the observations come from a different population than the rest, the resulting regression equation may be very misleading. The best-fitting straight line describing two populations jointly—joining two swarms of points, as it were—may be very different from the best-fitting straight line within either population separately.

EXAMPLE 5.12 Students trying out for college volleyball teams took a series of fitness tests. Subsequently, a Jump Index and a Sprint Index were calculated for each candidate to indicate performance in two areas of importance in volleyball. The results are shown in Figure 5.7 for 20 candidates, 13 for the men's team and 7 for the women's team.

Notice that when all 20 candidates are considered as a single group, the slope of the regression line is positive ($b = .51$). However, this seems to be a misrepresentation of the relationship between Jump Index and Sprint Index due to the mixing of the two populations (males and females), because when we look at the two groups separately, the relationship looks very different. For males, the relationship between the Sprint Index and the Jump Index is small but negative ($b = -.19$), and for females, it is small and positive ($b = .13$). When the two populations are considered jointly, the picture of the relationship between the Sprint Index and the Jump Index is distorted by the

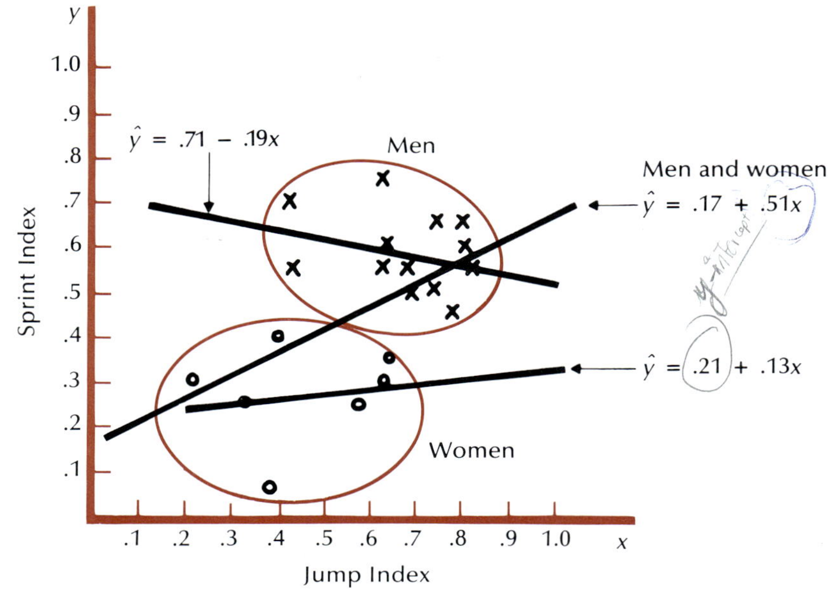

Figure 5.7 Scattergram of Sprint Index as a function of Jump Index for 20 volleyball candidates.

differences between the two groups, resulting in a description of the relationship that is correct for neither group.

(Source: Student project by H. Young.)

In Example 5.12, there were 13 observations from one population and 7 from another. The same kind of distortion can occur even if there is only one observation that is very different from the rest. Such **outliers** can often be detected in a scattergram, and a decision can then be made as to whether the deviant observation is from the same population as the rest.

Comment 6, Concerning Nonlinear Relationships. The formulas for b and a in Expressions (5.9) and (5.10) will produce correct values of b and a for any set of paired data, *whether or not it is appropriate* to try to fit a straight line to the data. Thus, the existence of an equation of the form $\hat{y} = bx + a$ should not be taken as evidence of a linear relationship between the variables x and y. (Example 5.10 presented data that could have been described better by a curvilinear function than by a linear function.) You should always plot the data to see whether a straight-line function is a reasonable choice for a description of the relationship. The next subsection introduces a more objective method of assessing how well the linear description $\hat{y} = bx + a$ fits the data. It is a good practice to include this method routinely as one component in any attempt to interpret a simple linear regression equation.

Strength of Linear Relationship

Before we can measure the strength of the linear relationship $\hat{y} = bx + a$, we must know how well we could estimate the values of the variable y if it had no relationship with x at all. In that case, where x_i provides no information about the value of y_i, our best guess as to the value of each and every y_i is the mean $\bar{y}$, yielding errors of estimation $y_i - \hat{y}_i$ equal to the deviations from the mean $y_i - \bar{y}$.

Now consider how the existence of a linear relationship between x and y can reduce our errors of estimation $y_i - \hat{y}_i$ from what they would be (namely, $y_i - \bar{y}$) if there were no linear relationship. If y is linearly related to x, then each deviation from the mean $y_i - \bar{y}$ can be partitioned into an error of estimation $y_i - \hat{y}_i$ and a portion $\hat{y}_i - \bar{y}$ due to regression (that is, due to the linear relationship), as illustrated in Figure 5.8.

Algebraically, we have

$$y_i - \bar{y} = (y_i - \hat{y}_i) + (\hat{y}_i - \bar{y}) \tag{5.12}$$

which says that each observation's total deviation from the mean consists of its error of estimation (error deviation) plus the deviation due to regression. Given an observation's total deviation from the mean, the larger the deviation due to regression (that is, the stronger the linear relationship), the smaller the error deviation.

Expression (5.12) gives the deviation for one observation. Such deviations $y_i - \bar{y}$ are positive for some observations and negative for others. Therefore, a

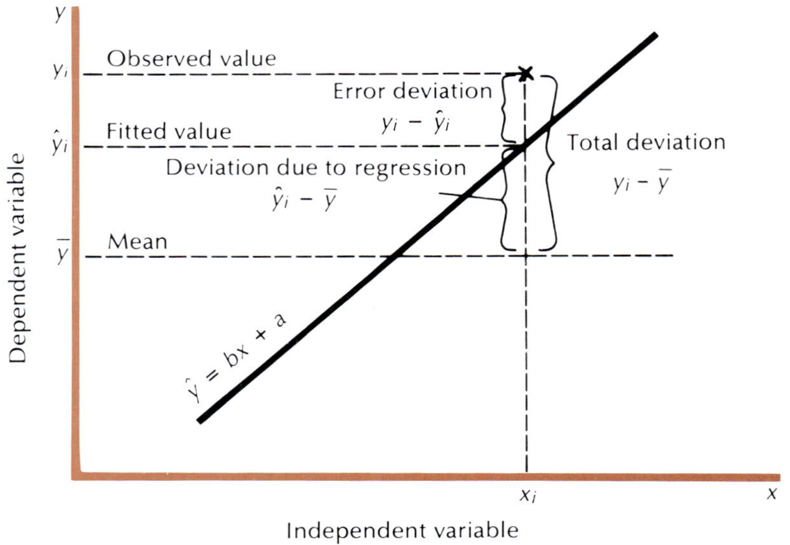

Figure 5.8
Stylized drawing to illustrate the partition of an observation's total deviation from the mean into error deviation plus deviation due to regression.

convenient measure of the total amount by which all the observations in the sample deviate from the mean is the sum of the squared deviations from the mean, which is usually called the **total sum of squares**:

$$SS_y = \sum (y_i - \bar{y})^2 \tag{5.13}$$

where the subscript y indicates that we are referring to variable y. A sum of squares is said to measure the **variation** in a set of data—that is, the extent to which certain values differ from each other—so SS_y is also called the **total variation in *y***, which is understood to refer to the variation of the observed values y_i around their mean $\bar{y}$. If all of the values y_i were equal and hence equal to the mean $\bar{y}$, the deviations $y_i - \bar{y}$ would all be zero and the total variation SS_y would likewise be zero. The more spread out the values of y_i are, the larger the magnitudes of the deviations $y_i - \bar{y}$, and consequently the larger the total variation SS_y.

Expression (5.12) and Figure 5.8 showed how a single observation's total deviation from the mean can be partitioned into error deviation and deviation due to regression. By squaring both sides of the equality in Expression (5.12) and summing over all the observations in the sample, we can show that the total variation in y (SS_y) is likewise partitioned into **residual error variation** (the **error sum of squares**, SS_{error}) and **variation due to the regression** (the **sum of squares due to regression**, SS_{regr}) as follows:

$$\sum (y_i - \bar{y})^2 = \sum (y_i - \hat{y}_i)^2 + \sum (\hat{y}_i - \bar{y})^2 \tag{5.14}$$

or

$$SS_y = SS_{error} + SS_{regr} \tag{5.15}$$

The partition of these sums for the entire sample is analogous to the partition of a single observation's deviation as depicted in Figure 5.8 and Expression (5.12). Given the total variation SS_y, the larger the variation due to regression SS_{regr} (that is, the stronger the linear relationship), the smaller the error variation SS_{error}.

Looking at Figure 5.8, we see that if the linear relationship is a strong one, such that the points all tend to fall very close to the least-squares line, then the error component will be relatively small and the variation due to regression SS_{regr} will be a large proportion of the total variation SS_y. On the other hand, if there is only a weak linear relationship, such that the points are widely scattered around the least-squares line, then the error component will be relatively large and the variation due to regression SS_{regr} will constitute a small proportion of the total variation SS_y. These facts are the basis for a measure of the strength of linear relationship, called the **coefficient of determination**, denoted by r^2:

> The **coefficient of determination** r^2 is the proportion of the total variation in y that can be accounted for by the linear relationship with x; that is, **(5.16)**

$$r^2 = \frac{SS_{regr}}{SS_y} = \frac{SS_{regr}}{SS_{regr} + SS_{error}}$$

$$= \frac{\sum (\text{deviation due to regression})^2}{\sum (\text{total deviation})^2} \qquad (5.17)$$

By examining Figure 5.8, you can see how the relative sizes of error deviation and deviation due to regression, when combined from all observations, determine the value of r^2.

The coefficient of determination takes on values from 0 (or 0%) when x and y are not linearly related, to 1 (or 100%) when there is a perfect straight-line relationship between x and y in the data. Thus, values of r^2 near 0 indicate that the data points do not fall in a well-defined straight-line shape, whereas values near 100% indicate that the points tend to fall close to the least-squares straight line.

EXAMPLE 5.13 The coefficient of determination for the data presented in Figure 5.5 is $r^2 = 58\%$. For the data in Figure 5.6, $r^2 = 76\%$. In Figure 5.7, $r^2 = 4\%$ for the women, $r^2 = 8\%$ for the men, and $r^2 = 26\%$ for the entire group considered jointly. You should reexamine these scattergrams to get an idea of how various coefficients of determination manifest themselves in real data.

Expression (5.17) is a definitional, not computational, formula for the coefficient of determination r^2. The usual way to compute r^2 is first to calculate its square root, r, by means of a formula given in the next section.

5.2 SIMPLE LINEAR CORRELATION

The value of r is of interest in its own right, and not just as the square root of the coefficient of determination r^2. Indeed, r is an important descriptive statistic known as the **Pearson product-moment correlation coefficient**, developed by Karl Pearson (see Biography 5.3). There are other coefficients of correlation, such as Spearman's rank correlation coefficient to be discussed in Section 16.2, but Pearson's r is used so much more frequently that when the term **correlation coefficient** is used, it is understood to be the Pearson product-moment correlation coefficient.

BIOGRAPHY 5.3 **Karl Pearson (1857–1936)**

Karl Pearson was born in London, England, on 27 March 1857. He studied at King's College, Cambridge, and later in Berlin and Heidelberg, Germany. He taught at

University College in London, where he was successively appointed as Goldsmid Professor of Applied Mathematics (1884–1911), Gresham Professor of Geometry (1891–1911), Galton Professor of Eugenics (1911–33), and professor emeritus (1933–36).

Pearson is considered to be among the leading contributors to the development of statistics, and he is sometimes referred to as the founder of the science of statistics. His influence is still present in many topics in statistics; for example, he coined the term *standard deviation* (Chapter 4); he was responsible for much of the theoretical development underlying correlation analysis (Chapters 5 and 13); he made significant contributions to the study of probability distributions (Chapters 7 and 8); and he invented the widely used chi-square test (Chapter 14).

Pearson married Maria Sharpe in 1890, and they had one son (Egon Sharpe Pearson, a prominent statistician in his own right) and two daughters. He died at Coldharbour, England, on 27 April 1936 at the age of 79.

(Source: *World Who's Who in Science*, 1968.)

Pearson's r is an index of the linear relationship between two variables, that is, an index of **simple linear correlation**. Like the simple linear regression equation, to which it is closely related, Pearson's r provides a description of linear relationship. Since r^2 is a proportion, ranging from a minimum of 0 to a maximum of 1, r is necessarily limited to the range from -1, which is called **perfect negative correlation**, to $+1$, which represents **perfect positive correlation**. Moreover, from what we know about the meaning of r^2, it follows that an r-value near 0 indicates a weak linear relationship and an r-value near either -1 or $+1$ indicates a strong linear relationship.

Calculation of the Correlation Coefficient

The coefficient r_{xy} of correlation between variables x and y may be defined as follows:

$$r_{xy} = \frac{\sum (x - \bar{x})(y - \bar{y})}{\sqrt{\sum (x - \bar{x})^2}\sqrt{\sum (y - \bar{y})^2}} \tag{5.18}$$

(The subscripts on r are dropped when the context makes clear which variables are involved.) Since the denominator is always positive, the numerator determines the sign of r. Notice that the numerator involves the product of the deviations from the means of the two variables for each observational unit. If the deviations for a given observational unit are both positive or both negative, the product will be positive and will contribute to a positive value of r. If one deviation is positive and the other negative for a given observational unit, the product will be negative and will contribute to a negative value of r. The sum (over all the observational units) of these products will be positive, as then will r, if there is a general tendency for relatively high values on one variable to be associated with relatively high values on the other variable, and for relatively low values on one variable to be associated with relatively low values on the other variable. In addition, the sum, and hence r, will be negative if relatively high values on one variable tend to be associated with relatively low values on the other variable. Let us illustrate with two simple artificial examples.

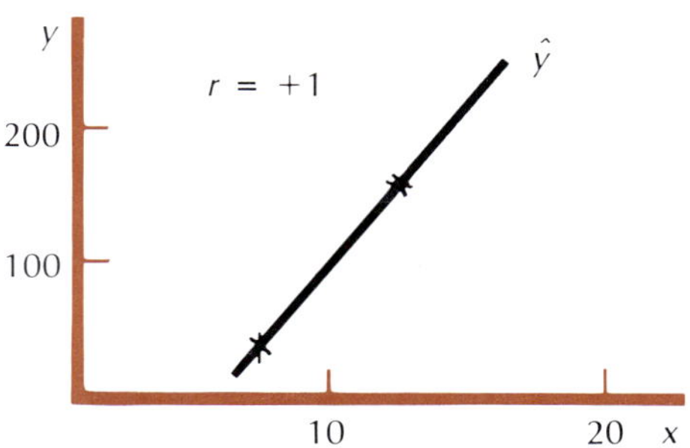

Figure 5.9 Positive correlation in a sample of $n =$ 2 observations.

EXAMPLE 5.14 Consider the following sample of $n = 2$ bivariate observations:

Observational unit 1: $x_1 = 7$ $y_1 = 40$

Observational unit 2: $x_2 = 13$ $y_2 = 160$

The means of the two variables are $\bar{x} = 10$ and $\bar{y} = 100$, and it is apparent that the first observational unit is below the mean on both variables and the second observational unit is above the mean on both variables. Therefore, the correlation is positive: As the value of one variable changes, the value of the other variable changes in the same direction. The correlation coefficient is $r = +1$ for these data, shown in Figure 5.9 with the regression line.

EXAMPLE 5.15 For the second example, consider another sample of $n = 2$ bivariate observations:

Observational unit 1: $x_1 = 8$ $y_1 = 170$

Observational unit 2: $x_2 = 12$ $y_2 = 30$

Again the means are $\bar{x} = 10$ and $\bar{y} = 100$, but this time the first observational unit is below the mean on x and above the mean on y, while the second observational unit is above the mean on x and below the mean on y. Therefore, the correlation is negative: As the value of one variable changes, the value of the other variable changes in the opposite direction. The correlation coefficient is $r = -1$ for these data, shown in Figure 5.10 with the regression line.

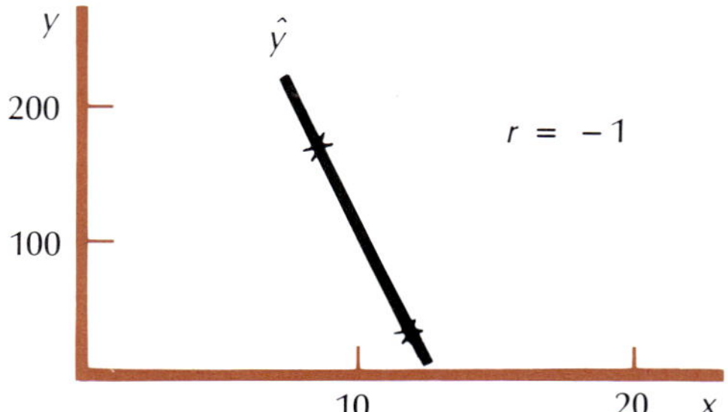

Figure 5.10 Negative correlation in a sample of $n = 2$ observations.

Algebraic manipulation analogous to that on pages 116–117 for the sample variance converts the definitional formula in Expression (5.18) to the usual **computational formula for the correlation coefficient:**

$$r_{xy} = \frac{\sum xy - \dfrac{\sum x \sum y}{n}}{\sqrt{\sum x^2 - \dfrac{(\sum x)^2}{n}}\sqrt{\sum y^2 - \dfrac{(\sum y)^2}{n}}} \qquad (5.19)$$

The correlation coefficient calculated by Expression (5.19) is usually reported to two decimal places, except that values that would round to ±1.00 are often reported with sufficient decimal places to indicate that they differ from ±1.00 (for example, .9997).

EXAMPLE 5.16 To illustrate the use of Expression (5.19), we calculate the coefficient of correlation between recalled birth weight (x) and reported adult weight (y) for the data of Example 5.5 from a sample of sixteen 18-year-old females. The scattergram and regression line are presented in Figure 5.5. The calculations for Expression (5.19) may be organized as follows.

x	y	xy	x^2	y^2
112	120	13,440	12,544	14,400
118	135	15,930	13,924	18,225
118	125	14,750	13,924	15,625
118	138	16,284	13,924	19,044
101	115	11,615	10,201	13,225
93	118	10,974	8,649	13,924
128	145	18,560	16,384	21,025
118	120	14,160	13,924	14,400
140	144	20,160	19,600	20,736
108	130	14,040	11,664	16,900
118	120	14,160	13,924	14,400
123	135	16,605	15,129	18,225
112	120	13,440	12,544	14,400
127	132	16,764	16,129	17,424
112	125	14,000	12,544	15,625
134	135	18,090	17,956	18,225
$\sum x = 1{,}880$	$\sum y = 2{,}057$	$\sum xy = 242{,}972$	$\sum x^2 = 222{,}964$	$\sum y^2 = 265{,}803$

Expression (5.19) yields the correlation coefficient:

$$r = \frac{242{,}972 - \dfrac{(1{,}880)(2{,}057)}{16}}{\sqrt{222{,}964 - \dfrac{1{,}880^2}{16}}\sqrt{265{,}803 - \dfrac{2{,}057^2}{16}}}$$

$$= \frac{1{,}274.5}{\sqrt{2{,}064}\sqrt{1{,}349.9375}} = .7635$$

The correlation between recalled birth weight and reported adult weight in this sample is $r = .76$.

Notice that the preliminary calculations are the same as for simple linear regression in Example 5.6, except for the inclusion of the y^2 column for correlation.

It should be apparent from both Expressions (5.18) and (5.19) that interchanging the roles of x and y has no effect on the correlation; that is,

$$r_{xy} = r_{yx} \tag{5.20}$$

Although correct identification of the dependent and independent variables as y and x, respectively, is critical in regression, it makes no difference in simple linear correlation.

Interpretation of the Correlation Coefficient

A little algebraic manipulation allows us to express the correlation coefficient as

$$r = b\frac{s_x}{s_y} \tag{5.21}$$

[You may find it instructive to use Expressions (4.13), (5.9), and (5.19) to demonstrate to yourself the truth of Expression (5.21).]

Careful examination of Expression (5.21) reveals several properties of the correlation coefficient relevant to its interpretation. The following comments outline these and other points regarding the meaning of a correlation coefficient. Because the interpretation of correlation is essentially an extension of the interpretation of regression, the numbering of the comments continues from the Comments 1–6 on regression on pages 157–160.

Comment 7, Concerning Linearity. Expression (5.21) shows that r is closely related to the slope b of the regression line $\hat{y} = bx + a$, and thus it emphasizes that r is an index of *linear* relationship—an index of the tendency of the points in the scattergram to form a single straight line. A correlation coefficient near 0 means not that the variables are not functionally related, but only that they have no strong *linear* relationship. They may still be related in a curvilinear, or nonlinear, fashion.

EXAMPLE 5.17 The following hypothetical data are plotted in Figure 5.11.

x:	1.5	1.6	1.7	1.8	2.0	2.2	2.5	3.0	3.8	4.2	5.0	5.5	5.8	6.0	6.2	6.3	6.4	6.5
y:	5.0	4.5	4.0	3.5	3.0	2.5	2.0	1.5	1.0	1.0	1.5	2.0	2.5	3.0	3.5	4.0	4.5	5.0

You may verify that the correlation between x and y is $r = 0$, and yet it seems clear that there is a systematic relationship between the variables. The explanation is that the relationship is nonlinear, and the value of $r = 0$ is an index of linear relationship. The correlation coefficient r does not detect the systematic nonlinear component in the relationship.

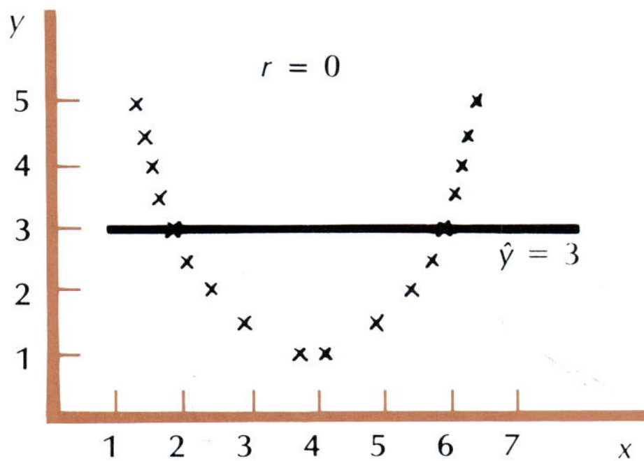

Figure 5.11 Scattergram showing y as a curvilinear function of x, with best-fitting linear regression line.

Example 5.17 illustrates why it is a good practice to plot the data in a scattergram before attempting an interpretation of a correlation coefficient. Comment 6 on page 160 makes much the same point in the context of regression.

Comment 8, Concerning the Direction of the Relationship. Expression (5.21) shows that the correlation coefficient r always has the same sign as the slope b of the regression line $\hat{y} = bx + a$. A negative slope implies a negative correlation (inverse relationship), and a positive slope implies a positive correlation (direct relationship).

EXAMPLE 5.18 The following are some pairs of variables, familiar from everyday experience, that tend to be positively correlated. Changes in one variable tend to be associated with changes in the same direction in the other variable:

a. size of house and amount of heating bill;
b. cost of long-distance phone call and distance called;
c. height and shoe size;
d. price of hardcover textbook and number of pages; and
e. engine size and fuel consumption.

The following are some pairs of variables that tend to be negatively correlated. Changes in one variable tend to be associated with changes in the opposite direction in the other variable:

f. score at darts and distance from board;
g. distance from student's home to college and frequency of visiting home;
h. height and number of strides to walk a fixed distance;
i. number of previous practices playing a musical composition and number of wrong notes played; and
j. engine size and gas mileage.

Comment 9, Concerning Measurement Units. Expression (5.21) indicates that the correlation coefficient r is a dimensionless quantity. The units attached to the slope b in the right-hand side of (5.21) are y-units per x-unit (rise over run); the units of s_x (the standard deviation of x) are x-units; and the units of s_y (the standard deviation of y) are y-units. Thus, the units in Expression (5.21) all cancel, leaving us with the dimensionless quantity r. In effect, r is a standardized form of the slope b. Just as we divided the deviation from the mean by the standard deviation to define standard scores in Section 4.4, pages 127–130, so we convert the "rise over run" of the regression equation to standard deviation units in order to obtain a standardized slope coefficient r. Consequently, the value of the correlation coefficient r is not altered by the choice of measurement units (for example, cm, km, or ft) for the variables. Conversely, rearranging the terms of Expression (5.21) as

$$b = r\frac{s_y}{s_x} \tag{5.22}$$

we may think of the slope as being an index of linear relationship converted into the original units of measurement. The correlation coefficient is an index of linear *relationship*—an index of the tendency of the cloud of points in the scattergram to appear as a single straight line—not an index of location or dispersion. The correlation coefficient for a given set of data is not affected by changes in location (sliding the point cloud left or right or up or down) or by changes in scale (linearly stretching or compressing one or both axes).

Comment 10, Concerning Strength of Linear Relationship. Correlation coefficients cannot be interpreted as proportions (except in certain special applications), and they are not percentages. For example, $r = .50$ is *not* twice as strong a correlation as $r = .25$; nor is it correct to write it as "50%." To interpret a value of r, it is usually useful to square it and interpret the coefficient of determination r^2 as the proportion of the variation that can be accounted for in one variable by its linear relationship with the other; r^2 is correctly interpreted as a proportion (or percentage), but r is not. As a very rough guideline, a correlation coefficient in the neighborhood of $r = \pm.30$ is usually considered to represent a weak linear relationship, accounting for less than 10% ($r^2 = 9\%$) of the variation in the dependent variable; a correlation coefficient in the neighborhood of $r = \pm.60$ might be considered to represent a moderate linear relationship, accounting for about one-third ($r^2 = 36\%$) of the variation in the dependent variable; and a correlation coefficient in the neighborhood of $r = \pm.90$ is often considered to represent a fairly strong linear relationship, accounting for more than 80% ($r^2 = 81\%$) of the variation in the dependent variable. However, no hard-and-fast rules can be stated as to what is a weak or strong correlation. For some applications, a linear relationship with $r = \pm.50$ may provide sufficient precision in the predictions or description required; in others, even a linear relationship with $r = \pm.95$ may not provide the precision required. In any case, the sign of the correlation coefficient r has no bearing on the strength of the relationship. For example, a correlation of $r = -.9$ provides evidence of just as strong a linear relationship as does a correlation of $r = +.9$. In both cases, the variables are highly correlated ($r^2 = 81\%$), but in the first case increases in one variable are accompanied by decreases in the other, while in the second case increases in one are accompanied by increases in the other.

EXAMPLE 5.19 Figure 5.12 shows scattergrams from several different sets of real data to illustrate the appearance of various degrees of correlation. There are several features to observe in each scattergram:

a. In general, the cloud of points tends to take on more of a straight-line pattern as the correlation coefficient r approaches either -1 or $+1$ (that is, as we move from the upper to the lower scattergrams).

b. The slope b of the regression line has the same sign as does the correlation coefficient r.

c. Changing the measurement units (for example, ft, in., or cm) on the axes (or in the calculations) would not change the shape of the cloud of points, and hence would not change the correlation coefficient.

d. A comparison of scattergrams (c) and (e) reveals that the point clouds look different, even though the data are identical. The two point clouds are in fact the same, but the scales on which they have been drawn make them look different. This comparison illustrates that intuitive judgments of correlation from a scattergram are not infallible. Conversely, scales on scattergrams should be chosen to give an accurate impression of the correlation or lack thereof in the data.

e. Break marks (//) are used to indicate breaks in the axes when the scales are not shown continuously from the origin (0, 0), and the regression line is never drawn across a break in either axis.

(Sources: Student projects by R. Schulz, L. Warman, S. Fiege, N. Daley, L. Nowochin, W. Vornbrock, B. Laskosky, M. Banack, and G. Rackette.)

Comment 11, Concerning the Observed Range of the Variables. The correlation that is manifest between two variables depends in part on the range of the variables observed. In general, the coefficient of correlation between two variables tends to be reduced when the range of one or both variables is reduced. Furthermore, the relationship may change as the range is changed; Comment 1 on page 157 concerning extrapolation in the context of regression applies to correlation as well.

EXAMPLE 5.20 Consider an investigation of the correlation between a person's age and speed of running one mile. From our everyday knowledge of normal growth and development, we would expect this correlation to be much higher if the age range investigated were, say, 5 to 20 years than if it were limited to 14 to 15 years. Even though running speed tends to increase substantially in the range from 5 to 20 years, the relationship may not show up if we look only at the change from 14 to 15 years.

In addition, a different age range may give a very different picture of the relationship between age and running speed. For example, the correlation is apt to be negative in the range from 30 to 70 years, even though it is positive in the range from 5 to 20 years.

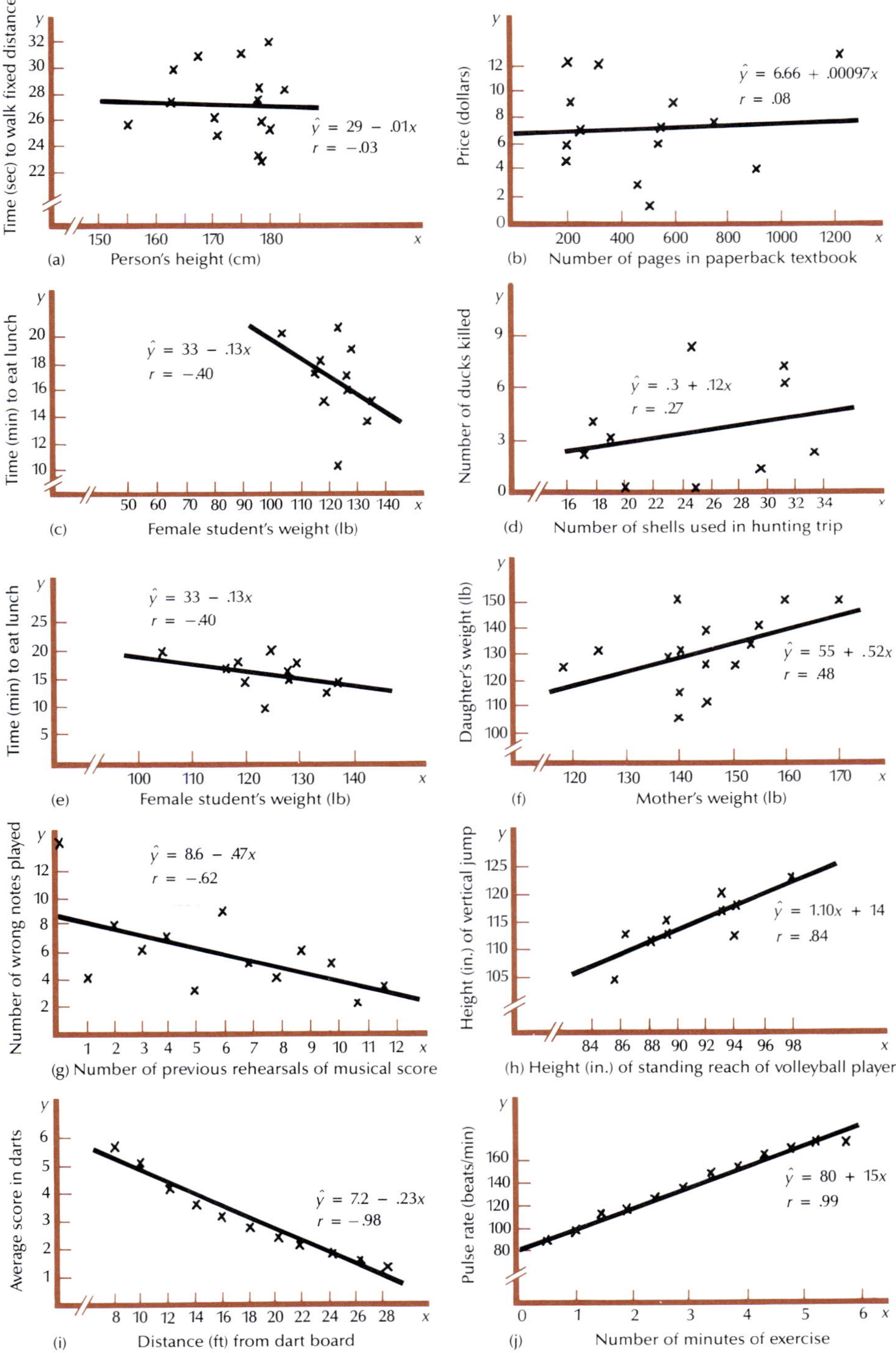

Figure 5.12 Scattergrams illustrating various degrees of linear correlation.

EXAMPLE 5.21 In Example 5.16, the independent variable x was birth weight, with observed values ranging from 93 to 140 ounces, and the correlation with young-adult weight was $r = .76$. If the sample had included only the 11 women with birth weights between 108 and 127 ounces, the correlation would have been only $r = .41$, as you may appreciate by mentally deleting the scattergram points for the two smallest and the three largest birth weights in Figure 5.5.

Similarly, if one or more observations come from a different population than the rest, the resulting correlation coefficient may be very misleading. This same point is discussed further in Comment 5 on page 159 in the context of regression.

EXAMPLE 5.22 In the data of Figure 5.7, the correlation coefficient calculated from the entire set of 20 observations is $r = .51$, whereas the relationship between the variables is more accurately represented by the correlation values $r = -.29$ for men and $r = .20$ for women.

Comment 12, Concerning Causality. Correlation does not indicate cause. No matter how high the correlation between two variables, the correlation coefficient never gives any grounds for saying that changes in one variable *cause* changes in another variable, but only that changes in one are associated with or are accompanied by changes in the other. Any judgment about causality must be based on what we know about the process we are investigating and about the ways the observations were made, not on the value of the correlation coefficient.

EXAMPLE 5.23 Very high correlations have been reported for the following pairs of variables:

a. teachers' salaries and alcohol sales;

b. the number of storks nesting in various European towns in the early 1900s and the number of human babies born in those towns in the same period;

c. the amount of rainfall on the prairies and the wheat yield from those areas;

d. the amount of time students study for a test and their mark on the test.

Do any of these correlations indicate that changes in the first variable are responsible for changes in the second variable?

In every case, the answer is no: The amount of correlation never indicates causation.

a. Both teachers' salaries and alcohol sales have increased over the years, along with the general standard of living and increasing human populations.

b. Larger towns had more places for storks to nest as well as more people to have babies.

c. It is our knowledge of wheat's need for moisture, not the amount of the correlation, that leads us to believe that low wheat yields result from insufficient moisture, rather than that insufficient moisture results from low wheat yields.

d. For any given student, additional study does indeed tend to result in higher test marks. However, an alternative explanation of the correlation would be that higher marks result from greater ability and that those with greater ability study more because they find it easier or more enjoyable.

In every case, the correlation merely indicates the degree of relationship or association, and we must rely on our knowledge of the way the world works to decide which of the various possible causal explanations is most reasonable.

EXAMPLE 5.24 "'My car won't start when I buy pistachio.'

"The manager of a Texas automobile dealership thought the woman who confronted him with this bizarre statement must be crazy. It seems that on hot summer days she would drive to a certain shop for ice cream to take home. It never failed, she said: the car would always start when she bought chocolate, vanilla or strawberry—but when she bought pistachio, she got stranded.

"The manager had to see this to believe it. He tried a chocolate trip, and the car worked fine. Vanilla or strawberry—no problem. Then came the trip for pistachio and, sure enough, the engine refused to start.

"It was an engineering troubleshooter whose insight solved the problem. He observed that chocolate, vanilla and strawberry were pre-packaged flavors, sold right out of the freezer. But take-home orders of pistachio were hand-packed at the shop. The time that it took to have the pistachio packed was just enough for the car to develop vapor lock in the summertime Texas heat. The woman wasn't crazy after all—her car *wouldn't* start when she bought pistachio."

(Source: Original source unknown. Reprinted in *Bulletin of the Greater New York Automobile Dealers Association* and subsequently in *Reader's Digest,* Canadian ed., May 1981, p. 69.)

STUDY MATERIAL FOR CHAPTER 5

Terms and Concepts

univariate methods
bivariate methods

Section 5.1

function
dependent variable
independent variable
slope
y-intercept
linear equation in two unknowns
regression analysis
regression fallacy
regression
simple linear regression

scatter diagram
scattergram
fitting by eye
method of least squares
least-squares line
fitted value
error of estimation
least-squares criterion
regression coefficient
regression constant
regression of y on x
regression of x on y
outlier
total sum of squares
variation
total variation in y
residual error variation
error sum of squares
variation due to the regression
sum of squares due to regression
coefficient of determination

Section 5.2

Pearson product-moment correlation coefficient
correlation coefficient
simple linear correlation
perfect negative correlation
perfect positive correlation

Key Formulas

Section 5.1

(5.4) $y = bx + a$

(5.7) $\hat{y} = bx + a$

(5.9) $$b = \frac{\sum xy - \dfrac{\sum x \sum y}{n}}{\sum x^2 - \dfrac{(\sum x)^2}{n}}$$

(5.10) $$a = \frac{\sum y - b \sum x}{n} = \bar{y} - b\bar{x}$$

(5.13) $$\text{SS}_y = \sum (y_i - \bar{y})^2$$

(5.15) $SS_y = SS_{error} + SS_{regr}$

(5.17) $r^2 = \dfrac{SS_{regr}}{SS_y}$

Section 5.2

(5.19) $$r_{xy} = \frac{\sum xy - \dfrac{\sum x \sum y}{n}}{\sqrt{\sum x^2 - \dfrac{(\sum x)^2}{n}}\sqrt{\sum y^2 - \dfrac{(\sum y)^2}{n}}}$$

(5.20) $r_{xy} = r_{yx}$

(5.21) $r = b\dfrac{s_x}{s_y}$

(5.22) $b = r\dfrac{s_y}{s_x}$

Exercises for Numerical Practice

Section 5.1

5/1. Four observational units yielded the following bivariate data.

x	y
3	5
1	9
4	6
2	7

a. Calculate the slope b of the linear regression equation $\hat{y} = bx + a$, describing y as a function of x.

b. Calculate the y-intercept a of the linear regression equation $\hat{y} = bx + a$, describing y as a function of x.

c. Write the linear regression equation describing y as a function of x. Draw the scattergram and the regression line.

5/2. Five observational units yielded the following bivariate data.

x	y
10	8
0	9
15	0
3	6
7	2

a. Calculate the slope b of the linear regression equation $\hat{y} = bx + a$, describing y as a function of x.

b. Calculate the y-intercept a of the linear regression equation $\hat{y} = bx + a$, describing y as a function of x.

c. Write the linear regression equation describing y as a function of x. Draw the scattergram and the regression line.

5/3. Calculate the linear regression equation describing y as a function of x for the following bivariate observations:

x:	5	8	3	7	0	3
y:	0	8	7	6	2	9

Draw the scattergram and the regression line.

5/4. Calculate the linear regression equation describing y as a function of x for the following bivariate observations:

x:	3	4	2	0	6	1	5
y:	6	2	4	2	9	6	1

Draw the scattergram and the regression line.

Section 5.2

5/5. Calculate the correlation coefficient r and the coefficient of determination r^2 for the data in Exercise 5/1.

5/6. Calculate the correlation coefficient r and the coefficient of determination r^2 for the data in Exercise 5/2.

5/7. Calculate r and r^2 for the data in Exercise 5/3.

5/8. Calculate r and r^2 for the data in Exercise 5/4.

5/9. In a sample of bivariate observations, the standard deviations of the variables x and y were $s_x = 524$ and $s_y = 10.2$, and the slope of the regression line describing y as a function of x was $b = -.0186$. Use Expression (5.21) to calculate the correlation coefficient r.

5/10. In a sample of bivariate observations, the standard deviations of the variables x and y were $s_x = 1.97$ and $s_y = 96.1$, and the slope of the regression line describing y as a function of x was $b = 24.8$. Use Expression (5.21) to calculate the correlation coefficient r.

5/11. A sample of bivariate observations was summarized as follows:

$$\bar{x} = 4.79, \quad s_x = 2.61, \quad \bar{y} = 63.6, \quad s_y = 40.2, \quad r = .92$$

Using Expressions (5.22) and (5.10), calculate the linear regression equation describing y as a function of x.

5/12. A sample of bivariate observations was summarized as follows:

$$\bar{x} = 237, \quad s_x = 88.4, \quad \bar{y} = 7.22, \quad s_y = 3.19, \quad r = -.77$$

Using Expressions (5.22) and (5.10), calculate the linear regression equation describing y as a function of x.

Problems for Applying the Concepts

5/13. Identify which variable is most reasonably designated as the dependent variable in each of the sets of variables (a) to (j) on page 146.

5/14. Identify which variable is more reasonably designated as the independent variable in each of the pairs of variables (a) to (j) in Example 5.18.

5/15. A flute player investigated the adage "practice makes perfect" by recording the number of incorrectly played notes in 13 successive rehearsals of the same six-page musical score, with the following results.

Number of previous rehearsals	Number of mistakes
0	13
1	4
2	8
3	6
4	7
5	3
6	9
7	5
8	4
9	6
10	5
11	2
12	3

a. Identify the dependent and independent variables and their values.
b. Calculate the simple linear regression equation to describe these data.
c. Calculate the correlation coefficient r.
d. Draw the scattergram, and comment on the appropriateness of using the methods of Parts (b) and (c).
e. What do the results of your analysis tell you about the data?

(Source: Student project by W. Vornbrock.)

5/16. A hunter recorded the following data from 10 hunting trips one fall.

Number of shells used	Number of ducks killed
20	0
25	0
17	2
30	1
32	7
25	8
19	3
34	2
18	4
32	6

a. Calculate the least-squares regression line to describe number of ducks killed as a function of number of shells used.
b. Calculate the correlation coefficient r.
c. Draw the scattergram, and comment on the appropriateness of using the methods of Parts (a) and (b).
d. What do the results of your analysis tell you about the data?

(Source: Student project by N. Daley.)

5/17. Ten different batches of dough were used in a study to investigate the effect of the amount of baking powder on the rising of baking-powder biscuits. Each batch of dough received a different amount of baking powder, while the rest of the ingredients remained the same. The dough was rolled to a standard thickness of .75 inch before baking, and the height of the biscuit was measured after baking, as follows.

Height of biscuit (in eighth-inch units)	Amount of baking powder (in half-teaspoon units)
6	0
8	1
10	2
11	3
12	4
11	5
10	6
10	7
10	8
10	9

a. Identify the dependent and independent variables and their values.

b. Calculate the simple linear regression equation to describe these data.

c. Calculate the correlation coefficient.

d. What would the correlation coefficient have been if the data had been recorded in inches and teaspoons rather than eighth-inches and half-teaspoons?

e. Draw the scattergram, and comment on the appropriateness of using the methods of Parts (b) and (c).

f. What do the results of your analysis tell you about the data?

(Source: Student project by W. Sjogren.)

5/18. The following data were collected in an investigation of the relationship between the time that a sugar cube takes to dissolve in a glass of water and the temperature of the water. The investigator heated a given quantity of water, measured its temperature, added a sugar cube, and, stirring gently, measured the time until the sugar was completely dissolved. The results of nine trials were as follows.

Trial number	Time (seconds)	Temperature (°C)
1	65	24
2	39	45
3	22	54
4	13	80
5	19	67
6	80	16
7	50	31
8	46	35
9	72	19

a. Identify the dependent and independent variables and their values.

b. Calculate the simple linear regression equation to describe these data.

c. Calculate the correlation coefficient.

d. What would the correlation coefficient have been if the data had been recorded in minutes and °F rather than seconds and °C?

e. Draw the scattergram, and comment on the appropriateness of using the methods of Parts (b) and (c).

f. What do the results of your analysis tell you about the data?

(Source: Student project by L. McKinney.)

5/19. Disposable personal income is that part of wages, salaries, interest earnings, and other forms of compensation that people may use either to save or to spend. It is one of the most closely watched components of the gross national product. The following table shows disposable personal income and personal consumption expenditures in the United States in billions of current dollars from 1965 to 1981.

Year	Disposable personal income	Personal consumption expenditures
1965	476	430
1967	548	490
1969	639	582
1971	752	672
1973	915	812
1975	1096	970
1977	1311	1206
1979	1642	1511
1981	2015	1858

a. Identify the dependent and independent variables and their values.

b. Calculate the simple linear regression equation to describe these data.

c. Calculate the correlation coefficient.

d. Draw the scattergram, and comment on the appropriateness of using the methods of Parts (b) and (c).

e. What do the results of your analysis tell you about the data?

(Source: U.S. President, *Economic Report of the President,* 1970 and 1982.)

5/20. A country's gross national product (GNP) is the total value of the goods and services produced by its residents in a given year. The following table shows the U.S. gross national product in constant 1972 dollars and the number of telephones in the United States for selected years from 1950 to 1979.

Year	Number of telephones (in millions)	GNP (in billions of 1972 dollars)
1950	39	535
1955	50	658
1960	66	737
1965	82	929
1970	105	1085
1975	130	1233
1979	153	1483

a. Calculate the least-squares regression line to describe number of telephones as a function of GNP.

b. Calculate the correlation coefficient.

c. Draw the scattergram, and comment on the appropriateness of using the methods of Parts (a) and (b).

d. What do the results of your analysis tell you about the data?

(Sources: U.S. Department of Commerce, Bureau of the Census, *Statistical Abstract of the United States 1981,* p. 560; U.S. President, *Economic Report of the President,* 1982, p. 234.)

5/21. In training for baseball school, a student athlete did an exercise known as step-ups, in which he would run up and down one step repeatedly. This exercise was used in an investigation of the effect of exercise on the athlete's pulse rate. Data were collected over a two-day period during which the athlete performed the exercise for 12 different lengths of time up to a maximum of 6 minutes, with rest periods of at least one hour between successive tests. The athlete's pulse rate in beats per minute was recorded immediately after each running of the exercise. The exercise duration had a mean of 3.25 and a standard deviation of 1.80 minutes, and the pulse rate had a mean of 127.1 and a standard deviation of 26.6 beats per minute. The linear regression equation was $\hat{y} = 79.6 + 14.6x$, with x representing the exercise duration and y the pulse rate.

a. Identify the dependent and independent variables.

b. What pulse rate does the regression equation predict if this athlete exercises for 4 minutes?

c. What do the slope and the y-intercept tell you about the relationship between the two variables?

d. Calculate the correlation coefficient r.

e. Is the linear regression equation a good description of the relationship in these data?

(Source: Student project by G. Rackette.)

5/22. The number of cards forgotten on each of 20 trials was recorded in a memory experiment in which varying numbers of playing cards (up to 20) were shown to an individual for 30 seconds, followed by immediate recall of as many cards as possible. The mean number of cards shown was 10.5 with a standard deviation of 5.9, and the mean number of cards forgotten was 2.1 with a standard deviation of 2.4. The relationship between the variables was summarized as $\hat{y} = .26x - .6$, with x representing the number of cards shown and y the number of cards forgotten.

a. Identify the dependent and independent variables.

b. What is the mean number of cards that would be forgotten, according to the equation, if 10 were shown?

c. What do the slope and the y-intercept tell you about the relationship between the two variables?

d. Calculate the correlation coefficient r.

e. Is the linear regression equation a good description of the relationship in these data?

(Source: Student project by J. Millang.)

5/23. A dart player threw 50 darts at a dart board from each of 11 different distances ranging from 8 to 28 feet, and he recorded the average score at each distance. The board was of the bull's-eye configuration with 10 points awarded for the bull's-eye and decreasing numbers of points for successive rings from the center. The distances chosen had a mean of 18.0 and a standard deviation of 6.6 feet, and the average scores had a mean of 3.15 and a standard deviation of 1.53 points. The correlation between distance and score was $r = -.98$.

a. Identify the dependent and independent variables.

b. Calculate the linear regression equation.

c. What do the slope and the y-intercept tell you about the relationship between the two variables?

d. Is the linear regression equation a good description of the relationship in these data?

(Source: Student project by M. Banack.)

5/24. A sample of 14 paperback college textbooks had a mean price of $7.14 with a standard deviation of $3.49, and a mean length of 497 pages with a standard deviation of 302 pages. The correlation between price and length was $r = .084$.

a. Identify the dependent and independent variables.

b. Calculate the linear regression equation.

c. What do the slope and the y-intercept tell you about the relationship between the two variables?

d. Is the linear regression equation a good description of the relationship in these data?

(Source: Student project by L. Warman.)

5/25. In the two sets of artificial data in Examples 5.14 and 5.15, the correlation coefficients were $r = +1$ and $r = -1$, respectively. Do samples of $n = 2$ observations always produce perfect correlations? Explain.

5/26. Data for the past 12 years from a women's clothing store revealed a correlation of $r = .81$ between annual sales revenue and the amount spent on advertising during the year. If $1000 is subtracted from each year's amount spent on advertising and the correlation coefficient is calculated for these modified data, the result is again $r = .81$. Explain.

5/27. "Regular attendance at worship is good for physical and spiritual health, according to a University of North Carolina professor. Dr. Berton Kaplan says research indicates that frequent churchgoers have lower blood pressure and are less liable to have strokes. A study in Israel shows that those who infrequently attend synagogue have a heart disease rate of 58 per thousand compared to 29 per thousand for the most orthodox." Does the correlation justify the conclusion? What alternative explanation can you offer?

(Source: *The Lutheran,* 2 September 1981, p. 19.)

5/28. A major cancer study at the University of Minnesota School of Public Health found that "eating cooked cereals like oatmeal is related to stomach cancer. Study participants who ate a lot of hot cereals . . . were about seven times more likely to die of stomach cancer than those who ate little of them." Does the correlation justify the conclusion that eating oatmeal increases the risk of stomach cancer? What alternative explanation can you offer?

(Source: *Bond,* Summer 1982, pp. 12–13.)

5/29. "A teacher noted that many students who had low scores on the midterm examination were much closer to class average on the final examination. This she attributed to her teaching skill. But she also noticed that several students who were exceptionally high on the midterm examination slumped noticeably on the final. This she attributed to slackening off due to overconfidence." What other explanation can you suggest?

(Source: Campbell, 1974, p. 194.)

5/30. "The instructors in a flight school adopted a policy of consistent positive reinforcement recommended by psychologists. They verbally reinforced each successful execution of a flight maneuver. After some experience with this training approach, the instructors claimed that contrary to psychological doctrine, high praise for good execution of complex maneuvers typically results in a decrement of performance on the next try. What should the psychologist say in response?"

(Source: Kahneman & Tversky, 1973, pp. 250–251.)

References

Bell, E. T. *Men of Mathematics*. New York: Simon and Schuster, 1937.

Campbell, S. K. *Flaws and Fallacies in Statistical Thinking*. Englewood Cliffs, NJ: Prentice-Hall, 1974.

Encyclopaedia Britannica, 1959, Vol. 13, pp. 876–77; 1977, Micropaedia Vol. VI, pp. 123–24.

Galton, F. *Hereditary Genius*. New York: Macmillan, 1869.

Kahneman, D., and Tversky, A. "On the Psychology of Prediction." *Psychological Review* 80 (1973): 237–51.

Newbrun, E. "Sugar and Dental Caries: A Review of Human Studies." *Science* 217 (1982): 418–23.

Stigler, S. *Springer Statistics Calendar 1981*. New York: Springer, 1980.

World Who's Who in Science. Chicago: Marquis Who's Who, 1968.

Further Reading

Campbell, S. K. *Flaws and Fallacies in Statistical Thinking*. Englewood Cliffs, NJ: Prentice-Hall, 1974. Chapter 13 gives an entertaining and informative introduction to regression and correlation.

Huff, D. *How to Lie with Statistics*. New York: Norton, 1954. Chapter 8 discusses several amusing and instructive examples concerning correlation and causation.

Reichmann, W. J. *Use and Abuse of Statistics.* London: Methuen, 1962. Chapter 10 discusses questions of cause and effect.

Tufte, E. R. *Data Analysis for Politics and Policy.* Englewood Cliffs, NJ: Prentice-Hall, 1974. This highly readable book provides an excellent introduction to the interpretation of regression analysis, written at a level accessible to the beginning student.

PART II

PROBABILITY THEORY

art I introduced some methods of descriptive statistics, the purpose of which is to describe or summarize empirical data in more manageable and informative ways. The methods included frequency distributions and various measures of location, dispersion, and relationship.

Descriptive statistics involves putting the data into a form different from that in which they were collected. Such reorganization of the data is often essential to understanding the data. It may reveal previously unknown features of the process being investigated, or it may suggest new insights or hypotheses to be tested by further investigation. But these benefits are achieved in descriptive statistics without adding anything to the data—except a different organization. In particular, there is no estimation, prediction, or inference about things unseen in descriptive statistics, but rather there is only description of what has been observed.

Often, however, the reason for collecting data is to answer a question that goes beyond the data at hand, a question for which only some of the relevant observations are available. The data then constitute a sample from which inferences about some underlying population are to be drawn.

EXAMPLE 6.0 In the following situations, researchers want to know what will happen in the general case, not just in the particular cases observed:

- A chemist performs tests on a sample of a new plastic in order to learn the properties of any sample of the plastic.
- A manufacturer puts a new shampoo on sale in a few test markets to estimate how well it would be received nationwide.
- An agricultural research station compares two depths of tillage on several plots of land in order to determine the optimal depth of tillage on any plot of comparable land.
- A polling organization interviews 2500 people in order to predict which candidate will win a national election.
- An accountant tests the adequacy of a firm's accounts receivable records by examining a small sample of the total number of accounts receivable.
- A consumer protection agency takes a random sample of 25 packages of sugar from grocery store shelves in order to determine whether the manufacturing process that filled the packages is working according to the required specifications.
- A psychologist studies the vision of 10 observers under different conditions of illumination to test a theory of human perception.
- A government agency contacts 50,000 households each month to determine the nation's monthly unemployment rate.
- Medical researchers assess the effectiveness of a new treatment for cancer in a group of volunteers.

Whenever one draws a conclusion that goes beyond the data actually observed, there is some uncertainty involved. Inferential statistics is, in large part, concerned with the question of how much uncertainty, and the mathematical concept that is basic to the measurement of uncertainty is probability.

Is probability theory a branch of statistics, along with descriptive statistics and inferential statistics? No, it is rather a branch of mathematics whose results are extensively applied in inferential statistics. Its role in statistics is analogous to that of other areas of mathematics, such as differential and integral calculus, which are not subdisciplines of statistics but which provide the mathematical tools with which statistical methods are built. Roughly speaking, we can say that probability theory deals with the nature and likelihood of the possible samples that could be drawn from a known population, whereas inferential statistics is concerned with what a known sample tells us about the (unknown) population from which it was drawn.

The next three chapters present some basic results in probability theory that will provide some necessary tools for inferential statistics.

CHAPTER 6

PROBABILITIES

EXAMPLE 6.1 Anthony S. Clancy of Dublin, Ireland, tells this story about himself: "I was born on the 7th day of the week, 7th day of the month, 7th month of the year, 7th year of the century, 7th child of a 7th child, and I have 7 brothers: That makes 7 sevens.

"On my 27th birthday, at a race meeting, when I looked at the race card to pick a winner in the 7th race, the horse numbered 7 was called Seventh Heaven, with a handicap of 7 stone. The odds were 7 to one. I put 7 shillings on this horse.

"It finished 7th."

(Source: Vaughan, 1979, pp. 130–131. Used with permission of the author.)

In addition to being a cornerstone of inferential statistics, probability has many interesting applications in everyday life. It is used in establishing the odds at horse races and other sporting events, assessing the effectiveness of new drugs, maintaining quality control in industry, marketing new products, monitoring the status of endangered species, forecasting the weather, and in numerous other areas.

The field of probability is a broad and fascinating one. This chapter will introduce the topic. Let us begin with a definition.

6.1 A DEFINITION OF PROBABILITY

A **probability** is a number that is assigned to an event to indicate the likelihood or the chance that the event will occur.

EXAMPLE 6.2 If the weather report says that there is a 70% chance of rain, we take that to mean that it is more likely that it will rain than that it will not, while a 50% chance of rain means that it is just as likely to rain as not.

EXAMPLE 6.3 If you are told that the probability that you will win a lottery prize is one in a million, you no doubt have some idea what that probability means—among other things, don't start spending the prize yet! The probability is a number, .000001, which is assigned to the event of your winning a prize and which indicates how (un)likely the event is.

A probability cannot be just any number. There is no such thing as a probability of −2 or of 150%. Let us consider a definition of probability that makes clear which numbers are eligible. For simplicity, we will speak in the context of a finite set of possible events:

> A **probability** is a real number $P(A)$ that is assigned to an event A and that has the following properties:
>
> **a.** $P(A)$ is neither less than 0 (with 0 indicating that the event A cannot possibly occur) nor greater than 1 (with 1 indicating that the event A occurs with certainty). **(6.1)**
>
> **b.** For any two mutually exclusive events A_1 and A_2, the probability that one or the other occurs is equal to the sum of their separate probabilities.

In symbols, Property (a) may be written

$$0 \leq P(A) \leq 1 \tag{6.2}$$

and Property (b) may be written

$$P(A_1 \text{ or } A_2) = P(A_1) + P(A_2) \quad \text{for mutually exclusive events } A_1, A_2 \tag{6.3}$$

Recall from Section 2.1 that the term **mutually exclusive** means that if one of the events occurs, the other one cannot occur; each excludes the other. In other words, the probability that two mutually exclusive events both occur equals zero:

$$P(A_1 \text{ and } A_2) = 0 \quad \text{for mutually exclusive events } A_1 \text{ and } A_2 \tag{6.4}$$

Thus, Property (b) in Expression (6.1) is analogous to requiring in the measurement of length that the length of two sticks placed end to end with no overlap must equal the sum of their separate lengths.

EXAMPLE 6.4 "It is estimated that 25 percent of Americans become depressed to the point of needing professional help at some point during their lives," and recovery is not certain, even with professional help. According to one study, the probability is about .50 that a depressed patient will recover and never suffer a relapse, while the probability is about .35 that a depressed patient will recover but subsequently suffer a relapse.

What is the probability that a depressed patient will at least recover from his current bout of depression? If we let A_1 represent recovery without relapse and A_2 represent recovery with relapse, we can use Expression (6.3) to calculate the total probability of recovery as

$$P(\text{recovery}) = P(A_1) + P(A_2) = .50 + .35 = .85$$

because recovery without relapse (A_1) and recovery with relapse (A_2) are mutually exclusive events.

What is the probability that a depressed patient will never recover? We know from Expression (6.1) that an event that is certain to occur has a probability of 1. Thus, the probability that a patient will either recover or not recover has a probability of 1: $P(\text{recovery or no recovery}) = 1$. However, because recovery and no recovery are mutually exclusive, we can write

$$P(\text{recovery or no recovery}) = P(\text{recovery}) + P(\text{no recovery})$$

and hence

$$.85 + P(\text{no recovery}) = 1$$

from which we see that

$$P(\text{no recovery}) = 1 - .85 = .15$$

Thus, the probability is 15% that a depressed patient will never recover.
(Source: Kolata, 1981.)

Probabilities may we written as simple fractions, decimals, or percentages. For example, the expressions $\frac{1}{4}$, .25, and 25% all specify the same probability. All three forms are acceptable; the choice of form is made on the basis of what is easiest to read in the particular context. For example, 1/64 is simpler than its decimal form .015625 or 15.625%, and 2/3 is more exact than the decimal form .67 or 67%. On the other hand, 1/7 and 3/17 are harder to compare with each other than are the decimal forms .14 and .18 (or 14% and 18%). Guidelines for how to round and how many digits to report are given in Appendix 1.

6.2 INTERPRETATION OF PROBABILITIES

It is not difficult to specify, as we did in Property (a) in Expression (6.1), what we mean by probabilities of 0 and 1. However, it is more difficult to agree on precisely what we mean by particular probability values between those extremes. We know that .25, for example, indicates a more likely event than does .24, but how is the probability .25 itself to be interpreted?

There are three major approaches to the interpretation of probabilities. These three approaches, to be discussed in the next three subsections, need not be viewed as competing theories. They may be regarded as different ways of looking at the same concept, with each interpretation contributing to our understanding.

Classical Probability and Random Selection

Historically, the study of probability has its roots in games of chance, where fairness demands that certain outcomes be equally likely. A fair die, for example, is one that is equally likely to expose any of its six faces when it is rolled; a fair deal from a deck of cards gives each card equal likelihood of being dealt; a fair coin is equally likely to show heads or tails when it is flipped. Limitation of applicability to situations with equally likely outcomes is the distinguishing characteristic of the **classical interpretation of probability**:

> If there are m equally likely outcomes of which m_1 are of the type that is of interest, then the probability of an outcome of that type, $P(A_1)$, is m_1/m; that is, **(6.5)**

$$P(A_1) = \frac{m_1}{m} \quad \text{for } m \text{ equally likely outcomes} \qquad \textbf{(6.6)}$$

the outcome you are interested in

equally likely outcomes.

EXAMPLE 6.5 A standard deck of playing cards consists of 13 cards (ace, two, three, four, five, six, seven, eight, nine, ten, jack, queen, king) of each of 4 suits (clubs, diamonds, hearts, spades) for a total of 52 cards.

What is the probability of drawing the ace of spades in a single draw from a well-shuffled deck of cards? There are $m = 52$ equally likely outcomes of which $m_1 = 1$ is of the type that is of interest. From Expression (6.6), we obtain the required probability as

$$P(\text{ace of spades}) = \frac{1}{52} = .019$$

What is the probability of drawing a diamond in a single draw from a well-shuffled deck of cards? There are $m = 52$ equally likely outcomes of which $m_1 = 13$ are of the type that is of interest. The required probability is

$$P(\text{diamond}) = \frac{13}{52} = .25$$

EXAMPLE 6.6 What is the probability of rolling a ten in a single roll of a pair of fair dice? There are $m = 36$ equally likely outcomes (6 possibilities for the first die, times 6 possibilities for the second die). Of these outcomes, $m_1 = 3$ yield a total of ten, namely, 6 on the first die and 4 on the second, 4 on the first and 6 on the second, and 5 on both dice. The probability of rolling a ten is therefore

$$\frac{3}{36} = \frac{1}{12} = .083$$

The classical interpretation applies not only to games of chance, but to any situation where the various outcomes are equally likely. One such situation that is of fundamental importance in statistical theory and practice is the case where the outcome is randomly selected.

The term **random** has a different meaning in statistics than it has colloquially: It does not mean haphazard (that is, without aim or plan); rather it means operating according to particular probabilities. In statistics, randomness is associated not with whim but with adherence to specific probabilities. As the term is used here, **random selection** from a set of elements means that each and every element in the set has an equal chance of being selected. We judge randomness not by looking at an outcome but only by knowing how the outcome was generated. An outcome may look haphazard even if the selection was not random, and the results of random selection may or may not look haphazard.

EXAMPLE 6.7 An introductory applied statistics class consists of 3 students specializing in agriculture, 2 in biology, 15 in business, 1 in economics, 2 in education, 2 in forestry, 1 in mathematics, 2 in nursing, 2 in pre-pharmacy, 5 in psychology, 2 in social welfare, and 3 in pre-veterinary medicine. What is the probability that a randomly selected student is specializing in nursing? The use of random selection ensures that each of the students in the class is equally likely to be selected. The total number of students in the class is 40, so there are $m = 40$ equally likely outcomes, of which $m_1 = 2$ are of the type that is of interest in the question. The required probability is therefore

$$P(\text{nursing}) = \frac{2}{40} = .05$$

Given that a nursing student has been selected (for example, called on to answer a question), was the selection random? We cannot tell from the outcome. By definition, any one of the 40 students was equally likely to have been chosen by random selection. Whether the selection was random depends on how it was done, not on the result.

Random selection requires the use of some inanimate device (for example, coins or dice) to guarantee randomness. One of the most useful such devices is a **table of random digits** (see Appendix 2), which consists of thousands of digits, each one independently generated such that in each position any of the 10 digits, 0, 1, 2, . . . , 9, is equally likely to appear. To use the table, start at a different place each time, either by picking a starting point haphazardly or by beginning where you left off after the previous use. You may go forward, backward, up, or down in the table, but, of course, you must blindly determine the starting point and direction of reading and not look for digits that "look right"! To obtain random numbers up to 9, single digits are used; for random numbers up to 99, pairs of digits are used, with 01, 02, . . . , 09 treated as 1, 2, . . . , 9; for numbers up to 999, digit triplets are used, with 001, 002, . . . , 099 treated as 1, 2, . . . , 99; and so on.

EXAMPLE 6.8 How would we use the table of random digits in Appendix 2 to select a student at random from a class of 40 students? Let us assume that we have a list of the names of the 40 students. This list may be in alphabetical order or any other order. We begin by deciding where to start in the table of random digits, without looking at the digits themselves while making this decision. Let us start with the digits in columns 55–56 of row 60 on the fourth page in Appendix 2. Then we read down until we find a pair of digits between 01 and 40 inclusive. (Any other starting point and direction of reading would also be satisfactory, as long as we choose them without looking at the digits involved.) The specified starting digits are 93; we pass these by because they are not between 01 and 40 inclusive. Reading down, we come next to 48, which we also pass, and 82 and then 03, which identifies the third student on the list as our random selection.

If a table of random digits is not used in making a random selection, some other inanimate device must be appropriately used to prevent any bias from entering that would favor some elements in the set over others. It is well established that human intuitions about randomness include many systematic biases; for example, in choosing a name "at random" from a list of students, we intuitively tend to avoid the first and last students on the list, making them less likely choices. Our intuitive choice is haphazard but not random, because not every element in the set has an equal chance of being selected. Drawing well-shuffled, equal-sized slips of paper from a container provides an alternative method of random selection, but it may be very time-consuming to create the required slips of paper, and there is some danger of insufficient shuffling.

The use of a table of random digits, or of the random-number generating function now available on some calculators, is often the easiest and safest method of random selection.

The use of random selection, in the sense of giving each element of a set an equal chance of being selected, enables us to apply the classical interpretation of probability to a wide range of problems in statistics.

The classical interpretation of probability cannot be applied to situations where the various outcomes are not equally likely. Consider, for example, the probabilities that a student will receive a grade of A, B, C, or Fail in a course; or the probabilities that a team will win, lose, or tie a game; or the probabilities that a cancer patient will survive for 1, 2, 3, or more years. In each example, the various outcomes are not equally likely to occur, so we need to consider other ways of interpreting probabilities that will handle such cases.

Frequentistic Probability and the Law of Large Numbers

A theorem established by Jakob Bernoulli (see Biography 6.1) provides the foundation for a second interpretation of probability.

BIOGRAPHY 6.1 Jakob (or Jacques or James) Bernoulli (1654–1705)

Jakob Bernoulli, sometimes called Jakob Bernoulli I, was born on 27 December 1654 in Basel, Switzerland. His ancestors were French Huguenots, Protestants who fled from Catholic persecution in Antwerp in 1583. The family settled in Basel, where members of successive generations were consistently highly successful in commerce, law, medicine, physics, and especially mathematics. There were eight famous mathematicians in three generations of the Bernoulli family:

Jakob I (1654–1705);
his brother Johann (or Jean or John) I (1667–1748);
their nephew Nikolaus I (1687–1759);
three sons of Johann I: Nikolaus II (1695–1726),
Daniel (1700–1782),
Johann II (1710–90);
two sons of Johann II: Johann III (1744–1807),
Jakob II (1759–89).

As a family, the Bernoulli mathematicians are best known for their development of differential and integral calculus and its applications.

Jakob Bernoulli I studied at the University of Basel and then traveled extensively in Europe beginning in 1676. In 1687 he was appointed professor of mathematics at the University of Basel. He developed differential and integral calculus from the foundations laid by Newton and Leibniz, coining the term *integral* in this connection. He studied the mathematics of catenaries (curves formed by chains or cords suspended between two points), achieving results with modern applications relating to power lines and suspension bridges. His work in probability theory, published posthumously in 1713, included the original version of the Law of Large Numbers, the binomial distribution, and other important contributions.

Jakob Bernoulli married Judith Stupanus in 1684, and they had one son and one daughter. He died in Basel on 16 August 1705 at the age of 50.

(Sources: Bell, 1937; *World Who's Who in Science,* 1968.)

Bernoulli's theorem, also known as the **Law of Large Numbers**, has been further developed and strengthened by later mathematicians, but the fundamental concept for the interpretation of probability was present from the beginning. Let us consider the following form of the Law of Large Numbers:

If

- n is the number of trials or repetitions of a situation,
- n_1 is the number of times that the outcome that is of interest occurs in those trials,
- $P(A_1)$ is the probability of that outcome in each individual trial, and **(6.7)**
- the number n of trials is made sufficiently large,

then the observed **proportion** n_1/n becomes arbitrarily close to the **probability** $P(A_1)$.

In symbols, this result may be written as

$$P(A_1) \simeq \frac{n_1}{n} \qquad \text{for large } n \tag{6.8}$$

Bernoulli

EXAMPLE 6.9 The Law of Large Numbers may be illustrated by repeated use of any device, such as coins, cards, dice, or a table of random digits, that generates outcomes with a constant known probability on each trial. Figure 6.1 was produced with the aid of a table of random digits. The probability of an even digit (0, 2, 4, 6, or 8) in such a table is .5, as is also the probability of an odd digit (1, 3, 5, 7, or 9).

Successive digits in the table, each classified as even or odd, were considered to be the outcomes of successive trials, like repeated flips of a fair coin with outcomes classified as heads or tails. The proportion of even digits was calculated from time to time as the number of trials increased. Figure 6.1 illustrates that the fluctuation in the observed proportion of evens diminishes as the number of trials increases, with the **proportion** gradually zeroing in on the **probability** (namely, .5) of an even digit in any one trial, as prescribed by the Law of Large Numbers in Expression (6.7). The proportion would continue to fluctuate if the number of trials were made even larger, but the fluctuations of the observed proportion around the true probability would continue to get smaller.

The Law of Large Numbers enables us to interpret a probability as the proportion of the time that an event will occur in the long run. This interpretation of probability as relative frequency in the long run is perhaps the most generally accepted and applicable interpretation. Its adherents are sometimes referred to as *frequentists,* and the relative-frequency view is called the **frequentistic interpretation of probability**. For example, if we know that

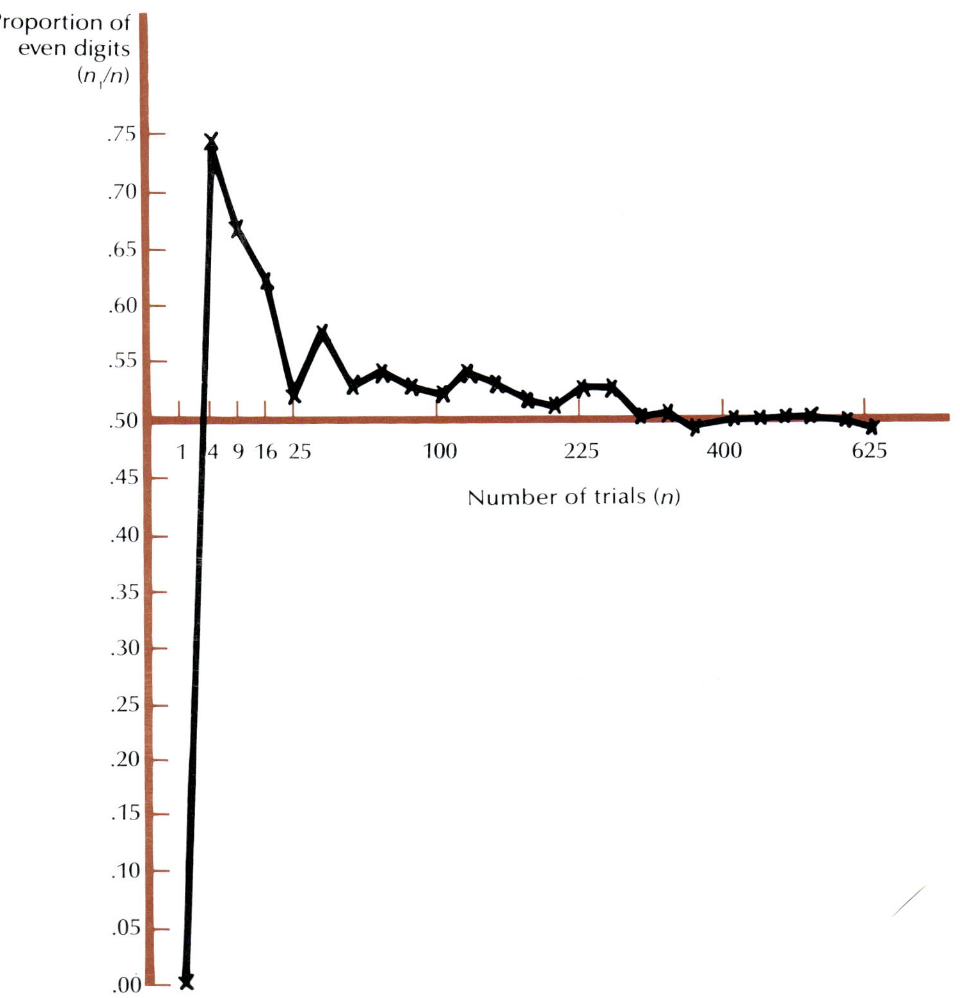

Figure 6.1
Proportion of even digits as a function of number of trials, illustrating the Law of Large Numbers. The horizontal axis is calibrated in equal square steps (1^2, 2^2, 3^2, and so on) to compress the upper end of the scale where changes are more gradual and hence do not require as fine a scale to be seen.

the probability of an event in a particular situation is .10, then we know that in the long run the event will occur in about 10% of such situations. The probability .10 does not mean that in 100 trials the event must occur exactly 10 times; nor does it mean even that the event must occur exactly 100 times in 1000 trials. Rather the Law of Large Numbers indicates that if a large enough number of trials were run, the event would occur very nearly 10% of the time. There is no specific number of trials that is "large enough"; it depends on how close to the true probability we want to insist on being and how sure we want to be about that closeness.

EXAMPLE 6.10 In Example 6.6, we saw that the probability of rolling a ten in a single roll of a pair of fair dice is $1/12 = .083$. What does this probability mean according to the frequentistic interpretation? It does *not* mean that a ten will turn up once in every 12 rolls or that the chances of rolling a ten are in any way improved by failing to roll a ten on 11 previous rolls. If the dice are fair, the probability .083 remains the same on each and every trial, regardless of what has just been rolled. According to the frequentistic view, the probability .083 means that if a pair of fair dice were rolled a large number of times, a ten would turn up approximately 8.3% of the time.

EXAMPLE 6.11 In Example 6.7, we found that the probability that a student randomly selected from a certain statistics class of 40 students would be in nursing was $2/40 = .05$. If the instructor calls on students randomly to answer questions in class, nursing students would be called on about 5% of the time, over a long series of questions. Either more or fewer than 2 of the first 40 questions (or of any 40 questions, for that matter) may be directed to a nursing student, but the probability .05 can be interpreted as 5% of the questions in the long run.

The Law of Large Numbers also provides us with a way of estimating probabilities. If there have been n trials or repetitions of a situation and the event A_1 has occurred in n_1 of the n trials, then the relative frequency n_1/n can be used as an estimate of the probability $P(A_1)$ of the event on each trial. The larger the number of trials, the better the estimate is apt to be, assuming that the situation is the same on each trial.

EXAMPLE 6.12 Suppose that a major league baseball team has a man on first base with nobody out. What is the probability that the team will score at least one run in the remainder of the inning?

The Law of Large Numbers allows us to estimate this probability on the basis of the proportion of times that runs have been scored in a large number of occurrences of this situation in the past. In major league games played in 1959 and 1960, the situation of a runner on first with no outs and no other runners on base occurred 1728 times. In 684 of those cases, the team scored at least one run in the remainder of the inning. We can estimate the probability of scoring at least one run in that situation as $684/1728 = .396$, or approximately 40%.

How good our probability estimate is depends not only on the number of trials observed (1728 in this case), but also on whether the situation in the observed trials is the same as the situation to which we want to apply the probability. For example, have there been any rule changes since 1960 or any changes in baseball managers' strategies that would change the probability so that the estimate from 1959 and 1960 would not be appropriate today?

(Source: Hooke, 1978.)

The difference between the classical and frequentistic interpretations relates to the fact that classical probability involves just one trial and, in particular, one with equally likely outcomes. Frequentistic probability bypasses the restriction of equiprobability by introducing the idea of repeating the selection process n times.

The reference in frequentistic probability to repetitions of a situation raises a problem unique to this interpretation. The problem is that many situations to which we may wish to attach probability statements do not seem to be repeatable and may indeed be unique.

EXAMPLE 6.13 The following questions deal with unique, nonrepeating situations:

- What is the probability that today's international tensions will lead to World War III?
- What is the probability that a certain political party will win the upcoming election?
- What is the probability that you will pass your first statistics course?

What is meant by "relative frequency in the long run" of a nonrepeatable event? There are two ways of handling this problem. The first is perfectly in accord with the frequentistic interpretation of probability: The frequentist can point out that *no* situation is ever repeated identically; repeated situations always differ at least in time or in place. What is relevant for the Law of Large Numbers is that the repeated situations be the same in whatever ways are critical to the probability of the outcome of interest.

EXAMPLE 6.14 The probability of rolling a ten with two dice is not affected by how recently the dice were last rolled, and probably not by whether the dice are rolled on the table or on the floor. Such changes do not prevent us from talking about a repetition of the situation, because they do not affect the relevant probability. However, filing some of the edges of the dice, or perhaps even just using different dice, may cause us to call it a different situation—different with respect to critical features for the question at hand.

The probability of a nonrepeatable event is interpreted in the frequentistic view by reference to a set of situations that are judged to be the same with respect to their critical features.

EXAMPLE 6.15 The following are frequentistic approaches to the questions in Example 6.13:

- The probability that today's international tensions will lead to World War III may be assessed with reference to similar sorts of tensions that were resolved with or without war in the past.

- The probability that a certain party will win an upcoming election may be based on how often the method of prediction being used has been correct in the past with the same amount of evidence (for example, public opinion polls).
- The probability that you will pass your first statistics course may be estimated either from the proportion of students who have passed your instructor's statistics courses in the past or, perhaps more accurately, from the proportion of students with your past academic record and apparent motivation who have passed.

Sometimes, of course, it is not at all apparent just what are the relevant and critical features of a situation. Yet that knowledge is essential if we are to use the Law of Large Numbers to estimate probabilities, since it determines the values of both n_1 and n in the proportion n_1/n. Coincidences sometimes seem more improbable than they are, either because we underestimate n_1 by failing to recognize how many things would surprise us (for example, we are surprised to run into an acquaintance during a visit to a distant city, but there are hundreds of other acquaintances that would have surprised us equally), or because we fail to recognize all the occasions on which the event could have occurred but did not (for example, we pay scant attention to the hundreds of strangers we see in the distant city). In any case, estimation of probabilities on the basis of the Law of Large Numbers requires careful consideration as to what are the relevant and critical features of the situation whose outcome is to be assigned a probability.

The second approach to probabilities of nonrepeatable events is to abandon the frequentist's attempt to define probability in terms of empirically observable events. One can point out, for example, that the frequentistic position loses some of its appeal as an objective interpretation when closer examination reveals that it is based on "situations that are *judged* to be the same with respect to their critical features" and that the decision as to which features are critical requires yet another judgment. An alternative approach is to bring the subjective element to the forefront.

* Subjective Probability and Odds

According to the **subjective interpretation of probability**, a probability is a measure of a person's degree of certainty about an event. This view, which has received considerable attention in recent years, is not as arbitrary as it might sound at first. It certainly does not imply that one simply picks a number out of the air and declares it to be the probability of some event. The problem of assigning a numerical value to one's degree of belief is a separate issue from the principle of subjective probability, namely, that probability can be viewed as a measure of one's degree of belief based on one's assessment of the relevant evidence and uncertainty. The subjective interpretation provides a way of understanding probabilities assigned to events where there are neither equally likely outcomes nor a large number of repetitions of essentially the same situation.

* This symbol identifies material that is not prerequisite for later topics in this book.

EXAMPLE 6.16 What does it mean if a business executive says that the subjective probability that a certain business venture will be a success is .75? It means that, on the basis of her knowledge of the situation and evaluation of the evidence, she feels that the venture is three times more likely to succeed (probability .75) than it is to fail (probability .25).

Similarly, a weather forecast of a 20% chance of rain may be interpreted as a subjective probability indicating that a weather expert who has examined the available data thinks it is four times more likely that it will not rain (probability .80) than that it will rain (probability .20).

Subjective probabilities are sometimes estimated by making use of the concept of odds. **Odds** are an alternative method of expressing any probability, whether subjective or not, but we introduce the concept here because of its applicability in assigning numerical values to subjective probabilities. If the probability of an event's occurrence is designated by p, and the probability of the event's nonoccurrence is designated by

$$q = 1 - p \tag{6.9}$$

then the *odds in favor of the event* are defined as the ratio of p to q. By convention, odds are expressed as the ratio of two positive integers, c to d, with no common factors. In other words, if

$$\frac{p}{q} = \frac{c}{d} \tag{6.10}$$

where c and d are positive integers with no common factors, then the odds in favor of the event are c to d. The corresponding *odds against the event* are d to c. Odds are conventionally stated as odds in favor if p is greater than q and as odds against if q is greater than p; that is, the larger integer is given first.

EXAMPLE 6.17 If the probability is $p = .92$ that you will pass your first statistics course, then the probability that you will not pass is $q = 1 - .92 = .08$, and the odds are

$$\frac{.92}{.08} = \frac{92}{8} = \frac{23}{2}$$

or 23 to 2 in your favor.

EXAMPLE 6.18 The probability of rolling a ten in one roll of a pair of fair dice is $p = 1/12$. The probability of not rolling a ten is therefore $q = 1 - 1/12 = 11/12$, and the odds are

$$\frac{11/12}{1/12} = \frac{11}{1}$$

or 11 to 1 against rolling a ten.

The equation $p/q = c/d$ in Expression (6.10) enables us to convert probabilities to odds. To perform the reverse operation, converting odds to probabilities, solve the equation for p as follows:

$$\frac{p}{1-p} = \frac{c}{d}$$

$$dp = c(1-p) = c - cp$$

$$cp + dp = c$$

$$p(c+d) = c$$

$$p = \frac{c}{c+d} \quad \text{for odds of } c \text{ to } d \tag{6.11}$$

Expression (6.11) allows us to calculate the probability of an event from the odds for or against the event.

EXAMPLE 6.19 The odds are approximately 21 to 20 that a human baby will be a boy. The probability of a boy is therefore

$$P(\text{boy}) = \frac{21}{21+20} = \frac{21}{41} = .51$$

The odds are approximately 13 to 1 that a human baby will become right-handed. The probability of right-handedness is therefore

$$P(\text{right}) = \frac{13}{13+1} = \frac{13}{14} = .93$$

Expression (6.11) can be used to estimate a person's subjective probability of an event if he is able to state what he considers to be odds for a fair bet regarding the event. In a **fair bet**, the person betting on the more likely outcome must put forward a correspondingly larger stake. This method is not without its difficulties. For example, the amount a person thinks he would be willing to bet may be influenced by any attraction or aversion he has to gambling, and his perception of the odds may change depending on whether the amounts wagered are in dollars or in thousands of dollars. However, the method is useful in many applications.

EXAMPLE 6.20 A baseball fan wants to bet \$5 against \$2 that the National League team will win the next World Series. Evidently, her subjective probability that the National League team will win is greater than $5/(5+2) = 5/7 = .71$.

EXAMPLE 6.21 A business executive is planning to replace some equipment. She can either discard the old equipment, at negligible cost, or try to sell it. If she tries to sell it and is unsuccessful, she will lose \$100,000 for salvage, storage, and advertising costs, but if she does sell it, she will make a net gain of \$300,000 on the sale. What is her subjective probability that a buyer can be found for the old equipment if she decides simply to discard it? She does not want to bet \$100,000 against \$300,000 that a buyer will be found. Evidently her subjective probability is less than

$$\frac{100{,}000}{100{,}000 + 300{,}000} = .25$$

On the other hand, her subjective probability could be higher than 25%, and other factors could dictate the decision; for example, the company's financial situation might be such that it cannot risk anything close to a 75% chance of a $100,000 loss.

It is worth reiterating that the three interpretations of probability—classical, frequentistic, and subjective—are not mutually exclusive. All are valid ways of thinking about probability, consistent with its definition in Expression (6.1). Like viewing a scene from a variety of perspectives, consideration of the various interpretations of probability provides a more complete picture.

6.3 BASIC RULES FOR CALCULATION OF PROBABILITIES

In discussing the three interpretations of probability, we treated the object of the probability as if it were always a simple statement; for example, $P(A)$ refers to the probability that "the event A occurs." We will call an event that is stated in this simple form a **simple event** and its probability a **simple probability**. In this section, we will consider probabilities of events represented by more complex statements and will introduce rules for calculating such probabilities from simple probabilities.

Conditional Probability

The frequentistic interpretation of probability makes it clear that probability is a relative thing: the number of times an event occurs in the long run, relative to the number of repetitions of a certain kind of situation. Sometimes, and especially when more than one context is possible, this relativity is made explicit, in which case the probability is called a *conditional probability*. The **conditional probability** $P(A_2 \mid A_1)$ is the probability that event A_2 occurs given that we are considering only those situations where event A_1 also occurs. The notation $P(A_2 \mid A_1)$ is read "the probability of A_2 given A_1."

EXAMPLE 6.22 In a study of shopping habits, shoppers were observed as they left a drugstore during a one-hour period from 4:00 to 5:00 P.M. The numbers of male and female shoppers were recorded and, for each one, whether a purchase had been made while in the drugstore. Such information is important for planning the size of the sales force needed at various times of the day. The results were as follows:

	Purchase	No purchase	Totals
Male	10	9	19
Female	37	13	50
Totals	47	22	69

Let us use these results to estimate the corresponding probabilities for this drugstore at the given time of day. We estimate the probability that a shopper is male as $P(\text{male}) = 19/69 = .28$; similarly, $P(\text{female}) = 50/69 = .72$. We estimate the probability that a shopper will make a purchase as $P(\text{purchase}) = 47/69 = .68$; similarly, $P(\text{no purchase}) = 22/69 = .32$. The preceding are simple probabilities.

We may also estimate various conditional probabilities. The probability that a male shopper will make a purchase is $P(\text{purchase} \mid \text{male}) = 10/19 = .53$, whereas the probability that a female shopper will make a purchase is $P(\text{purchase} \mid \text{female}) = 37/50 = .74$. The probability that a purchaser is a male is $P(\text{male} \mid \text{purchase}) = 10/47 = .21$, whereas the probability that a nonpurchaser is a male is $P(\text{male} \mid \text{no purchase}) = 9/22 = .41$.

Be sure to notice the distinction between terms such as $P(\text{purchase} \mid \text{male})$ and $P(\text{male} \mid \text{purchase})$. The first deals with the probability of a purchase but limits consideration to males, whereas the second deals with the probability that a shopper is a male but limits consideration to shoppers who made purchases. The event to the left of the vertical line is the one whose probability is being assessed; the event to the right of the line is the event or condition to which consideration is limited.

(Source: Student project by L. Jorgensen.)

A conditional probability $P(A_2 \mid A_1)$ can be viewed as the number of times that the two events A_1 and A_2 occur together, relative to the total number of times that the condition A_1 occurs (with or without A_2), in the long run. The event "A_1 and A_2 occur together" is called a **joint event**, and the probability of a joint event is called a **joint probability**, written $P(A_1 \text{ and } A_2)$. Notice that the notation $P(A_1 \text{ and } A_2)$ refers to the joint event where A_1 and A_2 occur together; the joint event does not include occasions on which only one of the simple events occurs. On the other hand, a simple probability includes all occurrences of the relevant event, whether in isolation or jointly with one or more other events. It follows that a joint probability can never be greater than the probability of any of its simple events; that is,

$$P(A_1 \text{ and } A_2) \leq P(A_1) \quad \text{and} \quad P(A_1 \text{ and } A_2) \leq P(A_2) \tag{6.12}$$

Thus, **conditional probability** may be written as the ratio of a joint probability to the simple probability of the condition:

$$\text{Conditional probability} = \frac{P(\text{joint event})}{P(\text{condition})}$$

or, more formally,

$$P(A_2 \mid A_1) = \frac{P(A_1 \text{ and } A_2)}{P(A_1)} \qquad \text{provided } P(A_1) \neq 0 \tag{6.13}$$

EXAMPLE 6.23 Let us verify Expression (6.13) in the data from Example 6.22. For example, the joint probability that a shopper is a female who made a purchase is $P(\text{purchase and female}) = 37/69$. According to Expression (6.13),

$$P(\text{purchase} \mid \text{female}) = \frac{P(\text{purchase and female})}{P(\text{female})} = \frac{37/69}{50/69} = \frac{37}{50}.$$

which agrees with the result obtained directly in Example 6.22. Similarly, the probability that a shopper who made no purchase was a female may be calculated either as

$$P(\text{female} \mid \text{no purchase}) = \frac{P(\text{female and no purchase})}{P(\text{no purchase})} = \frac{13/69}{22/69} = \frac{13}{22}$$

or directly from the no-purchase column of the table in Example 6.22 as

$$P(\text{female} \mid \text{no purchase}) = \frac{13}{22}$$

In practice, we use the latter (direct) method when we have access to the complete table of the form given in Example 6.22. We would use Expression (6.13) if we had only the simple and joint probabilities available.

Joint Probability and Rules for Multiplication

Rearranging the terms of Expression (6.13), we can calculate the joint probability of two events by multiplying the simple probability of one event times the conditional probability of the other event given the first. This result is the **general rule for the multiplication of probabilities:**

$$P(A_1 \text{ and } A_2) = P(A_1)P(A_2 \mid A_1) \tag{6.14}$$

Since it is arbitrary which event is called A_1 and which A_2, we could have written the general multiplication rule as

$$P(A_1 \text{ and } A_2) = P(A_2)P(A_1 \mid A_2) \tag{6.15}$$

These equations again show that the probability of the joint occurrence of two events A_1 and A_2 is always less than (or equal to) the simple probabilities $P(A_1)$ and $P(A_2)$, as stated in Expression (6.12). In particular, the joint probability is equal to a fraction of the simple probability of one event, the fraction being the conditional probability of the other event given the first.

Repeated application of the general rule in Equation (6.14) allows us to extend the rule to handle the joint occurrence of three or more events. For example, for three events,

$$P(A_1 \text{ and } A_2 \text{ and } A_3) = P(A_1)P(A_2 \mid A_1)P[A_3 \mid (A_1 \text{ and } A_2)] \tag{6.16}$$

EXAMPLE 6.24 What is the probability that two cards dealt from a well-shuffled standard deck will both be queens? Let Q_1 represent the event that the first is a queen and Q_2 the event that the second is a queen. Using Expression (6.14), we have

$$P(Q_1 \text{ and } Q_2) = P(Q_1)P(Q_2 \mid Q_1) = \frac{4}{52} \cdot \frac{3}{51} = \frac{1}{221} = .0045$$

What is the probability that three cards dealt from a well-shuffled standard deck will all be kings? Expression (6.16) yields

$$P(K_1 \text{ and } K_2 \text{ and } K_3) = \frac{4}{52} \cdot \frac{3}{51} \cdot \frac{2}{50} = \frac{1}{5525} = .00018$$

We can readily extend Expression (6.16) to determine, say, the probability that four cards dealt from a well-shuffled standard deck will all be aces:

$$P(A_1 \text{ and } A_2 \text{ and } A_3 \text{ and } A_4) = \frac{4}{52} \cdot \frac{3}{51} \cdot \frac{2}{50} \cdot \frac{1}{49} = \frac{1}{270{,}725} = .0000037$$

What is the probability that four cards dealt from a well-shuffled standard deck will result in this hand, in any order: six of clubs, jack of hearts, seven of diamonds, and three of clubs? Let B_1 represent the event that the first card dealt is one of those four, B_2 the event that the second card is one of those four, and so on. The required probability is

$$P(B_1 \text{ and } B_2 \text{ and } B_3 \text{ and } B_4) = \frac{4}{52} \cdot \frac{3}{51} \cdot \frac{2}{50} \cdot \frac{1}{49} = \frac{1}{270{,}725} = .0000037$$

The fact that this latter probability is exactly the same as the probability of four aces is worth noting, because we are much more likely to be surprised at being dealt four aces than at being dealt a four-card hand consisting of the three and six of clubs, the jack of hearts, and the seven of diamonds. Coincidences, such as being dealt a four-ace hand, are often no more improbable than other occurrences; they are just more interesting. For example, we watch as the car's odometer turns from 49999.9 to 50000.0, but we pay no special attention to the change from, say, 53761.1 to 53761.2, even though each event may occur just once in the life of the car. Thus, hundreds of events go by without attracting our attention because we fail to notice their uniqueness. Then, when a more easily remembered configuration occurs (such as four aces), we tend to think of it as improbable, failing to recognize that it is merely our inability to distinguish the uniqueness of other events that makes them seem more likely.

EXAMPLE 6.25 Insurance companies make use of mortality tables, which show the probability that a person of a given age will die within one year, based on records of deaths in a given country or region. Here is part of a mortality table for white females in the United States.

Age	Death rate per 1000
67	16.0
68	17.4
69	19.1

Using $P(\text{not } 68 \mid 67)$ to represent the conditional probability that a 67-year-old will not live to be 68, we can rewrite the preceding death rates as

$$P(\text{not } 68 \mid 67) = .0160$$
$$P(\text{not } 69 \mid 68) = .0174$$
$$P(\text{not } 70 \mid 69) = .0191$$

What is the probability that a 67-year-old white female in the United States will live to be at least 70? The probability that she will live to age 68 is

$$P(68 \mid 67) = 1 - P(\text{not } 68 \mid 67) = 1 - .0160 = .9840$$

Similarly,

$$P(69 \mid 68) = 1 - .0174 = .9826$$
$$P(70 \mid 69) = 1 - .0191 = .9809$$

From Expression (6.16), we can calculate the probability that such a 67-year-old will live to age 70 as

$$P(70 \mid 67) = P(68 \mid 67)P(69 \mid 68)P(70 \mid 69) = (.9840)(.9826)(.9809) = .9484$$

or approximately 95%.

(Source: *Information Please Almanac 1982,* pp. 780–781.)

The general multiplication rule in Expressions (6.14), (6.15), and (6.16) reduces to a simpler form in the special case of **independent events**. Two events are said to be **independent** if the occurrence of one has no effect on the probability of the other; that is,

$$P(A_1 \mid A_2) = P(A_1) \qquad \text{for independent events } A_1 \text{ and } A_2 \tag{6.17}$$

and

$$P(A_2 \mid A_1) = P(A_2) \qquad \text{for independent events } A_1 \text{ and } A_2 \tag{6.18}$$

In other words, if two events are independent, knowledge as to whether one of the events has occurred leaves us none the wiser as to whether the other event will occur. If events A_1 and A_2 are independent, so that Expressions (6.17) and (6.18) apply, then Expressions (6.14) and (6.15) both simplify to the following **special multiplication rule for independent events:**

$$P(A_1 \text{ and } A_2) = P(A_1)P(A_2) \qquad \text{for independent events } A_1 \text{ and } A_2 \tag{6.19}$$

Expression (6.19) says that the probability of the joint occurrence of two independent events is equal to the product of their simple probabilities. The extension to more than two independent events is straightforward, a special case of Expression (6.16):

$$P(A_1 \text{ and } A_2 \text{ and} \ldots \text{and } A_k) = P(A_1)P(A_2)\cdots P(A_k) \quad \text{for independent events } A_1, A_2, \ldots, \text{ and } A_k \tag{6.20}$$

EXAMPLE 6.26 The probability that a newborn human baby is a boy is approximately .51. What is the probability that all four singleton babies born in a certain hospital in one day are boys? It is probably safe to assume that the events are independent (the sex of one baby is in no way dependent on the sex of another of the babies), so the required joint probability is obtained as the product of the simple probabilities:

$$P(B_1 \text{ and } B_2 \text{ and } B_3 \text{ and } B_4) = (.51)(.51)(.51)(.51) = .068$$

or 6.8%.

What is the probability that the next baby born in the hospital is also a boy? Since the events are independent, the probability remains .51:

$$P(\text{boy} \mid 4 \text{ boys}) = P(\text{boy}) = .51$$

Although the probability of five consecutive male births is only .035, the probability of the fifth one is .51 regardless of the outcome of the previous four births.

Example 6.26 illustrates a common misunderstanding concerning independent events. The error of thinking that "chance" somehow makes up for "past excesses"—for example, thinking that a girl is more likely because the previous four births in the hospital were all boys—is known as the **gambler's fallacy**. The gambler's fallacy treats independent events as if they were not independent.

Disjunctive Probability and Rules for Addition

A **disjunctive event** is a complex event characterized as the occurrence of one or both of two simple events. The probability of a disjunctive event is referred to as a **disjunctive probability**; for example, the disjunctive probability $P(A_1 \text{ or } A_2)$ is the probability that either event A_1 or event A_2 (or both) occurs. Disjunctive probabilities may be calculated by means of the **general rule for the addition of probabilities**, according to which the disjunctive probability of two events is equal to the sum of their simple probabilities, minus their joint probability; that is,

$$P(A_1 \text{ or } A_2) = P(A_1) + P(A_2) - P(A_1 \text{ and } A_2) \quad \text{for any events } A_1 \text{ and } A_2 \tag{6.21}$$

EXAMPLE 6.27 Let us verify Expression (6.21) using the data of Example 6.22 concerning shoppers in a drugstore. What is the probability that a shopper is either male or a purchaser? We can see from the table of Example 6.22 that there were $10 + 9 + 37 = 56$ such shoppers out of 69, so our probability estimate is $P(\text{male or purchase}) = 56/69$.

We would use Expression (6.21) to compute this probability if the entire table of data were not available. If in place of the entire table we knew only the simple and joint probabilities, we could calculate

$$P(\text{male or purchase}) = P(\text{male}) + P(\text{purchase}) - P(\text{male and purchase})$$
$$= \frac{19}{69} + \frac{47}{69} - \frac{10}{69} = \frac{56}{69}$$

We subtract the term 10/69 because the male purchasers were counted twice—both as males (in the 19/69) and as purchasers (in the 47/69)—and they should be counted only once.

The general addition rule reduces to a simpler form in the special case of **mutually exclusive** events. Recall that mutually exclusive events are ones whose conditional and joint probabilities are zero; that is,

$$P(A_1 \mid A_2) = 0 \qquad \text{for mutually exclusive events } A_1 \text{ and } A_2 \tag{6.22}$$

and

$$P(A_1 \text{ and } A_2) = 0 \qquad \text{for mutually exclusive events } A_1 \text{ and } A_2 \tag{6.23}$$

[You should compare Expressions (6.22) and (6.23) with (6.17) and (6.19), respectively, to clarify the distinction between mutual exclusion and independence.] In the calculation of disjunctive probabilities for mutually exclusive events, the joint-probability terms vanish [by virtue of Expression (6.23)], yielding the **special addition rule for mutually exclusive events:**

$$P(A_1 \text{ or } A_2 \text{ or } \ldots \text{ or } A_k) = P(A_1) + P(A_2) + \cdots + P(A_k) \quad \text{for mutually exclusive events } A_1, A_2, \ldots, A_k \tag{6.24}$$

This special rule is, in fact, simply an extension of Property (b) in Expression (6.1), which was illustrated in Example 6.4.

EXAMPLE 6.28 The proportions of students born in the various months of the year were approximately as follows in one investigation:

Jan.	Feb.	March	April	May	June	July	Aug.	Sept.	Oct.	Nov.	Dec.
.09	.08	.10	.09	.08	.10	.09	.07	.08	.07	.08	.07

What is the probability that a randomly selected student was born in the first six months of the year? Since the six events (being born in January, being born in February, and so on) are mutually exclusive, the disjunctive probability is the sum of the simple probabilities:

$$P(\text{Jan. or Feb. or March or April or May or June}) = .09 + .08 + .10 + .09 + .08 + .10 = .54$$

or 54%. (It should be noted that the monthly distribution of birthdays is different in different parts of the world and even in different parts of the same country.)

(Source: Camrose Lutheran College, 1981–82 admissions records.)

* 6.4 VENN DIAGRAMS

In 1880 the British logician John Venn devised a method for the graphical representation of events and relationships between events. In the context of probability theory, a **Venn diagram** uses the following:

a. circles, or sometimes rectangles, to represent various classes of events;

b. overlapping of the circles to represent the possibility of simultaneous or joint occurrence of events; and

c. areas of the graph to represent probabilities of occurrence, although the areas usually are not drawn to scale.

Parts (a)–(e) of Figure 6.2 illustrate the basic elements of Venn diagrams. The entire set of possible events, or possible outcomes of a situation, is called the **sample space**, usually designated by S. Since the sample space is defined to include all possible outcomes, the probability that the outcome of any given trial is from the sample space is necessarily equal to 1; that is,

$$P(S) = 1 \tag{6.25}$$

The sample space is usually represented in a Venn diagram by a rectangle, as in Part (a) of Figure 6.2. An event A in the sample space is represented by a circle within the rectangle S, as shown in Part (b). By considering the total area of the rectangle S to be 1, we may think of the area of the circle A as

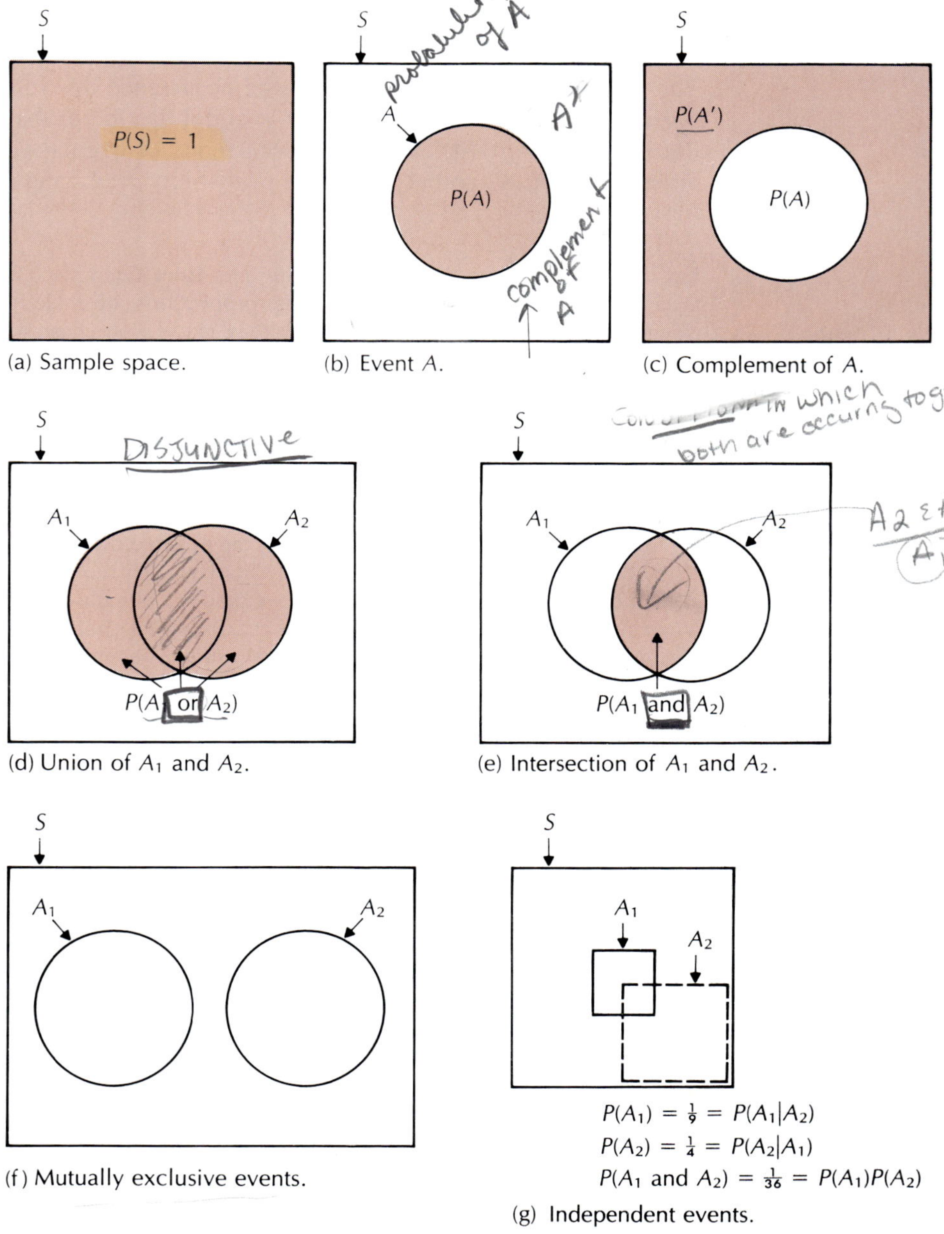

S

$P(S) = 1$

(a) Sample space.

S

A

$P(A)$

(b) Event A.

S

$P(A')$

$P(A)$

(c) Complement of A.

S

A_1 A_2

$P(A_1 \text{ or } A_2)$

(d) Union of A_1 and A_2.

S

A_1 A_2

$P(A_1 \text{ and } A_2)$

(e) Intersection of A_1 and A_2.

S

A_1 A_2

(f) Mutually exclusive events.

S

A_1 A_2

$P(A_1) = \frac{1}{9} = P(A_1|A_2)$

$P(A_2) = \frac{1}{4} = P(A_2|A_1)$

$P(A_1 \text{ and } A_2) = \frac{1}{36} = P(A_1)P(A_2)$

(g) Independent events.

Figure 6.2 Basic concepts illustrated by Venn diagrams. (a) Sample space; (b) event A; (c) complement of A; (d) union of A_1 and A_2; (e) intersection of A_1 and A_2; (f) mutually exclusive events; (g) independent events.

representing the probability of event A. The portion of the sample space not included in event A is called the **complement** of event A, and its probability is represented by the area within S but outside A, as shown by the shading in Part (c). The symbol A' (read "A-prime") designates the complement of A. The disjunctive event consisting of the occurrence of event A_1 or event A_2 or both is called the **union** of A_1 and A_2, and it has the probability $P(A_1 \text{ or } A_2)$ as illustrated in Part (d). The joint occurrence of events A_1 and A_2, called the **intersection** of A_1 and A_2, has the probability $P(A_1 \text{ and } A_2)$ as illustrated in Part (e).

EXAMPLE 6.29 Let us illustrate Venn diagrams with the data from Example 6.22. Observations made at a drugstore revealed the following proportions of male and female shoppers, categorized according to whether or not they made a purchase while in the drugstore.

	Purchase	No purchase	Totals
Male	.14	.13	.27
Female	.54	.19	.73
Totals	.68	.32	1.00

If one shopper is selected at random, the possible outcomes will be as shown in the Venn diagram in Figure 6.3, where M is the event that the selected shopper is male and B is the event that the selected shopper made a purchase. You should examine the Venn diagram to see how it represents the following probabilities associated with the various possible outcomes of the proposed random selection. From the Venn diagram, we can determine the probability that the selected shopper is:

a. male, $P(M) = .13 + .14 = .27$;

b. female, $P(M') = .19 + .54 = .73$;

c. a purchaser, $P(B) = .14 + .54 = .68$;

d. a nonpurchaser, $P(B') = .19 + .13 = .32$;

e. a male or a purchaser, $P(M \text{ or } B) = .13 + .14 + .54 = .81$;

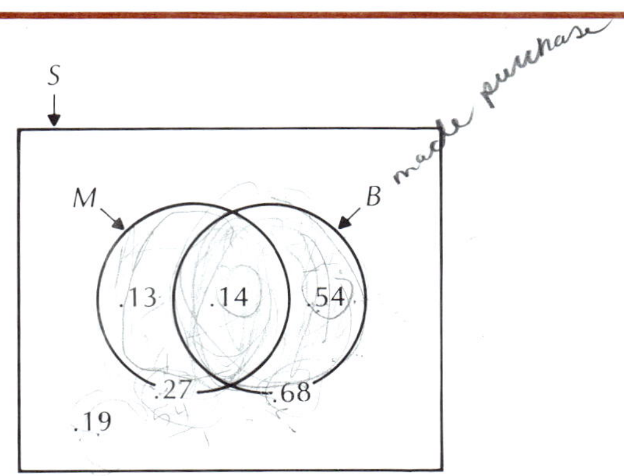

Figure 6.3 Venn diagram for the drugstore-shopper data. As usual for Venn diagrams, the areas are not drawn to scale.

f. a female or a purchaser, $P(M' \text{ or } B) = .19 + .14 + .54 = .87$;
g. a male or a nonpurchaser, $P(M \text{ or } B') = .13 + .14 + .19 = .46$;
h. a female or a nonpurchaser, $P(M' \text{ or } B') = .19 + .13 + .54 = .86$;
i. a male purchaser, $P(M \text{ and } B) = .14$;
j. a female purchaser, $P(M' \text{ and } B) = .54$;
k. a male nonpurchaser, $P(M \text{ and } B') = .13$;
l. a female nonpurchaser, $P(M' \text{ and } B') = .19$.

It is often helpful to sketch a Venn diagram as a first step in solving a probability question. The benefit is lost, however, if care is not taken in recording numerical values in the diagram. In particular, simple probabilities, which apply to an entire circle, should be written on the rim of the circle, whereas joint probabilities, which apply to an undivided segment of a circle, should be written in that segment. See Figure 6.3 for an example.

Illustration of Addition Rules

Venn diagrams may enhance your understanding of the rules for the addition of probabilities as presented in Section 6.3, pages 205–206. Probabilities are added when we want to find a disjunctive probability—the probability of one event *or* another. The key word for the use of addition is *or*. You should examine Part (d) in Figure 6.2 to verify the general rule for the addition of probabilities in Expression (6.21).

EXAMPLE 6.30 From the Venn diagram in Figure 6.3, we find that the probability that a randomly selected shopper is a male (M) or a purchaser (B) is

$$\begin{aligned} P(M \text{ or } B) &= P(M) + P(B) - P(M \text{ and } B) \\ &= (.13 + .14) + (.14 + .54) - .14 \\ &= .27 + .68 - .14 \\ &= .81 \end{aligned}$$

Similarly, you should examine Part (f) in Figure 6.2 to verify Expressions (6.22), (6.23), and (6.24) concerning the special case of mutually exclusive events.

Illustration of Multiplication Rules

Venn diagrams may also be used to illustrate the rules for the multiplication of probabilities as presented in Section 6.3, pages 202–205. Probabilities are multiplied when we want to find a joint probability—the probability of the joint occurrence of one event *and* another. The key word for the use of multiplication is *and*. You should examine Part (e) in Figure 6.2 to verify the general rule for the multiplication of probabilities in Expression (6.14). In terms of the Venn diagram in Part (e), the conditional probability of A_2 given A_1 is the shaded area, expressed as a proportion of the A_1 circle. Multiplying this proportion [Expression (6.13)] times the area of the A_1 circle, representing $P(A_1)$, yields the shaded area, representing the joint probability [Expression (6.14)].

EXAMPLE 6.31 Let us refer again to the Venn diagram in Figure 6.3. The probability that a randomly selected shopper made a purchase (B), given that he is male (M), is

$$P(B \mid M) = \frac{.14}{.13 + .14} = \frac{.14}{.27}$$

Now let us apply Expression (6.14) to determine the probability that a randomly selected shopper is a male purchaser:

$$\begin{aligned} P(M \text{ and } B) &= P(M)P(B \mid M) \\ &= (.27)\frac{.14}{.27} = .14 \end{aligned}$$

Recall from Section 6.3 that two events A_1 and A_2 are **independent** of each other if the simple probability of A_2 equals its conditional probability given A_1, as in Expression (6.18). Independence is not easily represented in a Venn diagram. To do so, we must draw the various areas in the diagram in their proper proportions, because independence relates to the relative sizes of areas. Part (g) of Figure 6.2 represents independent events in a quasi–Venn diagram. In terms of Part (g), the right-hand side of Expression (6.18), that is, $P(A_2)$, refers to the proportion of the sample space S that is included in A_2; and the left-hand side, that is, $P(A_2 \mid A_1)$, refers to the proportion of the square A_1 that is included in A_2. In the diagram these two proportions are equal: One-quarter of the area of S is occupied by A_2, and one-quarter of the area of A_1 is occupied jointly with A_2. Hence—and this is the important implication of independence—limiting our consideration to A_1 does not alter the probability of A_2, which is why we say in this case that A_1 and A_2 are independent of each other.

The multiplication rule takes on a particularly simple form in the special case of independent events. You should examine Part (g) of Figure 6.2 to verify the special multiplication rule for independent events [Expression (6.19)].

* 6.5 MARGINAL PROBABILITY

Sometimes a simple probability can be calculated as the sum of several joint probabilities. The simple probability of an event that can occur jointly with each of the members of a set of mutually exclusive and exhaustive events is sometimes called a **marginal probability**.

EXAMPLE 6.32 Consider the probability that a certain student who is applying for admission to graduate school will successfully complete her graduate degree program. She is applying to two graduate schools, A and B, but she prefers the more prestigious school, A. The probability that a student with her academic preparation will complete the degree is about .60 if she is at school A and .80 if she is at school B. If she is accepted at neither school (let us call this alternative C), the probability that she will complete the degree is 0. The probability that she will be accepted at school A, in which case she will go there, is .20. The probability that she will go to school B is .70, and the probability that she will be accepted at neither school is .10.

Representing the alternative schooling possibilities by A, B, and C, and the event that she completes the degree by the symbol D, we have

$$
\begin{aligned}
P(A) &= .20 \qquad & P(D \mid A) &= .60 \\
P(B) &= .70 \qquad & P(D \mid B) &= .80 \\
P(C) &= .10 \qquad & P(D \mid C) &= 0.00
\end{aligned}
$$

These probabilities may be represented in a table as follows:

	School A	School B	No school (C)
Degree (D)	(.2)(.6) = .12	(.7)(.8) = .56	(.1)(0.0) = 0.00
No degree (D')			
	.20	.70	.10

The entries in the top three cells are joint probabilities calculated by means of Expression (6.14):

$$
\begin{aligned}
P(A \text{ and } D) &= P(A)P(D \mid A) = (.20)(.60) &= .12 \\
P(B \text{ and } D) &= P(B)P(D \mid B) = (.70)(.80) &= .56 \\
P(C \text{ and } D) &= P(C)P(D \mid C) = (.10)(0.00) &= 0.00
\end{aligned}
$$

Now we can fill in the other three cells by subtraction and obtain the probabilities in the right-hand margin of the table by adding the joint probabilities in each row.

	A	B	C	Marginal probabilities
D	.12	.56	0.00	.68
D'	.08	.14	.10	.32
Marginal probabilities	.20	.70	.10	1.00

The simple probabilities $P(D) = .68$ and $P(D') = .32$ are marginal probabilities, equal to the sum of the joint probabilities in the appropriate row of the table. They indicate that the probability that the student will successfully complete the degree for which she is applying is .68. The simple probabilities $P(A) = .20$, $P(B) = .70$, and $P(C) = .10$ are also marginal probabilities. They are equal to the sum of the joint probabilities in the appropriate column of the table.

The marginal probability of an event A that can occur in k ways, namely, in combination with any of k mutually exclusive and exhaustive events B_j (for $j = 1, 2, \ldots, k$), is equal to the sum of the k joint probabilities $P(A \text{ and } B_j)$.

That is,

$$P(A) = P(A \text{ and } B_1) + P(A \text{ and } B_2) + \cdots + P(A \text{ and } B_k)$$
$$= \sum_{j=1}^{k} P(A \text{ and } B_j) \qquad \textbf{(6.26)}$$

for mutually exclusive and exhaustive events B_j

Expression (6.26) says that the marginal probability of an event—that is, its simple probability—is equal to the sum of the probabilities of the various mutually exclusive ways in which it can occur. Substituting Expression (6.14) into Expression (6.26), we may also write the marginal probability of event A as follows:

$$P(A) = P(B_1)P(A \mid B_1) + P(B_2)P(A \mid B_2) + \cdots + P(B_k)P(A \mid B_k)$$
$$= \sum_{j=1}^{k} P(B_j)P(A \mid B_j) \qquad \textbf{(6.27)}$$

for mutually exclusive and exhaustive events B_j

EXAMPLE 6.33 Example 6.32 presented simple probabilities of three schooling alternatives for a certain student, along with the conditional probability that she would successfully complete a degree if she followed each alternative. Expression (6.27) allows us to use these simple and conditional probabilities to calculate the marginal probability that she will successfully complete a degree. In the notation of Example 6.32, we have

$$P(D) = P(A)P(D \mid A) + P(B)P(D \mid B) + P(C)P(D \mid C)$$
$$= (.20)(.60) + (.70)(.80) + (.10)(0.00) = .68$$

To illustrate the use of Expression (6.26), let us rewrite the probability that she will obtain the degree as

$$P(D) = P(A \text{ and } D) + P(B \text{ and } D) + P(C \text{ and } D)$$
$$= .12 + .56 + 0.00 = .68$$

and write the probability that she will not obtain the degree as

$$P(D') = P(A \text{ and } D') + P(B \text{ and } D') + P(C \text{ and } D')$$
$$= .08 + .14 + .10 = .32$$

* 6.6 BAYES' RULE

According to Expression (6.13), the conditional probability of an event, say, B_1, given event A, is equal to the joint probability of the two events, divided by the probability of the condition; that is,

$$P(B_1 \mid A) = \frac{P(A \text{ and } B_1)}{P(A)} \qquad \text{provided } P(A) \neq 0 \qquad \textbf{(6.28)}$$

If we rewrite the numerator of Expression (6.28) using the general multiplication rule in Expression (6.15) and substitute the formula for marginal probability, Expression (6.27), in the denominator, we obtain an alternative formula for conditional probability known as **Bayes' rule**:

$$P(B_1 \mid A) = \frac{P(B_1)P(A \mid B_1)}{P(B_1)P(A \mid B_1) + P(B_2)P(A \mid B_2) + \cdots + P(B_k)P(A \mid B_k)} \tag{6.29}$$

for mutually exclusive and exhaustive events $B_1, B_2, \ldots, B_k$

This method of expressing a conditional probability is named after its discoverer, Thomas Bayes, a Presbyterian clergyman who lived in Britain from 1702 to 1761. Bayes' rule is especially useful for revising the probability of an event B_1 in the light of new evidence (event A).

EXAMPLE 6.34 Example 6.32 presented simple probabilities of three schooling alternatives (A, B, and C) for a certain student, along with the conditional probabilities that she would successfully complete a degree (event D) if she followed each alternative:

$$P(A) = .20 \qquad P(D \mid A) = .60$$
$$P(B) = .70 \qquad P(D \mid B) = .80$$
$$P(C) = .10 \qquad P(D \mid C) = 0.00$$

If we now learn that she obtained a degree, what is the probability that she went to school A? According to Bayes' rule, the probability that she went to school A, given that she obtained a degree, is

$$\begin{aligned} P(A \mid D) &= \frac{P(A)P(D \mid A)}{P(A)P(D \mid A) + P(B)P(D \mid B) + P(C)P(D \mid C)} \\ &= \frac{(.20)(.60)}{(.20)(.60) + (.70)(.80) + (.10)(0.00)} \\ &= \frac{.12}{.68} = .18 \end{aligned}$$

In light of the new evidence (the fact that she obtained a degree), we revise our prior probability that she went to school A (namely, .20) down to .18.

Combining Expressions (6.28) and (6.29) as

$$P(B_1 \mid A) = \frac{P(B_1)P(A \mid B_1)}{P(A)} \tag{6.30}$$

we can see that Bayes' rule expresses the conditional probability of B_1 given A, $P(B_1 \mid A)$, as a function of the simple probability of B_1, $P(B_1)$. $P(B_1)$ is the probability of B_1 in the absence of information about A; it is called the **prior probability**. The conditional probability $P(B_1 \mid A)$ is the probability of B_1 as revised in light of the information about A; $P(B_1 \mid A)$ is called the **posterior probability**. Bayes' rule describes how to revise the prior probability in light of additional information to yield the posterior probability. (In Example 6.34, the prior probability of school *A* was .20, and the posterior probability was .18.)

6.7 SOME TIPS FOR ANSWERING PROBABILITY QUESTIONS

This chapter has introduced principles that enable us to answer questions about four types of probability:

1. *Simple probability*. What is the probability of event *A*?
2. *Conditional probability*. What is the probability of event *A*, *if* consideration is limited to situations where event *B* also occurs?
3. *Joint probability*. What is the probability that events *A and B* both occur?
4. *Disjunctive probability*. What is the probability that either event *A or* event *B* occurs?

The words *if, and,* and *or* distinguish among the different types of probability. The fact that these key words are all little words should serve as a reminder that probability questions must be read very carefully. The beginner should not become discouraged if correct answers to probability questions do not leap from the page upon first reading. Even probability experts have to take time to read a question carefully, or to listen carefully, to sort out precisely which probability is required. Experience provides guidance as to what to watch for in a probability question, but it never removes the need for careful thinking.

The first step in answering questions about probability is to identify the events and to translate the question into the format of one of the four questions above. Sometimes this step will involve breaking the question down into parts, each part having the form of one of the above questions. In any case, once we have made the *and*'s, *or*'s and *if*'s explicit, our method of arriving at an answer depends on the information available:

a. If there are a finite number of equally likely outcomes under consideration and we know the number of outcomes of each type, the required probability of the outcome of interest—whether it be a simple, conditional, joint, or disjunctive event—is simply the number of outcomes of interest divided by the number of outcomes under consideration. (See Examples 6.5, 6.6, and 6.7.)

b. If we know the frequencies, or numbers of occurrences, of various possible events, the required probability of the event of interest—whether it be a simple, conditional, joint, or disjunctive event—is simply the number of times that the event of interest occurred, divided by the number of relevant occasions on which it could have occurred. (See Examples 6.12, 6.22, and the first paragraph of Example 6.27.)

* c. If we know the odds corresponding to the required probability—whether it be a simple, conditional, joint, or disjunctive probability—we use Expression (6.11) to obtain the required probability from the odds. (See Examples 6.19, 6.20, and 6.21.)

d. If we know the appropriate input probabilities for the relevant formula, we obtain the required probability using Expression (6.13) for a conditional event, (6.14) for a joint event, or (6.21) for a disjunctive event. (See Examples 6.25, 6.26, and 6.28.)

* e. If we know the joint or conditional probabilities relating to an exhaustive set of mutually exclusive ways in which a particular event can occur, we obtain the simple probability of that event as a marginal probability using Expression (6.26) or (6.27). (See Example 6.33.)

* f. If we are given a conditional probability $P(A \mid B)$ and require the reverse conditional probability $P(B \mid A)$, we are likely to need Bayes' rule in Expression (6.29) or (6.30). (See Example 6.34.)

STUDY MATERIAL FOR CHAPTER 6

Terms and Concepts

Section 6.1

probability
mutually exclusive events

Section 6.2

classical interpretation of probability
random
random selection
table of random digits
Bernoulli's theorem
Law of Large Numbers
proportion
frequentistic interpretation of probability
* subjective interpretation of probability
* odds
* fair bet

Section 6.3

simple event
simple probability
conditional probability
joint event
joint probability
general rule for the multiplication of probabilities
independent events

special multiplication rule for independent events
gambler's fallacy
disjunctive event
disjunctive probability
general rule for the addition of probabilities
special addition rule for mutually exclusive events

* Section 6.4

Venn diagram
sample space
complement
union
intersection

* Section 6.5

marginal probability

* Section 6.6

Bayes' rule
prior probability
posterior probability

Key Formulas

Section 6.2

(6.6) $P(A_1) = \frac{m_1}{m}$ for m equally likely outcomes

(6.8) $P(A_1) = \frac{n_1}{n}$ for large n

* **(6.9)** $q = 1 - p$

* **(6.11)** $p = \frac{c}{c + d}$ for odds of c to d

Section 6.3

(6.13) $P(A_2 \mid A_1) = \frac{P(A_1 \text{ and } A_2)}{P(A_1)}$ provided $P(A_1) \neq 0$

(6.14) $P(A_1 \text{ and } A_2) = P(A_1)P(A_2 \mid A_1)$

(6.17) $P(A_1 \mid A_2) = P(A_1)$ for independent events A_1 and A_2

(6.19) $P(A_1 \text{ and } A_2) = P(A_1)P(A_2)$ for independent events A_1 and A_2

(6.21) $P(A_1 \text{ or } A_2) = P(A_1) + P(A_2) - P(A_1 \text{ and } A_2)$

(6.22) $P(A_1 \mid A_2) = 0$ for mutually exclusive events A_1 and A_2

(6.23) $P(A_1 \text{ and } A_2) = 0$ for mutually exclusive events A_1 and A_2

(6.24) $P(A_1 \text{ or } A_2 \text{ or } \cdots \text{ or } A_k) = P(A_1) + P(A_2) + \cdots + P(A_k)$
for mutually exclusive events $A_1, A_2, \ldots, A_k$

* Section 6.5

(6.26) $P(A) = P(A \text{ and } B_1) + P(A \text{ and } B_2) + \cdots + P(A \text{ and } B_k)$
for mutually exclusive and exhaustive events B_j

(6.27) $P(A) = P(B_1)P(A \mid B_1) + P(B_2)P(A \mid B_2) + \cdots + P(B_k)P(A \mid B_k)$
for mutually exclusive and exhaustive events B_j

* Section 6.6

(6.29) $$P(B_1 \mid A) = \frac{P(B_1)P(A \mid B_1)}{P(B_1)P(A \mid B_1) + P(B_2)P(A \mid B_2) + \cdots + P(B_k)P(A \mid B_k)}$$
for mutually exclusive and exhaustive events B_j

Exercises for Numerical Practice

Section 6.1

6/1. Which of the following cannot be probabilities?

$-.13$, .25, 17%, 98, 300%, .7%, 3/4, 4/3, 2.67, 0

6/2. Which of the following cannot be probabilities?

112/117, 6/5, 99%, -10%, 110%, .001, $-\frac{1}{2}$, 100, 1.75, 1

Section 6.2

* **6/3.** Express each of the following probabilities as odds:

a. $P(A) = .7$ **b.** $P(B) = .55$ **c.** $P(C) = .1$ **d.** $P(D) = .15$

* **6/4.** Express each of the following probabilities as odds:

a. $P(A) = .99$ **b.** $P(B) = .001$ **c.** $P(C) = .4$ **d.** $P(D) = 5/8$

* **6/5.** Calculate $P(A)$ from the odds in each case:

a. 2 to 1 that A will occur. **b.** 2 to 1 that A will not occur.
c. 13 to 12 that A will occur. **d.** 13 to 12 that A will not occur.

* **6/6.** Calculate $P(A)$ from the odds in each case:

a. 5 to 1 that A will occur. **b.** 5 to 1 that A will not occur.
c. 11 to 7 that A will occur. **d.** 11 to 7 that A will not occur.

Section 6.3

6/7. Given that $P(A) = .62$, $P(B) = .46$, $P(A \text{ and } B) = .11$, calculate:

a. $P(A \mid B)$ **b.** $P(B \mid A)$

6/8. Given that $P(C) = .40$, $P(D) = .24$, $P(C \text{ and } D) = .18$, calculate:

a. $P(C \mid D)$ **b.** $P(D \mid C)$

6/9. Given that $P(E) = .75$, $P(E \mid F) = .60$, $P(F \mid E) = .10$, calculate:

a. $P(E \text{ and } F)$ **b.** $P(F)$

6/10. Given that $P(G) = .30$, $P(G \mid H) = .35$, $P(H \mid G) = .90$, calculate:

a. $P(G \text{ and } H)$ **b.** $P(H)$

6/11. Given that A and B are independent events with probabilities $P(A) = .2$ and

$P(B) = .4$, calculate:

a. $P(A \mid B)$ **b.** $P(B \mid A)$ **c.** $P(A \text{ and } B)$ **d.** $P(A \text{ or } B)$

6/12. Given that C and D are independent events with probabilities $P(C) = .7$ and $P(D) = .9$, calculate:

a. $P(C \mid D)$ **b.** $P(D \mid C)$ **c.** $P(C \text{ and } D)$ **d.** $P(C \text{ or } D)$

6/13. Given that E and F are mutually exclusive events with probabilities $P(E) = .25$ and $P(F) = .45$, calculate:

a. $P(E \mid F)$ **b.** $P(F \mid E)$ **c.** $P(E \text{ and } F)$ **d.** $P(E \text{ or } F)$

6/14. Given that G and H are mutually exclusive events with probabilities $P(G) = .17$ and $P(H) = .52$, calculate:

a. $P(G \mid H)$ **b.** $P(H \mid G)$ **c.** $P(G \text{ and } H)$ **d.** $P(G \text{ or } H)$

* Section 6.4

6/15. Given the Venn diagram of Figure 6.4, calculate each item (a) through (p).

a. $P(A)$ **b.** $P(B)$ **c.** $P(A')$
d. $P(B')$ **e.** $P(A \text{ or } B)$ **f.** $P(A \text{ or } B')$
g. $P(A' \text{ or } B)$ **h.** $P(A' \text{ or } B')$ **i.** $P(A \text{ and } B)$
j. $P(A \text{ and } B')$ **k.** $P(A' \text{ and } B)$ **l.** $P(A' \text{ and } B')$
m. $P(A \mid B)$ **n.** $P(A \mid B')$ **o.** $P(A' \mid B)$
p. $P(A' \mid B')$

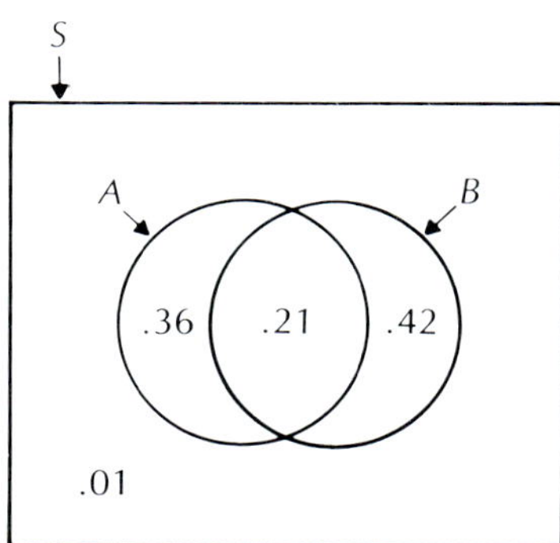

Figure 6.4

6/16. Given the Venn diagram of Figure 6.5, calculate each item (a) through (p) from Exercise 6/15.

* Section 6.5

6/17. Events B_1, B_2, and B_3 are mutually exclusive and exhaustive, with probabilities $P(B_1) = .15$, $P(B_2) = .28$, and $P(B_3) = .57$. Given that $P(A \mid B_1) = .80$, $P(A \mid B_2) = .20$, and $P(A \mid B_3) = .01$, calculate $P(A)$.

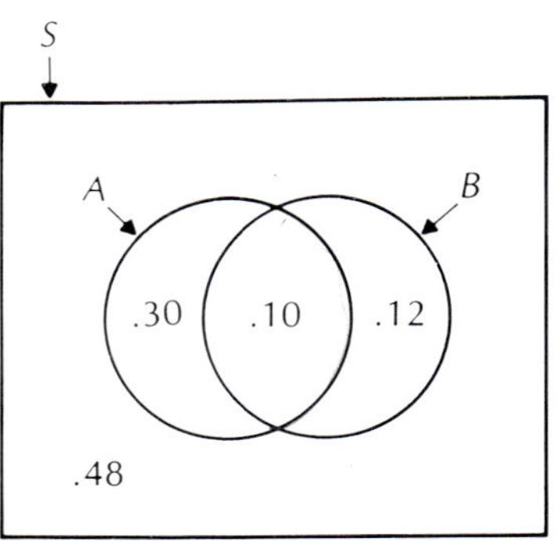

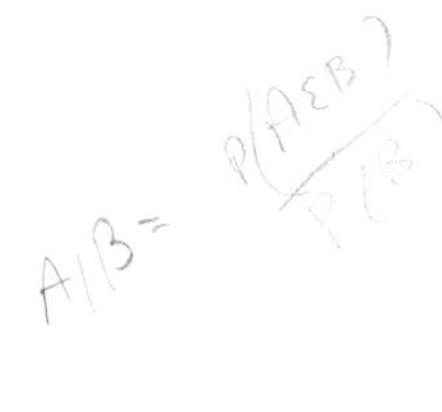

Figure 6.5

6/18. Events D_1, D_2, and D_3 are mutually exclusive and exhaustive, with probabilities $P(D_1) = .71$, $P(D_2) = .03$, and $P(D_3) = .26$. Given that $P(C \mid D_1) = .60$, $P(C \mid D_2) = .90$, and $P(C \mid D_3) = .70$, calculate $P(C)$.

* Section 6.6

6/19. Given the probabilities in Exercise 6/17, calculate:

a. $P(B_1 \mid A)$ **b.** $P(B_2 \mid A)$ **c.** $P(B_3 \mid A)$

6/20. Given the probabilities in Exercise 6/18, calculate:

a. $P(D_1 \mid C)$ **b.** $P(D_2 \mid C)$ **c.** $P(D_3 \mid C)$

Problems for Applying the Concepts

6/21. Most gasoline pumps have an automatic shut-off that turns off the pump when the car's gas tank is nearly full. The attendant then usually pumps a little more gas and stops when the amount owing on the pump is an easy number to work with, preferably so many dollars and no cents. If the pump is read to the nearest cent, what is the probability that the automatic shut-off stops the pump:

a. at an exact dollar amount (and no cents)?

b. at an amount ending in 47 cents?

c. at an amount that could be paid exactly with dollar bills and/or quarters?

6/22. What is the probability of rolling an eight in a single roll of a pair of fair dice?

6/23. With reference to the pizza bills in Problem 2/11 (Chapter 2), what is the probability that a randomly selected bill would be less than $10.00?

6/24. With reference to the income tax returns in Problem 2/23 (Chapter 2), what is the probability that a randomly selected return would show an adjusted gross income of at least $100,000?

6/25. Coding the letters of the alphabet as $A = 1$, $B = 2$, . . . , $Z = 26$, use the table of random digits in Appendix 2 to select a letter. Start at column 00 of row 01, and read from left to right.

6/26. Use the table of random digits to select one of the pages (numbered 1–220) up to this point in this book. Start at column 00 of row 94 in Appendix 2, and read from left to right.

* **6/27.** According to the exploration manager of a gas company, the probability that a particular drilling operation will result in a commercially viable gas discovery is 5/6. Express his subjective probability in the form of odds.

* **6/28.** According to a bank economist, the probability that the U.S. Federal Reserve system will tighten monetary policy is .75. Express her subjective probability in the form of odds.

* **6/29.** In 1980, 20% of all bachelor's degrees awarded in the United States were in business and commerce, whereas the corresponding figure in Canada was 10%. Express in the form of odds the probability that a randomly selected graduate was in business and commerce:

a. for the United States data.

b. for the Canadian data.

* **6/30.** Express the following probabilities as odds:

a. The probability of being hospitalized for a mental illness at least once in one's lifetime is approximately 1/8.

b. The probability of suffering depression sufficiently serious to require professional help at least once in one's lifetime is approximately 1/4.

* **6/31.** The odds that a student who begins work toward a university degree will not complete the degree are currently about 19 to 6 for men and 7 to 3 for women.

a. What is the probability of completion for men?

b. What is the probability of completion for women?

* **6/32.** Down's syndrome is a form of congenital mental retardation. For a woman 41 years of age when her child is born, the odds against having a child with Down's syndrome are 96 to 1 if the father is also 41 years old, but the odds are 82 to 1 if the father is 56 years old. For such a woman, what is the probability of giving birth to a Down's syndrome child:

a. if the father is 41 years old?

b. if the father is 56 years old?

* **6/33.** A hockey fan thinks that it would be fair to bet $4 to $3 that the New York Islanders would defeat the Edmonton Oilers in a hockey game. What is his subjective probability that the Oilers would win?

* **6/34.** A football fan thinks that it would be fair to bet $11 to $4 that the Dallas Cowboys would defeat the Minnesota Vikings in a football game. What is her subjective probability that the Vikings would win?

6/35. Calving records from one farm for the years 1978 through 1981 yielded the following numbers of male and female calves delivered at various times of the day.

	Day	Evening	Night	Totals
Bulls	29	5	17	51
Heifers	17	9	16	42
Totals	46	14	33	93

Assuming random selection in each item (a) through (g), what is the probability that

a. a calf is a heifer?
b. a calf either is a heifer or is delivered during the day?
c. a calf delivered during the day is a bull?
d. a calf is a bull delivered in the evening?
e. a calf either is a bull or is delivered at night?
f. a calf is born during the night?
g. a heifer is delivered in the evening?

For each item (a) through (g), calculate the probability and state whether it is simple, conditional, joint, or disjunctive.
(Source: Student project by J. Millang.)

6/36. Members of a group of 25 college students were classified according to hair and eye color as follows.

	Red hair	Blond hair	Brown hair	Totals
Blue eyes	3	6	2	11
Brown eyes	5	4	5	14
Totals	8	10	7	25

Assuming random selection in each item (a) through (g), what is the probability that

a. a group member has red hair?
b. a blue-eyed group member has blond hair?
c. a group member has red hair and blue eyes?
d. a group member has either brown hair or brown eyes?
e. a group member has blue eyes?
f. a group member with brown hair has brown eyes?
g. a group member has either blond hair or blue eyes?

For each item (a) through (g), calculate the probability and state whether it is simple, conditional, joint, or disjunctive.
(Source: Student project by L. Nowochin.)

6/37. A supermarket manager conducted an investigation of cashiers' accuracy prior to installing a new computerized checkout system. Errors were classified as coding errors (wrong item or price) and counting errors (missing or extra items). Based on the results of the study, the manager estimated that the probabilities that a customer's grocery bill will include coding errors, counting errors, or both types of errors are .72, .05, and .04, respectively.

a. If a grocery bill has a counting error, what is the probability that it will also have a coding error?
b. If a grocery bill has a coding error, what is the probability that it will also have a counting error?
c. What is the probability that a grocery bill will have at least one of these two types of errors?
d. What is the probability that a grocery bill will have no errors of either type?

6/38. A restaurant serves two kinds of pie: apple and cherry. The probabilities that on any one day the restaurant will run out of apple pie, cherry pie, or both are .52, .41, and .37, respectively.

- **a.** If the restaurant runs out of apple pie, what is the probability that it will also run out of cherry pie?
- **b.** If the restaurant runs out of cherry pie, what is the probability that it will also run out of apple pie?
- **c.** What is the probability that the restaurant will run out of at least one kind of pie?
- **d.** What is the probability that the restaurant will not run out of either kind of pie?

6/39. The following is part of a mortality table for the United States.

Age	Death rate per 1000
81	80.9
82	86.9
83	93.1
84	99.4

- **a.** What is the probability that an 81-year-old will live to be at least 84?
- **b.** What is the probability that an 81-year-old will live to be 83 but die before age 84?

(Source: *Information Please Almanac 1982,* pp. 780–81.)

6/40. What is the probability that a hand of five cards dealt from a well-shuffled standard deck contains five cards of the same suit?

6/41. The leading causes of death in North America are cardiovascular diseases, cancer, and accidents, with probabilities of approximately .50, .21, and .05, respectively, that death will be due to these causes.

- **a.** What is the probability that an individual's cause of death will be one of the three leading causes?
- **b.** What is the probability that an individual's cause of death will be any cause other than an accident?

(Source: *Information Please Almanac 1982,* p. 776.)

6/42. The proportions of babies born in Monroe County, New York (Rochester and area), in the various months of the year were approximately as follows for the years 1941 through 1968:

Jan.	Feb.	March	April	May	June	July	Aug.	Sept.	Oct.	Nov.	Dec.
.079	.075	.085	.081	.086	.083	.087	.087	.087	.086	.081	.083

What is the probability that a randomly selected native of Monroe County was born in the last six months of the year?
(Source: Knapp, 1982.)

6/43. With reference to the data in Problems 6/41 and 6/42:

- **a.** What is the probability that a Monroe County native is born in February and (eventually) dies of cancer?

b. What is the probability that a Monroe County native born in February will die of cancer?

c. What is the probability that a Monroe County native who dies of cancer was born in February?

6/44. What is the probability of rolling seven, eight, and nine in that order in three successive rolls of a pair of fair dice?

6/45. All students living in a college residence complex were classified according to whether they never wore visual corrective lenses, wore glasses at least occasionally but never contact lenses, or wore contact lenses at least occasionally. The probability that a student was male was .468, that a student wore contacts was .152, that a student wore no corrective lenses was .521, that a student was female and wore contacts was .125, that a male student wore no corrective lenses was .591, and that a female student wore glasses was .305.

a. What was the probability that a student was male and wore no corrective lenses?

b. What was the probability that a student with contact lenses was female?

c. What was the probability that a student was female or wore contacts?

(Source: Student project by K. Ritchie.)

6/46. With reference to Problem 6/45:

a. What was the probability that a student was female and wore glasses?

b. What was the probability that a male student wore contact lenses?

c. What was the probability that a student was male or wore no corrective lenses?

* **6/47.** Let H represent the event that a customer at a fast-food outlet orders a hamburger, and M the event that a customer orders a glass of milk. Given that $P(H) = .60$, $P(M) = .30$, and $P(M \mid H) = .25$, use a Venn diagram to determine each of the following:

a. $P(H \text{ and } M)$

b. $P(H \mid M)$

c. $P(H \text{ or } M)$

d. $P(H')$

e. $P(H' \text{ and } M')$

f. Are H and M mutually exclusive? How do you know?

g. Are H and M independent? How do you know?

* **6/48.** Let C represent the event that a college student takes a computing science course and Y the event that a student takes a physics course. Given that $P(C) = .25$, $P(Y) = .10$, and $P(C \mid Y) = .90$, use a Venn diagram to determine each of the following:

a. $P(C \text{ and } Y)$

b. $P(Y \mid C)$

c. $P(C \text{ or } Y)$

d. $P(C \text{ and } Y')$

e. $P(C \text{ or } Y')$

f. Are C and Y mutually exclusive? How do you know?

g. Are C and Y independent? How do you know?

* **6/49.** The probabilities that the radio announcer, the technician, or both will arrive on time for an early morning radio program are .80, .92, and .75, respectively. Use a Venn diagram to determine the probability that:

a. at least one of them arrives on time;

b. neither of them arrives on time.

* **6/50.** The probabilities that a college student subscribes to a news magazine, a sports magazine, or both are .09, .04, and .01, respectively. Use a Venn diagram to determine the probability that a student subscribes to:

a. either a news or a sports magazine;

b. neither a news nor a sports magazine.

* **6/51.** A baseball player has a batting average (that is, probability of getting a hit) of .302 against right-handed pitchers and of .184 against left-handed pitchers. If he faces right-handed pitchers 28% of the time, what is his overall batting average?

* **6/52.** The probability that a recently transplanted shrub will survive is .80 if it rains tomorrow and .30 otherwise. If the probability of rain tomorrow is .10, what is the probability that the shrub will survive?

* **6/53.** An inspector for a fruit wholesaler rejects 12% of all shipments of peaches inspected. Past experience has shown that this inspector rejects 95% of all bad shipments and that 10% of all shipments have ultimately proven unsatisfactory. What is the probability that one of the inspector's rejected shipments is actually bad?

* **6/54.** For people in a certain population, the probability of developing lung cancer is .12. If 40% of the population are smokers, and 90% of those who develop lung cancer are smokers, what is the probability that a smoker in this population will develop lung cancer?

* **6/55.** Twenty percent of the patients given a certain medical lab test have a serious contagious disease. The lab test for the disease is not infallible: 4% of those with the disease are not detected by the test, and 7% of those without the disease are incorrectly identified as having it. What is the probability that a person has the disease, if the test result indicates he does?

* **6/56.** Seventy percent of beginning college students require additional instruction in basic skills of English composition. Of those who require additional instruction, 91% fail a college entrance examination on composition skills, while 42% of those who do not require additional instruction also fail the test. What is the probability that a student who fails the test actually needs additional instruction in composition skills?

* **6/57.** A television manufacturer has three assembly plants. Records show that 1% of the sets shipped from plant A turn out to be defective, as compared with 2% of those from plant B and 4% of those from plant C. In all, 20% of the manufacturer's total output comes from plant A, 50% from plant B, and 30% from plant C. If a customer finds that her television set is defective, what is the probability that it came from plant C?

* **6/58.** There are three used-car dealers in a small town. In the past two years, 98% of dealer A's customers, 77% of dealer B's customers, and 84% of dealer C's customers have been satisfied with their purchase. Dealer A has had the smallest number of customers, only half as many as dealer B and only one-quarter as many as dealer C. What is the probability that a resident who bought a used car locally within the last two years, and is satisfied with it, bought it from dealer B?

6/59. A man who won a million dollars in a lottery was interviewed by a newspaper reporter. Discuss the following two statements from the report of the interview:

- **a.** The million-dollar winner "said he should have realized his luck was getting better because he won $50 a few months ago in [a different lottery]."
- **b.** The million-dollar winner planned to keep buying lottery tickets "but I'll only buy one or two tickets, like I always do. It doesn't matter if you've got one ticket or a million . . . . If you have the right number, you're going to win."

(Source: *The Montreal Star,* 6 December 1976, p. 1.)

6/60. "I've never won a lottery or sweepstakes, but it would be foolish to stop entering them now, because losing so many times brings me that much closer to winning." Discuss this statement.

6/61. "Whenever I look for a specific book in the library, it always seems to be the one book that's missing from the shelf. I must be jinxed, because the probability of that happening by chance is very small." Is the probability of this phenomenon very small? Discuss.

6/62. Was the event that was described in Example 6.1 a particularly unlikely event? Discuss.

References

Bell, E. T. *Men of Mathematics.* New York: Simon and Schuster, 1937.

Hooke, R. "Statistics, Sports, and Some Other Things." In *Statistics: A Guide to the Unknown,* 2nd ed., edited by J. M. Tanur *et al.* San Francisco: Holden-Day, 1978.

Information Please Almanac 1982. New York: Simon and Schuster, 1981.

Knapp, T. R. "The Birthday Problem: Some Empirical Data and Some Approximations." *Teaching Statistics* 4 (1982): 10–14.

Kolata, G. B. "Clinical Trial of Psychotherapies is Under Way." *Science* 212 (1981): 432–33.

Vaughan, A. *Incredible Coincidence: The Baffling World of Synchronicity.* New York: Lippincott, 1979. Case number 100, pp. 130–31.

World Who's Who in Science. Chicago: Marquis Who's Who, 1968.

Further Reading

Campbell, S. K. *Flaws and Fallacies in Statistical Thinking.* Englewood Cliffs, NJ: Prentice-Hall, 1974. Chapter 11 deals with probability, using a variety of examples to clarify some common misunderstandings.

Kotz, S., and Stroup, D. F. *Educated Guessing: How to Cope in an Uncertain World.* New York: Marcel Dekker, 1983. This introductory book attempts to bridge the gap between intuitive meanings and precise mathematical formulations of probability and to show its use in real-world applications. "Before visiting a casino," the authors add, "we suggest a careful reading of this book."

Rowntree, D. *Probability without Tears*. New York: Scribner, 1984. This book offers a painless introduction to probability for nonmathematicians.

Weaver, W. *Lady Luck: The Theory of Probability*. Garden City, NY: Doubleday, 1963. True to its dust jacket, this book is written "with wit and clarity." It provides a fascinating introduction to probability theory for the nonmathematician. Chapters 1 through 6 deal with basic concepts and illustrations. Chapter 13 is an excellent discussion of "rare events, coincidences, and surprising occurrences," complete with anecdotes.

CHAPTER 7

PROBABILITY DISTRIBUTIONS

EXAMPLE 7.1 Sometimes we talk on the phone with someone, such as a prospective employer, or hear about someone, such as a professor whose course we plan to take, and then form an impression of what that person looks like, only to find when we eventually meet him that he is much taller or shorter or fatter or thinner than we had expected. What probabilities are associated with various possible heights or with various possible weights?

Distribution of height of 8585 adult males born in the British Isles.

Height in inches	Observed distribution		Theoretical description based on the Gaussian distribution	
	Frequency	Proportion	Probability	Frequency
57–57$\frac{7}{8}$	2	.0002	.0001	1
58–58$\frac{7}{8}$	4	.0005	.0003	3
59–59$\frac{7}{8}$	14	.002	.001	10
60–60$\frac{7}{8}$	41	.005	.004	33
61–61$\frac{7}{8}$	83	.010	.010	86
62–62$\frac{7}{8}$	169	.020	.023	198
63–63$\frac{7}{8}$	394	.046	.046	392
64–64$\frac{7}{8}$	669	.078	.078	668
65–65$\frac{7}{8}$	990	.115	.114	978
66–66$\frac{7}{8}$	1223	.142	.143	1231
67–67$\frac{7}{8}$	1329	.155	.155	1331
68–68$\frac{7}{8}$	1230	.143	.144	1238
69–69$\frac{7}{8}$	1063	.124	.115	990
70–70$\frac{7}{8}$	646	.075	.079	681
71–71$\frac{7}{8}$	392	.046	.047	402
72–72$\frac{7}{8}$	202	.024	.024	205
73–73$\frac{7}{8}$	79	.009	.010	89
74–74$\frac{7}{8}$	32	.004	.004	34
75–75$\frac{7}{8}$	16	.002	.001	11
76–76$\frac{7}{8}$	5	.0006	.0004	3
77–77$\frac{7}{8}$	2	.0002	.0001	1
Total	8585	1.000	1.000	8585

The final report of the Anthropometric Committee to the British Association for the Advancement of Science in 1883 tabulated heights of 8585 adult males born in the British Isles. The distribution of height is very regular and is well approximated by probabilities obtained from the familiar bell-shaped curve known as the **Gaussian distribution**, to be discussed in Section 7.4. The accompanying table shows the observed distribution of height, together with the theoretical approximation. On the whole, the theoretical distribution provides a good description of the empirical data.

Based on methods to be introduced in this chapter, the theoretical distribution may be used as a mathematical model of the real-world distribution of height, allowing us to determine the probabilities associated with various possible heights.

(Source: Kendall & Stuart, 1969, p. 8.)

In Chapter 2 we saw that a frequency distribution of a variable is a description of the frequencies with which various values of the variable or categories of observation occur. Chapter 6 introduced some ways of calculating and interpreting the probability of an event or category of observation. Now this chapter brings together the themes of distributions and of probabilities as we consider how probabilities associated with a set of events, or with the possible values of a variable, can be organized into coherent descriptions of the variable.

7.1 BASIC CONCEPTS IN THE STUDY OF PROBABILITY DISTRIBUTIONS

This section outlines some fundamental concepts pertaining to probability distributions in general, and then the remainder of the chapter introduces some specific probability distributions of particular importance in statistics.

The Nature of Probability Distributions

A **probability distribution** of a variable is a description of the probabilities with which various values of the variable or categories of observation occur.

EXAMPLE 7.2 Let us consider the probabilities of obtaining various numbers of heads in three flips of a fair coin. There are 8 equally likely outcomes: HHH, HHT, HTH, THH, HTT, THT, TTH, and TTT, where H = heads and T = tails. Of these 8 possible outcomes, 1 consists of three heads, 3 include exactly two heads, 3 include exactly one head, and 1 includes no heads. From the classical interpretation of probability, we know that the probabilities associated with the four possible numbers of heads are 1/8, 3/8, 3/8, and 1/8, respectively; this constitutes the probability distribution. The distribution is shown in tabular form as follows:

Probability distribution of number of heads in three flips of a fair coin

Number of heads	Probability
0	1/8
1	3/8
2	3/8
3	1/8
Total	1.0

EXAMPLE 7.3 The following table is a probability distribution showing the probability associated with each of seven categories of scores on an intelligence test.

Probability distribution of IQ as measured by the Stanford-Binet test.

IQ (and verbal classification)		Probability
140 or more	(Gifted)	.01
120–139	(Superior)	.11
110–119	(High average)	.18
90–109	(Average)	.46
80– 89	(Low average)	.15
70– 79	(Borderline)	.06
Below 70	(Mentally retarded)	.03
Total		1.00

(Source: Terman & Merrill, 1937.)

Probability distributions are closely related to percentage frequency distributions, introduced in Section 2.4. A percentage distribution shows, for a particular set of data that have been collected, the percentages (or proportions) of observations that occurred in various categories. If the set of data constitutes a population (that is, all the observations relevant to the question being asked), then the percentage distribution is identical to the probability distribution of random selections from that population. The percentage distribution of sample data, on the other hand, may be considered to be an estimate of the probability distribution of the corresponding population.

Whether its data constitute a population or a sample, a percentage frequency distribution describes a set of observations that have already been made. A probability distribution derived from actual observations is called an **empirical probability distribution**. A probability distribution derived mathematically—for example, from the rules of probability rather than from empirical observations—is called a **theoretical probability distribution**. These concepts are illustrated in the preceding two examples: Example 7.3 presents an empirical probability distribution of IQ, obtained by measuring IQ in a large group of individuals, and Example 7.2 shows a theoretical probability distribution for coin flipping, obtained without flipping a single coin.

For reasons of availability and mathematical convenience, there will be times when we will want to use a theoretical probability distribution in place of an empirical distribution. We must be aware, however, that the conditions underlying any given theoretical distribution are probably never met perfectly in real-world data, and hence theoretical probability distributions should be regarded as approximations when applied to empirical data. (For example, the table of Example 7.1 shows an empirical probability distribution in the column headed "Proportion," along with the corresponding theoretical probability distribution in the column headed "Probability." The correspondence between the two is unusually good but, inevitably, less than perfect.) Consequently, before using a theoretical distribution as a substitute for an empirical distribution, we must be sure that the theoretical distribution yields probabilities that are sufficiently accurate for the problem at hand.

Random Variables

The term *random variable* is used in two different but closely related senses. In applied statistics, any *variable* whose values may be described by a probability distribution is called a **random variable**. This terminology is consistent with our previous definition of *random* in Section 6.2, which implied adherence to specific probabilities. A random variable may be thought of as a variable whose values depend on a probabilistic element described by a particular probability distribution. Except where otherwise specified, we follow this usage in this book.

In a more strictly mathematical sense, a **random variable** is a *function* that assigns a numerical value to each outcome in an experiment whose results depend to some extent on chance. The numerical value assigned is one that reflects the dimension of interest in the experiment. Although in this sense the random variable is the *function* (or rule) that assigns numerical values to probabilistic outcomes, it is common to refer to the observed *variable* as the random variable, as discussed in the previous paragraph. In either case, the observed values in an investigation are the values of a random variable.

EXAMPLE 7.4 In Example 7.2, the random variable, in the first sense, is number of heads. There are, of course, many other features we could record to indicate the outcome of three flips of a fair coin; for example, the time required to flip the three coins, the number of rotations of the coins in the air, and the distance between the points on the floor where they landed. In the more mathematical sense, the random variable is the function, or set of instructions, that dictates which feature of the outcome is relevant and what numerical value should be recorded for each outcome. In Example 7.2, the random variable directs us to ignore time, number of rotations, and so on, and to assign the numerical value 0, 1, 2, or 3, depending on the number of heads showing on the three flipped coins.

In Example 7.3, the random variable, in the strict sense, is the procedure that assigns an IQ score to an individual. In the more general sense, the random variable is IQ score, which takes on different values for different individuals.

In Example 7.1, the random variable, in the strict sense, is the procedure that assigned a numerical value (namely, height in inches) to each of the 8585 men in the investigation. More loosely, we say that the random variable is height.

Discrete Variables and Continuous Variables

Variables may be classified on the basis of their possible values as being either discrete or continuous. A variable is said to be **discrete** if each value of the variable on the real-number line is separated from the next larger or smaller value of the variable by real numbers that are not possible values of the variable. The most common examples of discrete variables are those whose values are limited to the integers (. . . , −2, −1, 0, 1, 2, 3, . . .). A variable is said to be **continuous** if its theoretically possible values include the infinite number of real numbers between two of its values; for example, between 3.6 and 3.7 are real numbers such as 3.61 and 3.62, and between those there are real numbers such as 3.613 and 3.614, and so on ad infinitum. In practice, of course, we do not record an infinite number of decimal places in our observations, but rather we round them to certain discrete values. Such rounding does not change the nature of the variable, which is determined by its theoretically possible values. The possible values of a **discrete variable** are disconnected points on the real number line, whereas the possible values of a **continuous variable** include all the values in an uninterrupted region of the real number line.

EXAMPLE 7.5 In Example 7.1, height in inches is a continuous variable, because between any two different heights, no matter how close together, there exist other possible heights, even though our measuring equipment may not be precise enough to detect the difference.

In Example 7.2, number of heads is a discrete variable, because there are no possible values of the variable between adjacent values, such as 2 and 3. Values such as 2.2 and 2.5 are not permissible values of the variable.

Sometimes specialized knowledge is necessary to recognize whether a variable is discrete or continuous. In Example 7.3, IQ is nowadays a discrete variable, because IQs can only be whole numbers when determined by modern IQ tests. Formerly, IQ (intelligence quotient) was calculated as a ratio (or quotient) of two numbers, in which case fractional results were possible, but that is no longer the case, and IQs now are always integers.

A probability distribution in which the random variable is a discrete variable is called a **discrete probability distribution**. A discrete probability distribution, such as that of number of heads presented in Example 7.2, shows the probability of occurrence of each discrete value of the variable.

A probability distribution in which the random variable is a **continuous** variable is called a **continuous probability distribution**. A continuous probability distribution, such as that of height in inches presented in Example 7.1, shows the probabilities with which a continuous variable takes a value within various intervals or categories of **possible** values.

Probability Functions

A convenient way of presenting a *discrete* probability distribution is by means of its **probability function**, which is a mathematical function giving the probability of occurrence of each value of a discrete random variable.

EXAMPLE 7.6 Consider the distribution of the number of dots ($x = 1, 2, 3, 4, 5,$ or 6) on the top surface after a single roll of a fair die. The probability function describing this distribution is

$$P(x) = \tfrac{1}{6} \qquad \text{for } x = 1, 2, \ldots, 6 \tag{7.1}$$

The probability function is pictured in Figure 7.1.

Expression (7.1) is a compact way of writing $P(1) = \frac{1}{6}$, $P(2) = \frac{1}{6}, \ldots,$ and $P(6) = \frac{1}{6}$. The random variable is number of dots on the top surface, and the probability function $P(x)$ gives the probability of occurrence of each of the six possible values of this discrete variable. For example, the probability of rolling a five is $P(5) = \frac{1}{6}$, and the probability of rolling an odd number is

$$P(1) + P(3) + P(5) = \tfrac{1}{6} + \tfrac{1}{6} + \tfrac{1}{6} = .5$$

Since the values of a probability function are probabilities, they must obey the requirements specified in the definition of probability. In particular, the following conditions must hold:

$$0 \leq P(x) \leq 1 \qquad \text{for all values of } x \tag{7.2}$$

$$\sum P(x) = 1 \tag{7.3}$$

$$P(x_i \text{ or } x_j) = P(x_i) + P(x_j) \qquad \text{for } x_i \neq x_j \tag{7.4}$$

You should verify that these conditions are met by the probability function $P(x)$ in Example 7.6.

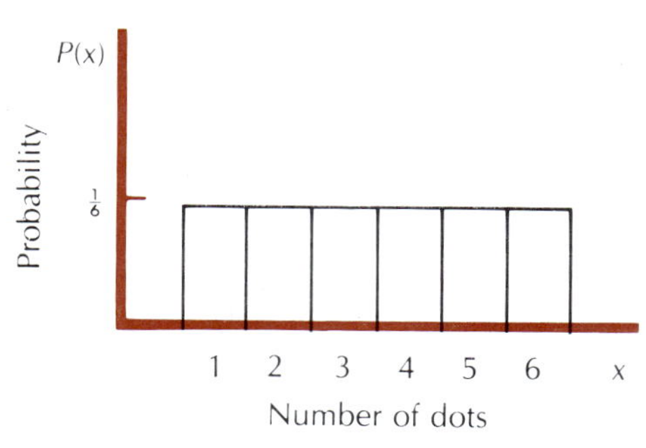

Figure 7.1 Histogram of the probability function $P(x) = 1/6$.

Probability Density Functions

A convenient way of representing a *continuous* probability distribution is by means of its **probability density function**. By definition, a continuous variable has an infinite number of possible values, and in fact between any two values are an infinite number of other possible values of the variable. It follows that any one particular value of a continuous variable occurs with a probability that is crudely represented as $1/\infty$ and that is usually interpreted as 0. What is of interest in the case of continuous probability distributions is the probability associated with an interval, rather than with a specific value of the variable. Probability functions, which are used to represent *discrete* probability distributions, give probabilities associated with each specific value of the random variable and therefore are not suitable for *continuous* variables, each specific value having a probability of 0 for such variables. Continuous probability distributions are represented instead by probability density functions.

We have previously met the concept of **density** in the discussion of the presentation of frequency distributions with unequal category intervals in Section 2.5. In the case of continuous probability distributions, we again have need of the concept of density, although this time for different reasons. This time it is the mathematical problems involved in having an infinite number of possible values of the random variable, each with a probability of 0, that lead us to abandon heights in favor of *areas* to represent probabilities.

A **probability density function** is a function of a continuous random variable such that probabilities are portrayed as areas, rather than heights, on the graph of the function. Specifically, the probability that on a given trial the random variable x will take on a value between any two values x_i and x_j is equal to the proportion that the area under the density function's curve between x_i and x_j is of the total area under the curve. In a probability density function (as opposed to a frequency density function), the units for density are chosen so that the total area under the curve of the probability density function is 1, and hence the probability of a value between x_i and x_j is simply the area under the curve between x_i and x_j. In Figure 7.2, the proportion of the area

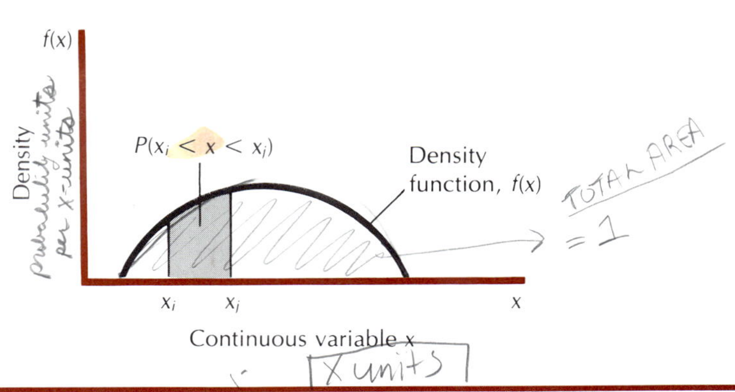

Figure 7.2 A probability density function, illustrating the probability that x takes a value between x_i and x_j.

under the curve that is shaded represents the probability that the value of the continuous random variable x occurring on a given trial will be between x_i and x_j.

A density function, which is conventionally designated by the lower-case symbol $f(x)$, gives the density, or concentration of probability, at any given value of the variable x. The units on the vertical axis in a graph of a probability density function may be thought of as "probability units per x-unit," so that multiplication of width times height (that is, "x-units" times "probability units per x-unit") yields area in probability units. In practice, we are usually not as interested in the density itself as in various areas (probabilities) under the density curve.

EXAMPLE 7.7 Consider a perfectly fair spinner that is free to stop at any point around a circle, and suppose that the circumference of the circle is calibrated in degrees from 0 to 360, with 0 and 360 being at the same point (say, due north). Each spin of the pointer yields a reading of the random variable—number of degrees clockwise from due north—which is not necessarily an integer. The probability density function describing the distribution of this continuous variable x is

$$f(x) = \tfrac{1}{360} \qquad \text{for } 0 < x < 360 \tag{7.5}$$

Notice that it is immaterial whether we write the qualification in Expression (7.5) as "less than" or "less than or equal to," because the probability of any specific value of a continuous random variable is 0. The density function in Expression (7.5) is pictured in Figure 7.3.

Such a function, in which the density takes on the same value at every value of x, is called a **uniform density**, and the corresponding probability distribution is called a **uniform distribution** or a **rectangular distribution**. (Example 7.6 gave an example of a discrete rectangular distribution.)

We can determine the probability that the spinner will stop in any specified region of the circle by calculating the relevant area under the density curve. For example, the

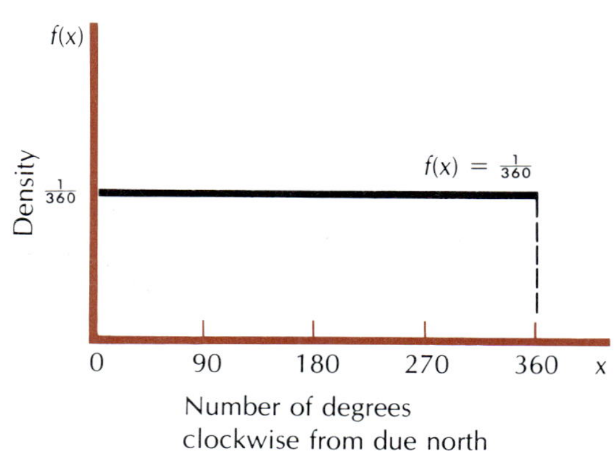

Figure 7.3 Probability density function describing the distribution of readings of a spinner.

probability that the spinner will stop in the quadrant bounded by 270 and 360 is

$$P(270 < x < 360) = (360 - 270)(\tfrac{1}{360}) = .25$$

The probability of obtaining a value between 90 and 126 is

$$P(90 < x < 126) = (126 - 90)(\tfrac{1}{360}) = .10$$

Finally, the total area under the curve is equal to 1, as required in any probability density function:

$$P(0 < x < 360) = (360 - 0)(\tfrac{1}{360}) = 1$$

You may find it instructive to reexamine Example 7.6 to see that probabilities may be viewed as proportions of total area in the discrete case, too.

Distribution Functions

A function showing the cumulative probability up to various values of the random variable is called the **distribution function**. The distribution function, which is designated by the upper-case symbol $F(x_i)$, gives the probability of obtaining a value of the random variable x less than or equal to x_i:

$$F(x_i) = P(x \leq x_i) \tag{7.6}$$

The values of $F(x_i)$ are therefore cumulative "or less" probabilities.

Since values of the distribution function $F(x_i)$ are probabilities, they are never less than zero:

$$F(x) \geq 0 \tag{7.7}$$

Also, when the variable x is at its maximum possible value $x_{\max}$, the value of $F(x)$ reaches its maximum of 1:

$$F(x_{\max}) = 1 \tag{7.8}$$

because the probability that the value of x is less than or equal to its maximum possible value is 1.

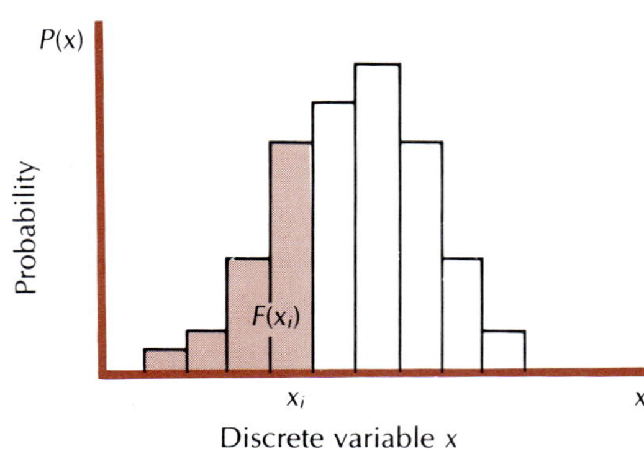

Figure 7.4 Graph showing the probability function $P(x)$ as a histogram and the value of the distribution function $F(x_i)$ as the shaded area.

Cumulative distributions were first introduced in Section 2.6, where we saw how such a function may be represented graphically by an ogive. In this subsection, we will not draw the ogive of the distribution function itself. Rather, we will consider how the distribution function $F(x)$ is portrayed by a graph of the corresponding probability function $P(x)$ or density function $f(x)$.

The relationship between distribution functions and probability functions of *discrete* random variables is illustrated in Figure 7.4. The value of the distribution function $F(x_i)$ is the shaded area of the histogram, corresponding to values of x up to and including x_i.

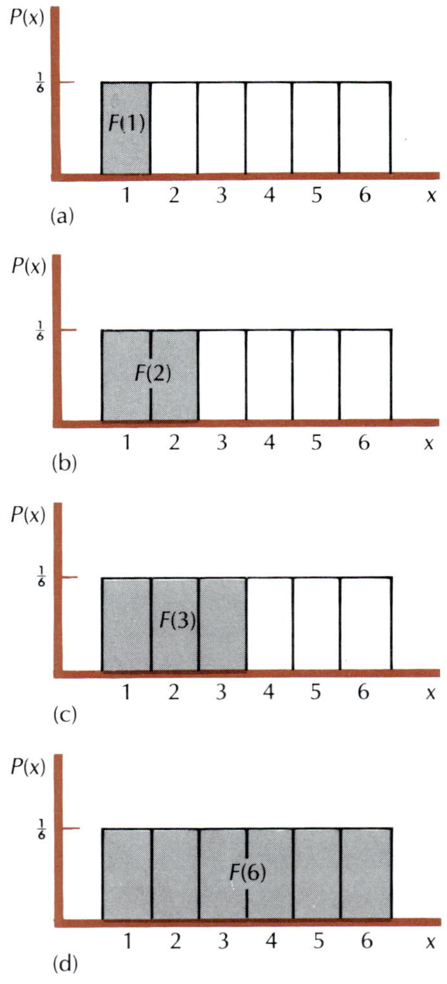

Figure 7.5

EXAMPLE 7.8 In Example 7.6, we considered the probability function

$$P(x) = \tfrac{1}{6} \qquad \text{for } x = 1, 2, \ldots, 6$$

where x is the number of dots on the top surface after a single roll of a fair die. The distribution function $F(x)$, or cumulative probability function, gives the probability of rolling a one or less as

$$F(1) = P(1) = \tfrac{1}{6}$$

the probability of rolling a two or less as

$$F(2) = P(1) + P(2) = \tfrac{1}{3}$$

the probability of rolling a three or less as

$$F(3) = P(1) + P(2) + P(3) = \tfrac{1}{2}$$

and so on up to the probability of rolling a six or less,

$$F(6) = P(1) + P(2) + \cdots + P(6) = 1$$

These probabilities are represented graphically by progressively shading the graph of the probability function from left to right, as illustrated in Figure 7.5.

The relationship between distribution functions and density functions of *continuous* random variables is illustrated by the graph of the probability density function in Figure 7.6. The value of the distribution function $F(x_i)$ is the area under the density-function curve to the left of the point x_i. Notice that as x_i moves across the graph from left to right, the shaded proportion of the area under the curve gradually increases from 0 to 1.

EXAMPLE 7.9 In Example 7.7, we considered the uniform probability density function

$$f(x) = \tfrac{1}{360} \qquad \text{for } 0 < x < 360$$

where x is the number of degrees clockwise from due north that a spinner comes to rest. The distribution function $F(x)$ gives the probability that x takes on a value between 0 and any selected value. For example, the probability that the spinner stops between 0 and 60 is

$$F(60) = (60 - 0)(\tfrac{1}{360}) = \tfrac{1}{6}$$

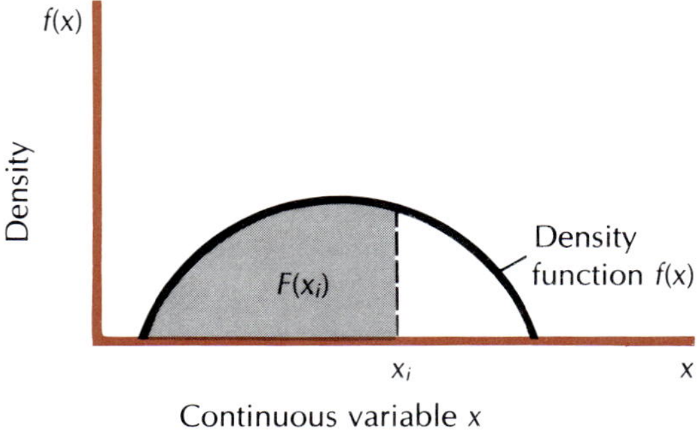

Figure 7.6 Graph showing the probability density function $f(x)$ as the curved line and the value of the distribution function $F(x_i)$ as the shaded area.

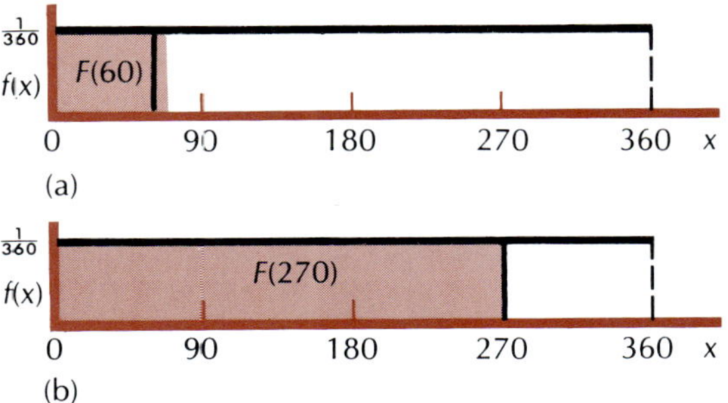

Figure 7.7

and the probability that x is less than 270 is

$$F(270) = (270 - 0)(\tfrac{1}{360}) = \tfrac{3}{4}$$

The value of $F(x_i)$ is represented graphically as the area under the probability density curve that lies to the left of the value x_i, as illustrated in Figure 7.7.

Thus, the term **distribution function** is used in connection with both discrete and continuous distributions. The terminology is summarized in Table 7.1.

TABLE 7.1 Functional representations of probability distributions.

Type of random variable	Probability distribution	Cumulative "or less" distribution
Discrete variable x	Probability function $P(x)$	Distribution function $F(x)$
Continuous variable x	Probability density function $f(x)$	Distribution function $F(x)$

Expected Values

A fundamental concept in probability theory is that of **mathematical expectation** or **expected value**. The **expected value** $E(x)$ of a variable x is the average value of the variable that would be obtained in the long run.

To calculate the expected value of a discrete random variable x, multiply each value x_i of the variable by its probability p_i of occurrence, and sum over all possible values of the variable:

$$\begin{aligned} E(x) &= \sum x_i p_i \\ &= x_1 p_1 + x_2 p_2 + \cdots + x_m p_m \end{aligned} \tag{7.9}$$

for m mutually exclusive and exhaustive values of x. An expected value is simply an average; in fact, you should recognize that an expected value is a weighted mean of the x's, where each value of x is weighted by its probability p_i and the sum of the weights $\sum p_i$ is equal to 1. (Weighted means were introduced in Section 3.2.) The expected value of a continuous random variable is defined in terms of integral calculus in a manner directly analogous to Expression (7.9).

EXAMPLE 7.10 A local grocery store advertised a promotional campaign whereby the amounts of shoppers' purchases were recorded on specially prepared "bonus dividend cards." After sufficient purchases had been recorded, the shopper could present the card to a store official who would break the seal on the card to reveal the amount of a cash prize to be awarded to the shopper, as shown in Figure 7.8. What is the expected value of the amount of prize on a card?

The amount of \$1,000 occurs with probability 6/36,000; the amount of \$250, with probability 18/36,000; and so on. The mathematical expectation of the amount of prize is therefore

$$\begin{aligned} E(\text{prize}) &= (1{,}000)(6/36{,}000) + (250)(18/36{,}000) + \cdots + (1)(34{,}710/36{,}000) \\ &= 1.6392 \end{aligned}$$

or approximately \$1.64. We speak of the expected amount of the prize as being \$1.64, even though no single individual can win exactly \$1.64. In fact, the expected value \$1.64 is the arithmetic mean of the amounts of the 36,000 prizes. The single most likely amount of prize to appear on an individual card is \$1.00, which is the mode, not the expected value, of the distribution.

(Source: *Camrose Booster,* 31 March 1981, p. 3. Used with permission of Camrose Booster, Ltd.)

every card is a winning card

here is the ratio of winners in 36,000 cards

\$1,000.00 Winners — 6 cards
\$250.00 Winners — 18 cards
\$100.00 Winners — 36 cards
\$20.00 — Winners 150 cards
\$10.00 Winners — 360 cards
\$5.00 Winners — 720 cards
\$1.00 Winners — 34,710 cards

total cash prizes available to be won \$59,010.

(if all cards redeemed)

Figure 7.8

The concept of mathematical expectation plays a central role in describing probability distributions. The **mean of a probability distribution** is defined as the expected value of the random variable:

$$\mu = E(x) \tag{7.10}$$

The **variance of a probability distribution** is defined as the expected value (or mean) of the squared deviations from the mean:

$$\sigma^2 = E(x - \mu)^2 \tag{7.11}$$

The **standard deviation of a probability distribution** is σ, the positive square root of the variance.

To illustrate how these definitions relate to our previous definitions of mean and variance for finite populations of N observations, consider a discrete random variable x with N values x_i, each occurring with probability $p_i = 1/N$. Using Expressions (7.10) and (7.9), we find the mean in this case to be

$$\mu = E(x) = \sum x_i p_i = \sum x_i \frac{1}{N} = \frac{\sum x_i}{N}$$

which was our previous definition of the population mean in Section 3.2. Using Expressions (7.11) and (7.9), we find the variance in this case (where $p_i = 1/N$ for all i) to be

$$\sigma^2 = E(x - \mu)^2 = \sum (x_i - \mu)^2 p_i = \sum (x_i - \mu)^2 \frac{1}{N} = \frac{\sum (x_i - \mu)^2}{N}$$

which was our previous definition of the population variance in Section 4.3. In fact, the expected-value definitions of mean and variance are the fundamental definitions of these concepts. They extend the application of these concepts beyond random variables as reflected in a finite set of actual observations, to random variables described in terms of theoretical probability distributions.

EXAMPLE 7.11 The table of Example 7.2 gives the probability distribution of the number of heads in three flips of a fair coin. What are the mean and the variance of this distribution?

Let the variable x represent number of heads. Using Expressions (7.9), (7.10), and (7.11), we find that the mean (or expected value) of the distribution is

$$\mu = E(x) = (0)(\tfrac{1}{8}) + (1)(\tfrac{3}{8}) + (2)(\tfrac{3}{8}) + (3)(\tfrac{1}{8}) = 1.5$$

and the variance of the distribution is

$$\begin{aligned}\sigma^2 &= E(x - \mu)^2 = E(x - 1.5)^2 \\ &= (-1.5)^2(\tfrac{1}{8}) + (-.5)^2(\tfrac{3}{8}) + (.5)^2(\tfrac{3}{8}) + (1.5)^2(\tfrac{1}{8}) = .75\end{aligned}$$

In other words, the mean number of heads in three flips of a fair coin is 1.5, and the standard deviation is $\sqrt{.75} \simeq .9$ heads.

7.2 THE BINOMIAL DISTRIBUTION

There are many discrete probability distributions, but perhaps the most important of them is the **binomial distribution**, discovered by Jakob Bernoulli and first published in 1713 (see Biography 6.1). Not only are there numerous applications of the binomial distribution in empirical problems, but also some knowledge of the binomial distribution is a prerequisite for certain further topics in statistics.

The Random Variable in a Binomial Distribution

The **binomial distribution** is the *theoretical* probability distribution of number of successes in a series of independent trials where the probability of success in each individual trial remains constant. In this context, the term **success** refers to the outcome, or category of observation, whose probability is to be determined; any other outcome is called a **failure**. Thus, the random variable in a binomial distribution is number of successes, and the binomial distribution shows the probability of $0, 1, 2, \ldots, n$ successes in n independent trials where the probability of success on each individual trial is the same (say, p) for every trial. Such a trial, with two mutually exclusive outcomes and a constant probability of success, is called a **Bernoulli trial**.

EXAMPLE 7.12 If we want to determine the probability of rolling exactly two sixes in three rolls of a fair die, then the random variable is number of sixes. Rolling a six would be a success, and rolling any other number would be a failure.

If we want to determine the probability that exactly two out of three particular 89-year-olds will die before their ninetieth birthday, then the random variable is number who die before age 90. In this context, dying before age 90 could be labeled a success, because it is the event whose probability is to be determined, and surviving to age 90 could be labeled a failure. The terms **success** and **failure** here imply nothing about the desirability of the outcomes. They are simply convenient labels that may be arbitrarily applied to either of the two categories of outcomes of a Bernoulli trial. The probability remains the same whether we talk about exactly two people dying out of three or, equivalently, exactly one person surviving out of three.

If we want to determine the probability of getting exactly four heads in five flips of the same coin, then the random variable is number of heads. A head would be a success, and a tail would be a failure.

The Binomial Probability Function

EXAMPLE 7.13 The probability that an American woman who has not married by age 28 will ever marry is approximately .70. Consider three unmarried 28-year-old women, and let us assume that whether one of the three marries or not has no influence on whether either of the others marries or not.

What is the probability that all three eventually marry? Since we have assumed that the three "trials" are independent, we can use Expression (6.20) to calculate the joint probability as

$$P(3 \text{ marry}) = (.70)(.70)(.70) = (.70)^3 = .343$$

What is the probability that two marry and one does not? There are three ways that this could happen: The first, or the second, or the third woman could be the one who does not marry. These three possibilities are mutually exclusive (given that only one does not marry), so the probability that two marry and one does not is [by Expression (6.24)] the sum of the probabilities of the three possibilities:

$$\begin{aligned} P(2 \text{ marry}) &= (.30)(.70)(.70) + (.70)(.30)(.70) + (.70)(.70)(.30) \\ &= 3(.70)^2(.30)^1 \\ &= .441 \end{aligned}$$

What is the probability that one marries and two do not? Again, there are three possibilities to consider:

$$\begin{aligned} P(1 \text{ marries}) &= (.70)(.30)(.30) + (.30)(.70)(.30) + (.30)(.30)(.70) \\ &= 3(.70)^1(.30)^2 \\ &= .189 \end{aligned}$$

Finally, what is the probability that none of the three ever marries? The probability that any one of the three does not marry is .30, so that if the three cases are independent we calculate

$$P(0 \text{ marry}) = (.30)(.30)(.30) = (.30)^3 = .027$$

The preceding four paragraphs answer the following type of question: What is the probability that n_1 of the three women marry and the other n_2 do not, if the probability of marriage is .70 for each of them and if it is unaffected by whether or not the others marry? The answer to this question may be written as

$$P(n_1 \text{ marry}) = \frac{3!}{n_1!\, n_2!}(.70)^{n_1}(.30)^{n_2} \quad \textbf{(7.12)}$$

where $n_1 + n_2 = 3$. (Recall that $3! = 3 \times 2 \times 1$ and is read "three-factorial.") The ratio of factorials after the equals sign is the number of possible ways of grouping 3 women into n_1 who marry and n_2 who do not, and this ratio is multiplied by the probability that a particular set of n_1 women marry and n_2 do not.

(Source: *Information Please Almanac 1982,* p. 768.)

Expression (7.12) is an example of a binomial probability function. In general, a **binomial probability function** gives the probability of exactly n_1 successes in n independent trials in which the probability of success on each individual trial remains the same on every trial. The general form of the binomial probability function may be written as

$$P_{\text{bin}}(n_1) = \frac{n!}{n_1!\, n_2!} p^{n_1} q^{n_2} \quad \textbf{(7.13)}$$

where $P_{\text{bin}}(n_1)$ is the probability of obtaining exactly n_1 successes (and hence n_2 failures) in $n = n_1 + n_2$ independent trials if on each trial the probability of success is p and the probability of failure is

$$q = 1 - p \tag{7.14}$$

Be careful not to confuse the three different probabilities in Expression (7.13):

- p is the probability of success on any one trial,
- $q = 1 - p$ is the probability of failure on any one trial, and
- $P_{\text{bin}}(n_1)$ is the probability of exactly n_1 successes and n_2 failures in a total of $n = n_1 + n_2$ independent trials.

In a typical problem, we know (or are given) the probability p, we compute the probability q as $1 - p$, and we use Expression (7.13) to find the probability $P_{\text{bin}}(n_1)$ for some particular value(s) of n_1 of interest.

The first term on the right-hand side of Expression (7.13) gives the number of ways of partitioning a set of n items into two sets, one having n_1 items (say, successes) and the other having n_2 items (say, failures). It is sometimes called the **number of combinations of n items taken n_1 at a time**, designated by

$$\binom{n}{n_1} = \frac{n!}{n_1!\,(n - n_1)!} = \frac{n!}{n_1!\,n_2!} \tag{7.15}$$

where $n = n_1 + n_2$. The term $p^{n_1}q^{n_2}$ in Expression (7.13) is the probability of n_1 successes and n_2 failures in any one particular order in n trials; for example, $p^3q^2 = pppqq = qqppp = pqpqp$. Thus, the binomial probability function in Expression (7.13) is the probability $p^{n_1}q^{n_2}$ of one particular pattern of n_1 successes and n_2 failures in n independent trials, multiplied by the number $n!/(n_1!\,n_2!)$ of different ways that n_1 successes and n_2 failures could occur in n trials.

EXAMPLE 7.14 Let us demonstrate the use of Expression (7.13) by repeating the calculations in Example 7.13. We assume that three unmarried women make independent decisions as to whether or not to marry, and that the probability of marriage is .70 for each of the three women. The respective probabilities that all, exactly two, exactly one, and none of the three eventually marry are as follows:

$$P_{\text{bin}}(3) = \frac{3!}{3!\,0!}(.70)^3(.30)^0 = (1)(.70)^3(1) = .343$$

$$P_{\text{bin}}(2) = \frac{3!}{2!\,1!}(.70)^2(.30)^1 = (3)(.70)^2(.30) = .441$$

$$P_{\text{bin}}(1) = \frac{3!}{1!\,2!}(.70)^1(.30)^2 = (3)(.70)(.30)^2 = .189$$

$$P_{\text{bin}}(0) = \frac{3!}{0!\,3!}(.70)^0(.30)^3 = (1)(1)(.30)^3 = .027.$$

The probability function is portrayed in Figure 7.9.

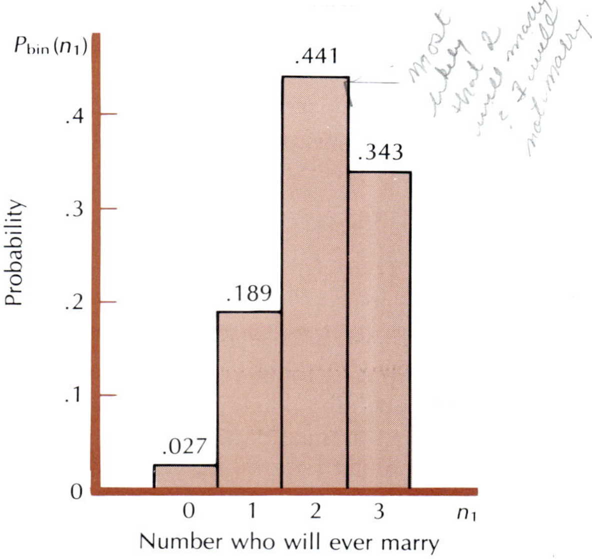

Figure 7.9 Histogram of the binomial probability function for $n = 3$ and $p = .70$.

According to the definition in Expression (7.6), the corresponding distribution function $F_{bin}(n_1)$ gives the probability of n_1 or fewer successes. In this case, the respective probabilities that none of the three marries, at most one of the three marries, two or fewer of the three marry, and three or fewer of the three marry are as follows:

$$F_{bin}(0) = P_{bin}(0) = .027$$
$$F_{bin}(1) = P_{bin}(0) + P_{bin}(1) = .216$$
$$F_{bin}(2) = P_{bin}(0) + P_{bin}(1) + P_{bin}(2) = .657$$
$$F_{bin}(3) = P_{bin}(0) + P_{bin}(1) + P_{bin}(2) + P_{bin}(3) = 1.000$$

The last line demonstrates that the sum of the four probabilities generated by the probability function $P_{bin}(n_1)$ is equal to 1, as it must be, because the four probabilities correspond to four mutually exclusive and exhaustive events: The probability that exactly none, or exactly one, or exactly two, or exactly three of three unmarried women will ever marry is 1.

The binomial distribution takes its name from the fact that the binomial probability function, as given in Expression (7.13), is the general form of any term in the binomial expansion

$$(p + q)^n = P_{bin}(0) + P_{bin}(1) + \cdots + P_{bin}(n)$$

which must equal 1, because $p + q = 1$. (Example 7.14 shows all the terms of the binomial expansion for $p = .7$, $q = .3$, and $n = 3$.) Thus, the "bi" in *binomial* indicates that the distribution is used in situations in which the outcomes may be considered as falling into one of *two* classes on each trial, success with probability p and failure with probability q.

Specifically, Example 7.14 illustrates that the binomial probability function given in Expression (7.13) answers a whole family of questions of the following form:

> What is the probability of obtaining exactly n_1 successes in n independent trials if the probability of success on each trial is p? **(7.16)**

Such probabilities could be determined by direct application of the principles of Chapter 6, as we did in Example 7.13, but the binomial probability function gives a general solution to all problems of this type.

EXAMPLE 7.15 Approximately 21% of North American deaths are due to cancer. What is the probability that 4 of the next 10 unrelated deaths reported in a certain county will be due to cancer? The question fits the pattern for the binomial distribution with $n = 10$, $n_1 = 4$, and $p = .21$. Applying Expression (7.13), we obtain

$$P_{\text{bin}}(4) = \frac{10!}{4!\,6!}(.21)^4(.79)^6 = .099$$

The required probability is approximately .10, which is the value of the relevant binomial probability function evaluated at $n_1 = 4$. By evaluating the function at other values of n_1, we obtain the entire binomial distribution for $n = 10$ and $p = .21$, which is shown in Figure 7.10.

The distribution function $F_{\text{bin}}(n_1)$ gives the cumulative probability that the number of successes will be n_1 or less. For example, the probability that no more than 2 of the next 10 unrelated deaths will be due to cancer is

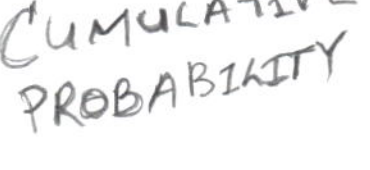

$$F_{\text{bin}}(2) = P_{\text{bin}}(0) + P_{\text{bin}}(1) + P_{\text{bin}}(2) = .095 + .252 + .301 = .648$$

You should verify that the distribution function equals 1 when evaluated at the

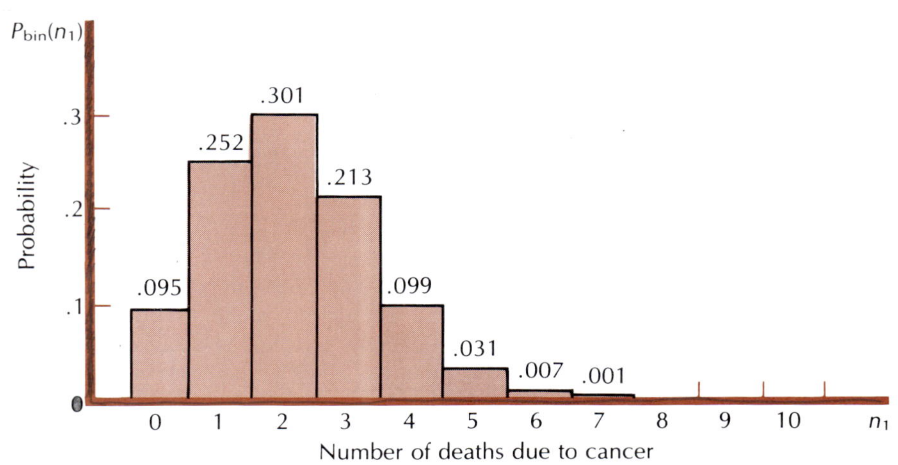

Figure 7.10 Histogram of the binomial probability function for $n = 10$ and $p = .21$.

maximum possible value of the random variable:

$$F_{\text{bin}}(10) = 1.000$$

It follows that the probability that more than 2 of the next 10 unrelated deaths will be due to cancer is given by

$$1 - F_{\text{bin}}(2) = 1.000 - .648 = .352$$

as you may verify in Figure 7.10.

(Source: *Information Please Almanac 1982,* p. 776.)

It is worth reiterating that the binomial distribution is a theoretical, rather than empirical, probability distribution. Thus, its probabilities are determined as a mathematical consequence of its stipulation of independent trials and a constant probability of success on each trial, rather than from empirical observations. Therefore, the question as to whether it is appropriate to use the binomial probability function to calculate probabilities in any particular application depends on the extent to which the **binomial-distribution requirements** of independent trials and constant probability are met. The binomial probability function may sometimes provide a reasonable approximation even if the requirements are not met perfectly. In some cases, we use it specifically to test whether it is reasonable to suppose that the binomial requirements apply; that is, we treat the theoretical distribution as if it were the population, and we evaluate the probability of drawing the observed sample from such a population.

EXAMPLE 7.16 Suppose that you flipped a coin six times and obtained heads exactly once. What is the probability of obtaining no more than one head in six fair flips of a fair coin? If this probability turns out to be very small, it would mean either that you witnessed a rather unusual event, or alternatively that one of the binomial requirements was not met—perhaps you flipped the coin in such a way that the outcome depended on which side was up when you picked up the coin.

Assuming fair flips of a fair coin, the required probability is given by the binomial distribution function with $n = 6$, $n_1 = 1$, and $p = .5$:

$$\begin{aligned} F_{\text{bin}}(1) &= P_{\text{bin}}(0) + P_{\text{bin}}(1) \\ &= \frac{6!}{0!\,6!}(.5)^0(.5)^6 + \frac{6!}{1!\,5!}(.5)^1(.5)^5 \\ &= \frac{1}{64} + \frac{6}{64} = \frac{7}{64} \simeq .11 \end{aligned}$$

The probability of obtaining 0 or 1 head in 6 flips of a fair coin is about 11%. Whether this probability is small enough to make us doubt the assumptions on which its calculation was based—namely, fair flips of a fair coin—is a matter of judgment. It is perhaps small enough that we would at least want to ask about the method of flipping. In any event, it is the probability of so few heads, *if* the binomial requirements were met. It can never tell us with certainty whether the binomial requirements were met—and it does *not* tell us the probability that they were met. Rather it tells us only what the probability would be *if* they were met. If we feel that this probability is unusually small, then we must examine the empirical situation in which the data were collected to help us decide whether (a) we have witnessed an unusual event or (b) the calculated probability does not apply to the empirical data because of violations of the binomial requirements.

Descriptions of the Binomial Distribution

Theoretical distributions such as the binomial may be described in terms of many of the same concepts as those introduced for empirical distributions in Chapters 2, 3, and 4. The following paragraphs review those concepts in the context of the binomial distribution.

Some of the preceding examples have illustrated that the binomial distribution may be portrayed graphically in the form of a *histogram*. Figure 7.11 shows more systematically how the overall shape of the binomial distribution depends on the values of both n and p. Comparing histograms in the same vertical column, notice the change in location as p changes for a given value of n; and comparing histograms in the same horizontal row, notice the change in dispersion as n changes for a given value of p.

A *measure of location* of a theoretical probability distribution indicates where the possible values of the random variable are located on the real-number line. We saw in Section 7.1, page 240, that the mean of a probability distribution is the expected value of the random variable. The **mean of a binomial distribution**, designated by μ_{bin}, is the expected value of the number of successes:

$$\mu_{\text{bin}} = E(n_1) = (0)P_{\text{bin}}(0) + (1)P_{\text{bin}}(1) + (2)P_{\text{bin}}(2) + \cdots + (n)P_{\text{bin}}(n)$$

which simplifies to

$$\mu_{\text{bin}} = np \tag{7.17}$$

Although the simplification is neither obvious nor easy, it is both algebraically exact and intuitively reasonable: The Law of Large Numbers allows us to interpret p as the long-run proportion of successes, and hence np is the mathematically expected number of successes in n trials.

EXAMPLE 7.17 In Example 7.13, we considered the probabilities that exactly 0, 1, 2, and 3 of three unmarried 28-year-old women would eventually marry, given that the probability that one such woman will marry is .70. The mean of this distribution is calculated from Expression (7.17) as

$$\mu_{\text{bin}} = (3)(.70) = 2.1$$

In other words, out of three unmarried 28-year-old women acting independently of each other, the expected number who will marry is 2.1. Thus, in the long run, we could expect a mean of 2.1 out of three such women to marry. Some sets of three such women would include more than 2.1 who will marry and some fewer than 2.1 who will marry, but the mean number over many sets of three unmarried 28-year-old women would be 2.1.

EXAMPLE 7.18 In Example 7.16, we considered the probability of obtaining no more than one head in six flips of a fair coin. The expected number of heads in six fair flips is

$$\mu_{\text{bin}} = (6)(.5) = 3.0$$

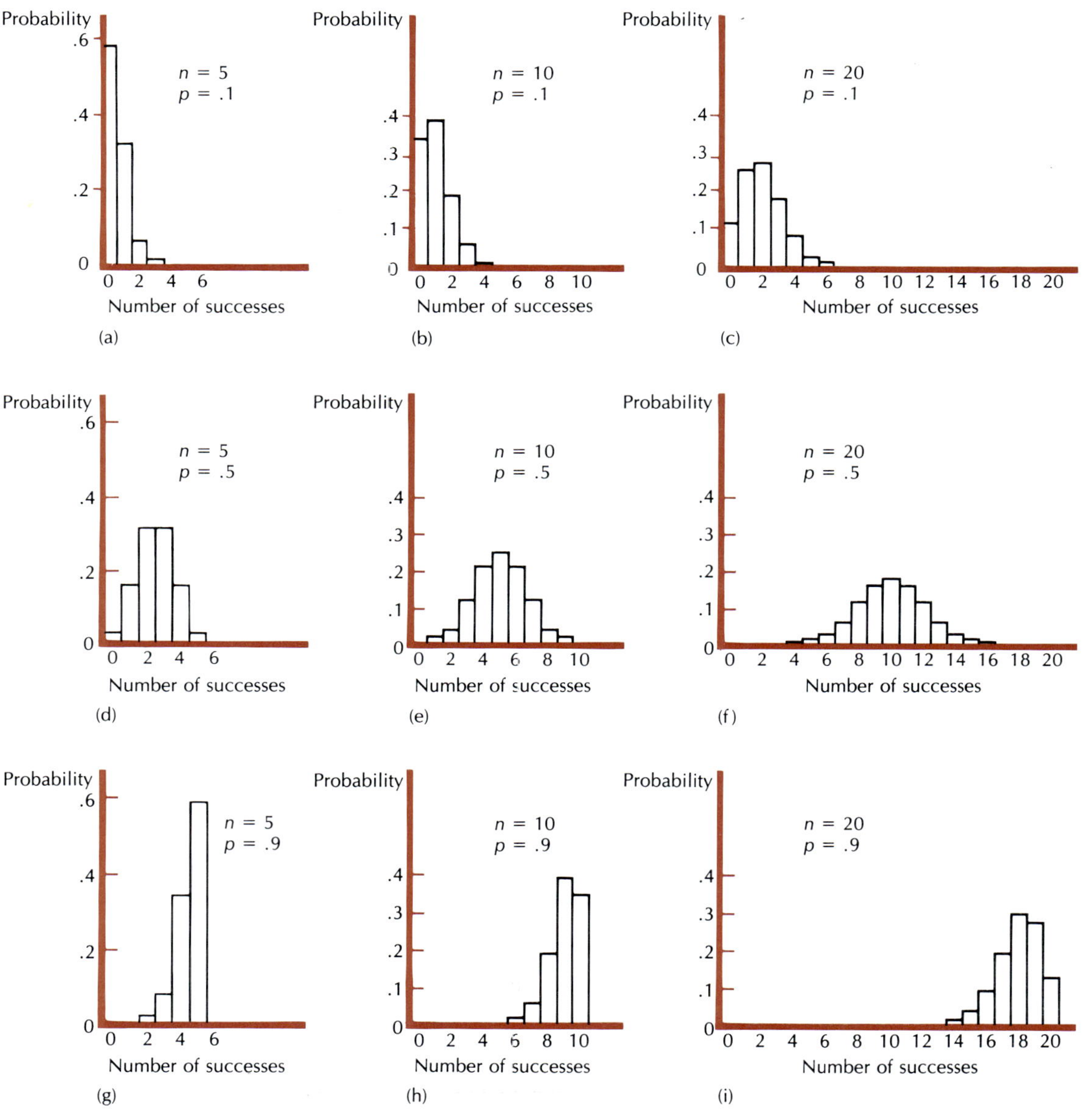

Figure 7.11
Histograms of nine binomial distributions.

In some sets of six fair flips we will obtain more than 3 heads and in some, fewer than 3 heads, but the mean of the distribution is 3.0 heads.

A *measure of dispersion* of a theoretical probability distribution indicates how spread out the possible values of the random variable tend to be along the real-number line. We saw in Section 7.1, page 240, that the variance of a probability distribution is the expected value of the squared deviations from the mean; in other words, the variance is the arithmetic mean of the squared deviations from the mean. The **variance of a binomial distribution** is designated by σ^2_{bin} and is defined by

$$\begin{aligned}\sigma^2_{\text{bin}} &= E(n_1 - np)^2 \\ &= (0 - np)^2 P_{\text{bin}}(0) + (1 - np)^2 P_{\text{bin}}(1) + \cdots + (n - np)^2 P_{\text{bin}}(n)\end{aligned}$$

which eventually simplifies to

$$\sigma^2_{\text{bin}} = npq \tag{7.18}$$

Hence, the **standard deviation of a binomial distribution** is

$$\sigma_{\text{bin}} = \sqrt{npq} \tag{7.19}$$

This standard deviation is interpreted as other standard deviations are: It is a kind of average amount of deviation from the mean, or a standard amount by which the actual number of successes differs from the expected number of successes.

EXAMPLE 7.19 Consider again the probability that exactly one out of three unmarried 28-year-old women will eventually marry, given that the probability that one such woman will marry is .70. We saw in Example 7.17 that the mean of this binomial distribution is $\mu_{\text{bin}} = 2.1$. Expression (7.19) gives the standard deviation as

$$\sigma_{\text{bin}} = \sqrt{(3)(.70)(.30)} = \sqrt{.63} = .79$$

Out of three unmarried 28-year-old women, the expected number who will ever marry is 2.1 women, and the standard or "average" amount by which the actual number will differ from the expected number is .8 women.

EXAMPLE 7.20 What is the probability of obtaining one head in six flips of a fair coin? In Example 7.18 we calculated the mean of this distribution to be $\mu_{\text{bin}} = 3.0$. From Expression (7.19), the standard deviation of the distribution is

$$\sigma_{\text{bin}} = \sqrt{(6)(.5)(.5)} = \sqrt{1.5} = 1.2$$

The expected number of heads in six flips of a fair coin is 3.0, but the actual number of heads in any six flips may well differ from the expected number, with 1.2 heads being a standard amount of deviation.

* 7.3 DISCRETE DISTRIBUTIONS RELATED TO THE BINOMIAL

In Section 7.1, a **discrete probability distribution** was defined as a description of a discrete random variable, a description that shows the probability of occurrence of each possible value of the variable. Section 7.2 introduced the binomial distribution as an example of a discrete probability distribution. Here we begin with a few words about the binomial distribution, to serve as a point of reference, and then we present three additional discrete distributions that have many important applications.

The **binomial distribution** is the theoretical probability distribution of number of successes (and number of failures) in a series of n independent trials where the probability of success on any one trial is the same for all trials. It is called *bi*nomial because there are just two categories of possible outcomes—success and failure—on each trial. The **binomial probability function** may be written as

$$P_{\text{bin}}(n_1) = \frac{n!}{n_1!\, n_2!} p_1^{n_1} p_2^{n_2} \tag{7.20}$$

where

n_1 and n_2 are the numbers of successes and failures, respectively;

$n = n_1 + n_2$ is the total number of trials;

p_1 and p_2 are the probabilities of success and failure, respectively, on each individual trial (with $p_1 + p_2 = 1$); and

$P_{\text{bin}}(n_1)$ is the probability of exactly n_1 successes (and n_2 failures) in n independent trials with the same probability p_1 of success on each trial.

The Multinomial Distribution

The **multinomial distribution** is a generalization of the binomial distribution to accommodate k categories of outcomes, where k is an integer two or more. Instead of the binomial's two categories, success and failure, we will refer to the multinomial's k categories simply as category 1, category 2, . . . , category k. The **multinomial probability function** is

$$P_{\text{mult}}(n_1, n_2, \ldots, n_k) = \frac{n!}{n_1!\, n_2! \cdots n_k!} p_1^{n_1} p_2^{n_2} \cdots p_k^{n_k} \tag{7.21}$$

where

$n_1, n_2, \ldots, n_k$ are the numbers of outcomes in categories $1, 2, \ldots, k$, respectively;

$n = n_1 + n_2 + \cdots + n_k$ is the total number of trials;

p_j is the probability that the outcome on any individual trial is in category j

* This symbol identifies material that is not prerequisite for later topics in this book.

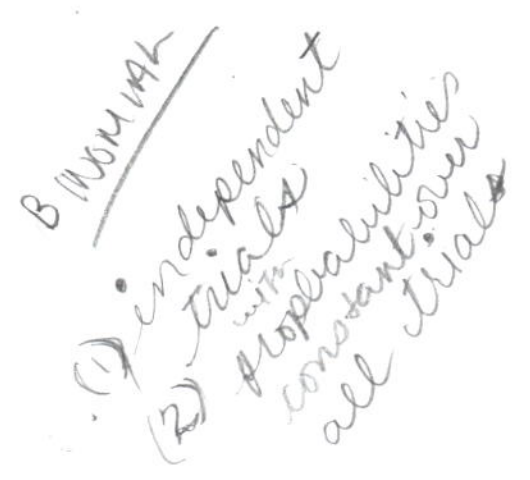

(for $j = 1, 2, \ldots, k$), with $\sum p_j = 1$; and

$P_{\text{mult}}(n_1, n_2, \ldots, n_k)$ is the probability of obtaining n_1 outcomes in category 1, n_2 outcomes in category 2, . . . , and n_k outcomes in category k in n independent trials, with the probabilities p_j constant over all trials.

EXAMPLE 7.21 In the game of craps, one player, called the *shooter,* rolls a pair of dice one or more times on each turn. On the first roll of each turn, a roll of seven or eleven is a winning roll (called a *natural*) for the shooter; a roll of two or three or twelve is a losing roll (called *craps*); and other outcomes are indeterminate, the winner depending on further rolls of the dice. Assuming fair dice, you may verify (see Example 6.6) that the probability of a natural (seven or eleven, which we will call category 1) is $p_1 = \frac{2}{9}$, the probability of craps (two or three or twelve, which we will call category 2) is $p_2 = \frac{1}{9}$, and the probability of any other outcome (category 3) is $p_3 = \frac{2}{3}$.

What is the probability of obtaining 3 naturals, no craps, and 2 indeterminate rolls on the first roll of 5 turns with a pair of fair dice? Expression (7.21) gives the required probability as

$$P_{\text{mult}}(3, 0, 2) = \frac{5!}{3!\,0!\,2!}\left(\frac{2}{9}\right)^3\left(\frac{1}{9}\right)^0\left(\frac{2}{3}\right)^2 = \frac{320}{6561} = .04877$$

or approximately 5%.

It should be noted that the binomial and multinomial distributions require that the probabilities associated with the various outcomes must remain constant from one trial to the next. Coin flipping and dice rolling are standard examples in which the probabilities do not change from trial to trial. On the other hand, if the outcomes consist of selections from a finite population (such as cards drawn from a deck) without replacement after every trial, the probability changes from trial to trial. In sampling from a finite population, the requirement of constant probability for binomial and multinomial distributions is met by the use of **sampling with replacement**—that is, each item selected is put back into the available pool of items before the next selection is made. A different distribution, introduced in the next section, is required for situations of sampling without replacement.

The Hypergeometric Distribution

The **hypergeometric distribution** is the theoretical probability distribution of various possible patterns of outcomes in random sampling without replacement from a finite population. In **sampling without replacement**, the pool of available items becomes progressively smaller as each selection is made, and hence the probabilities of the various categories of outcomes do not remain constant from trial to trial.

The hypergeometric distribution answers the same kind of question when there is sampling *without* replacement as the binomial and multinomial distributions answer when there is sampling *with* replacement: What is the probability of obtaining n_1 outcomes in category 1, n_2 outcomes in category

2, . . . , and n_k outcomes in category k in n trials? Specifically, the **hypergeometric probability function**,

$$P_{\text{hyper}}(n_1, n_2, \ldots, n_k) = \frac{\binom{N_1}{n_1}\binom{N_2}{n_2}\cdots\binom{N_k}{n_k}}{\binom{N}{n}} \tag{7.22}$$

gives the probability of obtaining

n_1 category-1 items out of N_1 such items;

n_2 category-2 items out of N_2 such items;

. . . ; and

n_k category-k items out of N_k such items,

in randomly selecting $n = \sum n_j$ items out of $N = \sum N_j$ items altogether, the k categories being mutually exclusive. The numerator of Expression (7.22) is the number of ways of obtaining the specified pattern of n items out of N, and the denominator is the total number of ways of choosing n items out of N.

EXAMPLE 7.22 Consider a wholesale food distributor that employs 30 salespersons, 12 of whom are men and 18 of whom are women. Five salespersons are to be selected at random to attend an important conference. What is the probability of selecting 2 men and 3 women?

We require the probability of selecting $n_1 = 2$ out of $N_1 = 12$ men and $n_2 = 3$ out of $N_2 = 18$ women. Expression (7.22) gives the probability as

$$P_{\text{hyper}}(2, 3) = \frac{\binom{12}{2}\binom{18}{3}}{\binom{30}{5}} = \frac{\dfrac{12!}{2!\,10!}\,\dfrac{18!}{3!\,15!}}{\dfrac{30!}{5!\,25!}}$$

$$= \frac{\dfrac{(12)(11)}{(2)(1)}\,\dfrac{(18)(17)(16)}{(3)(2)(1)}}{\dfrac{(30)(29)(28)(27)(26)}{(5)(4)(3)(2)(1)}} = .3779$$

or approximately 38%.

EXAMPLE 7.23 One important application of the hypergeometric distribution is in quality-control cases such as the following: A certain type of calculator is inspected in lots of 20 calculators. An inspector randomly selects 3 of the calculators from each lot for a thorough inspection in order to decide whether to pass the lot or withhold it for further testing. We can use the hypergeometric distribution to calculate the probabilities that this inspection procedure will detect various numbers of defective calculators in the lots of 20.

For example, what is the probability that exactly 1 of the 3 sampled calculators will be defective if there is a total of 4 defective calculators in the whole lot? We require the probability of obtaining $n_1 = 1$ out of $N_1 = 4$ defective calculators and $n_2 = 2$ out of

$N_2 = 16$ good calculators. Expression (7.22) gives the probability as

$$P_{\text{hyper}}(1, 2) = \frac{\binom{4}{1}\binom{16}{2}}{\binom{20}{3}} = \frac{\frac{4!}{1!\,3!}\frac{16!}{2!\,14!}}{\frac{20!}{3!\,17!}}$$

$$= \frac{\frac{(4)}{(1)}\frac{(16)(15)}{(1)(2)}}{\frac{(20)(19)(18)}{(1)(2)(3)}} = \frac{8}{19} = .4211$$

or approximately 42%.

What is the probability that this inspection procedure will pass a lot of 20 that includes 4 defective calculators? For the lot to pass, we must have $n_1 = 0$. The probability is

$$P_{\text{hyper}}(0, 3) = \frac{\binom{4}{0}\binom{16}{3}}{\binom{20}{3}} = \frac{\frac{4!}{0!\,4!}\frac{16!}{3!\,13!}}{\frac{20!}{3!\,17!}}$$

$$= \frac{(1)\frac{(16)(15)(14)}{(1)(2)(3)}}{\frac{(20)(19)(18)}{(1)(2)(3)}} = \frac{28}{57} = .4912$$

or approximately 49%. Almost half of the lots containing 4 defective calculators will escape detection by this inspection procedure.

EXAMPLE 7.24 The game of Scrabble includes 100 wooden tiles, each imprinted with a single letter on one side. The distribution of letters is as follows:

12 E's

9 each of A, I;

8 O's;

6 each of N, R, T;

4 each of D, L, S, U;

3 G's;

2 each of B, C, F, H, M, P, V, W, Y, blank; and

1 each of J, K, Q, X, Z.

If 6 letters are randomly selected from the 100 tiles, what is the probability of obtaining the letters to spell LETTER (with no blanks)? We require the probability of obtaining

$n_1 = 1$ out of $N_1 = 4$ L's;

$n_2 = 2$ out of $N_2 = 12$ E's;

$n_3 = 2$ out of $N_3 = 6$ T's;

$n_4 = 1$ out of $N_4 = 6$ R's; and

$n_5 = 0$ out of $N_5 = 72$ other tiles.

According to Expression (7.22), the required probability is

$$P_{\text{hyper}}(1, 2, 2, 1, 0) = \frac{\binom{4}{1}\binom{12}{2}\binom{6}{2}\binom{6}{1}\binom{72}{0}}{\binom{100}{6}} = \frac{9}{451{,}535} = .000020 \tag{7.23}$$

or approximately 1/50,000.

As the population size N becomes large, the calculations for the hypergeometric distribution can become tedious. Fortunately, the multinomial distribution provides a reasonably good approximation to the hypergeometric, provided that the sample size n is less than about 5% of the population size N. This usage is known as the **multinomial approximation to the hypergeometric distribution**.

EXAMPLE 7.25 Let us calculate the multinomial approximation for the problem posed in Example 7.24. The sample size $n = 6$ is 6% of the population size $N = 100$, so the accuracy of the approximation in this case may be somewhat questionable according to our guideline of "less than about 5%." Let us see whether it is.

We require the probability of obtaining $n_1 = 1$ L, $n_2 = 2$ E's, $n_3 = 2$ T's, $n_4 = 1$ R, and $n_5 = 72$ other tiles, where the probability of obtaining each letter can be approximated by the proportion of that letter in the population of 100 tiles: $p_1 = .04$, $p_2 = .12$, $p_3 = .06$, $p_4 = .06$, and $p_5 = .72$, respectively. To use the multinomial approximation to the hypergeometric distribution, we calculate the required probability from Expression (7.21) as

$$P_{\text{mult}}(1, 2, 2, 1, 0) = \frac{6!}{1!\,2!\,2!\,1!\,0!}(.04)^1(.12)^2(.06)^2(.06)^1(.72)^0 = .000022$$

Even though the sample size was slightly larger than recommended by the guideline for the use of the multinomial approximation, the result is a reasonable approximation of the exact probability (.000020) obtained in Expression (7.23).

The Poisson Distribution

Section 7.3 began with the binomial distribution, which gives the probability of exactly n_1 successes in n independent trials with a constant probability p of success on each trial. Next, we considered the possibility of more than two categories of outcomes (the multinomial distribution), and then we removed the requirement of a constant probability from trial to trial by allowing sampling without replacement (the hypergeometric distribution). Now we return to the basic binomial situation with two categories of outcomes (say, success and failure), n independent trials, and a constant probability p of success on every trial, and consider the **Poisson distribution**, which may be applied when n is large and p is small, even if their exact values are unknown.

The **Poisson distribution** is the theoretical probability distribution of the number of successes, or occurrences of any given category of event, in an arbitrary interval of time or space, where there is a small but constant probability of occurrence at each moment in the interval. In effect, each "moment" in the interval is a trial, but it is not necessary that we know the number n of trials (moments), only that it is large, nor the probability p of success in each moment, only that it is small and constant from moment to moment regardless of the outcome in any other moment. That is, the trials (moments) are independent of each other. The **Poisson probability function**, named after the French mathematician Siméon D. Poisson (1781–1840), is

$$P_{\text{Poisson}}(n_1) = \frac{\lambda^{n_1} e^{-\lambda}}{n_1!} \tag{7.24}$$

where

e is the constant 2.71828...;

λ (the lower-case Greek letter lambda) is the mean number of occurrences per unit time or per unit space; and

$P_{\text{Poisson}}(n_1)$ is the probability of exactly n_1 occurrences in a unit of time or space.

Appendix 9 gives values of e raised to selected powers useful in calculating Expression (7.24), for the convenience of readers whose calculators do not have the e^x function.

EXAMPLE 7.26 A highway gas station averages 1.3 customers every five minutes during the day. What is the probability that five minutes would go by with no customers arriving?

To apply the Poisson distribution, we do not need to know precisely how many "moments" there are in five minutes—that is, how many discrete occasions there are when a vehicle might have arrived at the gas station. However, it is reasonable to assume that there are a large number of such moments and that the probability that a customer will arrive in each moment is small and relatively unaffected by what happened in other moments. Thus, we can use the Poisson distribution with the mean $\lambda = 1.3$ customers per five minutes, to calculate the probability of obtaining $n_1 = 0$ customers in a particular five-minute period:

$$P_{\text{Poisson}}(0) = \frac{(1.3)^0 e^{-1.3}}{0!} = .273$$

where the value of $e^{-1.3}$ may be obtained either from Appendix 9 or by use of a pocket calculator. The probability that the gas station would receive no customers in a particular five-minute period is about 27%.

The probability of exactly one customer in a given five-minute period is

$$P_{\text{Poisson}}(1) = \frac{(1.3)^1 e^{-1.3}}{1!} = .354$$

or approximately 35%.

Similarly, we can work out the probability of any number of customers in a given five-minute period as shown in the following table.

n_1	$P_{\text{Poisson}}(n_1)$
0	.273
1	.354
2	.230
3	.100
4	.032
5	.008
6	.002

The probabilities continue to get smaller as n_1 continues to increase. The distribution is portrayed in Figure 7.12.

The Poisson distribution has been successfully applied to a wide variety of problems, such as the following:

a. *Medical diagnosis.* What is the distribution of numbers of cells in normal blood counts?

b. *Radioactive disintegration.* What pattern describes the emission of particles by a radioactive substance?

c. *Traffic safety.* What is the probability of x or more accidents in the next month on a certain stretch of highway, given its present speed limit?

d. *Telephone systems.* What is the probability that equipment capable of handling x calls simultaneously will be adequate for a given community or company?

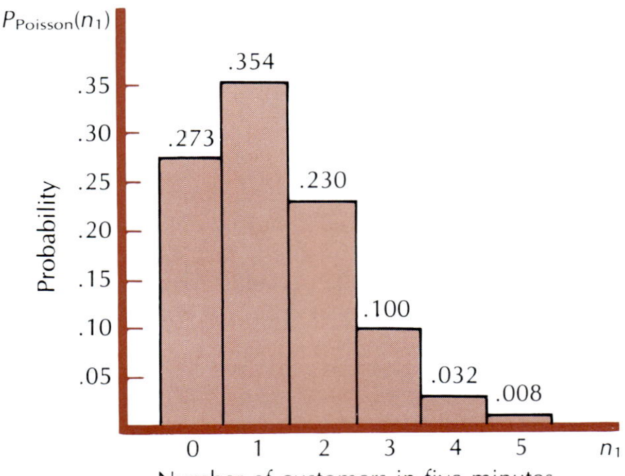

Figure 7.12 Histogram of the Poisson distribution with $\lambda = 1.3$.

e. *Commercial inventory control.* What is the probability that a supply of x units of an item will be adequate to meet the demand for the item, and what is the probability that x units will be an oversupply resulting in losses due to perishability and/or storage costs of the surplus?

It is interesting to note that independent random events occurring with constant probability, as in the case of the Poisson distribution, often have the appearance of clustering, or "coming in bunches." Gas-station attendants, clerks in shoe stores, and homeowners replacing burned-out light bulbs have probably all sometimes felt that "it never rains but it pours," a phenomenon characteristic of independent random events occurring according to a Poisson process. If such events were consistently evenly spaced, we would have good reason to doubt their randomness. Of course, the presence of clustering is not a proof of randomness. Sometimes the reason that "it never rains but it pours" is that the events do *not* have constant probability—people *are* more likely to be on the road, or in the bank, and so on, at certain times of the day. However, independent random events occurring with constant probability also exhibit clustering.

In addition to its use in describing the distribution of independent random events in which the actual number of discrete moments or trials is unknown, the Poisson distribution may also be used as an **approximation to the binomial distribution** when the number n of trials is large—at least about 30—and the probability p of success on each trial is small—less than about 10%. To use this approximation, we approximate the mean λ of the Poisson distribution by the mean of the binomial distribution:

$$\mu_{\text{bin}} = np \tag{7.25}$$

EXAMPLE 7.27 If 1.4% of new university students quit school in the first month of the academic year, what is the probability that exactly 5 students in a freshman class of 250 will quit in the first month?

The question could be answered by the binomial distribution with $n = 250$ and $p = .014$, but since n is larger than 30 and p is less than .10, we may use the Poisson approximation. We first calculate the expected number of withdrawals as

$$\mu_{\text{bin}} = (250)(.014) = 3.5$$

The probability of exactly 5 withdrawals may then be obtained via Expression (7.24) as

$$P_{\text{Poisson}}(5) = \frac{(3.5)^5 e^{-3.5}}{5!} = .132169$$

or approximately 13%.

What is the probability of 2 or more withdrawals in the first month? This probability may be calculated as

$$P(\text{2 or more withdrawals}) = 1.00 - P(\text{0 withdrawals}) - P(\text{1 withdrawal})$$

Since

$$P_{\text{Poisson}}(0) = \frac{(3.5)^0 e^{-3.5}}{0!} = .030197$$

and

$$P_{\text{Poisson}}(1) = \frac{(3.5)^1 e^{-3.5}}{1!} = .105691$$

the required probability is

$$P(\text{2 or more withdrawals}) = 1.00 - .030197 - .105691 = .86$$

7.4 THE NORMAL DISTRIBUTION

As we proceed through this book, we will meet several theoretical probability distributions of continuous random variables. One of these, called the **normal distribution**, has been central to a substantial portion of the theoretical developments in inferential statistics during the past two centuries. Like any other theoretical probability distribution, the normal distribution is a mathematical fiction that may or may not provide an adequate approximation to actual observations of empirical variables. However, it would be hard to overstate its theoretical importance. Its density function was first published in 1733 by Abraham de Moivre (see Biography 7.2), and it was studied further by Pierre Laplace (see Biography 8.1) and especially by Carl Gauss, as Biography 7.1 describes.

BIOGRAPHY 7.1 Carl Friedrich Gauss (1777–1855)

Carl F. Gauss was born Johann Friedrich Carl Gauss on 30 April 1777 in Brunswick (Braunschweig), Germany. His father, who made his living as a bricklayer, gardener, and canal tender, did all he could to keep his son from pursuing higher education rather than a trade as a bricklayer or gardener. His mother, a stonemason's daughter who had learned to read a little but not to write, and his mother's younger brother encouraged the young Gauss to develop his obvious genius. By two years of age, Gauss had taught himself reading and basic arithmetic. Once while watching his father calculate a weekly payroll, the not-yet-three-year-old Gauss noticed an error and gave the correction, subsequently verified by his father.

In his first class in arithmetic at age nine, Gauss correctly solved in an instant a problem that the rest of the boys in the class toiled at unsuccessfully for an hour. The teacher was duly impressed, bought Gauss the best arithmetic text he could find, and encouraged him to study further with the teacher's more mathematically inclined assistant. In 1791, through arrangements initiated by this assistant, the 14-year-old Gauss was introduced to Karl Wilhelm Ferdinand, the Duke of Brunswick, who thereafter provided financial support for Gauss's education and research until the Duke's death in 1806 in the invasion by Napoleon. Gauss studied at Brunswick's Caroline College from 1792 to 1795, where he mastered Greek and Latin as quickly as mathematics. Indeed, until 30 March 1796 he seriously considered specializing in language and literature, but on that date Gauss became the first ever to discover a ruler-and-compass construction of a regular 17-sided polygon, thus solving an age-old problem that had been attempted by centuries of mathematicians. He continued his studies at the University of Göttingen from 1795 to 1798, was awarded a doctorate in mathematics by the University of Helmstadt in 1799, and then pursued further research in mathematical astronomy in Brunswick under the Duke's patronage until he was named director of the observatory at the University of Göttingen in 1807 after the Duke's death. He remained in that position for almost 48 years.

In 1795, at the age of 18, Gauss had devised the method of least squares, thus anticipating its first publication by Legendre by at least 10 years (see Biography 5.2). During the course of his further development of the method of least squares, Gauss studied the properties of the so-called normal distribution. Although the normal probability density function had earlier appeared in the work of de Moivre (see Biography 7.2) and Laplace (see Biography 8.1) as an approximation for other distributions, Gauss was the first to consider it as a probability distribution in its own right and the first to explore its properties. It came to be known as the Gaussian Law of Error, a reference to his use of it to describe the distribution of errors of estimation in the method of least squares.

It would be difficult to exaggerate Gauss's scientific contributions. In addition to revolutionizing the whole of mathematics by his insistence on and provision of rigorous proofs, he is known for his work in astronomy, geodesy, and physics. He devised methods for the accurate prediction of the orbits of planets and comets and for the measurement of the intensity of the earth's magnetism. He invented the electric telegraph and a device for measuring magnetic field intensities, and he made improvements in measuring instruments used in astronomy and geodesy. According to the mathematical historian E. T. Bell, "Archimedes, Newton, and Gauss, these three, are in a class by themselves among the great mathematicians, and it is not for ordinary mortals to attempt to range them in order of merit."

Gauss married Johanne Osthof on 9 October 1805, but she died on 11 October 1809. Left with three young children, Gauss remarried on 4 August 1810 and had three more children. Gauss's early interest in languages remained his lifelong hobby, and he read widely in several European languages, including English, as well as in classical Greek and Latin. At age 62, he began teaching himself Russian—for relaxation, he said—and was reading, writing, and speaking the language fluently within two years. He died in Göttingen on 23 February 1855 at the age of 77.

(Sources: Bell, 1937; Eisenhart, 1978; *World Who's Who in Science,* 1968.)

The term **normal distribution** arose from the observation that the errors in repeated measurements of the same physical quantity tended to exhibit considerable regularity in their frequency distributions. This pattern of errors, referred to as the *normal curve of errors,* closely resembled the bell-shaped curve traced by de Moivre's density function, and his mathematical function came to be known as the *normal distribution* by virtue of its similarity to the normally observed pattern of errors.

It must be emphasized, however, that there is nothing particularly normal about the normal distribution; nor is there anything abnormal about other distributions. For example, it is just as normal for some variables to follow the binomial distribution as it is for others to follow the normal distribution. The term *normal distribution* is an historical accident—or even a misnomer—in the same way that the term *regression* is: In both cases the terminology arose from the limited perspective of an early application. The normal distribution is also called the **Gaussian distribution**, and although this label is not as prone to misinterpretation as the term *normal distribution,* neither is it yet as widely used. In this book, the terms *normal* and *Gaussian* will be used interchangeably. In contexts where *normal* seems liable to be incorrectly taken in its everyday sense of regular or typical, we will use the term *Gaussian* in order to emphasize that we are simply referring to the shape of a particular theoretical distribution and not assessing its prevalence or appropriateness. Regardless of its name, the normal distribution is a theoretical distribution that exists only as a mathematical formula and not (exactly) in nature.

There are at least two good reasons for the dominant position that the normal or Gaussian distribution has enjoyed in inferential statistics. First, there are some empirical variables besides measurement errors whose distributions are approximately Gaussian. Second, and more important, the normal distribution has certain convenient mathematical properties such that it is useful even when the empirical variables are clearly not Gaussian (see, for example, the Central Limit Theorem in Section 8.4).

The Normal Probability Density Function

The normal distribution may be represented by a bell-shaped curve. However, there are many shapes of bells, and only some of them may be said to have a Gaussian shape. The equation of the line that traces a Gaussian bell is the **normal probability density function**:

$$f_{\text{Gauss}}(x) = \frac{1}{\sigma\sqrt{2\pi}} e^{-(x-\mu)^2/2\sigma^2} \tag{7.26}$$

where

$f_{\text{Gauss}}(x)$ is the probability density of a continuous random variable x;

$\mu = E(x)$ is a location parameter (measure of location), namely, the mean or expected value of x;

$\sigma^2 = E(x - \mu)^2$ is a dispersion parameter (measure of dispersion), namely, the variance or squared standard deviation of x;

$e = 2.71828\ldots$ is a mathematical constant, best known as the base of natural logarithms; and

$\pi = 3.14159\ldots$ is another mathematical constant.

Although the equation may look rather formidable at first glance, the calculation is not difficult with the aid of a hand-held calculator. (A tabulation of values of e^a is provided in Appendix 9 for various values of a.) In fact, we will not have much occasion to calculate numerical values of the density function $f_{\text{Gauss}}(x)$, but you may find it instructive to do so at least once to convince yourself that a sketch of the normal curve is neither arbitrary nor magic but rather results from substituting values of x into Expression (7.26) for given values of the mean μ and standard deviation σ and plotting the results.

EXAMPLE 7.28 What do Gaussian distributions look like? Figure 7.13 shows the graphs of the normal probability density function, given in Expression (7.26), for selected values of the mean

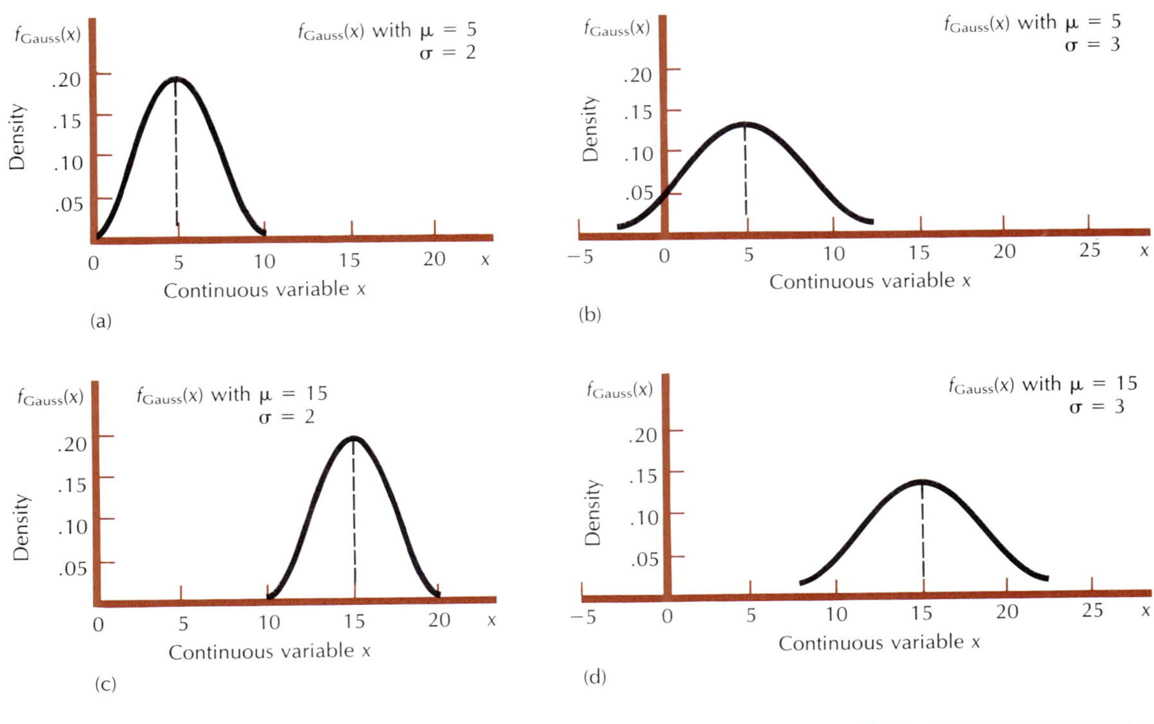

Figure 7.13
Normal probability density functions for selected values of μ and σ.

μ and the standard deviation σ. Notice in the figure that changing the mean merely shifts the whole distribution to the left or right along the x-axis; that is, it alters its location. Changing the standard deviation changes how spread out the distribution is over the x-axis; the accompanying change in height of the distribution is a by-product of changing the dispersion, because the total area under a probability density function is by definition equal to 1 (see Section 7.1, pages 233–235).

The random variable x in a Gaussian distribution is continuous from minus-infinity to plus-infinity. Thus, we can substitute any real number x, positive or negative or zero, into the density function in Expression (7.26) to calculate the associated value of $f_{\text{Gauss}}(x)$ for a given mean and standard deviation. Examination of (or, if you prefer, experimentation with) Expression (7.26) reveals that whether x is positive or negative, large or small, $f_{\text{Gauss}}(x)$ is always positive; that is, the curve never dips below, nor even touches, the horizontal axis, as illustrated in Figure 7.13.

Likewise, we can see from Expression (7.26) that regardless of the value of x, the exponent of e is never positive, and hence the value of $f_{\text{Gauss}}(x)$ attains its maximum when $x = \mu$ (in which case the exponent of e is 0). This implies that in a Gaussian distribution the mode equals the mean.

The normal distribution is said to be completely specified by two parameters, namely, its mean μ and its standard deviation σ. Thus, if a variable is known to be Gaussian, knowledge of its mean and its standard deviation is sufficient to enable us to graph its density function and, as we shall see shortly, to determine the probability of obtaining various sorts of values of the random variable.

The Standard Normal Distribution

In practice, we rarely need to use Expression (7.26) directly. Generally it is not density (height of the curve at a particular point) that is of interest in elementary statistical applications but rather probability (area under a segment of the curve), and tables of the relevant areas are readily available. There are an infinite number of different normal distributions that could be tabulated, one for each combination of possible values of the mean μ and the standard deviation σ. However, any value of a variable x with known mean and standard deviation can be converted to a *standard score* or *z-score* (see Section 4.4):

$$z_i = \frac{x_i - \mu}{\sigma} \tag{7.27}$$

It follows that tabulations of the normal distribution in terms of standard scores are sufficient for any combination of mean and standard deviation.

As we saw in Section 4.4, a set of observations expressed in standard units necessarily has a mean of 0 and a standard deviation of 1. Hence, the normal distribution tabulated in terms of standard scores—which is called the **standard normal distribution**—has the mean $\mu = 0$ and the standard deviation $\sigma = 1$. In this context, the standard scores z_i are often referred to as **standard normal deviates**, since a z-score is simply the number of standard deviation units by which an observation deviates from the mean.

When the random variable is expressed in standard units, Expression (7.26) simplifies to the *standard normal density function*:

$$f_{\text{Gauss}}(z) = \frac{1}{\sqrt{2\pi}} e^{-z^2/2} \simeq (.3989)e^{-z^2/2} \tag{7.28}$$

which is graphed in Figure 7.14.

An important property of any Gaussian distribution is its symmetry about the mean. This symmetry results from the squaring of z in Expression (7.28), such that for any value of z, both $+z$ and $-z$ yield the same value of $f_{\text{Gauss}}(z)$, and it implies that each point to the left of the mean is the mirror image of the corresponding point to the right of the mean. Thus, 50% of the area under the curve is to the right of the mean, and 50% is to the left of the mean. The value

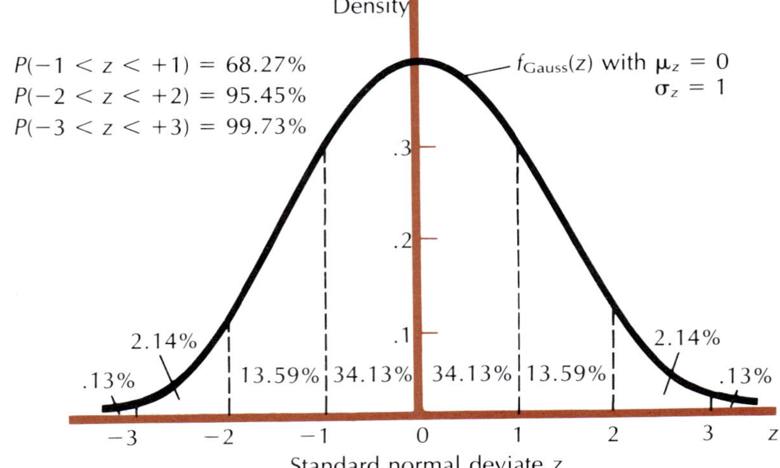

Figure 7.14 The standard normal density function, showing selected percentage areas (rounded to two decimal places) under the density curve.

that partitions a distribution into equal halves is, by definition, the *median*, so we know that in a Gaussian distribution the median equals the mean, which, as noted previously, also equals the *mode*. In other words,

$$\text{Mean} = \text{median} = \text{mode} \qquad \text{for any Gaussian distribution} \tag{7.29}$$

As a consequence of the symmetry of the normal distribution, the probability that an observation randomly selected from a normal distribution is greater than the mean of the distribution is .50; similarly, the probability that such an observation is less than the mean is .50.

The probabilities required in most applications of the normal distribution can be obtained through the use of the tabulation of the standard normal distribution in Appendix 3. Appendix 3 lists various values of a standard normal deviate z and gives two probabilities (that is, two areas under the density curve) associated with each z-value. First, the value $F_{\text{Gauss}}(z_A)$ of the distribution function (see Section 7.1, pages 235–238) is the probability that a randomly selected standard normal deviate will be less than some specified value z_A. In symbols,

$$F_{\text{Gauss}}(z_A) = P(z < z_A) \tag{7.30}$$

Second, the upper-case A identifying the upper-tail area represents the probability that a randomly selected standard normal deviate will be greater than z_A. In symbols,

$$A = P(z > z_A) \tag{7.31}$$

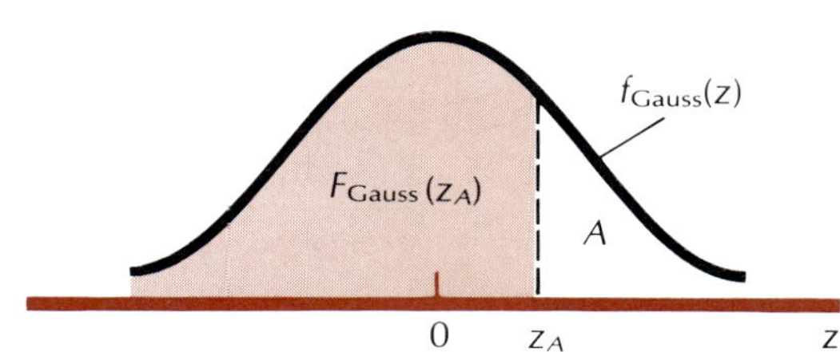

Figure 7.15 Standard normal distribution, with the shaded area representing the probability of a standard normal deviate less than z_A and the unshaded area representing the probability of a standard normal deviate greater than z_A.

In other words, the notation z_A stands for a z-value that cuts off an upper-tail area of A, leaving a lower-tail area of

$$F_{\text{Gauss}}(z_A) = 1 - A \tag{7.32}$$

These **areas of the standard normal distribution**, $F_{\text{Gauss}}(z_A)$ and A, shown in Figure 7.15, represent the probabilities specified in Expressions (7.30) and (7.31), respectively. Appendix 3 is a tabulation of such areas under the standard normal density curve. [Be careful to distinguish between the *distribution function* $F(z)$, which gives the area to the left of a given value of z, and the *density function* $f(z)$, which gives the height of the curve at a given value of z.]

EXAMPLE 7.29 What is the probability that a randomly selected standard normal deviate will be less than 2.00? From Appendix 3, we find that

$$P(z < 2.00) = F_{\text{Gauss}}(2.00) = .9772$$

which is the required probability, as shown in Figure 7.16(a).

What is the probability that an observation randomly selected from a Gaussian

Figure 7.16

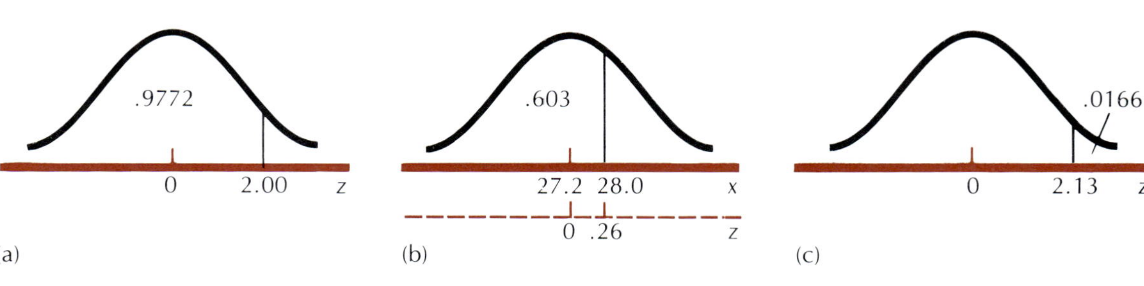

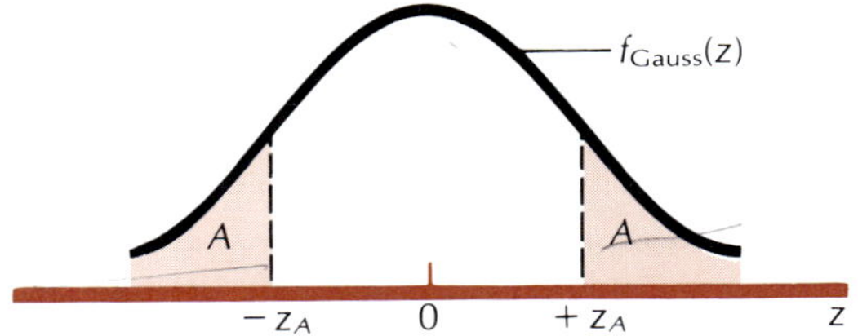

Figure 7.17 Standard normal distribution, with equal tail areas shaded.

distribution with a mean of 27.2 and a standard deviation of 3.1 will be less than 28.0? A value of 28.0 in the given distribution corresponds to a value of

$$z = \frac{28.0 - 27.2}{3.1} = .26$$

in a standard normal distribution according to Expression (7.27). The required probability, illustrated in Figure 7.16(b), is

$$P(z < .26) = F_{Gauss}(.26) = .603$$

What is the probability that a randomly selected standard normal deviate will be greater than 2.13? From Appendix 3, the required probability is the upper-tail area A corresponding to a z-value of 2.13, as shown in Figure 7.16(c):

$$P(z > 2.13) = .0166$$

In answering questions such as those in this example, you will often find it useful to draw a rough sketch similar to those in Figure 7.16 to help you visualize the area (probability) that is being calculated.

We can use the symmetry of the normal distribution to calculate the value of the distribution function for a negative value of z. If z_A cuts off an upper-tail area of A, then by symmetry we know that $-z_A$ cuts off a *lower-tail* area of A. Since values of the distribution function $F_{Gauss}(z)$ represent lower-tail areas, we have

$$F_{Gauss}(-z_A) = A \tag{7.33}$$

as illustrated in Figure 7.17.

EXAMPLE 7.30 What is the probability that a randomly selected standard normal deviate will be less than −1.37? The question asks about a left-hand tail area as illustrated in Figure 7.17. Because of the symmetry of the Gaussian distribution, we know that

$$P(z < -1.37) = P(z > 1.37)$$

Thus, we can obtain the required probability by looking up the corresponding upper-tail area A in Appendix 3,

$$P(z < -1.37) = .085$$

as shown in Figure 7.18(a).

What is the probability that a randomly selected standard normal deviate will be

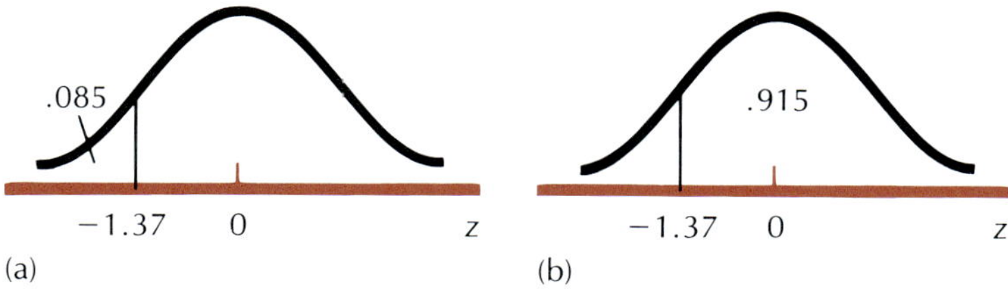

Figure 7.18

greater than −1.37? This question is the complement of the previous one. By symmetry, we know that the area to the right of −1.37 must equal the area to the left of +1.37, and the latter area is given in Appendix 3. Thus, the probability is

$$P(z > -1.37) = .915$$

as illustrated in Figure 7.18(b).

To determine the probability that a standard normal deviate has a value between any two values z_i and z_j, we obtain the distribution-function values $F_{\text{Gauss}}(z_i)$ and $F_{\text{Gauss}}(z_j)$ and subtract the smaller from the larger:

$$P(z_i < z < z_j) = F_{\text{Gauss}}(z_j) - F_{\text{Gauss}}(z_i) \qquad \text{for } z_i < z_j \tag{7.34}$$

This relationship is illustrated in Figure 7.19.

EXAMPLE 7.31 What is the probability that a randomly selected standard normal deviate will be between .50 and 2.50? We begin by drawing a rough sketch similar to that in Figure 7.19, labeling $z = 0$ as a reference point and $z = .50$ and $z = 2.50$, to help visualize the area we are calculating. Using Expression (7.34) and the table in Appendix 3, we find that

$$\begin{aligned} P(.50 < z < 2.50) &= F_{\text{Gauss}}(2.50) - F_{\text{Gauss}}(.50) \\ &= .9938 - .691 \\ &= .303 \end{aligned}$$

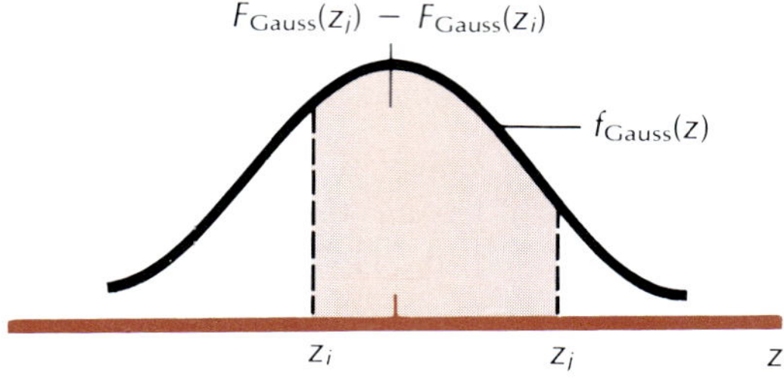

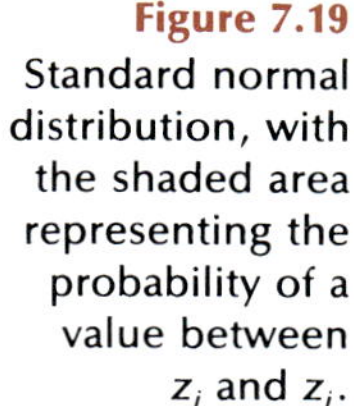

Figure 7.19 Standard normal distribution, with the shaded area representing the probability of a value between z_i and z_j.

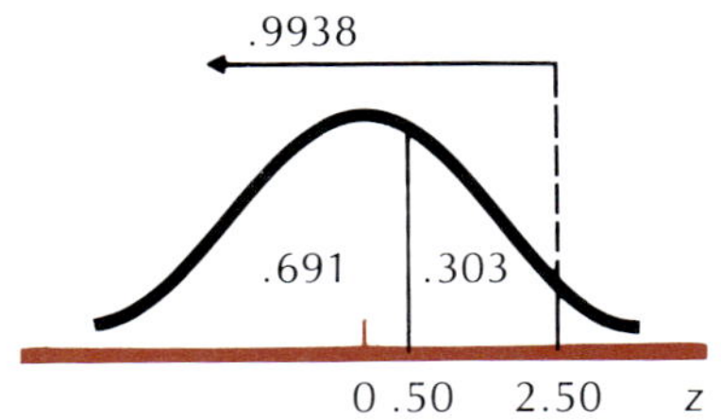

Figure 7.20

as shown in Figure 7.20. (Since the value .691 has been rounded to three decimal places, the answer is reliable to no more than three decimal places. Guidelines for how to round and how many digits to report are given in Appendix 1.)

EXAMPLE 7.32 Consider a population such as that in Example 7.1, in which adult male heights are approximately normally distributed with a mean of 67.46 inches and a standard deviation of 2.56 inches. What is the probability that a randomly selected observation from such a distribution will be between 66.00 and 72.00 inches?

We first convert the lower and upper critical values 66.00 and 72.00 to standard units:

$$z_{\text{lower}} = \frac{66.00 - 67.46}{2.56} = -.57$$

and

$$z_{\text{upper}} = \frac{72.00 - 67.46}{2.56} = 1.77$$

As shown in Figure 7.21, we label these values on a rough sketch of a normal

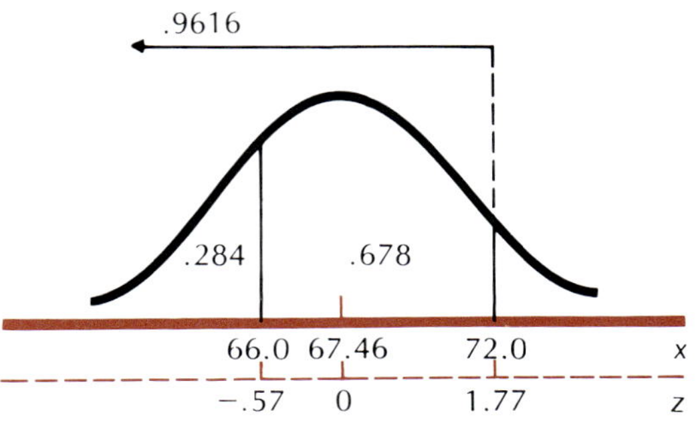

Figure 7.21

distribution to visualize the relevant area. We can then see that the required probability is obtained via Expressions (7.33) and (7.34) as

$$\begin{aligned} P(-.57 < z < 1.77) &= F_{\text{Gauss}}(1.77) - F_{\text{Gauss}}(-.57) \\ &= .9616 - .284 \\ &= .678 \end{aligned}$$

(The fourth decimal place in the answer would not be meaningful, because of the rounding of the input value .284.) We conclude that approximately 67.8% of the heights in the given population would be between 66.00 and 72.00 inches.

You may find it instructive to use the method of this example to verify the three probabilities given at the left-hand side of Figure 7.14.

Sometimes it is necessary to determine the value of the standard normal deviate associated with a given cumulative probability or tail area. We do so by entering the table in Appendix 3 with the appropriate value of $F_{\text{Gauss}}(z)$ or of A, and then reading off the corresponding value of z.

EXAMPLE 7.33 What z-score exceeds 67% of the observations in a standard normal distribution? In other words, given $F_{\text{Gauss}}(z) = .67$, what is the value of z? From the table in Appendix 3, we find that the required standard normal deviate is $z = .44$, as shown in Figure 7.22(a).

What z-score cuts off the top 15% of a standard normal distribution? In other words, given $A = .15$, what is the value of z? From Appendix 3 we read the required value of $z = 1.04$, as shown in Figure 7.22(b).

EXAMPLE 7.34 In a population such as that in Example 7.1, in which heights of adult males are approximately normally distributed with a mean of 67.46 inches and a standard deviation of 2.56 inches, what height separates the shortest 10% of the population from the tallest 90%? In other words, given $F_{\text{Gauss}}(z) = .10$, what is the value of z? Using Expression (7.33) and the table in Appendix 3, we find that the required z-score is -1.28, as illustrated in Figure 7.23. Solving for x in Expression (7.27), we obtain a height of

$$x = 67.46 - 1.28(2.56) = 64.18 \text{ inches}$$

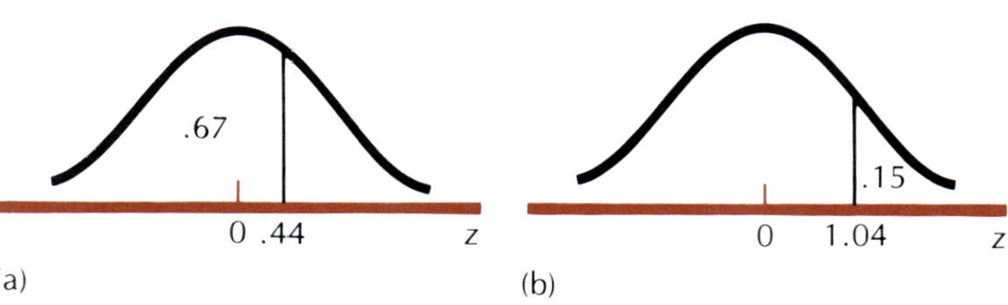

Figure 7.22

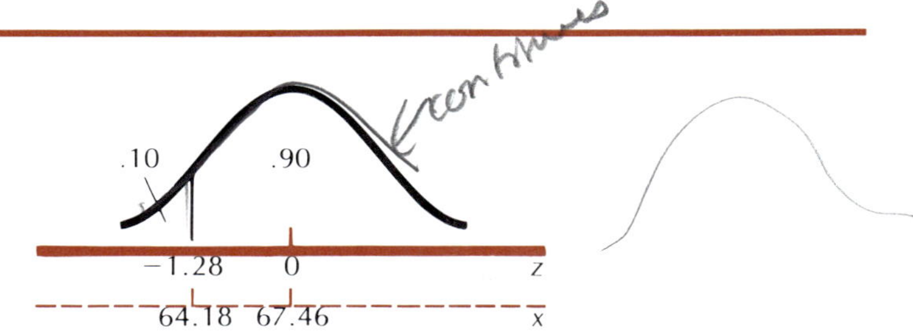

Figure 7.23

The Gaussian Approximation to Discrete Distributions

Recall from Section 7.1 that a discrete variable is one whose possible values are disconnected points on the real-number line, such as number of heads in 10 flips of a fair coin. A continuous variable, on the other hand, is one whose possible values include all the real numbers in an uninterrupted region between two of its values, such as a person's height.

The normal or Gaussian distribution is a continuous distribution: Its random variable can take on any value on the real-number line. Continuous variables observed in empirical investigations are probably never perfectly Gaussian, but the Gaussian distribution provides adequate approximations in many situations. The Gaussian distribution may also provide a satisfactory approximation in the case of certain discrete variables, if an adjustment, called the **continuity correction**, is made to compensate for the fact that a continuous distribution, which assigns probabilities (totaling 1.0) to every interval on the real-number line, is being used to approximate a discrete distribution, which "uses up" its total probability of 1.0 on a series of disconnected points.

The continuity correction is a convenient and reasonably accurate way of associating the disconnected values of a discrete variable with connected intervals of a continuous variable, so that the probabilities associated with those intervals can in turn be associated with the disconnected values of the discrete variable. The **continuity correction** consists in treating each value of the discrete variable as if it covered the interval from halfway down to the next smaller discrete value to halfway up to the next larger discrete value. In the terminology of Chapter 2, the continuity correction consists in treating each value of the discrete variable as a category covering the interval from one category boundary to the next.

EXAMPLE 7.35 Over a period of six months, a smoker kept a record of the number of cigarettes she smoked per day. Her daily consumption was approximately normally distributed with a mean of 18.7 cigarettes and a standard deviation of 2.4 cigarettes. What is the probability that she will smoke 21 or more cigarettes on a particular day?

To obtain a probability from the normal distribution in this situation, we must use the continuity correction, because number of cigarettes is a discrete variable. As shown in Figure 7.24(a), we treat the discrete value 21 as a category with boundaries 20.5 and

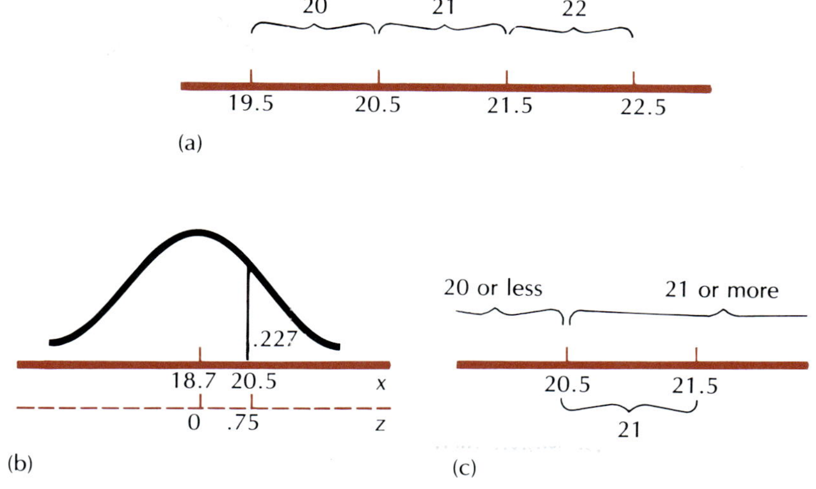

Figure 7.24

21.5, so that "21 or more" is interpreted to mean "20.5 or more." Converting 20.5 to standard units, we must find the probability that a standard normal deviate is greater than

$$z = \frac{20.5 - 18.7}{2.4} = .75$$

The required probability is found in Appendix 3 to be

$$P(z > .75) = .227$$

or approximately 23%, as shown in Figure 7.24(b).

In this example, the events "21 or more" and "20 or less" are mutually exclusive and exhaustive. Thus, we must have

$$P(21 \text{ or more}) + P(20 \text{ or less}) = 1$$

This requirement is met by using the continuity correction: The discrete value 21 is considered to be a category with boundaries 20.5 and 21.5, and the discrete value 20 is considered to be a category with boundaries 19.5 and 20.5. Thus, "21 or more" is taken as "20.5 or more" and "20 or less" is taken as "20.5 or less," so that these two exhaustive events account for the entire area under the normal density curve, as illustrated in Figure 7.24(c).

One important discrete distribution that may be approximated by the normal distribution is the binomial distribution. In fact, it was in this context that the normal distribution was first proposed by Abraham de Moivre (see Biography 7.2) and studied by Pierre Laplace (see Biography 8.1).

BIOGRAPHY 7.2 **Abraham de Moivre (1667–1754)**

Abraham de Moivre was born in Vitry, Champagne, France, on 26 May 1667. He studied Greek at the University of Sedan in France from 1678 to 1682, logic at Namur in Belgium from 1682 to 1683, and mathematics and physics at Collège d'Harcourt in Paris from 1684 to 1685. After the 1685 revocation of the Edict of Nantes, which had granted toleration to Protestants, de Moivre first went into seclusion in Paris and then fled to England in 1688 at the age of 21 in order to escape religious persecution. He settled in London and, according to H. L. Seal, made his living "by tutoring noblemen's sons and by advising gamblers and speculators who dealt in annuities, which were a popular form of investment in the first half of the eighteenth century."

De Moivre wrote several books, one of which was an important text in probability theory that included an original treatment of the general rules for the addition and multiplication of probabilities (see Section 6.3). A second important textbook dealt with the calculation of annuities and led to his recognition as the founder of the science of life contingencies. Yet another far-reaching contribution came on 12 November 1733. On that date, de Moivre introduced the normal density function as an approximation to the binomial distribution.

De Moivre died in London on 27 November 1754 at the age of 87.

(Sources: Seal, 1978; Stigler, 1982; *World Who's Who in Science,* 1968.)

The binomial probability gives the probability of exactly n_1 successes in n independent trials in which the probability of success on each individual trial is p. However, numerical evaluation of the function can be tedious, especially as n becomes large. Although number of successes is a discrete variable, it can be shown that as n becomes large, the binomial distribution for a given value of p approaches the shape of the normal distribution with mean $\mu = np$ and standard deviation $\sigma = \sqrt{npq}$. Thus, we can use the normal distribution with the continuity correction to obtain binomial probabilities if n is large enough. The result is called the **Gaussian approximation to the binomial distribution.** There is no magic value of n to define precisely what values of n are "large enough," but experience suggests that the following is a reasonable **rule of thumb for use of the Gaussian approximation**:

The Gaussian approximation to the binomial distribution may be used when both np and nq are greater than about 5. **(7.35)**

EXAMPLE 7.36 As noted in Example 7.15, approximately 21% of North American deaths are due to cancer. What is the probability that 7 or more of the next 20 unrelated deaths reported in a certain county will be due to cancer?

The required probability may be obtained by repeated application of the binomial probability function to evaluate

$$P(7 \text{ or more}) = P_{\text{bin}}(7) + P_{\text{bin}}(8) + \cdots + P_{\text{bin}}(20)$$

which, after considerable calculation, turns out to equal .107.

Alternatively, we may consider the Gaussian approximation to the binomial distribution with mean

$$\mu_{\text{bin}} = (20)(.21) = 4.2$$

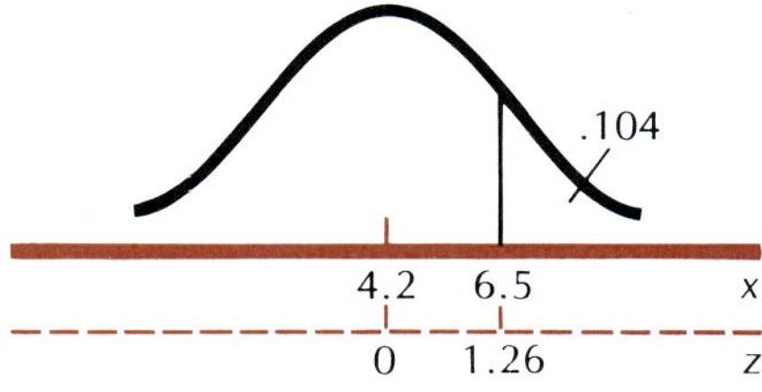

Figure 7.25

and standard deviation

$$\sigma_{\text{bin}} = \sqrt{(20)(.21)(.79)} = 1.8$$

Is n large enough to use the approximation? Not really. The value of np is only 4.2, which is less than the value of 5 suggested as a minimum by the rule of thumb in Expression (7.35). Let us use the approximation anyway for illustration, recognizing that the answer may be rather approximate because of the small value of np.

For the Gaussian approximation, we treat the discrete value 7 as a category with boundaries 6.5 and 7.5. The event "7 or more" is therefore interpreted as "6.5 or more," and we require the z-score corresponding to 6.5 deaths:

$$z = \frac{6.5 - 4.2}{1.82} = 1.26$$

From Appendix 3, we find that the probability of a standard normal deviate greater than 1.26 is

$$P(z > 1.26) = .104$$

as illustrated in Figure 7.25. Thus, the probability of 7 or more deaths due to cancer out of the next 20 deaths is .104, according to the Gaussian approximation to the binomial.

Notice that the Gaussian approximation of .104 is very close to the exact binomial value of .107, even though the rule of thumb in Expression (7.35) did not sanction its use, because of the small value of np. This result illustrates that the rule of thumb should be taken not as a fixed cutoff for the use of the Gaussian approximation but rather as a warning that the approximation gradually becomes less satisfactory as either np or nq drops below about 5.

It should be emphasized that the continuity correction is not required for all applications of the normal distribution, but only for those in which the normal distribution, which is continuous, is being used as an approximation for the distribution of a discrete variable (for example, one whose values are limited to the integers). No continuity correction is used for continuous variables, where precision is limited by rounding rather than by the nature of the variables.

STUDY MATERIAL FOR CHAPTER 7

Terms and Concepts

Section 7.1

probability distribution
empirical probability distribution
theoretical probability distribution
random variable
discrete variable
continuous variable
discrete probability distribution
continuous probability distribution
probability function
probability density function
uniform density
uniform distribution
rectangular distribution
distribution function
mathematical expectation
expected value
mean of a probability distribution
variance of a probability distribution
standard deviation of a probability distribution

Section 7.2

binomial distribution
success
failure
Bernoulli trial
binomial probability function
number of combinations of n items taken n_1 at a time
question answered by the binomial probability function
binomial-distribution requirements
mean of a binomial distribution
variance of a binomial distribution
standard deviation of a binomial distribution

* Section 7.3

multinomial distribution
multinomial probability function

sampling with replacement
hypergeometric distribution
sampling without replacement
hypergeometric probability function
multinomial approximation to the hypergeometric distribution
Poisson distribution
Poisson probability function
Poisson approximation to the binomial distribution

Section 7.4

normal distribution
Gaussian distribution
normal probability density function
standard normal distribution
standard normal deviate
areas of the standard normal distribution
continuity correction
Gaussian approximation to the binomial distribution
rule of thumb for use of the Gaussian approximation

Key Formulas

Section 7.1

(7.9) $E(x) = \sum x_i p_i$

(7.10) $\mu = E(x)$

(7.11) $\sigma^2 = E(x - \mu)^2$

Section 7.2

(7.13) $P_{\text{bin}}(n_1) = \dfrac{n!}{n_1!\, n_2!} p^{n_1} q^{n_2}$

(7.14) $q = 1 - p$

(7.17) $\mu_{\text{bin}} = np$

(7.19) $\sigma_{\text{bin}} = \sqrt{npq}$

* Section 7.3

(7.21) $P_{\text{mult}}(n_1, n_2, \ldots, n_k) = \dfrac{n!}{n_1!\, n_2! \cdots n_k!} p_1^{n_1} p_2^{n_2} \cdots p_k^{n_k}$

(7.22) $P_{\text{hyper}}(n_1, n_2, \ldots, n_k) = \dfrac{\binom{N_1}{n_1}\binom{N_2}{n_2}\cdots\binom{N_k}{n_k}}{\binom{N}{n}}$

(7.24) $P_{\text{Poisson}}(n_1) = \dfrac{\lambda^{n_1} e^{-\lambda}}{n_1!}$

Section 7.4

(7.27) $$z_i = \frac{x_i - \mu}{\sigma}$$

Exercises for Numerical Practice

Section 7.1

7/1. A random variable x has possible values $x = 1, 2, 3, 4$. Indicate whether or not each of the following is a probability function for such a variable, and if not, why not:

a. $P(1) = .05$ $P(2) = .35$ $P(3) = .55$ $P(4) = .15$
b. $P(1) = 1/8$ $P(2) = 1/2$ $P(3) = 1/4$ $P(4) = 1/16$
c. $P(1) = 13\%$ $P(2) = 69\%$ $P(3) = 1\%$ $P(4) = 17\%$
d. $P(1) = 51$ $P(2) = 12$ $P(3) = 34$ $P(4) = 3$
e. $P(1) = .66$ $P(2) = .32$ $P(3) = -.19$ $P(4) = .21$

7/2. Indicate whether or not each of the following is a probability function, and if not, why not:

a. $P(x) = x/4$, for $x = 1, 2, 3, 4$
b. $P(x) = x/10$, for $x = 0, 1, 2, 3, 4$
c. $P(x) = 1/3$, for $x = 0, 1, 2, 3$
d. $P(x) = x/2$, for $x = -1, 0, 1, 2$
e. $P(x) = x^2$, for $x = -.7, -.6, \sqrt{.15}$
f. $P(x) = 1/5$, for $x = 3, 4, 5$

7/3. The probabilities associated with the four possible values of a random variable x are as follows:

$$P(1) = .29, \quad P(2) = .53, \quad P(3) = .02, \quad P(4) = .16$$

Calculate $F(1)$, $F(2)$, $F(3)$, and $F(4)$.

7/4. The probabilities associated with the five possible values of a random variable x are given by $P(x) = x/10$, for $x = 0, 1, 2, 3, 4$. Calculate $F(0)$, $F(1)$, $F(2)$, $F(3)$, and $F(4)$.

7/5. Calculate the expected value of the variable in Exercise 7/3.

7/6. Calculate the expected value of the variable in Exercise 7/4.

7/7. Given the probability distribution defined in Exercise 7/3, calculate the following:

a. the mean.
b. the variance.
c. the standard deviation.

7/8. Given the probability distribution defined in Exercise 7/4, calculate the following:

a. the mean.
b. the variance.
c. the standard deviation.

7/9. A discrete probability distribution is defined by

$$P(x) = (x^2 - x)/20 \qquad \text{for } x = 0, 1, 2, 3, 4$$

Calculate the following:

a. the mean.
b. the variance.
c. the standard deviation.

7/10. A discrete probability distribution is defined by

$$P(x) = (x^2 + x + 12)/100 \qquad \text{for } x = 0, 1, 2, 3, 4$$

Calculate the following:

a. the mean.
b. the variance.
c. the standard deviation.

Section 7.2

7/11. For a binomial distribution with $n = 4$ trials and probability $p = 1/3$ for each trial, calculate the following:

a. $P_{\text{bin}}(0)$ **b.** $P_{\text{bin}}(1)$ **c.** $P_{\text{bin}}(2)$
d. $P_{\text{bin}}(3)$ **e.** $P_{\text{bin}}(4)$ **f.** $F_{\text{bin}}(3)$

7/12. For a binomial distribution with $n = 3$ trials and probability $p = .4$ for each trial, calculate the following:

a. $P_{\text{bin}}(0)$ **b.** $P_{\text{bin}}(1)$ **c.** $P_{\text{bin}}(2)$
d. $P_{\text{bin}}(3)$ **e.** $F_{\text{bin}}(2)$ **f.** $F_{\text{bin}}(3)$

7/13. What is the probability of exactly

a. 0 successes in 6 Bernoulli trials with $p = \frac{1}{8}$?
b. 6 successes in 9 Bernoulli trials with $p = \frac{2}{3}$?

7/14. What is the probability of exactly

a. 5 successes in 5 Bernoulli trials with $p = \frac{5}{6}$?
b. 2 successes in 7 Bernoulli trials with $p = \frac{1}{4}$?

7/15. For a binomial distribution with $n = 900$ and $p = .8$, calculate the following:

a. the mean.
b. the standard deviation.

7/16. For a binomial distribution with $n = 400$ and $p = .9$, calculate the following:

a. the mean.
b. the standard deviation.

7/17. For a binomial distribution where there are 1600 trials and the probability of success on each trial is .02, calculate the following:

a. the mean.
b. the standard deviation.

7/18. For a binomial distribution where there are 25 trials and the probability of success on each trial is .36, calculate the following:

a. the mean.
b. the standard deviation.

* Section 7.3

7/19. Evaluate the multinomial probability function

a. where $n_1 = 1$, $n_2 = 3$, $n_3 = 1$, and $p_1 = .4$, $p_2 = .5$, $p_3 = .1$.
b. where $n_1 = 4$, $n_2 = 1$, $n_3 = 0$, $n_4 = 2$, and $p_1 = .3$, $p_2 = .2$, $p_3 = .1$, $p_4 = .4$.

7/20. Evaluate the multinomial probability function

a. where $n_1 = 0$, $n_2 = 4$, $n_3 = 2$, and $p_1 = .1$, $p_2 = .3$, $p_3 = .6$.
b. where $n_1 = 2$, $n_2 = 0$, $n_3 = 5$, $n_4 = 0$, and $p_1 = .40$, $p_2 = .03$, $p_3 = .50$, $p_4 = .07$.

7/21. Evaluate the hypergeometric probability function

a. where $n_1 = 3$, $n_2 = 1$, and $N_1 = 12$, $N_2 = 3$.
b. where $n_1 = 1$, $n_2 = 5$, $n_3 = 0$, and $N_1 = 2$, $N_2 = 7$, $N_3 = 1$.

7/22. Evaluate the hypergeometric probability function

a. where $n_1 = 0$, $n_2 = 4$, and $N_1 = 4$, $N_2 = 5$.
b. where $n_1 = 2$, $n_2 = 0$, $n_3 = 6$, and $N_1 = 2$, $N_2 = 2$, $N_3 = 8$.

7/23. What is the probability of $n_1 = 3$ successes in a Poisson distribution

a. with a mean of 2.0?
b. with a mean of 4.0?

7/24. What is the probability of $n_1 = 4$ successes in a Poisson distribution

a. with a mean of 3.0?
b. with a mean of 2.5?

Section 7.4

7/25. Find the following areas of the standard normal distribution:

a. $P(z < 2.21)$
b. $P(z > .15)$
c. $P(1.08 < z < 2.40)$
d. $P(|z| > .89)$
e. $P(|z| < 1.68)$
f. $P(z < -1.38)$
g. $P(z > -.13)$
h. $P(-1.61 < z < -1.43)$
i. $P(|z| > 2.42)$
j. $P(|z| < 3.33)$

7/26. Find the following areas of the standard normal distribution:

a. $P(z < 1.98)$
b. $P(z > 1.80)$
c. $P(.19 < z < 2.54)$
d. $P(|z| > 1.93)$
e. $P(|z| < .50)$
f. $P(z < -.27)$
g. $P(z > -1.77)$
h. $P(-2.24 < z < -.64)$
i. $P(|z| > 1.14)$
j. $P(|z| < 3.50)$

* **7/27.** What answer would the Bienaymé-Chebyshev inequality [Expression (4.18)] give to Exercise 7/25(e)? Why does it differ from the original answer?

* **7/28.** What answer would the Bienaymé-Chebyshev inequality [Expression (4.18)] give to Exercise 7/26(d)? Why does it differ from the original answer?

7/29. Use Appendix 3 to find the values of the following standard normal deviates, where the subscript indicates the upper-tail area:

a. $z_{.09}$ **b.** $z_{.0001}$ **c.** $z_{.005}$ **d.** $z_{.50}$
e. $z_{.20}$ **f.** $z_{.997}$ **g.** $z_{.99}$ **h.** $z_{.9994}$

7/30. Use Appendix 3 to find the values of the following standard normal deviates, where the subscript indicates the upper-tail area:

a. $z_{.004}$ **b.** $z_{.10}$ **c.** $z_{.0003}$ **d.** $z_{.75}$
e. $z_{.9995}$ **f.** $z_{.05}$ **g.** $z_{.999}$ **h.** $z_{.97}$

7/31. What z-score in a standard normal distribution

a. exceeds 97.5% of the observations?
b. exceeds 20% of the observations?
c. is exceeded by 5% of the observations?
d. is exceeded by 75% of the observations?

7/32. What z-score in a standard normal distribution

a. exceeds 99.5% of the observations?
b. exceeds 2% of the observations?
c. is exceeded by 1% of the observations?
d. is exceeded by 50% of the observations?

7/33. In a normal distribution with a mean of 32.1 and a standard deviation of 9.0, what proportion of the area is

a. greater than 35.0?
b. greater than 5.0?
c. less than 25.0?
d. less than 45.0?
e. between 20.0 and 40.0?

7/34. In a normal distribution with a mean of 484 and a standard deviation of 121, what proportion of the area is

a. greater than 750?
b. greater than 450?
c. less than 400?
d. less than 600?
e. between 350 and 650?

Problems for Applying the Concepts

7/35. The following variables were referred to in the problems for Chapter 3. Identify each as discrete or continuous:

a. percentage butterfat content of milk.
b. amount of waitress's tip, in cents.
c. average hay yield of a field, in bales per acre.
d. number of letters received.

e. maximum daily temperature.

f. distance home for college students.

7/36. The following variables were referred to in the problems for Chapter 4. Identify each as discrete or continuous:

a. number of posters in a college residence room.

b. number of seconds to run 100 meters.

c. resting pulse rate, calculated as the average number of beats per minute.

d. potato mass in grams.

e. crop yield in bushels per acre.

f. breakfast attendance at a cafeteria.

7/37. A car dealer sells no used cars on 60% of her working days, just one used car on 35% of her working days, and two used cars on the remaining 5% of her working days. What is the mathematically expected number of used cars sold per day?

7/38. A private music teacher has no students absent in 55% of his teaching weeks, exactly one student absent in 30% of the weeks, two absences in 10% of the weeks, and three absences in the remaining 5% of the weeks. What is the mathematically expected number of absences per week?

7/39. The University of Alberta uses a 9-point grading system. The following table shows the verbal labels associated with each of the nine possible grades, along with the probability that a first-year student will obtain each grade in a given course.

Grade	Approximate percentage of first-year students' grades
9 (Excellent)	4%
8 (Excellent)	13%
7 (Good)	22%
6 (Good)	24%
5 (Pass)	18%
4 (Pass)	11%
3 (Fail)	4%
2 (Fail)	3%
1 (Fail)	1%

a. Identify the random variable, and state whether it is discrete or continuous.

b. The table shows values of which of the following: probability function, probability density function, distribution function?

c. What is the expected value of the random variable?

d. Calculate the mean, variance, and standard deviation of this probability distribution.

e. Find the numerical value of $F(3)$, and state in words what it represents.

f. Find the numerical value of $1 - F(7)$, and state in words what it represents.

(Source: University of Alberta, *Registration Procedures,* Winter Session 1979–80.)

7/40. The following table shows the cash prizes offered in a magazine publisher's promotional sweepstakes, in which one entry had the stated probabilities of winning the various prizes:

Amount of prize	Probability of winning
$50,000	.000 000 4
10,000	.000 000 4
5,000	.000 000 4
2,500	.000 000 4
500	.000 004 1
100	.000 010 3
10	.001 239 5
0	.998 744 5

a. Identify the random variable, and state whether it is discrete or continuous.

b. The table shows values of which of the following: probability function, probability density function, distribution function?

c. What is the expected value of the random variable?

d. Find the numerical value of $F(10)$, and state in words what it represents.

e. Find the numerical value of $1 - F(500)$, and state in words what it represents.

(Source: Reader's Digest Association (Canada), $105,000 "Cash Bonanza" Sweepstakes advertisement mailing, fall 1982.)

7/41. Calculate the mean and the standard deviation of the distribution of the number of dots appearing on the top surface in the process of rolling a single fair die. Is the calculated standard deviation consistent with the rule of thumb in Expression (4.17)?

7/42. Calculate the mean and the standard deviation of the distribution of the total number of dots appearing on the top surfaces in the process of rolling a pair of fair dice. Is the calculated standard deviation consistent with the rule of thumb in Expression (4.17)?

7/43. In 1980, 10% of all bachelor's degrees awarded in Canada were in business and commerce. Determine the probability that of three randomly selected 1980 graduates, the number in business and commerce is

a. 0 **b.** 1 **c.** 2 **d.** 3 **e.** 2 or fewer

7/44. In 1980, 20% of all bachelor's degrees awarded in the United States were in business and commerce. Determine the probability that of four randomly selected 1980 graduates, the number in business and commerce is

a. 0 **b.** 1 **c.** 2 **d.** 3 **e.** 4 **f.** 2 or fewer

7/45. The probability that an American man who has not married by age 40 will ever marry is approximately .42. Consider eight randomly selected unmarried 40-year-old men.

a. What is the probability that exactly two of the eight will ever marry?

b. What is the probability that no more than two of the eight will ever marry?

c. Calculate the mean and the standard deviation of the distribution of the number who will marry.

7/46. The probability that an American who has not married by age 19 will ever marry is approximately 96%. Consider 12 randomly selected unmarried 19-year-olds.

a. What is the probability that exactly 10 of the 12 will ever marry?

b. What is the probability that at least 10 of the 12 will ever marry?

c. Calculate the mean and the standard deviation of the distribution of the number who will marry.

7/47. A space vehicle has five independently functioning computers, all five of which must be operating successfully before liftoff will take place. The probability of malfunction is .005 for each computer.

a. What is the probability that computer malfunction will delay liftoff?

b. Calculate the mean and the standard deviation of the number of computers malfunctioning.

7/48. A manufacturer's records indicate that 10% of its television picture tubes burn out before the guarantee has expired. A particular dealer has sold eight such tubes. Assuming that the binomial-distribution requirements are met, calculate the following:

a. the probability that all eight tubes will outlive their guarantee.

b. the mean and the standard deviation of the number of picture tubes burning out prematurely.

* **7/49.** An instructor is making up a multiple-choice test consisting of 10 questions, each with four alternatives (A, B, C, and D), one of which is the correct answer. She uses a table of random digits to determine in which position (A, B, C, or D) to put the correct answers, in such a way that each position has equal probability. What is the probability that the correct answers will turn out to consist of 5 A's, 3 B's, 2 C's, and 0 D's?

* **7/50.** With respect to rolling a pair of dice, let us use the term *double* to refer to an outcome in which both dice show the same value, and the term *natural* to refer to an outcome of seven or eleven. What is the probability of obtaining two doubles, one natural, and four other outcomes in seven rolls of a pair of fair dice?

* **7/51.** There are four suits in a standard deck of 52 playing cards. Each of the four suits consists of 13 cards as follows: 1 ace, 9 numbered cards, and 3 face cards. What is the probability of receiving 2 aces, no numbered cards, and 3 face cards in 5 cards dealt fairly from a well-shuffled deck?

* **7/52.** A small college has 5 faculty members teaching French, 3 teaching Spanish, and 2 teaching German. A committee of 5 members is formed to investigate the possibility of including overseas studies in the college's language program. If the committee members were chosen randomly from the 10 members of the language departments, what is the probability that the committee would consist of 1 French instructor, 2 Spanish instructors, and 2 German instructors?

* **7/53.** Consider the distribution of letters in 100 Scrabble tiles as shown in Example 7.24. Three letters are randomly selected from the 100 tiles. Counting Y as a consonant, calculate the probability of drawing one vowel, two consonants, and no blanks according to

a. the exact method.

b. the multinomial approximation.

* **7/54.** A company that manufactures machine bolts inspects the bolts in lots of 144. Four bolts are randomly selected from each lot and checked against specifications. If any of the four fails the inspection, the entire lot is withheld for examination. If there are 12 defective bolts in the whole lot, what is the probability that no defective bolts will be in the sample

a. according to the exact method?
b. according to the binomial approximation?

If there are 24 defective bolts in the whole lot, what is the probability that exactly one defective bolt will be in the sample

c. according to the exact method?
d. according to the binomial approximation?

* **7/55.** A supermarket sells an average of .8 10-kg bags of rolled oats per week (that is, 8 bags in 10 weeks, on the average). Assume that the number of bags sold per week is a Poisson variable.

a. What is the probability that a week will go by with no bags sold?
b. What is the probability that exactly one bag will be sold in a particular week?
c. How many bags must the store have on hand at the start of each week to ensure that the chance of the demand exceeding the supply is less than 5%?
d. What is the probability that the demand for 10-kg bags of rolled oats in a particular week will exceed the supply if, at the start of each week, the store stocks the number of bags recommended in Part (c)?

* **7/56.** Assume that the daily number of reported traffic accidents on weekdays in a certain city is a Poisson variable with a mean of 5.3 accidents per day. What is the probability that on a particular weekday there will be

a. no reported accidents?
b. exactly one reported accident?
c. exactly two reported accidents?
d. exactly three reported accidents?
e. four or more reported accidents? [*Hint:* The answer to Part (e) can be obtained by subtraction.]

* **7/57.** The probability that a randomly selected person was born on 29 February is about .00068. Use the Poisson approximation to the binomial distribution to determine the probability that in a college of 1000 students, the number of students with their birthday on 29 February would be

a. zero.
b. one.
c. two.
d. three or more. [*Hint:* The answer to Part (d) can be obtained by subtraction.]

* **7/58.** The probability that a randomly selected person was born on 31 May is about .0027. Use the Poisson approximation to the binomial distribution to determine the probability that in a college of 1000 students, the number of students with their birthday on 31 May would be

a. zero.

b. one.

c. two.

d. three or more. [*Hint:* The answer to Part (d) can be obtained by subtraction.]

7/59. A manufacturing company produces home freezers with compressors having lifetimes that are approximately normally distributed with a mean of 12.0 years and a standard deviation of 3.0 years. In order to assess the costs associated with various alternative lengths of warranty that it might offer, the company needs to know how many freezers can be expected to last various lengths of time. In what proportion of the freezers will the compressor last

a. less than 2.0 years?

b. less than 3.0 years?

c. more than 5.0 years?

d. more than 10.0 years?

7/60. Amounts on the receipts issued by one of the regular (not express) cash registers in a supermarket are approximately normally distributed with a mean of $49 and a standard deviation of $18. What proportion of the receipts are

a. more than $100.00?

b. more than $25.00?

c. between $30.00 and $60.00?

d. less than $10.00?

7/61. Stanford-Binet IQ scores are approximately normally distributed with a mean of 100.0 and a standard deviation of 16.0 in the general population. Treating IQ score as a discrete variable, determine what proportion of individuals will be classified as follows:

a. as gifted, if gifted is defined by an IQ of 140 or more.

b. as superior, if superior is defined by an IQ between 120 and 139 inclusive.

7/62. Stanford-Binet IQ scores are approximately normally distributed with a mean of 100.0 and a standard deviation of 16.0 in the general population. Treating IQ score as a discrete variable, determine what proportion of individuals will be classified as follows:

a. as mentally retarded, if mentally retarded is defined by an IQ below 70.

b. as average, if average is defined by an IQ between 90 and 109 inclusive.

7/63. Use the Gaussian approximation to the binomial distribution to find the probability of obtaining 60 or more heads in 100 flips of a fair coin.

7/64. Use the Gaussian approximation to the binomial distribution to find the probability of getting more than 15 sixes in 75 rolls of a fair die.

7/65. The probability that a male student who begins work on a university degree will complete the degree is approximately .24. The following questions refer to the Gaussian approximation to the binomial distribution.

a. What is the probability that 20 or more out of 50 incoming male students will complete their degrees?

b. What is the probability that fewer than 15 out of 50 incoming male students will complete their degrees?

c. Was the Gaussian approximation appropriate here according to the rule of thumb in Expression (7.35)?

7/66. The probability that a female student who begins work on a university degree will complete the degree is approximately .31. The following questions refer to the Gaussian approximation to the binomial distribution.

a. What is the probability that 10 or fewer out of 40 incoming female students will complete their degrees?

b. What is the probability that more than 25 out of 40 incoming female students will complete their degrees?

c. Was the Gaussian approximation appropriate here according to the rule of thumb in Expression (7.35)?

References

Bell, E. T. *Men of Mathematics.* New York: Simon and Schuster, 1937.

Eisenhart, C. "Gauss, Carl Friedrich." *International Encyclopedia of Statistics.* Edited by W. H. Kruskal and J. M. Tanur. New York: The Free Press, 1978.

Information Please Almanac 1982. New York: Simon and Schuster, 1981.

Kendall, M. G., and Stuart, A. *The Advanced Theory of Statistics.* Vol. 1: *Distribution Theory,* 3rd ed. New York: Hafner, 1969.

Seal, H. L. "Moivre, Abraham de." *International Encyclopedia of Statistics.* Edited by W. H. Kruskal and J. M. Tanur. New York: The Free Press, 1978.

Stigler, S. M. "A Modest Proposal: A New Standard for the Normal." *American Statistician* 36 (1982): 137–38.

Terman, L. M., and Merrill, M. A. *Measuring Intelligence.* Boston: Houghton Mifflin, 1937.

World Who's Who in Science. Chicago: Marquis Who's Who, 1968.

Further Reading

Burr, I. W. *Elementary Statistical Quality Control.* New York: Marcel Dekker, 1978. This book describes applications of probability theory in problems relating to quality control through inspection of samples of a product.

Weaver, W. *Lady Luck: The Theory of Probability.* Garden City, NY: Doubleday, 1963. This book provides lucid discussions of mathematical expectation in Chapter 7, the binomial distribution in Chapter 10, and other probability distributions in Chapter 12.

CHAPTER 8

SAMPLING DISTRIBUTIONS

EXAMPLE 8.1 "Until recently, historical data about the age of menarche [first menstruation] were probably of interest only to a few specialists. Much of the current publicity about the historical age of menarche has to do with the belief that in American girls menarcheal age has dropped from 17 years in the 19th century to 12.5 years today. This supposed trend has been regarded as the most significant factor in increasing teenage sexuality and the 'growing problem of teenage pregnancy.' Stories about it have appeared in almost every kind of popular journal from *Newsweek* to *The Nation,* and in scholarly literature of anthropology, psychology, child development, nursing, and other fields. . . . Notions about the magnitude of the change in menarcheal age are based upon misinformation. There has been some change, but very much less than has been assumed.

"In the United States in the present day the mean age at normal menarche is estimated to be 12.3 years (some would say 12.5 or 12.6) and the range from 9 to 17 years. Individual differences are attributed to differences in general health and nutrition, heredity, psychosocial development, and a number of other factors. . . . It is important to emphasize the broad range of menarcheal age, because in the past each clinician contributing to the subject did so, with a few exceptions, on the basis of a very small sample . . . .

"For the 19th century the data . . . suggest that the age 17 is derived from an isolated report on a small sample, and that it is well within the range of a standard deviation and should not be taken as an average.

"Undoubtedly there has been some drop in menarcheal age in the United States since the 19th century, to under 13 in the 1980's. Women in most European countries still have menarche sometime after their 13th birthday. This difference is an additional reason for caution in using 19th-century European data to generalize about the changing age of menarche here [in the United States]."

(Source: Bullough, 1981. Copyright 1981 by the American Association for the Advancement of Science. Used with permission of the AAAS.)

Statistics is the science of answering questions on the basis of empirical data. Often, as in the preceding example, the question is such that it is impossible to obtain all of the relevant observations (the population), and a conclusion must be drawn on the basis of a sample of observations.

Sampling refers to the process of selecting a sample from a population. From the sample we calculate *statistics* (numerical summaries of the sample), which are used as *estimates* of the corresponding *parameters* (numerical summaries of the population). Since our purpose is to answer a particular question, interest is centered on the population of observations relevant to that question; the sample is of interest only insofar as it provides useful information about that population. Therefore, it is important to have a good sample, one that provides *accurate* and reasonably *precise* estimates of the parameters.

Let us digress for a moment to clarify the distinction between **accuracy** and **precision**:

Accuracy refers to the lack of bias, or lack of systematic error, in an estimate. **(8.1)**

Precision refers to the lack of variability, or lack of random error, in an estimate. **(8.2)**

EXAMPLE 8.2 The second and third paragraphs of Example 8.1 are concerned with the question of **precision**. They point out that both the sample size and the inherent variability in the data affect the precision of an estimate. The use of a very small sample to estimate average menarcheal age may introduce substantial random error into the estimate, because an estimate from a very small sample may, just by chance, be considerably too low or too high. An estimate from a larger sample is apt to be closer to the true value of the parameter and hence is said to be more precise. Quite apart from the matter of sample size, the mere existence of substantial individual differences makes any estimate less precise than it would be if menarcheal age showed less variability or dispersion.

The question of **accuracy** is the concern of the final paragraph of Example 8.1. The problem is that the use of European data to draw conclusions about United States women may introduce systematic error into the estimate of average menarcheal age. That is to say, the European parameter may be different from the American parameter, so that no matter how precisely we were to estimate it, we would still be estimating the wrong thing, namely, the European average rather than the American average. An estimate with a systematic tendency to be too high or too low is said to lack accuracy or to be a biased estimate.

Both accuracy and precision are important characteristics to strive for in our estimates. Perhaps less obviously, it is also important to be able to assess and to state the degree of precision associated with the estimates. In addition to introducing some basic considerations in sampling, this chapter deals with the problem of describing the chances that the resulting estimates will be close to the true values of the parameters.

8.1 BASIC CONCEPTS IN THE STUDY OF SAMPLING DISTRIBUTIONS

Simple Random Sampling

The reason for any form of **sampling**—for selecting a sample to learn about the corresponding population—is that it would be too expensive, too time-consuming, or just plain impossible to get the information we need by observing every member of the population of relevant observations. If we keep in mind that it is the population and not the sample that we are actually trying to learn about, we will recognize that any statistics (descriptions of samples) are meaningless unless we know enough about how the sample was selected so that we can judge the extent to which it is apt to be representative of the population.

EXAMPLE 8.3 According to a radio news report in July 1982, 48% of senior business executives answering a survey thought the economy would get worse before the end of the year, as compared with only 14% who thought it would get better; the rest thought it would stay about the same. The news report did not say what population of executives had been sampled, and yet the appropriate conclusions to draw depend critically on what kind of representation there was in the sample: What sizes of companies? What industries? What regions of the country? What political affiliations? Unless we know how the sample was selected, the statistics (48%, 14%) do not tell us very much.

There are several techniques that attempt to ensure that the sample is properly representative of the population, but the technique on which most methods in elementary inferential statistics are based is **simple random sampling**. We have already seen in Section 6.2 that the term *random* does not mean "haphazard" or "without a plan" in statistics but rather implies adherence to certain probabilities. In simple random sampling, these probabilities are equal for all members of the population. There are more complex kinds of random sampling in which the probabilities are known but not all equal, but the equiprobable (or "simple") case is so basic to statistical theory that the shorter term **random sampling** is often used to mean simple random sampling.

> A **random sample** is one in which every member of the population of relevant observations has equal probability of being selected. More precisely, a **simple random sample** of n observational units is one that is chosen in such a way that every possible combination of n observational units that could be drawn from the population has equal probability of being selected. **(8.3)**

Most statistical theory is based on the notion of random samples, not only because the equal-probability feature of simple random sampling makes the mathematics of determining precision more manageable, but more importantly because the use of an inanimate device for selection means that no characteristic of the observational unit can surreptitiously make it more or less likely than other units to be selected. Such prejudice in selection would introduce an unknown but systematic bias, causing the sample statistics to be inaccurate as estimates of the population parameters of interest.

EXAMPLE 8.4 A particularly well known example of unintentionally biased sampling occurred in the *Literary Digest* poll to predict the winner of the 1936 U.S. presidential election. Based on a sample of 2.3 million voters—an unusually large sample by today's standards—the *Literary Digest* predicted an easy victory for Alfred M. Landon. Instead, Franklin D. Roosevelt won by a landslide.

Hindsight revealed the flaws in the poll. For example, one characteristic that made some voters more likely than others to be in the sample was relative affluence. The sample had been obtained from *Literary Digest* subscription lists, telephone directories, and automobile registrations, all of which favored the well-to-do. If the relatively more affluent voters were more likely to vote for Landon, and if the relatively more affluent voters were overrepresented in the sample, then the poll would have a bias in favor of Landon.

But the *Literary Digest* sample had another serious flaw, this one without so many *if*'s. The *Literary Digest* had mailed "straw-vote" ballot forms to 10 million voters, and then relied on the voters to indicate their preference and return the forms. Only about 2.3 million ballots were returned. Even if the original 10 million voters had been randomly selected (which they were not), the final sample of 2.3 million was not random. The final sample contained 2.3 million voters who were sufficiently interested in registering their opinion in a straw vote that they took the trouble to mail back a response; it omitted the 7.7 million who showed less interest in the straw vote. As it turned out, the anti-Roosevelt protest vote was vastly overrepresented in the final sample, while the pro-Roosevelt majority were more apt to be content to pass up the straw vote and wait for the actual election. It is now known that mail questionnaires are notoriously unreliable indicators of public opinion, because of their reliance on voluntary response.

(Source: Bryson, 1976.)

Whenever observations are obtained from those who volunteer to take part in an investigation, there is a serious danger of systematic bias in the results. Those who are willing to volunteer may very well differ from nonvolunteers in ways that are relevant to the dependent variable. The advantage of random sampling is that it guarantees that observational units possessing a certain type of characteristic are no more likely to be selected than are any other observational units.

EXAMPLE 8.5 In a letter published in *Chemical & Engineering News*, an organic chemist wrote: "I recently became a father for the fifth time and I now have four daughters. I have made an admittedly small survey of the organic faculty at [the University of] Minnesota, and my former coworkers, who are practicing organic chemists. My sample of 30 have produced 29 male and 49 female children. In order to discover whether this remarkable sex bias is significant, I am writing to request data from chemists, stating their specialty (organic, physical, analytical, etc.), and the sex of their children. I will report on the results of this survey in due course. I predict that the theoretical chemists, who are not exposed to 'unnatural' odors, will have the same ratio of males and females as the general population."

Based on responses from 920 chemists, the subsequent data still showed the organic chemists to have a greater excess of daughters than other chemists. Is the hypothesis supported that the fumes to which organic chemists are exposed make them more likely to have daughters than sons? The data cannot answer the question because the sample was not random: Certain chemists would be more likely than others to respond to the request for information. As the original letter writer explained, "I may be receiving a biased sample, since the chemists with an excess of daughters will be eager to support my hypothesis that somehow the 'unnatural odors' are more lethal to the sperm carrying the Y chromosome."

(Sources: Leete, 1981a, 1981b. Used with permission of E. Leete.)

Genuine random sampling requires prior identification of every observational unit in the population, followed by the use of some inanimate device to select observational units without prejudice. The discussion of random selection in Section 6.2 introduced the use of a table of random digits for this purpose. (You may wish to review Example 6.8 at this point.) Procedures such as interviewing people who just happen to come along, selecting an assortment of observational units that "look random," and making use of the only observational units that are available do not lead to random samples, because not every observational unit in the population has an equal chance of being selected. The seriousness of the bias thus created depends on the extent to which the characteristic that makes an observational unit more or less likely to be selected is in any way related to the variables under investigation. Such relationships are often difficult or impossible to foresee, and random sampling provides one of the best safeguards against systematic bias in the sample.

Because of practical and sometimes ethical considerations, it is often impossible to meet the requirements of random sampling in practice, but that does not keep people from collecting data, doing research, and drawing inferences. It does make it harder to draw general conclusions from the data, but investigators can do their best to obtain a random sample, report how the sample was in fact obtained, and then leave it to the reader to decide if the results apply only to the particular sample or to a larger population.

The Nature of Sampling Distributions

Sampling involves the selection of a sample from which statistics are calculated to estimate the corresponding parameters of the population. Once an estimate has been obtained, it is reasonable to ask how close the estimate is to the true value. What is the probability that the estimate differs from the unknown parameter by less than so-and-so many units? Or what is the probability that the estimate is more than such-and-such a distance from the parameter? If the estimate is based on a random sample, then questions such as these can be answered through knowledge of the sampling distribution of the estimate.

A **sampling distribution** of a statistic is a description of the probabilities with which various values, or sets of values, of the statistic occur. In other words, a sampling distribution is a particular kind of probability distribution, namely, a probability distribution in which the random variable is itself a statistic (numerical description of a sample). Thus, we may inquire about the sampling distribution of the sample mean, or the sampling distribution of the sample standard deviation, or the sampling distribution of the sample median, or the sampling distribution of any other statistic. In each case, the sampling distribution relates the possible values of the statistic to their probabilities of occurrence.

At this point you may object: How can we have a *distribution* of something like a sample mean, where in practice we have just one sample whose mean has just one value? How can we have a *distribution* of the sample mean, or of any other statistic? Perhaps it is partly the terminology that is confusing here. A distribution of the sample mean is actually a distribution of the *possible* values of the sample mean. While it is true that a given sample provides only one sample mean, it must be remembered that the given sample occurred by chance—the observational units were selected randomly—and other samples could just as easily have occurred, each one of which would have yielded its own sample mean. In this sense, a sampling distribution is a theoretical distribution, although we could verify it empirically if we were to go to the trouble of collecting numerous samples from the sample population.

The basic concept of a sampling distribution of a statistic may be clarified if we review the appearance of the same concept in another theoretical probability distribution, namely, the binomial distribution of number of successes in n trials. In practice, there is just one sequence of n trials, and we record just one value of n_1, the number of successes. (For example, we may flip a coin $n = 6$ times and obtain $n_1 = 1$ head as in Example 7.16.) When we talk about the distribution of number of successes in n trials, we are talking about a theoretical distribution relating the *possible* values of n_1 (only one of which actually occurred) to their probabilities of occurrence. Similarly, for the sampling distribution of a statistic such as the sample mean, there is in practice just one sample, and we record just one value of the sample mean $\bar{x}$. When we talk about the distribution of the sample mean, we are talking about a theoretical distribution relating the *possible* values of $\bar{x}$ (only one of which actually occurred) to their probabilities of occurrence.

In the case of the binomial, it would be possible to repeat the whole sequence of n trials, recording a separate value of n_1 for each **replication** (as such a repetition is called). Plotting the frequency with which each value of n_1 occurred in a large number of replications would yield an empirical distribution of number of successes that should closely resemble the theoretical binomial distribution.

Similarly, in the case of the sample mean, it would be possible to collect repeated samples from the same population, recording a separate value of $\bar{x}$ for each replication. Plotting the frequency with which each value of $\bar{x}$ occurred would yield an **empirical sampling distribution** of sample means that should closely resemble the **theoretical sampling distribution** of the sample mean. The next section gives examples.

8.2 EMPIRICAL SAMPLING DISTRIBUTIONS

All of the sampling distributions that we will be referring to in the subsequent chapters of this book will be theoretical sampling distributions. After the present chapter, we will have no further occasion to tabulate or plot an empirical sampling distribution, since the exact theoretical sampling distributions are known for the inferential statistics we will be studying.

Why then do we take time in this section to consider empirical sampling distributions, which involve the artificial situation of drawing repeated samples from the same population? The reason is that it is instructive to do so: The notion that the observed sample in an investigation is only one of many samples that might have been observed is fundamentally important. Examination of empirical sampling distributions can teach us something about what is meant by "the sampling distribution of a statistic," and theoretical sampling distributions of various statistics form the foundation for virtually all methods of inferential statistics.

An **empirical sampling distribution** shows the frequencies with which various values of a given statistic occurred in the drawing of repeated samples of the same size n from the same population. This artificial situation of drawing repeated samples provides a concrete model of the abstract concept underlying theoretical sampling distributions.

EXAMPLE 8.6 A table of random digits was used to obtain 100 samples of $n = 5$ observations each from a discrete rectangular distribution. Selections were made according to the probability function

$$P(x) = \tfrac{1}{9} \qquad \text{for } x = 1, 2, \ldots, 9 \tag{8.4}$$

by recording successive nonzero digits from the table until 100 samples of $n = 5$ digits had been obtained. This procedure is an example of **simulation** by **Monte Carlo methods**, in which simulated data are generated randomly from known populations, either to model an empirical situation or, as in this case, to study the behavior of statistical tools or descriptions.

The following four statistics were calculated for each of the 100 samples: mean $\bar{x}$, median $\tilde{x}$, standard deviation s, and variance s^2. The graph in Column (1) of Figure 8.1(a) is a histogram of the rectangular population from which the samples were drawn. The other four histograms in Column (1) of Figure 8.1 show empirical sampling distributions of the 100 values of the four statistics.

Consider first the empirical sampling distribution of the means of the 100 samples from the rectangular population [Figure 8.1(b), Column (1)]. Notice that the mean of the 100 sample means, designated by $\hat{\mu}_{\bar{x}} = 4.95$, is very close to the population mean $\mu = 5$. (The subscript $\bar{x}$ in the symbol $\hat{\mu}_{\bar{x}}$ indicates that we are dealing with a distribution of sample means; that is, the variable in this distribution is the sample mean, which takes on different values for different samples. The symbol $\mu_{\bar{x}}$ refers to the mean of the theoretical sampling distribution of the sample means, and the "hat" in $\hat{\mu}_{\bar{x}}$ indicates an empirical estimate, rather than the actual value, of $\mu_{\bar{x}}$.) Notice, too, that the standard deviation of the 100 sample means, designated by $\hat{\sigma}_{\bar{x}} = 1.24$, is considerably smaller than the population standard deviation $\sigma = 2.58$: The standard deviation of a sample mean is smaller than the standard deviation of an individual observation. This tells us that sample means $\bar{x}$ tend to deviate from the true mean μ by a smaller amount than the individual observations x_i do. In other words, the sample mean has greater precision (less random error) than an individual observation. This feature can be seen in the histograms, too. For example, 23% + 28% + 22% = 73%

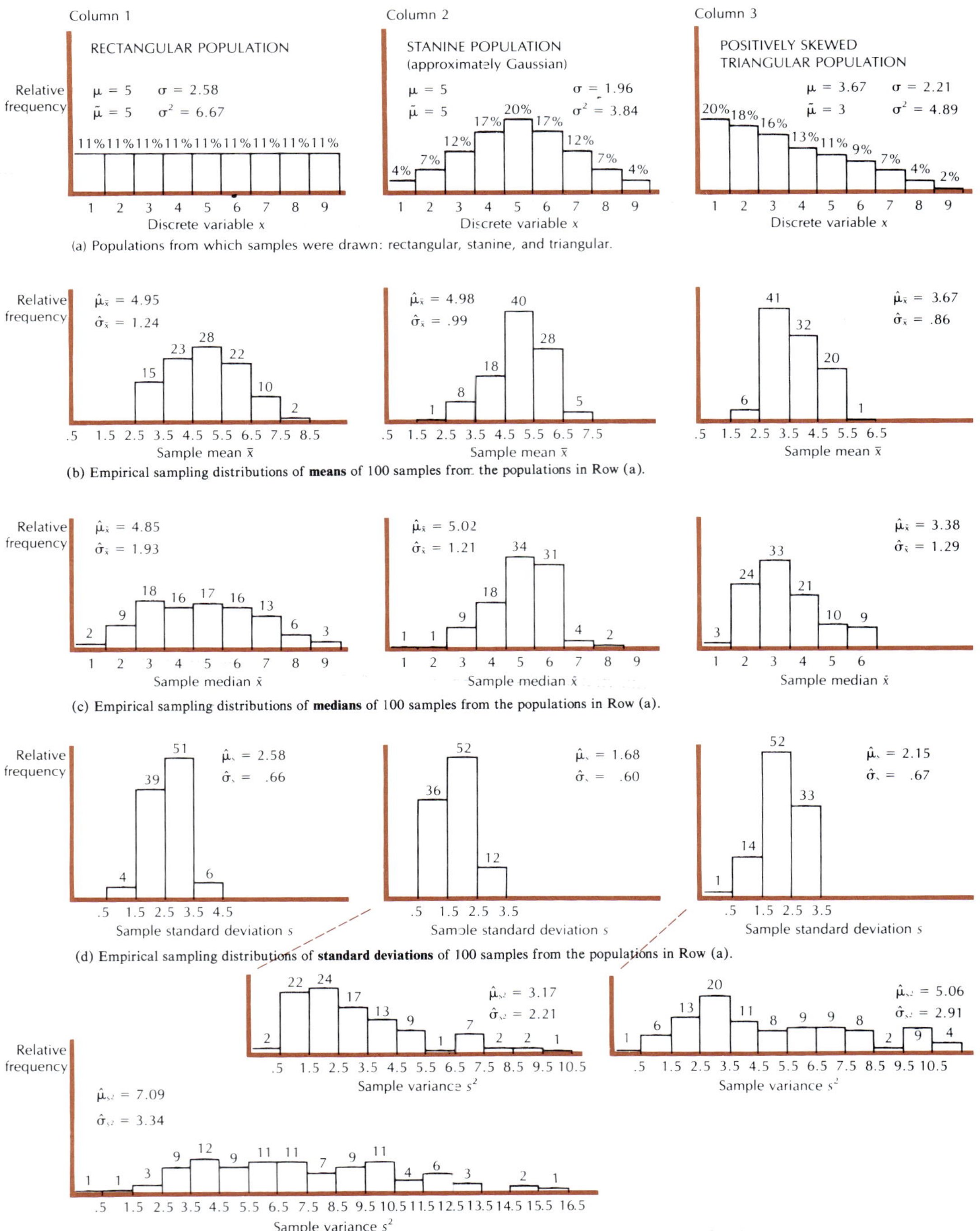

Figure 8.1
Histograms of three populations, each one accompanied by four empirical sampling distributions based on 100 samples of $n = 5$ observations.

of the sample means differed from the true mean μ by less than 1.5, whereas only 33% of the individual observations were so close.

Moving down Column (1) to Figure 8.1(c), which shows the empirical sampling distribution of the medians of the 100 samples, we find that the mean of the 100 medians, designated by $\hat{\mu}_{\tilde{x}} = 4.85$, is again very close to the population median $\tilde{\mu} = 5$. The standard deviation of the 100 medians, denoted by $\hat{\sigma}_{\tilde{x}} = 1.93$, is notably larger than the standard deviation of the 100 means, $\hat{\sigma}_{\bar{x}} = 1.24$, indicating that sample medians fluctuate more from sample to sample than do sample means. In other words, the sample mean has greater precision than the sample median, in random samples from a rectangular population. This fact can also be seen in the histograms, in that the empirical sampling distribution of the median is spread out over a wider range than that of the mean. For example, only 16% + 17% + 16% = 49% of the sample medians were within 1.5 of the true value of the parameter $\mu = \tilde{\mu} = 5$, whereas 73% of the sample means were that close.

The histogram in Figure 8.1(d), Column (1), shows the empirical sampling distribution of the standard deviations of the 100 samples. The mean of these 100 sample standard deviations, $\hat{\mu}_s = 2.58$, is very close indeed to the true value of the standard deviation of the parent population, $\sigma = 2.58$. Of course, not all of the samples produced an estimated standard deviation of exactly 2.58—some were higher and some were lower—but the mean of these estimates, $\hat{\mu}_s$, was 2.58. The amount by which they tended to be too high or too low on the average is reflected in the standard deviation of the 100 sample standard deviations, $\hat{\sigma}_s = .66$.

The histogram in Figure 8.1(e), Column (1), shows the empirical sampling distribution of the variances of the 100 samples from the rectangular population. The fact that the units of variance are squared units accounts in part for the different appearance of this distribution. The mean of the 100 sample variances, $\hat{\mu}_{s^2} = 7.09$, is reasonably close to the population variance $\sigma^2 = 6.67$, of which each of the sample variances is an estimate. The irregularities in the shape of the histogram are due to the fact that it is based on only 100 samples. These irregularities would tend to smooth out in an empirical distribution based on a very large number of samples or in the exact theoretical sampling distribution, which is, in essence, based on an infinite number of samples.

EXAMPLE 8.7 A table of random digits was used to obtain 100 samples of $n = 5$ observations each from an approximately Gaussian distribution of a discrete variable. The distribution was based on the **stanine scale**, which we obtain by partitioning a standard normal distribution into nine regions with category boundaries at standard normal deviates of $z = -\infty, -1.75, -1.25, -.75, -.25, .25, .75, 1.25, 1.75, +\infty$, and labeling the nine categories with the discrete variable values $1, 2, \ldots, 9$. The term **stanine** is shortened from "standard nine." Each of the nine values of a stanine variable occurs with the probability given for the associated region of the standard normal distribution; these probabilities are shown in the histogram of the stanine population in Column (2) of Figure 8.1(a).

Selections were made by considering successive pairs of random digits and recording a value of 1 for random-digit pairs 00–03, 2 for 04–10, 3 for 11–22, 4 for 23–39, 5 for 40–59, 6 for 60–76, 7 for 77–88, 8 for 89–95, and 9 for 96–99. This Monte Carlo method ensured that the value 1 occurred with probability 4%, 2 with probability 7%, and so on, as required for a stanine population. Four statistics (mean, median, standard deviation, and variance) were calculated for each of the 100 samples obtained from the stanine population, and their histograms are shown in Column (2) of Figure 8.1, Parts (b)–(e).

The graphs in Parts (b) and (c) of Column (2) show histograms of the empirical sampling distributions of the 100 sample means and the 100 sample medians, respectively, from the stanine population. The means of both of these sampling distributions, $\hat{\mu}_{\bar{x}} = 4.98$ and $\hat{\mu}_{\tilde{x}} = 5.02$, are very close to the true value of the parameter ($\mu = \tilde{\mu} = 5$) being estimated by each sample mean and each sample median. As was the case for the rectangular population in Example 8.6, the standard deviation of the 100 sample means, $\hat{\sigma}_{\bar{x}} = .99$, is smaller than the standard deviation of the 100 sample medians, $\hat{\sigma}_{\tilde{x}} = 1.21$, again indicating that the mean is a more precise estimate than the median.

The graphs in Parts (d) and (e) of Column (2) portray the empirical sampling distributions of the standard deviations and variances, respectively, of the 100 samples from the stanine population. The means of these estimates, $\hat{\mu}_s = 1.68$ and $\hat{\mu}_{s^2} = 3.17$, are slightly smaller than the parameters being estimated, $\sigma = 1.96$ and $\sigma^2 = 3.84$, respectively, but these differences merely point out the reason that in practice we use theoretical rather than empirical sampling distributions: An empirical sampling distribution may, just by "the luck of the draw," yield values that are smaller or larger than typical, especially when based on a relatively small number of samples. Notice the positive skewness in the distribution of variances; we will see later that this is typical of sampling distributions of variances of samples drawn from approximately Gaussian populations.

EXAMPLE 8.8 A table of random digits was used to obtain 100 samples of $n = 5$ observations each from a positively skewed distribution of a discrete variable. Specifically, the population was a discrete triangular distribution with the probability function

$$P(x) = \frac{(10 - x)}{45} \qquad \text{for } x = 1, 2, \ldots, 9 \tag{8.5}$$

In general, a **triangular distribution** is one in which, at any point within the range of the random variable x, a given increase in x changes the associated probability by a fixed amount. In the case at hand, each unit increase in x subtracts 1/45 from the probability.

The Monte Carlo method to simulate samples from a distribution having the probability function in Expression (8.5) was as follows: Selections were made by considering successive pairs of random digits and recording a value of 1 for random-digit pairs 11–19, 2 for 22–29, 3 for 33–39, 4 for 44–49, 5 for 55–59, 6 for 66–69, 7 for 77–79, 8 for 88–89, and 9 for 99 and ignoring all other pairs. After 100 samples of $n = 5$ observations each had been recorded, empirical sampling distributions were tabulated for four statistics (mean, median, standard deviation, and variance), as portrayed in Column (3) of Figure 8.1, Parts (b)–(e).

Each of the empirical sampling distributions shows the distribution of the 100 values of one of the four statistics. In every case, the mean of the sampling distribution of the statistic (3.67, 3.38, 2.15, and 5.06) is reasonably close to the true value of the parameter being estimated (3.67, 3, 2.21, and 4.89, respectively). Once again, the standard deviation of sample means, $\hat{\sigma}_{\bar{x}} = .86$, is smaller than the standard deviation of sample medians, $\hat{\sigma}_{\tilde{x}} = 1.29$; that is, the sample mean has greater precision than the sample median.

By comparing across Columns (1), (2), and (3) of Figure 8.1, you may also gain some impression of how the behavior of each of the statistics changes as a function of the population from which the sample is drawn. For example, sample variances vary most widely in the samples from a rectangular population (where observations in both tails occur with substantial probability), least widely in the samples from an approximately Gaussian population (where observations in the tails are rather improbable), and by an intermediate amount in the samples from a triangular population (where observations are probable in one tail but not in the other).

8.3 RELATIONSHIP AMONG POPULATIONS, SAMPLES, AND SAMPLING DISTRIBUTIONS

The previous section illustrated sampling distributions of four different statistics: mean $\bar{x}$, median $\tilde{x}$, standard deviation s, and variance s^2. It also showed that each sampling distribution may be characterized by *its* mean and standard deviation, so we have $\mu_{\bar{x}}$ and $\sigma_{\bar{x}}$ for the sampling distribution of the sample mean $\bar{x}$, and μ_s and σ_s for the sampling distribution of the sample standard deviation s, and so on. In earlier chapters we described *populations* in terms of their mean μ and standard deviation σ, and *samples* in terms of their mean $\bar{x}$ and standard deviation s, so now we have a third kind of mean and standard deviation, namely, the mean and standard deviation of a **sampling distribution**. In this section we distinguish among these three kinds of means and standard deviations.

There are just three cases to consider: populations, samples, and sampling distributions. We can, of course, have a population or sample of observations of any of a large number of different variables, and we can have a sampling distribution of any of several different statistics. However, the general nature of the relationship among populations, samples, and sampling distributions is the same regardless of which variable we consider in the population and sample and regardless of which statistic we consider in the sampling distribution. For simplicity, let us use the random variable x in our discussion, understanding that the discussion applies to other variables as well, and let us consider the sampling distribution of the mean $\bar{x}$, understanding that we could have phrased the discussion in terms of any other statistic (such as $\tilde{x}$, or s, or s^2) instead.

Corresponding to the three cases—populations, samples, and sampling distributions—there are three kinds of means to keep straight, as illustrated in the three columns of Figure 8.2. The *population,* consisting of the entire set of N observations (possibly an infinite number) relevant to the question at hand, has a mean of μ (or μ_x). Random selection of a subset of observations from the population yields a *sample* of n observations with a mean of $\bar{x}$. The sample mean $\bar{x}$ may itself be thought of as a random selection from the **theoretical sampling distribution** of the $\binom{N}{n}$ possible values of $\bar{x}$, whose expected value or mean is $\mu_{\bar{x}}$. Notice that the horizontal axis of the population distribution in Figure 8.2 is labeled x, whereas that of the sampling distribution is labeled $\bar{x}$. This difference emphasizes the fact that a sampling distribution is a probability distribution whose random variable is a statistic, in this case the sample mean.

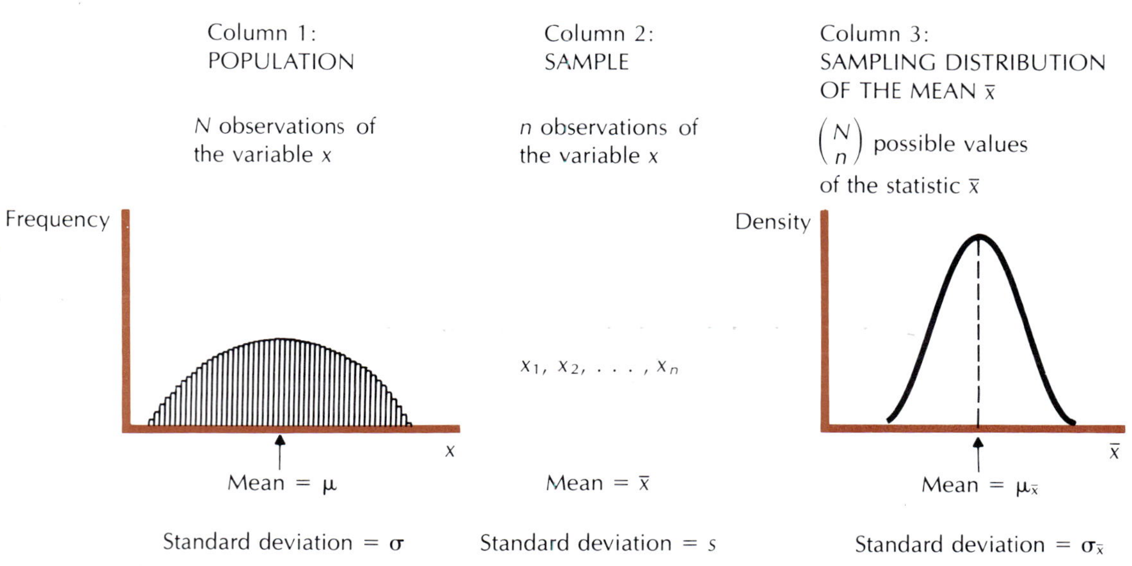

Figure 8.2
Illustration of the relationship among populations, samples, and the sampling distribution of the mean.

Similarly, there are three kinds of standard deviations to keep straight. The three are illustrated in Figure 8.2, again for the case in which the variable is x and the statistic in the sampling distribution is the sample mean $\bar{x}$. The standard deviation of the population is designated by σ (or σ_x), the standard deviation of the sample is denoted by s (or s_x), and the standard deviation of the sampling distribution of the mean $\bar{x}$ (that is, the standard deviation of the possible values of $\bar{x}$) is designated by $\sigma_{\bar{x}}$.

The standard deviation of a sampling distribution is often referred to as a **standard error**. Recall that a standard deviation is a kind of average amount by which observations deviate from their mean. In the case of a sampling distribution, the "observations" are estimates (statistics), and the mean of the distribution is the parameter they are estimating, so the deviation of an estimate (such as $\bar{x}$) from the parameter (such as μ) is rightly called an *error*, and the standard deviation is therefore the standard error. The **standard error of the mean**, denoted by $\sigma_{\bar{x}}$, indicates a kind of average amount by which the sample mean $\bar{x}$ is in error, or deviates from the population mean μ. It is a measure of the variability of the estimate $\bar{x}$.

Thus, a population may be described by its mean μ and standard deviation σ, a sample may be described by its mean $\bar{x}$ and standard deviation s, and a sampling distribution may be described by its mean (the expected value of the statistic) and standard deviation (the standard error of the statistic).

8.4 THEORETICAL SAMPLING DISTRIBUTION OF THE MEAN

In Section 8.2, we used Monte Carlo methods to generate repeated samples from the same population so that we could calculate statistics from each sample and examine their empirical sampling distributions. This repeated sampling and tabulation would not be necessary if we knew enough about the relevant sampling distribution in advance. This section presents the theoretical sampling distribution of the mean to illustrate how it is possible to know the approximate sampling distribution of a statistic in situations in which we have only one sample and only one value of the statistic.

The **theoretical sampling distribution of the mean** is known by virtue of results obtained through theoretical, rather than empirical, results in statistics. These results, describing the theoretical sampling distribution, can be proved to be the mathematical consequence of certain conditions. We will consider the results in three separate statements, corresponding to three slightly different sets of conditions. All three statements require the following condition:

> A random sample of n observations with a mean of $\bar{x}$ is drawn from a population of N observations with a mean of μ and a standard deviation of σ. **(8.6)**

In brief, Condition (8.6) stipulates that the sample must be random and that the population must have a finite standard deviation. In practice, the critically important component of this condition is random sampling.

Mean and Standard Deviation of the Sampling Distribution

The first of our three statements gives formulas for the mean and the standard deviation of the sampling distribution of the mean:

> If Condition (8.6) concerning random sampling holds, then the sampling distribution of the mean has a mean equal to the population mean μ and a standard deviation equal to $\sigma/\sqrt{n}$. **(8.7)**

That is,

$$\mu_{\bar{x}} = \mu \tag{8.8}$$

$$\sigma_{\bar{x}} = \frac{\sigma}{\sqrt{n}} \tag{8.9}$$

The formula for the standard error of the mean, $\sigma_{\bar{x}}$, in Expression (8.9) actually requires that the population size N be infinite if it is to hold exactly, but in practice the formula may be used as long as the population size N is large relative to the sample size n. We can make the formula accommodate finite populations simply by multiplying by $\sqrt{1 - n/N}$, which is called the **finite-population correction factor**. This adjustment reduces the calculated result slightly. However, it makes little difference unless the proportion of the population included in the sample, called the **sampling fraction** n/N, is about

10% or higher. We will not consider such cases here. In any event, the standard error of the mean in Expression (8.9) can be interpreted like any other standard deviation, as discussed in Section 4.4.

EXAMPLE 8.9 The quality-control staff for a certain manufacturer of electrical products have detailed records concerning the variability in the quality of their products. For example, the process being used to manufacture special-purpose 75-watt light bulbs is known to yield a standard deviation of 150 hours in bulb lifetimes. If the staff were to estimate the mean lifetime of the 75-watt bulbs produced by this process on the basis of data from a random sample of 36 bulbs, how large an error should they anticipate in the estimate?

The mean lifetime to be estimated is the unknown value μ, and the estimate of μ is the mean $\bar{x}$ of the 36 observed lifetimes. Any difference between the sample mean $\bar{x}$ and the population mean μ is an error in the estimate. Expression (8.9) indicates that a standard amount of error in a sample mean in this case is $\sigma_{\bar{x}} = 150/\sqrt{36} = 25$ hours. In other words, the standard error of the mean is 25 hours. Some sample means based on 36 observations would differ from the true value by more than 25 hours and some by less, but 25 hours is a standard amount of error in this type of sample mean.

We can make more specific statements by using the Bienaymé-Chebyshev inequality (Section 4.4, pages 122–126). For example, three standard errors would be 75 hours, so the probability that the estimate $\bar{x}$ is in error by 75 hours or more is at most $1/3^2 = 11\%$. Similarly, five standard errors would be 125 hours; the probability that the estimate $\bar{x}$ is "off" by 125 hours or more is at most $1/5^2 = 4\%$.

Notice that the staff would have to quadruple the sample size from $n = 36$ to $n = 144$ in order to reduce the standard error by one-half, from $150/\sqrt{36} = 25$ hours to $150/\sqrt{144} = 12.5$ hours.

Section 4.4 introduced z-scores to express individual observations in terms of how many standard deviation units they were above or below the mean:

$$z_i = \frac{x_i - \mu}{\sigma}$$

Similarly, we will now have occasion to convert sample means $\bar{x}$ to standard scores, expressing them in terms of how many standard errors they are above or below the population mean μ:

$$z_{\bar{x}} = \frac{\bar{x} - \mu}{\sigma_{\bar{x}}} \tag{8.10}$$

EXAMPLE 8.10 Over the years, a statistics instructor found that students' marks on her final examinations, which were of comparable difficulty, had a mean of 61.3% and a standard deviation of 15.1 percentage points. How unusual would it be for her current class of 43 students to obtain a mean mark of 68.0% on the final exam?

Here the population mean is $\mu = 61.3$, the population standard deviation is $\sigma = 15.1$, and the sample mean is $\bar{x} = 68.0$. If the current students' marks on the final exam can be regarded as a random sample from the same population as the final marks of previous students, then the standard error of the sample mean is

$$\sigma_{\bar{x}} = \frac{15.1}{\sqrt{43}} = 2.303$$

or approximately 2.3 percentage points. Thus, the sample mean is

$$z_{\bar{x}} = \frac{\bar{x} - \mu}{\sigma_{\bar{x}}} = \frac{68.0 - 61.3}{2.303} = 2.91$$

standard errors away from the population mean. One standard error away would be "standard"—some sample means would differ by more than that and some by less. Two standard errors away from the population mean is somewhat unusual—that's twice the standard amount of deviation. Qualitatively, we can say that a deviation of 2.91 standard errors is rather unusual, being 2.91 times the standard amount of deviation for random samples of this size. Using the Bienaymé-Chebyshev inequality, we can add that such deviant sample means occur with a probability of at most $1/2.91^2 = 12\%$.

Tucked away in Expression (8.7) are two very important bits of information, one about the accuracy of the sample mean $\bar{x}$ as an estimate of the population mean μ and the other about its precision. Concerning **accuracy**, or freedom from systematic error, Expression (8.7) says that the sample mean $\bar{x}$ is an unbiased estimate of the population mean μ; that is, $\mu_{\bar{x}} = \mu$. We have noted previously that this is a virtue of random sampling, that it avoids any systematic tendency for the sample mean to be too large or too small. This does not mean that the mean of a random sample is without error, but that its error is random, not systematic.

Concerning **precision,** or freedom from random error, Expression (8.7) says that the sample mean is subject to less variability than is the random variable in the population. Specifically, as stated in Expression (8.9), the variability is reduced by a factor of one divided by the square root of the sample size:

$$\sigma_{\bar{x}} = \sigma\left(\frac{1}{\sqrt{n}}\right)$$

If the samples were of size $n = 1$, then each sample mean would be the value of that one observation, and the standard error of the mean would equal the population standard deviation σ. As the sample size n is increased, the value of $\sigma_{\bar{x}} = \sigma/\sqrt{n}$ decreases, which is to say that the standard error becomes progressively smaller the larger the sample. As a measure of the variability or fluctuations in the sample mean, the standard error reflects the precision of the sample mean as an estimate of the population mean. Thus, increasing the sample size improves the precision of the estimate, because increasing the sample size reduces the standard error.

A rather interesting implication of Expression (8.9) concerning the precision of the sample mean is that for a given amount of dispersion in the population (as measured by σ), the precision of the estimate $\bar{x}$ depends heavily on sample size n and hardly at all on population size N (as long as the population is at least 10 times larger than the sample). Since random sampling excludes systematic bias and ensures that each of the n observations provides an additional independent item of information, it follows that precision depends on the number of observations that *are* in the sample rather than on the number of potential observations that *are not* in the sample.

EXAMPLE 8.11 Given that the standard deviation of the heights of North American women is approximately 2.5 inches, how large a random sample would we need to estimate, with a standard error of .1 inch, the mean height of women in a city in which 100,000 women live? Applying Expression (8.9), we have $.1 = 2.5/\sqrt{n}$, from which we determine that we would need a random sample of $n = (2.5/.1)^2 = 625$ women. Notice that the number of women in the city (100,000) does not enter into the calculation, as long as it is more than 10 times the sample size (which it is). How large a random sample would we need to estimate, with a standard error of .1 inch, the mean height of 50,000,000 North American women? The answer is again 625.

Samples from Gaussian Populations

The second of our three statements tells us more about the sampling distribution of the mean for the special case where the population of observations is a normal distribution. As noted in Section 7.4, the normal distribution is a theoretical rather than empirical distribution. Nonetheless, it provides a satisfactory approximation for many empirical variables; Example 7.1 provided an illustration concerning the distribution of adult heights. The second of our three statements is useful in such cases:

> If Condition (8.6) concerning random sampling holds, and if the population is Gaussian, then the sampling distribution of the mean follows the Gaussian distribution with a mean equal to the population mean μ and a standard deviation equal to $\sigma/\sqrt{n}$. **(8.11)**

The only new element in Expression (8.11) is that for Gaussian populations, the sampling distribution of the mean is likewise Gaussian. This fact allows us to make much more exact probability statements than are possible when we do not know the shape of the sampling distribution. In particular, it allows us to go to tables of the standard normal distribution function (rather than to the Bienaymé–Chebyshev inequality) to find the probability that our estimate is in error by more than any amount we may wish to specify.

EXAMPLE 8.12 Example 8.9 described a situation in which a light-bulb manufacturing process resulted in a standard deviation of 150 hours in bulb lifetimes. If the quality-control staff used a random sample of 36 bulbs to estimate the mean lifetime of 75-watt bulbs produced by this process, the standard error of the sample mean would be $150/\sqrt{36} = 25$ hours [Expression (8.9)]. According to the Bienaymé-Chebyshev inequality (Section 4.4, pages 122–126), the probability that the estimate $\bar{x}$ is in error by 75 hours or more is at most $1/3^2 = 11\%$. The Bienaymé-Chebyshev inequality neither requires nor uses information about the shape of the sampling distribution.

In fact, the lifetimes of light bulbs produced by the process in question are approximately normally distributed. It follows from Expression (8.11) that the sampling distribution of sample mean lifetimes is also Gaussian. Thus, we can go to the table of the standard normal distribution in Appendix 3 to find the probabilities associated with various possible values of the sample mean $\bar{x}$.

What is the probability that the mean $\bar{x}$ of a random sample of 36 bulbs is off the true parameter value μ by 75 hours or more? An error of $\bar{x} - \mu = \pm 75$ hours is

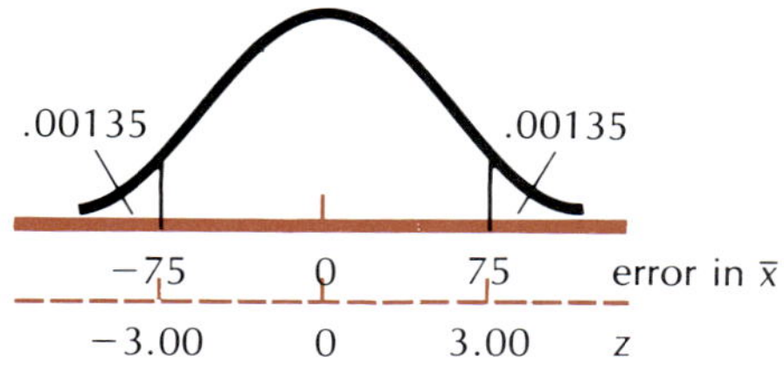

Figure 8.3

$z_{\bar{x}} = \pm 75/25 = \pm 3.00$ standard errors away from the true value [Expression (8.10)]. From Appendix 3, we find that the probability that a standard normal deviate exceeds $z = 3.00$ is .00135. By symmetry, the probability that a standard normal deviate is less than $z = -3.00$ is also .00135, as illustrated in Figure 8.3. Therefore, the probability of an error of 75 hours or more is $2 \times .00135 = .00270$, or .27%.

The answer in the previous paragraph demonstrates the value of knowing the shape of the sampling distribution. Without Expression (8.11), which gave us the shape of the sampling distribution for this problem, we had to resort to the Bienaymé-Chebyshev inequality for an answer. The answer it gave was "at most 11%." Knowing that the population was approximately Gaussian, and hence [by Expression (8.11)] that the sampling distribution was Gaussian, we found the required probability to be .27%. The Bienaymé-Chebyshev answer is not wrong, but the latter answer is much more specific.

In samples from Gaussian populations, the standard error of the mean is smaller than the standard error of the median or of the mode, which is to say that the mean is a more precise estimate. For example, the standard error of the median, $\sigma_{\tilde{x}}$, is approximately 1.25 times the standard error of the mean, $\sigma_{\bar{x}}$, in samples from Gaussian populations; this relationship is illustrated in Column (2) of Figure 8.1. The greater precision of the mean, relative to other measures of location, in samples from Gaussian populations is one of the reasons that the mean has been the most widely used kind of average in statistical inference.

The Central Limit Theorem

The first of our three statements about the sampling distribution of the mean, Expression (8.7), told us its mean and its standard deviation. The second statement, Expression (8.11), told us the shape of the sampling distribution of the mean for one particular case, namely, when we are sampling from a Gaussian population. Our third and final statement tells us about the shape of the sampling distribution of the mean even when we do not know the shape of the population distribution. The third statement is the **Central Limit Theorem**, originally introduced by Pierre Laplace (see Biography 8.1) and subsequently refined and extended by more recent mathematicians.

BIOGRAPHY 8.1 Pierre Simon de Laplace (1749–1827)

Pierre Laplace was born on 28 March 1749 in the village of Beaumont-en-Auge, Normandy, France. He studied at a military school in Beaumont and then from 1765 to 1767 in Caen. In 1768, at the age of 19, he became a professor of mathematics at the École Militaire in Paris and subsequently at the École Normale. Meanwhile, in 1783 he became an artillery inspector, and in 1785 he examined a 16-year-old named Napoleon Bonaparte, who later became, among other things, his colleague at the French Academy of Sciences. In 1799, Napoleon named Laplace minister of the interior and, later in the same year, a senator and chancellor of the senate. A shrewd politician, Laplace survived both the French Revolution and the restoration of the monarchy. Indeed, he held various government appointments—in the republic, the empire, and the monarchy—changing his political loyalties to suit whoever was in power at the time. All the while, he pursued his mathematical research.

Sometimes called the Newton of France, Laplace devoted most of his research career to mathematical astronomy. During the course of this work, he made significant contributions to probability theory, publishing an influential treatise on the topic in 1812. He extended de Moivre's use (see Biography 7.2) of the normal density function as a means of approximating certain probabilities, suggested tabulating the distribution to aid in such approximations, and he was the first to discover the Central Limit Theorem. In addition to the applications of his research in astronomy, he is known for his contributions to the study of tides, capillary action, electricity, speed of sound, specific heat, and combustion.

Laplace's writing is known not only for its content but also for two less desirable features. One was his frequent use of the phrase "Il est aisé à voir [It is easy to see] . . ." to introduce mathematical results that were not at all easy to see. The other was his failure to give credit to others whose work he used. "To call a spade a spade," wrote mathematical historian E. T. Bell, "Laplace stole outrageously, right and left, wherever he could lay his hands on anything of his contemporaries and predecessors which he could use . . . . Laplace need not have been so ungenerous. His own colossal contributions to the dynamics of the solar system easily overshadow the works of others whom he ignores."

Laplace retired to his country estate near Paris and died on 5 March 1827, shortly before his 78th birthday.

(Sources: Bell, 1937; Fréchet & Stigler, 1978; *World Who's Who in Science,* 1968.)

The **Central Limit Theorem** is a remarkable result of tremendous importance for the field of inferential statistics. It may be stated as follows:

> If Condition (8.6) concerning random sampling holds, and if the sample size n is large enough, then the sampling distribution of the mean is approximately Gaussian with a mean equal to the population mean μ and a standard deviation equal to $\sigma/\sqrt{n}$. **(8.12)**

The Central Limit Theorem allows us to use tables of the standard normal distribution function to obtain probabilities of various amounts of error in the estimation of the population mean, regardless of the shape of the population distribution. Actually, the theorem does not promise that the sampling distribution will be exactly Gaussian until the sample size n reaches infinity, but the sampling distribution becomes close enough to the Gaussian distribution for much smaller sample sizes. How large is large enough depends on the shape of the population distribution. For example, from Expression (8.11) we know that *any* sample size (as small as $n = 1$) is large enough to assure the truth of the theorem's conclusion in the case of a Gaussian population. In the case of other symmetric or nearly symmetric population distributions, samples as small as $n = 5$ may be large enough to invoke the Central Limit Theorem. The generally accepted **rule of thumb for the Central Limit Theorem** is as follows:

> For almost any population distribution likely to be encountered in practice, a sample size of $n = 30$ is apt to be large enough to apply the results of the Central Limit Theorem with reasonable accuracy. **(8.13)**

EXAMPLE 8.13 Let us consider again the data from the statistics instructor whose students over the years obtained a mean mark on her final exams of 61.3% and a standard deviation of 15.1 percentage points. How unusual would it be for her current class of 43 students to obtain a mean mark as high as 68.0% on the final exam?

In Example 8.10, we saw that the standard error of the mean of a random sample of 43 observations would be $\sigma_{\bar{x}} = 15.1/\sqrt{43} = 2.303$ and that a sample mean of 68.0 would be $z = (68.0 - 61.3)/2.303 = 2.91$ standard errors above the population mean. The Central Limit Theorem in Expression (8.12) enables us to obtain the required probability from the normal distribution. From Appendix 3, we find $F_{\text{Gauss}}(2.91) = .99819$ and the corresponding upper-tail area $A = .00181$, as illustrated in Figure 8.4.

In other words, the probability that a sample consisting of 43 marks, randomly selected from a population with a mean of 61.3% and a standard deviation of 15.1 percentage points, would have a mean of 68.0% or more is approximately .18%. Thus, a class average of 68.0% would be rather unusual, and if it were to occur, we might wonder whether the 43 marks really were from the same population as the previous marks. Was the current final exam as difficult as previous ones, for example, or was the course taught differently?

The answer of .18% in the previous paragraph illustrates the advantage of the Central Limit Theorem over the Bienaymé-Chebyshev inequality, which produced an answer of "at most 12%." The answer obtained from the Bienaymé-Chebyshev inequality is correct but is not nearly as specific as that obtained when we know the shape of the sampling distribution of the mean, thanks to the Central Limit Theorem.

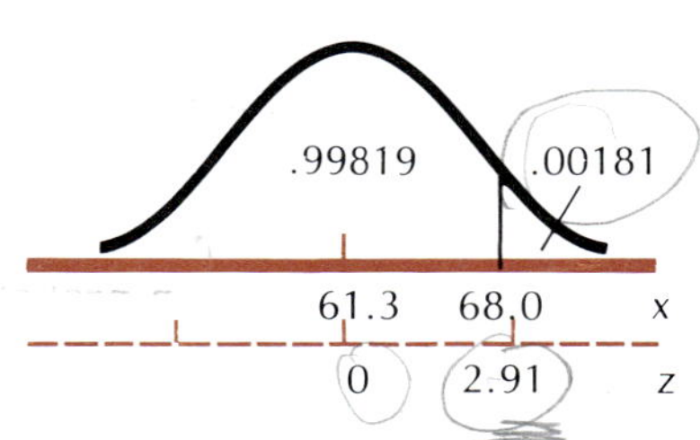

Figure 8.4

* 8.5 A CASE IN POINT: SAMPLE SURVEYS

Many of the ideas in this chapter have direct application in sample surveys. A **sample survey** is a method of ascertaining the characteristics of a well-defined population on the basis of information from the members of an appropriately chosen sample. The alternative to a sample survey would involve the much more costly and time-consuming procedure of obtaining information from every single member of the population, as in a complete census.

Sample surveys form the basis for a variety of widely publicized reports, including monthly employment and unemployment rates, the Consumer Price Index, the Nielsen ratings of television viewing, and public opinion polls, such as those of the Gallup organization. In addition, numerous businesses conduct sample surveys to find out what features people like and dislike about their products, and government agencies at all levels, from federal to municipal, conduct sample surveys to determine which of their programs are meeting the needs they were designed for and what improvements or new programs should be introduced.

Accuracy and **precision** are the basic statistical concerns in sample surveys. Problems with accuracy are generally the more serious, since precise answers to the wrong questions (precise but inaccurate results) are apt to be less useful than approximate answers to the right questions (imprecise but accurate results). The following paragraphs deal with questions of accuracy and precision in three phases of a sample survey: sampling the population, asking the question, and evaluating the results.

Sampling the Population

A vitally important component in planning a sample survey is the sampling procedure. The first step here is to define clearly the relevant population. The selection of respondents must be specifically planned according to whether the population is all adults in the country, all students in a particular school, all senior citizens in a given city, all retail food stores in a specified region, or some other well-defined set of individuals or institutions.

Once the population has been defined, it is possible to devise a sampling procedure that will permit scientifically valid inferences to be drawn about the population on the basis of the sample data. We met one such procedure, called **simple random sampling**, in Section 8.1. In simple random sampling (without replacement), every member of the population of relevant observational units not yet selected for the sample has equal probability of being selected. Selection of such a sample requires the use of some inanimate device, such as a table of random digits, to ensure that no characteristic of the observational unit can surreptitiously make it more or less likely than other units to be selected.

A good sampling procedure enables us to calculate and report the margin

* This symbol identifies material that is not prerequisite for later topics in this book.

of error in the resulting estimates. Consider, for example, a simple random sample of n observations selected from a population of size N. If the survey is concerned with a numerically measured variable (such as how much money a family spends on food in a week) where the data will be summarized by a mean, the *standard error of the mean of a simple random sample* can be estimated by a simple formula based on Expression (8.9). If the survey yields categorized data that will be summarized by proportions, such as the proportions of individuals favoring various political candidates, the *standard error of the proportion observed in a simple random sample* can be estimated by a simple formula to be introduced in Chapter 14. It must be emphasized that these particular formulas—the ones we will meet in this book—give correct estimates of standard errors *for the case of simple random sampling only*; they will give inappropriate estimates for any other type of sampling. The important point here is that the use of simple random sampling enables us to determine the **sampling error**, or the standard amount of error (in our estimates) that results from observing only a sample rather than the whole population.

Sometimes it is possible to improve upon simple random sampling by moving to more complex forms of random sampling. These forms of random sampling also permit calculation of standard errors, although the methods are beyond the scope of this book. A more complex form is an improvement if a given sample size n yields greater precision—that is, if the standard error for the more complex sampling plan is smaller than that for a simple random sample of the same size. One important alternative to simple random sampling is **stratified random sampling**. Stratified random sampling will result in improved precision if the population can be partitioned into two or more relatively homogeneous subpopulations, called **strata** (plural of **stratum**), such that the observational units within any given stratum tend to be more similar to each other than to observational units in different strata. In stratified random sampling, a sample is randomly selected from each stratum separately, and these subsamples together constitute the total sample of size n.

EXAMPLE 8.14 Consider a four-year college with 500 first-year, 400 second-year, 300 third-year, and 200 fourth-year students, for a total of $N = 1400$ students. Suppose that we want to estimate the average age of students at the college using a sampling fraction of 1%. Since the population size is $N = 1400$, a sampling fraction of 1% implies a sample size of $n = (.01)(1400) = 14$ students.

In simple random sampling, we would use a table of random digits to select the student identification numbers of 14 of the 1400 students. Of course, it could happen that no fourth-year students would be included in a simple random sample, or that, say, first-year students would be underrepresented. Such results, which could lead to a relatively large error in our estimate, could occur just by chance in a simple random sample.

To prevent some of the misrepresentations that are possible in a simple random sample, we could treat first-year, second-year, third-year, and fourth-year students as constituting four strata and use **stratified random sampling**. This method involves specifying in advance how many of the 14 students in the entire sample we will select from each stratum. There are several ways in which these numbers may be specified. For example, if the number selected from each stratum is proportional to the size of the stratum, the procedure is called stratified random sampling with **proportional allocation**. Using this procedure, we apply the sampling fraction to each stratum separately, so that we take 1% of the 500 first-year students, 1% of the 400 second-year students, and so on, yielding a sample of 5 first-year, 4 second-year, 3 third-year, and 2 fourth-year students, for a total sample size of 14. In an alternative procedure known as stratified random sampling with **optimum allocation**, the variability within each stratum is also taken into account, in addition to the size of the stratum, and relatively more observational units are selected from strata with greater variance. In any case, once we have specified how many of the 14 students in the entire sample we will select from each stratum, then we use a table of random digits to select the student identification numbers of the specified numbers of first-year, second-year, third-year, and fourth-year students, for a total sample size of 14.

If stratified random sampling eliminates some of the larger errors that might occur by chance under simple random sampling, such as those due to not including any first-year students or not including any fourth-year students, then the resulting estimates will be more precise. In other words, the standard errors in such cases of stratified random sampling will be smaller than those for simple random sampling.

Earlier we referred to sampling procedures that permit scientifically valid inferences and enable us to calculate and report the size of the standard error of the resulting estimates. The general category of sampling procedures with these desirable characteristics is known as **probability sampling**, and it includes simple random sampling, stratified random sampling, and a variety of other applications of random sampling. Recall that the term *random* does not mean "haphazard" or "without a plan" in statistics; rather it implies adherence to certain probabilities. In simple random sampling, these probabilities are equal for all members of the entire population; in stratified random sampling, the probabilities are equal for all members of a given stratum but not necessarily equal for different strata. In general, **probability sampling** is any sampling procedure in which every observational unit in the entire population has a known (but not zero) probability of being included in the sample. The probabilities need not all be equal, but they must all be known and they must not be zero. A sample survey must use some form of probability sampling if it is to be taken seriously as a scientifically valid survey.

The alternative to probability sampling is known as **nonprobability sampling**, or **nonrandom sampling**, in which observational units are selected by nonrandom methods. The use of nonrandom methods introduces the possibility of systematic bias in the estimates and precludes determination of the size of the sampling error in our estimates. That is, no formulas exist for estimating standard errors of nonrandom samples, and use of the random-sample formulas (such as those presented in this book) will yield incorrect estimates for the nonrandom case. Examples of nonprobability sampling include the following:

> **convenience sampling**, in which observational units that are more readily available are included in the sample and less convenient units have no chance of being included;

judgment sampling, in which someone, perhaps an expert, selects what appears to be a representative sample; and

quota sampling, in which hired technicians, such as interviewers, are required to collect data from a certain number of observational units in specified categories (for example, five unemployed male teenagers, five female senior citizens, and so on) but are free to select any observational unit satisfying the guidelines.

Results from all forms of nonprobability sampling are susceptible to serious biases, because the value of the variable of interest may be different in those (unobserved) observational units possessing whatever characteristic it was that made them unlikely to be selected. In other words, nonprobability sampling omits certain ill-defined subpopulations from possible inclusion in the sample, leaving us no way of estimating the size of sampling errors. Examples 8.4 and 8.5 illustrate the dangers of nonprobability sampling.

One particular form of nonprobability sampling warrants special mention: **Systematic sampling** involves selecting every 10th, or every 100th, or in general every mth observational unit from a sequential array of the members of the population (for example, every 10th name on a list, or every 100th product from an assembly line). Such a sample is not a random sample even if the starting point is selected randomly. However, it may sometimes be treated as a random sample, provided that there are no hidden periodicities in the underlying process. Any regularly recurring, or periodic, pattern in the underlying process may lead to serious bias in the resulting estimates. (For example, every 10th house on a street may be a corner house; every 100th product off an assembly line may be at the start of a work shift.) The use of systematic sampling should raise a mental warning flag to be alert for possible periodicities. Probability sampling is safer, although sometimes less convenient.

Asking the Question

Whether the data in a sample survey are collected by personal interviews, by telephone, or by mail, the way the question is phrased is a critical consideration. Even apparently minor changes in the wording of a question can make a substantial difference in the answers received. Some words are interpreted differently by different people—for example, does "unemployed" include those on temporary layoff, on strike, and on holiday?—so survey questions should be specific and should avoid technical terms. Furthermore, questionnaires should always be pretested in a subsample of the relevant population to check for previously overlooked ambiguities and misinterpretations. It is not sufficient that particular words are used correctly on the questionnaire; what matters is that the way the words are interpreted by the respondents corresponds to what the surveyor intended to ask.

Similarly, interviewers must be carefully trained to understand the importance of the manner in which questions are put to the respondents, not only in terms of the precise wording but also in terms of their delivery. Tone of voice, rate of speaking, and even gestures provide subtle but powerful cues that can unintentionally bias the results of a survey conducted by inadequately trained interviewers, especially with respect to attitudinal questions such as political preferences.

Evaluating the Results

When designed and conducted properly, sample surveys can provide highly precise and accurate information about large populations on the basis of relatively small samples. The following list identifies some of the major considerations that are relevant in the evaluation of sample surveys and their results. The first item pertains to the question of **precision**; the other five items pertain primarily to the question of **accuracy**.

1. *Sampling error.* The report of a sample survey should include a statement of the sampling error for each estimate. In succeeding chapters we will consider several forms that such a statement may take, but they are all based on the relevant standard error.
2. *Sampling procedure.* Serious distortion can result from the use of nonprobability sampling methods such as convenience sampling (including the use of volunteer respondents), judgment sampling, and quota sampling. (News reporters sometimes report that they obtained information about public opinion from a "random" sample of people at a mall or on a downtown street. They mean a haphazard, convenience sample.)
3. *Missing data.* Frequently data cannot be obtained from all of the observational units selected for inclusion in the sample. Respondents who refuse to answer a question, or who are never at home, or who cannot be located may very well differ from other respondents with respect to the variable of interest. Thus, even if probability sampling is used to identify the original sample, a low response rate destroys the randomness and reduces the ultimate sample of cooperating respondents to a convenience sample (namely, those in the originally identified sample that were available for observation). Such distortions in the sample create a serious problem. It is difficult to overemphasize the importance of making every effort to obtain data from all of the observational units in the originally identified random sample. The percentage of nonrespondents in the original sample should be reported in the results of the survey.
4. *Respondent errors.* Survey results can be rendered meaningless if respondents do not provide correct information, whether because of lack of knowledge, misinterpretation of the question, or deliberate misrepresentation, perhaps out of fear or embarrassment. Examples 1.2, 1.3, and 1.4 provide illustrations.
5. *Wording of the question.* The wording of the question is the primary stimulus to which the respondent responds, and it must therefore be such as to elicit appropriate information. Whether intentionally or

unintentionally, the wording may draw respondents' attention to certain aspects of the topic and bias their answers accordingly. (In Example 14.1, is the higher approval rate in 1976 due to the nearness of the games or to the omission of cost information in the question that year?)

6. *Interviewer errors.* Interviewers may unwittingly reinforce respondents for certain kinds of answers or misinterpret their answers for classification on the response form.

STUDY MATERIAL FOR CHAPTER 8

Terms and Concepts

sampling
accuracy
precision

Section 8.1

simple random sampling
random sampling
random sample
simple random sample
sampling distribution
replication
empirical sampling distribution
theoretical sampling distribution

Section 8.2

simulation
Monte Carlo methods
stanine scale
triangular distribution

Section 8.3

standard error
standard error of the mean

Section 8.4

theoretical sampling distribution of the mean
finite-population correction factor
sampling fraction
Central Limit Theorem
rule of thumb for the Central Limit Theorem

* Section 8.5

sample survey
sampling error
stratified random sampling
stratum/strata
proportional allocation
optimum allocation
probability sampling
nonprobability sampling
nonrandom sampling
convenience sampling
judgment sampling
quota sampling
systematic sampling

Key Formulas

Section 8.4

(8.9) $$\sigma_{\bar{x}} = \frac{\sigma}{\sqrt{n}}$$

(8.10) $$z_{\bar{x}} = \frac{\bar{x} - \mu}{\sigma_{\bar{x}}}$$

Exercises for Numerical Practice

Section 8.4

8/1. A random sample of 16 observations with a mean of 61.0 is drawn from a Gaussian population with a mean of 56.9 and a standard deviation of 23.1. Calculate the following:

- **a.** the mean and the standard deviation of the sampling distribution of the sample mean.
- **b.** the standard error of the mean.
- **c.** the number of standard errors that the sample mean is away from the population mean.
- **d.** the probability of getting a sample mean of 61.0 or more in this situation.

8/2. A random sample of 9 observations with a mean of 3.19 is drawn from a Gaussian population with a mean of 3.64 and a standard deviation of .87. Calculate the following:

- **a.** the mean and the standard deviation of the sampling distribution of the sample mean.
- **b.** the standard error of the mean.
- **c.** the number of standard errors that the sample mean is away from the population mean.
- **d.** the probability of getting a sample mean of 3.19 or less in this situation.

8/3. A random sample of 12 observations is drawn from a Gaussian population with a standard deviation of 42.6. Calculate the following:

- **a.** the standard error of the mean.
- **b.** the probability that the sample mean will differ from the population mean by no more than 10.0.

8/4. A random sample of 20 observations is drawn from a Gaussian population with a standard deviation of 677. Calculate the following:

- **a.** the standard error of the mean.
- **b.** the probability that the sample mean will differ from the population mean by no more than 50.

8/5. A random sample of 64 observations with a mean of 144 is drawn from a large population with a mean of 156 and a standard deviation of 88. Calculate the following:

- **a.** the mean and the standard deviation of the sampling distribution of the sample mean.
- **b.** the standard error of the mean.
- **c.** the number of standard errors that the sample mean is away from the population mean.
- **d.** the probability of getting a sample mean of 144 or less in this situation.

8/6. A random sample of 100 observations with a mean of 7.50 is drawn from a large population with a mean of 7.24 and a standard deviation of 1.27. Calculate the following:

- **a.** the mean and the standard deviation of the sampling distribution of the sample mean.
- **b.** the standard error of the mean.
- **c.** the number of standard errors that the sample mean is away from the population mean.
- **d.** the probability of getting a sample mean of 7.50 or more in this situation.

8/7. A random sample of 50 observations is drawn from a large population with a standard deviation of 3.05. Calculate the following:

- **a.** the standard error of the mean.
- **b.** the probability that the sample mean will differ from the population mean by more than 1.00.

8/8. A random sample of 75 observations is drawn from a large population with a standard deviation of 52.0. Calculate the following:

- **a.** the standard error of the mean.
- **b.** the probability that the sample mean will differ from the population mean by more than 20.0.

Problems for Applying the Concepts

8/9. What effect would each of the following have on the accuracy and the precision of the resulting estimate of the mean of the variable?

a. Accidentally omitting unnumbered pages (for example, color plates or prefatory material) in counting total numbers of pages in books.

b. Recording attitudes as 1 (pro) or 0 (con) rather than on, say, a seven-point scale (1, 2, . . . , 7).

c. Reporting age in years in the usual way (for example, a person says she is 19 from her nineteenth birthday until the day before her twentieth birthday) rather than to the *nearest* year.

8/10. What effect would each of the following have on the accuracy and the precision of the resulting estimate of the mean of the variable?

a. Rounding actual family incomes to the nearest $1000.

b. Determining mass on a balance that, unknown to you, reads 10 grams too low.

c. Accepting verbal reports of a student's study time for a test.

8/11. American adult IQ scores on the Wechsler Adult Intelligence Scale are known to be approximately normally distributed with a mean of 100.0 and a standard deviation of 15.0.

a. What are the mean and the standard deviation of the sampling distribution of means of random samples of

i. n = 9 observations?
ii. n = 36 observations?

b. How large a random sample would we need to estimate the mean with a standard error of

i. 1.0 IQ points?
ii. 2.0 IQ points?

c. What is the probability that the mean IQ score of a random sample of 16 American adults would be

i. 110.0 or higher?
ii. 95.0 or lower?

d. What is the probability that the mean IQ score of 4 randomly selected American adults would differ from the true mean of 100.0 by

i. 15.0 IQ points or more?
ii. 12.0 IQ points or more?

8/12. Family income in a certain region is known to be approximately normally distributed with a mean of $25,000 and a standard deviation of $6,000.

a. What are the mean and the standard deviation of the sampling distribution of means of random samples of

i. n = 100 observations?
ii. n = 400 observations?

b. How large a random sample would we need to estimate the mean with a standard error of $500?

c. What is the probability that the mean income of a random sample of 80 families would be

i. $27,000 or more?
ii. $24,000 or less?

d. What is the probability that the mean income of 150 randomly selected families would differ from the population mean income by

i. \$500 or less?
ii. \$200 or less?

8/13. For adult females, the red blood cell count obtained in a medical blood test has a mean of 4,500,000 per mm^3 and a standard deviation of 350,000 per mm^3.

a. What are the mean and the standard deviation of the sampling distribution of means of random samples of

i. n = 200 observations?
ii. n = 50 observations?

b. How large a random sample would we need to estimate the mean with a standard error of

i. 40,000 per mm^3?
ii. 120,000 per mm^3?

c. What is the probability that the mean red blood cell count of a random sample of 40 women would be

i. 4,550,000 per mm^3 or higher?
ii. 4,400,000 per mm^3 or lower?

d. What is the probability that the mean red blood cell count of 75 randomly selected women would differ from the true mean by

i. 25,000 per mm^3 or less?
ii. 100,000 per mm^3 or less?

8/14. The actual weight of coffee in a 1-pound package produced by a food-processing firm is approximately normally distributed with a mean of 16.00 ounces and a standard deviation of .05 ounce.

a. What are the mean and the standard deviation of the sampling distribution of means of random samples of

i. n = 18 packages?
ii. n = 60 packages?

b. How large a random sample would we need to estimate the population mean weight with a standard error of .0075 ounce?

c. What is the probability that the mean weight of a random sample of 35 packages would be

i. 15.99 ounces or less?
ii. 16.02 ounces or more?

d. What is the probability that the mean weight of 10 randomly selected packages would differ from the population mean by

i. .01 ounce or more?
ii. .05 ounce or more?

8/15. Problem 7/39 (Chapter 7) presented a grade distribution with a mean of 5.91 and a standard deviation of 1.67. If students' grades could be regarded as a random sample from the given distribution, what is the probability that

a. a class of 40 students would have a mean grade of 6.00 or higher?
b. a class of 400 students would have a mean grade of 6.00 or higher?

8/16. The mean length of human pregnancies is 266 days with a standard deviation of 16 days.

a. What is the probability that the mean length of pregnancy of a random sample of 30 pregnant women would be 272.0 days or more?

b. What is the probability that the mean length of pregnancy of a random sample of 90 pregnant women would be 272.0 days or more?

8/17. Consider the discrete rectangular distribution with the probability function

$$P(x) = .1 \qquad \text{for } x = 0, 1, 2, \ldots, 9$$

a. Calculate the mean, the median, and the standard deviation of this population.

b. Calculate the mean and the standard deviation of the theoretical sampling distribution of means of random samples of $n = 5$ observations drawn from this population.

c. Describe a Monte Carlo procedure for obtaining 25 random samples of $n = 5$ observations from the stated population. Record the observations resulting from the procedure.

d. Calculate the mean of each of the 25 samples, and present the 25 sample means in a histogram with category boundaries of $-.5, .5, 1.5, \ldots, 9.5$.

e. Calculate the mean and the standard deviation of the 25 sample means. Compare these two values with the mean and the standard deviation computed in Part (b).

f. Calculate the median of each of the 25 samples, and present the 25 sample medians in a histogram as in Part (d).

g. Calculate the mean and the standard deviation of the 25 sample medians. Compare this mean with the population median in Part (a), and compare this standard deviation with that from Part (e). What does the latter comparison indicate about sample means and sample medians?

8/18. Consider the discrete triangular distribution with the probability function

$$P(x) = (x + 1)/55 \qquad \text{for } x = 0, 1, 2, \ldots, 9$$

a. Calculate the mean, the median, and the standard deviation of this population.

b. Calculate the mean and the standard deviation of the theoretical sampling distribution of means of random samples of $n = 5$ observations drawn from this population.

c. Describe a Monte Carlo procedure for obtaining 20 random samples of $n = 5$ observations from the stated population. Record the observations resulting from the procedure.

d. Calculate the mean of each of the 20 samples, and present the 20 sample means in a histogram with category boundaries of $-.5, .5, 1.5, \ldots, 9.5$.

e. Calculate the mean and the standard deviation of the 20 sample means. Compare these two values with the mean and the standard deviation computed in Part (b).

f. Calculate the median of each of the 20 samples, and present the 20 sample medians in a histogram as in Part (d).

g. Calculate the mean and the standard deviation of the 20 sample medians. Compare this mean with the population median in Part (a), and compare this standard deviation with that from Part (e). What does the latter comparison indicate about sample means and sample medians?

References

Bell, E. T. *Men of Mathematics*. New York: Simon and Schuster, 1937.

Bryson, M. C. "The Literary Digest Poll: Making of a Statistical Myth." *American Statistician* 30 (1976): 184–85.

Bullough, V. L. "Age at Menarche: A Misunderstanding." *Science* 213 (1981): 365–66.

Fréchet, M., and Stigler, S. M. "Laplace, Pierre Simon de." *International Encyclopedia of Statistics*. Edited by W. H. Kruskal and J. M. Tanur. New York: The Free Press, 1978.

Leete, E. "Children of Organic Chemists." *Chemical & Engineering News,* 28 September 1981, p. 57. (a).

Leete, E. Personal communication to J. O. Olson and K. A. Olson, 1 December 1981. (b).

World Who's Who in Science. Chicago: Marquis Who's Who, 1968.

Further Reading

Campbell, S. K. *Flaws and Fallacies in Statistical Thinking*. Englewood Cliffs, NJ: Prentice-Hall, 1974. Chapter 12 discusses sampling, and Chapters 2, 3, and 4 provide a variety of interesting illustrations of the importance of having adequate information about the entire data-collection procedure if statistical results are to be meaningful.

Ferber, R., Sheatsley, P., Turner, A., and Waksberg, J. *What Is a Survey*? Washington, DC: American Statistical Association, 1980. This excellent booklet describes in nontechnical language the planning, execution, and interpretation of sample surveys. Single copies are available free upon request from the American Statistical Association, 806 Fifteenth Street, N.W., Washington, DC 20005.

Huff, D. *How to Lie with Statistics*. New York: Norton, 1954. Chapter 10 offers good, sound advice on "how to talk back to a statistic," indicating what information must be provided about sampling and other aspects of data collection if statistical results are to be meaningful.

Weaver, W. *Lady Luck: The Theory of Probability*. Garden City, NY: Doubleday, 1963. Besides introducing the basic ideas of sampling, Chapter 14 includes a clear account of the relationship between probability and statistics.

Wheeler, M. *Lies, Damn Lies, and Statistics: The Manipulation of Public Opinion in America*. New York: Liveright, 1976. This nontechnical book gives insights into some of the potential pitfalls in sample surveys, illustrated by a journalistic account of public opinion polling in the 1972 U.S. presidential election campaign.

Williams, W. H. *A Sampler on Sampling*. New York: Wiley, 1978. This book gives a more thorough treatment of several topics introduced in this chapter.

PART III

INFERENTIAL STATISTICS

EXAMPLE 9.0 "In the space of one hundred and seventy-six years the Lower Mississippi has shortened itself two hundred and forty-two miles. That is an average of a trifle over one mile and a third per year. Therefore, any calm person, who is not blind or idiotic, can see that in the Old Oölitic Silurian Period, just a million years ago next November, the Lower Mississippi River was upward of one million three hundred thousand miles long, and stuck out over the Gulf of Mexico like a fishing-rod. And by the same token any person can see that seven hundred and forty-two years from now the Lower Mississippi will be only a mile and three-quarters long, and Cairo [Illinois] and New Orleans will have joined their streets together, and be plodding comfortably along under a single mayor and a mutual board of aldermen. There is something fascinating about science. One gets such wholesale returns of conjecture out of such a trifling investment of fact."

(Source: Mark Twain, *Life on the Mississippi*, 1883.)

Inferential statistics is the branch of the science of statistics dealing with methods for drawing appropriate inferences about populations on the basis of samples drawn from those populations. In studying descriptive statistics in Part I, we considered ways of describing numerical data, especially by using a single numerical value to summarize a certain aspect of the data. Such a numerical value was termed a *statistic* in the case of sample data or a *parameter* in the case of population data. Now, in inferential statistics, we will apply the principles of probability theory from Part II to the descriptive statistics from Part I to see what the various sample statistics can tell us about the populations from which the samples were drawn.

By its very nature, inferential statistics is vitally important in science, in business, and in everyday life. It assists us in knowing what we can and cannot conclude when we have only a sample of all the relevant observations that would be required to answer an empirical question with certainty.

CHAPTER 9

OVERVIEW OF THE LOGIC OF CLASSICAL STATISTICAL INFERENCE

EXAMPLE 9.1 "It all started on a hot July day in 1946 in one of those new home freezing units. In his laboratory at the General Electric Company, Vincent Schaefer serendipitously discovered that a bit of Dry Ice could create a virtual snowstorm in a freezer that until then had contained only a fog of cold water. If it worked in a freezer, many reasoned, it should work in clouds to tame storms and make rain.

". . . The hazard for [weather-modification] researchers during the past 35 years has not been so much the weather itself as their repeated failures to prove early claims that they could change the weather. . . . Today, only a single set of experiments, which were conducted in Israel, appears to have confirmed an increase in precipitation after cloud seeding. The results of a few other experiments seem encouraging but hardly convincing. Both the one apparent success and the failures demonstrate that weather-modification experiments require statistical rigor as well as some idea of how clouds work if researchers are to overcome the confounding natural variability of the weather.

". . . The first of two Israeli weather modification experiments ran from 1961 to 1967 under the direction of three researchers from the Hebrew University of Jerusalem . . . . They wanted to determine whether seeding wintertime clouds with fine particles of silver iodide would increase rainfall over northern Israel. They assumed that many Israeli clouds could yield more rain if silver iodide were added to them to promote the formation of ice particles from water droplets cooled below the freezing point, the first step in the precipitation process. The Israeli I experiment appeared to have succeeded. Rainfall, according to a number of independent analyses, increased about 15 percent in the target areas after cloud seeding. That is a respectable amount in rainmaking circles . . . .

"The primary purpose of the second, 1969–1975 Israeli experiment was to see if seeding would enhance rainfall over the drainage system that supplies water to the Sea of Galilee . . . . As it turned out, Israeli II also served as a confirmatory experiment; that is, one in which specific hypotheses are tested and a strict design is adhered to throughout in order to confirm an apparent effect of a preceding, exploratory experiment. Only in the late 1970s did the concepts of exploratory and confirmatory experiments become formally accepted among weather-modification researchers, and then only at the insistence of statisticians. . . .

"The strict design of Israeli II . . . seems to have paid off. According to recent analyses by the experimenters, precipitation increased 13 percent in the target area as a whole and 18 percent in the smaller catchment area. The probability that [such a large increase would result] from a chance distribution of particularly rainy days was 2.8 percent for the whole target area and 1.7 percent for the catchment area."

The observational units in Israeli II were days with suitable clouds for seeding in the given region. The sample consisted of 388 such days; the population is all such days that might ever occur. Methods of inferential statistics were used to draw conclusions about the true effect of this type of cloud seeding in general (that is, conclusions about the population) on the basis of a limited set of observations (that is, on the basis of a sample).

(Source: Kerr, 1982. Copyright 1982 by the American Association for the Advancement of Science. Used with permission of the AAAS.)

Inferential statistics includes an impressive array of tools for drawing conclusions about populations on the basis of sample data, but for purposes of our discussion they can all be organized according to their use of one of two basic sorts of logical procedure: estimation and hypothesis testing. This chapter briefly outlines the logic of each of these two general procedures before we attempt a more detailed coverage in the context of some particular methods in subsequent chapters. The main purpose of this chapter is not to give you an exhaustive tour of the realm of inferential statistics but rather to show you a map, as it were, of the area, to help you organize the material and recognize some important distinctions among different types of inferences.

9.1 ESTIMATION

Estimation refers to the procedure of obtaining an approximate numerical value, called the **estimate**, of a population parameter through the analysis of sample data. Any characteristic of a population that can be described in terms of a parameter can be estimated on the basis of data from an appropriate sample. That is, there are methods for estimating population means, variances, standard deviations, regression coefficients, dependent-variable means at given independent-variable values, correlation coefficients, proportions, and various other population values. Thus, methods of estimation can be categorized according to the parameter being estimated: estimation of population means, estimation of population variances, and so on.

For each estimable parameter, two types of estimation can be distinguished according to whether the estimate is presented in the form of a single numerical value or an interval of values. The resulting estimates are called **point estimates** and **interval estimates**, respectively.

A good estimate should be both accurate (free from systematic error) and precise (relatively free of random error), and it should be reported in such a way that the reader gains a correct impression of its accuracy and its precision. As discussed in Section 8.1, pages 286–289, considerations related to *accuracy* require the provision of information about how the sample was selected. Now we will see that the inclusion of information relevant to *precision* distinguishes interval estimates from point estimates.

Point Estimates

The simplest form of statistical inference is characterized by the use of a single statistic to estimate a parameter.

> A **point estimate** is a single numerical value calculated from sample data and taken to be indicative of the value of the population parameter. **(9.1)**

EXAMPLE 9.2 How many hours of sleep do college students get on a typical weeknight? A random sample of 22 male students living in college residences were asked about their average length of weeknight sleep. Their reported weeknight sleep durations had a mean of 7.05 hours and a standard deviation of .86 hour.

The sample mean, $\bar{x} = 7.05$, is a point estimate of the mean μ of weeknight sleep durations that would be reported by all male students living in college residences. The sample standard deviation, $s = .86$, is a point estimate of the standard deviation σ of weeknight sleep durations that would be reported by all members of that population. How closely the reported sleep durations correspond to the actual sleep durations is, of course, an open question.

(Source: Student project by D. Stewart.)

EXAMPLE 9.3 In the experiment identified as Israeli I in Example 9.1, rainfall was found to have increased by about 15% in the target areas after cloud seeding. The observed value of 15% is a point estimate of the true effect of this type of cloud seeding; that is, it is a point estimate of the percentage increase in rainfall that would result if this type of seeding were used in the indefinitely large population of suitable days that might ever occur.

EXAMPLE 9.4 According to a radio news report in July 1982, 48% of senior business executives answering a survey thought the economy would get worse before the end of the year, as compared with only 14% who thought it would get better; the rest thought it would stay about the same. The values 48% and 14% are point estimates of the corresponding proportions of business executives holding those views in the entire population from which the sample was drawn. As noted in Example 8.3, the point estimates do not tell us very much if we do not know what population of business executives was sampled.

The chief drawback of point estimates is that they provide no information about their precision. As we know from our discussion of sampling distributions in Chapter 8, estimates are rarely exactly correct; rather, the possible values of the estimate are distributed around the true value of the parameter, according to the sampling distribution. Nonetheless, point estimates sometimes leave the impression of exactness, perhaps—ironically—because they are completely silent about their margin of error. We can partially overcome this drawback by reporting, along with the point estimate, sufficient information to enable the reader to judge its precision.

EXAMPLE 9.5 In Example 9.4, we have no information about the precision of the estimate that 48% of senior business executives thought the economy would get worse. As a general rule, precision improves with increasing sample size, but the report gives no information as to how many business executives were surveyed. Does the value of 48% represent 10 out of 21, or 48 out of 100, or perhaps 480 out of 1000? Results from a sample of 21 are not nearly as precise as those from a sample of 1000. Information about sample sizes is an important ingredient in specifying the precision of an estimate.

A direct measure of the precision of the mean of a random sample of n observations is given by its standard error [Expression (8.9)]:

$$\sigma_{\bar{x}} = \frac{\sigma}{\sqrt{n}}$$

which is a standard or "average" amount by which the mean of a random sample differs from the population mean of which it is an estimate. In practice, we generally do not know the value of the population standard deviation σ, and so we cannot calculate the actual value of the standard error of the mean. Instead we use the **estimated standard error of the mean:**

$$s_{\bar{x}} = \frac{s}{\sqrt{n}} \quad (9.2)$$

where s is the sample standard deviation and n is the sample size. The estimated standard error provides information about the precision of a random sample mean: The smaller the standard error, the more precise is the sample mean as an estimate of the population mean.

EXAMPLE 9.6 In Example 9.2, the mean number of hours of sleep in a weeknight reported by 22 male students living in college residences was 7.05 hours. The fact that the estimate was based on 22 observations is relevant to precision. Other things being equal, such an estimate is more precise than one based on, say, 2 observations, but not as precise as one based on, say, 200 observations.

The fact that the sample standard deviation was .86 hour is also relevant to precision. We know that .86 hour was a standard amount of deviation from the mean and that not too many of the observations in the sample would be more than about two standard deviations—that is, 1.72 hours—away from the sample mean.

An even more direct measure of the precision of the sample mean is given by its estimated standard error:

$$s_{\bar{x}} = \frac{.86}{\sqrt{22}} = .18 \text{ hour}$$

Thus, we have an estimate of 7.05 hours with an estimated standard error of .18 hour.

Point estimates are widely used and unquestionably useful, but they should be accompanied by collateral information relevant to accuracy (for example, how the sample was chosen, how the data were collected) and precision (for example, sample size, standard error).

One way of presenting collateral information with a point estimate is sufficiently common, especially in the natural sciences, that it warrants special mention. The practice is to report the point estimate in the form

point estimate ± estimated standard error of the point estimate **(9.3)**

to indicate how precise the point estimate is. Notice in the following examples that reporting a point estimate in this form requires a statement as to whether the value given is "± standard error" or "± standard deviation," because the alternative format

mean ± standard deviation **(9.4)**

is sometimes used if the purpose is simply to describe a set of numerical values and their variability.

EXAMPLE 9.7 In the investigation described in Example 9.6 to determine how much sleep male college students get, the average reported sleep duration was 7.05 ± .18 hours (mean ± standard error). Here we wanted to indicate the precision of the point estimate, so we used Expression (9.3).

EXAMPLE 9.8 The report of a study comparing the diets of individuals with and without essential hypertension (high blood pressure) gave the daily "total caloric intakes (mean ± standard error)" of the two groups as 1790 ± 108 kcal for the hypertensives and 1945 ± 107 kcal for the control subjects.

Here again the form given in Expression (9.3) was used to indicate the precision of the point estimates.

(Source: McCarron *et al.*, 1982.)

EXAMPLE 9.9 In a study of how sleep disturbances can be minimized for shift workers who rotate between day, evening, and night shifts, the researchers described their subjects as follows: "We compared 85 male rotating shift workers, aged 19 to 68 (mean ± standard deviation, 31.4 ± 10.0), with a control group of 68 male non-rotating day and swing shift workers with comparable jobs, aged 19 to 56 (mean, 27.3 ± 8.2), at the Great Salt Lake Minerals and Chemicals Corporation in Ogden, Utah."

Here the purpose was to describe the individuals and the variability among them, so Expression (9.4) was used.

(Source: Czeisler *et al.*, 1982.)

Interval Estimates

Even if collateral information relevant to precision is provided with a point estimate, some statistical sophistication is still required to translate that information into an appreciation of the precision—or imprecision—of the estimate. Estimation in terms of intervals rather than points incorporates the collateral information relevant to precision into the estimate itself and thus imposes fewer demands on the recipient of the information.

An **interval estimate** is a set of adjacent numerical values (or an interval on the real-number line) determined from sample data and accompanied by a statement of the probability that such an interval includes the population parameter. **(9.5)**

EXAMPLE 9.10 Here is an interval estimate for the data in Example 9.6 concerning college students' amount of sleep: We can be 95% confident that the population mean of weeknight sleep durations reported by male students living in college residences is between 6.67 and 7.43 hours. In Chapter 10, we will study how such values are calculated from the sample data; for now, we simply introduce the concept of the interval estimate.

In the type of interval estimate that we will study in this book, the interval is usually called a **confidence interval** (6.67 to 7.43, in the Example 9.10), the endpoints of the interval are the **confidence limits** (6.67 and 7.43, in the example), and the accompanying probability is the **confidence level** or **degree of confidence**. The most commonly used confidence levels are 95% and 99%.

The **confidence level** is probably best interpreted as the success rate of the method of estimation, in this sense: If we used this method of estimation with many different random samples from the same population, then in about 95% of those samples, the 95% confidence interval would include the population value. Note in particular that the 95% confidence interval does *not* imply that the population value (the parameter) moves around and is in the interval 95% of the time. The parameter is a single fixed number; it does not vary. The variability is in the sample and hence in the confidence limits or endpoints, which are determined from the sample data and are therefore liable to random error. Different samples yield different confidence limits (different interval estimates) just as they yield different sample means (different point estimates). Thus, the confidence limits would vary from sample to sample, if repeated samples were taken from the same population, but they are established by a method that assures us that the limits include the parameter with the stated level of confidence. In brief, the parameter is an unknown constant; the confidence limits are variable (subject to error) but are characterized by a success rate given by the confidence level.

9.2 HYPOTHESIS TESTING

The second general type of statistical inference, after estimation, is **hypothesis testing** or **significance testing**, which is used to determine whether an observed (sample) difference provides enough evidence for us to believe that there is a corresponding difference in the underlying population. Although we will discuss hypothesis testing further in Section 11.4 after we have actually used some significance tests, it will be helpful to introduce the basic ideas here.

A **significance test** is a procedure for evaluating whether

a. the difference between a sample value and another value (either a constant or another statistic) should be attributed simply to random error ("chance"), in which case the sample difference is said to be **not statistically significant**, **(9.6)**

or alternatively whether

b. the difference is large enough for us to conclude that the corresponding population values are different, in which case the sample difference is said to be **statistically significant**.

The designations *significant* and *not significant* are applied to sample values only, never to population values.

It is important not to confuse statistical significance with practical importance. A difference may be statistically significant (that is, not due to chance) and yet be so small or due to such a commonplace cause as to be of no practical importance. Some knowledge of the area of application is needed to judge the importance of a statistically significant finding.

Significance tests may be categorized according to which feature of the population is being tested. There are tests about population means, variances (or standard deviations), regression coefficients, dependent-variable means at given independent-variable values, correlation coefficients, proportions, and various other features of populations.

EXAMPLE 9.11 In the study of hypertension described in Example 9.8, a significance test was performed to test the difference between the mean amount of calcium consumed by hypertensive subjects and that consumed by control subjects. The results showed that "compared to 44 normotensive controls, 46 subjects with essential hypertension reported significantly less daily calcium ingestion (668 ± 55 milligrams [mean ± standard error] compared to 886 ± 89 milligrams)."

The sample means, $\bar{x}_{\text{hyper}} = 668$ mg and $\bar{x}_{\text{control}} = 886$ mg, were estimates of the true calcium ingestion means, μ_{hyper} and μ_{control}, in the underlying populations of hypertensive and normal individuals. The fact that the sample difference was "significant" indicates that it is a larger difference than we would expect to find between means of random samples from populations with the same mean; that is, it is a larger difference than we would expect to occur if random error were the only reason for the difference. If random error is not the only reason for the difference, what else is responsible? If hypertension was the only relevant characteristic distinguishing the hypertensive subjects from the control subjects, the obvious conclusion is that hypertensives tend to consume less calcium than do other individuals: $\mu_{\text{hyper}} < \mu_{\text{control}}$.

It is a separate question as to whether the statistically significant difference in reported calcium consumption between hypertensive and other individuals is of any practical importance. It is the researchers' expertise in the health sciences that enabled them to conclude: "The data suggest that inadequate calcium intake may be a previously unrecognized factor in the development of hypertension."

(Source: McCarron *et al.*, 1982.)

EXAMPLE 9.12 Traffic researchers tested the influence of the popular tranquilizer diazepam (Valium) on a person's highway driving performance. They found that the variability, as measured by the variance, of the left–right position of the vehicle on the road was significantly greater when the driver was under the influence of a 10-milligram dosage of diazepam than when he had received either 0 or 5 milligrams of the drug. The sample variances s_0^2, s_5^2, and s_{10}^2 for the three dosages were estimates of the true variability σ_0^2,

σ_5^2, and σ_{10}^2 in left–right position produced by the three dosages. Drivers occasionally drifted across their lane boundaries whether under the influence of diazepam or not, but the significance test showed that the increased variability in left–right position under the influence of diazepam was larger than one would expect by chance, so that "potentially dangerous impairment [due to diazepam] was inferred from the reactions of some subjects."

(Source: O'Hanlon *et al.*, 1982.)

EXAMPLE 9.13 Are male students more likely than female students to wear glasses when their photograph is being taken? A random sample of 63 photographs taken for student identification cards was examined to investigate this question. It was found that 33% of the men and 15% of the women in the sample wore glasses for the picture, but the difference between these sample proportions was not statistically significant. A difference as large as that observed in this sample could reasonably be expected to occur just by chance even if males and females are equally likely to wear glasses for a photograph. The sample proportions $\hat{p}_m = .33$ and $\hat{p}_f = .15$ were estimates of the true proportions p_m and p_f of glasses-wearers in the relevant populations of male and female college students.

The absence of a significant difference between $\hat{p}_m$ and $\hat{p}_f$ does not prove that the population proportions p_m and p_f are the same—that males and females are equally likely to wear glasses for a photograph. For example, a larger sample would have given us more precise estimates of the population proportions and might have led to a different conclusion. All we can say is that the sample does not provide enough evidence to rule out that hypothesis. Even if the difference had been statistically significant, we would not know whether it occurred because women have better eyesight, or because women are more likely to wear contact lenses, or because women with glasses are more likely to remove them for a photograph.

(Source: Student project by M. Brink.)

The concept of **significance testing** or **hypothesis testing** requires a clear distinction between samples and populations, and in the terminology appropriate to each. You may find it helpful to think in terms of two distinct realms of knowledge. One realm is the realm of observation, consisting of incomplete evidence, samples, sample values, significance or nonsignificance, and other things we can observe or calculate from our observations. The other realm is the realm of the unknown, consisting of the underlying reality, the true state of affairs, populations, parameters, and other things about which we form hypotheses and draw inferences, because we cannot observe them directly.

Hypothesis testing involves assessing to what extent the sample data are consistent with two competing hypotheses about the underlying population. The sample data are the known but incomplete evidence; the hypotheses are competing claims about the true but unknown state of reality. In what follows, we will use the general term *sample data* or *sample difference* to refer to the evidence collected in the realm of observation, and the general term *population values* to refer to parameters or constants in the realm of the unknown underlying reality.

Two Hypotheses

The two competing hypotheses in hypothesis testing are known as the **null hypothesis** and the **alternative hypothesis**. Both hypotheses are *always* about unknown population values, never about sample data.

The **null hypothesis** is a claim that certain population values are *not* different from each other. (For example, the null hypothesis in Example 9.11 would propose that $\mu_{\text{hyper}} = \mu_{\text{control}}$; in Example 9.12, that $\sigma_0^2 = \sigma_5^2 = \sigma_{10}^2$; and in Example 9.13, that $p_m = p_f$, or that the population proportions of male and female students who wear glasses for a photograph are equal.) This constitutes a specific hypothesis that is set up as a working assumption because it allows us to calculate certain relevant probabilities. In effect, we evaluate the conditional probability of obtaining a sample difference at least as large as the one actually observed, given the null hypothesis. [In Example 9.13, we would evaluate the probability that we would find a sample difference as large as 33% versus 15% *if* the sample had come from a population in which the proportions of male and female students who wear glasses for a photograph are equal; that is, we evaluate P(such a large sample difference | null hypothesis).]

If we find that the sample data are sufficiently unlikely to have occurred in the situation proposed by the null hypothesis, we reject the null hypothesis, which says that the population values are equal, and conclude instead that the population values are different from each other. This is the case in which the sample difference is said to be **statistically significant** [Expression (9.6)].

If we find that the observed sample data are reasonably likely to occur in the situation proposed by the null hypothesis, we do not reject the null hypothesis, and we attribute the sample difference to chance. This is the case in which the sample difference is said to be **not statistically significant** [Expression (9.6)]. (In Example 9.13, the probability that we would obtain sample proportions as different from each other as 33% and 15%, if the population proportions were actually equal, turns out to be about .14. An event that would occur by chance 14 times in 100 is not especially unusual, so we conclude that the sample difference could reasonably be due to chance—it is not statistically significant—and the sample is consistent with the null hypothesis.)

Notice that hypothesis testing is carried out like a game of "let's pretend": Let us suppose, the reasoning goes, that the null hypothesis is true; then we can see what the chances would be of getting a sample as unusual as the one that actually occurred. Hypothesis testing does not tell us how likely it is that the null hypothesis is true, but rather how likely our sample data would be *if the null hypothesis were true.* That is, hypothesis testing does not evaluate P(null hypothesis), but rather P(such a large sample difference | null hypothesis). (In Example 9.13, P[such a large sample difference | null hypothesis] = .14.) From that information, we make a judgment as to whether the null hypothesis is sufficiently inconsistent with the sample data for us to reject the null hypothesis in favor of an alternative hypothesis. The **alternative hypothesis** says that the relevant population values are indeed different from each other. We will examine methods for evaluating such probabilities and guidelines for making such judgments in subsequent chapters.

Type I and Type II Errors

The result of a significance test is not and cannot be absolute certainty, for we are dealing with sample data—incomplete information, by definition. We may therefore make the error of rejecting the null hypothesis of no difference when in fact the population values are not different; this is called a **Type I error**. Or we may make the opposite error of failing to reject the null hypothesis of no difference when the population values are in fact different from each other; this is called a **Type II error**.

Moreover, our conclusion will depend in part on what we judge to be a "sufficiently unlikely" outcome for us to reject the null hypothesis. In practice, we usually define "sufficiently unlikely" in terms of the Type I error rate that we are willing to risk—that is, how willing we are to proclaim that we have found a population difference when in fact none exists—and the maximum acceptable Type I error rate is conventionally 5%. Thus, if sample differences as large as the one observed in our sample would occur less than 5% of the time in the situation proposed by the null hypothesis, it is conventional to reject the null hypothesis, running a 5% risk of a Type I error. (In Example 9.13, sample differences as large as the one observed would occur 14% of the time if the null hypothesis were true. Conventionally, we require a sample difference that would occur less than 5% of the time under the null hypothesis before we reject the null hypothesis, so in this case we do not reject the null hypothesis.)

The reader may find that the following analogy clarifies the nature of hypothesis testing. The analogy likens a significance test to a criminal court trial:

- **a.** The null hypothesis of equality corresponds to the assumption of innocence in criminal proceedings.
- **b.** The researcher typically seeks to demonstrate that the null hypothesis should be rejected, thus playing the role of the prosecutor. (If the researcher hopes to uphold the null hypothesis, it is generally more appropriate to use interval estimation than hypothesis testing.)
- **c.** The sample data constitute the evidence.
- **d.** The evidence is evaluated to see if it is consistent with the null hypothesis (assumption of innocence).
- **e.** Statistical significance corresponds to a verdict of guilty; lack of statistical significance, to a verdict of not guilty.
- **f.** A Type I error is represented by the conviction of an innocent person.
- **g.** A Type II error is represented by the acquittal of a guilty person.
- **h.** Both the significance test and the criminal trial involve a judgment about the true state of affairs on the basis of less than complete information.

In summary, a conclusion that a sample difference is statistically significant is a rejection of the null hypothesis. It is either a correct rejection or a Type I error. A conclusion that a sample difference is not statistically significant is a decision not to reject the null hypothesis. It is either a correct nonrejection or a Type II error.

Independent Variables and Dependent Variables

In Chapter 5, we saw that a *dependent* variable is one that is described in terms of, or as a function of, another variable, called the *independent* variable. Let us introduce some further terminology concerning the variables in hypothesis testing.

Hypothesis testing is widely used in comparing experimental treatments or conditions. In this context, the dimension along which the treatments vary is called the **independent variable**, or sometimes the **treatment variable**, and the different treatments themselves are the **levels of the independent variable** or, especially if they are numerically defined, the **values of the independent variable**. The term **treatment** itself is used quite broadly in statistics to refer to any experimental condition being investigated (whether or not any actual "treatment" is administered); that is, each of the levels of any independent variable may be referred to as a treatment.

The random variable that is observed is called the **dependent variable**, or sometimes the **response variable**, and the numerical observations recorded are the **values of the dependent variable**. This terminology is entirely consistent with the use of these terms in connection with regression and correlation in Chapter 5: The dependent (response) variable is described as a function of the independent (treatment) variable.

EXAMPLE 9.14 In Example 9.11, the independent variable was blood pressure (determined according to a standard procedure), and the "treatments" (levels of the independent variable) were normal blood pressure and high blood pressure. The dependent variable was calcium consumption in a 24-hour period, as determined by a computer analysis of verbal reports of food eaten.

The 44 values of the dependent variable for the normal subjects had a mean of 886 mg, and the 46 values of the dependent variable for the hypertensive subjects had a mean of 668 mg. The study found a significant difference in reported calcium consumption (the dependent variable) as a function of blood pressure (the independent variable). In other words, the null hypothesis, which said that the population means of calcium intake were the same for normal and hypertensive individuals, was rejected; the population mean for hypertensives is lower than that for normals. However, we cannot know for certain whether this conclusion is correct or whether it is a Type I error.

EXAMPLE 9.15 In Example 9.12, the independent (treatment) variable was amount or dosage of diazepam, and the treatments (levels of the independent variable) were 0 mg, 5 mg, and 10 mg. The dependent (response) variable was the left–right position of the vehicle on the road, and its values were successive distances between the midline of the vehicle and the right-hand boundary of the traffic lane. The left–right position of the vehicle was studied as a function of the amount of diazepam administered to the driver.

EXAMPLE 9.16 In Example 9.13, the independent variable was sex of the student, and the "treatments" (levels of the independent variable) were male and female. The dependent variable was whether or not the subject wore glasses for a photograph, and the values of the dependent variable were 0 for those who did not wear glasses and 1 for those who did. Presence or absence of eyeglasses was studied as a function of the sex of the student.

The data were found to be consistent with the null hypothesis, which said that sex of the student is irrelevant to whether or not the student wears glasses for a photograph. The failure to reject the null hypothesis may correctly reflect the absence of a difference in the population, or it may be a Type II error.

Even if the treatments have identical effects in terms of the relevant dependent variable, any summary statistics (such as sample means) for the different samples to which the different treatments are applied would not likely be exactly equal. Some difference would be expected just by chance—that is, because of random error. Thus, researchers find themselves looking at differences among samples (for example, 668 mg calcium for hypertensives versus 886 mg for controls) and needing to decide whether those sample differences should be attributed to chance or, alternatively, if they are too large to be dismissed as random error and should rather be attributed to something else, such as the effects of the treatments.

Hypothesis testing helps the researcher make the decision as to whether the evidence from the sample is sufficient to rule out chance as the sole explanation for sample differences. If chance can be ruled out, and if the experiment was planned and conducted carefully to avoid any systematic bias, then the researcher concludes that the treatments in fact have different effects "in the population." The population in such cases is often thought of as the hypothetical population of all possible observational units to whom the treatments might possibly have been administered, or simply as the underlying **true effects of the treatments**, which are manifest only imperfectly in any given sample. (This notion of the true effect of a treatment is illustrated in Example 9.3.)

* 9.3 CLASSICAL VERSUS BAYESIAN INFERENCE

Statistical inference refers to the use of information from samples to draw conclusions about population parameters. The most widely used approach to accomplish this goal is **classical inference**, which is the approach that has been introduced in this chapter and around which the rest of this book revolves. However, it is not the only approach. To give a better idea of what

* This symbol identifies material that is not prerequisite for later topics in this book.

assumptions characterize the classical approach, this section outlines how it differs from an alternative approach known as **Bayesian inference**.

Classical inference derives from the classical and frequentistic interpretations of probability (see Section 6.2). A statistic, such as a sample mean, takes on various values with certain probabilities, different samples yielding different values of the sample mean. According to the frequentistic interpretation of probability, the probability that the sample mean takes on a value in a given interval is the relative frequency with which sample means would be observed in that interval in the long run in repeated random samples from the same population. The corresponding parameter, the population mean, does not vary. The parameter's value is unknown, but it is a constant value: The frequency with which a parameter takes on various values is zero for all values except the one value that it always has! Given that probability is interpreted as "relative frequency in the long run" in classical inference, it does not make sense in classical inference to talk about the probability (relative frequency) with which a parameter equals a certain value, because a parameter *always* equals the same value.

Bayesian inference, on the other hand, is based on the subjective, rather than the frequentistic, interpretation of probability. According to the subjective interpretation (see Section 6.2), probability is a measure of one's certainty about an event. In this view, it makes sense to talk about the probability that a parameter equals a certain value, because probability here does not imply relative frequency in the long run but rather one's degree of belief.

The fundamental distinction between classical and Bayesian inference is that classical inference treats parameters as constants, whereas Bayesian inference treats them as random variables. Thus, classical inference does not allow us to assign probabilities to various possible values of a parameter, whereas Bayesian inference does. According to classical inference, the value of a parameter is just part of the way things are—a "state of nature"—and claims about states of nature are either true or false. In classical inference, uncertainty about the state of nature does not make a claim "probably true" or "probably false," because that would imply, according to the frequentistic interpretation of probability, that it is sometimes true and sometimes false. Bayesian inference, on the other hand, allows us to attach probabilities to various possible states of nature, or various possible values of parameters. Thus, Bayesian inference allows us to make the kind of probability statement that answers our intuitive question: What is the probability—that is, how certain can we be—that the hypothesized value of the parameter is correct? Such a question is not addressed by classical methods of inference, which use instead the logic of Sections 9.1 and 9.2.

Bayesian inference is so named because of the central role played by Bayes' rule (Section 6.6) in this approach to statistical inference. However, it should be noted that Bayes' rule itself is acceptable in both the classical and the Bayesian schools. The two approaches differ not in their acceptance of Bayes' rule, but in their acceptance of its use in Bayesian inference to revise subjective probabilities assigned to possible states of nature or values of parameters.

STUDY MATERIAL FOR CHAPTER 9

Terms and Concepts

Section 9.1

estimation
estimate
point estimate
estimated standard error of the mean
point estimate ± estimated standard error of the point estimate
mean ± standard deviation
interval estimate
confidence interval
confidence limits
confidence level

Section 9.2

hypothesis testing
significance testing
significance test
not statistically significant
statistically significant
null hypothesis
alternative hypothesis
Type I error
Type II error
independent variable
treatment variable
level of the independent variable
value of the independent variable
treatment
dependent variable
response variable
value of the dependent variable
true effect of a treatment

* Section 9.3

classical inference
Bayesian inference

Key Formula

Section 9.1

(9.2) $$s_{\bar{x}} = \frac{s}{\sqrt{n}}$$

Exercises for Numerical Practice

Section 9.1

9/1. Given $\bar{x} = 183.7$, $s = 13.1$, and $n = 100$, calculate $s_{\bar{x}}$.

9/2. Given $\bar{x} = 7.89$, $s = 2.40$, and $n = 400$, calculate $s_{\bar{x}}$.

9/3. A random sample of 34 observations has a mean of 66.4 and a standard deviation of 5.1. Calculate the estimated standard error of the mean.

9/4. A random sample of 96 observations has a mean of 1.39 and a standard deviation of .76. Calculate the estimated standard error of the mean.

Problems for Applying the Concepts

9/5. Based on findings in a random sample of 36 light bulbs, we can be 99% confident that the mean lifetime of the entire population of such bulbs is between 1018 and 1146 hours.

a. What is the confidence interval?

b. What are the confidence limits?

c. What is the confidence level?

9/6. Based on the findings in a random sample of 20 Holstein cows, we can be 95% confident that the mean butterfat content of milk in the relevant population of such cows is between 3.68% and 4.08%.

a. What is the confidence interval?

b. What are the confidence limits?

c. What is the confidence level?

(Source: Student project by S. Bailer.)

9/7. A farmer recorded the hay yield (in bales per acre) from two cuttings per year in the same fields. Nine first-cutting yields over the years had a mean of 119 and a standard deviation of 38 bales per acre, and seven second-cutting yields had a mean of 47 and a standard deviation of 15 bales per acre. A significance test revealed that first-cutting yields were significantly more variable than second-cutting yields.

a. Identify in words and in symbols the parameters that were estimated.

b. Identify the estimates in words and in symbols.

c. Give the sample means in the form of Expression (9.3).

d. Identify the dependent and independent variables, and state the levels of the independent variable.

e. In the significance test whose results are given in the description of this investigation, the null hypothesis was a statement about which of the following: population means, sample means, population standard deviations, sample standard deviations, population proportions, sample proportions?

f. State the null hypothesis (in words) for the significance test referred to above.

g. Was the null hypothesis rejected? How do you know?

h. Which type of error (I or II), if either, might have been made in the significance test reported above, and which type of error, if either, could not possibly have been made?

(Source: Student project by S. Bailer.)

9/8. In a sample of 42 college students, 85% of the females and 50% of the males reported owning pet dogs or cats. The difference was statistically significant.

a. What parameters were estimated?

b. What were the estimates?

c. Identify the dependent and independent variables, and state the levels of the independent variable.

d. In the significance test whose results are given in the description of this investigation, the null hypothesis was a statement about which of the following: population means, sample means, population standard deviations, sample standard deviations, population proportions, sample proportions?

e. State the null hypothesis (in words) for the significance test referred to above.

f. Was the null hypothesis rejected? How do you know?

g. Which type of error (I or II), if either, might have been made in the investigation as reported above, and which type of error, if either, could not possibly have been made?

(Source: Student project by C. Rokos.)

9/9. In a study of the distribution of various species of trees, counts were made of the numbers of trees (over a certain height) in a 400-square-foot area adjacent to a river near Fort McMurray in northern Alberta, Canada, and in a comparable riverbank area near Camrose in central Alberta. A total of 126 trees were counted, of which 73% of the Fort McMurray sample and 69% of the Camrose sample were willows. The difference was not statistically significant.

a. What parameters were estimated?

b. What were the estimates?

c. Identify the dependent and independent variables, and state the levels of the independent variable.

d. In the significance test whose results are given in the description of this investigation, the null hypothesis was a statement about which of the following: population means, sample means, population standard deviations, sample standard deviations, population proportions, sample proportions?

e. State the null hypothesis (in words) for the significance test referred to above.

f. Was the null hypothesis rejected? How do you know?

g. Which type of error (I or II), if either, might have been made in the investigation as reported above, and which type of error, if either, could not possibly have been made?

(Source: Student project by C. Morris.)

9/10. Twenty residence rooms, 10 of males and 10 of females, were randomly selected on a college campus, and the number of posters displayed in each room was counted. The females' rooms were found to have significantly more posters (a mean of 13.1 and a standard deviation of 7.5 posters) than the males' rooms (a mean of 5.3 and a standard deviation of 5.3 posters).

a. Identify in words and in symbols the parameters that were estimated.

b. Identify the estimates in words and in symbols.

c. Give the sample means in the form of Expression (9.3).

d. Identify the dependent and independent variables, and state the levels of the independent variable.

e. In the significance test whose results are given in the description of this investigation, the null hypothesis was a statement about which of the following: population means, sample means, population standard deviations, sample standard deviations, population proportions, sample proportions?

f. State the null hypothesis (in words) for the significance test referred to above.

g. Was the null hypothesis rejected? How do you know?

h. Which type of error (I or II), if either, might have been made in the significance test reported above, and which type of error, if either, could not possibly have been made?

(Source: Student project by J. Kruschel.)

References

Czeisler, C. A., Moore-Ede, M. C., and Coleman, R. M. "Rotating Shift Work Schedules That Disrupt Sleep Are Improved by Applying Circadian Principles." *Science* 217 (1982): 460–63.

Kerr, R. A. "Cloud Seeding: One Success in 35 Years." *Science* 217 (1982): 519–21.

McCarron, D. A., Morris, C. D., and Cole, C. "Dietary Calcium in Human Hypertension." *Science* 217 (1982): 267–69.

O'Hanlon, J. F., Haak, T. W., Blaauw, G. J., and Riemersma, J. B. J. "Diazepam Impairs Lateral Position Control in Highway Driving." *Science* 217 (1982): 79–81.

Further Reading

Campbell, S. K. *Flaws and Fallacies in Statistical Thinking*. Englewood Cliffs, NJ: Prentice-Hall, 1974. Chapter 12 provides an easy-reading introduction to the topic of statistical inference.

Huff, D. *How to Lie with Statistics*. New York: Norton, 1954. Chapter 4 is a brief introduction to standard errors.

Winkler, R. L. *An Introduction to Bayesian Inference and Decision*. New York: Holt, Rinehart & Winston, 1972. This book is recommended for readers who wish to pursue the topic of Bayesian inference.

CHAPTER 10

ESTIMATION OF MEANS

EXAMPLE 10.1 How long would a person sleep at a time if he did not have to get up each day to go to class or to go to work, and in fact if he had no way of knowing what time of day it was? Results from sleep-research volunteers living alone on just such a "self-scheduled routine in an environment free of time cues" have "contradicted the intuitive assumption that the length of sleep is determined by the length of prior wakefulness."

On the basis of such studies, a group of sleep researchers has found that sleep duration "depends on when subjects go to sleep, rather than how long they have been awake beforehand." It has long been known that human body temperature fluctuates systematically about 1°C each day, from a minimum that typically occurs in the middle of the night to a maximum that is typically in the middle of the day. Length of sleep until "natural" waking seems to depend on how close to the minimum of one's temperature cycle one is at the beginning of sleep.

For example, for one subject living without time cues for 3.5 months, the mean length of sleep (after a waking day of normal length or longer) was 7.8 hours if he chose to go to sleep within about 2 hours of his temperature-cycle minimum, but it was 14.4 hours if he chose to go to sleep within 2 hours of his temperature-cycle maximum. These values are estimates of the true mean sleep duration for this subject when he goes to sleep at the times indicated and is allowed to sleep undisturbed. Each value is a point estimate based on the mean of a sample of sleep durations observed under the specified conditions. The corresponding 95% confidence intervals for this subject's true mean sleep duration are approximately 7.2 to 8.4 hours for sleep that starts near the temperature minimum and 13.4 to 15.4 hours for sleep that starts near the temperature maximum.

As noted by the researchers, "these findings have major implications for understanding the timing of human sleep and may also help explain the sleep–wake patterns in shift workers and in certain clinical sleep disorders."

(Source: Czeisler *et al.*, 1980.)

Often one of the first things we want to know about a random variable is its typical or average value, and so it is that many empirical questions in science, business, and everyday life call for a measure of central location. As in the preceding example, the arithmetic mean is widely used for this purpose. In general, if we have a random sample of observed values of the variable, we can use the sample mean $\bar{x}$ for purposes of estimation and hypothesis testing concerning the population mean μ. This chapter discusses the estimation of population means, and Chapter 11 introduces significance tests about population means in the one-sample and two-sample cases.

10.1 POINT ESTIMATES OF μ

There are numerous statistics that we might consider in attempting to estimate the population mean μ of a random variable x. The sample mean, median, and mode are three possibilities, and literally dozens of others have been proposed. However, the most commonly used **point estimate of the population mean μ** is the sample mean $\bar{x} = \sum x/n$, which, as we saw in Section 8.4, is an unbiased estimate of μ.

EXAMPLE 10.2 In Example 10.1, the sample mean, 7.8 hours, was calculated from a sample of sleep episodes that began near the minimum of the subject's temperature cycle. It is a point estimate of that subject's true mean length of sleep episodes that begin near his temperature minimum. That is, it is an estimate of the mean length of the hypothetical population of his possible sleep episodes beginning near his temperature minimum. Similarly, the sample mean, 14.4 hours, is a point estimate of the subject's true mean length of sleep episodes that begin near his temperature maximum.

10.2 INTERVAL ESTIMATES OF μ WHEN σ IS KNOWN

One limitation of any point estimate is that it does not include information about its precision. This limitation can be avoided by the use of an interval estimate instead. It is not difficult to set up such an interval for the population mean by making use of what we know about the sampling distribution of the sample mean $\bar{x}$ (see Section 8.4).

EXAMPLE 10.3 The quality-control staff for a certain manufacturer of electrical products have detailed records concerning the variability in the quality of their products. For example, the process being used to manufacture special-purpose 75-watt light bulbs is known to yield a standard deviation of 150 hours in bulb lifetimes. To estimate the mean lifetime of the 75-watt bulbs being produced by this process, the staff tested a random sample of 36 bulbs and found a sample mean lifetime of $\bar{x} = 1082$ hours. What is the probability that the sample mean would be within 50 hours of the population mean that the staff were trying to estimate?

From the Central Limit Theorem in Expression (8.12), we know that the sampling distribution of the sample mean is approximately Gaussian and that the standard error of the mean is

$$\sigma_{\bar{x}} = \frac{\sigma}{\sqrt{n}}$$

or in this case

$$\sigma_{\bar{x}} = \frac{150}{\sqrt{36}} = 25 \text{ hours}$$

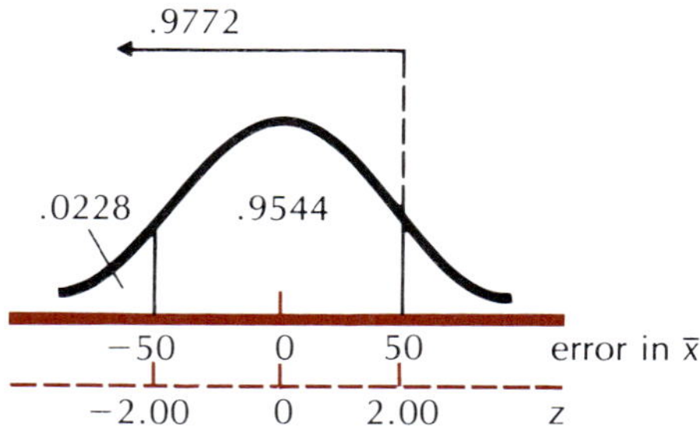

Figure 10.1

A sample mean that is 50 hours away from the population mean would be, according to Expression (8.10),

$$z_{\bar{x}} = \frac{50}{25} = 2.00 \text{ standard errors}$$

away from the population mean. In the table of the standard normal distribution in Appendix 3, we find $F_{\text{Gauss}}(2.00) = .9772$ and $F_{\text{Gauss}}(-2.00) = .0228$, from which we can determine that the probability that a sample mean would be less than 2.00 standard errors away from the population mean is $.9772 - .0228 = .9544$ (see Figure 10.1).

Therefore the probability is .9544 that such a sample mean is within 50 hours of the population mean. Since the sample mean is given as 1082 hours, we can say that we are 95.44% confident that the interval from $1082 - 50 = 1032$ to $1082 + 50 = 1132$ hours includes the population mean. We have thus calculated the 95.44% confidence interval for the population mean as 1032 to 1132 hours.

Example 10.3 illustrates how a confidence interval may be obtained in one specific set of data. The following paragraphs derive a general expression for confidence intervals for the population mean μ when the population standard deviation σ is known.

The Sampling Distribution of z_{obs}

The Central Limit Theorem, given in Expression (8.12), indicates that for a random sample from a population with mean μ and standard deviation σ, the sampling distribution of the mean $\bar{x}$ is Gaussian with mean μ and standard deviation

$$\sigma_{\bar{x}} = \frac{\sigma}{\sqrt{n}} \qquad \textbf{(10.1)}$$

if the sample size n is large enough. Expressing $\bar{x}$ in standard units as a z-score, let us relabel $z_{\bar{x}}$ from Expression (8.10) as z_{obs}:

$$z_{\text{obs}} = \frac{\bar{x} - \mu}{\sigma_{\bar{x}}} \qquad \textbf{(10.2)}$$

The subscript on z_{obs} (read "z-observed") indicates that the value of z is based on observations, summarized by $\bar{x}$, rather than on tabulated values of the standard normal distribution, where we also use the symbol z. We have, in fact, already used Expression (10.2) in Example 8.10, where z_{obs} was called $z_{\bar{x}}$. It is simply an application of the definition of a standard score (Section 4.4) and is interpreted as the number of standard errors that the sample mean $\bar{x}$ is away from the population mean μ. A negative value indicates that the sample mean is below the population mean, and a positive value indicates that the sample mean is above the population mean.

If the sampling distribution of the sample mean $\bar{x}$ is Gaussian, which it is under the Central Limit Theorem, then the **sampling distribution of z_{obs}** is also Gaussian: Subtraction of the constant μ from $\bar{x}$ merely changes the location (but not the shape) of the sampling distribution, and division by the constant $\sigma_{\bar{x}}$ merely changes the scale of measurement from x-units to standard (z) units. The sampling distribution of z_{obs} in Expression (10.2) is therefore the standard normal distribution.

It follows that if the sampling distribution of $\bar{x}$, and hence of z_{obs}, is Gaussian, then z_{obs} falls in the central $1 - \alpha$ area of the standard normal distribution with probability $1 - \alpha$ (where α is alpha, the lower-case first letter of the Greek alphabet). The symbol α is used to refer to the total tail area of a probability distribution, and hence $1 - \alpha$ is the remaining central area.

EXAMPLE 10.4 If $\alpha = .05$, then $1 - \alpha = 1.00 - .05 = .95$. If the sampling distribution of $\bar{x}$ is Gaussian, then z_{obs} falls in the central 95% of the standard normal distribution with probability .95. Conversely, we can state that z_{obs} falls in the extreme 5% of the standard normal distribution with probability $\alpha = .05$, and since the Gaussian distribution is symmetric, we can add that z_{obs} falls in the lower tail with probability $\alpha/2 = .025$ and in the upper tail with probability $\alpha/2 = .025$ (see Figure 10.2).

Let us use the notation z_A introduced in Section 7.4 to refer to the value of z that cuts off an upper-tail area A of the standard normal distribution. Then we can write that z_{obs} falls in the central $1 - \alpha$ area of the standard normal distribution with probability $1 - \alpha$ as follows:

$$-z_{\alpha/2} < z_{obs} < z_{\alpha/2} \quad \text{with probability } 1 - \alpha \tag{10.3}$$

The values $-z_{\alpha/2}$ and $z_{\alpha/2}$ cut off $\alpha/2$ of the area in the left and right tails,

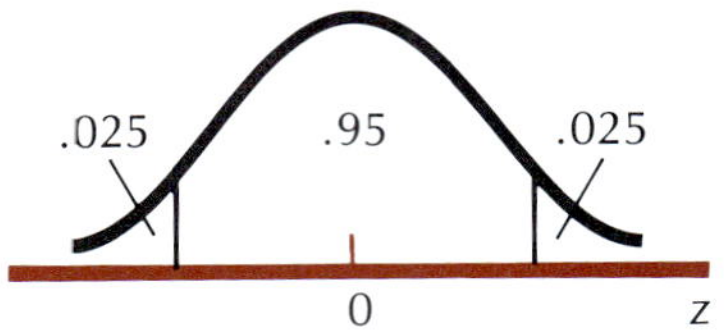

Figure 10.2

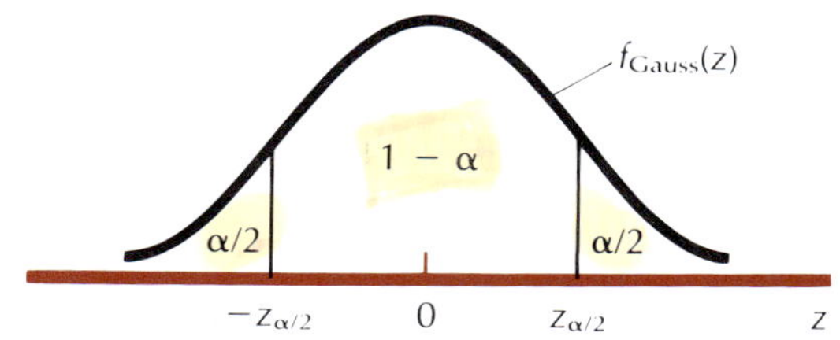

Figure 10.3 Illustration of the notation $z_{\alpha/2}$ to denote the z-score cutting off an area of $\alpha/2$ in the upper tail of the standard normal distribution.

respectively, of the standard normal distribution, leaving a central area of $1 - \alpha$, as illustrated in Figure 10.3.

EXAMPLE 10.5 If $\alpha = .05$, then $\alpha/2 = .025$ and $1 - \alpha = .95$. What is the value of $z_{.025}$? Looking for the upper-tail area $A = .025$ in Appendix 3, we find that $z_{.025} = 1.96$. You should try to visualize this answer in terms of Figure 10.3: z-scores of −1.96 and 1.96 cut off $\alpha/2 = 2.5\%$ of the area in each tail of the distribution, leaving a central area of $1 - \alpha = 95\%$, as shown in Figure 10.4(a).

If $\alpha = .01$, then $\alpha/2 = .005$ and $1 - \alpha = .99$. What is the value of $z_{.005}$? Looking for the upper-tail area $A = .005$ in Appendix 3, we find that $z_{.005} = 2.58$. Substituting these values into Figure 10.3, we see that standard normal deviates of −2.58 and 2.58 cut off $\alpha/2 = .005 = \frac{1}{2}\%$ of the area in each tail of the distribution, leaving a central area of $1 - \alpha = 99\%$, as shown in Figure 10.4(b).

The Margin of Error of the Estimate

The fact that the sampling distribution of z_{obs} is Gaussian is summarized in Expression (10.3), which says that z_{obs} falls between the standard normal deviates $-z_{\alpha/2}$ and $z_{\alpha/2}$ with probability $1 - \alpha$. Substituting Expression (10.2) into (10.3), we have

$$-z_{\alpha/2} < \frac{\bar{x} - \mu}{\sigma_{\bar{x}}} < z_{\alpha/2} \qquad \text{with probability } 1 - \alpha \tag{10.4}$$

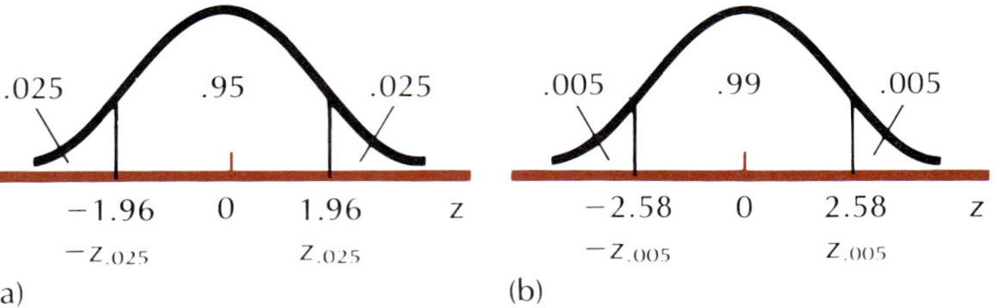

Figure 10.4

Multiplication by the standard error $\sigma_{\bar{x}}$ yields

$$-z_{\alpha/2}\sigma_{\bar{x}} < (\bar{x} - \mu) < z_{\alpha/2}\sigma_{\bar{x}} \qquad \text{with probability } 1 - \alpha \tag{10.5}$$

Expression (10.5) says that the probability is $1 - \alpha$ that the sample mean $\bar{x}$ will not differ from the population mean μ by more than $z_{\alpha/2}$ times the standard error of the mean.

EXAMPLE 10.6 Given the results obtained in Example 10.5, Expression (10.5) says, for example, that

$$-1.96\sigma_{\bar{x}} < (\bar{x} - \mu) < 1.96\sigma_{\bar{x}} \qquad \text{with probability } 95\%$$

and

$$-2.58\sigma_{\bar{x}} < (\bar{x} - \mu) < 2.58\sigma_{\bar{x}} \qquad \text{with probability } 99\%$$

The probability is 95% that $\bar{x}$ is less than 1.96 standard errors away from μ, and the probability is 99% that $\bar{x}$ is less than 2.58 standard errors away from μ.

Since the difference $\bar{x} - \mu$ is the error of the sample mean $\bar{x}$ as an estimate of the population mean μ, Expression (10.5) gives us a way of determining the probabilities associated with various amounts of error in the sample mean. In this context, the value $1 - \alpha$, represented by the central area in Figure 10.3, is called the **confidence level**. Let us use the symbol $e_{\alpha/2}$ to represent the **margin of error of an estimate at the $1 - \alpha$ confidence level**. In the case at hand, where we are estimating the population mean μ when the population standard deviation σ is known, Expression (10.5) tells us that, at the $1 - \alpha$ confidence level, the **margin of error of $\bar{x}$ when σ is known** is $z_{\alpha/2}$ times the standard error of the mean:

$$e_{\alpha/2} = z_{\alpha/2}\sigma_{\bar{x}} \tag{10.6}$$

We may rewrite Expression (10.5) as

$$-e_{\alpha/2} < \bar{x} - \mu < e_{\alpha/2} \qquad \text{with probability } 1 - \alpha \tag{10.7}$$

which says that the probability is $1 - \alpha$ that the sample mean $\bar{x}$ will differ from the true mean value μ by less than the margin of error $e_{\alpha/2}$.

EXAMPLE 10.7 Consider again the light-bulb manufacturing process in Example 10.3. Given that the true standard deviation of bulb lifetimes is 150 hours, what statement can we make with 95% confidence concerning the error in the mean lifetime, $\bar{x} = 1082$ hours, of a random sample of $n = 36$ bulbs?

The question is asking for the value of $e_{\alpha/2}$ in Expression (10.7) when the confidence level is $1 - \alpha = .95$. If $1 - \alpha = .95$, then $\alpha = .05$ and $\alpha/2 = .025$. Expression (10.6) gives the margin of error at the $1 - \alpha = .95$ confidence level as

$$e_{.025} = z_{.025}\sigma_{\bar{x}}$$

The value of $z_{.025} = 1.96$ is found in Appendix 3 (as demonstrated in Example 10.5), and the value of the standard error $\sigma_{\bar{x}}$ was calculated in Example 10.3 to be $150/\sqrt{36} = 25$ hours. Thus we have

$$e_{.025} = 1.96(25) = 49 \text{ hours}$$

We conclude that we can be 95% confident that the sample mean, $\bar{x} = 1082$ hours, does not differ from the population mean μ by more than 49 hours.

For the sample specified in the preceding paragraph, what statement can we make with 99% confidence concerning the error in the sample mean? For this statement, the confidence level is $1 - \alpha = .99$, and hence $\alpha = .01$ and $\alpha/2 = .005$. From the table of the standard normal distribution, we find $z_{.005} = 2.58$ (as demonstrated in Example 10.5), so we can use Expression (10.6) to calculate the margin of error at the 99% confidence level as

$$e_{.005} = z_{.005}\sigma_{\bar{x}} = 2.58(25) = 64.5 \text{ hours}$$

We conclude that we can be 99% confident that the sample mean, $\bar{x} = 1082$ hours, is within 64.5 hours of the unknown population mean μ.

Confidence Intervals for μ

To continue our derivation of a general expression of the confidence interval for the population mean, we note that multiplying both sides of an inequality by a negative number changes the direction of the inequality. (For example, if $x < 5$, then $-x > -5$.) Multiplication of Expression (10.5) by -1 yields

$$z_{\alpha/2}\sigma_{\bar{x}} > (\mu - \bar{x}) > -z_{\alpha/2}\sigma_{\bar{x}} \qquad \text{with probability } 1 - \alpha$$

which may be rewritten, with the order of the terms reversed, as

$$-z_{\alpha/2}\sigma_{\bar{x}} < (\mu - \bar{x}) < z_{\alpha/2}\sigma_{\bar{x}} \qquad \text{with probability } 1 - \alpha$$

Finally, addition of $\bar{x}$ yields the **1 − α confidence interval for the population mean μ when σ is known and the sampling distribution of the sample mean $\bar{x}$ is Gaussian:**

$$\bar{x} - z_{\alpha/2}\sigma_{\bar{x}} < \mu < \bar{x} + z_{\alpha/2}\sigma_{\bar{x}} \qquad \text{with confidence level } 1 - \alpha \tag{10.8}$$

In the terminology introduced in Section 9.1, the **1 − α confidence limits for μ when σ is known** are

$$\bar{x} \pm z_{\alpha/2}\sigma_{\bar{x}} \tag{10.9}$$

which may also be written as

$$\bar{x} \pm e_{\alpha/2} \tag{10.10}$$

and the confidence level is $1 - \alpha$. For example, the 95% confidence limits are $\bar{x} \pm 1.96\sigma_{\bar{x}}$, and the 99% confidence limits are $\bar{x} \pm 2.58\sigma_{\bar{x}}$.

EXAMPLE 10.8 In Example 10.7, the sample mean was $\bar{x} = 1082$ hours, and its margin of error at the 95% confidence level was $e_{.025} = 49$ hours. That is, we can be 95% confident that the sample mean of 1082 hours is within 49 hours of the population mean. Therefore, the 95% confidence limits may be stated in the form of Expression (10.10) as 1082 ± 49 hours. Since $1082 - 49 = 1033$ and $1082 + 49 = 1131$, the 95% confidence interval for the population mean μ is 1033 to 1131 hours. This is an interval estimate of the population mean.

We also saw in Example 10.7 that we can be 99% confident that the sample mean of 1082 hours is within $e_{.005} = 64.5$ hours of the population mean μ. Therefore, the 99%

confidence limits are 1082 ± 64.5 hours, and the 99% confidence interval for the mean μ is 1017.5 to 1146.5 hours.

Understanding the Confidence Interval

The general expression of the confidence interval in Expression (10.8) may look rather formidable at first glance, but if we take time to read it carefully, we will see that looks can be deceiving.

Notice that the confidence limits, $\bar{x} \pm e_{\alpha/2}$, are centered around the sample mean $\bar{x}$. The sample mean is, after all, an unbiased estimate of the population mean μ and hence is our "best guess." We obtain the confidence limits from this "best guess," the sample mean, by subtracting and adding, respectively, the appropriate margin of error, $e_{\alpha/2} = z_{\alpha/2}\sigma_{\bar{x}}$.

The margin of error, $e_{\alpha/2} = z_{\alpha/2}\sigma_{\bar{x}}$, is stated as a certain number $z_{\alpha/2}$ of standard errors $\sigma_{\bar{x}}$. The width of this margin of error depends on how confident we want to be in the resulting interval, as follows: The quantity $z_{\alpha/2}$ states how many standard errors the interval extends above and below the sample mean, and its subscript $\alpha/2$, which indicates which value of z is called for, is in turn determined by the chosen confidence level $1 - \alpha$. For example, in a 95% confidence interval the margin of error is $z_{.025} = 1.96$ standard errors, whereas if we want to be 99% confident that the confidence limits capture the population mean, we must be prepared to accept a wider margin of error, namely, $z_{.005} = 2.58$ standard errors.

In other words, Expression (10.8) says that the 95% confidence limits for the population mean are 1.96 standard errors above and below the sample mean, and the 99% confidence limits are 2.58 standard errors above and below the sample mean. Any desired confidence level may be used in Expression (10.8), but 95% and 99% are the most common ones.

As noted in Section 9.1, the confidence level in expressions such as (10.8) should be interpreted as the success rate of this particular method of estimation. The population mean μ is a constant. What is subject to variability is the sample mean $\bar{x}$ and hence the confidence limits, which are calculated from the sample mean as $\bar{x} \pm e_{\alpha/2}$. Different samples would have different sample means, resulting in correspondingly different confidence limits. The confidence level $1 - \alpha$ indicates the theoretical success rate; for example, 95% of the theoretically possible samples would have 95% confidence intervals that include the population mean μ.

It is *not* appropriate to express a calculated confidence interval, such as that in the first paragraph of Example 10.8, in the form $P(1033 < \mu < 1131) = .95$. The reason is that 1033, μ, and 1131 are all constants; there is no random variable in such a statement. The development of confidence intervals in this section began with probability terminology, and that is appropriate as long as our statement includes the random variable $\bar{x}$. However, it is not meaningful according to the classical understanding of probability after we have inserted a specific value of $\bar{x}$. Thus, the appropriate terminology for reporting confidence intervals is *confidence level*, not *probability*.

Size of Sample Required for a Given Margin of Error

In planning an investigation to obtain an interval estimate of a population mean, how should we decide how large a sample size n is required? The required sample size depends in part on the desired degree of precision, and sometimes it is necessary to operate by trial and error in this regard. However, if we can state what margin of error we will tolerate at our desired confidence level, and if we have a reasonably good idea as to the value of the population standard deviation σ, then we can use Expression (10.6) to determine the required sample size n.

If the standard error $\sigma_{\bar{x}}$ is expressed as $\sigma/\sqrt{n}$, Expression (10.6) becomes

$$e_{\alpha/2} = \frac{z_{\alpha/2}\sigma}{\sqrt{n}} \tag{10.11}$$

which says that the margin of error $e_{\alpha/2}$ in the sample mean at the $1 - \alpha$ confidence level is $z_{\alpha/2}$ times the standard error. Rearranging terms in Expression (10.11) to solve for sample size, we obtain the **size of sample required for a given margin of error in estimating μ:**

$$n = \left(\frac{z_{\alpha/2}\sigma}{e_{\alpha/2}}\right)^2 \tag{10.12}$$

If the resulting value is not an integer, the next larger integer should be taken for the required sample size. Even if we have only a rough idea as to the value of the standard deviation σ, we can still use Expression (10.12) to obtain a correspondingly rough idea of the required sample size.

EXAMPLE 10.9 Consider again the light-bulb manufacturing process in Example 10.3. Given that the true standard deviation of bulb lifetimes is 150 hours, how large a random sample would the quality-control staff need in order to be 95% confident that the sample mean would be within 20 hours of the population mean? Expression (10.12) gives the answer as

$$n = \left(\frac{(1.96)(150)}{20}\right)^2 = 216.09$$

which we interpret as a required sample size of 217 (the next larger integer).

How large a random sample would the staff need in order to be 99% confident that the sample mean is off by no more than 20 hours? Again we apply Expression (10.12), obtaining

$$n = \left(\frac{(2.58)(150)}{20}\right)^2 = 374.4225$$

which we interpret as a required sample size of 375 (the next larger integer). To the extent that the population standard deviation σ used in Expression (10.12) is an approximation, the prescribed sample size will likewise be an approximation, but it at least provides useful information to aid our judgment as to how large a sample to observe.

10.3 INTERVAL ESTIMATES OF μ WHEN σ IS ESTIMATED BY s

A limitation of the confidence interval in Expression (10.8) is that it requires knowledge of the population standard deviation σ. In practical applications, the population standard deviation σ is typically unknown, and the only information available is the sample data, from which statistics such as the sample mean $\bar{x}$ and the sample standard deviation s may be calculated.

The entire derivation of Expression (10.8) was based on the fact that if the sampling distribution of the sample mean $\bar{x}$ is Gaussian, then the sampling distribution of

$$z_{\text{obs}} = \frac{\bar{x} - \mu}{\sigma_{\bar{x}}} \tag{10.13}$$

is also Gaussian, because the subtraction of μ and the division by $\sigma_{\bar{x}}$ have no effect on the shape of the distribution. However, if we use the sample standard deviation s to estimate the population standard deviation σ and hence use

$$s_{\bar{x}} = \frac{s}{\sqrt{n}} \tag{10.14}$$

to estimate the standard error $\sigma_{\bar{x}}$ so that we have

$$\frac{\bar{x} - \mu}{s_{\bar{x}}} \tag{10.15}$$

in place of the ratio in Expression (10.13), then the shape of the distribution is changed. The shape is changed because, unlike the parameter $\sigma_{\bar{x}}$, the estimate $s_{\bar{x}}$ is not a constant but is subject to sampling error, which sometimes makes it too large and sometimes too small.

Degrees of Freedom

The distribution of the ratio in Expression (10.15) is in fact a whole family of distributions, because its exact shape depends, in effect, on how good our estimate $s_{\bar{x}}$ of the standard error is. The goodness of the estimate depends in turn upon the number of independent pieces of information used in calculating the estimated standard error, a quantity known as the number of degrees of freedom. The **number of degrees of freedom** of a statistic refers to the number of "free-to-vary" or unrestricted values entering into the calculation of the statistic. The usual application of the concept is in measuring the variability or dispersion of a set of sample values, in which case the number of degrees of freedom is the number of values whose variability is being measured minus the number of sample-based restrictions on those values.

EXAMPLE 10.10 You can obtain an intuitive idea—and that is really all you need at this stage—of the concept of degrees of freedom by considering a series of three mental illustrations.

1. For the first illustration, pick three numbers, any three numbers, and write them down. You have 3 degrees of freedom in this case, because you are free to choose each and every one of the three numbers without restriction. In general, in picking n numbers without restriction, we have n degrees of freedom.
2. For the second illustration, pick three numbers such that their mean is 10.0. You have only 2 degrees of freedom in this case, because although you are free to choose the first two numbers without restriction, once you have chosen them, the third number is determined by the restriction that the mean of the three numbers must be 10.0. Where did the other degree of freedom go? I used it up, as it were, in specifying that the mean must be 10.0, a number that I chose without restriction—with 1 degree of freedom. In general, in picking n numbers whose mean must be $\bar{x}$, we have $n - 1$ degrees of freedom, the other 1 degree of freedom being associated with the mean $\bar{x}$.
3. For the third illustration, pick three numbers whose mean is 10.0 and whose range is 12.0. Here your first number is considered to be free to vary, even though you are limited to numbers that are not too far away from the mean so that the eventual range will not become too large. But once you have picked the first number, then the other two numbers are determined by the two restrictions: If the first number is the smallest of the three, say, x, then the second number must be $x + 12.0$ to produce the specified range, and the other number is determined by the restriction that the mean must be 10.0. Analogous constraints apply if the first number happens to be the largest or the middle one. In any case, you have only 1 degree of freedom, because although you are free to choose the first number (from a set of permissible numbers), once you have chosen it, the remaining two numbers are determined by the specified restrictions, each of which uses up 1 degree of freedom. In general, in picking n numbers subject to two independent restrictions, we have $n - 2$ degrees of freedom.

EXAMPLE 10.11 Now let us consider a specific example. The sample standard deviation

$$s = \sqrt{\frac{\sum (x - \bar{x})^2}{n - 1}}$$

which we met in Section 4.3, has $n - 1$ degrees of freedom. It measures the dispersion or variability of n sample values, but it measures their variability around a quantity (the sample mean $\bar{x}$) whose value is calculated from those same sample values, leaving $n - 1$ degrees of freedom for variation around the mean.

Alternatively, we may think of the standard deviation s in terms of the deviations $x - \bar{x}$ from the sample mean, of which there are n. However, only $n - 1$ of the deviations are unrestricted or free to vary, because by definition deviations from the mean must sum to zero [see Expression (3.6)]; this imposes one restriction on the n deviations, leaving $n - 1$ degrees of freedom.

The exact distribution of the ratio in Expression (10.15) depends on the value of $n - 1$, which is the number of degrees of freedom associated with the sample standard deviation s. As the value of $n - 1$ becomes larger, the estimated standard deviation s becomes more dependable, because it is based on a greater number of observations.

Student's *t* Distribution

To give a name to the ratio in Expression (10.15), let us define the *t*-statistic

$$t_{\text{obs}} = \frac{\bar{x} - \mu}{s_{\bar{x}}} \tag{10.16}$$

where the subscript on t_{obs} (read "*t*-observed") indicates that the value of t is based on observations, summarized by $\bar{x}$, rather than on a tabulation of theoretical values. When the population standard deviation σ is unknown and hence must be estimated by the sample standard deviation s, the ratio t_{obs} plays a role analogous to that played by z_{obs} for the case where the population standard deviation σ is known (which was discussed in Section 10.2). The British statistician W. S. Gosset (see Biography 10.1), writing under the pen name Student, studied the sampling distribution of t_{obs} for situations where the sampling distribution of the sample mean $\bar{x}$ is Gaussian. The resulting family of distributions is referred to collectively as **Student's *t* distribution**, which may be defined as the sampling distribution of the ratio of (a) the error in a point estimate having a Gaussian sampling distribution, divided by (b) the estimated standard error of the estimate:

$$t = \frac{\text{(point estimate having a Gaussian sampling distribution) minus (parameter)}}{\text{estimated standard error of the point estimate}} \tag{10.17}$$

Expression (10.16) is the special case of Expression (10.17) that occurs when the point estimate is a sample mean $\bar{x}$ with a Gaussian sampling distribution.

BIOGRAPHY 10.1

William Sealy Gosset (1876–1937)

William Gosset was born 13 June 1876 in Canterbury, England. He won scholarships to study first at Winchester College and then at New College, Oxford, where he earned degrees in mathematics and chemistry. Immediately afterwards, in 1899, he accepted a position on the scientific staff of the Guinness Brewery in Dublin, Ireland. He remained with the firm throughout his life, including a year's further study with Karl Pearson (see Biography 5.3) in London from 1906 to 1907. In 1935 he was put in charge of the Guinness Brewery in London.

Gosset's contributions to statistics resulted from his application of statistical methods to the experimental and industrial problems that he encountered in his work. J. O. Irwin observed that Gosset saw "the need for adequate methods to deal with small samples in examining the relations between the quality of the raw materials of beer, such as barley and hops, the conditions of production, and the finished article. The importance of controlling the quality of barley ultimately led him to study the design of agricultural field trials." Thus, he developed small-sample theory related to means, variances, and correlation coefficients, including the introduction in 1908 of Student's *t* distribution for small-sample inferences about means. All of Gosset's scientific articles except one were published under the pen name Student, in compliance with a policy of his employer.

Gosset married Marjory Surtees Phillpotts in 1906, and they had one son and two daughters. He died in Beaconsfield, England, on 16 October 1937 at the age of 61.

(Sources: Irwin, 1978; *World Who's Who in Science,* 1968.)

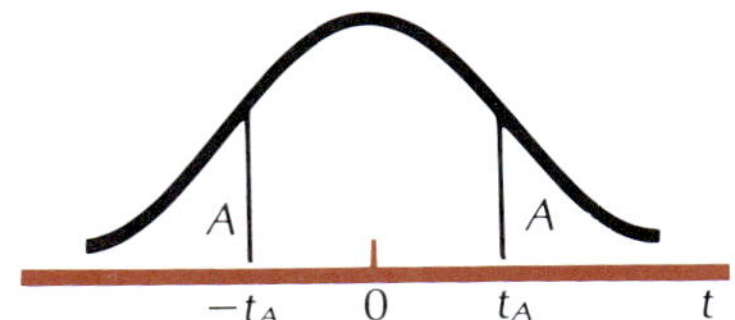

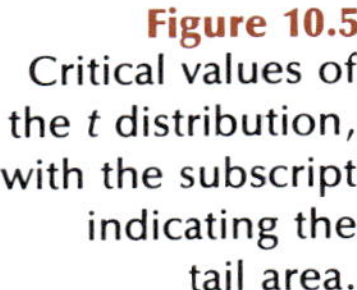
Figure 10.5 Critical values of the t distribution, with the subscript indicating the tail area.

The important result of Gosset's work for our present purposes is that if the sampling distribution of the mean $\bar{x}$ is Gaussian, as it is under the Central Limit Theorem, then

$$t_{\text{obs}} = \frac{\bar{x} - \mu}{s/\sqrt{n}} \tag{10.18}$$

follows the t distribution with $n - 1$ degrees of freedom (df). In other words, the **sampling distribution of t_{obs}** is the t distribution with $n - 1$ degrees of freedom. A wavy line is used as the symbol for "is distributed as," and a parenthetical value is attached to the symbol t to indicate which t distribution, so the expression

$$t_{\text{obs}} \sim t(n - 1 \text{ df}) \tag{10.19}$$

says that the statistic t_{obs} follows the t distribution with $n - 1$ degrees of freedom.

Certain critical values of the t distribution for various numbers of degrees of freedom are tabulated in Appendix 4, where t_A is the value of t that cuts off the upper-tail area A under the t probability density curve (see Figure 10.5). The area to the right of t_A is A, and therefore the area to the left of t_A is $1 - A$. The t distribution is symmetric with a mean of 0, so the area to the left of $-t_A$ equals the area to the right of t_A. One practical implication of this fact is that the critical values that cut off an area of $\alpha/2$ in each tail are $-t_{\alpha/2}$ and $t_{\alpha/2}$, giving a total tail area of α.

EXAMPLE 10.12 The critical value that cuts off 5% of the area in the right-hand tail of the t distribution with 17 degrees of freedom is

$$t_{.05}(17 \text{ df}) = 1.740$$

as given in Appendix 4 and illustrated in Figure 10.6(a).

The critical value that cuts off 1% of the area in the left-hand tail of the t distribution with 21 degrees of freedom is

$$t_{.99}(21 \text{ df}) = -t_{.01}(21 \text{ df}) = -2.518$$

as given in Appendix 4 and illustrated in Figure 10.6(b).

The critical values that cut off 2.5% of the area in each tail of the t distribution

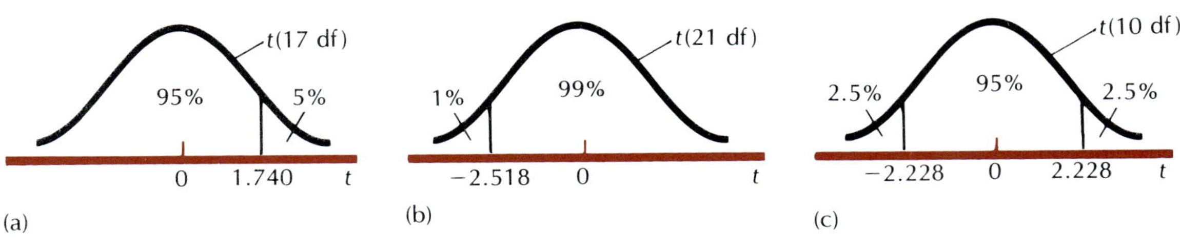

Figure 10.6

with 10 degrees of freedom are

$$\pm t_{.025}(10 \text{ df}) = \pm 2.228$$

as given in Appendix 4 and illustrated in Figure 10.6(c).

What do the t distributions look like? We have already noted that the t distribution is symmetric with a mean of 0. The standard deviation of the t distribution is slightly greater than that of the standard normal distribution, which is 1.00. (For example, the standard deviation of t with 5 degrees of freedom turns out to be 1.29.) As the number of degrees of freedom increases, the standard deviation of the t distribution approaches 1.00 fairly rapidly. Thus, the t distribution has the same mean and, except for very small numbers of degrees of freedom, almost the same standard deviation as the standard normal distribution.

Also like the Gaussian distribution, the t distribution is continuous from minus-infinity to plus-infinity. However, the t distribution has slightly heavier tails than the normal distribution; that is, the t distribution is leptokurtic (see Section 2.7). The leptokurtosis diminishes as the number of degrees of freedom increases, until for infinite degrees of freedom, the t distribution is identical to the standard normal distribution. That is,

$$\boldsymbol{t(\infty\ \textbf{df})} \sim \text{standard normal distribution} \tag{10.20}$$

as illustrated in Figure 10.7.

For most practical purposes, the number of degrees of freedom does not have to become very large before the t distribution is virtually indistinguishable from the standard normal distribution. You can verify this statement by examining the bottom few rows of critical values of t in Appendix 4. The values in the bottom row of the table (infinite degrees of freedom) are identical to the corresponding critical values of the standard normal distribution (Appendix 3), by virtue of Expression (10.20). For example,

$$t_{.025}(\infty \text{ df}) = z_{.025} = 1.960$$

The immediately preceding rows (for 40, 60, and 120 degrees of freedom) are

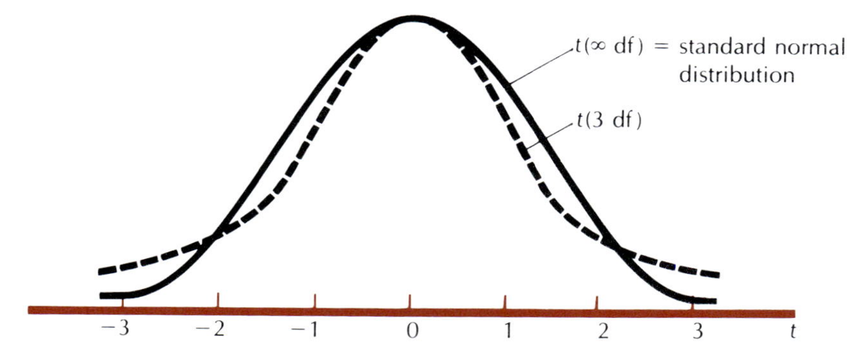

Figure 10.7 Student's t distributions with 3 and with infinite degrees of freedom.

not very different from the bottom row; after the number of degrees of freedom is up to 30 or 40, the t distribution changes relatively little with increasing degrees of freedom. Thus, if the exact number of degrees of freedom we require for a particular problem is not given in Appendix 4, we may approximate, either by interpolating between the t-values of the nearest numbers of degrees of freedom that are given or, in many cases, by using the t-value for the next smaller number of degrees of freedom that is given (that is, the next larger t-value). It turns out that use of the next larger value of t will cause our estimates to err, if anything, in the direction of caution.

EXAMPLE 10.13 The critical value that cuts off 1% of the area in the right-hand tail of the t distribution with 35 degrees of freedom is

$$t_{.01}(35 \text{ df}) \simeq 2.44 \qquad \text{(or, at most, 2.46)}$$

The first value is obtained by interpolation and the second from the line for 30 degrees of freedom. The first value, 2.44, represents an attempt to determine the exact value, while the second value sets an upper limit that the exact value cannot exceed. The latter kind of approximation will be all we need in many applications.

The critical value that cuts off .5% of the area in the right-hand tail of the t distribution with 200 degrees of freedom is

$$t_{.005}(200 \text{ df}) \simeq 2.59 \qquad \text{(or, at most, 2.62)}$$

The first value is obtained by interpolation and the second from the line for 120 degrees of freedom.

In the past, when tables of the t distribution were less available, many authors recommended comparing t_{obs} to the standard normal distribution for sample sizes n greater than 30. This rough approximation is not necessary when tabulations for degrees of freedom between 30 and infinity are available, as in Appendix 4. Whether or not you use this approximation, the important point is that the conceptual distinction between z_{obs} and t_{obs} is based not on sample size but on whether the population standard deviation σ is known, as in z_{obs}, or must be estimated by the sample standard deviation s, as in t_{obs}.

The Margin of Error of the Estimate

We have seen that the statistic t_{obs} as defined in Expression (10.18) follows the t distribution with $n - 1$ degrees of freedom if the sampling distribution of the mean $\bar{x}$ is Gaussian. It follows that t_{obs} falls in the central $1 - \alpha$ area of the t distribution with probability $1 - \alpha$; that is,

$$-t_{\alpha/2} < t_{obs} < t_{\alpha/2} \qquad \text{with probability } 1 - \alpha \tag{10.21}$$

where $\pm t_{\alpha/2}$ are the critical values of the t distribution with appropriate degrees of freedom. Substituting Expression (10.16) into (10.21) and rearranging terms as in Section 10.2, we find that

$$-t_{\alpha/2}s_{\bar{x}} < (\bar{x} - \mu) < t_{\alpha/2}s_{\bar{x}} \qquad \text{with probability } 1 - \alpha \tag{10.22}$$

Expression (10.22) says that the probability is $1 - \alpha$ that the sample mean $\bar{x}$ will not differ from the population mean μ by more than $t_{\alpha/2}$ times the estimated standard error of the mean. As before, let us use the symbol $e_{\alpha/2}$ to represent the margin of error of an estimate at the $1 - \alpha$ confidence level. Then in this case, where we are estimating the population mean μ without knowledge of the population standard deviation σ, the **margin of error of $\bar{x}$ when σ is estimated by s** is

$$e_{\alpha/2} = t_{\alpha/2}s_{\bar{x}} \qquad \text{with confidence level } 1 - \alpha \tag{10.23}$$

EXAMPLE 10.14 Tests for butterfat content were performed on milk from a random sample of 20 Holstein dairy cows. The sample mean was 3.88%, and the standard deviation was .42 percentage point. What statement can we make with 95% confidence concerning the error in the sample mean as an estimate of the mean of the relevant population of Holstein cows?

Expression (10.23) gives the margin of error at the $1 - \alpha = .95$ confidence level as

$$e_{.025} = t_{.025}s_{\bar{x}}$$

The relevant t distribution in this case has $n - 1 = 20 - 1 = 19$ degrees of freedom, and the value of $t_{.025}(19\,\text{df}) = 2.093$ is found in Appendix 4. The value of $s_{\bar{x}}$ is calculated via Expression (10.14) as

$$s_{\bar{x}} = \frac{s}{\sqrt{n}} = \frac{.42}{\sqrt{20}} = .0939 \text{ percentage point}$$

Using these values, we calculate

$$e_{.025} = (2.093)(.0939) = .197$$

or approximately .20 percentage point. We conclude that we can be 95% confident that the sample mean, $\bar{x} = 3.88\%$, is within .20 percentage point of the population mean of which it is an estimate.

For the same sample, what statement can we make with 99% confidence concerning the error in the sample mean? The margin of error at the 99% confidence level is

$$e_{.005} = t_{.005}s_{\bar{x}}$$

Appendix 4 gives the value of $t_{.005}$(19 df) as 2.861, so we have

$$e_{.005} = (2.861)(.0939) = .269$$

Therefore, we can be 99% confident that the sample mean, $\bar{x} = 3.88\%$, does not differ from the population mean by more than .27 percentage point.

(Source: Student project by S. Bailer.)

Confidence Intervals for μ

Using the same procedure as for the case with σ known (Section 10.2), we may rearrange the terms of Expression (10.22) to yield the **1 − α confidence interval for the population mean μ when σ is estimated by *s* and the sampling distribution of $\bar{x}$ is Gaussian:**

$$\bar{x} - t_{\alpha/2}s_{\bar{x}} < \mu < \bar{x} + t_{\alpha/2}s_{\bar{x}} \qquad \text{with confidence level } 1 - \alpha \tag{10.24}$$

Therefore, the **1 − α confidence limits for μ when σ is estimated by *s*** are

$$\bar{x} \pm t_{\alpha/2}s_{\bar{x}} \tag{10.25}$$

which Expression (10.23) allows us to write as

$$\bar{x} \pm e_{\alpha/2} \tag{10.26}$$

EXAMPLE 10.15 In Example 10.14, the mean butterfat content for a random sample of 20 Holstein cows was 3.88%, and the margin of error at the 95% confidence level was .20 percentage point. The 95% confidence limits are therefore 3.88 ± .20, and the 95% confidence interval is 3.68% to 4.08%. We can be 95% confident that this interval includes the population mean.

The margin of error at the 99% confidence level was .27 percentage point. In other words, the 99% confidence limits are 3.88 ± .27, or 3.61% and 4.15%. We can be 99% confident that the population mean is between 3.61% and 4.15%.

Notice that the confidence interval for the population mean takes the same general form in the present case, where the population standard deviation σ is estimated by the sample standard deviation s [Expression (10.24)], as in the previous case, where σ is known [Expression (10.8)], the general form being

$$\bar{x} - e_{\alpha/2} < \mu < \bar{x} + e_{\alpha/2} \qquad \text{with confidence level } 1 - \alpha \tag{10.27}$$

The confidence limits are again centered around the sample mean $\bar{x}$ and are again of the form $\bar{x} \pm e_{\alpha/2}$, that is, the sample mean plus and minus the appropriate margin of error, which in this case is $e_{\alpha/2} = t_{\alpha/2}s_{\bar{x}}$. In other words, the confidence interval again estimates the population mean μ in terms of an interval that is centered around the sample mean $\bar{x}$ and that provides for a specified margin of error.

The derivation of the confidence interval involved the assumption that the sampling distribution of the sample mean $\bar{x}$ is Gaussian. Fortunately, the method is not excessively sensitive to departures from this **assumption of normality**; the confidence intervals are approximately correct in spite of moderate departures from normality. The method is therefore said to be relatively **robust** against nonnormality.

10.4 SUMMARY CONCERNING INTERVAL ESTIMATES OF μ

The results of Sections 10.2 and 10.3 concerning interval estimates of the population mean μ can be organized into the following summary statement, using the results in Section 8.4 about the sampling distribution of the mean $\bar{x}$. (Symbols not specifically defined in the statement have their conventional meanings.)

If a random sample is either

- large enough, or
- drawn from a Gaussian population,

then

a. the sampling distribution of $\bar{x}$ is Gaussian, and hence

b. the sampling distribution of $z_{\text{obs}} = \dfrac{\bar{x} - \mu}{\sigma_{\bar{x}}}$ is Gaussian, and

c. the sampling distribution of $t_{\text{obs}} = \dfrac{\bar{x} - \mu}{s_{\bar{x}}}$ is $t(n - 1 \text{ df})$, and

d. the $1 - \alpha$ confidence limits for μ are $\bar{x} \pm e_{\alpha/2}$, where

$$e_{\alpha/2} = z_{\alpha/2}\sigma_{\bar{x}} \qquad \text{if } \sigma \text{ is known, or}$$

$$e_{\alpha/2} = t_{\alpha/2}s_{\bar{x}} \qquad \text{if } \sigma \text{ is estimated by } s$$

STUDY MATERIAL FOR CHAPTER 10

Terms and Concepts

Section 10.1

point estimate of μ

Section 10.2

sampling distribution of z_{obs}
confidence level
margin of error of an estimate at the $1 - \alpha$ confidence level
margin of error of $\bar{x}$ when σ is known

$1 - \alpha$ confidence interval for μ when σ is known
$1 - \alpha$ confidence limits for μ when σ is known
size of sample required for a given margin of error in estimating μ

Section 10.3

number of degrees of freedom
Student's t distribution
sampling distribution of t_{obs}
$t(\infty \text{ df})$
margin of error of $\bar{x}$ when σ is estimated by s
$1 - \alpha$ confidence interval for μ when σ is estimated by s
$1 - \alpha$ confidence limits for μ when σ is estimated by s
assumption of normality
robust

Key Formulas

Section 10.2

(10.2) $$z_{obs} = \frac{\bar{x} - \mu}{\sigma_{\bar{x}}}$$

(10.6) $$e_{\alpha/2} = z_{\alpha/2}\sigma_{\bar{x}}$$

(10.8) $$\bar{x} - z_{\alpha/2}\sigma_{\bar{x}} < \mu < \bar{x} + z_{\alpha/2}\sigma_{\bar{x}}$$ with confidence level $1 - \alpha$

(10.12) $$n = \left(\frac{z_{\alpha/2}\sigma}{e_{\alpha/2}}\right)^2$$

Section 10.3

(10.14) $$s_{\bar{x}} = \frac{s}{\sqrt{n}}$$

(10.16) $$t_{obs} = \frac{\bar{x} - \mu}{s_{\bar{x}}}$$

(10.17) $$t = \frac{(\text{point estimate having a Gaussian sampling distribution}) - (\text{parameter})}{\text{estimated standard error of the point estimate}}$$

(10.23) $$e_{\alpha/2} = t_{\alpha/2}s_{\bar{x}}$$

(10.24) $$\bar{x} - t_{\alpha/2}s_{\bar{x}} < \mu < \bar{x} + t_{\alpha/2}s_{\bar{x}}$$ with confidence level $1 - \alpha$

Exercises for Numerical Practice

Section 10.2

10/1. Use the table in Appendix 3 to find the value of each of the following:

a. $z_{.25}$ **b.** $z_{.025}$ **c.** $z_{.01}$ **d.** $z_{.90}$ **e.** $z_{.98}$

10/2. Use the table in Appendix 3 to find the value of each of the following:

a. $z_{.20}$ **b.** $z_{.05}$ **c.** $z_{.005}$ **d.** $z_{.85}$ **e.** $z_{.999}$

10/3. Using the 95% confidence level, calculate the margin of error in the mean of a random sample of 45 observations from a population whose standard deviation is 9.15.

10/4. Using the 99% confidence level, calculate the margin of error in the mean of a random sample of 60 observations from a population whose standard deviation is 76.4.

10/5. A random sample of 50 observations with a mean of 33.81 is drawn from a population with a standard deviation of 2.60. Calculate the 98% confidence limits for the population mean.

10/6. A random sample of 75 observations with a mean of 1.01 is drawn from a population with a standard deviation of .18. Calculate the 99.9% confidence limits for the population mean.

10/7. How large a random sample do we need from a population with a standard deviation of 247 in order to be 99% confident that the sample mean is within 100 of the population mean?

10/8. How large a random sample do we need from a population with a standard deviation of .35 in order to be 95% confident that the sample mean is within .05 of the population mean?

Section 10.3

10/9. Use the table in Appendix 4 to find the value of each of the following:

a. $t_{.025}(5\text{ df})$ **b.** $t_{.995}(15\text{ df})$ **c.** $t_{.005}(49\text{ df})$ **d.** $z_{.025}$
e. $t_{.99}(8\text{ df})$ **f.** $t_{.05}(37\text{ df})$ **g.** $t_{.0005}(124\text{ df})$ **h.** $z_{.01}$

10/10. Use the table in Appendix 4 to find the value of each of the following:

a. $t_{.005}(12\text{ df})$ **b.** $t_{.95}(28\text{ df})$ **c.** $t_{.975}(3\text{ df})$ **d.** $z_{.05}$
e. $t_{.0005}(99\text{ df})$ **f.** $t_{.01}(55\text{ df})$ **g.** $t_{.025}(149\text{ df})$ **h.** $z_{.005}$

10/11. A random sample of 19 observations from a Gaussian population has a mean of 42.3 and a standard deviation of 10.8. Calculate the margin of error of the mean at the 90% confidence level.

10/12. A random sample of 27 observations from a Gaussian population has a mean of 738 and a standard deviation of 123. Calculate the margin of error of the mean at the 98% confidence level.

10/13. A random sample of 17 observations from a Gaussian population has a mean of 396 and a standard deviation of 92. Calculate the 99% confidence limits for the population mean.

10/14. A random sample of 8 observations from a Gaussian population has a mean of 55.1 and a standard deviation of 6.9. Calculate the 95% confidence limits for the population mean.

Problems for Applying the Concepts

10/15. A manufacturing process produces ball bearings whose diameters are known to have a standard deviation of approximately .12 mm.

a. What statement can we make with 95% confidence concerning the error in the mean of a random sample of 64 observations?

b. What are the 99% confidence limits for the population mean diameter of bearings produced by this process, as determined from a random sample of 64 observations with a mean of 10.023 mm?

c. What statement can we make with 99% confidence concerning the error in the mean of a random sample of 40 observations?

d. What are the 95% confidence limits for the population mean diameter of bearings produced by this process, as determined from a random sample of 40 observations with a mean of 9.971 mm?

e. How large a random sample would we need in order to be 90% confident that the sample mean is within .01 mm of the population mean?

f. How large a random sample would we need in order to be 99% confident that the sample mean is within .01 mm of the population mean?

10/16. The number of deliveries per day on the maternity ward of a small hospital is known to have a standard deviation of 1.4. The mean number of births per day in a sample of 30 days was 2.1.

a. What statement can we make with 95% confidence concerning the error in the sample mean?

b. What are the 95% confidence limits for the true mean number of births per day in the hospital?

c. What statement can we make with 99% confidence concerning the error in the sample mean?

d. What is the 99% confidence interval for the true mean number of births per day in the hospital?

e. What would the 99% confidence interval be if it were based on a sample of 91 days with a mean of 1.9 births per day?

f. How large a sample of days would we need in order to be 95% confident that the sample mean does not differ from the true mean by more than .2?

g. How large a sample of days would we need in order to be 98% confident that the sample mean does not differ from the true mean by more than .2?

10/17. A farmer weighed a sample of 9 healthy, egg-producing hens. The sample had a mean mass of 1134 grams and a standard deviation of 124 grams.

a. What statement can we make with 95% confidence concerning the error in the sample mean?

b. What are the 95% confidence limits for the mean mass of the population of hens from which this sample was selected?

c. What is the 99% confidence interval for the mean mass of the relevant population of hens?

(Source: Student project by P. Spitzer.)

10/18. In an investigation of physical fitness of college students, resting heart rates were recorded for a sample of 15 female students. The sample had a mean of 71.1 beats per minute and a standard deviation of 6.4 beats per minute.

a. What statement can we make with 95% confidence concerning the error in the sample mean?

b. What are the 95% confidence limits for the mean resting heart rate of the relevant population of female students?

c. What is the 99% confidence interval for the mean resting heart rate of the relevant population of female students?

(Source: Student project by A. Graham.)

10/19. A golfer recorded the following scores (numbers of strokes) in seven rounds on a nine-hole course:

43, 39, 42, 41, 44, 37, 41

Construct a 95% confidence interval for the golfer's true ability (that is, for the mean of the hypothetical population of scores he might have obtained) on that course at that time.

(Source: Student project by V. Lindstrand.)

10/20. To help him in planning his time, a printer in a local printing office kept detailed records of the orders he printed. From an extensive file of completed orders for 500 copies of a standard 2″-by-3½″ uncoated business card in black ink, he selected a random sample of 6 orders. The time required to ink up, get a proper impression in the correct place on the card, and run 500 copies of these 6 orders had been recorded to the nearest five minutes as follows:

40, 35, 40, 30, 35, 35

Construct a 99% confidence interval for the true mean time required to complete this type of order.

(Source: Student project by D. Hutchinson.)

10/21. To estimate the average amount purchased by female shoppers, a small-town grocer recorded the dollar amounts of a random sample of eight women's cash-register receipts:

46, 22, 50, 61, 22, 4, 80, 56

Construct a 98% confidence interval for the mean amount purchased in the relevant population of female shoppers.

(Source: Student project by D. Paulgaard.)

10/22. The daily sales in dollars at a small store were recorded as follows for a random sample of seven days:

923.41, 614.93, 586.17, 819.33, 956.07, 1022.47, 1144.72

Construct a 90% confidence interval for the true mean daily sales under the economic conditions at the time of this investigation.

(Source: Student project by J. Voigt.)

References

Czeisler, C. A., Weitzman, E. D., Moore-Ede, M. C., Zimmerman, J. C., and Knauer, R. S. "Human Sleep: Its Duration and Organization Depend on Its Circadian Phase." *Science* 210 (1980): 1264–67.

Irwin, J. O. "Gosset, William Sealy." *International Encyclopedia of Statistics*. Edited by W. H. Kruskal and J. M. Tanur. New York: The Free Press, 1978.

World Who's Who in Science. Chicago: Marquis Who's Who, 1968.

Further Reading

Moroney, M. J. *Facts from Figures.* Harmondsworth, Middlesex: Penguin, 1951. Chapter 14 introduces confidence intervals under the heading "how to be precise though vague."

Walker, H. M. "Degrees of Freedom." *Journal of Educational Psychology* 31 (1940): 253–69. This article presents both an intuitive and a geometrical explanation of the concept of degrees of freedom. It has been reprinted in several collections, including the following: Haber, A.; Runyon, R. P.; and Badia, P., eds. *Readings in Statistics.* Reading, MA: Addison-Wesley, 1970.

CHAPTER 11

TESTING HYPOTHESES ABOUT MEANS

EXAMPLE 11.1 There has been considerable controversy in recent years about differences in intellectual abilities between males and females. Results of many tests have shown that on the average females score higher than males on intellectual tasks requiring verbal skills while males tend to score higher on visual-spatial tasks, although the extensive overlap between the score distributions precludes drawing conclusions about any given individual. The controversy has revolved chiefly around the question as to whether these differences are primarily social in origin, reflecting the way children are brought up, or primarily biological in origin, reflecting brain differences.

It is known that the left side of the human brain is usually more specialized for some functions, such as verbal ability, and that the right side of the brain tends to be more specialized for other functions, such as visual-spatial ability; this pattern of specialization is called *cerebral lateralization*. The major band of fibers through which information passes between the two halves of the brain is called the *corpus callosum*. Cerebral lateralization and the exchange of information between the cerebral hemispheres are currently topics of substantial research activity, but few sex differences in brain structure have been established.

In one important study, 14 normal human brains, 9 male and 5 female, were examined in connection with autopsies performed at the Dallas Forensic Institute in Dallas, Texas. Researchers measured the width of the splenium, which is the posterior portion of the corpus callosum, and found that the mean for the male sample was 1.14 cm as compared with 1.64 cm for the female sample.

Of course, there is some natural variation among individuals in any bodily measurement (the standard deviations were reported as .174 and .182 cm for males and females, respectively), and so the researchers wanted to determine whether the observed sample difference reflected a true difference between males and females or whether it should be attributed to chance variation among individuals. Using a procedure to be introduced in this chapter, the researchers found that the probability was less than .001 that such a large difference would have occurred in random samples if the male and female populations actually had the same mean. In other words, the sample difference was statistically significant: It was larger than could reasonably be attributed to chance alone and therefore lent support to the hypothesis that the true mean width of the splenium is different for males and females.

The researchers noted that their finding "is of pragmatic interest to the forensic scientist. In addition, it has wide-ranging implications for students of human evolution and comparative neuroanatomy, as well as for neuropsychologists in search of an anatomical basis for possible gender differences in the degree of cerebral lateralization."

(Source: Lacoste-Utamsing & Holloway, 1982.)

As outlined in Section 9.2, the general objective of hypothesis testing is to determine whether an observed (sample) difference provides enough evidence for us to conclude that there is a corresponding difference in the underlying population. We start with a working assumption, called the **null hypothesis** and designated by H_0, that there is no difference between the quantities of interest in the population. This assumption enables us to determine how likely or unlikely our actual sample would be in that case. The **alternative hypothesis**, designated by H_A, states that there is a difference between the quantities of interest in the population.

If our sample is very unlikely to have occurred under the null-hypothesis assumption of no population difference, we reject the null hypothesis in favor of the alternative hypothesis. Still, we acknowledge that even very unlikely events occur sometimes and so our rejection may be a mistake—a Type I error. On the other hand, if our sample is consistent with what would be expected under the null hypothesis, we conclude that we do not have enough evidence to reject the null hypothesis. Notice that we cannot prove that the null hypothesis is true: Even if the sample data are consistent with the null hypothesis of no population difference, there may actually be a difference in the population, a difference with which the sample data are also consistent. In that case, our failure to reject the null hypothesis would be a Type II error. In any case, the results of hypothesis testing depend on the determination of how likely or unlikely our sample data would be if the null hypothesis were true, and to that problem we now turn.

The summary statement in Section 10.4 concerning interval estimates of the population mean μ provides the theoretical basis for our discussion of testing hypotheses about means. We will consider three types of tests about means in Sections 11.1, 11.2, and 11.3:

1. one-sample *t* test for a specified value of the mean;
2. independent-samples *t* test of the difference between two means; and
3. paired-samples *t* test of the difference between two means.

11.1 ONE-SAMPLE *t* TEST FOR A SPECIFIED VALUE OF THE MEAN

Let us begin with an example to illustrate the nature of the first type of test; then we will look at the procedure in general terms.

EXAMPLE 11.2 The Parker Brothers game of Boggle requires players to list, within three minutes, as many words as they can find in a scrambled assortment of letters. The game includes a three-minute, hourglass-type timer.

For this example, I used a stopwatch to measure the amount of time required for the sand to run through a Boggle timer. In a sample of 10 trials, the timer had a mean of 158.2 seconds and a standard deviation of 7.7 seconds. Are these results consistent with the notion that the observed times were a random sample from a population with a mean of $\mu = 180$ seconds (three minutes)?

The null hypothesis to be tested here is that the population mean is 180 seconds; in symbols,

$$H_0: \mu = 180 \tag{11.1}$$

The alternative hypothesis is that the population mean is not equal to 180 seconds; in symbols,

$$H_A: \mu \neq 180$$

We want to see if the sample provides enough evidence for us to reject the null hypothesis in favor of the alternative hypothesis.

We know from Chapter 10 (see the summary statement in Section 10.4) that the sampling distribution of

$$t_{\text{obs}} = \frac{\bar{x} - \mu}{s/\sqrt{n}}$$

is the t distribution with $n - 1$ degrees of freedom. Therefore, if the null hypothesis in Expression (11.1) is true, then

$$t_{\text{obs}} = \frac{158.2 - 180}{7.7/\sqrt{10}} = -8.95$$

may be regarded as a random selection from a t distribution with $n - 1 = 9$ degrees of freedom. From the table in Appendix 4, we find that critical values of ± 4.781 cut off an area of .0005 in each tail of the t distribution with 9 degrees of freedom, for a total tail area of .001, or .1% (see Figure 11.1). That is, a value of t_{obs} more extreme than ± 4.781 would occur only .1% of the time if the null hypothesis were true. Since our observed value -8.95 is more extreme than the critical values, we reject the null hypothesis, acknowledging that our procedure would lead to a Type I error .1% of the time if the null hypothesis were in fact true.

We conclude that the sample mean of 158.2 seconds is significantly different from the hypothesized value of 180 seconds, testing at the .001 significance level. The data support the view that the population mean μ is less than 180 seconds.

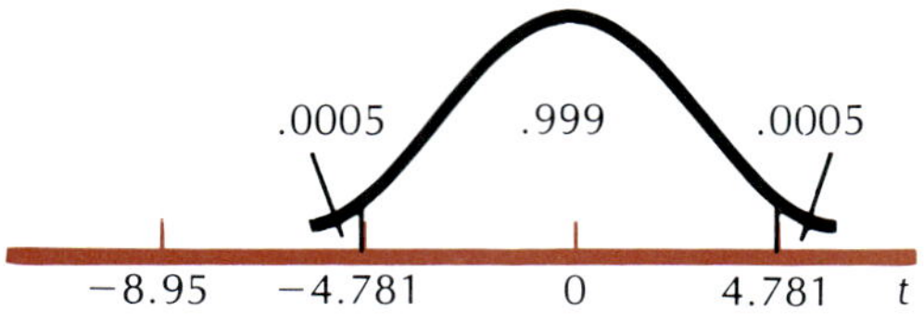

Figure 11.1

Hypotheses

If we have data from a single random sample, as in Example 11.2, it may be of interest to test whether the sample mean differs significantly (in the statistical sense—that is, more than by chance) from a specified value, say, μ_0, that had previously been conjectured to be the population mean. In this situation, the general form of the null hypothesis is

$$H_0\colon \mu = \mu_0 \tag{11.2}$$

That is, the population mean μ is hypothesized to be equal to the specified value μ_0. Notice that the symbol μ refers to the unknown mean of the population, whereas μ_0 refers to a specific known value (such as 180 in Example 11.2); μ is an unknown constant, whereas μ_0 is a known constant. What is not known about μ_0 (or about 180 in the example) is whether it is the correct value of the unknown population mean μ. The null hypothesis says, in effect, "Let us suppose that the unknown constant μ equals this specific known constant."

The usual alternative hypothesis is

$$H_A\colon \mu \neq \mu_0 \tag{11.3}$$

This alternative hypothesis does not specify an individual value for the population mean μ; rather it says that μ is not equal to the value specified in the null hypothesis. An hypothesis, such as that in Expression (11.2), that states an exact equality is called a **simple hypothesis**, whereas one, such as that in Expression (11.3), that involves some sort of inequality is called a **composite hypothesis**, because it may be thought of as being a composite or combination of simple hypotheses. In classical hypothesis testing, the null hypothesis is typically a simple hypothesis, or at least may be viewed as such, and the alternative hypothesis is typically a composite hypothesis.

The Test Statistic

From the summary statement in Section 10.4, we know that for random samples whose means follow a Gaussian sampling distribution, the statistic

$$t_{\text{obs}} = \frac{\bar{x} - \mu}{s/\sqrt{n}} \tag{11.4}$$

follows the t distribution with $n - 1$ degrees of freedom. If the null hypothesis specified in Expression (11.2) is true, then we can substitute $\mu = \mu_0$ in Expression (11.4) to obtain the **test statistic** for the **one-sample t test for a specified value of the mean**:

$$t_{\text{obs}} = \frac{\bar{x} - \mu_0}{s/\sqrt{n}} \tag{11.5}$$

which is distributed according to $t(n - 1\ \text{df})$ if the null hypothesis is true. Since μ_0 is the constant given in the null hypothesis, and $\bar{x}$, s, and n can all be determined from the sample data, Expression (11.5) allows us to calculate the numerical value of the test statistic t_{obs}. (Guidelines for how to round and how many digits to report are given in Appendix 1.)

The Decision Rule

Once we have a numerical value of t_{obs} contingent on the truth of the null hypothesis, we can compare t_{obs} to tabled values of the t distribution to see what the chances would be of getting a value of t at least that large if the null hypothesis were true. Looking at Figure 11.2, we can see that the farther a value of t_{obs} is away from the mean of the t distribution (that is, away from 0), the less likely it is.

But how unlikely must the value of the test statistic t_{obs} be for us to decide that we do not think such an unlikely event happened to us in this particular sample? In other words, how small must the conditional probability

$$P = P(\text{such a large sample difference} \mid H_0) \tag{11.6}$$

be for us to conclude that the null hypothesis should be rejected? The particular conditional probability specified in Expression (11.6) is called the ***P*-value** of the significance test: It is the probability of such a large difference between the quantities of interest in the sample if the null hypothesis is true. Graphically, it is the total tail area (in both tails) more extreme than the **observed value of the test statistic** (see Figure 11.2).

Recall from Section 9.2 that a significance test is carried out like a game of "let's pretend": Let us suppose, the logic says, that the null hypothesis of no difference between the quantities of interest in the population is true. If so, a sample difference at least as large as the one we obtained would occur with probability P. If P is very small, indicating a very large difference between the quantities of interest in the sample, we will reject the null hypothesis. To decide how small is "very small," we specify the maximum probability, denoted by α, with which we are willing to risk making a Type I error—incorrectly rejecting a true null hypothesis. If P is less than α, we reject the null hypothesis, and if P is greater than α, we do not reject the null

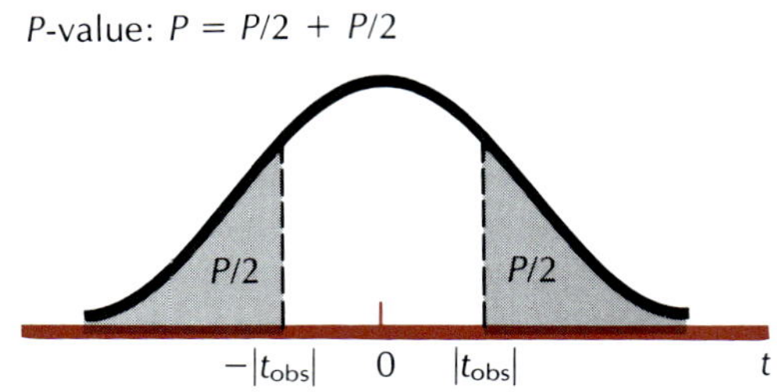

Figure 11.2 The sampling distribution of t_{obs} if the null hypothesis is true, showing the P-value as the total tail area more extreme than a particular **observed value** of t_{obs}.

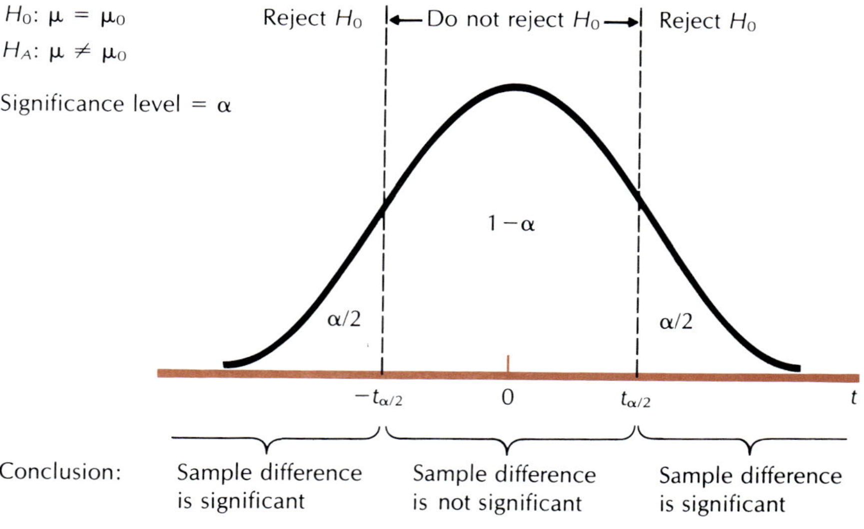

Figure 11.3 The sampling distribution of t_{obs} if the null hypothesis is true, with **critical values** ($\pm t_{\alpha/2}$) shown for the decision about the null hypothesis.

hypothesis. The value α is called the **significance level** of the test, and it determines the Type I error rate—the total tail area beyond the **critical values** in the null-hypothesis distribution, as shown in Figure 11.3.

In practice, all we have to do is to compare the numerical value of t_{obs} to the **critical values**, $\pm t_{\alpha/2}$, which separate a total tail area equal to the significance level α from the rest of the distribution, as shown in Figure 11.3. If t_{obs} is in the **rejection region**, beyond the critical value in one of the tails of the t distribution, we reject the null hypothesis and conclude that the mean μ of the population from which our sample was drawn is not equal to the specified value μ_0. On the other hand, if t_{obs} is in the **nonrejection region**, between the critical values, then we do not reject the null hypothesis and we conclude that we do not have enough evidence to say that the population mean μ differs from the hypothesized value μ_0.

In either case, the conclusion should always be qualified by an indication of the significance level employed. This is often done by means of the ***P*-value inequality:**

$$P < \alpha \qquad \text{for a result that is statistically significant } (H_0 \text{ rejected}) \tag{11.7}$$

or

$$P > \alpha \qquad \text{for a result that is not statistically significant } (H_0 \text{ not rejected}) \tag{11.8}$$

The P-value inequality is written (with the numerical value of α in place of the symbol α) at the end of the statement of the significance-test results.

Alternatively, it may simply be reported that the result was or was not significant at the α level.

Conventionally, the maximum acceptable value of α is .05, which is to say that we are willing in the long run to make Type I errors (incorrect rejections) on 5% of the occasions when the null hypothesis we are testing is true. Of course, we do not know which occasions those are. On each occasion, all we know is how large the sample difference is. Rejection of the null hypothesis because the P-value is less than .05 implies a judgment that an event (the sample outcome) so extreme as to happen less than 5% of the time did not happen to us on this particular occasion; rather, we conclude, the sample was a reflection of the alternative hypothesis. For a more stringent test, the significance level α may be reduced, say, to .01. Rejection of the null hypothesis because the P-value is less than .01 implies a judgment that an event so extreme as to happen less than 1% of the time did not happen to us on this particular occasion; rather, we conclude, the sample was a reflection of the alternative hypothesis.

There is nothing magical about the 5% and 1% significance levels. Any P-value obtained may be reported as a summary of the evidence provided by the sample against the null hypothesis. The conventional value of $\alpha = .05$ can be treated as a guideline for interpretation when relevant experience or other information suggests nothing to the contrary.

EXAMPLE 11.3 Let us review Example 11.2 in light of the preceding discussion. In that example, we tested the null hypothesis that the true mean time μ required for the sand to run through a certain hourglass-type timer was $\mu_0 = 180$ seconds as claimed by the manufacturer. Given a sample of $n = 10$ observations with a mean of $\bar{x} = 158.2$ seconds and a standard deviation of $s = 7.7$ seconds, we calculated the value of the test statistic in Expression (11.5) as

$$t_{\text{obs}} = \frac{158.2 - 180}{7.7/\sqrt{10}} = -8.95$$

We compared the observed value of the test statistic, -8.95, with the critical values, $\pm t_{.0005}(9 \text{ df}) = \pm 4.781$, of the sampling distribution of t_{obs} under the null hypothesis (see Figure 11.4). Using the decision rule illustrated in Figure 11.3, we rejected the null hypothesis at the .001 significance level. The evidence, evaluated at the .001 significance level, suggests that the timer's true mean μ is less than the hypothesized

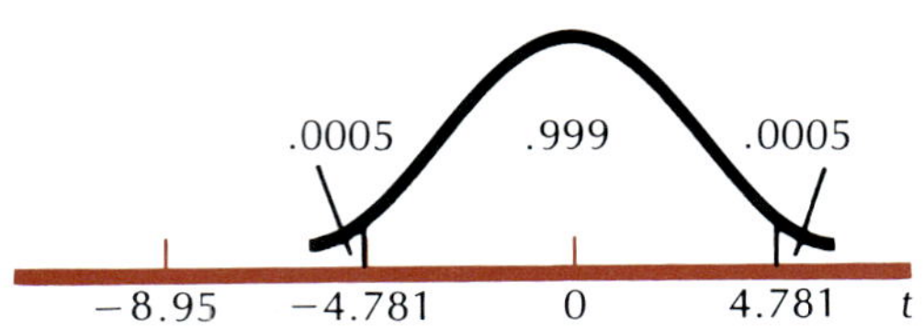

Figure 11.4

value of 180 seconds. Another way of reporting this conclusion would be to say that the sample mean was significantly less than the hypothesized value of 180 seconds ($P < .001$).

It is instructive to consider the 99.9% confidence interval for the true mean μ as determined from this sample. As given in the summary statement in Section 10.4, the 99.9% confidence limits are

$$158.2 \pm (4.781)(7.7/\sqrt{10}) = 158.2 \pm 11.6$$

and hence the 99.9% confidence interval is 146.6 to 169.8 seconds. Notice that the interval does not include the hypothesized value of 180 seconds. A statement that the sample mean is significantly different from 180 seconds at the .1% significance level is equivalent to saying that the 99.9% confidence interval does not include the value 180 seconds. These are alternative ways of saying that the natural sampling variability in the sample mean 158.2 would rarely (less than .1% of the time) be sufficient to account for such a large difference from 180 seconds, if 180 seconds were the true population mean.

EXAMPLE 11.4 Some theorists have suggested that babies who are heavier at birth have a better chance of developing high intelligence than those with lower birth weights. (Others have dubbed this the "fat-chance" theory.) If the theory is correct, then we might expect to find that individuals who are pursuing higher education had higher birth weights than the general population.

A sample of 16 female college students reported their birth weights. The recalled birth weights had a mean of 3.331 kg and a standard deviation of .333 kg. Official community health records show the mean female live birth weight to be 3.25 kg for the region from which the college draws its students. Let us test the null hypothesis that the college population from which these weights were sampled had a mean μ of $\mu_0 = 3.25$ kg, against the alternative hypothesis that the population mean μ for female college students is not equal to the female regional average of $\mu_0 = 3.25$ kg. In symbols,

$$H_0\colon\ \mu = 3.25$$

$$H_A\colon\ \mu \neq 3.25$$

If the null hypothesis H_0 is true, the observed value of the test statistic, as given by Expression (11.5) is

$$t_{\text{obs}} = \frac{3.331 - 3.25}{.333/\sqrt{16}} = .97$$

If the null hypothesis is true, $t_{\text{obs}} = .97$ may be regarded as a random selection from the t distribution with $n - 1 = 16 - 1 = 15$ degrees of freedom, whose .05-level critical values are $\pm t_{.025}(15\text{ df}) = \pm 2.131$ (see Figure 11.5). Since the observed value t_{obs} falls between the critical values, we cannot reject the null hypothesis at the .05 significance level (see Figure 11.3).

The data are consistent with the notion that these college students came from a population whose mean birth weight was 3.25 kg; the observed sample mean was not significantly different from the hypothesized value of 3.25 kg ($P > .05$). Of course, it could be that the students' recollections of their birth weights were inaccurate. But even if the recollections were perfectly accurate, we have not proved that the null hypothesis is true; perhaps, for example, a larger sample would have given us a sufficiently precise estimate of college-student birth weights so that we could have distinguished it from the hypothesized value of 3.25 kg. A sample difference that is not statistically significant

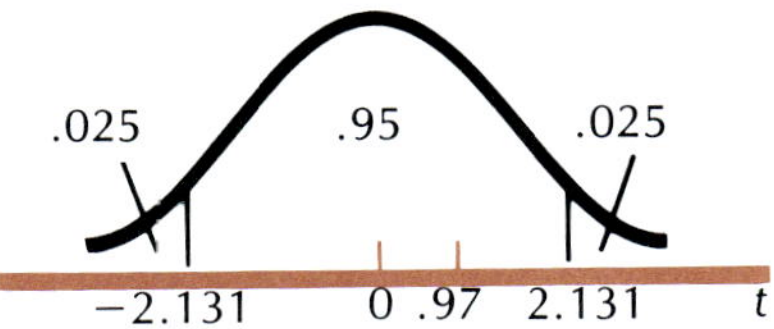

Figure 11.5

implies not that there is no difference in the population, but only that the sample does not provide enough evidence to rule out the possibility of no population difference.

This conclusion is spelled out more specifically in the 95% confidence interval for the mean of the population from which the observations in this example were sampled. Using information provided in the summary statement in Section 10.4, we find that the 95% confidence limits are

$$3.331 \pm (2.131)(.333/\sqrt{16}) = 3.331 \pm .177$$

and hence the 95% confidence interval is 3.154 to 3.508 kg, which includes the hypothesized value of 3.25 kg as well as a range of other values. The data do not rule out the hypothesized value of 3.25 kg for the population mean, but neither are they inconsistent, at the 95% confidence level, with any other value between 3.154 and 3.508 kg.

A statement that the sample mean is not significantly different from 3.25 kg ($P > .05$) is equivalent to saying that the 95% confidence interval includes the value 3.25 kg. Confidence intervals and significance tests differ in their units of measurement: In confidence intervals we express the variability associated with the sample mean in the original units of measurement (kilograms, in this case), whereas in significance testing we evaluate the variability associated with the sample mean in t-units (number of estimated standard errors away from an hypothesized value).

(Source: Student project by C. Hochhausen.)

One-Sided Alternatives

So far, the alternative hypotheses H_A that we have considered have been of the form given in Expression (11.3): $\mu \neq \mu_0$. This is indeed the usual form, but there are certain circumstances in which one of the following two types of alternative hypotheses may replace that in Expression (11.3):

$$H_A: \mu < \mu_0 \tag{11.9}$$

or

$$H_A: \mu > \mu_0 \tag{11.10}$$

Notice that only one of these two possible alternative hypotheses would be specified in any given problem; if both were of interest, the alternative hypothesis in Expression (11.3) would be used. The hypotheses in Expressions (11.9) and (11.10) are called **one-sided alternatives** or **directional alternatives**, because they indicate that a difference is sought in one direction only.

Hypotheses in the form of Expression (11.3) ($\mu \neq \mu_0$) are called **two-sided alternatives** or **nondirectional alternatives**. A one-sided alternative is appropriate in the following circumstance only:

> A difference in the direction opposite to that specified by the one-sided alternative is of no more interest than no difference at all. **(11.11)**

In effect, this is equivalent to treating the null hypothesis as the composite $\mu \geq \mu_0$ or $\mu \leq \mu_0$ for the alternatives in Expressions (11.9) and (11.10), respectively, but for conceptual simplicity we will treat all null hypotheses as simple hypotheses, imposing the restriction in Expression (11.11) on the use of one-sided alternatives.

EXAMPLE 11.5 In Example 11.4, we wanted to see whether individuals pursuing higher education had a higher mean birth weight μ than the general regional average of 3.25 kg. However, we did not use a one-sided alternative ($\mu > 3.25$), because it would also be of interest if the college students' birth weights had been significantly *below* the regional average. Even though we were perhaps expecting or "hoping for" a difference in the positive direction, which would support the "fat-chance" theory, a difference in the opposite direction would have implications different from those of a finding of no difference—perhaps favoring a "slim-chance" theory. Therefore, a two-sided alternative was used.

The calculation of the test statistic t_{obs} is the same whether the alternative hypothesis is one-sided or two-sided. However, with a one-sided alternative there is only one critical value, rather than two, and the entire rejection region of area α is in one tail of the distribution. Figure 11.6 illustrates the decision rule for the two types of one-sided alternatives. A test with a rejection region in only one tail of the relevant sampling distribution is called a **one-tailed test** (Figure 11.6); one with rejection regions in both tails is called a **two-tailed test** (Figure 11.3).

One-sided alternatives are sometimes misused because, if used incorrectly, they make it "easier" to obtain a statistically significant result. A positive value of t_{obs}, for example, does not have to be as large to exceed t_α [see Figure 11.6(b)] as it does to exceed $t_{\alpha/2}$ (which must be to the right of t_α in order to divide the α-area tail in half). However, if the reported significance level is to be meaningful, it is important that one-sided alternatives be used only in the circumstance specified in Expression (11.11), and in particular that one-sided alternatives never be proposed on the basis of inspection of the sample data. Experience in a particular area of application makes it easier to determine when one-sided alternatives are appropriate, but the beginning student is best advised to guard against overuse.

Since directional alternatives are the exception rather than the rule, their use is stated explicitly in reports of the results of any significance tests in which they are used, for example, "$P < .05$, one-tailed." The reader of the results can then determine that a difference that is significant at the α level (say, .05) against a one-sided alternative would be significant at the 2α level (for

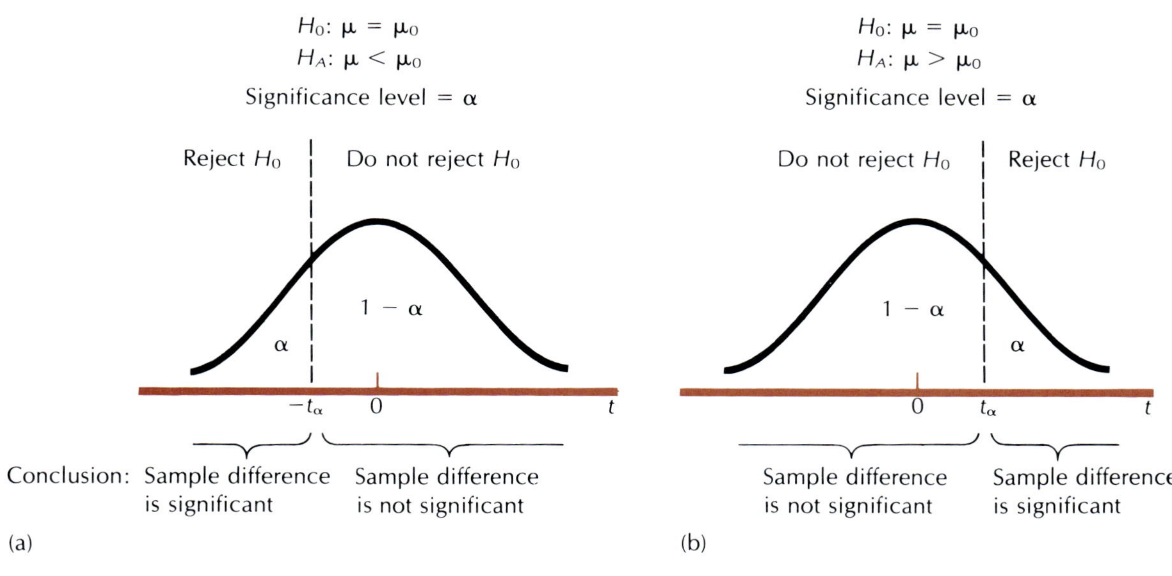

Figure 11.6
Sampling distributions of t_{obs} if the null hypothesis is true. Distribution (a) shows the critical value for the decision about the null hypothesis against a one-sided "less than" alternative; distribution (b) shows the critical value for a one-sided "greater than" alternative.

example, .10) against a two-sided alternative. Conversely, a difference significant at the α level (say, .05) against a two-sided alternative would be significant at the $\alpha/2$ level (for example, .025) against a one-sided alternative in the appropriate direction.

EXAMPLE 11.6 In a search for weaknesses in his golfing, an amateur golfer keeps track of how many of his strokes are putts. Although he recognizes that there will be some variability from round to round, he considers his putting to be satisfactory if he is convinced that his true mean putting ability is less than 18 putts per round on a nine-hole course; if not, he devotes more practice time to putting. In seven rounds of golf on a nine-hole course, he had a mean of 15.3 putts per round and a standard deviation of 1.6 putts per round. Do these data provide sufficient evidence to convince the golfer that his true mean putting performance is satisfactory according to his criterion?

In this case, the golfer is interested in knowing about a difference in only one direction, namely, whether his true mean putting ability μ is less than 18 putts per round. We use a one-sided alternative here, because a difference in the other direction (more than 18 putts per round) would be treated in the same way as a finding of no difference; namely, more practice time would be devoted to putting. In symbols, we have

$$H_0: \ \mu = 18$$

$$H_A: \ \mu < 18$$

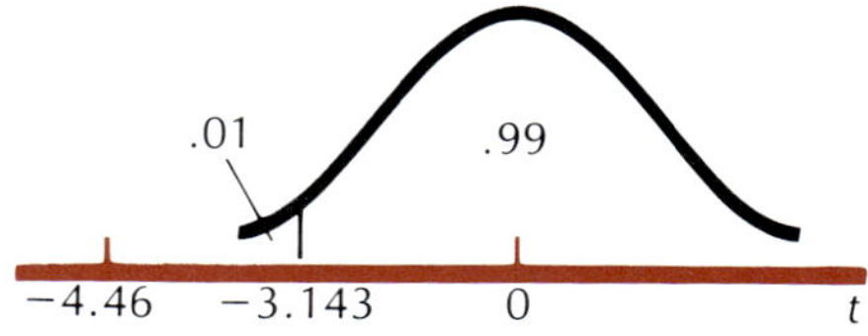

Figure 11.7

If the null hypothesis is true, the value of the test statistic, given by Expression (11.5), is

$$t_{\text{obs}} = \frac{15.3 - 18}{1.6/\sqrt{7}} = -4.46$$

which may be regarded as a random selection from a t distribution with $n - 1 = 6$ degrees of freedom. The critical value cutting off 1% of the area in the left tail of the distribution is $-t_{.01}(6\text{ df}) = -3.143$, as shown in Figure 11.7. Using the decision rule illustrated in Figure 11.6(a), we reject the null hypothesis at the .01 significance level, because $t_{\text{obs}} = -4.46$ is in the rejection region beyond the critical value -3.143. We conclude that the mean number of putts in the sample of seven rounds is significantly less than 18 putts per round ($P < .01$, one-tailed).

(Source: Student project by V. Lindstrand.)

Summary of the Test Procedure

The **one-sample *t* test for a specified value of the mean** may be summarized as follows:

Assumptions.

a. Random sample.

b. Normality of the sampling distribution of the mean.

Null hypothesis. $H_0\colon \mu = \mu_0$

	Nondirectional test	Left-directional test	Right-directional test
Alternative hypothesis.	$H_A\colon \mu \neq \mu_0$	$H_A\colon \mu < \mu_0$	$H_A\colon \mu > \mu_0$
Critical values.	$\pm t_{\alpha/2}$	$-t_\alpha$	t_α

Step 1. Calculate the point estimate $\bar{x}$ to see what sample value is being tested.

Step 2. Calculate the null-hypothesis value of the test statistic:

$$t_{\text{obs}} = \frac{\bar{x} - \mu_0}{s/\sqrt{n}}$$

Step 3. Look up the critical value(s) in the t distribution with $n - 1$ degrees of freedom. (If σ is known and used in place of s in Step 2, it is considered to have infinite degrees of freedom.)

Step 4. If the observed value t_{obs} is at or beyond the critical value in the tail of the distribution, conclude that the sample mean is significantly different from the specified value μ_0 ($P < \alpha$); otherwise, conclude that the observed difference is not statistically significant ($P > \alpha$).

11.2 INDEPENDENT-SAMPLES *t* TEST OF THE DIFFERENCE BETWEEN TWO MEANS

The t test described in Section 11.1 is intended for situations in which we have a single sample of n observations on the basis of which we want to draw an inference about the mean of the population from which the sample was taken. This section introduces another t test, this one for two samples, one with n_1 observations and the other with n_2 observations, on the basis of which we want to learn something about the means, μ_1 and μ_2, of the two populations from which the two samples were drawn. The type of situation in which such a test would be used is illustrated in Example 11.1 and in the following example.

EXAMPLE 11.7 The operations research department of a large food-processing company undertook a project to determine whether orange juice made from a new variety of faster-growing oranges tastes different from orange juice made from the usual variety of oranges purchased by the company.

Twenty-nine individuals who had been randomly assigned to one of two treatment conditions were each given a glass of orange juice to taste and were asked to rate it on a scale from 1 (very poor) to 10 (outstanding). The two treatment conditions corresponded to the two varieties of oranges: $n_1 = 15$ individuals rated juice made from the old variety (variety 1), and $n_2 = 14$ individuals rated juice made from the new variety (variety 2). The sample means for the two varieties were $\bar{x}_1 = 7.5$ and $\bar{x}_2 = 6.1$, which are point estimates of the corresponding population mean ratings μ_1 and μ_2 for the two varieties. The null hypothesis to be tested is that the two population means are equal—that there is no systematic tendency for the two varieties to be rated differently—and the alternative hypothesis is that the true mean ratings are different for variety 1 and variety 2; in symbols,

$$H_0: \mu_1 = \mu_2$$
$$H_A: \mu_1 \neq \mu_2$$

The one-sample t test described in Section 11.1 is based on the fact that the ratio

$$t = \frac{\text{(point estimate having a Gaussian sampling distribution) minus (parameter)}}{\text{estimated standard error of the point estimate}} \quad \textbf{(11.12)}$$

follows the t distribution having the number of degrees of freedom associated with the estimated standard error. The same general principle forms the basis for testing whether the difference between the means of two **independent random samples** is statistically reliable or whether it should be attributed to chance. (Two samples are **independent** if the selection of observational units for one sample has no bearing on which observational units are in the other sample.) To apply this general principle in the present case, we must identify the quantities that play the roles of the various terms in Expression (11.12): the relevant parameter, its point estimate, and the estimated standard error of the point estimate. We will identify these quantities in the next four subsections and then bring them together in the fifth.

Among other applications, the two-sample t test introduced in this section is frequently used in research involving the comparison of an experimental group and a control group. As noted in Section 9.2, the two groups correspond to the two levels of the independent variable (the treatment variable), and the observations are the values of the dependent variable (the response variable). Although it is customary in the context of correlation and regression to use the symbol x for the independent variable and the symbol y for the dependent variable, the convention in the context of t tests on means, where generally no symbol is needed for the independent variable, is to use the symbol x for the dependent variable. We will follow this convention here, but you should recognize that in testing for differences between means of the variable x, we are testing whether the value of the variable x depends on the level of the independent variable that distinguishes the two groups. (In Example 11.7, the independent variable is the variety of oranges, and the dependent variable is the rating of taste.)

Hypotheses

The null hypothesis for the t test based on two independent samples is, as usual, an assumption of no difference in the population. Specifically, the mean μ_1 of the population from which one sample was drawn is hypothesized to be equal to the mean μ_2 of the population from which the second sample was drawn; in symbols,

$$H_0\colon\ \mu_1 = \mu_2 \tag{11.13}$$

The alternative hypothesis is typically the general two-sided composite hypothesis:

$$H_A\colon\ \mu_1 \neq \mu_2 \tag{11.14}$$

However, one-sided alternatives analogous to those in Expressions (11.9) and (11.10) are also possible and may be used if a difference in the opposite direction would be of no more interest than no difference at all.

The Parameter and Its Estimate

The null hypothesis in Expression (11.13) says that the difference between the parameters is zero, as demonstrated by subtracting μ_2 from both sides of Expression (11.13):

$$H_0\colon\ \mu_1 - \mu_2 = 0 \tag{11.15}$$

In effect, the difference $\mu_1 - \mu_2$ is the parameter that we wish to estimate on the basis of our two independent random samples.

If the two sample means are $\bar{x}_1$ and $\bar{x}_2$, the sample difference $\bar{x}_1 - \bar{x}_2$ is an unbiased estimate of the population difference $\mu_1 - \mu_2$, and the sampling distribution of the difference $\bar{x}_1 - \bar{x}_2$ is Gaussian if the sampling distributions of $\bar{x}_1$ and $\bar{x}_2$ are both Gaussian (for example, under the Central Limit Theorem).

The Standard Error of the Estimate

What is the standard error $\sigma_{\bar{x}_1-\bar{x}_2}$ of the difference between two sample means? In other words, what is the standard deviation of the sampling distribution of the difference between two sample means? We will answer this question in terms of the variance $\sigma^2_{\bar{x}_1-\bar{x}_2}$ of the sampling distribution, that is, the square of the required standard error.

It can be shown that the variance of the difference between two independent random variables is equal to the sum of their separate variances. It follows that for the difference between means of two independent random samples, the square of the required standard error is the sum of the two squared standard errors:

$$\sigma^2_{\bar{x}_1-\bar{x}_2} = \sigma^2_{\bar{x}_1} + \sigma^2_{\bar{x}_2} \qquad \text{for two independent random samples} \qquad \textbf{(11.16)}$$

Given that there are n_1 observations in the first sample and n_2 in the second, Expression (11.16) becomes, by application of Expression (8.9),

$$\sigma^2_{\bar{x}_1-\bar{x}_2} = \frac{\sigma_1^2}{n_1} + \frac{\sigma_2^2}{n_2} \qquad \text{for two independent random samples} \qquad \textbf{(11.17)}$$

If we can assume that the two populations being sampled have equal dispersion ($\sigma_1 = \sigma_2 = \sigma$, say), then Expression (11.17) simplifies to

$$\sigma^2_{\bar{x}_1-\bar{x}_2} = \sigma^2\left(\frac{1}{n_1} + \frac{1}{n_2}\right) \qquad \begin{array}{l}\text{for two independent random samples}\\ \text{from populations with equal variances}\end{array} \qquad \textbf{(11.18)}$$

Taking square roots, we obtain the *standard error of the difference between the means of two independent samples from populations with equal variances*:

$$\sigma_{\bar{x}_1-\bar{x}_2} = \sqrt{\sigma^2\left(\frac{1}{n_1} + \frac{1}{n_2}\right)} \qquad \textbf{(11.19)}$$

The Estimated Standard Error of the Estimate

To obtain an estimate of the standard error in Expression (11.19), we need somehow to pool the variance estimates s_1^2 and s_2^2 from the two samples into a single estimate of the common population variance σ^2. We can easily obtain this estimate by taking a fresh look at the definitional formula for the sample variance:

$$s^2 = \frac{\sum (x - \bar{x})^2}{n - 1} \qquad \textbf{(11.20)}$$

The sample variance, or the variance estimate, is the sum of the squared deviations from the mean, divided by the number of degrees of freedom. As noted in Section 5.1, the phrase *sum of squared deviations from the mean* is often shortened to **sum of squares** and is symbolized by SS. The number of degrees of freedom is denoted by df. Thus, the variance estimate in Expression (11.20) may be written as a sum of squares divided by its number of degrees of freedom:

$$s^2 = \frac{\text{SS}}{\text{df}} \tag{11.21}$$

When we have two independent samples and require a **pooled estimate of the common population variance σ^2**, we add the sum of squares SS_1 from the first sample and the sum of squares SS_2 from the second sample to obtain the total sum of squares. Then we divide by the total of the numbers of degrees of freedom from the two samples, df_1 and df_2:

$$s^2_{\text{pooled}} = \frac{\text{SS}_1 + \text{SS}_2}{\text{df}_1 + \text{df}_2} \tag{11.22}$$

$$= \frac{\sum (x_1 - \bar{x}_1)^2 + \sum (x_2 - \bar{x}_2)^2}{(n_1 - 1) + (n_2 - 1)} \quad \text{by the definition of SS} \tag{11.23}$$

$$= \frac{\sum x_1^2 - \dfrac{(\sum x_1)^2}{n_1} + \sum x_2^2 - \dfrac{(\sum x_2)^2}{n_2}}{n_1 + n_2 - 2} \quad \text{by the derivation on pages 116–117} \tag{11.24}$$

$$= \frac{(n_1 - 1)s_1^2 + (n_2 - 1)s_2^2}{n_1 + n_2 - 2} \quad \text{by rearranging Expression (11.21)} \tag{11.25}$$

In Expressions (11.23) and (11.24), the subscript on the variable x indicates whether the x-values are taken from sample 1 or sample 2. Expressions (11.22) through (11.25) are all algebraically equivalent, but (11.23) is perhaps the most transparent. It shows that $\text{SS}_1 + \text{SS}_2$ is the sum of the squared deviations of the observations from their own sample mean. Expressions (11.24) and (11.25) generally are preferable for computational purposes.

The denominator of these expressions is more straightforward:

$$\text{df}_1 + \text{df}_2 = (n_1 - 1) + (n_2 - 1) = n_1 + n_2 - 2 \tag{11.26}$$

Notice that there is a total of $n_1 + n_2$ observations in the two samples combined, and there are two sample-based restrictions imposed in the calculation of s^2_{pooled}, namely, the two sample means. Thus, there is a total of $n_1 + n_2 - 2$ degrees of freedom.

Therefore, the estimate of the standard error in Expression (11.19) is

$$s_{\bar{x}_1 - \bar{x}_2} = \sqrt{s^2_{\text{pooled}}\left(\frac{1}{n_1} + \frac{1}{n_2}\right)} \tag{11.27}$$

with $n_1 + n_2 - 2$ degrees of freedom.

EXAMPLE 11.8 Example 11.7 described an investigation into the taste of orange juice produced from two varieties of oranges. The sample of $n_1 = 15$ ratings for variety 1 had a mean of 7.5 and a standard deviation of 1.3, and the sample of $n_2 = 14$ ratings for variety 2 had a mean of 6.1 and a standard deviation of 1.5. The estimate of the population difference $\mu_1 - \mu_2$ is the sample difference

$$\bar{x}_1 - \bar{x}_2 = 7.5 - 6.1 = 1.4$$

How precise is this estimate? We may estimate its standard error from Expression (11.27), using Expression (11.25) to obtain

$$s_{\bar{x}_1-\bar{x}_2} = \sqrt{\frac{(15-1)(1.3)^2 + (14-1)(1.5)^2}{15+14-2}\left(\frac{1}{15}+\frac{1}{14}\right)} = .5202$$

Thus, the estimated mean difference between ratings for varieties 1 and 2 is 1.4, with a standard error of approximately .52. (Recall that the standard error of an estimate is a representative or standard amount of error in such an estimate.)

The Test Statistic

We began our development of this test procedure by noting that we would be using the general principle that the ratio

$$t = \frac{\text{(point estimate having a Gaussian sampling distribution) minus (parameter)}}{\text{estimated standard error of the point estimate}} \qquad \textbf{(11.28)}$$

follows the t distribution having the number of degrees of freedom associated with the estimated standard error. Since then, we have seen that for testing whether the difference between the means of two independent random samples is statistically significant,

a. the relevant parameter is $\mu_1 - \mu_2$;

b. its point estimate $\bar{x}_1 - \bar{x}_2$ has a Gaussian sampling distribution under the terms of the Central Limit Theorem; and

c. the estimated standard error of the point estimate is given by Expression (11.27), having $n_1 + n_2 - 2$ degrees of freedom, provided that the two population variances are equal.

Putting these facts together in Expression (11.28), we have the result that

$$t_{\text{obs}} = \frac{(\bar{x}_1 - \bar{x}_2) - (\mu_1 - \mu_2)}{s_{\bar{x}_1-\bar{x}_2}} \qquad \textbf{(11.29)}$$

follows the t distribution with $n_1 + n_2 - 2$ degrees of freedom. Assuming for the moment that the null hypothesis given in Expression (11.13) is true, we can substitute the null-hypothesis value of $\mu_1 - \mu_2 = 0$ into Expression (11.29) to obtain the **test statistic** for the **independent-samples *t* test of the difference between two means**:

$$t_{\text{obs}} = \frac{\bar{x}_1 - \bar{x}_2}{\sqrt{s^2_{\text{pooled}}\left(\frac{1}{n_1}+\frac{1}{n_2}\right)}} \qquad \textbf{(11.30)}$$

which is distributed as $\boldsymbol{t(n_1 + n_2 - 2\ \mathrm{df})}$ if the null hypothesis is true. Expression (11.30) indicates how far apart $\bar{x}_1$ and $\bar{x}_2$ are, in estimated-standard-error units.

If the null hypothesis is true, the test statistic may be regarded as a random selection from the t distribution with $n_1 + n_2 - 2$ degrees of freedom; but if it is not true, the test statistic will be more likely than specified by the t distribution to take on an extreme value. In order to evaluate these two possibilities, the numerical value of the test statistic t_{obs} may be compared with the critical values $\pm t_{\alpha/2}(n_1 + n_2 - 2\ \mathrm{df})$ for a two-tailed test at the α significance level. As illustrated for the one-sample test in Figure 11.3, if the value of t_{obs} falls in one of the tails of the sampling distribution, the sample difference is judged to be statistically significant (that is, not just due to chance), and the null hypothesis is rejected. Otherwise, the sample difference is not statistically significant and the null hypothesis cannot be rejected. This decision rule is easily modified to test against a one-sided alternative hypothesis, as discussed in Section 11.1.

It should be noted that the standard error of the difference between two independent sample means may also be used to obtain a **confidence interval for the population mean difference $\boldsymbol{\mu_1 - \mu_2}$**:

$$(\bar{x}_1 - \bar{x}_2) - t_{\alpha/2}s_{\bar{x}_1-\bar{x}_2} < \mu_1 - \mu_2 < (\bar{x}_1 - \bar{x}_2) + t_{\alpha/2}s_{\bar{x}_1-\bar{x}_2} \quad \text{with confidence level } 1 - \alpha \qquad \textbf{(11.31)}$$

This interval estimate has the same form as those discussed previously, in that the confidence limits are

$$\text{(point estimate)} \pm \text{(margin of error)}$$

Confidence intervals and significance tests provide alternative ways of looking at the same data: We use a confidence interval to make a statement concerning what the parameter is equal to, whereas we use a significance test to make a statement concerning what it is not equal to.

EXAMPLE 11.9 In the investigation described in Example 11.8, the sample of $n_1 = 15$ ratings for variety 1 had a mean of 7.5 and a standard deviation of 1.3, and the sample of $n_2 = 14$ ratings for variety 2 had a mean of 6.1 and a standard deviation of 1.5. The estimate of the population difference $\mu_1 - \mu_2$ is the sample difference $\bar{x}_1 - \bar{x}_2 = 7.5 - 6.1 = 1.4$.

If the null hypothesis that $\mu_1 - \mu_2 = 0$ is true, then the test statistic in Expression (11.30),

$$t_{\mathrm{obs}} = \frac{7.5 - 6.1}{\sqrt{\dfrac{(15-1)(1.3)^2 + (14-1)(1.5)^2}{15 + 14 - 2}\left(\dfrac{1}{15} + \dfrac{1}{14}\right)}} = \frac{1.4}{.5202} = 2.69$$

may be regarded as a random selection from the t distribution with $n_1 + n_2 - 2 = 27$ degrees of freedom, whose .02-level critical values are $\pm t_{.01}(27\ \mathrm{df}) = \pm 2.473$, as shown in Figure 11.8. The observed value 2.69 indicates that the sample mean difference of 1.4 is a difference of 2.69 estimated standard errors. Since the observed value 2.69 is beyond the critical value 2.473, we reject the null hypothesis at the 2% significance

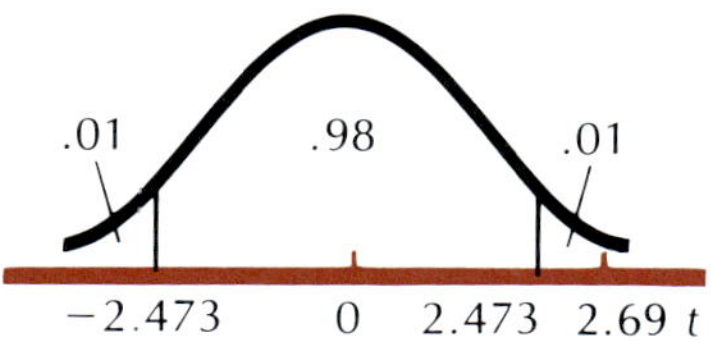

Figure 11.8

level. The mean rating of juice made from the usual variety of oranges was significantly higher than that for the new variety of oranges ($P < .02$).

The 98% confidence limits for the mean difference in the population are

$$(7.5 - 6.1) \pm (2.473)(.5202) = 1.4 \pm 1.286$$

yielding a 98% confidence interval of .1 to 2.7 rating points in favor of the usual variety. The fact that the 98% interval does not include 0 (which was the difference hypothesized in the null hypothesis) indicates that the sample difference is statistically significant at the 2% significance level.

EXAMPLE 11.10 Do unmarried men and women have different ideas as to what is an appropriate age for a person to get married? Twenty unmarried college students (aged 17–19 years), 10 males and 10 females, were asked their opinion as to what, in general, is an appropriate age for a person to get married. Their responses were as follows:

Males:	26,	22,	23,	25,	27,	25,	23,	24,	23,	28
Females:	25,	23,	25,	23,	30,	23,	27,	26,	23,	19

The sample mean for males, $\bar{x}_1 = 24.6$, is slightly larger than that for females, $\bar{x}_2 = 24.4$; the difference is $24.6 - 24.4 = .2$ year. Is this difference statistically significant? That is, should the data lead one to believe that male and female students have different ideas as to an appropriate age for marriage?

To answer this question, we calculate the value of the t test statistic in Expression (11.30). As shown in Expressions (11.22), (11.23), and (11.24), this calculation requires pooling the variation (SS) of the male observations around their mean,

$$SS_1 = \sum x_1^2 - \frac{(\sum x_1)^2}{n_1} = (26^2 + 22^2 + \cdots + 28^2) - \frac{246^2}{10} = 34.4$$

with the variation of the female observations around their mean,

$$SS_2 = \sum x_2^2 - \frac{(\sum x_2)^2}{n_2} = (25^2 + 23^2 + \cdots + 19^2) - \frac{244^2}{10} = 78.4$$

to obtain a pooled variance estimate as follows:

$$s^2_{\text{pooled}} = \frac{SS_1 + SS_2}{df_1 + df_2} = \frac{34.4 + 78.4}{(10 - 1) + (10 - 1)} = \frac{112.8}{18}$$

Substituting into Expression (11.30), we find that the observed value of the test statistic

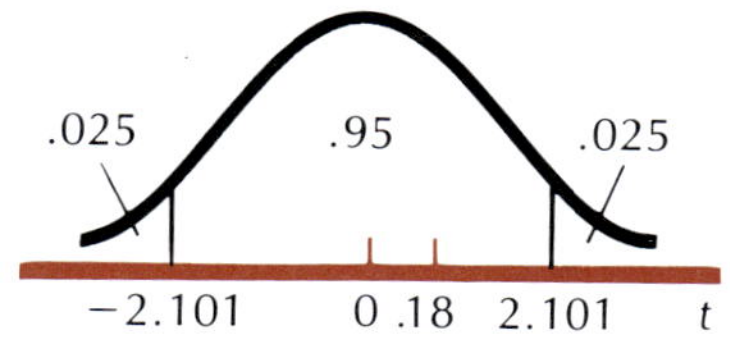

Figure 11.9

is

$$t_{obs} = \frac{24.6 - 24.4}{\sqrt{\frac{112.8}{18}\left(\frac{1}{10} + \frac{1}{10}\right)}} = \frac{.2}{1.12} = .18$$

The critical values of the t distribution with $n_1 + n_2 - 2 = 18$ degrees of freedom are $\pm t_{.025}(18\text{ df}) = \pm 2.101$ at the 5% significance level (see Figure 11.9).

Since the observed value of the test statistic $t_{obs} = .18$ falls between the critical values, ±2.101, we cannot reject the null hypothesis of no population difference, and so we conclude that the difference observed in the sample is not statistically significant ($P > .05$). The observed difference is well within the bounds of natural sampling variability, and the data provide no support for the alternative hypothesis that men and women have different ideas as to an appropriate age for marriage.

It should be noted that the pooled variation $SS_1 + SS_2$ could alternatively have been obtained from the sample standard deviations $s_1 = 1.95505$ and $s_2 = 2.95146$ via Expression (11.25):

$$SS_1 + SS_2 = (10 - 1)(1.95505^2) + (10 - 1)(2.95146^2) = 112.8$$

but sufficient decimal places must be retained for computational purposes so that the final answer will not suffer from rounding error. If the original observations are available, Expression (11.24) is the preferred computational formula for $SS_1 + SS_2$.

From Expression (11.31), the 95% confidence limits for the mean difference in the population are

$$(24.6 - 24.4) \pm (2.101)(1.12) = .2 \pm 2.4 \text{ years}$$

yielding a 95% confidence interval of −2.2 to +2.6 years. The fact that the interval includes 0 (which was the difference hypothesized in the null hypothesis) indicates that the sample difference is not statistically significant. At the 95% confidence level, the sample data are consistent with a true mean difference of 0 years, or of any number of years between −2.2 and +2.6.

(Source: Student project by H. Asfeldt.)

Summary of the Test Procedure

The **independent-samples *t* test of the difference between two means** may be summarized as follows:

Assumptions.

a. Independent random samples.

b. Normality of the sampling distributions of both sample means.

c. Equal population variances.

Null hypothesis. H_0: $\mu_1 = \mu_2$

	Nondirectional test	Left-directional test	Right-directional test
Alternative hypothesis.	H_A: $\mu_1 \neq \mu_2$	H_A: $\mu_1 < \mu_2$	H_A: $\mu_1 > \mu_2$
Critical values.	$\pm t_{\alpha/2}$	$-t_\alpha$	t_α

Step 1. Calculate the point estimates $\bar{x}_1$ and $\bar{x}_2$ to see what sample difference is being tested.

Step 2. Calculate the null-hypothesis value of the test statistic:

$$t_{\text{obs}} = \frac{\bar{x}_1 - \bar{x}_2}{\sqrt{s^2_{\text{pooled}}\left(\frac{1}{n_1} + \frac{1}{n_2}\right)}}$$

Step 3. Look up the critical value(s) in the t distribution with $n_1 + n_2 - 2$ degrees of freedom.

Step 4. If the observed value t_{obs} is at or beyond the critical value in the tail of the distribution, conclude that the sample mean difference is statistically significant ($P < \alpha$); otherwise, conclude that the observed difference is not statistically significant ($P > \alpha$).

Violations of Assumptions

In the derivation of the test statistic in Expression (11.30), it was assumed that the two populations being sampled have equal dispersion, $\sigma_1 = \sigma_2$. This is called the **assumption of homogeneity of variance**. How can we tell if the assumption is met, and what difference does it make? We will introduce a test for equality of dispersion in Chapter 12, but in practice, small or moderate violations of the assumption do not make much difference in the results of the t test, especially if the samples are of equal size or if the larger sample is the more variable one. The t test therefore is said to be moderately robust against **heterogeneity of variance** (violation of the equal-variance assumption). In case of severe departures from equal variances, there is an alternative test procedure available, a reference for which is given under Further Reading at the end of this chapter.

We should also note that the independent-samples t test, like the other two t tests presented in this chapter (Sections 11.1 and 11.3), involves an **assumption of normality** of the sampling distribution of the sample mean. If the sample size n is large enough, the Central Limit Theorem assures us that this condition will be met regardless of the shape of the population distribution being sampled. As we saw in Section 8.4, the Central Limit Theorem can be applied for samples as small as $n = 1$ if the population distribution is approximately Gaussian, and for samples as small as about $n = 5$ as long as the population distribution is approximately symmetric. In any case, all three of the *t* tests presented in this chapter are fairly robust against violations of the normality assumption, in that such violations generally do not have a serious effect on the results of the *t* test.

A much more critical assumption is the **assumption of random sampling**. In order to call on the Central Limit Theorem as we have just done, we must assume that the observations are randomly sampled from the underlying population(s); see Expression (8.6). Violations of this assumption have serious consequences for all three *t* tests presented in this chapter. Without random sampling, the *t* test statistic and the corresponding *P*-value are mere descriptions of the observed sample and cannot be used for formal inference about the underlying population(s). Conversely, the inferences drawn from *t* tests apply only to the population from which the observations were randomly sampled.

11.3 PAIRED-SAMPLES *t* TEST OF THE DIFFERENCE BETWEEN TWO MEANS

Paired samples are two samples in which (a) the observational units come in pairs that are highly similar in ways related to the dependent variable and (b) one member of each pair is observed in one sample and the other member in the other sample.

EXAMPLE 11.11 A standard example of paired samples comes in research involving identical twins. The twins are very similar in many respects, so when one twin is assigned to one sample to receive one treatment or level of the independent variable and the other twin is assigned to the other sample to receive the other level of the independent variable, any effects of the treatments will be more easily seen in the twins' responses than they would be in the responses of two unrelated individuals (whose responses are more likely to differ even if the treatment has no effect).

A much more common example of paired samples occurs in before-after studies, in which the same observational units are observed both before and after the administration of the treatment being studied. Each observational unit is said to serve as its own control in this case.

A third example occurs when matched pairs of observational units are formed. For example, in agricultural research two adjacent plots of land may be considered as one pair, since adjacent land is likely to be much more similar than widely separated plots of land.

It is important to recognize the distinction between **paired samples**, discussed in this section, and **independent samples**, which were discussed in Section 11.2. In independent samples, the observational units in one sample have no relationship with the observational units in the other sample. In paired samples, each observational unit is in some sense linked with one particular observational unit in the other sample. Paired-samples data can, in fact, be thought of as bivariate observations in which the pair, rather than a member of the pair, is the observational unit.

The use of paired samples will increase the precision (reduce the standard error) of the comparison of two means if the similarity within pairs is sufficiently greater than the similarity between different pairs. To the extent that the pairing results in a positive correlation between the observations in the two samples, the standard error of the difference between the means will be reduced, which is to say that the precision of the estimated difference will be improved.

Difference Scores

Because of the correlation induced by pairing, we cannot test for differences between paired samples by means of the method outlined for independent samples in Section 11.2. The method for paired samples involves calculating a difference score for each pair and then using the one-sample t test (Section 11.1), with pairs (not members of pairs) as the observational units and with difference score as the random variable.

Let x_{1i} be the observation of the member of the ith pair in sample 1, and let x_{2i} be the observation of the other member of the ith pair, that is, the one in sample 2. (In this notation, the subscripts $2i$ are read as "2, i" and not as "2 times i.") We define the **difference score** for the ith pair as

$$d_i = x_{1i} - x_{2i} \qquad \text{for } i = 1, 2, \ldots, n \tag{11.32}$$

where a negative value of d_i indicates that the sample 1 observation was smaller than the sample 2 observation.

EXAMPLE 11.12 Six university students with above-average grades took part in a study of reading ability. Each student took two reading tests, one in relatively quiet surroundings and one with a radio on in the same room. The reading speed in words per minute was determined for each test, as recorded in the first two columns of the following table. The observations under the two conditions (radio off and radio on) constitute paired samples, and these are converted into a single sample of difference scores, also in words per minute, in the right-hand column of the table.

	Radio off, x_1	Radio on, x_2	Difference score, d
Individual 1	125	109	16
Individual 2	347	278	69
Individual 3	265	275	−10
Individual 4	195	191	4
Individual 5	535	416	119
Individual 6	235	250	−15

Is the difference between the radio-off mean, $\bar{x}_1 = 283.7$, and the radio-on mean, $\bar{x}_2 = 253.2$, statistically significant? Notice that this question is the same as asking whether the mean of the difference scores, $\bar{d} = 30.5$, is significantly different from 0. The difference between the two sample means,

$$\bar{x}_1 - \bar{x}_2 = 283.7 - 253.2 = 30.5 \text{ words per minute}$$

is necessarily equal to the mean of the difference scores, $\bar{d} = 30.5$ words per minute, by virtue of Characteristic 2 of summation in Section 1.6.

(Source: Student project by A. Warnke.)

Once the paired-samples data have been converted to difference scores, we can use the one-sample t test to evaluate whether the mean $\bar{d}$ of the difference scores is significantly different from 0. The test follows the same procedure as that in Section 11.1, with the variable d substituted for the variable x throughout.

The Test Statistic

The null hypothesis is that there is no difference between the means of the two populations with respect to the dependent variable. Equivalently, the null hypothesis says that the mean difference in the population is equal to zero; that is,

$$H_0\text{: } \mu_d = 0$$

where μ_d is the mean difference in the population. Either a two-sided or a one-sided alternative hypothesis may be tested, as described in Section 11.1.

Once again applying the principle that the ratio

$$t = \frac{\text{(point estimate having a Gaussian sampling distribution) minus (parameter)}}{\text{estimated standard error of the point estimate}}$$

follows the t distribution, we know that

$$t_{\text{obs}} = \frac{\bar{d} - \mu_d}{s_{\bar{d}}} \tag{11.33}$$

is distributed according to $t(n - 1 \text{ df})$ if the sampling distribution of $\bar{d}$ is Gaussian, where

$$s_{\bar{d}} = \frac{s_d}{\sqrt{n}} \tag{11.34}$$

is the estimated standard error of the mean difference score and s_d is the standard deviation of the difference scores. Substitution of the hypothesized value $\mu_d = 0$ into Expression (11.33) yields the **test statistic** for the **paired-samples t test of the difference between two means**:

$$t_{\text{obs}} = \frac{\bar{d}}{s_d/\sqrt{n}} \tag{11.35}$$

which follows the t distribution with **$n - 1$ degrees of freedom** if the null hypothesis is true.

If the null hypothesis is true, the test statistic may be regarded as a random selection from the t distribution with $n - 1$ degrees of freedom; but if it is not true, the test statistic t_{obs} will be more likely than specified by the t distribution to take on an extreme value. We assess statistical significance by comparing the numerical value of t_{obs} to critical values from the t distribution, as illustrated for two-sided alternatives in Figure 11.3 and for one-sided alternatives in Figure 11.6.

Again, it should be noted that an **interval estimate of the population difference** is readily obtained from the above quantities; specifically, the **confidence interval for the mean difference μ_d** is given by

$$\bar{d} - t_{\alpha/2}s_{\bar{d}} < \mu_d < \bar{d} + t_{\alpha/2}s_{\bar{d}} \qquad \text{with confidence level } 1 - \alpha \tag{11.36}$$

The confidence limits in this case have the familiar form,

$$\text{(point estimate)} \pm \text{(margin of error)}$$

EXAMPLE 11.13 In the data of Example 11.12, the difference in reading speeds with the radio off and with the radio on had a mean of $\bar{d} = 30.5$ words per minute (with higher speeds for the radio-off condition) and a standard deviation of $s_d = 52.797$ words per minute for $n = 6$ university students with above-average grades. From Expression (11.35), the observed value of the test statistic is

$$t_{obs} = \frac{30.5}{52.797/\sqrt{6}} = \frac{30.5}{21.554} = 1.42$$

If the null hypothesis of no population difference is true, t_{obs} may be regarded as a random selection from the t distribution with $n - 1 = 5$ degrees of freedom, whose critical values are $\pm t_{.025}(5 \text{ df}) = \pm 2.571$ at the 5% significance level. Since the observed value of the test statistic $t_{obs} = 1.42$ falls between the critical values into the central region of the t distribution, as shown in Figure 11.10, we cannot reject the null hypothesis. The evidence from the sample is insufficient to conclude that above-average students read at different speeds with the radio on and with the radio off ($P > .05$).

The preceding result does not demonstrate that the radio has no effect; it merely fails to demonstrate an effect. The 95% confidence limits for the true mean difference between the conditions are

$$30.5 \pm (2.571)(21.554) = 30.5 \pm 55.4 \text{ words per minute}$$

yielding a 95% confidence interval of −24.9 to 85.9 words per minute. At the 95% confidence level, the data are compatible with various possibilities: Turning off the radio may have no effect on reading speed (on the average); or it may reduce reading speed by as much as 24.9 words per minute on the average; or it may increase reading speed by as much as 85.9 words per minute on the average. We would need a more

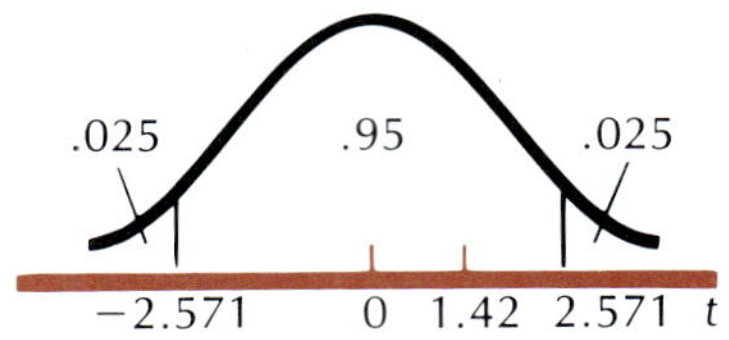

Figure 11.10

precise estimate—one with a smaller standard error—to distinguish among these possibilities. A larger sample would be an obvious first step.

It is worth noting the gain in precision that was achieved in this investigation by the use of paired rather than independent samples. The estimated standard error of the difference between the means $\bar{x}_1 - \bar{x}_2$, calculated according to Expression (11.27), which ignores the pairing of observations, is

$$s_{\bar{x}_1 - \bar{x}_2} = 71.9 \text{ words per minute}$$

On the other hand, the estimated standard error of the mean difference score $\bar{d}$, calculated according to Expression (11.34), which takes account of the pairing of observations, is

$$s_{\bar{d}} = 21.6 \text{ words per minute}$$

In other words, a representative or standard amount of error in the estimate based on independent samples would be 71.9 words per minute, while a representative or standard amount of error in the estimate based on these paired samples was only 21.6 words per minute. These values illustrate that the use of paired samples can lead to improved precision (reduced standard errors) in estimating differences.

Summary of the Test Procedure

The **paired-samples *t* test of the difference between two means** may be summarized as follows:

Assumptions.

a. Random sample of difference scores.

b. Normality of the sampling distribution of the mean difference.

Null hypothesis. H_0: $\mu_d = 0$

	Nondirectional test	Left-directional test	Right-directional test
Alternative hypothesis.	H_A: $\mu_d \neq 0$	H_A: $\mu_d < 0$	H_A: $\mu_d > 0$
Critical values.	$\pm t_{\alpha/2}$	$-t_\alpha$	t_α

Step 1. Calculate a difference score, $d_i = x_{1i} - x_{2i}$, for each of the n pairs, as well as the point estimate $\bar{d}$ to see what sample difference is being tested.

Step 2. Calculate the null-hypothesis value of the test statistic,

$$t_{\text{obs}} = \frac{\bar{d}}{s_d/\sqrt{n}}$$

Step 3. Look up the critical value(s) in the t distribution with $n - 1$ degrees of freedom.

Step 4. If the observed value t_{obs} is at or beyond the critical value in the tail of the distribution, conclude that the sample mean difference is statistically significant ($P < \alpha$); otherwise, conclude that the observed difference is not statistically significant ($P > \alpha$).

11.4 OUTCOMES OF HYPOTHESIS TESTING

Now that we have seen some specific examples of significance tests, let us take a closer look at the logic underlying the hypothesis-testing procedure, in order to gain a better understanding of what its results mean.

The Purpose of Hypothesis Testing

The purpose of hypothesis testing is to draw conclusions on the basis of sample data about the unknown parameters of the relevant population(s). In particular, the purpose is to evaluate whether the sample evidence is sufficient to conclude that there is a difference with respect to a certain feature in the population—either a difference between a parameter and a specified constant or a difference between two (or more) parameters.

Hypothesis-testing investigations usually are designed so that the outcome of interest is associated with demonstration of a difference (rejection of the null hypothesis) rather than with demonstration of equality (retention of the null hypothesis). Failure to reject the null hypothesis does not prove the equality stated in the null hypothesis. Indeed, examination of the various test statistics reveals that a small sample size n or a high degree of dispersion s in the observations, perhaps due to measurement error or other random error that could be reduced by better experimental technique, make statistical significance (rejection of the null hypothesis) less likely to be achieved even if the alternative hypothesis is true. Investigations designed to demonstrate a population equality are more likely to use confidence intervals, which seek to specify the true value of a parameter within a specified margin of error, rather than hypothesis testing, which is more suited to demonstrating population differences.

In terms of its underlying logic, hypothesis testing begins with a null hypothesis, which is a statement of the working assumption of no difference between the population values in question, and an alternative hypothesis, which is a statement of the possible population difference, evidence of which is to be judged from the sample data. The hypotheses in significance testing are always statements about parameters (population values, such as μ or σ), not about statistics (sample values, such as $\bar{x}$ or s). It is unnecessary to make assumptions (hypotheses) about what the sample values might be, since we can calculate them exactly. The sample values are known. It is the population values that are unknown, and so in hypothesis testing we propose statements about the possible values of those parameters and then test those hypotheses.

Four Possible Outcomes

A significance test leads to one of two conclusions: "reject the null hypothesis" or "do not reject the null hypothesis." Whichever conclusion is drawn may be either correct or incorrect, depending on the actual but unknown state of the

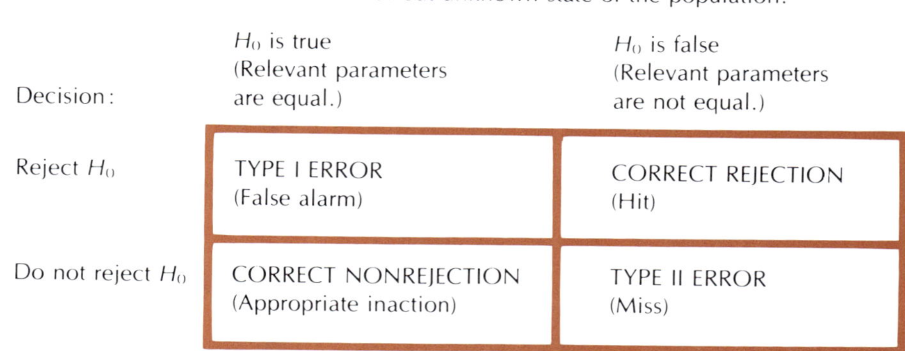

Actual but unknown state of the population:
H_0 is true
(Relevant parameters are equal.)
H_0 is false
(Relevant parameters are not equal.)
Decision:
Reject H_0
TYPE I ERROR
(False alarm)
CORRECT REJECTION
(Hit)
Do not reject H_0
CORRECT NONREJECTION
(Appropriate inaction)
TYPE II ERROR
(Miss)

Figure 11.11 Four possible outcomes of a significance test.

population. Thus, there are four possible outcomes of a significance test, as illustrated in Figure 11.11.

If the null hypothesis is actually true, then the decision to reject it is a mistake known as a **Type I error** or, in the more colorful terminology of signal-detection experiments, a false alarm (because a difference was claimed that did not exist), while the decision not to reject the null hypothesis is a **correct nonrejection** or appropriate inaction (because no difference was claimed, nor did one exist). On the other hand, if the null hypothesis is actually false, then the decision to reject it is a **correct rejection** or a hit (because the difference was detected), while the decision not to reject the null hypothesis is a mistake known as a **Type II error** or a miss (because the difference was not detected).

Two Possible Sampling Distributions

To illustrate the probabilities associated with these four possible outcomes, let us consider the sampling distribution of the test statistic t_{obs}. Expressions (11.5), (11.29), and (11.33) show that t_{obs} is simply the difference between a point estimate and the null-hypothesis value of the parameter being estimated, expressed in estimated-standard-error units.

If the null hypothesis is true (and if the assumptions listed in the summary of each test procedure are met), then the sampling distribution of the test statistic t_{obs} is the t distribution, which has a mean of 0; that is, the expected value of the difference between the point estimate and the null-hypothesis value of the parameter is 0 if the null hypothesis is true. In any given sample, the point estimate may be larger or smaller than the null-hypothesis value even if the null hypothesis is true, but the expected value of the difference is 0, as illustrated in the top half of Figure 11.12.

On the other hand, if the null hypothesis is false, then the mean of the

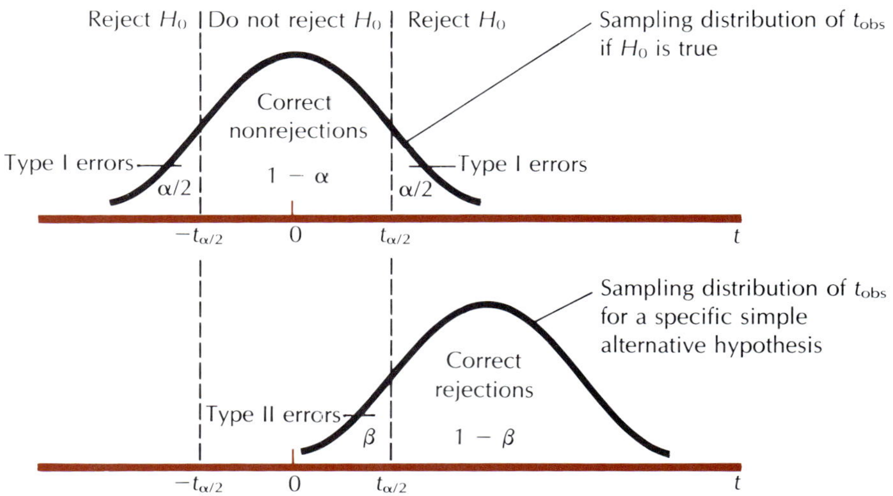

Figure 11.12 Schematic diagram illustrating possible sampling distributions in a significance test.

sampling distribution of the test statistic t_{obs} is no longer equal to 0. The sampling distribution is shifted to the left if the actual value of the parameter is less than the null-hypothesis value, or it is shifted to the right if the actual value of the parameter is greater than the null-hypothesis value. The usual composite alternative hypothesis does not specify how far the sampling distribution is shifted, but it includes an infinite number of simple alternatives corresponding to all the possible positions to which the sampling distribution could be shifted from its null-hypothesis position. In the lower half of Figure 11.12, one of the possible simple alternatives has been arbitrarily selected for illustration, and the sampling distribution of t_{obs} that would apply if that particular simple alternative hypothesis were true has been drawn.

In any particular application, we do not know if we are sampling from the distribution represented in the upper half of Figure 11.12, or from the distribution in the lower half of the figure, or from one of the other possible alternative distributions that could have been drawn on the lower line. Indeed, that is precisely what we want to find out: Which hypothesis is true?

The hypothesis-testing procedure involves focusing on the sampling distribution of the test statistic under the null hypothesis, illustrated in the upper half of Figure 11.12. Because the null hypothesis is a simple hypothesis, it allows us to calculate specific probabilities associated with certain outcomes. Critical values, represented by the vertical lines in Figure 11.12, are established on the basis of the null-hypothesis distribution, such that values of the test statistic beyond the critical values will be considered sufficient evidence to warrant rejection of the null hypothesis. Values of the test statistic t_{obs} between the critical values will be deemed sufficiently consistent with the null hypothesis that it cannot be rejected.

Probabilities of the Four Outcomes

The critical values, established on the basis of the null-hypothesis distribution (the top half of Figure 11.12), define four kinds of areas in the sampling distributions—two in the null-hypothesis distribution and two in the alternative-hypothesis distribution. The four areas, labeled in Figure 11.12 with the names of the four possible outcomes of hypothesis testing, represent the probabilities of those outcomes. (Recall that areas of probability distributions represent probabilities.) These are all conditional probabilities, the condition specifying which of the two illustrated distributions is assumed to be correct—the null-hypothesis distribution or an alternative-hypothesis distribution.

The symbols for the **probabilities of the four outcomes of hypothesis testing** involve the first two letters of the Greek alphabet: lower-case alpha (α) and beta (β). The following terminology is used:

$P(\text{Type I error} \mid H_0) = \alpha =$ the **significance level of the test** **(11.37)**

$P(\text{correct nonrejection} \mid \mathrm{H}_0) = 1 - \alpha =$ the **confidence level of the interval estimate** of the parameter **(11.38)**

$P(\text{Type II error} \mid \text{a specific simple } H_A) = \beta$ **(11.39)**

$P(\text{correct rejection} \mid \text{a specific simple } H_A) = 1 - \beta =$ the **power of the test** to detect the population difference stated in the given simple H_A **(11.40)**

Notice that the term *significance level* is used in the context of significance tests and not for confidence intervals, whereas the term *confidence level* is used in the context of confidence intervals and not for significance tests.

Reducing Error Probabilities

Naturally, what we would like to do in significance testing is to arrange things so that our probability of being correct is very large and that of making either kind of error is very small. Unfortunately, these goals are somewhat incompatible. To reduce Type I errors, we move the critical values farther apart, thus reducing the tail areas of the null-hypothesis distribution (see Figure 11.12). But as we do that, we are simultaneously increasing the Type II error region by enlarging the tail of the alternative-hypothesis distribution. Conversely, if we move the critical values closer together to increase the power of the test and to reduce the probability of a Type II error under the alternative-hypothesis distribution, we simultaneously increase the probability of a Type I error. As we have seen, the usual solution to this dilemma is to concentrate on Type I errors, setting a significance level α to control the likelihood of claiming that a population difference exists when in fact it does not. A high Type II error rate for some alternative hypotheses may result, but they will tend to be the alternatives that are not very different from the null hypothesis anyway.

In spite of these trade-offs between Type I and Type II errors, there are some ways of simultaneously reducing the probabilities of both types of errors. One way is to increase the sample size n. Notice in the formulas for the test statistics t_{obs} that as n increases, the value of t_{obs} also increases. This increase is due to the fact first introduced in Chapter 8 that the mean becomes a more precise estimate (that is, its standard error decreases) as the sample size n is increased. A second, possibly more difficult, way to minimize both types of errors is to find ways to reduce the measurement errors or other errors in the observations. This approach may involve using a more precise measuring instrument, or it may involve careful selection of observational units, as in the use of paired rather than independent samples. (Example 11.13 illustrates the latter sort of improvement in precision.) In whatever way it may be accomplished, reducing the variability of the observations reduces the standard error of the estimate and hence reduces the spread or dispersion of the sampling distributions (illustrated in Figure 11.12) so that they overlap less. Reducing variability causes the sampling distributions to "pull in their tails," as it were, leaving less tail area in the error regions. Thus, increasing sample size and reducing the variability of the observations are two ways of improving the power of a test without making Type I errors more likely.

Reporting and Interpreting Results

Reports of the results of a significance test should always include the symbol for the test statistic, its number of degrees of freedom, its numerical value, and information about the significance level. We will consider two ways of giving the latter information, both of which involve the P-value, first introduced in Section 11.1.

Information about the significance level is usually given in the form of the *P-value inequality* for the most informative α level. For example, if P is less than .01 ($P < .01$), then P is necessarily also less than .05 ($P < .05$). On the other hand, knowing that P is less than .05 does not indicate whether P is also less than .01 or not; hence, "$P < .01$" is more informative than "$P < .05$" if both are true. Similarly, "$P > .05$" is more informative than "$P > .01$" if both are true. Of the α levels for which critical values of the test statistic are available, the most informative level for the particular result at hand is usually reported.

EXAMPLE 11.14 In technical reports, the results of significance tests are often reported in the style of the following examples:

a. "The fertilized trees were significantly taller than the untreated trees, $t(13) = 3.27$, $P < .01$."

b. "The children significantly underestimated the length of a kilometer, $t(13) = 2.27$, $P < .05$."

c. "The mean weight loss of the group was not statistically significant, $t(13) = 1.27$, $P > .10$."

Notice that it is the *observed* value of the test statistic that is reported, not the critical value. Hence there is no ambiguity in abbreviating the notation for the observed value from "$t_{obs}(13 \text{ df})$" to "$t(13)$," as is the usual practice in technical reports.

An alternative way of giving information about the significance level is to report the actual P-value of the observed test statistic.

EXAMPLE 11.15 If the observed value of the test statistic in an investigation with a two-sided alternative hypothesis is found to be $t_{obs}(498 \text{ df}) = 2.12$, we can see from the t table in Appendix 4 that $P < .05$. However, since $t(498 \text{ df})$ is almost identical to the standard normal distribution, we can obtain a more exact P-value from Appendix 3 by finding the tail area beyond ± 2.12 in a standard normal distribution, namely, $P = .034$. Thus, the observed difference could be reported as statistically significant, $t(498) = 2.12$, $P = .034$.

The exact P-value is, of course, more informative than the P-value inequality, but generally the tables that would enable us to determine the P-value are not readily available (the Gaussian distribution is an exception). However, the reporting of P-values is rapidly gaining in popularity because many standard computer programs for hypothesis testing, and even some electronic calculators, provide exact P-values. Reporting actual P-values enables readers to assess the reported results according to their own personal standard for α, even if their standard differs from that of the writer of the report.

Finally, four points about results of significance tests should be noted in an attempt to prevent some common mistakes.

Comment 1. The probability P that appears in P-value inequalities such as $P < .05$ and in exact P-values such as $P = .034$ is *not* the probability that the null hypothesis is true. Nor is it the probability of any other hypothesis. Indeed, classical statistical inference does not allow the assignment of a probability to an hypothesis, because hypotheses are about parameters, and parameters are constants, not random variables. As stated in Expression (11.6), the P-value is the probability of obtaining a certain kind of sample data *if* the null hypothesis were true. In every significance test, the observed value of the test statistic represents the position of the sample outcome *in the null-hypothesis sampling distribution*; it shows how far the sample outcome is from the mean of the sampling distribution that would exist if the null hypothesis were true. Recall that in deriving the formulas for the test statistics t_{obs}, we had to employ a value of the parameter obtained from the null hypothesis. Classical hypothesis testing assumes the truth of the null hypothesis as a working hypothesis in order to determine how consistent the sample data are with that hypothesis. It evaluates the probability of the sample outcome, not of the hypothesis.

Comment 2. Statistical significance is a property pertaining to sample values, which are subject to variability, and never to population parameters, which are constants. For example, sample means may be said to be significantly different or not, but population means may not be so designated. Population means are either equal or unequal, without qualification as to "statistical significance" since they are invariable quantities that do not depend on which observations happen to be sampled. Unequal sample means, on the other hand, may or may not be "significantly" different—that is, so different that the vagaries of random sampling cannot reasonably account for their difference.

Comment 3. As noted in Section 9.2, statistical significance does not imply practical importance.

EXAMPLE 11.16 A difference of one percentage point in the mean grades of large samples of students from two cities may be statistically significant but too small to be of any practical importance.

EXAMPLE 11.17 The fact that mean reading ability is significantly higher in a sample of Grade 8 students than in a sample of Grade 1 students has an obvious explanation that diminishes the practical importance of this statistically significant finding.

Statistical significance should draw attention to differences that are judged to be greater than expected by chance, in order that reasons for their occurrence may be sought. Some knowledge of the area of application is generally needed to identify such reasons and to assess the importance of the finding in the light of those reasons. The interpretation of the results may well be facilitated by examining interval estimates of the relevant parameters as an adjunct to hypothesis testing, in order to gain a better understanding of how precise the results are.

Comment 4. Statistical significance does not rule out the possibility of poorly collected or contaminated or inappropriate data. The P-value is a descriptive statistic that indicates the probability of obtaining, through random selection from a population described by the null hypothesis, a sample outcome at least as extreme as the one observed. "Significantly" small P-values may occur

a. by chance (Type I error);

b. because the null hypothesis is false (correct rejection);

c. because of problems related to data collection, such as systematic departures from randomness in sample selection (see Sections 1.1 and 8.1); or

d. because of violations of assumptions underlying the particular significance test (see the final subsection in Section 11.2).

The P-value does not single out any one of these possibilities. Careful thought about the investigation and its results is required to draw the correct conclusion from a statistically significant finding.

STUDY MATERIAL FOR CHAPTER 11

Terms and Concepts

null hypothesis
alternative hypothesis

Section 11.1

simple hypothesis
composite hypothesis
test statistic
one-sample t test for a specified value of the mean
degrees of freedom for the one-sample t test
P-value
observed value of a test statistic
significance level
critical value
rejection region
nonrejection region
P-value inequality
one-sided alternative
directional alternative
two-sided alternative
nondirectional alternative
one-tailed test
two-tailed test

Section 11.2

independent random samples
sum of squares
pooled estimate of the common population variance σ^2
independent-samples t test of the difference between two means
degrees of freedom for the independent-samples t test
confidence interval for the mean difference $\mu_1 - \mu_2$
assumption of homogeneity of variance
heterogeneity of variance
assumption of normality
assumption of random sampling

Section 11.3

paired samples
difference score
paired-samples t test of the difference between two means
degrees of freedom for the paired-samples t test
confidence interval for the mean difference μ_d

Section 11.4

Type I error
correct nonrejection
correct rejection
Type II error
probabilities of the four outcomes of hypothesis testing
significance level of a test
confidence level of an interval estimate
power of a test

Key Formulas

Sections 11.1, 11.2, and 11.3

See Table 11.1.

Section 11.2

(11.21) $$s^2 = \frac{\text{SS}}{\text{df}}$$

(11.22) $$s^2_{\text{pooled}} = \frac{\text{SS}_1 + \text{SS}_2}{\text{df}_1 + \text{df}_2}$$

TABLE 11.1 Basic notation and formulas for inferences about one or two means.

	One sample (Section 11.1)	Two independent samples (Section 11.2)	Paired samples (Section 11.3)
Random variable	x	x	$d_i = x_{1i} - x_{2i}$
Parameter	μ	$\mu_1 - \mu_2$	μ_d
Null hypothesis	H_0: $\mu = \mu_0$	H_0: $\mu_1 - \mu_2 = 0$	H_0: $\mu_d = 0$
Point estimate	$\bar{x}$	$\bar{x}_1 - \bar{x}_2$	$\bar{d}$
Estimated standard error	$s_{\bar{x}} = \frac{s}{\sqrt{n}}$	$s_{\bar{x}_1-\bar{x}_2} = \sqrt{s^2_{\text{pooled}}\left(\frac{1}{n_1} + \frac{1}{n_2}\right)}$	$s_{\bar{d}} = \frac{s_d}{\sqrt{n}}$
Test statistic	$t_{\text{obs}} = \frac{\bar{x} - \mu_0}{s_{\bar{x}}}$	$t_{\text{obs}} = \frac{\bar{x}_1 - \bar{x}_2}{s_{\bar{x}_1-\bar{x}_2}}$	$t_{\text{obs}} = \frac{\bar{d}}{s_{\bar{d}}}$
Sampling distribution	$t(n - 1 \text{ df})$	$t(n_1 + n_2 - 2 \text{ df})$	$t(n - 1 \text{ df})$
Margin of error	$e_{\alpha/2} = t_{\alpha/2}s_{\bar{x}}$	$e_{\alpha/2} = t_{\alpha/2}s_{\bar{x}_1-\bar{x}_2}$	$e_{\alpha/2} = t_{\alpha/2}s_{\bar{d}}$
Confidence limits	$\bar{x} \pm e_{\alpha/2}$	$(\bar{x}_1 - \bar{x}_2) \pm e_{\alpha/2}$	$\bar{d} \pm e_{\alpha/2}$

Exercises for Numerical Practice

Section 11.1

11/1. A random sample of 21 observations from a Gaussian population had a mean of 95.3 and a standard deviation of 6.7.

a. Test at the 5% significance level the claim that the population mean is 100.0.

b. Construct the 95% confidence interval for the population mean.

11/2. A random sample of 14 observations from a Gaussian population had a mean of 10.21 and a standard deviation of 1.68.

a. Test at the 1% significance level the claim that the population mean is 12.00.

b. Construct the 99% confidence interval for the population mean.

11/3. The following six observations were randomly selected from an approximately Gaussian population:

$$5, \quad 8, \quad 3, \quad 5, \quad 0, \quad 2$$

Test at the 1% significance level whether the population mean is less than 5.0.

11/4. The following seven observations were randomly selected from an approximately Gaussian population:

$$6, \quad 5, \quad 9, \quad 7, \quad 4, \quad 5, \quad 6$$

Test at the 5% significance level whether the population mean is greater than 4.0.

Section 11.2

11/5. Two random samples from approximately normal distributions were summarized as follows:

$$n_1 = 13, \quad \bar{x}_1 = 146, \quad s_1 = 20$$
$$n_2 = 11, \quad \bar{x}_2 = 140, \quad s_2 = 19$$

a. Making use of Expression (11.25), determine whether the two sample means differed significantly at the 10% level.

b. Construct the 90% confidence interval for the population mean difference.

11/6. Two random samples from approximately normal distributions were summarized as follows:

$$n_1 = 9, \quad \bar{x}_1 = 61.2, \quad s_1 = 3.5$$
$$n_2 = 14, \quad \bar{x}_2 = 62.5, \quad s_2 = 3.7$$

a. Making use of Expression (11.25), determine whether the two sample means differed significantly at the 5% level.

b. Construct the 95% confidence interval for the population mean difference.

11/7. The following are random samples from approximately Gaussian populations:

Sample 1: 2, 0, 4, 0

Sample 2: 7, 8, 10, 9, 8, 7, 9

a. Test the significance of the difference between the sample means at the .001 level.

b. Construct the 99.9% confidence interval for the population mean difference.

11/8. The following are random samples from approximately Gaussian populations:

Sample 1: 3, 2, 0, 2, 1, 0, 4, 3

Sample 2: 5, 8, 4, 6, 7

a. Test the significance of the difference between the sample means at the .01 level.

b. Construct the 99% confidence interval for the population mean difference.

Section 11.3

11/9. Four randomly selected pairs of observations of a normally distributed variable yielded the following paired samples.

	Sample 1	Sample 2
Pair 1	55	50
Pair 2	31	29
Pair 3	6	3
Pair 4	19	13

a. Test the significance of the difference between the sample means at the .05 level.

b. Construct the 95% confidence interval for the population mean difference.

11/10. Six randomly selected pairs of observations of a normally distributed variable yielded the following paired samples.

	Sample 1	Sample 2
Pair 1	43	39
Pair 2	51	52
Pair 3	92	89
Pair 4	14	14
Pair 5	7	2
Pair 6	83	77

a. Test the significance of the difference between the sample means at the .05 level.

b. Construct the 95% confidence interval for the population mean difference.

Problems for Applying the Concepts

11/11. A sample of 17 female college students reported the total amount of their summer earnings from May through August 1980. The mean of the 17 amounts was $1612, and the standard deviation was $1097. Test at the 1% significance level whether the observations may be viewed as a random sample from a population with a mean of $2500.

a. Write the null and alternative hypotheses in symbols and in words.

b. State the results of the significance test.

c. Which of the following might describe your statement in Part (b): Type I error, Type II error, correct rejection, correct nonrejection?

d. Construct a 99% confidence interval for the mean summer earnings of the relevant population of female college students.

(Source: Student project by C. Hay.)

11/12. Tests for butterfat content were performed on milk from a sample of 20 Holstein dairy cows. The sample mean was 3.88%, and the standard deviation was .42 percentage point. Test at the 5% significance level whether the data provide sufficient evidence to conclude that the population from which the sample was taken had a mean less than 4.00%.

a. Write the null and alternative hypotheses in symbols and in words.

b. State the critical value for the test.

c. State the results of the significance test.

d. Which of the following might describe your statement in Part (c): Type I error, Type II error, correct rejection, correct nonrejection?

(Source: Student project by S. Bailer.)

11/13. The resting pulse rate was counted (on the neck, just below the jawbone) in a sample of 10 mature light horses. The numbers of beats per minute were as follows:

38, 38, 40, 45, 28, 30, 27, 41, 29, 26

Perform a significance test to determine whether these data contradict the claim that the normal pulse rate is 37 beats per minute for such horses.

a. Write the null and alternative hypotheses in symbols and in words.

b. State the results of the significance test.

c. Which of the following might describe your statement in Part (b): Type I error, Type II error, correct rejection, correct nonrejection?

d. Construct a 95% confidence interval for the mean pulse rate of the relevant population of light horses.

(Source: Student project by H. Hoyland.)

11/14. Many North American colleges and universities require foreign students to achieve a certain standard on the Educational Testing Service "Test of English as a Foreign Language" (TOEFL) as an admission requirement. The following TOEFL scores were obtained by a sample of 9 Hong Kong students seeking admission to North American universities:

495, 525, 570, 610, 552, 497, 593, 505, 551

Perform a significance test to determine whether the data provide sufficient evidence to conclude that the population from which the sample was selected had a mean TOEFL score greater than 500.

a. Write the null and alternative hypotheses in symbols and in words.

b. State the results of the significance test.

c. Which of the following might describe your statement in Part (b): Type I error, Type II error, correct rejection, correct nonrejection?

(Source: Student project by B. Dung.)

11/15. In an area where both mule deer and white-tailed deer are hunted, dressed weights were obtained for a sample of 5 three-point bucks of each species. The sample of mule deer had a mean of 141 pounds and a standard deviation of 16.7 pounds, while the white-tailed sample had a mean of 132 pounds and a standard deviation of 17.5 pounds. Is the observed difference between the means statistically significant?

a. Identify the dependent and independent variables.

b. Write the null and alternative hypotheses in symbols and in words.

c. State the results of the significance test.

d. Construct a 95% confidence interval for the true difference in mean dressed weight between three-point bucks of the two species in this area.

(Source: Student project by B. Reil.)

11/16. Thirty-seven students living in college residences were asked about their average length of weeknight sleep. The weeknight sleep durations reported by the 22 males in the sample had a mean of 7.05 hours and a standard deviation of .86 hour, while the durations reported by the 15 females in the sample had a mean of 7.20 hours and a standard deviation of 1.01 hours. Is the observed difference between the means statistically significant?

a. Identify the dependent and independent variables.

b. Write the null and alternative hypotheses in symbols and in words.

c. State the results of the significance test.

d. Construct a 95% confidence interval for the true difference in mean reported sleep duration between males and females in the relevant population of college students.

(Source: Student project by D. Stewart.)

11/17. A sample of 12 university students were interviewed in the second month of the academic year concerning their study habits during the previous month. Those who were in their first year of studies reported their average number of hours of study per week as

21, 20, 14, 15, 17, 12

Those who were in their second or third year of studies reported their average weekly study times in hours as

30, 28, 30, 23, 10, 29

Test whether the sample means for first-year and later-year students differ significantly.

a. Identify the dependent and independent variables.

b. Write the relevant hypotheses in symbols and in words.

c. State the results of the significance test.

d. Construct a 95% confidence interval for the true difference in mean reported study times between first-year and later-year students in the relevant population.

(Source: Student project by I. Servold.)

11/18. The starting five players on a men's intercollegiate basketball team and the starting five on a women's team each attempted 30 foul shots at the same basket from the same position. The numbers of successful throws were as follows:

Men:	27,	28,	26,	27,	24
Women:	26,	20,	27,	21,	20

Perform a significance test to determine whether the sample means for men and women differ significantly or whether the difference can reasonably be attributed to chance.

a. Identify the dependent and independent variables.

b. Write the relevant hypotheses in symbols and in words.

c. State the results of the significance test.

d. Construct a 95% confidence interval for the true difference in mean foul-shot ability between these two teams.

(Source: Student project by R. Marshall.)

11/19. Example 11.1 described a neuroanatomical study that used a significance test to compare mean splenial width in adult men and women.

a. Identify the dependent and independent variables.

b. Write the relevant hypotheses in symbols and in words.

c. Calculate the observed value of the test statistic, and state the results of the significance test.

d. Construct a 99.9% confidence interval for the population mean difference in splenial width between men and women.

11/20. An arbitrator in a labor dispute recorded average hourly wages in 8 randomly selected firms in each of two comparable industries.

Industry A:	\$6.87,	6.90,	5.67,	6.14,	8.14,	7.25,	6.75,	8.10
Industry B:	\$6.80,	7.25,	7.15,	6.25,	9.25,	9.60,	8.90,	8.55

Perform a significance test to determine whether the sample mean wages differed significantly between the two industries.

a. Identify the dependent and independent variables.

b. Write the relevant hypotheses in symbols and in words.

c. State the results of the significance test.

d. Construct a 95% confidence interval for the population mean difference in hourly wages between the two industries.

11/21. In a study of the effects of caffeine, the resting pulse rates of 8 college students were recorded both immediately before and five minutes after they each drank a cup of tea. The pulse rates in beats per minute were as follows:

	Before	After
Student 1	69	75
Student 2	76	76
Student 3	70	71
Student 4	67	70
Student 5	72	78
Student 6	64	64
Student 7	68	70
Student 8	70	77

Perform a significance test to determine whether there was a significant change in mean pulse rates.

a. Identify the dependent and independent variables.

b. Write the relevant hypotheses in symbols and in words.

c. State the results of the significance test.

d. Construct a 95% confidence interval for the true mean change in pulse rates under the conditions of this experiment.

(Source: Student project by B. Weir.)

11/22. Five female students took part in a study to determine whether knowledge that one is being timed has any effect on the number of typing errors that one makes. Each student was instructed to type a certain 78-word passage once to familiarize herself with the typewriter, and then a second time while being timed with a stopwatch. Subsequently, the numbers of words typed incorrectly on the two trials were counted and recorded as follows.

	Untimed trial	Timed trial
Individual 1	11	21
Individual 2	9	13
Individual 3	6	7
Individual 4	3	3
Individual 5	8	13

Perform a significance test to determine whether the mean number of errors was affected by the subject's knowledge of being timed.

a. Identify the dependent and independent variables.

b. Write the relevant hypotheses in symbols and in words.

c. State the results of the significance test.

d. Construct a 95% confidence interval for the true mean effect of knowledge of being timed, in the population of student typists from which the sample was selected.

(Source: Student project by L. Mills.)

11/23. A realty company wanted to know if there was a significant difference between the appraised values of houses according to two different appraisers. Making sure that neither appraiser was aware of the procedure, the company selected 9 houses at random and had them appraised by the two appraisers, with the following results (in $1000s):

House:	1	2	3	4	5	6	7	8	9
Appraiser A:	$90	72	65	115	165	100	91	59	41
Appraiser B:	$80	70	68	110	145	98	87	55	38

Perform a significance test to compare the mean appraised values by the two appraisers.

a. Identify the dependent and independent variables.
b. Write the relevant hypotheses in symbols and in words.
c. State the results of the significance test.
d. Construct a 95% confidence interval for the true mean difference in appraised values between the two appraisers.

11/24. A manufacturer of personal computers is faced with severe competition from foreign suppliers. To meet this competition, the company devised a new training program to increase worker productivity. Before going to the expense of training all workers, the company decided to test the new program on a random sample of six workers assembling one particular component, to see if the program actually has a significant impact on productivity. The numbers of components assembled per day by each worker before and after the training program were as follows:

Worker:	A	B	C	D	E	F
Before:	34	38	33	35	34	36
After:	39	41	36	36	40	38

Perform a significance test to determine whether productivity was significantly greater after the training program than before.

a. Identify the dependent and independent variables.
b. Write the relevant hypotheses in symbols and in words.
c. State the results of the significance test.

11/25. The reported weights (in pounds) of a sample of 15 female college students and their mothers are shown on p. 400.

Perform a significance test to determine whether these data provide enough evidence for us to conclude that, on the average, mothers are more than 10 pounds heavier than their college-student daughters. [*Hint*: Substitute the null-hypothesis value of the parameter into Expression (11.33.)]

a. Identify the dependent and independent variables.
b. Write the null and alternative hypotheses in symbols and in words.
c. State the results of the significance test.
d. Construct a 95% confidence interval for the true mean difference in weight between mothers and their college-student daughters.

(Source: Student project by L. Nowochin.)

Daughter	Mother
125	145
123	118
110	145
150	140
128	138
105	140
130	140
150	170
133	153
138	145
150	160
115	140
140	155
125	150
130	125

11/26. According to folk wisdom, the distance that a person can reach horizontally, measured from fingertip to fingertip, is equal to that person's height. The following data (in centimeters) were obtained by measuring horizontal arm reach and height in a sample of 20 young adults.

Males		Females	
Arm reach	Height	Arm reach	Height
183	178	174	173
187	184	163	163
183	183	163	165
178	178	171	174
201	187	166	169
185	183	151	152
187	182	161	164
187	179	164	163
186	180	152	152
186	180	168	169

a. Perform a significance test to evaluate the folk-wisdom claim as it applies to males, and report the results.

b. Construct a 95% confidence interval for the true mean difference between arm reach and height of males.

c. Perform a significance test to evaluate the folk-wisdom claim as it applies to females, and report the results.

d. Construct a 95% confidence interval for the true mean difference between arm reach and height of females.

(Source: Student project by A. Kautz.)

11/27. A report concerning an experiment in Israel that demonstrated a significant increase in rainfall due to cloud seeding included the following statement of the results of the statistical analysis:

"The probability that the increase resulted from a chance distribution of particularly rainy days was 2.8 percent for the whole target area and 1.7 percent for the [smaller] catchment area. That compares with the approximate significance level of 40 per cent for the recently analyzed Florida Area Cumulus Experiment (FACE-2), an unsuccessful attempt to confirm FACE-1."

In what way does this statement misinterpret classical hypothesis testing, and how could the statement be corrected?

(Source: Kerr, 1982. Copyright 1982 by the American Association for the Advancement of Science. Used with permission of the AAAS.)

References

Kerr, R. A. "Cloud Seeding: One Success in 35 Years." *Science* 217 (1982): 519–21.

Lacoste-Utamsing, C. de, and Holloway, R. L. "Sexual Dimorphism in the Human Corpus Callosum." *Science* 216 (1982): 1431–32.

Further Reading

Snedecor, G. W., and Cochran, W. G. *Statistical Methods.* 7th ed. Ames, IA: Iowa State University Press, 1980. This text is an excellent reference concerning methods of inferential statistics. Various topics introduced in the present chapter are presented more thoroughly in Snedecor and Cochran's Chapters 5 and 6, including (on pages 95–98) a test comparing the means of two independent samples from populations with unequal variances.

CHAPTER 12

INFERENCES ABOUT STANDARD DEVIATIONS AND VARIANCES

EXAMPLE 12.1 In the midst of an unusually cold winter, a newspaper article reported that "leading U.S. climatologists believe the unusual Arctic cold that has besieged that nation's eastern half provides new evidence that North America's weather—and perhaps the world's—is becoming more erratic.

"The scientists say this apparent increase in the weather's variability seasonally and yearly has important implications for global food supplies. More erratic weather is likely to reflect in wider fluctuations in the yields of such staple crops as wheat, corn and soybeans.

"The ups and downs of harvests, in turn, add instabilities not only to consumer prices but also to the economies of those nations that have come increasing[ly] to depend on the U.S. as the world's leading exporter of cereal grains, livestock feed, and rice.

"Scientists disagree about whether it is becoming warmer or cooler. But there is something of a consensus among climatologists that a 14- to 16-year period of unusually mild and constant weather is probably over.

" 'Looking back over the past 100 years, there really is no period that compares in the evenness of weather,' says Dr. Robert Shaw, a professor of agricultural climatology at Iowa State University.

"Shaw said it's almost impossible to tell whether or not this winter is part of a general cooling trend. But the record cold does, he noted, seem in keeping with a return to more variable weather."

(Source: *The Montreal Star,* 24 January 1977. Used with permission of Thomson Newspaper Publishers.)

Population means, discussed in Chapters 10 and 11, are not the only parameters that can be estimated and about which hypotheses can be tested. As illustrated in Example 12.1, sometimes it is of interest to estimate variability or to detect changes in measures of dispersion. In this chapter we introduce methods for drawing inferences about the variability or dispersion of populations of observations. The measures of dispersion to be considered are the standard deviation σ and its square, the variance σ^2. These inferential methods are, unfortunately, quite limited to data from normal populations. Nonetheless, an introduction to inferences about dispersion is important, not only in its own right but also as background for material in Chapters 13, 14, and 15.

12.1 POINT ESTIMATES OF σ AND σ²

The usual **point estimate of the population variance σ^2** is the sample variance s^2, whose definitional formula is

$$s^2 = \frac{\sum (x - \bar{x})^2}{n - 1} \tag{12.1}$$

For a random sample from a given population, the sample variance s^2 is an **unbiased estimate** of the population variance σ^2. The fact that the estimate is unbiased implies that the expected value of s^2, or the mean of its sampling distribution, is equal to the population variance σ^2. In other words, the fact that

$$E(s^2) = \sigma^2 \tag{12.2}$$

indicates that s^2 is an unbiased estimate of σ^2.

Thus, it seems natural to use the sample standard deviation s, which is the positive square root of s^2, as a **point estimate of the population standard deviation σ**, and indeed this is the usual practice even though s is *not* an unbiased estimate of σ. An estimate is unbiased if its expected value equals the parameter that it is being used to estimate—if the average of all possible estimates, as it were, equals the parameter. However, it is a simple arithmetical fact that if the mean of a set of s^2-values equals the constant σ^2, then the mean of the corresponding set of s-values does not generally equal σ. Thus, s is a **biased estimate** of σ, because

$$E(s) \neq \sigma \tag{12.3}$$

However, its bias is not so serious as to preclude its usefulness in many practical applications.

EXAMPLE 12.2 Twenty unmarried female university students, aged 17 to 21 years, took part in a study concerning family size. The students filled out a brief questionnaire in which they reported the number of children they would like to have in their future families, as follows:

3, 3, 4, 2, 2, 2, 3, 4, 3, 3, 2, 0, 5, 2, 4, 5, 5, 4, 3, 3

The observed values of desired number of children had a mean of

$$\bar{x} = \frac{\sum x}{n} = \frac{62}{20} = 3.10 \text{ children}$$

a variance of

$$s^2 = \frac{\sum x^2 - \dfrac{(\sum x)^2}{n}}{n - 1} = \frac{222 - \dfrac{62^2}{20}}{19} = 1.5684$$

and a standard deviation of

$$s = \sqrt{1.5684} = 1.2524$$

or approximately 1.25 children.

From our discussion in Section 10.1, we know that the sample mean, $\bar{x} = 3.10$, is an unbiased estimate of the true mean desired family size that would be reported by the relevant population of female university students. Now we can add that the sample variance, $s^2 = 1.57$, is an unbiased point estimate of the variance σ^2 of the relevant population of desired family sizes, and the sample standard deviation, $s = 1.25$, is a slightly biased, but nonetheless useful, point estimate of the standard deviation σ of the relevant population of desired family sizes. With respect to the relevant population, we can estimate that the number of children desired by unmarried female university students in their future families has a mean (±standard deviation) of 3.10 ± 1.25 children.

(Source: Student project by T. Kopperud.)

12.2 INTERVAL ESTIMATES OF σ AND σ²

It is an important general principle that we can attach probability statements to inferences about a parameter if we know the sampling distribution of a statistic that relates the point estimate to the estimated parameter. We saw this to be true in the case of inferences about population means: The statistic

$$t_{\text{obs}} = \frac{\bar{x} - \mu}{s_{\bar{x}}} \tag{12.4}$$

which relates the point estimate $\bar{x}$ to the parameter μ, follows the t distribution, provided that the sampling distribution of $\bar{x}$ is Gaussian. Knowledge of the sampling distribution of t_{obs} allowed us to attach probability statements to our interval estimates of μ in Chapter 10 and to our significance-test conclusions about population means in Chapter 11.

The Chi-Square Distribution

The same general principle applies also in the case of inferences about population standard deviations and variances. In this case, the relevant sampling distribution is not the t distribution but a distribution called the **chi-square distribution**, or the **χ^2 distribution**. (The lower-case Greek letter χ is called "chi," which is pronounced "kie" to rhyme with sky and pi, π.) In particular, if a random sample of n observations with variance s^2 is drawn from a Gaussian population with variance σ^2, the chi-square statistic

$$\chi^2_{\text{obs}} = \frac{(n - 1)s^2}{\sigma^2} \tag{12.5}$$

follows the chi-square distribution with $n - 1$ degrees of freedom. Notice that, like the t statistic in Expression (12.4), the chi-square statistic in Expression (12.5) relates a point estimate to the parameter being estimated.

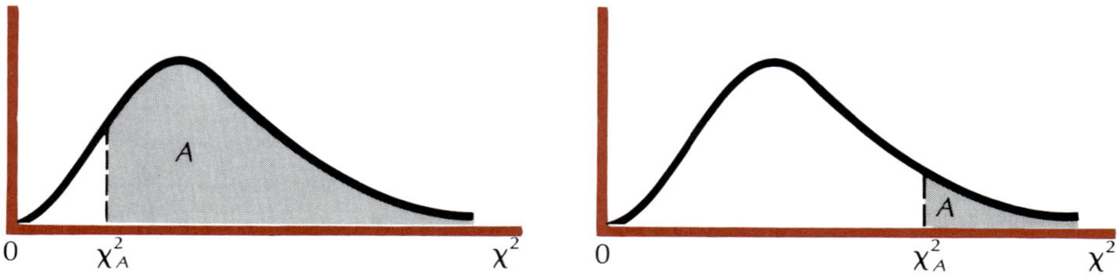

Figure 12.1 Critical values of the chi-square distribution, with the subscript indicating the upper-tail area.

Knowledge of the chi-square distribution will enable us to attach probability statements to certain inferences about standard deviations and variances.

Certain critical values of the chi-square distribution are tabulated in Appendix 5, where χ^2_A is the value of χ^2 such that A of the area under the chi-square probability density function is to the right of χ^2_A; equivalently, the area to the left of χ^2_A is $1 - A$ (see Figure 12.1). We will use the notation $\chi^2_A(n - 1\text{ df})$ to refer to the value of χ^2 that cuts off the proportion A of the area in the upper tail of the chi-square distribution with $n - 1$ degrees of freedom.

Unlike the tables of the Gaussian and the t distributions in Appendixes 3 and 4, the chi-square table in Appendix 5 includes separate tabulations for the left and right tails of the distribution. Separate tabulation of the two tails is required because the chi-square distribution is not symmetric but is skewed right (has positive skewness). Therefore, it is necessary to look up chi-square values for the left tail separately from those for the right tail. Critical values in the left tail are *not* obtained by attaching a minus sign to critical values in the right tail.

EXAMPLE 12.3 From the table in Appendix 5, we find that

$$\chi^2_{.025}(10\text{ df}) = 20.483$$

cuts off a right-tail area of 2.5%, as shown in Figure 12.2(a), while

$$\chi^2_{.975}(10\text{ df}) = 3.247$$

cuts off a left-tail area of 2.5%, as shown in Figure 12.2(b).

As another example, you should verify in Appendix 5 that

$$\chi^2_{.01}(21\text{ df}) = 38.932$$

cuts off an upper-tail area of 1%, as shown in Figure 12.2(c), while

$$\chi^2_{.99}(21\text{ df}) = 8.897$$

cuts off a lower-tail area of 1%, as shown in Figure 12.2(d).

The chi-square distribution, illustrated in Figure 12.3, is a continuous distribution, with the continuous random variable χ^2 taking values from 0 to

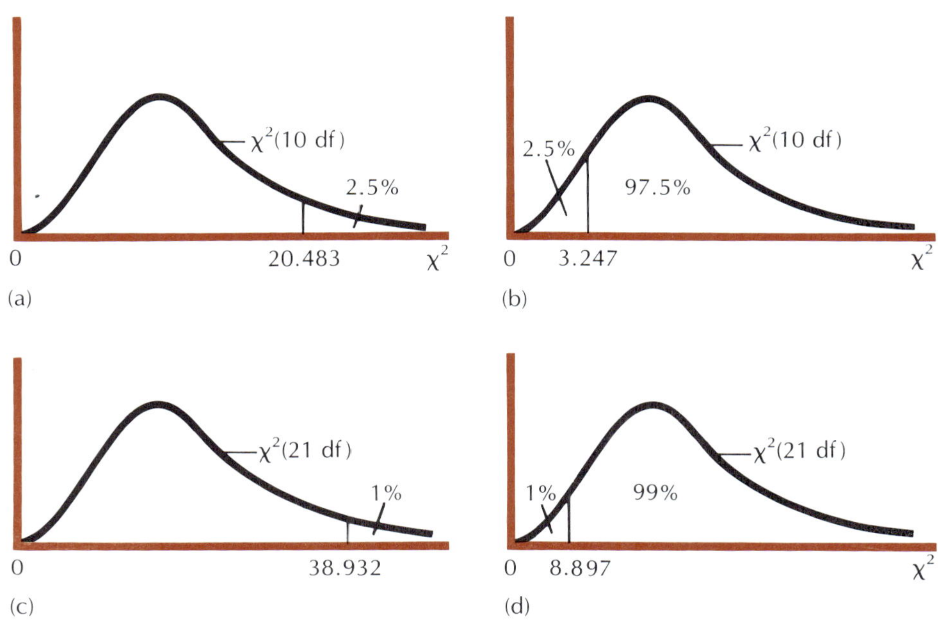

Figure 12.2

Figure 12.3
Chi-square distributions with various numbers of degrees of freedom.

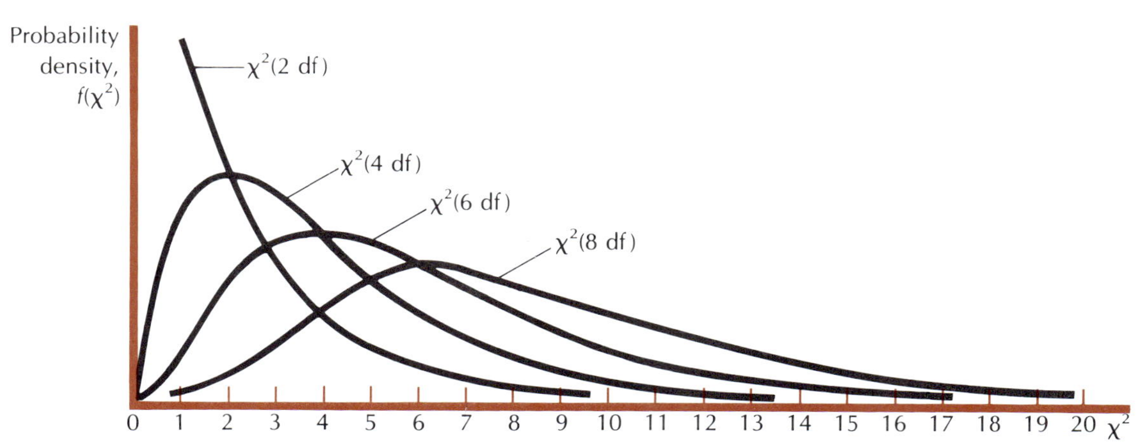

plus-infinity. As might be expected for a squared quantity, the value of χ^2 cannot be negative.

As the number of degrees of freedom increases, the chi-square distribution becomes more symmetric, gradually approaching the Gaussian bell shape. By the time the number of degrees of freedom reaches 30, the chi-square distribution is sufficiently close to the Gaussian shape that for most practical applications we no longer need separate tables of the chi-square distribution. Rather, the chi-square table in Appendix 5 gives a formula that translates standard normal deviates from the table in Appendix 3 into chi-square critical values, for use when there are more than 30 degrees of freedom associated with the chi-square statistic.

For sketching the chi-square distribution, it is useful to know that its mode is 2 less than the number of degrees of freedom (df); that is, mode = df − 2, provided df > 2. Because it is a positively skewed unimodal distribution, the mean is greater than the mode; specifically, the mean of the chi-square distribution is equal to its number of degrees of freedom. Finally, as illustrated in Figure 12.3, the chi-square distribution becomes more spread out—its dispersion increases—as the number of degrees of freedom becomes larger.

The chi-square distribution is closely related to the normal distribution. One instance of this relationship is easily demonstrated as follows: If we randomly select z-scores from a standard normal population, then z^2 has a chi-square distribution with 1 degree of freedom; that is,

$$z^2 \sim \chi^2(1 \text{ df}) \tag{12.6}$$

EXAMPLE 12.4 From the table of the standard normal distribution in Appendix 3, we know that 5% of a distribution of standard normal deviates would have absolute values greater than 1.96. Thus, 5% of a distribution of squared standard normal deviates would have values greater than $1.96^2 = 3.84$. From the table of the chi-square distribution in Appendix 5, we find that 5% of the chi-square distribution with 1 degree of freedom has chi-square values greater than

$$\chi^2_{.05}(1 \text{ df}) = 3.84$$

as shown in Figure 12.4(a).

Similarly, the standard normal critical values

$$\pm z_{.005} = \pm 2.576$$

cut off a total tail area of 1% of the standard normal distribution, and the one-degree-of-freedom chi-square critical value

$$\chi^2_{.01}(1 \text{ df}) = 2.576^2 = 6.64$$

cuts off a right-tail area of 1% in the chi-square distribution with 1 degree of freedom, as shown in Figure 12.4(b).

Confidence Intervals for σ and σ^2

Confidence intervals can be derived from the chi-square statistic in Expression (12.5) in a manner analogous to that used with the z and t statistics in Chapter

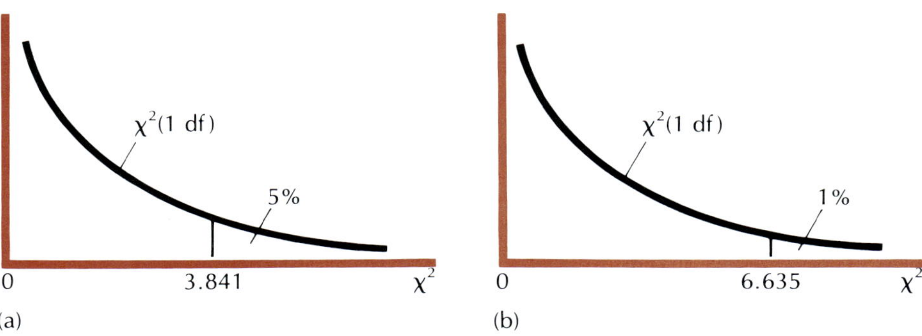

Figure 12.4

10. If

$$\chi^2_{\text{obs}} = \frac{(n-1)s^2}{\sigma^2} \tag{12.7}$$

follows the chi-square distribution with $n - 1$ degrees of freedom, then χ^2_{obs} falls in the central $1 - \alpha$ region of $\chi^2(n - 1 \text{ df})$ with probability $1 - \alpha$; that is,

$$\chi^2_{1-\alpha/2} < \chi^2_{\text{obs}} < \chi^2_{\alpha/2} \qquad \text{with probability } 1 - \alpha \tag{12.8}$$

Substitution of Expression (12.7) into (12.8) yields

$$\chi^2_{1-\alpha/2} < \frac{(n-1)s^2}{\sigma^2} < \chi^2_{\alpha/2} \qquad \text{with probability } 1 - \alpha \tag{12.9}$$

whose terms can be algebraically rearranged into the form of the **1 − α confidence interval for the population variance σ² for random samples from a Gaussian population:**

$$\frac{(n-1)s^2}{\chi^2_{\alpha/2}} < \sigma^2 < \frac{(n-1)s^2}{\chi^2_{1-\alpha/2}} \qquad \text{with confidence level } 1 - \alpha \tag{12.10}$$

Taking square roots, we obtain the **1 − α confidence interval for the population standard deviation σ for random samples from a Gaussian population:**

$$\sqrt{\frac{(n-1)s^2}{\chi^2_{\alpha/2}}} < \sigma < \sqrt{\frac{(n-1)s^2}{\chi^2_{1-\alpha/2}}} \qquad \text{with confidence level } 1 - \alpha \tag{12.11}$$

Notice that the confidence limits in Expressions (12.10) and (12.11) are based on the point estimate s^2. Different samples would produce different values of s^2, and hence different confidence limits, but a confidence level of, say, 95% tells us that 95% of the random samples that might be drawn from the population would yield 95% confidence limits that would capture the unknown (constant) value of the parameter.

The confidence limits are obtained from the point estimate s^2, not by adding and subtracting a certain margin of error, but by multiplying the point estimate by an appropriate error factor. The error is multiplicative for inferences about variances and standard deviations because the chi-square statistic relates the point estimate to the parameter in the form of a ratio (s^2/σ^2), while the error is additive for inferences about means because the t statistic relates the point estimate to the parameter in the form of a difference $(\bar{x} - \mu)$. However, in both cases the confidence interval uses the point estimate as a sort of "best-guess" value from which lower and upper confidence limits are determined according to the sampling distribution of a statistic (t or χ^2) that relates the point estimate to the parameter being estimated.

EXAMPLE 12.5 In Example 12.2, we presented data concerning the number of children desired by $n = 20$ unmarried female university students in their future families. The mean desired number was $\bar{x} = 3.10$, the variance was $s^2 = 1.5684$, and the standard deviation was $s = 1.2524$.

A 95% confidence interval for the population variance σ^2 may be calculated from Expression (12.10) as follows, using critical values from the chi-square distribution with $n - 1 = 19$ degrees of freedom:

$$\frac{(20 - 1)(1.5684)}{32.852} < \sigma^2 < \frac{(20 - 1)(1.5684)}{8.907} \qquad \text{with 95\% confidence}$$

$$.91 < \sigma^2 < 3.35 \qquad \text{with 95\% confidence}$$

Taking square roots, as indicated in Expression (12.11), we obtain the 95% confidence interval for the population standard deviation σ:

$$.95 < \sigma < 1.83 \qquad \text{with 95\% confidence}$$

Thus, we can be 95% confident that the true standard deviation of desired number of children in the population that was sampled is between .95 and 1.83 children.

Unfortunately, inferences about standard deviations and variances are not particularly robust against departures from the assumption of normality, an assumption [introduced in connection with Expression (12.5)] that is required in the preceding derivation of confidence intervals. Caution should therefore be exercised in interpreting such confidence intervals if, as is often the case in practice, it is not known that the population from which the observations were sampled is approximately Gaussian. However, a confidence interval can provide some indication of the precision of the point estimate, even if the confidence level must be qualified as being subject to error due to violation of the normality assumption.

12.3 ONE-SAMPLE CHI-SQUARE TEST FOR A SPECIFIED VALUE OF THE VARIANCE

The chi-square statistic that enables us to establish confidence intervals for the population variance can also be used to test whether a sample variance differs significantly from a specified value (say, σ_0^2) that had previously been postulated to be the population variance.

The null hypothesis for such a test states that the variance σ^2 of the population being sampled is equal to the specified value σ_0^2:

$$H_0\colon \sigma^2 = \sigma_0^2 \tag{12.12}$$

The symbol σ^2 refers to an unknown constant; σ_0^2 is a known constant whose numerical value appears in place of the symbol in any given application. The null hypothesis implies the equality of the corresponding square roots, $\sigma = \sigma_0$, and indeed the test may be considered a test about the standard deviation, with the null hypothesis written accordingly. The alternative hypothesis is usually two-sided:

$$H_A\colon \sigma^2 \neq \sigma_0^2 \tag{12.13}$$

However, it may be one-sided:

$$H_A\colon \sigma^2 < \sigma_0^2 \tag{12.14}$$

or

$$H_A\colon \sigma^2 > \sigma_0^2 \tag{12.15}$$

if a difference in the opposite direction is of no more interest than a finding of no difference.

Substituting the null-hypothesis value from Expression (12.12) into the chi-square statistic in Expression (12.5), we obtain the **test statistic** for a **one-sample chi-square test for a specified value of the variance:**

$$\chi^2_{\text{obs}} = \frac{(n - 1)s^2}{\sigma_0^2} \tag{12.16}$$

which is distributed according to $\boldsymbol{\chi^2(n - 1\,\text{df})}$ if the null hypothesis is true. Note that it is the variance, or square of the standard deviation, that appears in the chi-square test statistic, regardless of whether the null hypothesis is phrased in terms of variances or standard deviations.

If the null hypothesis is true, the test statistic in Expression (12.16) may be regarded as a random selection from the chi-square distribution with $n - 1$ degrees of freedom; but if it is not true, the test statistic will be more likely than specified by the chi-square distribution to take on an extreme value. To test at the α significance level, we compare the numerical value of χ^2_{obs} to the critical values $\chi^2_{1-\alpha/2}(n - 1\text{ df})$ and $\chi^2_{\alpha/2}(n - 1\text{ df})$ for a test against a two-sided alternative, or to the critical value $\chi^2_{1-\alpha}(n - 1\text{ df})$ for a one-sided "less than" alternative, or to the critical value $\chi^2_{\alpha}(n - 1\text{ df})$ for a one-sided "greater than" alternative, as illustrated in Figure 12.5.

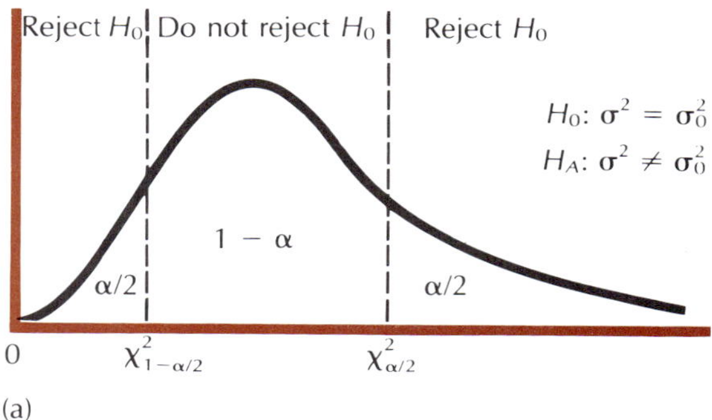

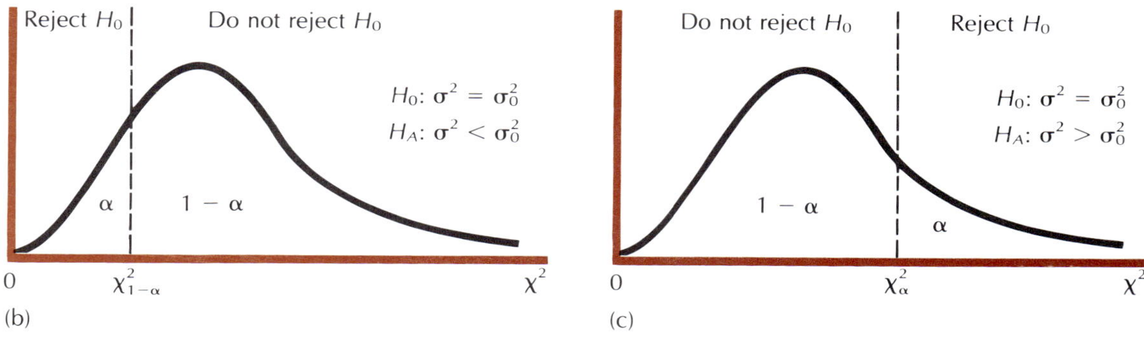

Figure 12.5
Sampling distributions of χ^2_{obs} if the null hypothesis is true, with critical values shown for the decision about the null hypothesis at the α significance level. Parts (a)–(c) correspond to three possible alternative hypotheses that may be proposed.

The one-sample chi-square test for a specified value of the variance is not particularly robust against violations of the assumption of normality of the underlying population, especially for small sample sizes. It is wise to interpret the results of this test cautiously—to recognize, for example, that the actual significance level may differ from the nominal significance level—unless there is some assurance that the population being sampled is reasonably well approximated by a normal distribution.

EXAMPLE 12.6 In Example 12.2, a sample of $n = 20$ unmarried female university students reported the desired number of children in their future families. The sample mean was $\bar{x} = 3.10$, and the standard deviation was $s = 1.25$ (rounded from 1.2524).

Does the sample provide enough evidence for us to conclude that the variability of desired family size in the students' generation differs from the variability of actual family size in their parents' generation, if the relevant standard deviation of actual

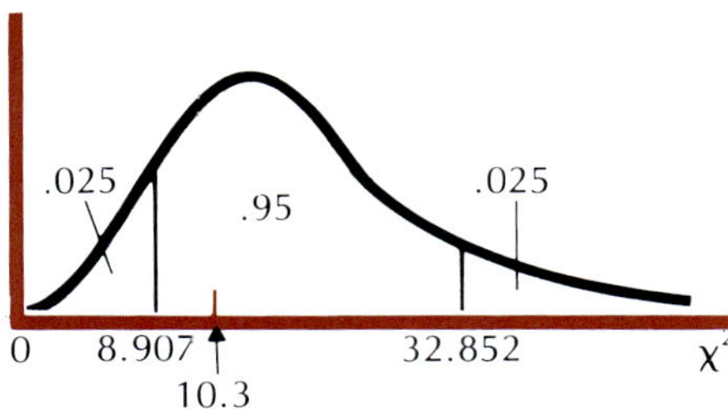

Figure 12.6

family size is known from census data to be 1.7 children? The null and alternative hypotheses implied by this question are

$$H_0: \sigma = 1.7$$

$$H_A: \sigma \neq 1.7$$

From Expression (12.16), the observed value of the test statistic is

$$\chi^2_{\text{obs}} = \frac{(20 - 1)(1.2524^2)}{1.7^2} = 10.3$$

If the null hypothesis is true, the test statistic χ^2_{obs} may be regarded as a random selection from the chi-square distribution with $n - 1 = 19$ degrees of freedom, whose .05-level critical values are $\chi^2_{.975}(19\text{ df}) = 8.907$ and $\chi^2_{.025}(19\text{ df}) = 32.852$, as shown in Figure 12.6.

Since $\chi^2_{\text{obs}} = 10.3$ falls between the critical values, we cannot reject the null hypothesis at the 5% significance level (see Figure 12.5). The evidence from the sample is insufficient to conclude that the variability of desired family size differs from that of actual family size in the preceding generation ($P > .05$).

Although the data are consistent with the null hypothesis, they do not prove it to be true. In fact, at the 95% confidence level, the data are consistent with any value of the population standard deviation in the 95% confidence interval from .95 to 1.83, which was calculated in Example 12.5. The confidence interval includes the hypothesized value of 1.7 as well as a range of other possible values.

Guidelines for how to round and how many digits to report are given in Appendix 1.

Summary of the Test Procedure

The **one-sample chi-square test for a specified value of the variance** may be summarized as follows:

Assumptions.

a. Random sample.

b. Normality of the underlying population.

Null hypothesis. H_0: $\sigma^2 = \sigma_0^2$

	Nondirectional test	Left-directional test	Right-directional test
Alternative hypothesis.	H_A: $\sigma^2 \neq \sigma_0^2$	H_A: $\sigma^2 < \sigma_0^2$	H_A: $\sigma^2 > \sigma_0^2$
Critical values.	$\chi^2_{1-\alpha/2}$ and $\chi^2_{\alpha/2}$	$\chi^2_{1-\alpha}$	χ^2_{α}

Step 1. Calculate the sample variance s^2 for computational purposes and the sample standard deviation s for interpretation.

Step 2. Calculate the null-hypothesis value of the test statistic:

$$\chi^2_{\text{obs}} = \frac{(n - 1)s^2}{\sigma_0^2}$$

Step 3. Look up the critical value(s) in the chi-square distribution with $n - 1$ degrees of freedom.

Step 4. If the observed value χ^2_{obs} is at or beyond the critical value in the tail of the distribution, conclude that the sample variance is significantly different from the specified value σ_0^2 ($P < \alpha$); otherwise, conclude that the observed difference is not statistically significant ($P > \alpha$).

12.4 INDEPENDENT-SAMPLES *F* TEST OF THE DIFFERENCE BETWEEN TWO VARIANCES

The test described in this section is used to determine whether two independent random samples provide sufficient evidence to conclude that the underlying populations differ with respect to dispersion—that is, to learn whether the observations in one population are more spread out than those in the other population. To say that the samples are independent implies that the choice of observational units for one sample has no bearing on which particular observational units are in the other sample. Each sample is selected independently of the other.

EXAMPLE 12.7 In an investigation of factors affecting retail prices of gasoline, prices of regular gasoline were recorded for two samples of gas stations in central Alberta. The study was limited to gas stations along major highways and, further, to stations where the price of regular gasoline was announced on a sign visible to passing motorists. One sample consisted of 9 stations in small towns, and the other consisted of 6 stations in a large city (Edmonton).

Everyday experience suggests that gasoline prices are often slightly higher in smaller towns, but the question of interest here is whether there is also a difference in variability. For example, the lower average price in larger cities may result either from lower prices by a few stations only, in which case variability might be greater, or from increased competitiveness, in which case variability might be lower than in smaller towns.

In the present data, the small-town stations had a mean price of 19.20 cents per liter with a standard deviation of .79 cent per liter (variance = .6225), while the large-city stations had a mean price of 19.70 cents per liter with a standard deviation of .13 cent per liter (variance = .016). The difference in the sample means turns out to be in the opposite direction from what might have been expected, and the variability, as indicated by the sample standard deviations, is greater in the small-town sample. Is this observed difference in variability statistically significant? Is the difference between the sample standard deviations large enough for us to conclude that the underlying populations differ with respect to dispersion? The following paragraphs provide a method that will enable us to answer such questions.

(Source: Student project by B. Swanson.)

The Test Statistic

In order to test whether two independent random samples come from populations with equal variances (or, equivalently, equal standard deviations), we entertain the null hypothesis,

$$H_0:\ \sigma_1^2 = \sigma_2^2 \tag{12.17}$$

where σ_1^2 and σ_2^2 are the true variances of the two underlying populations. The alternative hypothesis is generally two-sided:

$$H_A:\ \sigma_1^2 \neq \sigma_2^2 \tag{12.18}$$

However, it may be one-sided:

$$H_A:\ \sigma_1^2 < \sigma_2^2 \tag{12.19}$$

or

$$H_A:\ \sigma_1^2 > \sigma_2^2 \tag{12.20}$$

if a difference in the opposite direction would be of no more interest than a finding of no difference.

For a test of the null hypothesis, we require a statistic (a) that relates the variance estimates s_1^2 and s_2^2 to the relevant parameters σ_1^2 and σ_2^2 and (b) whose sampling distribution under the null hypothesis is known. Such information will allow us to determine how likely or unlikely we would be to obtain sample values s_1^2 and s_2^2 as different as those actually obtained in two given samples if the null hypothesis of no population difference were true.

Thanks to the work of R. A. Fisher (see Biography 12.1), we know that for independent random samples of size n_1 and n_2 from Gaussian populations, the ratio

$$F_{\text{obs}} = \frac{s_1^2/\sigma_1^2}{s_2^2/\sigma_2^2} \tag{12.21}$$

follows the ***F* distribution** with $n_1 - 1$ degrees of freedom for the numerator and $n_2 - 1$ degrees of freedom for the denominator. Substituting the null-hypothesis equality from Expression (12.17) into (12.21), we obtain the **test statistic** for the **independent-samples *F* test of the difference between two variances:**

$$F_{\text{obs}} = \frac{s_1^2}{s_2^2} \tag{12.22}$$

which follows the F distribution with **$n_1 - 1$ degrees of freedom for the numerator and $n_2 - 1$ degrees of freedom for the denominator**, if the null hypothesis is true. In view of Expression (12.22), the ***F* distribution**, sometimes called the **variance-ratio distribution**, may be thought of as the sampling distribution of the ratio of two independent estimates of the same Gaussian-population variance.

BIOGRAPHY 12.1 Ronald Aylmer Fisher (1890–1962)

R. A. Fisher was born on 17 February 1890 in East Finchley, near London, England. His mathematical interests and ability appeared early in his schooling. In 1909, he won a scholarship to Gonville and Caius College, Cambridge, where he first earned the B.A. degree with a specialization in optics in 1912 and then continued for a year's graduate study in physics. He next worked for an investment company for two years. His extreme nearsightedness kept him out of the army during World War I, and instead he taught mathematics and physics in various public schools from 1915 to 1919.

Meanwhile, he had published several important research papers in mathematical statistics and statistical genetics. These led to two offers of professional appointments, one as statistician at an agricultural research institution known as Rothamsted Experimental Station and the other from Karl Pearson (see Biography 5.3) at University College in London. Fisher had felt slighted by Pearson's treatment of his earlier work, and he went instead to Rothamsted for 14 highly productive years from 1919 to 1933. (The friction between these two leading statisticians of the twentieth century, Pearson and Fisher, developed into a well-known, lifelong feud. They are said to have been in agreement on only two occasions, one in 1915 and the other in 1932!) Fisher succeeded Pearson as the Galton Professor of Eugenics at University College from 1933 to 1943, and then held the Arthur Balfour Chair of Genetics at Cambridge from 1943 until his retirement in 1957. Among numerous other honors, Fisher was knighted in 1952.

Fisher made significant contributions in both statistics and genetics. One of his early papers dealt with the sampling distribution of the correlation coefficient (see Chapter 13), and he recognized that when the sampling distribution of an estimate is known, it is possible to test the significance of its difference from any specified value of the parameter. Thus, he, along with Pearson, Gosset (see Biography 10.1), and others, was among the pioneers of significance testing; it was Fisher who coined the term *null hypothesis*. Fisher is perhaps best known for his many contributions to experimental design, his recognition of the importance of randomization to eliminate bias in estimates, and his development of the analysis of variance (see Chapter 15), one of the most versatile and widely used procedures in statistical inference. It was in connection with this latter work that he derived and tabulated the variance–ratio distribution, later named the F distribution in his honor by George Snedecor.

Fisher married Ruth Eileen Guinness in 1917, and they had eight children. After his retirement in 1957, Fisher became a research fellow at the Commonwealth Scientific and Industrial Research Organisation in Adelaide, Australia, where he died in 1962.

(Sources: Bartlett, 1978; Neyman, 1967; *World Who's Who in Science,* 1968.)

The *F* Distribution

Let us devote a few paragraphs to the general nature of the F distribution before we present the procedure for testing the null hypothesis in Expression (12.17). Dividing both sides of Expression (12.17) by σ_2^2 shows that the null hypothesis implies that

$$\frac{\sigma_1^2}{\sigma_2^2} = 1 \tag{12.23}$$

This ratio of population variances is estimated by the ratio of the corresponding sample variances, which is the test statistic in Expression (12.22):

$$F_{\text{obs}} = \frac{s_1^2}{s_2^2} \tag{12.24}$$

If the first population variance σ_1^2 is actually larger, the sample variance ratio in Expression (12.24) will tend on the average to be greater than 1. If the second population variance σ_2^2 is actually larger, the sample variance ratio will tend on the average to be less than 1. Even if the population variances are exactly equal (as proposed by the null hypothesis), the sample variance ratio in Expression (12.24) will probably not be exactly equal to 1, but it should not be too far from it. The F distribution assigns probabilities to the various sorts of sample variance ratios that might occur when the population variance ratio is 1.

Certain critical values of the F distribution are tabulated in Appendix 6. Notice that we require two numbers of degrees of freedom, one for the variance estimate in the numerator and one for the variance estimate in the denominator, in order to find the appropriate critical value of F. It is customary to write the numerator degrees of freedom first; for example, s_1^2/s_2^2 is distributed according to $F(n_1 - 1, n_2 - 1 \text{ df})$.

The notation F_A is used to designate the value of F such that the proportion A of the area under the F probability density function is to the right of F_A. Equivalently, the area to the left of F_A is $1 - A$. (See Figure 12.7.)

Only upper-tail values of the F distribution are provided in the tables, because any lower-tail value is equal to the reciprocal of the corresponding upper-tail value with the numbers of degrees of freedom interchanged; that is,

$$F_{1-A}(q, r \text{ df}) = \frac{1}{F_A(r, q \text{ df})} \tag{12.25}$$

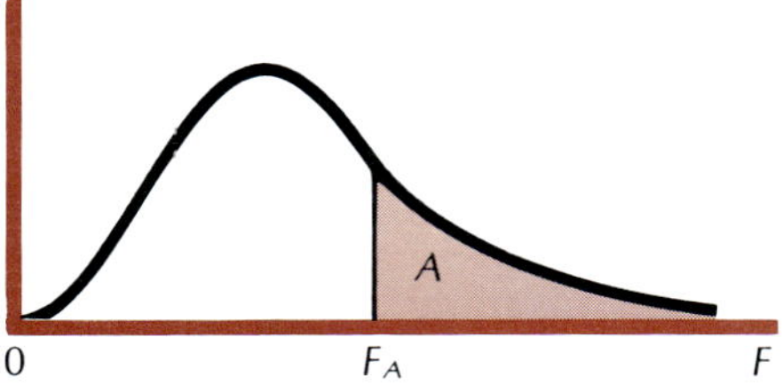

Figure 12.7 Critical value of the F distribution, with the subscript indicating the upper-tail area.

where q and r represent numbers of degrees of freedom. In practice, it is usually possible to avoid any need for Expression (12.25): Since it is arbitrary which sample is called sample 1 and which sample 2, they should be labeled so that the *larger* variance estimate appears in the numerator. This ensures that the value of F_{obs} always appears in the right-hand half of the distribution, and hence that the upper-tail values of the distribution are the only ones needed.

EXAMPLE 12.8 What critical value of the F distribution with 12 degrees of freedom for the numerator and 20 degrees of freedom for the denominator cuts off an upper-tail area of 5%? The value is found in Appendix 6:

$$F_{.05}(12, 20 \text{ df}) = 2.28$$

as shown in Figure 12.8(a).

What critical value of the F distribution with 10 degrees of freedom for the numerator and 3 degrees of freedom for the denominator cuts off an upper-tail area of 1%? From Appendix 6, you should verify that

$$F_{.01}(10, 3 \text{ df}) = 27.2$$

as shown in Part (b) of the figure.

To illustrate the use of Expression (12.25), let us find the critical value of the F distribution with 9 and 15 degrees of freedom that cuts off a lower-tail area of 5%.

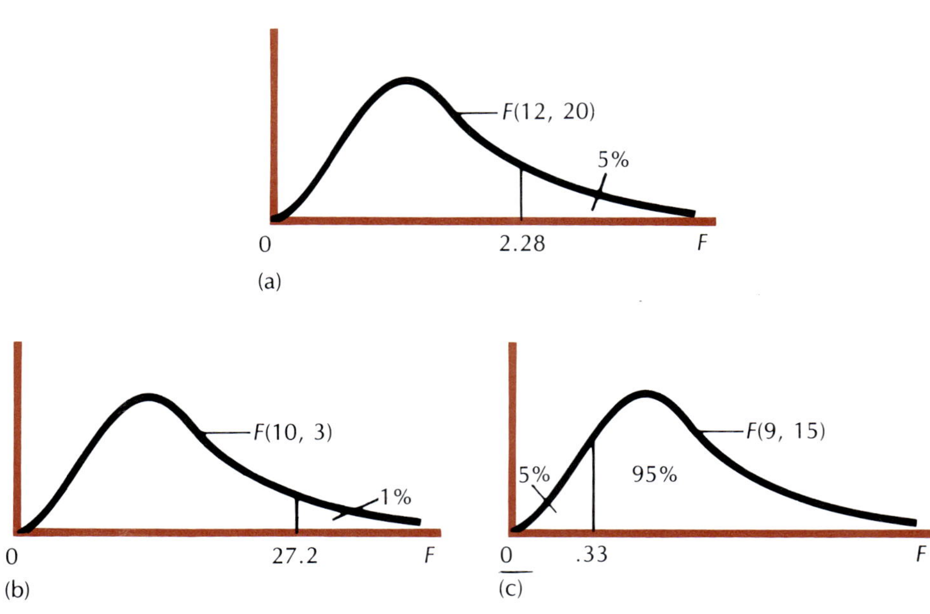

Figure 12.8

Using Expression (12.25) and Appendix 6, we find

$$F_{.95}(9,\ 15\text{ df}) = \frac{1}{F_{.05}(15,\ 9\text{ df})} = \frac{1}{3.01} = .33$$

as shown in Figure 12.8(c).

If the exact numbers of degrees of freedom we require for a particular problem are not given in Appendix 6, we may approximate, either by interpolating between the F-values for the nearest numbers of degrees of freedom that are given or, in many cases, by using the F-value for the next smaller number of degrees of freedom that is given (that is, the next larger F-value). It turns out that the use of the next larger value of F will cause our conclusions to err, if anything, in the direction of caution.

EXAMPLE 12.9 As shown in Figure 12.9(a), the critical value that cuts off an upper-tail area of 1% in the F distribution with 21 and 16 degrees of freedom is

$$F_{.01}(21,\ 16\text{ df}) \simeq 3.24 \qquad \text{(or, at most, 3.26)}$$

the first value being obtained by interpolation and the second from $F(20,\ 16\text{ df})$.

As shown in Figure 12.9(b), the critical value that cuts off an upper-tail area of 5% in the F distribution with 11 and 35 degrees of freedom is

$$F_{.05}(11,\ 35\text{ df}) \simeq 2.08 \qquad \text{(or, at most, 2.16)}$$

the first value being obtained by interpolation and the second from $F(10,\ 30\text{ df})$.

As illustrated in Figure 12.10, the F distribution is a continuous distribution, with the continuous random variable F taking on values from 0 to plus-infinity. The value of F cannot be negative, since it is the ratio of two positive (squared) quantities. With finite degrees of freedom, the mode of the F distribution is always slightly less than 1.00, and the mean is always slightly greater than 1.00. Accordingly, the F distribution is not symmetric but is

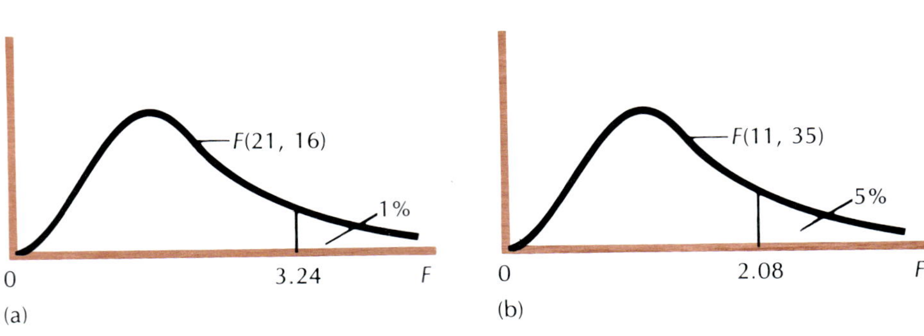

Figure 12.9

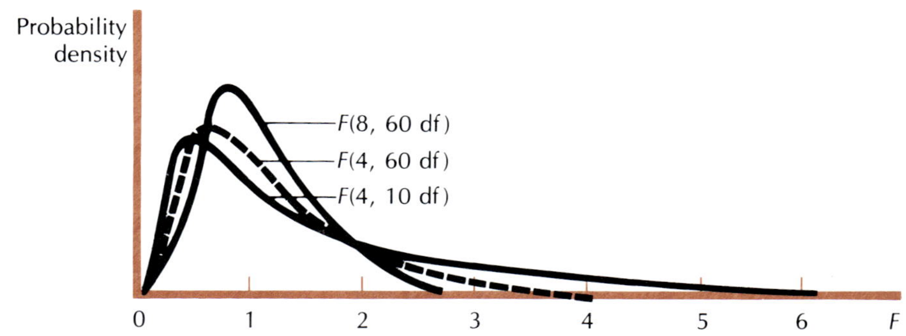

Figure 12.10 *F* distributions for various numbers of degrees of freedom.

skewed right. However, it gradually approaches the shape of the Gaussian distribution as both numbers of degrees of freedom increase.

The *F* distribution is closely related to other distributions that we have studied. One point of correspondence that we will meet again in Chapter 15 is that an *F*-value with 1 and, say, *r* degrees of freedom equals the square of a *t* statistic with *r* degrees of freedom:

$$F_{\alpha}(1, r \text{ df}) = [t_{\alpha/2}(r \text{ df})]^2 \qquad (12.26)$$

EXAMPLE 12.10 In Appendix 6, we find that

$$F_{.05}(1, 13 \text{ df}) = 4.67$$

as shown in Figure 12.11. Appendix 4 shows that

$$t_{.025}(13 \text{ df}) = 2.160$$

whose square is $2.160^2 = 4.67$, as promised by Expression (12.26).

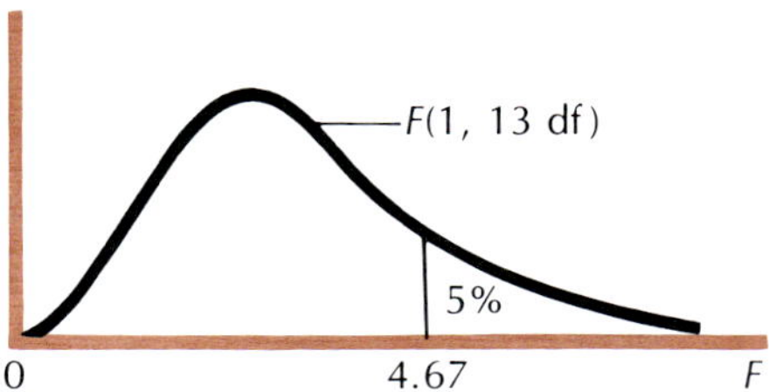

Figure 12.11

The Decision Rule

Returning now to the problem of testing the null hypothesis in Expression (12.17), we may compare the numerical value of the test statistic

$$F_{\text{obs}} = \frac{s_1^2}{s_2^2} \tag{12.27}$$

with the critical value(s) from the table of the F distribution in Appendix 6. If the null hypothesis is true, then the test statistic F_{obs} may be regarded as a random selection from the F distribution with appropriate degrees of freedom; but if it is not true, then the test statistic will be more likely than specified by the F distribution to take on an extreme value. In a test against the two-sided alternative at the α significance level, the critical values are $F_{1-\alpha/2}(n_1 - 1, n_2 - 1 \text{ df})$ and $F_{\alpha/2}(n_1 - 1, n_2 - 1 \text{ df})$. However, if we remember to put the larger estimate of the variance in the numerator of the variance ratio, we will require only the latter critical value, which is the one tabulated in Appendix 6. Similarly, when a one-sided alternative is required, appropriate labeling of the samples, so that the alternative hypothesis is in the form of Expression (12.20) rather than (12.19), can ensure that the critical value is the tabled value $F_\alpha(n_1 - 1, n_2 - 1 \text{ df})$, rather than a lower-tail value that must be calculated via Expression (12.25). The rule for deciding whether or not to reject the null hypothesis is illustrated for two-sided and one-sided alternatives in Figure 12.12.

Figure 12.12
Sampling distribution of F_{obs} if the null hypothesis is true, with critical values shown for the decision about the null hypothesis at the α significance level. Part (a) illustrates a two-sided alternative, and Part (b) illustrates a one-sided alternative.

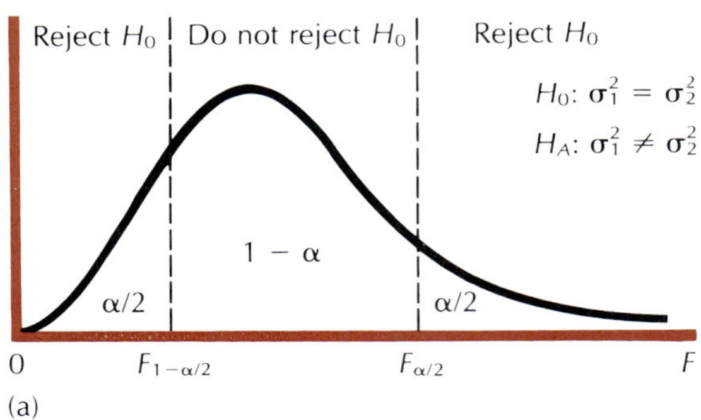

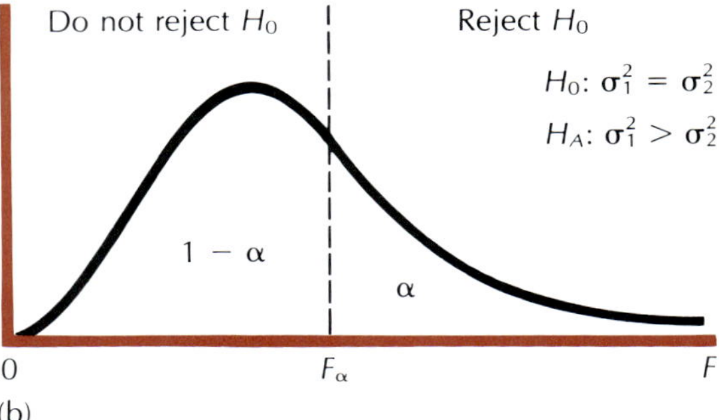

EXAMPLE 12.11 Example 12.7 presented data concerning gasoline prices from two samples of gas stations. The data may be summarized as follows.

	Large-city sample	Small-town sample
Sample size	6	9
Mean (cents per liter)	19.70	19.20
Variance estimate	.0160	.6225
Sample standard deviation	.13	.79

Is the difference in dispersion between the two samples statistically significant?

We could label either sample as sample 1, but to ensure that we need only the upper-tail critical value, which is the one tabulated in Appendix 6, and that we do not need to determine a lower-tail critical value via Expression (12.25), we put the larger variance estimate, .6225, in the numerator of the test statistic in Expression (12.27):

$$F_{\text{obs}} = \frac{.6225}{.0160} = 38.9$$

The numerator .6225 is based on $9 - 1 = 8$ degrees of freedom, and the denominator .0160 is based on $6 - 1 = 5$ degrees of freedom. If the null hypothesis is true, $F_{\text{obs}} = 38.9$ may be regarded as a random selection from the $F(8, 5\text{ df})$ distribution, whose upper-tail critical value is $F_{.01}(8, 5\text{ df}) = 10.3$ for a two-tailed test at the 2% significance level. Since $F_{\text{obs}} = 38.9$ is in the tail region beyond the critical value (see Figures 12.7 and 12.13), we reject the null hypothesis and conclude that there is greater variability in the small-town gasoline prices than in the large-city prices ($P < .02$), at least in the populations that were sampled in this investigation.

It is worth noting that the relevant populations in this case not only were defined by geographical location but also were limited to prices posted on the particular day of the investigation. The investigation did not sample different days, and there is no statistical basis in the data for assessing how typical that day's prices were of the usual pricing pattern in that area.

Guidelines for how to round and how many digits to report are given in Appendix 1. However, results from this test should not be reported without first giving due consideration to the cautionary remarks in "Violations of Assumptions," following the summary of the test procedure.

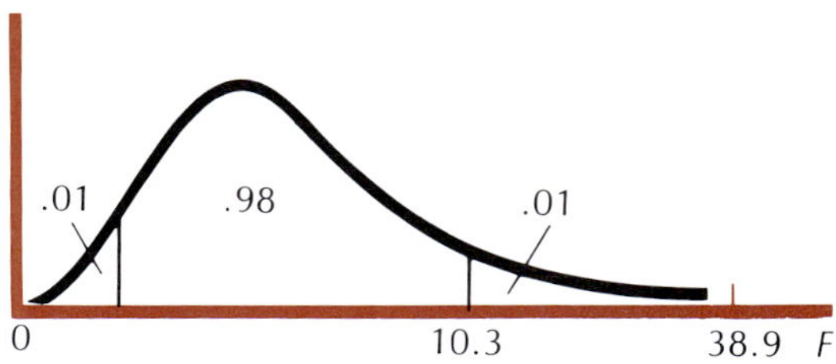

Figure 12.13

Summary of the Test Procedure

The **independent-samples F test of the difference between two variances** may be summarized as follows:

Assumptions.

a. Independent random samples.

b. Normality of the underlying populations.

Null hypothesis. H_0: $\sigma_1^2 = \sigma_2^2$

	Nondirectional test	**Directional test**
Alternative hypothesis.	H_A: $\sigma_1^2 \neq \sigma_2^2$	H_A: $\sigma_1^2 > \sigma_2^2$
Critical values.	$F_{\alpha/2}$	F_α

Step 1. Calculate the sample variances s^2 for computational purposes and the standard deviations s for interpretation. If the alternative hypothesis is two-sided, label as sample 1 the sample whose variance estimate s^2 is larger. If the alternative is one-sided, label as sample 1 the sample whose population variance is hypothesized to be larger.

Step 2. Calculate the null-hypothesis value of the test statistic:

$$F_{\text{obs}} = \frac{s_1^2}{s_2^2}$$

Step 3. Look up the critical value in the F distribution with $n_1 - 1$ and $n_2 - 1$ degrees of freedom.

Step 4. If the observed value F_{obs} is greater than or equal to the critical value, conclude that the variance of population 1 is greater than the variance of population 2 ($P < \alpha$); otherwise, conclude that the observed difference between the sample variances is not statistically significant ($P > \alpha$).

Violations of Assumptions

Recall that one of the assumptions underlying the independent-samples t test of the difference between two means was equality of the two population variances. You may be tempted to use the preceding F test to check on the assumption in order to see if the t test may legitimately by used. Such a practice is *not* recommended as a general rule, because this particular F test, like other methods discussed in this chapter to draw inferences about population variances, is rather sensitive to its own assumption of an underlying Gaussian population. In fact, the t test is considerably more robust against violations of the homogeneity-of-variance assumption than the F test of the difference between two variances is against **violations of the normality assumption**. Statistician G. E. P. Box has likened the preliminary use of the F test on variances, as a check on whether it is safe to use the usual t test on means, to going out in a rowboat in order to see if the water is calm enough for an ocean liner! The results of the F test of the difference between two variances should be interpreted with caution if there is a danger that the populations being sampled are not approximately Gaussian. (We will meet other, more robust, applications of the F distribution in Chapters 13 and 15.)

The test presented in this section is for variances estimated from two independent samples. In the case of paired samples, a different test—a reference for which is given under Further Reading at the end of the chapter—is required.

STUDY MATERIAL FOR CHAPTER 12

Terms and Concepts

Section 12.1

point estimate of σ^2
unbiased estimate
point estimate of σ
biased estimate

Section 12.2

chi-square distribution
$1 - \alpha$ confidence interval for σ^2
$1 - \alpha$ confidence interval for σ

Section 12.3

one-sample χ^2 test for a specified value of the variance
degrees of freedom for the one-sample χ^2 test

Section 12.4

F distribution
independent-samples F test of the difference between two variances
degrees of freedom for the independent-samples F test on variances
variance-ratio distribution
violation of the normality assumption in the independent-samples F test

Key Formulas

Section 12.2

(12.5) $$\chi^2_{\text{obs}} = \frac{(n-1)s^2}{\sigma^2}$$

(12.10) $$\frac{(n-1)s^2}{\chi^2_{\alpha/2}} < \sigma^2 < \frac{(n-1)s^2}{\chi^2_{1-\alpha/2}} \quad \text{with confidence level } 1 - \alpha$$

Section 12.3

(12.16) $$\chi^2_{\text{obs}} = \frac{(n-1)s^2}{\sigma_0^2}$$

Section 12.4

(12.22) $$F_{\text{obs}} = \frac{s_1^2}{s_2^2}$$

Exercises for Numerical Practice

Section 12.2

12/1. Use the table in Appendix 5 to find the value of each of the following:

a. $\chi^2_{.05}(24\text{ df})$ **b.** $\chi^2_{.01}(8\text{ df})$ **c.** $\chi^2_{.005}(16\text{ df})$ **d.** $\chi^2_{.025}(25\text{ df})$
e. $\chi^2_{.975}(13\text{ df})$ **f.** $\chi^2_{.995}(20\text{ df})$ **g.** $\chi^2_{.95}(30\text{ df})$ **h.** $\chi^2_{.99}(2\text{ df})$

12/2. Use the table in Appendix 5 to find the value of each of the following:

a. $\chi^2_{.99}(18\text{ df})$ **b.** $\chi^2_{.005}(3\text{ df})$ **c.** $\chi^2_{.05}(12\text{ df})$ **d.** $\chi^2_{.95}(28\text{ df})$
e. $\chi^2_{.025}(1\text{ df})$ **f.** $\chi^2_{.975}(1\text{ df})$ **g.** $\chi^2_{.995}(22\text{ df})$ **h.** $\chi^2_{.01}(15\text{ df})$

12/3. A random sample of 20 observations from a normal distribution has a standard deviation of 21.6. Calculate the 95% confidence interval for the population standard deviation.

12/4. A random sample of 27 observations from a normal distribution has a standard deviation of 6.12. Calculate the 95% confidence interval for the population standard deviation.

12/5. A random sample of 24 observations from a normal distribution has a variance of 984. Calculate the 99% confidence interval for the population standard deviation.

12/6. A random sample of 30 observations from a normal distribution has a variance of 769. Calculate the 99% confidence interval for the population standard deviation.

Section 12.3

12/7. Using the .05 significance level, test the claim that the population standard deviation is 10.0 if

a. a random sample of 13 observations from a Gaussian population has a standard deviation of 5.6.

b. a random sample of 19 observations from a Gaussian population has a standard deviation of 13.0.

12/8. Using the .01 significance level, test the claim that the population standard deviation is 2.00 if

a. a random sample of 15 observations from a Gaussian population has a standard deviation of 1.13.

b. a random sample of 18 observations from a Gaussian population has a standard deviation of 2.92.

Section 12.4

12/9. Use the tables in Appendix 6 to find the value of the following:

a. $F_{.05}(4, 12\text{ df})$ **b.** $F_{.01}(24, 15\text{ df})$
c. $F_{.025}(15, 8\text{ df})$ **d.** $F_{.10}(1, 30\text{ df})$

12/10. Use the tables in Appendix 6 to find the value of the following:

a. $F_{.01}(2, 4\text{ df})$ **b.** $F_{.025}(10, 6\text{ df})$
c. $F_{.10}(20, 7\text{ df})$ **d.** $F_{.05}(3, 5\text{ df})$

12/11. Consider two independent random samples, one having 10 observations with a standard deviation of 13.2 and the other having 5 observations with a standard deviation of 26.8. Assuming normality, test at the .05 significance level whether the underlying populations have equal variances.

12/12. Consider two independent random samples, one having 16 observations with a standard deviation of 3.76 and the other having 10 observations with a standard deviation of 6.28. Assuming normality, test at the .05 significance level whether the underlying populations have equal standard deviations.

Problems for Applying the Concepts

12/13. A sample of 16 female college students reported their birth weights. The recalled birth weights had a mean of 3.331 kg and a standard deviation of .333 kg. Official community health records show the mean (±standard deviation) of female live birth weights to be 3.25 ± .60 kg for the region from which the college draws its students. Perform a test of significance to determine whether the sample data contradict the claim that recalled birth weights of female college students have the same dispersion as the distribution of actual birth weights in the region.

a. Write the null and alternative hypotheses in symbols and in words.

b. State the results of the significance test.

c. Construct a 98% confidence interval for the true standard deviation of the relevant population of recalled birth weights of female college students.

(Source: Student project by C. Hochhausen.)

12/14. Tests for butterfat content were performed on milk from a sample of 20 Holstein dairy cows. The sample mean was 3.88%, and the standard deviation was .42 percentage point. Perform a significance test to determine whether the sample data contradict the claim that the true variance of butterfat content for such cows is .12.

a. Write the null and alternative hypotheses in symbols and in words.

b. State the results of the significance test.

c. Construct a 95% confidence interval for the true variance of butterfat content for such cows.

(Source: Student project by S. Bailer.)

12/15. The resting pulse rate was counted in a sample of 10 mature light horses. The numbers of beats per minute were as follows:

38, 38, 40, 45, 28, 30, 27, 41, 29, 26

Perform a significance test to determine whether these data contradict the claim that the resting pulse rate for such horses has a variance of 30.0.

a. Write the null and alternative hypotheses in symbols and in words.

b. State the results of the significance test.

c. Construct a 95% confidence interval for the true variance of resting pulse rate for such horses.

(Source: Student project by H. Hoyland.)

12/16. Many North American colleges and universities require foreign students to achieve a certain standard on the Educational Testing Service "Test of English as a Foreign Language" (TOEFL) as an admission requirement. The following TOEFL scores were obtained by a sample of 9 Hong Kong students seeking admission to North American universities:

495, 525, 570, 610, 552, 497, 593, 505, 551

Perform a significance test to determine whether these data contradict the claim that TOEFL scores have a standard deviation of 100.0 in the population from which the sample was selected.

a. Write the null and alternative hypotheses in symbols and in words.

b. State the results of the significance test.

c. Construct a 98% confidence interval for the true standard deviation of TOEFL scores in the relevant population.

(Source: Student project by B. Dung.)

12/17. The price per page was recorded for each book in a sample of 25 college textbooks purchased in 1979–80: 11 hardcover texts had a mean price per page of 2.50 cents with a standard deviation of 1.41 cents, and 14 paperback texts had a mean price per page of 2.04 cents with a standard deviation of 1.62 cents. Perform a significance test to determine whether these data contradict the claim that the variability in price per page is the same for hardcover and paperback textbooks.

a. Identify the dependent and independent variables.

b. Write the relevant hypotheses in symbols and in words.

c. State the results of the significance test.

d. Which of the following might describe your statement in Part (c): Type I error, Type II error, correct rejection, correct nonrejection?

(Source: Student project by L. Warman.)

12/18. Thirty-seven students living in college residences were asked about their average length of weeknight sleep. The weeknight sleep durations reported by the 22 males in the sample had a mean of 7.05 hours and a standard deviation of .86 hour, while the durations reported by the 15 females in the sample had a mean of 7.20 hours and a standard deviation of 1.01 hours. Is the observed difference between the standard deviations statistically significant?

a. Identify the dependent and independent variables.

b. Write the relevant hypotheses in symbols and in words.

c. State the results of the significance test.

d. Which of the following might describe your statement in Part (c): Type I error, Type II error, correct rejection, correct nonrejection?

(Source: Student project by D. Stewart.)

12/19. The average lengths of songs on 10 country record albums and 11 rock albums were found to be the following numbers of minutes:

Country: 2.96, 3.12, 3.33, 3.12, 3.04, 3.23, 4.07, 2.90, 3.73, 3.52

Rock: 4.06, 4.17, 3.39, 4.06, 6.01, 3.85, 4.83, 4.10, 3.45, 4.88, 3.05

Are these data consistent with the notion that the dispersion of average song lengths is the same for country albums and for rock albums?

a. Identify the dependent and independent variables.
b. Write the relevant hypotheses in symbols and in words.
c. State the results of the significance test.
d. Which of the following might describe your statement in Part (c): Type I error, Type II error, correct rejection, correct nonrejection?

(Source: Student project by K. E. George.)

12/20. According to folk wisdom, the distance that a person can reach horizontally, measured from fingertip to fingertip, is equal to that person's height. The following differences (in centimeters) were obtained by subtracting height from arm reach in a sample of 20 young adults:

Males:	5,	3,	0,	0,	14,	2,	5,	8,	6,	6
Females:	1,	0,	−2,	−3,	−3,	−1,	−3,	1,	0,	−1

Perform a significance test to evaluate whether the difference between arm reach and height is equally variable in males and females.

a. Identify the dependent and independent variables.
b. Write the relevant hypotheses in symbols and in words.
c. State the results of the significance test.
d. Which of the following might describe your statement in Part (c): Type I error, Type II error, correct rejection, correct nonrejection?

(Source: Student project by A. Kautz.)

12/21. Two suppliers are competing for a contract to supply spacecraft components that must be produced to a very tight specification. Inspection of random samples of a particular component from both suppliers yielded the following results concerning the thickness of the component: Eight specimens from supplier A had a standard deviation of .007 mm, and six specimens from supplier B had a standard deviation of .017 mm. Was there a significant difference in variability between the two suppliers?

12/22. Two brands of radial tires were compared by a consumer research agency. The average tread-life was very similar for the two brands, but the standard deviation of tread-life was 1990 km in a random sample of five brand A tires and 4030 km in a random sample of four brand B tires. Was there a significant difference in variability of tread-life between the two brands?

References

Bartlett, M. S. "Fisher, R. A." *International Encyclopedia of Statistics.* Edited by W. H. Kruskal and J. M. Tanur. New York: The Free Press, 1978.

Neyman, J. "R. A. Fisher (1890–1962): An Appreciation." *Science* 156 (1967): 1456–60.

World Who's Who in Science. Chicago: Marquis Who's Who, 1968.

Further Reading

Snedecor, G. W., and Cochran, W. G. *Statistical Methods.* 7th ed. Ames, IA: Iowa State University Press, 1980. Topics introduced in the present chapter are discussed in Chapters 5 and 6, and a test of the difference between two variances based on paired samples is given on pages 190–91.

CHAPTER 13

INFERENCES ABOUT CORRELATION AND REGRESSION

EXAMPLE 13.1 "Among the various series of long-range [weather] forecasts produced in the United States, the oldest and perhaps most widely quoted is that of the *Old Farmer's Almanac*. Unfortunately, recent systematic evaluations of long-range forecasts . . . have omitted the Almanac from their lists of forecasters. In the interest of completeness and in response to the high regard expressed by some local farmers for the Almanac forecasts, we present here a systematic evaluation of the accuracy of these forecasts over a five-year period. In view of the potentially high economic stakes associated with long-range forecasts, verification statistics represent an essential adjunct to any series of such forecasts. . . .

"While forecasting methodologies are not our main concern here, we note that the Almanac's forecasting methodology is not documented with the utmost of professional rigor. The extent of the documentation is the statement that 'Our weather forecasts are determined both by the use of a secret weather forecasting formula devised by the founder of this almanac in 1792 and by the most modern scientific calculations based on solar activity.' Forecasts are issued for sixteen geographical regions of the U.S. The forecasts for each region consist of a general textual summary of weather for the 12-month forecast period (November through October), forecast departures from normal temperature and precipitation for each of the 12 months, and a short 1–6 word summary of the forecast weather for 2–7 day periods throughout the year. The monthly temperature and precipitation forecasts, which are the only elements of the Almanac forecasts amenable to objective verification, form the basis for the following analysis."

Using 60 pairs of observations for each weather station (one pair for each of the 60 months in the five-year period from November 1975 through October 1980), the researchers calculated the correlations between forecast departure from normal temperature and actual departure from normal temperature in the 16 regions, as well as the correlations between forecast departure from normal precipitation and actual departure from normal precipitation in the 16 regions. Correlations near zero would result from random guessing, while large positive correlations would indicate successful forecasting. This chapter introduces methods for determining whether an observed relationship is strong enough to convince us that it should not be attributed to chance alone but rather represents a genuine relationship between the variables.

In the case at hand, the researchers found that although 10 of the 16 temperature correlations were positive and 10 of the 16 precipitation correlations were positive, "only one of each set of 16 correlations exceeds the [5%] significance level, $r = 0.255$. (One of 20 values should exceed the [5%] level by chance.) . . . The largest fractions of variance described by the monthly forecasts are $(0.27)^2 \simeq 0.07$ for temperature in

Region 3 [around Washington, D.C.] and $(0.41)^2 \simeq 0.17$ for precipitation in Region 16 [California]." Averaging over all 16 regions for the five-year period, the mean correlation between forecast and actual departures from normal temperature was .02, and that between forecast and actual departures from normal precipitation was .04. Overall, the Almanac's forecasts accounted (linearly) for substantially less than 1% of the variation in actual departures from normal temperature and precipitation. In other words, simply predicting that each and every month would have normal temperature and precipitation would have given virtually the same success as was achieved by the Almanac over this five-year period.

(Source: Walsh & Allen, 1981. Used with permission of Heldref Publications.)

Chapter 5 introduced methods of regression and correlation as ways of describing relationships between two variables, namely, one dependent variable y as a straight-line function ($y = bx + a$) of one independent variable x. Now in this chapter we move from descriptive to inferential methods as we consider what such a relationship, observed in a sample, tells us about the underlying relationship in the population. Or, more generally, what does the observed relationship in one particular set of data tell us about the true relationship between the variables?

13.1 INFERENCES ABOUT SIMPLE LINEAR CORRELATION

Simple linear correlation between two variables refers to the extent to which a given increase in the value of one variable tends to be associated with a constant change in the value of the other variable. It may be thought of as the extent to which two variables exhibit a straight-line relationship with each other. As in Example 13.1, samples consisting of pairs of observations of two variables are frequently employed to draw inferences about the true correlation between the variables in the underlying population. The inferences may be in the form of point estimation, interval estimation, or hypothesis testing.

Point Estimates of the Population Correlation Coefficient

The Pearson product-moment correlation coefficient r was presented in Section 5.2 as a descriptive statistic summarizing the degree and direction of linear relationship between two variables for a sample of observations. In addition to its role as a purely descriptive statistic, the correlation coefficient calculated for a particular sample of a population is a point estimate of what the correlation coefficient would be if it were calculated for the entire population. That is, the sample correlation coefficient r is a **point estimate of the corresponding population correlation coefficient** ρ. (ρ is the lower-case Greek letter rho; be careful not to confuse the Greek letter ρ with the Roman letter p in your reading and writing.) The computational formula for r is given in Expression (5.19).

EXAMPLE 13.2 In Example 13.1, the calculated correlations between forecast and actual departures from normal weather during a particular five-year period were point estimates of the true correlation in general between Almanac forecasts and actual weather measurements.

As usual, the point estimate gives us no information about its precision—how close it is to the actual population value. With any sample correlation coefficient, say, $r = .50$, we know (by definition) that the corresponding population value ρ must be between -1.0 and $+1.0$, but we do not know from the point estimate how close the estimate r is likely to be to the true value ρ nor even how likely the true value is to be positive.

* Confidence Intervals for the Population Correlation Coefficient

To attach probability statements to inferences about the population correlation coefficient ρ, we need to know the sampling distribution of a statistic that relates the estimate r to the parameter ρ. Unfortunately, the sampling distribution of r itself is quite complicated. For one thing, being bounded by a minimum of -1.0 and a maximum of $+1.0$, it is asymmetric unless the underlying correlation ρ in the population being sampled is equal to zero, and although the sampling distribution of r approaches normality as the sample size increases, it does so very slowly for nonzero values of ρ. However, in 1915, R. A. Fisher (see Biography 12.1) devised a statistic with the required properties for inferences about the true correlation ρ.

The statistic that Fisher invented transforms correlation values r into values $z(r)$ that are approximately normally distributed, and so it is known as **Fisher's $z(r)$ transformation**. Fisher considered the case of two variables x and y that follow the **bivariate normal distribution**, which may be thought of as a two-dimensional version of the Gaussian distribution introduced in Section 7.4. Detailed consideration of the bivariate normal distribution is beyond the scope of this book, but, among other things, it implies that each of the variables separately follows a Gaussian distribution. Fisher showed that if r is the correlation coefficient in a sample of size n for two variables x and y that follow the bivariate normal distribution with a population correlation coefficient of ρ, then the sampling distribution of the statistic

$$z(r) = \frac{1}{2}\ln\frac{1+r}{1-r} \tag{13.1}$$

is approximately Gaussian with a mean of approximately

$$z(\rho) = \frac{1}{2}\ln\frac{1+\rho}{1-\rho} \tag{13.2}$$

and a standard error of approximately

$$\sigma_{z(r)} = \frac{1}{\sqrt{n-3}} \qquad \text{for } n > 3 \tag{13.3}$$

* This symbol identifies material that is not prerequisite for later topics in this book.

where ln indicates the natural logarithm, or logarithm to the base $e = 2.71828\ldots$. Expressions (13.1) and (13.2) are easily computed on any of several models of handheld calculators, but for readers whose calculators lack the natural logarithm function (ln), values of Expressions (13.1) and (13.2) are tabulated in Appendix 7. In using the table, note that negative values of r yield negative values of $z(r)$:

$$z(-r) = -z(r)$$

EXAMPLE 13.3 Using either Expression (13.1) or Appendix 7, we find that Fisher's $z(r)$ transformation converts a sample correlation of $r = .31$ to a $z(r)$-value of $z(.31) = .321$, and it converts a sample correlation of $r = -.986$ to a $z(r)$-value of $z(-.986) = -2.477$. Similarly, it converts a population correlation of $\rho = -.89$ to a $z(\rho)$ value of $z(-.89) = -1.422$, and a population correlation of $\rho = .914$ to $z(.914) = 1.551$.

If the statistic $z(r)$ is normally distributed with mean $z(\rho)$ and standard deviation $\sigma_{z(r)}$ as given in Expressions (13.2) and (13.3), as it is under the assumption of bivariate normality that Fisher studied, then the statistic

$$z_{\text{obs}} = \frac{z(r) - z(\rho)}{\sigma_{z(r)}} \tag{13.4}$$

must have the standard normal distribution. [The statistic in Expression (13.4) has the same form as z_{obs} in Section 10.2; namely, the difference between a point estimate having a Gaussian sampling distribution and the corresponding parameter, divided by the standard error of the point estimate.] It follows directly that z_{obs} falls in the central $1 - \alpha$ region of the standard normal distribution with probability $1 - \alpha$; that is,

$$-z_{\alpha/2} < z_{\text{obs}} < z_{\alpha/2} \qquad \text{with probability } 1 - \alpha \tag{13.5}$$

or, equivalently,

$$-z_{\alpha/2} < \frac{z(r) - z(\rho)}{\sigma_{z(r)}} < z_{\alpha/2} \qquad \text{with probability } 1 - \alpha \tag{13.6}$$

Rearranging terms as we did in the derivations of confidence intervals for means in Chapter 10, we obtain the $1 - \alpha$ confidence interval for $z(\rho)$:

$$z(r) - z_{\alpha/2}\sigma_{z(r)} < z(\rho) < z(r) + z_{\alpha/2}\sigma_{z(r)} \qquad \text{with confidence level } 1 - \alpha \tag{13.7}$$

Notice that the confidence limits for $z(\rho)$ are of the general form

(point estimate) ± (margin of error)

Specifically, the $1 - \alpha$ confidence limits for $z(\rho)$ are

$$z(r) \pm e_{\alpha/2} \tag{13.8}$$

where

$$e_{\alpha/2} = z_{\alpha/2}\sigma_{z(r)} = \frac{z_{\alpha/2}}{\sqrt{n-3}} \tag{13.9}$$

is the margin of error at the $1 - \alpha$ confidence level.

The lower and upper confidence limits for $z(\rho)$, namely,

$$z(r_{\text{lower}}) = z(r) - e_{\alpha/2} \tag{13.10}$$

and

$$z(r_{\text{upper}}) = z(r) + e_{\alpha/2} \tag{13.11}$$

are then converted from the $z(r)$-scale back into the r-scale by substituting their values into the reverse $z(r)$ transformation [obtained by solving Expression (13.1) for r]:

$$r = \frac{e^{2z(r)} - 1}{e^{2z(r)} + 1} \tag{13.12}$$

[If your calculator lacks the exponential function e^x, you may obtain values of Expression (13.12) by reading from $z(r)$ to r in Appendix 7.] Thus, $z(r_{\text{lower}})$ and $z(r_{\text{upper}})$, which are the confidence limits for $z(\rho)$, are back-transformed to r_{lower} and r_{upper}, respectively, which are the confidence limits for ρ. We thus obtain the **$1 - \alpha$ confidence interval for ρ in a bivariate normal population:**

$$r_{\text{lower}} < \rho < r_{\text{upper}} \qquad \text{with confidence level } 1 - \alpha \tag{13.13}$$

EXAMPLE 13.4 For each of 10 gasoline-burning car models, the car's engine size (in cubic inches of displacement) and its gasoline mileage (in miles per gallon) in highway driving were obtained from the April 1980 issue of *Consumer Reports*. The correlation between engine size and gas mileage was $r = -.71$, the negative value indicating that cars with smaller engines tended to go farther on a gallon of gasoline. The sample value $r = -.71$ is a point estimate of the true correlation ρ between engine size and gas mileage in car models sold in North America.

To obtain an interval estimate of the true correlation ρ, we first transform $r = -.71$ to $z(-.71) = -.887$ using Expression (13.1) or Appendix 7. The 95% confidence limits for $z(\rho)$ are then obtained from Expressions (13.8) and (13.9) as

$$z(r) \pm \frac{z_{.025}}{\sqrt{n-3}} = -.887 \pm \frac{1.96}{\sqrt{10-3}} = -.887 \pm .741$$

or -1.628 and $-.146$. These are the lower and upper confidence limits on the $z(r)$-scale. Now we need only back-transform them to the r-scale using Expression (13.12) or Appendix 7 to obtain $r_{\text{lower}} = -.93$ and $r_{\text{upper}} = -.15$. Thus, we can be 95% confident that the true correlation ρ between engine size and gas mileage lies between $-.93$ and $-.15$.

We can also be 95% confident that between $(-.15)^2 = 2\%$ and $(-.93)^2 = 86\%$ of the variation in gas mileage can be accounted for by the linear relationship with engine size. A larger sample would give us a more precise estimate of ρ and hence narrower confidence intervals.

(Source: Student project by C. Kooyman.)

Examination of Expressions (13.3) and (13.9) reveals that the *precision* of the interval estimates produced by the preceding method depends largely on the sample size n: the larger the sample, the smaller the standard error and hence the narrower the confidence interval.

There are two questions that arise concerning the *accuracy* of the confidence limits produced by the method described in this section. The first is, To what extent does the stated sampling distribution of $z(r)$ depend on its assumption that the variables x and y are (bivariate) normally distributed? The method is in fact reasonably robust, at least against mild departures from normality.

The second question is, What is the error introduced by the fact that the stated sampling distribution of $z(r)$ is only approximate in terms of its mean, standard deviation, and Gaussian density function? The actual error introduced into values of z_{obs} by these factors depends on the true value of ρ as well as on the sample size, but the error is apt to be in the neighborhood of $1/(2n)$ or less—for example, only $\frac{1}{100} = .01$ or less for samples of size $n = 50$, but $\frac{1}{20} = .05$ or less for samples of size $n = 10$. Consequently, the confidence intervals described in this section should be regarded as approximate, especially for markedly non-Gaussian variables and for small samples (less than 30, perhaps). There is, for example, rarely any justification for reporting more than two decimal places of r in the confidence limits, since the method's exactness does not extend beyond the second decimal place unless the sample is very large. However, the accuracy is adequate for many practical applications.

Statistical Significance of the Sample Correlation Coefficient

In Chapter 5, we considered the coefficient of determination r^2 as a measure of the strength of linear relationship *in a sample.* It must be recognized that r^2 is a descriptive statistic, and the fact that a *sample* produces an r^2 value of, say, 90% does not imply that the variables have a strong linear relationship in the underlying *population.* Even if there is no linear relationship in the underlying population, a sample from that population may exhibit some degree of linear relationship just by chance. As we saw in the previous section, smaller samples give less precise estimates of the population correlation. In the extreme case, if there are just $n = 2$ observations in the sample, the coefficient of determination is always $r^2 = 100\%$, but this merely indicates that a straight line can always be found to fit any two points exactly and tells us nothing about the strength of the linear relationship between the variables in the underlying population.

EXAMPLE 13.5 Consider the following five samples consisting of $n = 2, 3, 4, 5,$ and 6 bivariate observations. The data in each sample were obtained by rolling a pair of dice, one red and one green, n times and recording the outcomes.

Sample 1		Sample 2		Sample 3		Sample 4		Sample 5	
Red	Green	Red	Green	Red	Green	Red	Green	Red	Green
1	6	4	2	1	2	3	2	3	2
2	1	1	4	3	6	5	6	1	4
		4	4	1	1	1	1	3	5
				6	6	4	6	3	4
						4	2	3	2
								3	4
$r = -1.00$		$r = -.50$		$r = .84$		$r = .77$		$r = -.20$	
$r^2 = 100\%$		$r^2 = 25\%$		$r^2 = 71\%$		$r^2 = 59\%$		$r^2 = 4\%$	

The population correlation between the outcomes of two fair dice is $\rho = 0$, and yet none of the correlations obtained in these small samples is 0. Some are positive and some are negative, but none is exactly 0. Even if we ignore the limiting case of $n = 2$ observations, up to 71% of the variation in the values observed on one die can be accounted for by the linear relationship with the values observed on the other die.

These results stem not from any bias or unfairness in the dice nor from any mysterious communication between the dice; they simply represent chance correlation. You may find it instructive to collect your own data on chance correlation by rolling a pair of dice to generate samples of 3 or 4 bivariate observations from which to calculate r. A sample correlation coefficient r will sometimes be larger and sometimes smaller than the population coefficient ρ that it is estimating, and the amount of the discrepancy depends to a considerable extent on the sample size n. We turn now to a method of determining how large the absolute value of r should be, in a sample of a given size, to convince us that it represents something other than mere chance correlation.

A basic question to ask about a sample correlation coefficient r is whether it is significantly different from the value of 0, which reflects the case of no linear relationship. A test of whether r differs from 0 by more than merely a chance amount is called a **test of the statistical significance of the correlation coefficient r**. The null hypothesis for the test is that there is no linear relationship between the variables in the population; in symbols,

$$H_0\colon\ \rho = 0 \qquad \textbf{(13.14)}$$

We may regard r itself as the *test statistic* to test the null hypothesis in Expression (13.14). When the null hypothesis in Expression (13.14) is true, the sampling distribution of r is symmetric with a mean of 0. Critical values of this null-hypothesis sampling distribution of r are given in Appendix 8 (based on the assumption that at least one of the variables is normally distributed). In the table in Appendix 8, r_A is the value of r such that A of the area of the sampling distribution of r is to the right of r_A; for example, $r_{.025}$ cuts off an upper-tail area of 2.5% of the sampling distribution of r when the population coefficient $\rho = 0$. Because of the symmetry of the distribution, $-r_{.025}$ cuts off a lower-tail area of 2.5%.

To enter the table in Appendix 8, we require the number of degrees of freedom associated with r. Recall from Section 10.3 that the number of degrees of freedom is the number of values whose variability is being measured minus the number of independent sample-based restrictions on those values. Because the value of r is based on the variability of the n sample points (as in a scattergram) around the best-fitting regression line $\hat{y} = bx + a$, where b and a are calculated from the same sample of n observations, the correlation coefficient has $\boldsymbol{n - 2}$ **degrees of freedom**, the 2 degrees of freedom being used for the determination of b and a.

To perform the **test of the significance of a correlation coefficient $\boldsymbol{r}$** based on a random sample of n observations, we simply compare the observed value of r with the critical values having $n - 2$ degrees of freedom in Appendix 8. If the observed value of r is at or beyond the critical value in the tail of the distribution, we reject the null hypothesis in Expression (13.14); otherwise, we conclude that r is not significantly different from 0. The null hypothesis is generally tested against a two-sided alternative, in which case the critical values are $\pm r_{\alpha/2}$ for a test at the α significance level, but one-tailed alternatives may be employed if a difference in the opposite direction would be of no more interest than a finding of no difference.

EXAMPLE 13.6 In Example 13.4, the correlation between engine size and gas mileage was reported to be $r = -.71$ in a sample of $n = 10$ car models sold in North America. Is r significantly different from 0?

Based on $n = 10$ bivariate observations, r has $n - 2 = 10 - 2 = 8$ degrees of freedom. In Appendix 8, the critical values with 8 degrees of freedom for a test at the .05 significance level are $\pm r_{.025}(8\text{ df}) = \pm.632$. Since the observed value of $r = -.71$ is in the tail area beyond the lower critical value, as shown in Figure 13.1, the sample provides enough evidence to conclude that engine size and gas mileage are negatively correlated in the relevant population of car models, $P < .05$. In other words, the sample value $r = -.71$ is significantly different from 0, testing at the 5% significance level.

The result of the significance test is consistent with the interval estimate obtained in Example 13.4: The 95% confidence interval, which was found to be from $-.93$ to $-.15$, does not include the value of 0, which was likewise rejected by the significance test.

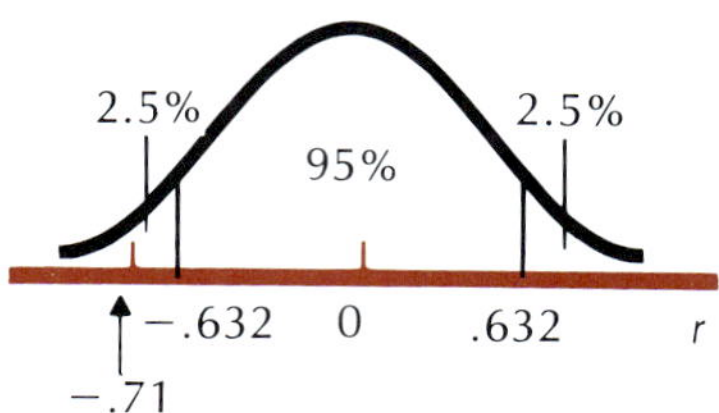

Figure 13.1

EXAMPLE 13.7 You may use Appendix 8 to verify that none of the five sample correlation coefficients for the dice outcomes in Example 13.5 is statistically significant at the .10 level.

It is worth noting, however, that if we were to continue generating repeated samples of any size n by rolling a pair of dice as in Example 13.5, approximately α of the resulting correlation coefficients r would be significant at the α level. These would be Type I errors. That is, after all, what α tells us: the Type I error rate when the null hypothesis is true. For example, when $\alpha = .05$, the probability is 5% that the correlation r in a random sample from a population with $\rho = 0$ will exceed the .05-level critical values. Thus, 5% of the sample values in the null-hypothesis sampling distribution can be expected to exceed the .05-level critical values just by chance.

The significance test in this section is reasonably robust against mild departures from the assumption that at least one of the variables is normally distributed. In practice, this test provides at least a rough indication of statistical significance even for non-Gaussian variables, in which case the stated significance level is taken to be only an approximation of the actual Type I error rate under the null hypothesis.

As noted in Chapter 5, the correlation coefficient r_{xy} is a standardized version of the slope coefficient b in the least-squares regression line $\hat{y} = bx + a$. Therefore, testing whether r is significantly different from 0 is exactly equivalent to testing whether the slope b in the regression equation is significantly different from 0. The next section introduces this equivalent test concerning slope, as well as a number of other inferences that are possible when we tackle the problem in terms of regression rather than correlation.

13.2 SIMPLE LINEAR REGRESSION ANALYSIS

Regression analysis refers to the study of the functional relationships among variables. In simple linear regression, one variable y is described as a straight-line function of another variable x by an equation of the form

$$\hat{y} = bx + a \tag{13.15}$$

where $\hat{y}$ is the fitted value of the dependent variable y as estimated on the basis of the independent variable x.

The difference $y_i - \hat{y}_i$ between an observed value y_i and the corresponding fitted value $\hat{y}_i$ obtained from the equation is an error of estimation, often called a **residual**. Because they represent the variation of the observations around the straight-line description, the residuals provide essential information in assessing the error in any inferences that may be drawn from the results. This section considers both significance testing and interval estimation, after a preliminary subsection to introduce some prerequisite terminology and notation.

Sums of Squares and Sums of Products

In Section 5.1, we met the term *sum of squares,* designated by SS_x for a variable x or SS_y for a variable y and used to refer to the sum of the squared deviations from the mean of a set of observations. The notation and formulas for such a sum of squares are included in Table 13.1. The table also shows that the *variance* of a sample of n observations (Section 4.3) can be expressed as the sum of squares divided by $n - 1$.

By analogy, the **sum of products** SP_{xy} for a sample of n observations on two variables x and y is the sum of the n products of corresponding deviations from the respective means:

$$SP_{xy} = \sum (x - \bar{x})(y - \bar{y}) \quad \textbf{(13.16)}$$

The **covariance** of x and y in a sample of n observations is obtained from the sum of products divided by $n - 1$:

$$\text{cov}\,(x, y) = \frac{SP_{xy}}{n - 1} \quad \textbf{(13.17)}$$

Although its numerical value is not directly interpretable in any simple way, covariance is a useful building block in measuring the way one variable varies with another.

Table 13.1 shows how the concepts of covariance and sum of products are analogous to the concepts of variance and sum of squares: In the special case where x and y are the same variable ($x = y$), the sum of products is the sum of squares ($SP_{xx} = SS_x$), and the covariance is then a measure of the way the variable "varies with itself," cov (x, x), which is usually called the sample variance, var (x). [The symbol var (x) is an alternative notation for s_x^2, in which the lower-case v on "var" distinguishes the sample value var (x) from the population value Var (x).]

The equations in Table 13.1 provide us with some simpler ways of writing the computational formulas for the regression coefficient b and the correlation

TABLE 13.1 Notation and formulas for sums of squares, sums of products, variance, and covariance.

Sums of squares and products			Variance and covariance	
Notation	Definitional formula	Computational formula	Notation	Formula
SS_x	$= \Sigma (x - \bar{x})^2$	$= \Sigma x^2 - \dfrac{(\Sigma x)^2}{n}$	var $(x) = s_x^2$	$= \dfrac{SS_x}{n - 1}$
SS_y	$= \Sigma (y - \bar{y})^2$	$= \Sigma y^2 - \dfrac{(\Sigma y)^2}{n}$	var $(y) = s_y^2$	$= \dfrac{SS_y}{n - 1}$
SP_{xy}	$= \Sigma (x - \bar{x})(y - \bar{y})$	$= \Sigma xy - \dfrac{\Sigma x \Sigma y}{n}$	cov (x, y)	$= \dfrac{SP_{xy}}{n - 1}$

coefficient r, as given in Expressions (5.9) and (5.19), respectively:

$$b = \frac{\sum xy - \dfrac{\sum x \sum y}{n}}{\sum x^2 - \dfrac{(\sum x)^2}{n}}$$

$$= \frac{\mathrm{SP}_{xy}}{\mathrm{SS}_x} = \frac{(n-1)\operatorname{cov}(x,y)}{(n-1)\operatorname{var}(x)} = \frac{\operatorname{cov}(x,y)}{\operatorname{var}(x)} = \frac{\operatorname{cov}(x,y)}{s_x^2} \tag{13.18}$$

$$r = \frac{\sum xy - \dfrac{\sum x \sum y}{n}}{\sqrt{\sum x^2 - \dfrac{(\sum x)^2}{n}}\sqrt{\sum y^2 - \dfrac{(\sum y)^2}{n}}}$$

$$= \frac{\mathrm{SP}_{xy}}{\sqrt{\mathrm{SS}_x}\sqrt{\mathrm{SS}_y}} = \frac{(n-1)\operatorname{cov}(x,y)}{\sqrt{(n-1)\operatorname{var}(x)}\sqrt{(n-1)\operatorname{var}(y)}}$$

$$= \frac{\operatorname{cov}(x,y)}{\sqrt{\operatorname{var}(x)}\sqrt{\operatorname{var}(y)}} = \frac{\operatorname{cov}(x,y)}{s_x s_y} \tag{13.19}$$

These expressions do not require any change in our method of computing values of b and r, but they should help you see the structure in what might otherwise seem to be rather intimidating computational formulas in Expressions (5.9) and (5.19).

Notice that the measurement units in the rightmost term of Expression (13.18) are x-units times y-units in the numerator and x-units-squared in the denominator, yielding y-units per x-unit, as required for the slope (rise/run). In Expression (13.19), the measurement units all cancel, since the correlation coefficient is the covariance expressed in standard deviation units. Attention to the appropriate units for b and r makes it easier to remember their formulas and their meanings.

EXAMPLE 13.8 Home gardeners soon learn that even for the same variety of seed, the number of days from the time that seeds are planted in the ground until the sprouts can first be seen breaking through the soil varies from year to year. What is the relationship between the germination time and the average daily (air) temperature after planting?

A gardener in central Alberta investigated this question for one particular variety of peas, using data from her gardening records over a five-year period. Eleven separate plantings of Lincoln-Homesteader peas were made during the five years, and these were the 11 observational units. The dependent variable y was the number of days from planting until sprouting, and the independent variable x was the mean maximum daily temperature (°C) over the seven consecutive days beginning with the day of planting. (Seed catalogs indicate a germination time of 6–8 days under ideal conditions.) The observations are recorded in the following table, along with some preliminary calculations for analyzing the relationship.

Observation number	Germination time (days) y	Mean maximum temperature (°C) x
1	11	14.8
2	12	17.4
3	10	15.6
4	8	20.9
5	11	20.1
6	12	17.1
7	9	19.9
8	8	21.4
9	11	13.5
10	8	21.4
11	14	13.4
	$\bar{y} =$ 10.363636	$\bar{x} =$ 17.772727
	$s_y =$ 1.963300	$s_x =$ 3.123489
	$\sum y =$ 114	$\sum x =$ 195.5
	$\sum y^2 =$ 1220	$\sum x^2 =$ 3572.13
	$SS_y =$ 38.545455	$SS_x =$ 97.561818
		$\sum xy =$ 1978.7
		$SP_{xy} =$ −47.390909

Extra digits are retained at this stage so that subsequent calculations will not suffer unduly from rounding errors. The germination times had a mean (±SD) of 10.4 (±2.0) days, and the mean maximum temperatures had a mean (±SD) of 17.8 (±3.1) °C.

Examination of the scattergram in Figure 13.2 suggests that a linear description of the relationship may be reasonable. To express germination time y as a linear function of temperature x, we use Expressions (13.18) and (5.10) to find

$$b = \frac{-47.390909}{97.561818} = -.48575262$$

and

$$a = \frac{114 - (-.48575262)(195.5)}{11} = 18.996785$$

from which we may write the regression equation as

$$\hat{y} = 19.0 - .49x$$

Expression (13.19) gives us the correlation coefficient as

$$r = \frac{-47.390909}{\sqrt{97.561818}\,\sqrt{38.545455}} = -.77280249$$

which would be reported as $r = -.77$. The linear equation, which accounts for $r^2 = 60\%$ of the variation in germination times, tells us that when the maximum temperatures after planting average 14°C, the mean germination time is about $19.0 - (.49)(14) = 12$ days, and that each increase of 1°C in the temperature variable is accompanied by an average decrease of about half a day in the germination time. The equation is plotted in the scattergram in Figure 13.2.

(Source: Student project by G. Hamilton.)

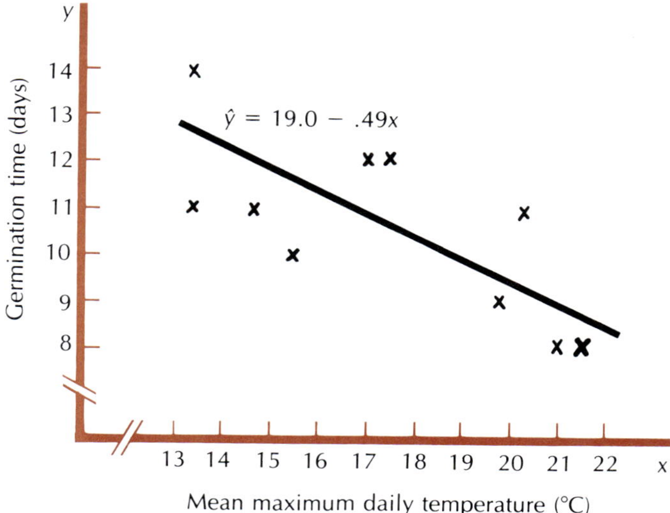

Figure 13.2 Scattergram of germination time as a function of mean maximum daily temperature for 11 plantings of peas. The least-squares linear function is also shown.

Statistical Significance of the Slope Coefficient *b*

Having calculated a regression equation $\hat{y} = bx + a$, we often want to know whether the relationship that has been observed between the variables x and y in the sample at hand is statistically significant or whether it should be attributed to chance. Is the slope coefficient b significantly different from zero? The value of b obtained in a particular sample may be considered to be an estimate of the corresponding slope coefficient β in the underlying population (where β is the lower-case Greek letter beta). Thus, the null hypothesis to be tested is that the true slope β of the straight line relating x and y in the population is zero; that is,

$$H_0\colon\ \beta = 0 \tag{13.20}$$

The null hypothesis says that x and y are not linearly related. The test of this null hypothesis is developed in this section.

As explained in Section 5.1, simple linear regression analysis involves the partition of the total variation in the dependent variable y,

$$\text{SS}_y = \sum (y_i - \bar{y})^2 \tag{13.21}$$

into two orthogonal (additive) parts. One part,

$$\text{SS}_{\text{regr.}} = \sum (\hat{y}_i - \bar{y})^2 \tag{13.22}$$

TABLE 13.2 ANOVA summary table for simple linear regression analysis.

Source of variation	SS	df	MS = SS/df	F
Regression on x (b)	$SS_{regr.}$	1	$MS_{regr.}$	$\frac{MS_{regr.}}{MS_{res.}}$
Residual	$SS_{res.}$	$n - 2$	$MS_{res.}$	—
Total	SS_y	$n - 1$		

is the variation that may be attributed to the linear relationship with the independent variable x. The other part,

$$SS_{res.} = \sum (y_i - \hat{y}_i)^2 \tag{13.23}$$

is the **residual variation** due to the errors of estimation, the portions of y that cannot be described as a straight-line function of x. It is sometimes designated by SS_{error}.

The partition of the total variation in y by simple linear regression analysis is summarized in Table 13.2. Such a table is called an **analysis of variance summary table**, often abbreviated to **ANOVA summary table**. In successive columns, the table shows the identifiable sources of variation in y along with their sums of squares (SS), degrees of freedom (df), mean squares (MS), and F ratios where applicable. The following paragraphs discuss these elements of the ANOVA summary table in turn.

Sum of squares. The sum-of-squares formulas in Expressions (13.21), (13.22), and (13.23) are definitional formulas that generally are not convenient for purposes of calculation. Algebraic manipulation leads to computational formulas that may be expressed in a variety of ways, as shown in Table 13.3. The formulas given in terms of x and y are the most direct and least likely to lead to rounding errors. If the formulas expressed in terms of b or r are used, it is important to remember that any premature rounding of b or r will limit the number of correct digits in the solution. It is safest to carry extra digits in all values entering the formulas and to round for reporting only.

TABLE 13.3 Sum-of-squares formulas for simple linear regression analysis.

SS	Definitional formula	Computational formulas		
		in terms of x, y	in terms of b	in terms of r
$SS_{regr.}$	$= \Sigma (\hat{y}_i - \bar{y})^2$	$= \frac{SP_{xy}^2}{SS_x}$	$= bSP_{xy}$	$= r^2 SS_y$
$SS_{res.}$	$= \Sigma (y_i - \hat{y}_i)^2$	$= SS_y - \frac{SP_{xy}^2}{SS_x}$	$= SS_y - bSP_{xy}$	$= (1 - r^2)SS_y$
SS_{total}	$= \Sigma (y_i - \bar{y})^2$	$= SS_y$	$= SS_y$	$= SS_y$

Degrees of freedom. Each source of variation has a certain number of degrees of freedom associated with it. There is one degree of freedom for the regression coefficient b, or for the difference between the regression line and the horizontal line through the mean of the y-values. There is also one degree of freedom for the y-intercept a, which is the same as the mean of the y-values if the slope is zero. If there are a total of n observations, there remain $n - 2$ degrees of freedom for residual variation of the n points around the regression line specified by the two sample-based values b and a.

Mean squares. The term **mean square** (MS) is shortened from "mean of the squared deviations from the mean." It is calculated by dividing a sum of squares (SS) by its number of degrees of freedom (df):

$$\text{MS} = \frac{\text{SS}}{\text{df}} \tag{13.24}$$

A mean square is a variance estimate. One example of a mean square is the sample variance introduced in Section 4.3.

The ANOVA summary table in Table 13.2 provides two estimates of the variance associated with the dependent variable y, each based on a different aspect of the data. First, the regression mean square,

$$\text{MS}_{\text{regr.}} = \frac{\text{SS}_{\text{regr.}}}{1} \tag{13.25}$$

reflects the difference of the slope b from a slope of zero. Second, the residual mean square,

$$\text{MS}_{\text{res.}} = \frac{\text{SS}_{\text{res.}}}{n - 2} \tag{13.26}$$

measures the variability of the individual points around the regression line.

F ratios. Recalling from Section 12.4 that the ratio of two independent estimates of the same Gaussian-population variance follows the F distribution, we can indicate the conditions under which the F ratio in Table 13.2 is the *test statistic* for the ***F* test of the significance of the regression coefficient *b***: Under the assumption that the errors of estimation are independent normal deviates with constant variance throughout the range of the independent variable, the F ratio

$$F_{\text{regr.}} = \frac{\text{MS}_{\text{regr.}}}{\text{MS}_{\text{res.}}} \tag{13.27}$$

may be evaluated against the critical value F_α $(1, n - 2 \text{ df})$ to test whether the linear relationship observed in the sample is statistically significant at the α level. This is a test of the null hypothesis in Expression (13.20), a test of whether the calculated regression coefficient b is significantly different from zero. Since the correlation coefficient r is simply a standardized form of the slope b, this is simultaneously a test of the null hypothesis in Expression (13.14) concerning the population correlation coefficient. Indeed, the F test in

Expression (13.27) is exactly equivalent to the two-tailed test of the significance of the correlation coefficient r (using Appendix 8) as presented in Section 13.1, pages 434–37.

The ANOVA summary table may also be used for descriptive purposes, without any assumptions about the underlying population. In particular, the **coefficient of determination**,

$$r^2 = \frac{SS_{regr.}}{SS_{total}} \tag{13.28}$$

is the proportion of the total variation in the dependent variable that can be accounted for by the linear relationship with the independent variable. The interpretation of r^2 was discussed in Sections 5.1 and 5.2.

EXAMPLE 13.9 Example 13.8 described an investigation of the germination time y of pea seeds as a function of air temperature x during the week after planting. The linear regression description of germination time,

$$\hat{y} = 19.0 - .49x$$

implies that the total variation in germination time (SS_y) has been partitioned into one portion due to the linear relationship with air temperature ($SS_{regr.}$) and a residual portion due to discrepancies from this relationship ($SS_{res.}$). Applying the computational formulas in terms of x and y from Table 13.3 to the data in Example 13.8, we calculate the sums of squares as

$$SS_{regr.} = \frac{(-47.390909)^2}{97.561818} = 23.0202584$$

$$SS_{res.} = 15.5251961$$

$$SS_y = 38.5454545$$

You should verify that the computational formulas in terms of b and in terms of r in Table 13.3 yield the same values.

For illustrative purposes, let us recompute these sums of squares from the definitional formulas in Table 13.3. We calculate the fitted values $\hat{y} = bx + a$ using the unrounded values of b and a as given in Example 13.8 and obtain results, recorded to four decimal places, as shown in the following table.

Observed y	Fitted $\hat{y}$	$\hat{y} - \bar{y}$	Residuals $y - \hat{y}$	$y - \bar{y}$
11	11.8076	1.4440	−.8076	.6364
12	10.5447	.1811	1.4553	1.6364
10	11.4190	1.0554	−1.4190	−.3636
8	8.8446	−1.5191	−.8446	−2.3636
11	9.2332	−1.1305	1.7668	.6364
12	10.6904	.3268	1.3096	1.6364
9	9.3303	−1.0333	−.3303	−1.3636
8	8.6017	−1.7620	−.6017	−2.3636
11	12.4391	2.0755	−1.4391	.6364
8	8.6017	−1.7620	−.6017	−2.3636
14	12.4877	2.1241	1.5123	3.6364
$\Sigma y = 114$	$\Sigma \hat{y} = 114.0000$	$\Sigma (\hat{y} - \bar{y}) = 0.0000$	$\Sigma (y - \hat{y}) = 0.0000$	$\Sigma (y - \bar{y}) = 0.0000$
$\bar{y} = 10.3636$	$\bar{\hat{y}} = 10.3636$	$\Sigma (\hat{y} - \bar{y})^2 = 23.0203$	$\Sigma (y - \hat{y})^2 = 15.5252$	$\Sigma (y - \bar{y})^2 = 38.5455$

Notice that the fitted values have the same mean as the observed values and that the last three columns have sums equal to zero and sums of squares equal to the previously calculated values of $SS_{regr.}$, $SS_{res.}$, and SS_y, respectively.

The ANOVA summary table for the linear regression analysis of germination times is as follows.

Source of variation	SS	df	MS	F	
Regression on temperature (b)	23.0202584	1	23.02	13.3	$P < .01$
Residual	15.5251961	9	1.73	—	
Total	38.5454545	10			

For convenience of reference in subsequent calculations, extra digits have been retained in the sums of squares. (Guidelines for how to round and how many digits to report are given in Appendix 1.)

From the ANOVA table, we can obtain the coefficient of determination via Expression (13.28):

$$r^2 = \frac{23.0202584}{38.5454545} = .59722369$$

or approximately 60%. This is the proportion of the total variation in germination time that can be accounted for by the linear relationship with temperature. You should verify that this value equals the square of the correlation coefficient calculated in Example 13.8.

Comparing the ANOVA-table F ratio of 13.3 with the critical value $F_{01}(1, 9 \text{ df}) = 10.6$, we find that the regression is significant at the 1% level, as shown in Figure 13.3(a). Thus, the slope of the regression line, $b = -.49$, is significantly different from zero, and hence there is a significant linear relationship between germination time and temperature. Equivalently, we can say that the F test means that the correlation coefficient, $r = -.77$, is significantly different from zero. We could have performed the identical significance test by referring r with 9 degrees of freedom to Appendix 8, where we find that an absolute r-value of .735 is required for significance at the 1% level, as shown in Figure 13.3(b).

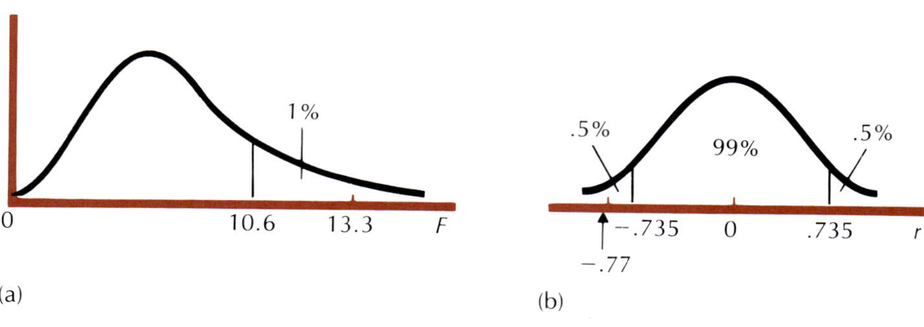

Figure 13.3

```
                                   SAS

                          ANALYSIS OF VARIANCE

                             SUM OF         MEAN
        SOURCE        DF    SQUARES        SQUARE       F VALUE       PROB>F

        MODEL          1  23.02025840   23.02025840      13.345       0.0053
        ERROR          9  15.52519615    1.72502179
        C TOTAL       10  38.54545455

            ROOT MSE       1.313401     R-SQUARE         0.5972
            DEP MEAN      10.36364      ADJ R-SQ         0.5525
            C.V.          12.67317

                          PARAMETER ESTIMATES

                       PARAMETER       STANDARD    T FOR HO:                  VARIABLE
VARIABLE    DF          ESTIMATE          ERROR  PARAMETER=0    PROB > !T!       LABEL

INTERCEP     1       18.99678526     2.39620867        7.928        0.0001  INTERCEPT
X            1       -0.48575262     0.13297113       -3.653        0.0053  TEMPERATURE


                                              PREDICT
                   OBS           ACTUAL         VALUE      RESIDUAL

                      1         11.0000       11.8076       -0.8076
                      2         12.0000       10.5447        1.4553
                      3         10.0000       11.4190       -1.4190
                      4          8.0000        8.8446       -0.8446
                      5         11.0000        9.2332        1.7668
                      6         12.0000       10.6904        1.3096
                      7          9.0000        9.3303       -0.3303
                      8          8.0000        8.6017       -0.6017
                      9         11.0000       12.4391       -1.4391
                     10          8.0000        8.6017       -0.6017
                     11         14.0000       12.4877        1.5123

                 SUM OF RESIDUALS                       5.55112E-14
                 SUM OF SQUARED RESIDUALS                   15.5252
```

Figure 13.4
SAS/STAT™ printout for Example 13.9.

Many computer programs are available to perform linear regression analysis. Output from one of the more popular statistical packages, known as the SAS/STAT™ software, is shown in Figure 13.4 for this example. (SAS/STAT is a trademark of SAS Institute Inc., Cary, NC, U.S.A.) Although we have not discussed all of the items in the SAS/STAT printout, you should be able to find the ANOVA summary table, the *P*-value for the *F* test, the coefficient of determination, the values of b and a, the observed values, the fitted values, and the residuals.

Summary of the test procedure. The **F test of the significance of the regression coefficient b** may be summarized as follows:

Assumptions.

a. Random sampling of y-values at any given x-value.

b. Independent normal distribution of y at every value of x.

c. Constant variance of y at every value of x.

d. Linear relationship between the true means of the y distributions and the corresponding x-values.

Null hypothesis. H_0: $\beta = 0$

Alternative hypothesis. H_A: $\beta \neq 0$

Step 1. Calculate the linear regression equation $\hat{y} = bx + a$.

Step 2. Calculate the values required to complete the ANOVA summary table in the format of Table 13.2.

Step 3. Look up the critical value $F_\alpha(1, n - 2\,\text{df})$.

Step 4. If the observed F ratio in the ANOVA table is greater than or equal to the critical value, conclude that the observed slope b is significantly different from zero ($P < \alpha$). If the F ratio is less than the critical value, conclude that the slope b is not significantly different from zero ($P > \alpha$).

Standard Errors and Confidence Intervals

This section deals with the precision of the information provided by the least-squares regression line. **Precision** refers to the extent to which a statistic is free of random error, and it can be measured in terms of the standard error of the statistic. The smaller the standard error of a statistic, the more precisely the statistic estimates the corresponding parameter, and hence the less random error there is in the estimate. We will meet standard errors corresponding to several different statistics in this section, each standard error providing information about the precision of the relevant statistic.

The information that a standard error provides about the precision of an estimate can often be summarized concisely in the form of a confidence interval with limits given by

$$\text{point estimate} \pm \text{margin of error}$$

where the margin of error is a certain number of standard errors. Under the assumption that the errors of estimation in regression analysis are independent Gaussian deviates with constant variance throughout the range of the independent variable, the various confidence intervals to be presented in this section all have limits given by

$$\text{point estimate} \pm (t_{\alpha/2} \text{ times the relevant standard error}) \quad \textbf{(13.29)}$$

where $t_{\alpha/2}$ is obtained from the t distribution with $n - 2$ degrees of freedom and, incidentally, is equal to the positive square root of the critical value $F_\alpha(1, n - 2 \text{ df})$ for the F test in Table 13.2.

The following five paragraphs present five different types of standard errors relevant to simple linear regression and, where appropriate, the confidence intervals that can be obtained from them. Guidelines for how to round and how many digits to report are given in Appendix 1.

1. Standard error of estimation: s_e. The positive square root of the residual mean square in Table 13.2 is called the **standard error of estimation**:

$$s_e = \sqrt{\text{MS}_{\text{res.}}} \quad \textbf{(13.30)}$$

It is an average or "standard" amount by which the errors of estimation $y_i - \hat{y}_i$ differ from their mean, which is zero. It may be interpreted as a typical amount by which the fitted values $\hat{y}$ differ from the observed values y for the values of x included in the sample. Although the standard error of estimation itself is interpreted most directly in terms of the particular x-values included in the sample, it is of considerable importance in regression analysis: First, it provides a convenient indication, in the measurement units of the dependent variable, of how close the fitted values tend to be to the observed values. Second, it forms the basis for certain other standard errors required for making inferences about the underlying population.

EXAMPLE 13.10 The residual mean square in the linear regression analysis of germination time as a function of temperature was reported in the ANOVA table in Example 13.9 as 1.73. Calculating from the unrounded data, we find that the standard error of estimation is

$$s_e = \sqrt{15.5251961/9} = 1.313401$$

or approximately 1.3 days. This tells us that the observed germination times y differ from the regression equation's predicted times $\hat{y}$ by about 1.3 days on the average, for the temperatures included in the sample. In other words, the standard error of estimation gives a rough idea as to the size of the residuals, or errors of estimation, $y - \hat{y}$: They are in the neighborhood of ± 1.3 days. You should examine the column of residuals shown in the table in Example 13.9 to see that this is the case.

The standard error of estimation answers the question, How close to the observed germination times in this sample are the estimates provided by the regression line? To answer, we could list all 11 errors of estimation (as tabulated in Example 13.9) or, more concisely, we can say that the *standard* error of estimation is about 1.3 days.

You should be able to find the standard error of estimation in the SAS/STAT printout in Figure 13.4.

2. Standard error of the slope: s_b. The **standard error of the estimated slope** ***b*** in the regression equation is computed as

$$s_b = \frac{s_e}{\sqrt{SS_x}} \tag{13.31}$$

This standard error measures the variability of the sample regression coefficient b, just as the standard error of the mean, $s_{\bar{x}}$, measures the variability of the sample mean $\bar{x}$. We can use the standard error of b to test whether b is significantly different from zero by evaluating the test statistic

$$t_{\text{obs}} = \frac{b}{s_b} \tag{13.32}$$

against the critical value $t_{\alpha/2}(n - 2\text{ df})$. This test is exactly equivalent to the F test in Table 13.2 since $t^2_{\text{obs}} = F_{\text{regr.}}$. Another use of the standard error of the slope b is in obtaining the **1 − α confidence interval for the population regression coefficient β:**

$$b - t_{\alpha/2}s_b < \beta < b + t_{\alpha/2}s_b \quad \text{with confidence level } 1 - \alpha \tag{13.33}$$

where $t_{\alpha/2}$ is obtained from the t distribution with $n - 2$ degrees of freedom.

EXAMPLE 13.11 The linear regression equation describing germination time as a function of temperature was given in Example 13.8 as $\hat{y} = 19.0 - .49x$, with a standard error of estimation given in Example 13.10 as $s_e = 1.3$. Using unrounded values from the earlier examples, we can calculate the standard error of the slope ($b = -.49$) as

$$s_b = \frac{1.313401}{\sqrt{97.561818}} = .132971$$

Given the value $t_{.025}(9\text{ df}) = 2.262$ from Appendix 4, Expression (13.33) gives the 95% confidence limits for the true slope (in the population) as

$$-.4858 \pm (2.262)(.1330) = -.4858 \pm .3008$$

or approximately $-.79$ and $-.18$. We can be 95% confident that the true reduction in germination time that accompanies a 1°C increase in mean maximum daily temperature is between .18 and .79 day.

The standard error of the slope may also be used to test whether the estimated slope ($b = -.4858$) is significantly different from zero, according to Expression (13.32):

$$t_{\text{obs}} = \frac{-.4858}{.1330} = -3.65$$

The critical values for a test at the 1% level are $\pm t_{.005}(9\text{ df}) = \pm 3.250$, and we conclude that the estimated slope is significantly different from zero at the 1% level. When there is only one independent variable (temperature, in this case), this test is equivalent to the F test of the regression, as reported in Example 13.9:

$$t^2_{\text{obs}} = (-3.65)^2 = 13.3 = F_{\text{regr.}}$$

You should be able to find the standard error of the slope and the t test statistic in the SAS/STAT printout in Figure 13.4.

3. Standard error of the y-intercept: s_a. The **standard error of the estimated y-intercept a** in the regression equation is calculated as

$$s_a = s_e \sqrt{\frac{\sum x^2}{n\text{SS}_x}} = s_e \sqrt{\frac{1}{n} + \frac{\bar{x}^2}{\text{SS}_x}} \tag{13.34}$$

This standard error measures the variability of the estimated y-intercept a, which is the estimated mean value of y when $x = 0$. The **$1 - \alpha$ confidence interval for the population y-intercept A** is

$$a - t_{\alpha/2}s_a < A < a + t_{\alpha/2}s_a \qquad \text{with confidence level } 1 - \alpha \tag{13.35}$$

where $t_{\alpha/2}$ is obtained from the t distribution with $n - 2$ degrees of freedom.

EXAMPLE 13.12 The linear regression equation describing germination time as a function of temperature was given in Example 13.8 as $\hat{y} = 19.0 - .49x$, with a standard error of estimation given in Example 13.10 as $s_e = 1.3$. Using unrounded values from the earlier examples, we find that the standard error of the y-intercept ($a = 19.0$) is

$$s_a = 1.313401 \sqrt{\frac{3572.13}{(11)(97.561818)}} = 2.3962$$

or approximately 2.4. We could use Expression (13.35) to obtain a confidence interval for the population y-intercept, that is, for the true mean germination time when the mean maximum daily temperature is 0°C. In this case, such an interval would not be meaningful. The investigation was limited to values of the temperature variable ranging from 13.4°C to 21.4°C, and it would be unwarranted and unwise to extrapolate the linear relationship observed in that range to a temperature of 0°C. If the daily maximum temperature were reduced, the seeds would at some point not germinate at all. There is no statistical basis for extrapolating beyond the observed range of temperatures.

You should be able to find the standard error of the y-intercept in the SAS/STAT printout in Figure 13.4.

4. Standard error of the mean $\hat{y}_0$ at x_0: $s_{\hat{y}_0}$. According to the regression equation, the value $\hat{y}_0 = bx_0 + a$ is an estimate of the true (population) mean of the dependent variable at a particular x-value x_0. The **standard error of the mean y-value at a given x-value** is given by

$$s_{\hat{y}_0} = s_e \sqrt{\frac{1}{n} + \frac{(x_0 - \bar{x})^2}{\text{SS}_x}} \tag{13.36}$$

This standard error measures the variability of $\hat{y}_0$ as an estimate of the population mean μ_{y_0} at x_0. Examination of the second term under the square-root sign in Expression (13.36) reveals that estimation of mean y-values is best (that is, most precise, having smallest standard error) when x_0 is near the mean of the x-values. Indeed, for estimation of the mean y-value right at $x_0 = \bar{x}$, the second term vanishes ($\bar{x} - \bar{x} = 0$), and the standard error of $\hat{y}_0$ in

Expression (13.36) reduces to the standard error of the sample y-mean, $s_e/\sqrt{n}$. This is the minimum possible value of $s_{\hat{y}_0}$. As we estimate mean y-values at values of x_0 progressively farther from $\bar{x}$, the standard error $s_{\hat{y}_0}$ becomes progressively larger—our estimates become less precise as we move away from the "center" of our observations. [Notice that Expression (13.34) is the special case of Expression (13.36) that occurs when $x_0 = 0$.] The **$1 - \alpha$ confidence interval for the population mean at x_0** is

$$\hat{y}_0 - t_{\alpha/2}s_{\hat{y}_0} < \mu_{y_0} < \hat{y}_0 + t_{\alpha/2}s_{\hat{y}_0} \qquad \text{with confidence level } 1 - \alpha \tag{13.37}$$

where $t_{\alpha/2}$ is obtained from the t distribution with $n - 2$ degrees of freedom.

EXAMPLE 13.13 The linear regression equation describing germination time as a function of temperature was given in Example 13.8 as $\hat{y} = 19.0 - .49x$, with a standard error of estimation given in Example 13.10 as $s_e = 1.3$. Using unrounded values from the earlier examples, we find that when the mean maximum daily temperature is, say, $x_0 = 20°\text{C}$, the standard error of the mean germination time,

$$\hat{y}_0 = 19.00 - (.4858)(20) = 9.28 \text{ days}$$

is

$$s_{\hat{y}_0} = 1.313401\sqrt{\frac{1}{11} + \frac{(20 - 17.772727)^2}{97.561818}} = .4945$$

or approximately half a day. The 95% confidence limits for the true mean germination time when the mean maximum daily temperature is 20°C are

$$9.28 \pm (2.262)(.4945) = 9.28 \pm 1.12$$

or approximately 8.2 and 10.4 days.

5. Standard error of the forecast mean at x_0: s_f. In addition to its use in estimating *population* mean y-values, the regression equation $\hat{y} = bx + a$ may be used for predicting mean y-values of *future samples* from the population. The use of the regression equation for genuine prediction, as distinguished from estimation or description of the population, is called **forecasting**. The forecast sample mean y-value at a particular x-value, say, x_0, is $\hat{y}_0 = bx_0 + a$, which is the same as our estimate of the population mean at x_0. However, the standard error for forecasting is larger than the standard error for estimating, because future sample means will themselves vary around the population mean. The standard error of a forecast mean takes into account not only the variability in the present sample, on which the forecast is based, but also the variability in the future sample whose mean is being predicted. The **standard error s_f of a forecast mean at an x-value of x_0** is

$$s_f = s_e\sqrt{\frac{1}{m} + \frac{1}{n} + \frac{(x_0 - \bar{x})^2}{\text{SS}_x}} \tag{13.38}$$

where m is the size of the future sample whose mean is being forecast and n is the size of the present sample on which the forecast is based. The standard error of a forecast sample mean in Expression (13.38) is the same as the standard error of an estimated population mean in Expression (13.36), except for the inclusion of the extra term $1/m$ based on the future sample size m. The contribution of this term to the size of the standard error is greatest when $m = 1$, that is, when we are forecasting the value of a single future observation. We cannot predict an individual observation as precisely as we can a mean of several observations. As the future sample size m increases, the contribution of this term gradually diminishes and finally vanishes for infinite m, that is, when the future "sample" is the entire population, in which case Expression (13.38) is indeed identical to Expression (13.36). The **$1 - \alpha$ confidence interval for the future sample mean $\bar{y}_{f_0}$ at x_0** is

$$\hat{y}_0 - t_{\alpha/2}s_f < \bar{y}_{f_0} < \hat{y}_0 + t_{\alpha/2}s_f \qquad \text{with confidence level } 1 - \alpha \qquad (13.39)$$

where $t_{\alpha/2}$ is obtained from the t distribution with $n - 2$ degrees of freedom, and $\bar{y}_{f_0}$ may refer either to an individual observation ($m = 1$) or to the mean of $m \geq 2$ observations.

EXAMPLE 13.14 Let us suppose that next spring, the gardener in Example 13.8 finds that the mean maximum daily temperature on the day of planting and the next six days is 20°C. What forecast can we make about the date of sprouting?

The forecast germination time is $\hat{y}_0 = 19.00 - (.4858)(20) = 9.28$, or approximately 9.3 days after planting. The standard error of the forecast mean of $m = 1$ observation at a temperature of $x_0 = 20$°C is

$$s_f = 1.313401\sqrt{\frac{1}{1} + \frac{1}{11} + \frac{(20 - 17.772727)^2}{97.561818}} = 1.4034$$

or approximately 1.4 days. The 95% confidence limits for the forecast are

$$9.28 \pm (2.262)(1.4034) = 9.28 \pm 3.17$$

or approximately 6.1 and 12.5 days. We can be 95% confident that the seeds will sprout between 6.1 and 12.5 days after planting, under the stated conditions. Notice that the confidence interval for a forecast sample value—in this case, a future sample of one observation only—is considerably wider than that for the population mean at the same temperature (see Example 13.13).

STUDY MATERIAL FOR CHAPTER 13

Terms and Concepts

Section 13.1

point estimate of ρ

* Fisher's $z(r)$ transformation
* bivariate normal distribution
* $1 - \alpha$ confidence interval for ρ

test of the statistical significance of the correlation coefficient r
degrees of freedom of the correlation coefficient r

Section 13.2

residual
sum of products
sample covariance
residual variation
ANOVA summary table for simple linear regression analysis
mean square
F test of the significance of the regression coefficient b
coefficient of determination
precision
standard error of estimation
standard error of the slope
confidence interval for the population regression coefficient β
standard error of the y-intercept
confidence interval for the population y-intercept
standard error of the mean y-value at a given x-value
confidence interval for the population mean y-value at a given x-value
forecasting
standard error of the forecast mean at a given x-value
confidence interval for the future sample mean at a given x-value

Key Formulas

Section 13.1

* **(13.3)** $$\sigma_{z(r)} = \frac{1}{\sqrt{n-3}} \quad \text{for } n > 3$$

* **(13.4)** $$z_{\text{obs}} = \frac{z(r) - z(\rho)}{\sigma_{z(r)}}$$

* **(13.9)** $$e_{\alpha/2} = z_{\alpha/2}\sigma_{z(r)}$$

* **(13.13)** $$r_{\text{lower}} < \rho < r_{\text{upper}} \quad \text{with confidence level } 1 - \alpha$$

Section 13.2

(13.15) $$\hat{y} = bx + a$$

(13.18) $$b = \frac{\text{SP}_{xy}}{\text{SS}_x} = \frac{(n-1)\,\text{cov}\,(x, y)}{(n-1)\,\text{var}\,(x)} = \frac{\text{cov}\,(x, y)}{\text{var}\,(x)} = \frac{\text{cov}\,(x, y)}{s_x^2}$$

(13.19) $$r = \frac{\text{SP}_{xy}}{\sqrt{\text{SS}_x}\sqrt{\text{SS}_y}} = \frac{(n-1)\,\text{cov}\,(x, y)}{\sqrt{(n-1)s_x^2}\sqrt{(n-1)s_y^2}} = \frac{\text{cov}\,(x, y)}{s_x s_y}$$

(13.24) $$\text{MS} = \frac{\text{SS}}{\text{df}}$$

(13.28) $r^2 = \dfrac{SS_{regr.}}{SS_{total}}$

(13.30) $s_e = \sqrt{MS_{res.}}$

(13.31) $s_b = \dfrac{s_e}{\sqrt{SS_x}}$

(13.32) $t_{obs} = \dfrac{b}{s_b}$

(13.33) $b - t_{\alpha/2}s_b < \beta < b + t_{\alpha/2}s_b$ with confidence level $1 - \alpha$

(13.34) $s_a = s_e\sqrt{\dfrac{1}{n} + \dfrac{\bar{x}^2}{SS_x}}$

(13.35) $a - t_{\alpha/2}s_a < A < a + t_{\alpha/2}s_a$ with confidence level $1 - \alpha$

(13.36) $s_{\hat{y}_0} = s_e\sqrt{\dfrac{1}{n} + \dfrac{(x_0 - \bar{x})^2}{SS_x}}$

(13.37) $\hat{y}_0 - t_{\alpha/2}s_{\hat{y}_0} < \mu_{y_0} < \hat{y}_0 + t_{\alpha/2}s_{\hat{y}_0}$ with confidence level $1 - \alpha$

(13.38) $s_f = s_e\sqrt{\dfrac{1}{m} + \dfrac{1}{n} + \dfrac{(x_0 - \bar{x})^2}{SS_x}}$

(13.39) $\hat{y}_0 - t_{\alpha/2}s_f < \bar{y}_{f_0} < \hat{y}_0 + t_{\alpha/2}s_f$ with confidence level $1 - \alpha$

Also see the formulas in Tables 13.1, 13.2, and 13.3.

Exercises for Numerical Practice

Section 13.1

* **13/1.** Find the value of $z(r)$ for each of the following values of r:

a. .36 **b.** −.87 **c.** .995 **d.** −.908 **e.** −.59 **f.** .80

* **13/2.** Find the value of $z(r)$ for each of the following values of r:

a. .965 **b.** .72 **c.** −.88 **d.** −.02 **e.** .70 **f.** −.924

* **13/3.** A random sample of 52 observations from a bivariate normal distribution yielded a value of $r = -.30$. Calculate the confidence limits for the population correlation coefficient

a. at the 95% confidence level.
b. at the 99% confidence level.

* **13/4.** A random sample of 28 observations from a bivariate normal distribution yielded a value of $r = .45$. Calculate the confidence limits for the population correlation coefficient

a. at the 95% confidence level.
b. at the 99% confidence level.

13/5. Given the data in Exercise 13/3:

a. How many degrees of freedom does the observed r-value have?
b. Is the correlation statistically significant at the 5% level?
c. Is the correlation statistically significant at the 1% level?

13/6. Given the data in Exercise 13/4:

a. How many degrees of freedom does the observed r-value have?

b. Is the correlation statistically significant at the 5% level?

c. Is the correlation statistically significant at the 1% level?

Section 13.2

13/7. The following random sample of six bivariate observations was recorded in order to describe a normally distributed variable y as a function of an independent variable x:

x:	0	1	3	6	7	9
y:	17	21	16	13	10	12

Calculate:

a. SS_x, SS_y, and SP_{xy}; s_x, s_y, and cov (x, y).

b. the regression equation describing y as a linear function of x.

c. the correlation coefficient r.

d. the ANOVA summary table, including the P-value inequality for the F test. (Use the computational formulas for the sums of squares.)

e. the coefficient of determination, using the sums of squares in the ANOVA table. Verify that it equals the square of your answer to Part (c).

f. the standard error of estimation.

g. the standard error of the slope and the 95% confidence limits for the true slope.

h. the standard error of the y-intercept and the 95% confidence limits for the population y-intercept.

i. the standard error of the mean value of y at $x = 4$ and the corresponding 95% confidence limits for the true mean.

j. the standard error of the forecast sample mean of 10 future observations at $x = 7$ and the corresponding 95% confidence limits for the forecast.

k. the standard error of the forecast value of one future observation at $x = 7$ and the corresponding 95% confidence limits for the forecast.

13/8. The following random sample of eight bivariate observations was recorded in order to describe a normally distributed variable y as a function of an independent variable x:

x:	0	2	5	8	11	15	16	19
y:	33	41	35	56	48	55	72	70

Perform the calculations itemized in Parts (a)–(k) of Exercise 13/7.

13/9. A random sample of 30 bivariate observations was recorded in order to describe a normally distributed variable y as a linear function of an independent variable x. The sum of squares due to the linear regression was found to be 571, and the residual sum of squares was 284.

a. Complete the ANOVA summary table, including the P-value inequality for the F test.

b. Calculate the coefficient of determination.

c. Calculate the correlation coefficient.

d. Calculate the standard error of estimation.

13/10. A random sample of 25 bivariate observations was recorded in order to describe a normally distributed variable y as a linear function of an independent variable x. The sum of squares due to the linear regression was found to be 3807, and the residual sum of squares was 3121. Perform the calculations itemized in Parts (a)–(d) of Exercise 13/9.

13/11. Given that $n = 27$, $s_y = 42.31$, and $r = -.2077$ in a simple linear regression analysis of y as a function of x:

a. Complete the ANOVA summary table, including the P-value inequality for the F test.

b. Calculate the standard error of estimation.

13/12. Given that $n = 18$, $s_y = 6.173$, and $r = -.3856$ in a simple linear regression analysis of y as a function of x:

a. Complete the ANOVA summary table, including the P-value inequality for the F test.

b. Calculate the standard error of estimation.

Problems for Applying the Concepts

13/13. Bivariate observations based on verbal report were obtained from 13 students haphazardly selected on a college campus. The correlation between grade point average and number of evenings spent in social drinking (at the bar) in the previous month was found to be $r = -.67$.

a. How many degrees of freedom does the observed r-value have?

b. Is the observed correlation greater than would be expected due to chance alone? Write the relevant null and alternative hypotheses in symbols and in words, and state the results of the significance test.

* **c.** Construct a 98% confidence interval for the true value of the correlation between these variables.

(Source: Student project by A. Knudson.)

13/14. A sample of 11 female college students agreed to take part in a study of eating habits. Each student was observed eating lunch in a cafeteria and was subsequently asked to record her weight. The correlation between time taken to eat lunch and reported weight was $r = -.40$.

a. How many degrees of freedom does the observed r-value have?

b. Is the observed correlation statistically significant? Write the relevant null and alternative hypotheses in symbols and in words, and state the results of the significance test.

* **c.** Construct a 95% confidence interval for the true value of the correlation between these variables.

(Source: Student project by S. Fiege.)

13/15. The results of a Highland dance competition were used to assign points to each of the 14 dancers in the highest class of 15-year-olds, with the number of points reflecting the degree of success in the competition. The correlation between this measure of placement in the competition and reported number of hours of practice per week was $r = .85$.

a. How many degrees of freedom does the observed r-value have?

b. Test the statistical significance of r, and state the results.

* **c.** Construct a 95% confidence interval for the true value of the correlation between these variables, and interpret it in terms of percentage of variation accounted for.

(Source: Student project by S. Lauber.)

13/16. Fifteen members of the Edmonton Oilers hockey team played at least 50 games in the 1981–82 season. The correlation between player's weight and total number of penalty minutes for the season was $r = .39$ for these 15 players.

a. How many degrees of freedom does the observed r-value have?

b. Test the statistical significance of r, and state the results.

* **c.** Construct a 95% confidence interval for the true value of the correlation between these variables, and interpret it in terms of percentage of variation accounted for.

(Source: Student project by B. Dixon.)

13/17. Records of one year's performance by seven different hockey players were randomly selected from the last ten years' files of a junior "A" hockey team (ages 15–20 years). The purpose of the study was to see whether scores on a preseason test of physical fitness, measured on a scale from 0 to 10, could be used to predict players' year-end point totals (goals plus assists). The following data were recorded.

Player	Physical fitness test	Year-end point total
1	3.0	26
2	7.5	72
3	9.0	120
4	7.0	65
5	8.5	83
6	4.0	26
7	6.5	30

Draw a scattergram of the data, and perform a linear regression analysis to obtain the following information:

a. the regression equation. Interpret the values of b and a.

b. the correlation coefficient.

c. the ANOVA summary table. Is the regression significant?

d. the coefficient of determination. What does it mean?

e. the standard error of estimation. What does it tell you?

f. the standard error of the slope and a 95% confidence interval for the true slope.

g. the standard error of the mean year-end point total for players with a fitness score of 4.5. Give the corresponding 95% confidence interval for the true mean.

h. the standard error of the forecast year-end point total for a player with a preseason fitness score of 8.0. Give the corresponding 95% confidence interval for the forecast.

(Source: Student project by R. Bishop.)

13/18. In an investigation of the relationship between blood–alcohol level and number of alcoholic drinks consumed prior to the testing, data were obtained from 10 men, aged 20–25 years, who had been charged with impaired driving, where impairment was legally defined by a blood–alcohol test reading over 80 mg of alcohol per 100 ml of blood. Each individual gave a retrospective verbal report of the number of alcoholic drinks he had consumed leading up to the impaired-driving charge, along with his measured blood–alcohol level (in mg of alcohol per 100 ml of blood) when charged.

Observation	Number of drinks	Blood–alcohol level
1	17	160
2	20	140
3	7	120
4	14	160
5	6	110
6	30	215
7	20	210
8	28	175
9	4	100
10	12	150

Draw a scattergram of the data, and perform a linear regression analysis of blood–alcohol level as a function of number of drinks, to obtain the following information:

a. the regression equation. Interpret the values of b and a.

b. the correlation coefficient.

c. the ANOVA summary table. Is the regression significant?

d. the coefficient of determination. What does it mean?

e. the standard error of estimation. What does it tell you?

f. the standard error of the slope and a 95% confidence interval for the true slope.

g. the standard error of the mean blood–alcohol level of young men charged with impaired driving after having 10 alcoholic drinks. Give the corresponding 95% confidence interval for the true mean.

h. the standard error of the forecast mean blood–alcohol level in a future sample of four young men charged with impaired driving after having had 10 drinks. Give the corresponding 95% confidence interval for the forecast.

(Source: Student project by J. McLean.)

13/19. Ten families were randomly selected from a certain metropolitan area. For each family, the annual family income and the annual expenditure on consumer durables (for example, appliances, furniture, and cars) were recorded as follows.

Family	Expenditure on durables ($)	Family income ($1000s)
1	6600	24
2	2100	9
3	4500	15
4	7000	35
5	6000	21
6	7200	24
7	3600	15
8	5400	30
9	4800	21
10	2600	12

Draw a scattergram of the data, and perform a linear regression analysis of annual expenditures on durables as a function of annual family income, to obtain the following information:

a. the regression equation. Interpret the values of b and a.

b. the correlation coefficient.

c. the ANOVA summary table. Is the regression significant?

d. the coefficient of determination. What does it mean?

e. the standard error of estimation. What does it tell you?

f. the standard error of the slope and the 99% confidence limits for the population slope.

g. the standard error of the mean expenditure on durables by families with an annual income of $25,000. Give the corresponding 99% confidence interval for the true mean.

h. the standard error of the forecast mean expenditure on durables by a future sample of five families with annual incomes of $25,000. Give the corresponding 99% confidence limits for the forecast.

13/20. To what extent can the asking price of houses in the real estate market be predicted from their size? Values of price and total floor area were recorded as follows for a sample of 13 houses advertised in a local newspaper.

House	Floor area (ft^2)	Price ($1000s)
1	1960	90
2	2000	119
3	2246	152
4	3000	220
5	1600	76
6	1200	76
7	1550	100
8	1900	113
9	2568	180
10	2400	150
11	2400	120
12	1100	67
13	2800	180

Draw a scattergram of the data, and perform a linear regression analysis of price as a function of floor area, to obtain the following information:

a. the regression equation. Interpret the values of b and a.

b. the correlation coefficient.

c. the ANOVA summary table. Is the regression significant?

d. the coefficient of determination. What does it mean?

e. the standard error of estimation. What does it tell you?

f. the standard error of the slope and the 99% confidence limits for the population slope.

g. the standard error of the mean asking price of houses with a floor area of 1500 square feet in the relevant population. Give the corresponding 99% confidence limits for the true mean.

h. the standard error of the forecast asking price of a house with a floor area of 1500 square feet. Give the corresponding 99% confidence limits for the forecast.

(Source: *The Democrat and Chronicle,* Rochester, NY, 15 December 1985.)

Reference

Walsh, J. E., and Allen, D. "Testing the Farmer's Almanac." *Weatherwise* (a publication of the Helen Dwight Reid Educational Foundation) 34 (1981): 212–15.

Further Reading

Johnson, P. O., and Jackson, R. W. B. *Modern Statistical Methods: Descriptive and Inductive.* Chicago: Rand McNally, 1959. Pages 353–56 discuss the problem of testing the significance of the difference between correlation coefficients calculated from two samples of bivariate observations, where the two samples are not independent of each other.

Kleinbaum, D. G., and Kupper, L. L. *Applied Regression Analysis and Other Multivariable Methods.* North Scituate, MA: Duxbury, 1978. This text offers a thorough survey of regression methods without assuming that the reader has been exposed to calculus or matrix algebra.

SAS Institute Inc. *SAS® User's Guide: Statistics, Version 5 Edition.* Cary, NC: SAS Institute Inc., 1985. This guide describes the use of the SAS/STAT computer programs, such as the one whose output was shown in Figure 13.4.

Snedecor, G. W., and Cochran, W. G. *Statistical Methods.* 7th ed. Ames, IA: Iowa State University Press, 1980. Chapters 9 and 10 provide concise but thorough coverage, extending the topics introduced in this chapter, including, for example, independent-samples tests of the difference between correlation coefficients.

CHAPTER 14

INFERENCES ABOUT PROPORTIONS

EXAMPLE 14.1 The results of public opinion polls are reported regularly in daily newspapers and on radio and television newscasts. Based on a sample from the relevant population, the polls provide information about the percentages of people favoring various political parties or candidates, supporting various causes, and holding various opinions about current social, political, and economic issues. In addition to their intrinsic interest, results from current polls may influence planning and decision making by government and business, while polls from previous years provide an historical record of public opinion.

For example, shortly before the 1976 summer Olympic Games were held in Montreal, the Gallup poll report shown in Figure 14.1 appeared in a Montreal newspaper. The report draws inferences about approving and disapproving proportions for an entire country (of 23,000,000 at that time) on the basis of a sample of 1017 people. Using methods to be introduced in the present chapter, it includes several point estimates, such as the value 68% in the headline, along with their margin of error (4 percentage points) and the confidence level (19 in 20, or 95%).

In principle, the methods in this chapter are very similar to the other inferential methods that have been presented in Chapters 10 through 13. Confidence intervals for population proportions take the same form as those for means and for regression parameters, and significance tests about proportions follow the general procedure for hypothesis testing with which we have become familiar in recent chapters.

There is, however, a fundamentally important difference about the inferences in this chapter. In the discussions of inferences about means, standard deviations, regression, and correlation, it was always assumed that the value of the observed random variable was a numerical measurement of some characteristic of the observational unit. It may have been an actual measurement, such as the height or pulse rate of each observational unit, or it may have been a less direct numerical index of some characteristic of the observational unit, such as the price of a product or the number of posters displayed in a student's room. In any case, the value of the random variable for each observational unit was always a number that could be located on the real-number line. In other words, the distinguishing feature of the kind of data that we have been dealing with is that each observational unit yields a

Gallup Poll

By Canadian Institute of Public Opinion

Olympic Games approved by 68%

NATIONALLY today, 68% of Canadians approve of having the 1976 summer Olympic Games in our country. Only 28% disapprove, while 10% qualify their answers or are undecided.

In spite of the fact that the estimated cost of holding the Games in Canada has more than doubled, from about $500 million dollars in 1973 to over one billion dollars today, approval level has risen from 54% three years ago to the current 68%.

Most of this increase has occurred outside of the province of Quebec, the site of the Games. In Quebec, 57% approved in 1973 compared to 61% now. In the rest of the country, the level of approval has risen from 53% three years ago to 71% today. It should be pointed out, however, that the question asked in the earlier study mentioned the expected cost. In the current question no mention was made of cost, and it would seem logical that Quebecers are much more aware of and concerned about the expected outlay than are other Canadians.

The current poll is based on personal, in-home interviews with 1,017 Canadian adults, 18 years and over, during the first week of January. The earlier study was based on 725 interviews. Both would be accurate within four percentage points, 19 in 20 times.

The questions were:

1976: **"As you probably know, the 1976 Olympic Games will be held in Montreal. On the whole do you approve or disapprove of holding these games in Canada?"**

1973: **"It has been estimated that the proposed summer Olympic Games, planned for Montreal in 1976, will cost about five hundred million dollars for the two weeks. Do you approve of holding these Olympic Games in Canada, or do you think they should be cancelled?"**

		Approve	**Disapprove**	**Qualified, Undecided**
1976 —	National	68%	22%	10%
	Quebec	61	31	8
	Rest of Canada	71	19	10
		Approve	**Cancelled**	**Undecided**
1973 —	National	54%	34%	12%
	Quebec	57	31	12
	Rest of Canada	53	35	12

Figure 14.1 Gallup poll reported in *The Montreal Star*, 2 March 1976. Used with permission of Thomson Newspaper Publishers.

numerical value (or perhaps two, in the case of paired data). A list of the raw data would show a numerical value (or perhaps two) for each observational unit.

In the data discussed in this chapter, observational units are not so much *measured* as *categorized* and *counted.* The only number that might be assigned to each observational unit is either 1 (meaning that it is in the category whose membership is being counted) or 0 (meaning that it is not). The raw data probably would show not a number *for* each observational unit, but rather the number *of* observational units in each of two or more categories. If a value were shown for each observational unit, it would be 1 or 0, or "yes" or "no," or "present" or "absent," or "approve" or "disapprove," or some other abbreviation for the label of a category.

The initial summaries of **numerically measured variables** are generally in terms of means, standard deviations, and/or correlations, but the initial summary of **categorized variables**, or **categorical data**, is apt to be in terms of proportions—the proportions of observational units in the various categories.

EXAMPLE 14.2 In Example 14.1, the observed variable is attitude toward holding the Olympics in Montreal, and the values of the variable are "approve," "disapprove," and "undecided." A list of the 1017 observations would consist of 1017 words: approximately 691 approve's, 224 disapprove's, and 102 undecided's.

The observational units (persons interviewed) do not *each* yield a numerical value, but rather each is counted as being in one of the three categories. Contrast this type of categorical data with a numerically measured variable, such as age, where a list of 1017 observations would consist of 1017 numerical values. We could calculate the mean and standard deviation from such a list of ages, but because the values of the attitude variable are not numerical, we cannot calculate statistics such as the mean and standard deviation to summarize them. Rather, the data are summarized by the proportions of individuals observed in the three categories: Nationally, 68% approved, 22% disapproved, and 10% were undecided.

14.1 POINT ESTIMATES OF p

There is no difficulty specifying an unbiased point estimate of a population proportion. The random-sample proportion $\hat{p}$ of observational units having a certain attribute is an unbiased **point estimate of the population proportion p** of observational units having that attribute. (The hat on $\hat{p}$, which is read "p-hat," is used to indicate that the value is an estimate.) If there are n observational units (or trials) in the sample and n_1 of them are of the type that is of interest, then the sample proportion is

$$\hat{p} = \frac{n_1}{n} \tag{14.1}$$

What is sometimes confusing is the variety of ways in which the sample proportion $\hat{p}$ may be described:

a. as a fraction, n_1/n, perhaps expressed in decimal form; or

b. as a percentage, $100(n_1/n)\%$; or

c. as frequencies, n_1 successes and n_2 ($= n - n_1$) failures (where the terms *success* and *failure* are defined as in Section 7.2); or

d. as a relative frequency, n_1 observational units out of a total of n; or

e. as a mean of a variable x whose only possible values are 0 and 1, so that if there are n_1 1s and n_2 0s, the sum of the xs is n_1 and the mean is $\bar{x} = \sum x/n = n_1/n = \hat{p}$.

These are all common alternative ways of reporting the same sample proportion. (Guidelines for how to round and how many digits to report are given in Appendix 1.) Notice in particular that sample proportions are often described by frequencies [alternatives (c) and (d)], and hence may not look like proportions at all.

The population proportion *p*, which is being estimated, is sometimes referred to as the **true proportion**, or the *true percentage*, or the *probability* of the given type of observational unit (probability of a success).

EXAMPLE 14.3 What is the probability that a woman who enters a drugstore will make a purchase there? This question asks for the value of a parameter—a population proportion. We could refer to the required parameter as the true proportion of female shoppers who make purchases, or as the true percentage of female shoppers who make purchases, or as the probability that a female shopper in a drugstore will make a purchase.

The required parameter may be estimated on the basis of sample data. For example, in a study of shopping habits, shoppers were observed as they left a drugstore during a one-hour period from 4:00 to 5:00 P.M. The numbers of male and female shoppers were recorded, along with the numbers of males and females who actually made purchases while in the drugstore. Any one of the following describes the resulting point estimate of the probability that a female shopper in the drugstore will make a purchase:

a. The proportion of purchasers among the female shoppers in the sample was .74.

b. 74% of the women in the sample made purchases.

c. The sample included 37 women who made purchases and 13 women who did not.

d. Of the 50 women in the sample, 37 made purchases.

e. If shoppers who made a purchase are given a score of 1 and those who did not make a purchase are given a score of 0, the mean number of female purchasers the drugstore served for each female shopper who entered the store was .74.

It is not so much the specific wording of these statements that concerns us here; the statements could be worded differently. Rather, notice the different forms in which the numerical information is given: .74, 74%, 37 and 13, 37 out of 50—these all describe the same point estimate. They all lead to the same conclusion about the parameter, namely, that the true proportion of female shoppers who make purchases is estimated to be .74, or that the probability that a female shopper in a drugstore will make a purchase is estimated to be 74%.

It is, of course, a separate question as to whether the estimate has any relevance to other drugstores and other times of day than those in which the data were collected.

(Source: Student project by L. Jorgensen.)

14.2 INTERVAL ESTIMATES OF p

The Sampling Distribution of the Sample Proportion

In order to attach probability statements to inferences about proportions, we need to know the sampling distribution of a statistic that relates the estimate $\hat{p}$ to the parameter p. If the estimate is based on a random sample of n observations from a population where the true proportion of successes is p, then, according to the definition of a random sample, the observations are independent, each having a probability p of being a success. Under these conditions, the sampling distribution of the number n_1 of successes is the binomial distribution. (Recall from Section 7.2 that the binomial distribution gives the probability of exactly n_1 successes and n_2 failures in $n = n_1 + n_2$ independent trials in each of which the probability of success is p and the probability of failure is $q = 1 - p$.)

As discussed on page 271, if both np and nq are greater than about 5, and if the appropriate continuity correction is used, the binomial distribution may be approximated by a Gaussian distribution with mean

$$\mu_{\text{bin}} = np \tag{14.2}$$

and standard deviation

$$\sigma_{\text{bin}} = \sqrt{npq} \tag{14.3}$$

In other words, the sampling distribution of the number n_1 of successes is approximately Gaussian. Dividing each element in the sampling distribution by the constant n rescales the distribution (changes the units on the horizontal axis) but leaves its Gaussian shape unchanged. Therefore, if the sampling distribution of the number n_1 of successes is approximately Gaussian with the mean and the standard deviation given in Expressions (14.2) and (14.3), it follows that the **sampling distribution of the sample proportion $\hat{p} = n_1/n$** is approximately Gaussian with mean

$$\mu_{\hat{p}} = \frac{np}{n} = p \tag{14.4}$$

and standard deviation

$$\sigma_{\hat{p}} = \frac{\sqrt{npq}}{n} = \sqrt{\frac{pq}{n}} \tag{14.5}$$

The standard deviation in Expression (14.5) is called the **standard error of a proportion**. Its formula in Expression (14.5) actually requires that the population size N be infinite if it is to hold exactly, but in practice the formula may be used as long as the population size N is large relative to the sample size n. (The formula can be made to accommodate finite populations simply by multiplying by the *finite-population correction factor* $\sqrt{1 - n/N}$. This adjustment reduces the calculated result slightly, but it makes little difference unless the sampling fraction n/N is about 10% or higher. We will not consider such cases here.)

If the sample proportion $\hat{p}$ follows the normal distribution with a mean of p [Expression (14.4)] and a standard deviation of $\sigma_{\hat{p}}$, then the statistic

$$z_{\text{obs}} = \frac{\hat{p} - p}{\sigma_{\hat{p}}} \qquad \textbf{(14.6)}$$

must follow the normal distribution with a mean of 0 and a standard deviation of 1, that is, the standard normal distribution. Expression (14.6) will enable us to attach probability statements to inferences about proportions, because z_{obs} relates the estimate $\hat{p}$ to the parameter p according to a known sampling distribution.

The result in the preceding paragraph is a **large-sample approximation to the sampling distribution of $\hat{p}$** that is acceptable, at least for rough work, provided that both np and nq are greater than about 10. If a continuity correction were made by subtracting $1/(2n)$ from the absolute value of the numerator, the approximation would be improved for smaller values of np and nq, even down to about 5. For simplicity, we will not use this continuity correction in this book. The important point is that the confidence intervals and significance tests presented in this chapter are based on large-sample approximations.

Confidence Intervals for *p*

To say that z_{obs} as defined in Expression (14.6) follows the standard normal distribution implies that the value of z_{obs} falls between the two-tailed α-level critical values $\pm z_{\alpha/2}$ with probability $1 - \alpha$; that is,

$$-z_{\alpha/2} < z_{\text{obs}} < z_{\alpha/2} \quad \text{with probability } 1 - \alpha \qquad \textbf{(14.7)}$$

Substituting Expression (14.6) into (14.7) and rearranging terms as we did in deriving confidence intervals for means in Chapter 10, we arrive at the **approximate $1 - \alpha$ confidence interval for the population proportion p**:

$$\hat{p} - z_{\alpha/2}\sigma_{\hat{p}} < p < \hat{p} + z_{\alpha/2}\sigma_{\hat{p}} \quad \text{with confidence level } 1 - \alpha \qquad \textbf{(14.8)}$$

The interval is only approximate because it is based on the Gaussian approximation to the binomial distribution, through Expression (14.6). The approximation improves as np and nq become larger. Therefore, the **approximate $1 - \alpha$ confidence limits for the true proportion p** are

$$\hat{p} \pm z_{\alpha/2}\sigma_{\hat{p}} \qquad \textbf{(14.9)}$$

Strictly, calculation of the standard error $\sigma_{\hat{p}}$ in Expression (14.5) requires knowledge of the population value p (and hence of $q = 1 - p$), but in practice the use of the sample values $\hat{p}$ and $\hat{q}$ provides a satisfactory approximation to the standard error:

$$\sigma_{\hat{p}} \approx \sqrt{\frac{\hat{p}\hat{q}}{n}} \qquad \textbf{(14.10)}$$

whenever $n\hat{p}$ and $n\hat{q}$ are large enough to apply the Gaussian approximation.

EXAMPLE 14.4 Example 14.3 described an investigation in which 37 out of 50 female shoppers in a drugstore made a purchase there, while the other 13 left without buying anything. The proportion of purchasers in the sample, $\hat{p} = 37/50 = .74$, is a point estimate of the probability that a female shopper in a drugstore will make a purchase.

The standard error of the sample proportion may be approximated by Expression (14.10) as

$$\sigma_{\hat{p}} \approx \sqrt{\hat{p}\hat{q}/n} = \sqrt{(.74)(.26)/50} = .0620$$

and the critical values of the standard normal distribution for a 95% confidence interval are $\pm z_{\alpha/2} = \pm 1.96$. Substituting these values into Expression (14.9), we obtain the approximate 95% confidence limits for the true proportion as

$$.74 \pm (1.96)(.0620) = .74 \pm .12$$

or .62 and .86. We can be 95% confident that the probability p is between about .62 and .86 that a female shopper in a drugstore will make a purchase.

The Margin of Error of the Estimate

Confidence limits for means, for regression parameters, and now for proportions take the general form

$$\text{(point estimate)} \pm \text{(margin of error)} \tag{14.11}$$

For an interval estimate of a proportion, the approximate $1 - \alpha$ confidence limits are

$$\hat{p} \pm e_{\alpha/2} \tag{14.12}$$

where

$$e_{\alpha/2} = z_{\alpha/2}\sigma_{\hat{p}} \tag{14.13}$$

is the **approximate margin of error of the sample proportion $\hat{p}$ at the $1 - \alpha$ confidence level**. The margin of error is only approximate because it is based on the Gaussian approximation to the binomial distribution, through Expression (14.6). The approximation improves as np and nq become larger. The margin of error may be reported directly in some applications, especially when a single margin of error applies to several point estimates.

When the point estimate is expressed as a percentage (and in other applications of percentages), care must be taken to **distinguish between percentages and percentage points. Percentage** refers to a proportion of an initial value; it implies multiplication of some initial value by the stated percentage. **Percentage points**, on the other hand, may be thought of as the units on the percent scale; they imply addition to or subtraction from an initial percentage value. Thus, the margin of error that is added to and subtracted from the point estimate is stated as a certain number of percentage points, not as a percentage (%).

EXAMPLE 14.5 To illustrate the distinction between percentages and percentage points, consider a situation where the unemployment rate rises from 8% to 10% of the labor force. That is an increase of 2 percentage points (*not* 2%) in the unemployment rate. An increase of 2 percentage points from an initial rate of 8% is an increase of (10% − 8%)/(8%) = 25% in the unemployment rate.

EXAMPLE 14.6 A good report of an opinion poll includes a statement of its margin of error. For example, the report of the Gallup poll in Example 14.1 includes the statement that the reported percentages "would be accurate within four percentage points, 19 in 20 times." The statement indicates that the margin of error at the 95% confidence level is 4 percentage points. Since the margin of error is added to and subtracted from the point estimate, the appropriate units are percentage points, not percent: 4 percentage points, not 4%. In symbols, we could write the margin of error as $e_{.025} = .04$. For example, the approximate 95% confidence limits for the point estimate (68%) in the headline of the Gallup report are $.68 \pm .04$, or 64% and 72%.

Substituting Expression (14.5) into (14.13), we may write the approximate margin of error of a sample proportion at the $1 - \alpha$ confidence level as

$$e_{\alpha/2} = z_{\alpha/2}\sqrt{\frac{pq}{n}} = z_{\alpha/2}\sqrt{\frac{p(1-p)}{n}} \tag{14.14}$$

Expression (14.14) shows that the margin of error depends not only on how confident we want to be $(1 - \alpha)$, but also on the true proportion p and on the sample size n.

However, there is an interesting characteristic of binomial variability that enables us to calculate the **maximum margin of error** for a given sample size and confidence level, regardless of the value of p or $\hat{p}$. For a given confidence level $1 - \alpha$ (which determines the value of $z_{\alpha/2}$) and a given sample size n, the margin of error in Expression (14.14) depends solely on the product pq, and since $pq = p(1 - p)$, the margin of error depends solely on the value of p. Notice in Table 14.1 what happens to the value of $pq = p(1 - p)$ as p increases: No matter what the value of p, the product pq can never exceed .25, which is the value of pq when $p = .5$.

If the product pq reaches its maximum possible value when $p = .5$, then the margin of error in Expression (14.14) must also be greater for a population proportion of $p = .5$ than it is for any other value of p, other things being equal. There is more variability in point estimates of true proportions in the neighborhood of .5 than in, for example, the neighborhood of $p = .1$ or $p = .9$.

TABLE 14.1

p	$q = 1 - p$	$pq = p(1 - p)$
.0	1.0	.00
.1	.9	.09
.2	.8	.16
.3	.7	.21
.4	.6	.24
.5	.5	.25
.6	.4	.24
.7	.3	.21
.8	.2	.16
.9	.1	.09
1.0	.0	.00

The implication of this result is that if we want to state a margin of error with a confidence level of *at least* $1 - \alpha$ no matter what the true value of p, we can simply "assume the worst" (that is, the greatest possible variability in the point estimate) by setting $p = .5$ so that $pq = .25$ in calculating the margin of error from Expression (14.14). In other words, the **maximum margin of error of a sample proportion at the $1 - \alpha$ confidence level** is

$$e_{\alpha/2} = z_{\alpha/2}\sqrt{\frac{(.5)(.5)}{n}} \qquad \textbf{(14.15)}$$

for a sample of n observations. This expression gives a margin of error of which we can be *at least* $1 - \alpha$ confident for *any* value of the true proportion p, and more confident for most values of p.

EXAMPLE 14.7 The 1973 Gallup poll reported in Example 14.1 was based on 725 observations. According to Expression (14.15), the resulting maximum margin of error of a sample proportion at the 95% confidence level is

$$e_{.025} = 1.96\sqrt{(.5)(.5)/725} = .0364$$

which is "within four percentage points" as claimed in the report.

Let us compare the maximum margin of error in this example with the margins of error that could be computed separately for each of the three national percentages for 1973: 54% approve, 34% cancel, and 12% undecided. Using Expressions (14.10) and (14.13), we obtain for these three sample proportions the following margins of error:

$$e_{.025} = 1.96\sqrt{(.54)(.46)/725} = .0363$$
$$e_{.025} = 1.96\sqrt{(.34)(.66)/725} = .0345$$
$$e_{.025} = 1.96\sqrt{(.12)(.88)/725} = .0237$$

respectively, at the 95% confidence level. Thus, the margins of error computed separately would be 3.6, 3.4, and 2.4 percentage points; these are not very different from the general claim of 3.6 percentage points that could be made on the basis of the maximum margin of error. Rounded to two decimal places, the 95% confidence limits may be given as $.54 \pm .04$, $.34 \pm .03$, and $.12 \pm .02$.

The 1976 Gallup poll reported in the same example was based on 1017 observations. The maximum margin of error of a sample proportion at the 95% confidence level in this case is

$$e_{.025} = 1.96\sqrt{(.5)(.5)/1017} = .0307$$

which could be reported as 3 percentage points. The report claims only that the error is "within four percentage points," a conservative claim that allows a single margin of error to be used for both the 1976 sample and the smaller 1973 sample. Again, the margins of error that could be computed separately for the three national percentages (68%, 22%, and 10%) for 1976 are only slightly smaller than the maximum margin of error of 3.1 percentage points; they are 2.9, 2.5, and 1.8 percentage points, respectively.

It should be noted that the margins of error that we have been discussing apply only to the national percentages in the poll in Example 14.1, because we have based them on the national sample sizes of 725 and 1017. The breakdown into regions (Quebec and the rest of Canada) involves smaller subsamples, and the number of persons interviewed in each region is not given. However, even if the total sample sizes of 725 and 1017 had been divided evenly between the two regions, the maximum margins of error at the 95% confidence level for the regional percentages would be 5.1 and 4.3 percentage points for 1973 and 1976, respectively. Also, if the samples were not divided equally between Quebec and the rest of Canada, the margin of error for some of the subsample proportions would be even greater. In any case, the claim that the results from both polls "would be accurate within four percentage points, 19 in 20 times" applies only to the national, not the regional, results. The margin of error for a regional proportion depends on the number of observations in that region. The regional sample sizes and margins of error should be reported to enable the reader to assess the precision of the regional estimates.

Size of Sample Required for a Given Margin of Error

The terms in Expression (14.14) can be manipulated algebraically to allow us to calculate the **size of random sample required to estimate a population proportion with a given margin of error and a confidence level of 1 − α**; specifically,

$$n = pq \frac{z^2_{\alpha/2}}{e^2_{\alpha/2}} \tag{14.16}$$

Since our goal in observing the proposed sample is to estimate the population proportion p, we do not know the value of p for use in Expression (14.16) where we are trying to decide how large a sample to observe. However, the fact that binomial variability can never exceed a certain amount, namely, that which occurs when $p = .5$, enables us to determine the maximum sample size that may be required. If we have prior information from which we know that the true proportion p cannot possibly be as high as .5 or as low as .5, we can substitute into Expression (14.16) the value of p closest to .5 of those values it might possibly take.

Thus, the rule for the use of Expression (14.16) is as follows:

> Consider what range of values the quantities p and q $(= 1 - p)$ might possibly take in the problem at hand, and assign to p and q the values from that range that are closest to .5. **(14.17)**

This rule implies setting $p = q = .5$ for the case where there is no prior information limiting the possible values of p, in which case Expression (14.16) becomes

$$n = .25 \frac{z^2_{\alpha/2}}{e^2_{\alpha/2}} \tag{14.18}$$

If the value resulting from the use of Expression (14.16) or (14.18) is not an integer, the result should be increased to the next larger integer as the prescribed sample size.

Notice that the size N of the population does not enter into the formula for determining the sample size n required to estimate a proportion with a given margin of error. As noted in connection with Expression (14.5), the effect of population size can be ignored as long as the population is at least 10 times larger than the sample.

EXAMPLE 14.8 How large a random sample would a polling organization require to estimate the true proportions of voters favoring the various national political parties, if it wants a maximum margin of error of 3 percentage points with 95% confidence? For $1 - \alpha = .95$ and hence $\alpha/2 = .025$, the size of random sample required is obtained from Expression (14.18) as

$$n = (.25)\frac{1.96^2}{.03^2} = 1067.11$$

or 1068 observations.

How large a random sample would be required for a maximum margin of error of 2 percentage points at the 95% confidence level? Expression (14.18) indicates a required sample size of $n = (.25)(1.96^2)/(.02^2) = 2401$ observations.

These specifications of required sample size apply whether one is surveying a city of 100,000 voters or a country of 100,000,000 voters. The size of the random sample, not the size of the population, is the main factor determining the precision of the estimates. Major polling organizations often consider that samples of 1500 are large enough to give adequate precision for practical purposes in national polls. A sample of 1500 observations yields a margin of error of less than 3 percentage points at the 95% confidence level. Of course, much larger total samples will be required if that degree of precision is desired for each particular city or region in the country.

EXAMPLE 14.9 How large a random sample would be required to estimate, with a maximum margin of error of 2 percentage points at the 99% confidence level, the true proportion of left-handed university students, if it is known that the true proportion p is between 3% and 10%? The possible values of p range from .03 to .10, and therefore the possible values of q range from .90 to .97. The maximum margin of error is associated with the values of p and q closest to .5. The values closest to .5 from the possible ranges in this case are $p = .10$ and $q = .90$. Using Expression (14.16) as interpreted by (14.17), with $1 - \alpha = .99$ and $\alpha/2 = .005$, we obtain a required sample size of

$$n = (.10)(.90)(2.576^2)/(.02^2) = 1493.05$$

or 1494 observations.

14.3 GENERAL PRINCIPLES OF CHI-SQUARE TESTS ABOUT PROPORTIONS

There are several ways that we might approach the problem of testing hypotheses about proportions. We could, for example, follow the approach that we used in introducing significance tests about means, where we obtained a t test statistic by substituting the null-hypothesis value into the t-statistic formula that we had previously used to set up confidence intervals. In the case of tests about proportions, that approach would lead us to a z test statistic through substitution of a null-hypothesis value into the z-statistic formula [in Expression (14.6)] on which the confidence intervals of Section 14.2 were based.

As noted in Section 12.2, the square of a standard normal deviate z is a chi-square variate with one degree of freedom; that is,

$$z^2 \sim \chi^2(1 \text{ df}) \tag{14.19}$$

Thus, a z test statistic has an equivalent chi-square test statistic with one degree of freedom.

In this chapter, we approach the problem of testing hypotheses about proportions in terms of chi-square statistics. The main reason for adopting this approach is that the z tests for proportions are limited to one-sample and two-sample cases, whereas the chi-square tests handle a larger class of problems. The chi-square tests with one degree of freedom are exactly equivalent to the z tests: Both are based on the Gaussian approximation to the binomial distribution and hence require sufficiently large samples (see Section 14.2). However, the same chi-square tests with more than one degree of freedom can also handle hypotheses about three or more proportions, while the z tests cannot. On the other hand, the z tests can test against one-sided as well as against two-sided alternative hypotheses in cases with one degree of freedom. However, if we require a test against a one-sided alternative, we can take the square root of the chi-square statistic as a standard normal deviate [see Expression (14.19)].

Section 14.1 indicated that a sample proportion may be described by the frequencies of the different categories of observational units observed in the sample. The number of observational units classified into the ith category is called the **observed frequency** of that category and is designated by the symbol o_i for values of i from 1 up to k, the total number of categories into which the n observational units are classified. The sum of the observed frequencies for all k categories is equal to the sample size:

$$\sum_{i=1}^{k} o_i = n \tag{14.20}$$

EXAMPLE 14.10 In Example 14.3, a sample of 50 female shoppers were observed in a drugstore in order to estimate the probability p that such a shopper will make a purchase; 37 of the women made purchases and 13 did not. The number of categories into which observational units were classified was $k = 2$ (purchase or no purchase), the observed frequencies were $o_1 = 37$ and $o_2 = 13$, and the sample size was $n = 50$.

EXAMPLE 14.11 In Example 14.2, a sample of 1017 interviewees were classified as 691 approving, 224 disapproving, and 102 undecided. The number of categories into which the observational units were classified was $k = 3$; the observed frequencies in the three categories were $o_1 = 691$, $o_2 = 224$, and $o_3 = 102$; and the sample size was $n = 1017$.

If we knew the true proportion p_i in the ith category, it would be easy to determine how many observational units to expect in the ith category in the sample. The expected number in the ith category is the true proportion p_i times the sample size n; in the two-category case, this is an application of the mean of the binomial distribution, $\mu_{\text{bin}} = np$. In practice, of course, we do not know the true proportions p_i, but we can work out the implications of various hypothesized values of the true proportions. Under a given set of hypothesized values p_{0i} (for $i = 1, 2, \ldots, k$) of the true proportions, the expected number of observations in the ith category is called the **expected frequency** of that category and is designated by the symbol e_i for values of i from 1 up to k. The term *expected* is used here in the sense of mathematical expectation, introduced in Section 7.1. It does not mean that the observed frequencies in any one sample are likely to equal the expected frequencies; rather, the expected frequencies are the average frequencies that would result in the long run if repeated samples of the same size were taken from the same population. In particular samples, the observed frequency o_i in the ith category will sometimes be less than the expected frequency e_i and sometimes greater.

We will be able to attach probability statements to inferences about the true proportions if we know the sampling distribution of a statistic that relates the observed frequencies o_i to the expected frequencies e_i. This sampling distribution will tell us, in effect, how likely it is for the observed frequencies o_i to differ by various amounts from the expected frequencies e_i. The statistic that we will consider is known as **Pearson's X^2 statistic**,

$$X^2 = \sum_{i=1}^{k} \frac{(o_i - e_i)^2}{e_i} \tag{14.21}$$

which is based on the difference between the observed and expected frequencies in each of k categories.

If the expected frequencies e_i correctly reflect the population proportions p_i, then, in a result due to Karl Pearson (see Biography 5.3), the **sampling distribution of Pearson's X^2 statistic** is approximated by the chi-square distribution with degrees of freedom equal to the number k of categories minus the number of independent sample-based restrictions used in the calculation of the expected frequencies. In symbols, the number of degrees of freedom for Pearson's X^2 statistic is

$$\text{df}_{X^2} = k - \text{sbr} \tag{14.22}$$

where sbr is the number of independent sample-based restrictions on the expected frequencies. For example, one restriction that is always imposed is that the expected frequencies must add up to n, the number of observational units in the sample; that is,

$$\sum_{i=1}^{k} e_i = n \tag{14.23}$$

If that is the only sample-based restriction on the expected frequencies, then there are $k - 1$ degrees of freedom. We will see that in certain applications there are fewer degrees of freedom. Notice, however, that the maximum number of degrees of freedom for this X^2 statistic is $k - 1$, not $n - 1$. The number of degrees of freedom depends on the number of categories, not on the number of observations.

The symbol χ^2 (lower-case chi squared) was formerly used in place of X^2 (capital ex squared) for Pearson's statistic in Expression (14.21), and you may encounter the earlier notation in some references. The modern notation emphasizes that the sampling distribution of Pearson's statistic is only approximated by the chi-square distribution, which was introduced in Chapter 12, and it may reduce confusion with the genuine chi-square statistic defined there.

The closeness with which the chi-square distribution approximates the sampling distribution of Pearson's X^2 statistic improves as the sample size n becomes larger. In practice, the following two statements constitute a reasonable **rule of thumb for the chi-square approximation to the sampling distribution of Pearson's X^2 statistic:**

a. If there is just one degree of freedom, the chi-square approximation is sufficiently accurate, at least for rough work, if none of the expected frequencies e_i is less than about 10.

b. If there is more than one degree of freedom, the chi-square approximation is sufficiently accurate if 80% or more of the expected frequencies e_i are at least 5 and none is less than 1. **(14.24)**

Statement (a) is explained in Section 14.2 in connection with the z-statistic in Expression (14.6), since the square of a standard normal deviate is a chi-square variate with one degree of freedom. As in that case, there exists a continuity correction that would improve the chi-square approximation and allow smaller values of e_i, down to about 5, but we will not use it in this book. Inclusion of the continuity correction would add $1/(2n)$ to the margin of error [Expression (14.13)] of the sample proportions. Hence its omission implies that we will claim slightly more precise knowledge of the true proportion than is warranted and will find significant differences among sample proportions slightly more often than we should, at least in small samples. (For example, if $n = 20$, the margin of error of a sample proportion should be increased by $1/(2 \times 20) = .025$ or $2\frac{1}{2}$ percentage points, whereas if $n = 100$, the margin of error should be increased by $1/(2 \times 100) = .005$ or half of one percentage point.) The second statement in the rule of thumb in Expression (14.24) extends the previous rule for the Gaussian approximation [Expression (7.35)] to cover situations with more than one degree of freedom. No continuity correction is required in such situations. A reference under Further Reading at the end of this chapter provides a guide to more exact tests for small samples.

Sections 14.4 and 14.5 describe the use of Pearson's X^2 statistic for significance tests about proportions in one-way and two-way classifications. Section 14.4 deals with one-way classifications, where we can test for specified values of proportions, and Section 14.5 deals with two-way classifications, where we can test the significance of observed differences between proportions.

14.4 CHI-SQUARE GOODNESS-OF-FIT TEST

Sample proportions may be expressed in terms of the observed frequencies o_i with which the different categories of observational units occur in the sample. Likewise, hypothesized population proportions may be expressed in terms of the expected frequencies e_i that would occur on the average in the long run if the hypothesized values were correct.

To test whether the true proportions of the various categories in a population are as postulated on the basis of prior information, we compare the expected frequencies e_i, as determined from the postulated proportions, to the observed frequencies o_i via Pearson's X^2 statistic in Expression (14.21). This procedure is called **Pearson's chi-square goodness-of-fit test**: It tests how well the hypothesized distribution of the observational units by categories fits or describes the actual distribution observed in the sample.

Hypotheses

As we will see when we look at some examples, the null hypothesis in a goodness-of-fit test does not always lend itself to such a brief symbolic statement as has been possible in the case of other tests that we have studied. Sometimes it will be easier just to state the null hypothesis in words.

If there are only two categories being recorded (in which case they may be referred to as *successes* and *failures*), then the null hypothesis is that the population proportion of successes is some specified value p_0 postulated on the basis of prior information:

$$H_0\colon\ p = p_0 \tag{14.25}$$

This null hypothesis for the two-category case implies that the population proportion of failures is $q_0 = 1 - p_0$, and hence may equivalently be written

$$H_0\colon\ p = p_0,\ q = q_0 \qquad \text{where } p_0 + q_0 = 1 \tag{14.26}$$

EXAMPLE 14.12 A sample of 90 college students were interviewed and classified as smokers or nonsmokers. There were 13 smokers and 77 nonsmokers. Are these data consistent with the claim that 25% of college students smoke?

The null hypothesis is that 25% of the relevant population of college students smoke, and the alternative hypothesis is that the true proportion of smokers is not equal to 25%. If p represents the true proportion of smokers in the relevant population, the hypotheses may be written as follows:

$$H_0\colon\ p = .25$$
$$H_A\colon\ p \neq .25$$

Using q to represent the true proportion of nonsmokers in the relevant population (that is, $q = 1 - p$), we can spell these hypotheses out more fully:

$$H_0\colon\ p = .25 \quad \text{and} \quad q = .75$$
$$H_A\colon\ p \neq .25 \quad \text{and} \quad q \neq .75$$

We will see shortly that these hypotheses can be evaluated by Pearson's goodness-of-fit test to determine how well the hypotheses fit the data.

(Source: Student project by D. Damkar.)

If there are three categories involved in a goodness-of-fit test, the null hypothesis states that the true proportions p_1, p_2, and p_3 of the population in the three categories (where $p_1 + p_2 + p_3 = 1$) are certain specified values p_{01}, p_{02}, and p_{03}, respectively:

$$H_0\colon\ p_1 = p_{01}, p_2 = p_{02}, p_3 = p_{03} \qquad \text{where } \sum_{i=1}^{3} p_{0i} = 1 \tag{14.27}$$

EXAMPLE 14.13 For this example, I flipped a pair of coins 50 times and obtained the following outcomes: Two tails occurred 15 times, one head and one tail occurred 22 times, and two heads occurred 13 times. Let us test whether it is reasonable to believe that these were fair coins, fairly tossed.

The null and alternative hypotheses in this case could be stated as follows:

$$\begin{aligned} &H_0\colon \text{ The coins were fair coins, fairly tossed} \\ &H_A\colon \text{ Not } H_0 \end{aligned} \tag{14.28}$$

To test the null hypothesis, we must determine what it implies about the probability of each of the three types of outcomes (no heads, one head, and two heads). If the coins are fair, the probability of heads is .5 for each coin. If they are fairly tossed, the results for the two coins will be independent of each other. Under these conditions (Bernoulli trials—see Section 7.2), the number of heads should follow the binomial distribution with $p = .5$ and $n = 2$ trials. Indeed, the null hypothesis in Expression (14.28) may be written more explicitly as

H_0: The number of heads follows the binomial distribution with $n = 2$ trials and $p = .5$ on each trial

Calculation of the binomial probabilities [via Expression (7.13)] gives us $P_{\text{bin}}(0\text{ heads}) = .25$, $P_{\text{bin}}(1\text{ head}) = .50$, and $P_{\text{bin}}(2\text{ heads}) = .25$. Alternatively, we could arrive at the same probabilities by recognizing that there are four equally likely outcomes (HH, HT, TH, and TT), of which one involves no heads, two involve one head, and one involves two heads. Thus, the null hypothesis in Expression (14.28) could be written in numerical form as

$$H_0\colon\ P(0\text{ heads}) = .25, P(1\text{ head}) = .50, P(2\text{ heads}) = .25$$

This hypothesis may be tested by Pearson's goodness-of-fit test, to be described shortly. The test will assess how well the hypothesized proportions fit the observed data.

In general, if there are k categories involved in a goodness-of-fit test, the null hypothesis specifies the proportion of observations in each of the k categories in the population:

$$H_0\colon\ p_i = p_{0i} \qquad \text{for } i = 1, 2, \ldots, k \qquad \text{where } \sum_{i=1}^{k} p_{0i} = 1 \tag{14.29}$$

Pearson's goodness-of-fit test is, in effect, a test for specified values of a set of proportions. The specified values of the proportions are the hypothesized values of the probabilities of occurrence of the various possible categories of observational units. As a preliminary to testing, we must first determine what probabilities of occurrence of the categories are implied by the null hypothesis. Examples 14.12 and 14.13 illustrate the process of translating a verbal null hypothesis into the numerical values of the proportions implied by the null hypothesis. There is no set procedure for accomplishing this translation; rather, it is a matter of working out the numerical implications of the verbal statement of the null hypothesis.

In every case, the alternative hypothesis that we will consider is the general nondirectional alternative that not all of the population proportions are as specified:

$$H_A\text{: Not } H_0 \qquad (14.30)$$

The alternative hypothesis does not propose that none be as specified; rather, it proposes only that not all be as specified.

The Test Statistic

To perform Pearson's goodness-of-fit test, we record the observed frequencies o_i for the k categories, such that every one of the n observational units in the sample is counted in one and only one category. The sum of the observed frequencies must equal the total sample size n, as noted in Expression (14.20).

Next, we determine the expected frequencies e_i by allocating n hypothetical observational units to the k categories in the manner specified by the null hypothesis. The expected frequency e_i in the ith category is the frequency that would be expected (mathematically) if the null hypothesis in Expression (14.29) were true. We calculate it by multiplying the sample size n times the hypothesized proportion p_{0i} in the ith category:

$$e_i = np_{0i} \qquad \text{for } i = 1, 2, \ldots, k \qquad (14.31)$$

The sum of the expected frequencies e_i is constrained by Expression (14.31) to equal the total sample size n, as indicated in Expression (14.23).

EXAMPLE 14.14 Let us use the investigation in Example 14.12 to illustrate. Under the null hypothesis that 25% of college students smoke, the expected frequency of smokers in a sample of 90 college students is 25% of 90, or

$$e_1 = (90)(.25) = 22.5 \text{ students}$$

The null hypothesis implies that 75% do not smoke, so the expected frequency of nonsmokers in the sample is 75% of 90, or

$$e_2 = (90)(.75) = 67.5 \text{ students}$$

if the null hypothesis is true.

The expected frequencies indicate how a sample of 90 students would be categorized theoretically by the null hypothesis. Next, we need a test statistic that compares the theoretical, expected frequencies with the actual, observed frequencies.

If any of the expected frequencies are too small according to the rule of thumb in Expression (14.24), it is sometimes possible to rectify the situation by combining adjacent categories and redefining k, the number of categories, accordingly.

The *test statistic* for **Pearson's chi-square test of goodness of fit** is Pearson's X^2 statistic:

$$X^2 = \sum \frac{(o_i - e_i)^2}{e_i} \qquad \textbf{(14.32)}$$

which compares the observed frequencies o_i to the frequencies e_i that would be expected if the null hypothesis were true. (Guidelines for how to round and how many digits to report are given in Appendix 1.)

If knowledge of the null hypothesis and the total sample size n is sufficient to calculate the expected frequencies e_i for the goodness-of-fit test, then Pearson's X^2 statistic has $k - 1$ degrees of freedom, one degree of freedom being used for the restriction that the sum of the k expected frequencies must equal n [Expression (14.23)]. Then, if the null hypothesis on the basis of which the expected frequencies were determined is true, the chi-square distribution with $k - 1$ degrees of freedom approximates the sampling distribution of Pearson's X^2 statistic in Expression (14.32).

The Decision Rule

If the null hypothesis is true, and if the sample size is large enough [see Expression (14.24)], the test statistic X^2 may be regarded as a random selection from the chi-square distribution with appropriate degrees of freedom. If the null hypothesis is not true, the observed frequencies o_i will tend to differ from the expected frequencies e_i, that is, from the frequencies that would be expected under the null hypothesis. Because these differences $o_i - e_i$ are squared in the calculation of the test statistic in Expression (14.32), the value of the test statistic X^2 will tend to be larger than specified by the chi-square distribution if the null hypothesis is not true.

Notice that whether an observed frequency o_i is larger or smaller than its expected frequency e_i, the difference contributes to a *larger* value of the test statistic X^2. Thus, the alternative hypothesis in a chi-square goodness-of-fit test is nondirectional or two-sided, even though differences in either direction contribute to a value of X^2 in the upper tail and the test is therefore one-tailed. This may seem puzzling at first—that a two-sided alternative is evaluated by a one-tailed test—but by squaring the discrepancies between o_i and e_i, we guarantee that large negative discrepancies as well as large positive discrepancies will all be in the upper tail of the distribution.

For a test at the α significance level, the critical value therefore is χ^2_α with appropriate degrees of freedom. If the test statistic X^2 is equal to or larger than the critical value χ^2_α with appropriate degrees of freedom, the null hypothesis upon which the expected frequencies were based is rejected at the α significance level. We conclude that the categories in the population from

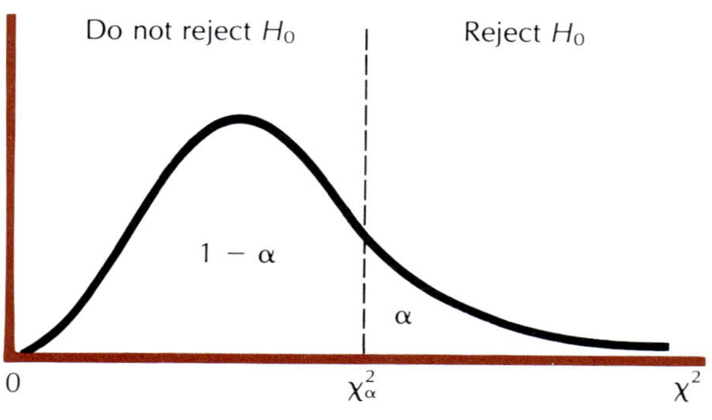

Figure 14.2 Approximate sampling distribution of the X^2 statistic for Pearson's test if the null hypothesis is true. The critical value is shown for the decision about the null hypothesis against the general nondirectional (two-sided) alternative at the α significance level.

which the sample came do not occur in the hypothesized proportions, according to an α-level test. On the other hand, if the observed value of the test statistic X^2 is less than the critical value χ^2_α, the data are consistent with the null hypothesis, testing at the α level. We conclude that the sample does not provide enough evidence to reject the null hypothesis. The decision rule is illustrated in Figure 14.2.

Because two-sided alternatives are tested in one tail, the Pearson X^2 statistic does not lend itself well to making tests against one-sided alternatives. However, if there is just one degree of freedom, the square root of a chi-square statistic [Expression (14.19)] may be compared to the standard normal distribution in Appendix 3 for testing one-sided alternatives.

EXAMPLE 14.15 Example 14.12 described an investigation in which 90 college students were found to include 13 smokers and 77 nonsmokers. To test whether these data are consistent with the claim that 25% of college students smoke, we saw that the relevant hypotheses are

$$H_0\colon\ p = .25$$
$$H_A\colon\ p \neq .25$$

where p is the true proportion of smokers in the relevant population of college students.

The point estimate of the true proportion of smokers is $\hat{p} = 13/90 = .14$, based on the present sample. Is this value significantly different from the hypothesized proportion .25?

We obtain the expected frequencies in this investigation by categorizing $n = 90$ hypothetical observational units (college students) according to the proportions specified by the null hypothesis. The observed frequencies o_i, the expected frequencies e_i, and the calculation of the test statistic X^2 in Expression (14.32) are summarized in the following table.

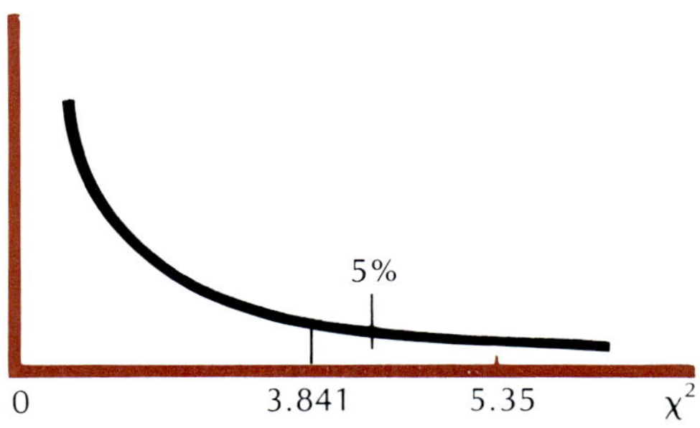

Figure 14.3

Category	o_i	e_i	$o_i - e_i$	$(o_i - e_i)^2/e_i$
Smoker	13	(.25)(90) = 22.5	−9.5	4.01
Nonsmoker	77	(.75)(90) = 67.5	9.5	1.34
	n = 90	n = 90		X^2 = 5.35

In this case, there were $k = 2$ categories, and the only information we required from the sample in order to calculate the expected frequencies was the sample size. Thus, there was just one sample-based restriction on the expected frequencies, namely, that they must add up to 90, and there was just $k - 1 = 2 - 1 = 1$ degree of freedom in the data.

The observed value of the test statistic, $X^2 = 5.35$, may be regarded as a random selection from the chi-square distribution with one degree of freedom, with a critical value of $\chi^2_{.05}(1\text{ df}) = 3.841$ for a goodness-of-fit test at the .05 significance level (see Figure 14.3). Since the observed value 5.35 is greater than the critical value 3.841, we reject the null hypothesis. Looking at the point estimate, which was $\hat{p} = 14\%$, we conclude that the true proportion of smokers in the relevant population of college students is less than the hypothesized value of 25% ($P < .05$).

It is often useful to supplement hypothesis testing with interval estimation. Using Expression (14.9), we find the 95% confidence limits for the true proportion to be $.144 \pm .073$, so we can be 95% confident that the true proportion of smokers in the relevant population is between approximately 7% and 22%.

EXAMPLE 14.16 Example 14.13 described an experiment in which a pair of coins was flipped 50 times, resulting in the following outcomes: No heads occurred 15 times, exactly one head occurred 22 times, and two heads occurred 13 times. The example also explained how the null hypothesis that these were fair coins, flipped fairly, implies that we are hypothesizing the following numerical values of the true probabilities of the three types of outcomes:

$$H_0\text{: } P(0\text{ heads}) = .25,\ P(1\text{ head}) = .50,\ P(2\text{ heads}) = .25$$

The alternative hypothesis is nondirectional:

$$H_A\text{: Not } H_0$$

The point estimates of the true proportions of 0-head, 1-head, and 2-head outcomes are 15/50 = .30, 22/50 = .44, and 13/50 = .26, respectively. Do these differ significantly from the hypothesized values of .25, .50, and .25?

We obtain the expected frequencies in this experiment by categorizing $n = 50$ hypothetical observational units (pairs of flips) according to the proportions specified by the null hypothesis, using Expression (14.31). The observed frequencies o_i, the expected frequencies e_i, and the calculation of the test statistic X^2 in Expression (14.32) are summarized in the following table.

Category	o_i	e_i	$o_i - e_i$	$(o_i - e_i)^2/e_i$
0 heads	15	(.25)(50) = 12.5	2.5	.50
1 head	22	(.50)(50) = 25.0	−3.0	.36
2 heads	13	(.25)(50) = 12.5	.5	.02
	$n = 50$	$n = 50$		$X^2 = .88$

In this case, there are $k = 3$ categories, and the only information we required from the sample in order to calculate the expected frequencies was the sample size. Thus, there was just one sample-based restriction on the expected frequencies, namely, that they must add up to 50, and the observed value of the test statistic, $X^2 = .88$, has $k - 1 = 3 - 1 = 2$ degrees of freedom.

The critical value for a test at the .05 level is therefore $\chi^2_{.05}(2\text{ df}) = 5.991$ (see Figure 14.4), and since the observed value of the test statistic is less than the critical value, we cannot reject the null hypothesis (see Figure 14.2).

The data do not provide significant evidence against the fairness of the coins ($P > .05$). The discrepancies between the observed frequencies and the mathematically expected frequencies under the null hypothesis are within the limits of what might reasonably be expected to occur by chance. In other words, the difference between (a) the observed proportions (.30, .44, and .26) and (b) the hypothesized proportions (.25, .50, and .25) is not statistically significant at the .05 level.

In Examples 14.15 and 14.16, the only thing we needed to know from the sample in order to calculate the expected frequencies was the sample size n. The use of the sample size n in the calculation of the expected frequencies to

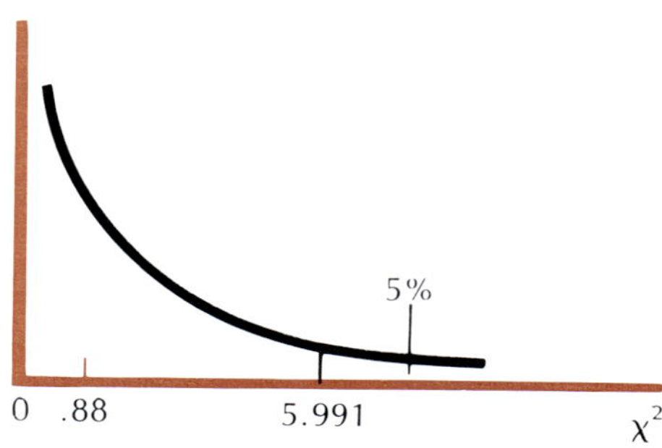

Figure 14.4

ensure that the expected frequencies would add up to the actual sample size n constituted the one sample-based restriction on the k expected frequencies, leaving us with $\boldsymbol{k - 1}$ **degrees of freedom**. Such is the case in many, but not all, applications of Pearson's test of goodness of fit.

When knowledge of the total sample size n (and the null hypothesis) must be supplemented by other sample-based restrictions in order to calculate the k expected frequencies e_i, the number of degrees of freedom is further reduced by one for each such restriction. In the context of testing goodness of fit, this situation most often arises in tests to determine whether the distribution of observations in a sample is consistent with an hypothesized population distribution, one or more parameters of which must be estimated from the sample in order to calculate the expected frequencies. We will not pursue such cases in this book.

Summary of the Test Procedure

Pearson's chi-square test of goodness of fit may be summarized as follows:

Assumptions.

a. Random sample of n observational units independently classified into k categories. (More precisely, the assumption is that the requirements for the multinomial distribution of Section 7.3 are met.)

b. Expected frequencies satisfying Expression (14.24).

Null hypothesis. H_0: $p_i = p_{0i}$, for $i = 1, 2, \ldots, k$, where $\sum p_{0i} = 1$

Alternative hypothesis. H_A: $p_i \neq p_{0i}$ for at least one category i

Step 1. Calculate the sample proportions $\hat{p}_i$ (for $i = 1, 2, \ldots, k$) in the k categories to see what sample values are being tested.

Step 2. Determine the population proportions p_{0i} (for $i = 1, 2, \ldots, k$) that would be in the k categories according to the null hypothesis.

Step 3. Calculate the k expected frequencies, $e_i = np_{0i}$ (for $i = 1, 2, \ldots, k$), that is, the frequencies expected under the null hypothesis.

Step 4. Calculate the null-hypothesis value of the test statistic:

$$X^2 = \sum \frac{(o_i - e_i)^2}{e_i}$$

Step 5. Look up the critical value χ^2_α in the chi-square distribution with k − sbr degrees of freedom. There are $k - 1$ degrees of freedom in the common case where the sample size is the only sample information used in the calculation of the expected frequencies.

Step 6. If the observed value X^2 is greater than or equal to the critical value, conclude that the set of sample proportions differs significantly from the set of proportions specified by the null hypothesis ($P < \alpha$); otherwise, conclude that the sample proportions are not significantly different from the specified values ($P > \alpha$).

14.5 CHI-SQUARE TEST FOR TWO-WAY TABLES

The procedure for testing the significance of differences between sample proportions is, in many ways, similar to that for testing for specified values of proportions. Again we will represent sample proportions in terms of observed frequencies; we will represent null-hypothesis population proportions in terms of expected frequencies; and we will assess the correspondence between the observed and expected frequencies by means of Pearson's X^2 statistic. The test in this case is called the **chi-square test for two-way tables**. The major distinction between this test and the one in the preceding section lies in the null hypothesis being tested and hence in the method of determining the frequencies to be expected under the null hypothesis.

The test in this section is applicable to two types of problems. Depending on how the data were collected, the frequency data in a two-way classification may be appropriate for testing either of two different null hypotheses corresponding to the two types of problems. However, both null hypotheses lead to the same method of determining the expected frequencies and to the same chi-square test for two-way tables.

Null Hypotheses

To test the null hypothesis that the proportion p_1 of successes in one population is equal to the proportion p_2 of successes in another population,

$$H_0\colon\ p_1 = p_2 \tag{14.33}$$

we collect data from a random sample from each population and record the number of successes and the number of failures in each sample. The sample data may be presented as observed frequencies in a 2-by-2 table showing the two samples cross-classified according to the two outcomes. Such a table usually includes the **marginal totals** (row totals and column totals) as well as the total sample size n. The two samples correspond to the levels of the independent variable, and the two outcomes are the values of the dependent variable.

EXAMPLE 14.17 In a study of the distribution of various species of trees, counts were made of the numbers of trees (over a certain height) in a 400-square-foot area adjacent to a river near Fort McMurray in northern Alberta and in a comparable riverbank area near Camrose in central Alberta. The independent variable in this case is region, and the dependent variable is species of tree. Is the true proportion of willows different in the two regions?

Let p_1 and p_2 represent the true proportions of willows in riverbank areas in the Fort McMurray and Camrose regions, respectively. The null hypothesis is that these population proportions are equal:

$$H_0\colon p_1 = p_2$$

The following 2-by-2 table shows the numbers of trees counted in the samples from the two populations.

		Region		
		Fort McMurray	Camrose	Totals
Species	Willow	52	38	90
	Other	19	17	36
	Totals	71	55	$n = 126$

The sample proportions, which are point estimates of the true proportions, are $\hat{p}_1 = 52/71 = .73$ and $\hat{p}_2 = 38/55 = .69$. We will see that the chi-square test for two-way tables can be used to test whether the difference between the sample proportions is statistically significant.

(Source: Student project by C. Morris.)

When there are more than two categories of outcomes and/or more than two populations, it becomes fairly cumbersome to write the null hypothesis in the symbolic notation of Expression (14.33). However, it is not hard to find a verbal generalization that expresses the null hypothesis clearly and concisely:

H_0: The ratio of the numbers of observational units in the various categories is the same in every population. **(14.34)**

Expression (14.34) is not limited to two categories of outcomes (successes and failures) or to just two populations; rather, it refers more generally to the various categories of outcomes in every population. Use of the chi-square test for two-way tables to test this type of null hypothesis is sometimes called the **chi-square test of homogeneity of populations.**

EXAMPLE 14.18 Consider the following investigation as an illustration of the use of more than two categories of the dependent variable. The 176 male and 200 female students living in a college residence complex were classified according to whether they never wore visual corrective lenses, wore glasses at least occasionally but never contact lenses, or wore contact lenses at least occasionally, as follows.

		Visual corrective			
		Glasses	Contacts	None	Totals
Gender of student	Male	62	10	104	176
	Female	61	47	92	200
	Totals	123	57	196	$n = 376$

Are these results compatible with the results that would be expected in random samples from male and female populations in which there is no difference between populations with respect to the true proportions wearing glasses, contact lenses, and no visual correctives? More simply, did the males and females differ significantly with respect to the wearing of visual corrective lenses?

The independent variable is gender, and the dependent variable is type of visual corrective, classified into three categories. We will see that the chi-square test for two-way tables can be used to test whether the observed differences provide enough evidence to conclude that the underlying populations of male and female students differ with respect to type of visual corrective. The null hypothesis to be tested [Expression (14.34)] is that the ratio of glasses wearers, contact wearers, and no-corrective wearers is the same in the male and female populations.

(Source: Student project by K. Ritchie.)

The null hypothesis in Expression (14.34) applies to any number of categories of the dependent variable and any number of populations (levels of the independent variable). Let us use the symbol k_A for the number of categories of variable A (say, the independent variable) and the symbol k_B for the number of categories or levels of variable B (say, the dependent variable). The sample data in the general case, then, are presented as observed frequencies in a k_A-by-k_B table. (In Example 14.18, the number of categories of gender is $k_A = 2$, and the number of categories of type of visual corrective is $k_B = 3$, so we have a 2-by-3 table.)

The null hypothesis in Expression (14.34) says, in effect, that the proportions of observational units in the various categories of the dependent variable do not depend on which population is being sampled. Knowledge of which population (level of the independent variable) an observational unit came from does not help, if the null hypothesis is true, in predicting which dependent-variable category it falls in. A chi-square test of this hypothesis requires independent random samples from the different populations, with a total of n observations altogether.

Exactly the same chi-square test for two-way tables may be used with the cross-classification of n observational units according to any two variables or types of attribute, say, variable A and variable B. That is, the same chi-square procedure may be used even if the researcher does not start with two or more identifiable populations (for example, males and females), but merely draws a random sample of n observational units from a single population and cross-classifies them according to two variables A and B. A general way of writing the null hypothesis for such tests is

$$H_0\text{: Variables } A \text{ and } B \text{ are independent of each other} \qquad (14.35)$$

where it is completely arbitrary which variable is labeled A and which one B. Accordingly, use of the chi-square test for two-way tables to test this type of null hypothesis is sometimes called the **chi-square test of independence**.

EXAMPLE 14.19 Is there any relationship between students' expressed interest in a course and their habitual time of arrival for lectures in that course? A sample of 50 students taking one or more mathematics courses filled out a rating scale regarding their interest in mathematics and reported their usual time of arrival for their mathematics lectures. The 50 students were subsequently classified in the following 3-by-3 table.

		Attitude toward mathematics			
		Negative	Neutral	Positive	Totals
Usual time of arrival for mathematics lectures	Early	7	10	10	27
	Just on time	3	7	4	14
	Late	3	4	2	9
	Totals	13	21	16	$n = 50$

Here the investigator did not determine the marginal totals for either variable; both sets of marginal totals were left to chance. (Contrast these data with the data in Example 14.18, where the investigator decided how many males and how many females to interview.) The chi-square test for two-way tables can be used to see if the variables are related.

Even though the investigator did not manipulate either variable, we may still consider one of the variables to be the independent variable and the other the dependent variable, just as we did in Chapter 5 when we talked about the relationship between numerically measured variables. In this example, time of arrival would probably be considered to be the dependent variable, and attitude toward mathematics, the independent variable. Thus, the sample proportions arriving early, just on time, and late are

$$7/13 = .54,\quad 3/13 = .23,\quad 3/13 = .23 \quad \text{for negative-attitude students}$$

$$10/21 = .48,\quad 7/21 = .33,\quad 4/21 = .19 \quad \text{for neutral-attitude students}$$

$$10/16 = .62,\quad 4/16 = .25,\quad 2/16 = .12 \quad \text{for positive-attitude students}$$

We can see a general trend that students with more positive attitudes are more likely to arrive early and less likely to arrive late than those with negative attitudes, but the significance of this difference remains to be tested. Testing whether these three sets of proportions are significantly different from each other is equivalent to testing whether the two variables (attitude and time of arrival) are independent of each other.

(Source: Student project by K. Ng.)

The two types of problem introduced in this section differ as follows: If a random sample from a single population is cross-classified on two variables A and B, then neither the row totals nor the column totals are fixed by the investigator, and the test is called a *test of independence of the two variables.* On the other hand, if independent random samples from populations identified by the levels of an independent variable A are cross-classified in terms of another variable B, then the variable-A totals (the sample sizes) are set by the investigator, and the test is called a *test of homogeneity of the populations.*

Although the underlying exact tests are fundamentally different, both types of problem lead to the same approximate chi-square test for two-way tables. Indeed, many authors refer rather loosely to both types of problem as *chi-square tests of independence.* Such writers note that both types of null hypothesis [Expressions (14.34) and (14.35)] imply that knowledge of the outcome on one variable does not help to predict the outcome on the other variable. In both cases, the chi-square test for two-way tables examines the relationship between two categorized variables and helps us judge whether the two categorized variables are related to each other or whether they are independent of each other.

EXAMPLE 14.20 In the context of Example 14.17, saying that the true proportion of riverbank willows in the Fort McMurray region is equal to that in the Camrose region is equivalent to saying that region (Fort McMurray vs. Camrose) and species of tree (willow vs. other) are unrelated, or are independent of each other. On the other hand, to say that the proportions of willows are different is to say that region and species of tree are related and that they are not independent of each other—the proportion of each species depends on the region. We use the chi-square test for two-way tables to help decide between these two possibilities.

General Form of the Contingency Table of Observed Frequencies

In general, if there are k_A categories of variable A and k_B categories or levels of variable B, the sample data are presented in what is sometimes called a k_A-by-k_B **contingency table**, which is a table whose cells show the observed frequencies, whose right-hand margin shows the frequency distribution of one variable, and whose bottom margin shows the frequency distribution of the other variable. Let us adopt the following notation:

o_{ab} is the observed frequency of the joint occurrence of the ath level of variable A and the bth level of variable B. (Double subscripts were introduced in Section 11.3.)

m_a is the marginal total of the observed frequencies at the ath level of variable A.

n_b is the marginal total of the observed frequencies at the bth level of variable B.

$n = \sum \sum o_{ab} = \sum m_a = \sum n_b$ is the total sample size, **(14.36)**

which may be obtained in any of the three ways indicated: by adding the observed frequencies in all the cells, or by adding the right-margin totals, or by adding the bottom-margin totals.

EXAMPLE 14.21 Notice that the **double summation** $\sum \sum$ in Expression (14.36) simply means "the sum of the sum of." A single summation sign means "the sum of" the various values of the variable that follows the sign. The variable that follows the first summation sign in a double summation is another summation, so we have the sum of a set of sums. Thus, we evaluate a double summation by systematically applying the rules for a single summation, first to the inside or right-hand summation sign, and then to the outer or left-hand summation sign.

For example, the expression $\sum \sum o_{ab}$ is an abbreviated version of

$$\sum_{a=1}^{k_A} \sum_{b=1}^{k_B} o_{ab}$$

which may be written as

$$\sum_{a=1}^{k_A} \left(\sum_{b=1}^{k_B} o_{ab} \right)$$

Expanding the single summation within parentheses, we have

$$\sum_{a=1}^{k_A} (o_{a1} + o_{a2} + \cdots + o_{ak_B})$$

Now we can expand the other single summation to obtain the sum of the various sums in parentheses:

$$(o_{11} + o_{12} + \cdots + o_{1k_B}) + (o_{21} + o_{22} + \cdots + o_{2k_B}) + \cdots + (o_{k_A1} + o_{k_A2} + \cdots + o_{k_Ak_B}).$$

The notation may look formidable at first, but the operations are quite straightforward: adding the values in columns and/or rows of a table. The following table illustrates the notation using the data (numbers of students) from Example 14.19. It is arbitrary which variable is labeled A and which one B. The result of the double summation is shown in the lower right-hand corner. You should verify the result by using the three methods listed in Expression (14.36).

		Variable B			
		Level 1	Level 2	Level 3	Totals
	Level 1	$o_{11} = 7$	$o_{12} = 10$	$o_{13} = 10$	$m_1 = \sum o_{1b} = 27$
Variable A	Level 2	$o_{21} = 3$	$o_{22} = 7$	$o_{23} = 4$	$m_2 = \sum o_{2b} = 14$
	Level 3	$o_{31} = 3$	$o_{32} = 4$	$o_{33} = 2$	$m_3 = \sum o_{3b} = 9$
	Totals	$n_1 = \sum o_{a1} = 13$	$n_2 = \sum o_{a2} = 21$	$n_3 = \sum o_{a3} = 16$	$n = \sum \sum o_{ab} = 50$

The general form of a k_A-by-k_B contingency table shows the observed frequencies o_{ab}, as illustrated in Table 14.2.

TABLE 14.2. The general form of a k_A-by-k_B contingency table.

		Variable B						Marginal totals for A
		Level 1	Level 2	$\cdots$	Level b	$\cdots$	Level k_B	
	Level 1	o_{11}	o_{12}	$\cdots$	o_{1b}	$\cdots$		m_1
	Level 2	o_{21}	o_{22}	$\cdots$	o_{2b}	$\cdots$		m_2
	$\vdots$	$\vdots$	$\vdots$		$\vdots$			$\vdots$
Variable A	Level a	o_{a1}	o_{a2}	$\cdots$	o_{ab}	$\cdots$		m_a
	$\vdots$	$\vdots$	$\vdots$		$\vdots$			$\vdots$
	Level k_A							
Marginal totals for B		n_1	n_2	$\cdots$	n_b	$\cdots$		n

Expected Frequencies

We have seen how observed frequencies o_{ab} may be presented in a contingency table. What are the corresponding frequencies e_{ab} that are to be expected if the null hypothesis that the variables are unrelated [Expression (14.34) or (14.35)] is true?

EXAMPLE 14.22 Let us illustrate with the data from Example 14.17 concerning species of riverbank trees in two regions. We determine the expected frequencies on the basis of the marginal totals, reprinted here for convenient reference.

		Region		
		Fort McMurray	Camrose	Totals
Species	Willow			90
	Other			36
	Totals	71	55	$n = 126$

The null hypothesis says that the relative numbers of trees of the different species are the same in both regions. If that hypothesis is true, then we would expect $\frac{90}{126}$ of the 71 trees in the Fort McMurray region to be willows ($71 \times \frac{90}{126} = 50.71$), and we would expect $\frac{90}{126}$ of the 55 trees in the Camrose region to be willows ($55 \times \frac{90}{126} = 39.29$). Similarly, if the null hypothesis is correct that the proportions are the same in the two regions, we would expect $\frac{36}{126}$ of the 71 trees in the Fort McMurray region to be other species ($71 \times \frac{36}{126} = 20.29$), and we would expect $\frac{36}{126}$ of the 55 trees in the Camrose region to be other species ($55 \times \frac{36}{126} = 15.71$). Notice that for each cell of the table, the expected frequency is simply the product of the corresponding marginal totals divided by n; for example, $(71 \times 90)/126$ for the first cell, and $(55 \times 36)/126$ for the last cell.

The following table shows the expected frequencies in parentheses together with the corresponding observed frequencies. Notice that the expected frequencies necessarily add up to the marginal totals (for example, $50.71 + 39.29 = 90$), because the marginal totals were used in the calculation of the expected frequencies.

		Region		
		Fort McMurray	Camrose	Totals
Species	Willow	52 (50.71)	38 (39.29)	90
	Other	19 (20.29)	17 (15.71)	36
	Totals	71	55	$n = 126$

In the general case (see Table 14.2), if the null hypothesis is true, variable B has no effect on the relative frequencies at the different levels of variable A; that is, the expected frequencies in each column are in the same proportions as the right-hand marginal totals. Thus the **expected frequencies e_{ab} in a contingency table** are obtained by allocating each column total n_b according to the proportions m_a/n indicated by the row totals (see Example 14.22) or, equivalently, by allocating each row total m_a according to the proportions n_b/n displayed by the column totals. Either way, as illustrated in Example 14.22, the expected frequency e_{ab} corresponding to the observed frequency o_{ab} is simply the product of the relevant marginal totals, divided by the total sample size n:

$$\text{Expected frequency} = \frac{\text{product of relevant marginal totals}}{\text{total sample size } n} \tag{14.37}$$

or, more explicitly,

$$e_{ab} = \frac{m_a n_b}{n} \tag{14.38}$$

This simple formula gives the frequency e_{ab} that would be mathematically expected if variables A and B were unrelated, as specified in the null hypotheses in Expressions (14.34) and (14.35). [For practice, you should use Expression (14.37) or (14.38) to verify the expected frequencies in Example 14.22.]

We obtain a partial check on the calculation of the expected frequencies by verifying that the expected frequencies in each row and each column add up to the corresponding marginal total. In the type of test being discussed here, the marginal totals of the expected frequencies must be the same as those of the observed frequencies.

The Test Statistic

Once we have the observed frequencies o_{ab} and the corresponding expected frequencies e_{ab} as implied by the null hypothesis that variables A and B are unrelated, then Pearson's X^2 statistic is calculated as the *test statistic* for the **chi-square test for two-way tables**. In effect, it compares the observed and expected frequencies.

We can calculate Pearson's statistic in the form previously given,

$$X^2 = \sum^{k} \frac{(o - e)^2}{e} \tag{14.39}$$

by taking the summation over all $k = k_A k_B$ cells of the k_A-by-k_B table. Expression (14.39) directs us to calculate $(o - e)^2/e$ for each cell of the table, and then to add the resulting values. (Guidelines for how to round and how many digits to report are given in Appendix 1.)

If the null hypothesis that the two variables are unrelated [Expression (14.34) or (14.35)] is true, and if the total sample size is large enough [see Expression (14.24)], then the sampling distribution of the X^2 statistic is approximated by the chi-square distribution with **$(k_A - 1)(k_B - 1)$ degrees of freedom**.

The number of degrees of freedom is $(k_A - 1)(k_B - 1)$ for the following reason: There are $k = k_A k_B$ categories (or observed frequencies) altogether, but there are several sample-based restrictions imposed in the calculation of the corresponding expected frequencies. Not only must the expected frequencies add up to the total sample size n (1 restriction), but, in addition, the k_A marginal totals for the levels of variable A are used in calculating the expected frequencies ($k_A - 1$ more restrictions, the last of the marginal totals being implied by the total-sample-size restriction), as are the k_B marginal totals for the levels of variable B ($k_B - 1$ more restrictions, the last of the marginal totals again being implied by the total-sample-size restriction). Thus, the number of categories minus the number of sample-based restrictions on the expected frequencies is

$$\begin{aligned} k - \text{sbr} &= k_A k_B - [1 + (k_A - 1) + (k_B - 1)] \\ &= k_A k_B - k_A - k_B + 1 \\ &= (k_A - 1)(k_B - 1) \end{aligned} \tag{14.40}$$

which is the number of degrees of freedom. Intuitively, we note that if we try to fill in frequencies in a k_A-by-k_B table with given marginal totals, we have freedom of choice for all but the last cell in each row and in each column; that is, $(k_A - 1)(k_B - 1)$ of the frequencies are free to vary, for a given set of marginal totals.

The Decision Rule

If the null hypothesis in Expression (14.34) or (14.35) is true, the test statistic in Expression (14.39) may be regarded as a random selection from the chi-square distribution with $(k_A - 1)(k_B - 1)$ degrees of freedom. If the null hypothesis is not true, the observed frequencies will tend to differ from the expected frequencies by more than a chance amount, and, because these differences $(o - e)$ are squared in the calculation of the test statistic, the value of the test statistic X^2 will tend to be *larger* than specified by the chi-square distribution. All $(o - e)$ differences, whether positive or negative, contribute to a larger value of the test statistic. Hence, the test is one-tailed, with the entire rejection region in the upper tail of the distribution, even though the alternative hypothesis is the nondirectional complement of Expression (14.34) or (14.35):

H_A: The ratio of the numbers of observational units in the various categories is not the same in every population **(14.41)**

or

H_A: Variables A and B are not independent of each other **(14.42)**

The critical value for the chi-square test for two-way tables at the α significance level is therefore $\chi^2_\alpha[(k_A - 1)(k_B - 1)\,\text{df}]$. The decision rule is illustrated in Figure 14.2.

As noted in Section 14.4, page 479, Pearson's X^2 statistic does not lend itself well to making tests against one-sided alternatives, because nondirectional or two-sided alternatives are tested by a one-tailed Pearson test. However, if there is just one degree of freedom, the square root of a chi-square statistic [see Expression (14.19)] may be compared to the standard normal distribution in Appendix 3 for testing one-sided alternatives.

EXAMPLE 14.23 Example 14.18 described an investigation in which 376 college students, 176 males and 200 females, were categorized according to the type of visual corrective they wore. Do males and females differ with respect to type of visual corrective in the relevant population of college students?

From the observed frequencies in the following table, we can determine that the sample proportions of glasses wearers, contact wearers, and no-corrective wearers were 35%, 6%, and 59% for the males, as compared to 30%, 24%, and 46%, respectively, for the females. We will use the chi-square test for two-way tables to determine whether the male sample proportions differed significantly from the female sample proportions.

		Visual corrective			
		Glasses	Contacts	None	Totals
Gender of student	Male	62 (57.57)	10 (26.68)	104 (91.74)	176
	Female	61 (65.43)	47 (30.32)	92 (104.26)	200
	Totals	123	57	196	$n = 376$

The 2-by-3 table above shows the expected frequencies in parentheses below the observed frequencies. The expected frequencies were obtained via Expression (14.37) or (14.38); for example, the expected frequency in the top left cell is (176)(123)/376 = 57.57, and the expected frequency in the top middle cell is (176)(57)/376 = 26.68. The rest of the expected frequencies may be obtained by the same procedure and verified by checking the marginal totals.

The test statistic is computed via Expression (14.39) as follows:

$$\begin{aligned} X^2 &= \frac{(62 - 57.57)^2}{57.57} + \frac{(10 - 26.68)^2}{26.68} + \frac{(104 - 91.74)^2}{91.74} \\ &\quad + \frac{(61 - 65.43)^2}{65.43} + \frac{(47 - 30.32)^2}{30.32} + \frac{(92 - 104.26)^2}{104.26} \\ &= .34 + 10.43 + 1.64 + .30 + 9.18 + 1.44 \\ &= 23.3 \end{aligned}$$

The associated number of degrees of freedom is $(k_A - 1)(k_B - 1) = (3 - 1)(2 - 1) = 2$ degrees of freedom. Since the observed value X^2 exceeds the critical value $\chi^2_{.01}(2\text{ df}) = 9.210$, the observed difference between males and females with respect to the wearing of visual correctives may be said to be statistically significant at the 1% level (see Figure 14.5). The type of visual corrective worn is related to gender ($P < .01$).

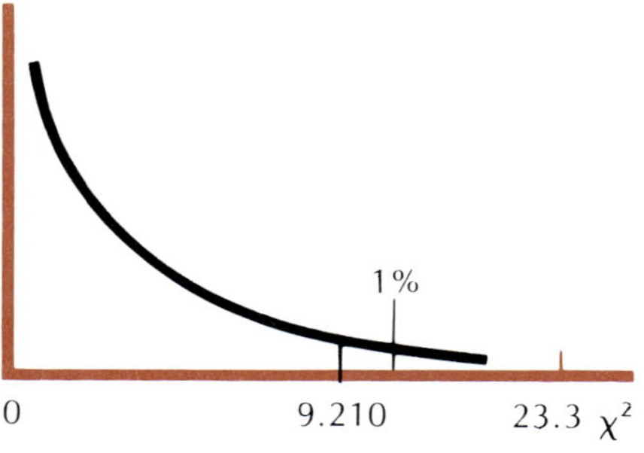

Figure 14.5

The two previously mentioned sets of sample proportions (35%, 6%, and 59% for males and 30%, 24%, and 46% for females) are significantly different from each other. From the separate terms in the calculation of X^2, we can see that the largest contribution to the value of X^2 comes from the cells for contact wearers, indicating that category as the location of greatest difference in the sample data. Female students in the sample were more likely to wear contact lenses than were male students.

Most of the major computer program packages for statistical analysis include the chi-square test for two-way tables. One such package is known as BMDP. Output from BMDP program 4F is shown in Figure 14.6 for this example. You should be able to recognize all of the items in the output by comparing it with the annotated outline of the calculations above.

EXAMPLE 14.24 Example 14.22 presented both the observed and the expected frequencies in a 2-by-2 table from a study concerning species of riverbank trees in two regions of Alberta. Example 14.17 noted that the sample proportions of willows were 73% and 69% in the two regions. Is this a statistically significant difference?

In a chi-square test for two-way tables, a 2-by-2 table has $(2 - 1)(2 - 1) = 1$ degree of freedom. The calculation of the test statistic X^2 may be summarized as follows.

Category	o_i	e_i	$o_i - e_i$	$(o_i - e_i)^2/e_i$
Willow/Fort McMurray	52	50.7	1.3	.033
Willow/Camrose	38	39.3	−1.3	.043
Other/Fort McMurray	19	20.3	−1.3	.083
Other/Camrose	17	15.7	1.3	.108
	n = 126	126		X^2 = .27

Since the observed value $X^2 = .27$ is less than the critical value $\chi^2_{.05}(1\text{ df}) = 3.841$, the null hypothesis cannot be rejected at the 5% significance level (see Figure 14.7). The sample proportions of willows are not significantly different in the two regions ($P > .05$).

Notice that this test did *not* test whether there are more willows than other species or whether there are more trees in one region than the other. Rather, it tested whether

***** OBSERVED FREQUENCY TABLE 1

GENDER	VISUAL			
	GLASSES	CONTACTS	NONE	TOTAL
MALE	62	10	104	176
FEMALE	61	47	92	200
TOTAL	123	57	196	376

ALL CASES HAD COMPLETE DATA FOR THIS TABLE.

***** PERCENTS OF ROW TOTALS -- TABLE 1

GENDER	VISUAL			
	GLASSES	CONTACTS	NONE	TOTAL
MALE	35. 2	5. 7	59. 1	100. 0
FEMALE	30. 5	23. 5	46. 0	100. 0
TOTAL	32. 7	15. 2	52. 1	100. 0

MINIMUM ESTIMATED EXPECTED VALUE IS 26. 68

STATISTIC	VALUE	D. F.	PROB.
PEARSON CHISQUARE	23. 323	2	0. 0000

***** EXPECTED VALUES -- TABLE 1

GENDER	VISUAL			
	GLASSES	CONTACTS	NONE	TOTAL
MALE	57. 6	26. 7	91. 7	176. 0
FEMALE	65. 4	30. 3	104. 3	200. 0
TOTAL	123. 0	57. 0	196. 0	376. 0

***** DIFFERENCES = OBSERVED - EXPECTED --- TABLE 1

GENDER	VISUAL			
	GLASSES	CONTACTS	NONE	TOTAL
MALE	4. 4	-16. 7	12. 3	0. 0
FEMALE	-4. 4	16. 7	-12. 3	0. 0
TOTAL	0. 0	0. 0	0. 0	0. 0

***** COMPONENTS OF CHI SQUARE = (OBS - EXP)**2 /EXP --- TABLE 1

GENDER	VISUAL			
	GLASSES	CONTACTS	NONE	TOTAL
MALE	0. 3	10. 4	1. 6	12. 4
FEMALE	0. 3	9. 2	1. 4	10. 9
TOTAL	0. 6	19. 6	3. 1	23. 3

Figure 14.6
BMDP4F printout for Example 14.23.

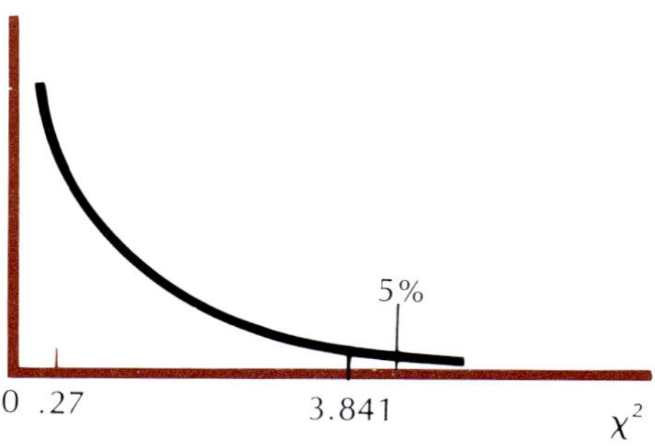

Figure 14.7

the *relative* numbers of trees of different species are the same in the two regions. In other words, it tested whether region and species are related or, on the other hand, whether they are independent of each other. It turned out that there was no significant evidence of relationship ($P > .05$).

It is important to recognize that, as noted in Example 14.24, the chi-square test for two-way tables does not test for differences between the levels of variable A or for differences between the levels of variable B, per se. Rather, it tests whether the proportions of observational units at the various levels of variable A remain unchanged regardless of which level of variable B is considered. Or, equivalently, it tests whether the proportions of observational units at the various levels of variable B remain unchanged regardless of which level of variable A is considered. It is, in effect, a test of the correlation between variable A and variable B: Are variables A and B related, or are they independent?

If the test statistic X^2 in the chi-square test for two-way tables is statistically significant, we usually want to know what the nature of the relationship is. Which levels of variable A are most likely to occur with which levels of variable B? To answer this question, it is often helpful to look at the sample proportions (which are point estimates of the true proportions) of observational units at the various levels of one variable, calculated separately for each level of the other variable, as illustrated in Examples 14.19, 14.23, and 14.24. The individual terms that are added in the calculation of X^2 may also be revealing. The cells that make the largest contribution to the value of X^2 are those in which the sample deviates most markedly from the hypothesized values, as illustrated in Example 14.23.

Again, we must remember that Pearson's X^2 statistic is only approximated by the chi-square distribution. The approximation is good provided that none of the *expected* frequencies is too small, as specified in Expression (14.24).

Summary of the Test Procedure

The **chi-square test for two-way tables** may be summarized as follows:

Assumptions.

a. Random sample of n observational units independently classified into $k_A k_B$ cells, or k_A independent random samples where the observational units in each sample are classified into k_B categories.

b. Expected frequencies satisfying Expression (14.24).

Null hypothesis. H_0: Variables A and B are independent of each other, or the ratio of the numbers of observational units in the various categories of variable B is the same for every category of variable A.

Alternative hypothesis. H_A: Not H_0.

Step 1. Calculate separately for each level of one variable the point estimates of the proportions in the various categories of the other variable, in order to see what sample proportions are being tested.

Step 2. Calculate the expected frequency, $e_{ab} = m_a n_b / n$, corresponding to each observed frequency o_{ab}.

Step 3. Calculate the null-hypothesis value of the test statistic:

$$X^2 = \sum \frac{(o - e)^2}{e}$$

Step 4. Look up the critical value $\chi^2_\alpha[(k_A - 1)(k_B - 1)\text{ df}]$.

Step 5. If the observed value X^2 is greater than or equal to the critical value, conclude that the proportions of observational units in the various levels of one variable are significantly different for different levels of the other variable or that the variables are related ($P < \alpha$). In this case, some understanding of the nature of the relationship can be gained by looking at the point estimates calculated in Step 1, as well as at the separate terms in the calculation of X^2 to see which cells contributed most to the value of X^2. If the observed value of the test statistic is less than the critical value, conclude that the differences between the proportions are not statistically significant or that the data are consistent with the idea that the variables are independent of each other ($P > \alpha$).

Relationship to Other Significance Tests

Table 14.3 shows the relationship of the chi-square test for two-way tables to other significance tests. Situations in which tests on proportions are appropriate are characterized by categorical data rather than numerically measured variables (see page 463). The chi-square test for two-way tables is suitable for testing for independence of two variables or for testing the significance of the difference between proportions in independent random samples. Either way,

TABLE 14.3 Summary of significance tests that have been presented for four different parameters, categorized into one-sample tests, independent-samples tests, and paired-samples tests. (Parenthetical references identify the chapter and section where each test is discussed.)

	Type of parameter under investigation			
Type of test	**Mean**	**Variance**	**Correlation**	**Proportions**
One-sample tests	One-sample t test (Sec. 11.1)	One-sample chi-square test (Sec. 12.3)	Test of significance of r (Sec. 13.1)	Chi-square goodness-of-fit test (Sec. 14.4)
Independent-samples tests	Independent-samples t test (Sec. 11.2)	Independent-samples F test (Sec. 12.4)	(See Further Reading in Chapter 13.)	Chi-square test for two-way tables (Sec. 14.5)
Paired-samples tests	Paired-samples t test (Sec. 11.3)	(See Further Reading in Chapter 12.)	(See Further Reading in Chapter 13.)	Paired-samples sign test (Sec. 16.5)

the test assumes that the observational units were selected independently of each other. The so-called sign test, presented in Section 16.5, may be used for testing the difference between proportions in paired samples.

STUDY MATERIAL FOR CHAPTER 14

Terms and Concepts

numerically measured variable
categorized variable
categorical data

Section 14.1

point estimate of p
true proportion

Section 14.2

sampling distribution of the sample proportion $\hat{p}$
standard error of a proportion
large-sample approximation to the sampling distribution of $\hat{p}$
approximate $1 - \alpha$ confidence interval for p
approximate $1 - \alpha$ confidence limits for p
approximate margin of error of $\hat{p}$
distinction between percentages and percentage points
maximum margin of error of $\hat{p}$
size of sample required for a given margin of error in estimating p

Section 14.3

observed frequency
expected frequency

Pearson's X^2 statistic
sampling distribution of Pearson's X^2 statistic
rule of thumb for the chi-square approximation to Pearson's X^2

Section 14.4

Pearson's chi-square goodness-of-fit test
degrees of freedom for chi-square goodness-of-fit test

Section 14.5

chi-square test for two-way tables
marginal total
chi-square test of homogeneity of populations
chi-square test of independence
contingency table
double summation
expected frequencies in a contingency table
degrees of freedom for chi-square test for two-way tables

Key Formulas

Section 14.1

(14.1) $$\hat{p} = \frac{n_1}{n}$$

Section 14.2

(14.5) $$\sigma_{\hat{p}} = \sqrt{\frac{pq}{n}}$$

(14.8) $$\hat{p} - z_{\alpha/2}\sigma_{\hat{p}} < p < \hat{p} + z_{\alpha/2}\sigma_{\hat{p}} \quad \text{with confidence level } 1 - \alpha$$

(14.13) $$e_{\alpha/2} = z_{\alpha/2}\sigma_{\hat{p}}$$

(14.16) $$n = pq\frac{z^2_{\alpha/2}}{e^2_{\alpha/2}}$$

Section 14.3

(14.21) $$X^2 = \sum \frac{(o_i - e_i)^2}{e_i}$$

Section 14.4

(14.31) $$e_i = np_{0i}$$

Section 14.5

(14.38) $$e_{ab} = \frac{m_a n_b}{n}$$

Exercises for Numerical Practice

Section 14.2

14/1. What are the mean and the standard deviation of the sampling distribution of $\hat{p}$ based on a random sample of 160 observations from a population in which $p = .80$?

14/2. What are the mean and the standard deviation of the sampling distribution of $\hat{p}$ based on a random sample of 1000 observations from a population in which $p = .40$?

14/3. Calculate the 99% confidence limits for the true proportion p if a random sample of 300 observations includes 99 successes.

14/4. Calculate the 99% confidence limits for the true proportion p if a random sample of 2000 observations includes 1740 successes.

14/5. Calculate the margin of error of the sample proportion at the 98% confidence level if a random sample consists of 900 successes and 300 failures.

14/6. Calculate the margin of error of the sample proportion at the 99.9% confidence level if a random sample consists of 75 successes and 425 failures.

14/7. Calculate the maximum margin of error of a sample proportion at the 95% confidence level based on a random sample of 3600 observations.

14/8. Calculate the maximum margin of error of a sample proportion at the 95% confidence level based on a random sample of 400 observations.

14/9. How large a random sample would be required to estimate a population proportion with a maximum margin of error of 6 percentage points at the 99.9% confidence level?

14/10. How large a random sample would be required to estimate a population proportion with a maximum margin of error of half of one percentage point at the 98% confidence level?

Section 14.4

14/11. A random sample of 200 observations consisted of 75, 70, 44, and 11 observations in four mutually exclusive categories. Use the chi-square goodness-of-fit test to test the claim that the true proportions in the four categories are 40%, 30%, 20%, and 10%, respectively.

14/12. A random sample of 50 observations consisted of 6, 13, 14, 12, and 5 observations in five mutually exclusive categories. Use the chi-square goodness-of-fit test to test the claim that the true proportions in the five categories are 10%, 20%, 30%, 25%, and 15%, respectively.

Section 14.5

14/13. Calculate the summations requested for the following values of o_{ab} and y_{ij}:

$$o_{11} = 9, \quad o_{12} = 3 \qquad y_{11} = 0, \quad y_{12} = 9, \quad y_{13} = 5$$
$$o_{21} = 6, \quad o_{22} = 9 \qquad y_{21} = 3, \quad y_{22} = 1, \quad y_{23} = 0$$
$$o_{31} = 5, \quad o_{32} = 4 \qquad y_{31} = 8, \quad y_{32} = 3, \quad y_{33} = 1$$

a. $\sum o_{a1}$ **b.** $\sum o_{2b}$ **c.** $\sum \sum o_{ab}$
d. $\sum y_{1j}$ **e.** $\sum y_{i2}$ **f.** $\sum \sum y_{ij}$

14/14. Calculate the summations requested for the following values of o_{ab} and x_{ij}:

$o_{11} = 7, \quad o_{12} = 2, \quad o_{13} = 8 \qquad x_{11} = 6, \quad x_{12} = 1, \quad x_{13} = 5, \quad x_{14} = 1$

$o_{21} = 5, \quad o_{22} = 2, \quad o_{23} = 1 \qquad x_{21} = 0, \quad x_{22} = 8, \quad x_{23} = 0, \quad x_{24} = 2$

a. $\sum o_{1b}$ **b.** $\sum o_{a3}$ **c.** $\sum \sum o_{ab}$
d. $\sum x_{i4}$ **e.** $\sum x_{2j}$ **f.** $\sum \sum x_{ij}$

14/15. A random sample 100 observations was cross-classified on variables A and B as follows:

	B_1	B_2	B_3	B_4
A_1	9	10	14	7
A_2	5	29	11	15

Perform the chi-square test for two-way tables to test whether there is a significant relationship between variables A and B.

14/16. A random sample of 200 observations yielded the following contingency table:

	B_1	B_2	B_3
A_1	69	47	34
A_2	31	13	6

Use the chi-square test for two-way tables to judge whether variables A and B are independent of each other.

Problems for Applying the Concepts

14/17. An accountant for a large department store randomly selected 100 accounts receivable and found that 8 of the accounts contained at least one error. Calculate the 95% confidence limits for the true proportion of accounts containing errors.

14/18. The research department of a seed company randomly selected 500 seeds of a certain variety of nasturtium for germination trials and found that 410 of the seeds germinated. Calculate the 99% confidence limits for the true germination rate of this variety of nasturtiums.

14/19. A magazine publishing company conducted a survey to estimate what proportions of domestic subscribers and of foreign subscribers read certain specific sections of the magazine. The publisher obtained data from a random sample of 3,000 of its 300,000 domestic subscribers and 300 of its 100,000 foreign subscribers and used the 98% confidence level.

a. What is the maximum margin of error in the proportions of the domestic sample who reported reading various sections?

b. What is the maximum margin of error in the proportions of the foreign sample who reported reading various sections?

c. What would happen to the margins of error in Parts (a) and (b) if the sample sizes were quadrupled?

14/20. A random sample of 1800 adults from across the country indicated their attitude toward unilateral nuclear disarmament (for, against, or undecided). The percentage choosing each alternative was reported for the nation as a whole and also separately for each of four regions that were equally represented in the total sample. At the 95% confidence level:

a. What is the maximum margin of error in the proportions of the national sample choosing each alternative?

b. What is the maximum margin of error in the regional sample proportions choosing each alternative?

c. What would happen to the margins of error in Parts (a) and (b) if the sample size were quadrupled?

14/21. How large a random sample is needed to estimate the true proportion of General Motors vehicles on a busy highway, with a maximum margin of error of 5 percentage points at the 95% confidence level?

14/22. How large a random sample is needed to estimate the true proportion of bus riders who are over 65 years of age, with a maximum margin of error of 10 percentage points at the 99% confidence level?

14/23. How large a random sample is needed to estimate a country's true unemployment rate to within $\frac{1}{4}$ of one percentage point at the 95% confidence level,

a. if the true rate is known to be between 3% and 15%?

b. if the true rate is known to be between 2% and 10%?

14/24. How large a random sample is needed to estimate, with a maximum margin of error of 2 percentage points at the 95% confidence level, the true proportion of Ontario residents whose mother tongue is English, if it is known to be between .75 and .95?

14/25. A table of random digits, such as that in Appendix 2, is constructed so that each digit is equally likely to appear in any position in the table. Before publication, such tables are subjected to various tests for errors in construction. The following data show the number of times each digit occurred in 500 successive digits of a table of random digits:

Digit:	0	1	2	3	4	5	6	7	8	9
Frequency:	42	47	43	52	43	59	49	54	55	56

Use Pearson's goodness-of-fit test to determine whether these data are consistent with the claim that the ten digits occurred with equal probability.

a. State the null hypothesis in symbols and in words.

b. Calculate the point estimates of the probabilities of the various digits.

c. What is the critical value of the test statistic for a test at the 5% significance level?

d. State the results of the significance test.

14/26. Roll a single die 60 times, recording the numbers of ones, twos, threes, fours, fives, and sixes that you get. Use the data to test whether the die is fair (that is, equally likely to turn up each of its six faces).

a. State the null hypothesis in symbols and in words.

b. What is the critical value of the test statistic for a test at the 5% significance level?

c. State the results of the significance test.

14/27. A normal-winged fruit fly was crossed with a vestigial-winged fruit fly in a genetics experiment. The offspring were found to consist of 75 normal-winged flies and 23 vestigial-winged flies. Are these data consistent with the claim that 25% of the offspring of this type of monohybrid cross are vestigial-winged?

a. State the null hypothesis in symbols and in words.

b. What is the point estimate of the true proportion of vestigial-winged offspring in such a cross?

c. State the results of the significance test.

d. Construct a 95% confidence interval for the true proportion of vestigial-winged offspring.

(Source: Student project by C. Merriman.)

14/28. In an experiment in genetics, 602 kernels of corn resulting from a dihybrid cross were categorized as being either yellow or purple and as either smooth or wrinkled, as follows:

333 purple and smooth,
117 purple and wrinkled,
111 yellow and smooth,
41 yellow and wrinkled.

According to Mendelian laws of inheritance, the proportions of these four types of kernels should be 9/16, 3/16, 3/16, and 1/16, respectively. Are the data consistent with the Mendelian prediction?

a. State the null hypothesis in words.

b. Write the hypothesized proportions and the estimated proportions in decimal form (to see how they compare).

c. State the results of the significance test.

d. Construct a 95% confidence interval for the true proportion of yellow-and-wrinkled kernels.

(Source: Student project by D. Latam.)

14/29. A pair of dice was rolled 360 times, with the following results:

Outcome:	2	3	4	5	6	7	8	9	10	11	12
Frequency:	4	16	24	31	42	58	54	45	39	28	19

Is there significant evidence that the dice are not fair?

a. State the null hypothesis in words.

b. Write the hypothesized proportions and the estimated proportions in decimal form (to see how they compare).

c. State the results of the significance test.

d. Construct a 95% confidence interval for the true proportion of elevens for this pair of dice.

14/30. The two left columns of the following table show the percentage distribution of family incomes in the United States in 1982.

Income	Percentage of households	Number of households in a given county
Under \$5,000	9.5%	18
\$5,000 but less than \$10,000	14.4	33
\$10,000 but less than \$15,000	13.6	30
\$15,000 but less than \$20,000	12.4	25
\$20,000 but less than \$25,000	10.8	23
\$25,000 but less than \$35,000	16.9	31
\$35,000 but less than \$50,000	13.4	23
\$50,000 or more	9.0	17
Total	100.0%	200

The right-hand column shows the distribution of family income in a random sample of 200 households from one particular county. Does this sample provide significant evidence that the county's income distribution is different from the national distribution?

a. State the null hypothesis in words.

b. Write the estimated county proportions as percentages (to see how they compare with the national percentages).

c. State the results of the significance test.

d. Construct a 95% confidence interval for the true proportion of county families with incomes of \$50,000 or more.

(Source: U.S. Department of Commerce, Bureau of the Census, *Statistical Abstract of the United States 1985,* page 444.)

14/31. Do the data in Example 14.19 provide significant evidence of a relationship between students' expressed interest in a course and their habitual time of arrival for lectures in that course?

a. State the null hypothesis in words.

b. State the results of the significance test.

14/32. Calving records from one farm for the years 1978 through 1981 yielded the following numbers of male and female calves delivered at various times of the day.

	Day	Evening	Night
Bulls	29	5	17
Heifers	17	9	16

Was there a significant difference between bulls and heifers with respect to time of delivery?

a. Calculate the estimated proportions in the various categories for bulls and for heifers.

b. State the results of the significance test.

c. Construct 95% confidence intervals for the true proportion of bulls born in the evening and for the true proportion of heifers born in the evening.

(Source: Student project by J. Millang.)

14/33. Do people living in different regions have different color preferences in the cars they drive? A sample of 107 cars were observed along a mountain highway near Cranbrook, B.C., and a sample of 137 cars were observed along a highway in the mixed-farming country near Camrose, Alberta. Vehicles larger than pickup trucks were omitted from the count, and two-tone vehicles were counted for the color of the body rather than the roof. The numbers of vehicles of each color were as follows.

	White	Yellow	Red or orange	Blue	Black	Gray	Green	Brown
Cranbrook	17	9	19	24	4	10	9	15
Camrose	10	5	17	24	6	16	24	35

Was there a significant difference between the two regions with respect to the proportions in which the various car colors were observed?

a. Calculate the estimated proportions in the various categories for each region.

b. State the results of the significance test.

(Source: Student project by C. Grue.)

14/34. In a study of shopping habits, shoppers were observed as they left a drugstore during a one-hour period from 4:00 to 5:00 P.M. The numbers of male and female shoppers were recorded, along with the numbers of males and females who actually made purchases while in the drugstore. The results were as follows.

	Purchase	No purchase
Males	10	9
Females	37	13

Is there a significant difference between the male and female shoppers with respect to the proportions who made purchases?

a. State the null hypothesis in symbols and in words.

b. Calculate the point estimates of the relevant true proportions.

c. State the results of the significance test.

(Student project by L. Jorgensen.)

14/35. In a sample of 530 college students, 117 of the 273 men and 71 of the 257 women owned a vehicle. Is the difference between these two proportions greater than would be expected by chance alone?

a. State the null hypothesis in symbols and in words.

b. Write the point estimates of the true proportions in decimal form.

c. State the results of the significance test.

d. Construct 95% confidence intervals for the true proportion of male students who own cars and for the true proportion of female students who own cars in the population from which this sample was selected.

(Source: Student project by D. Parkin.)

14/36. Three randomly selected samples consisting of 200 popcorn kernels each were stored for six hours at three different temperatures: −18°C, 8°C, and 22°C. Each sample was then taken directly from its respective storage place and put immediately into an oiled and heated frying pan for popping in the conventional fashion. The three samples (−18°C, 8°C, and 22°C) yielded 26, 4, and 2 unpopped kernels, respectively. Are the sample differences statistically significant?

a. State the null hypothesis in symbols and in words.

b. What are the point estimates of the relevant true proportions?

c. State the results of the significance test.

(Source: Student project by K. Dziwenka.)

14/37. Random samples of students and of staff at a college were asked to specify which of two candidates for the U.S. presidency they preferred. Which statistical method could be used for testing whether students and staff differed significantly in their preferences?

14/38. A horse owner kept track of the number of shoe replacements for each of 11 work horses in the past year. Which statistical method could be used to test whether the average number of shoes required by these 11 horses differed significantly from 12 shoes per horse?

14/39. A duck hunter sometimes went hunting by himself and sometimes with one or more friends. Each time, he recorded the size of the hunting party and the number of ducks brought home. Which statistical method could be used to test the significance of the relationship between the size of the hunting party and the number of ducks brought home?

14/40. A sample of 200 college students were cross-classified according to hair color and eye color. Which statistical method could be used to test the significance of the relationship between the two variables?

14/41. A sample of 10 individuals took the same physical fitness test both before and after taking a course in TaeKwon-Do (Korean art of self-defense). Which statistical method could be used for testing whether there was a significant change in the average fitness score?

14/42. Twenty female volunteers were randomly assigned to one of two conditions in a study of the effects of caffeine. Subjects in one condition drank a cup of coffee, and those in the other condition drank a cup of tea. The change in resting pulse rate that accompanied (within five minutes) the drinking of the tea or coffee was recorded for each subject. Which statistical method could be used to test whether there was significantly greater variability in the pulse-rate changes for one of the beverages?

14/43. A sample of 15 students who classified themselves as smokers reported the number of cigarettes they each smoked in a typical day. Which statistical method could be used to test whether the average number of cigarettes reported by these students differed significantly from 20 cigarettes?

14/44. The numbers of students pictured with and without eyeglasses were counted separately for males and females in a high school yearbook. Which statistical method could be used to test whether males and females differed significantly with respect to wearing glasses for their yearbook photograph?

14/45. A commercial greenhouse compared two methods of germination of flower seeds by planting a dozen flats according to each method and counting the number of marketable seedlings obtained from each flat. Which statistical method could be used to test whether the two methods differed significantly in the variability of the number of seedlings produced?

14/46. Eight male college students reported their height to the nearest tenth of an inch for subsequent verification by actual measurement. Which statistical method could be used to test whether the students had a significant bias to say that they were taller than they actually were?

14/47. As part of a study comparing Catholic and Protestant life-styles, samples of students claiming Catholic or Protestant religious backgrounds reported on a questionnaire as to whether they now "attend church regularly" or "do not attend church regularly." Which statistical method could be used to determine whether reported church attendance differed significantly between Catholics and Protestants?

14/48. As part of a study comparing Catholic and Protestant life-styles, samples of students claiming Catholic or Protestant religious backgrounds reported the size of the family (that is, the number of children) in which they grew up. Which statistical method could be used for testing whether average family size is different in the relevant Catholic and Protestant populations?

14/49. A hatchery received a shipment of eggs from each of three different suppliers. Each batch was incubated for three weeks under the same conditions, and the number of live chicks that hatched from each shipment was recorded, along with the number of eggs that failed to hatch from the shipment. Which statistical method could be used to test whether the three suppliers differed significantly with respect to the percentage of eggs successfully hatched?

14/50. Ten houses with natural-gas heating were randomly selected from the same neighborhood. The total floor area of each house was measured (in square feet), and the amount of natural gas consumed in the previous month was recorded. Which statistical method could be used to test the significance of the relationship between floor area and natural gas consumption?

14/51. Ten subjects were randomly assigned to each of two conditions in a memory experiment. The subjects in one condition were each given three minutes to study a photograph of a busy fairground scene, while those in the other condition each had three minutes to study a paragraph describing the same scene in detail. Subjects in both conditions were then allowed up to 20 minutes to write down as many items from the scene as they could remember. Which statistical method could be used to test whether the average number of items recalled differed significantly between the two conditions?

14/52. A large aircraft manufacturer has records of the number of airplanes ordered each month for the past eight months. Which statistical method could be used to see if it is reasonable to suppose that the observed distribution might have been generated by an underlying discrete uniform probability distribution?

Further Reading

Campbell, S. K. *Flaws and Fallacies in Statistical Thinking.* Englewood Cliffs, NJ: Prentice-Hall, 1974. Chapter 8 discusses common errors in the use of percentages, including the confusion of percentages and percentage points.

Dixon, W. J., et al., eds. *BMDP Statistical Software Manual, 1985 Reprinting.* Berkeley: University of California Press, 1985. BMDP is an extensive package of statistical programs. One example of its output was shown in Figure 14.6.

Hill, M., ed. *BMDP User's Digest, 3rd edition.* Los Angeles: BMDP Statistical Software, Inc., 1984.

Reichmann, W. J. *Use and Abuse of Statistics.* London: Methuen, 1962. Chapter 16 provides a good account of the meaning of the standard error of a proportion.

Snedecor, G. W., and Cochran, W. G. *Statistical Methods.* 7th ed. Ames, IA: Iowa State University Press, 1980. Chapters 7 and 11 deal with the analysis of count data, including appropriate uses of continuity corrections to improve the results from large-sample approximation methods.

Wheeler, M. *Lies, Damn Lies, and Statistics.* New York: Liveright, 1976. Journalistic rather than scientific in style, this book attempts "to demystify and demythologize the polling process, to demonstrate in non-technical terms just what [public opinion] polls can and cannot do." It succeeds.

CHAPTER 15

INTRODUCTION TO THE ANALYSIS OF VARIANCE

EXAMPLE 15.1 "The National Institute of Mental Health (NIMH) is conducting a 3-year, $3.4-million, multi-institutional clinical trial comparing drugs to two forms of psychotherapy in the treatment of depression. . . . The project, which was initiated for purely scientific reasons, has taken on political importance as well. Insurance companies and the government are increasingly concerned about whether to reimburse for psychotherapy and, if so, whether all forms of psychotherapy should be equally reimbursable. . . . Although many members of the professional community are convinced that psychotherapy works, a number of legislators and legal experts are awaiting more rigorous evidence than has been provided to date.

". . . The NIMH chose to study depression therapies in part because depression is a major public health problem and in part because several specific forms of therapy seemed ripe for testing.

"For the past year, 27 therapists have been in training to participate in the study by learning one of two psychotherapies or by learning a drug treatment regimen using [an antidepressant known as] imipramine . . . or a placebo. Next year 144 patients whose depression has been carefully documented will be randomly assigned to one of these four treatments. In addition, 36 patients will be assigned to a 'treatment as usual' group. They will be referred to experienced therapists, who will treat them as they treat their other depressed patients. The efficacy of the therapies will be determined both by asking the patients and the therapists whether the therapy relieved the patients' symptoms of depression and by conducting an independent clinical evaluation."

Thus, relief from depression will be measured as the dependent variable, yielding what, for simplicity, we might call an "improvement rating" for each of the depressed patients in the five conditions. The five conditions represent the five levels of the independent variable:

a. cognitive behavioral therapy,

b. interpersonal psychotherapy,

c. imipramine drug therapy,

d. the placebo drug treatment as a control condition, and

e. treatment as usual, as a second control condition.

The mean improvement rating may be calculated for each of the five conditions, and the researchers will then want to know whether the differences among the means represent true differences in the effectiveness of the treatments, or whether the observed differences can reasonably be attributed to chance. This chapter introduces a statistical method for answering such questions where there are means from several conditions to be compared at once.

(Source: Kolata, 1981. Copyright 1981 by the American Association for the Advancement of Science. Used with permission of the AAAS.)

In Chapter 11, we studied one-sample and two-sample t tests. These t tests are suitable for testing hypotheses about the value of a single population mean or about the equality of two population means. However, there are many situations, such as that illustrated in Example 15.1, where one would like to compare three or more means at a time.

If the goal is to compare three or more means from one investigation, is it appropriate to use the t test repeatedly to compare the means two at a time in all possible pairs? As a rule, such multiple t tests are not appropriate, because the probability associated with the result of the t test is based on the notion that the observed value of the test statistic t_{obs} is a single random selection from the t distribution. As we draw repeatedly from the sampling distribution of t_{obs}, the probability increases that one of the values of t_{obs} will be beyond the critical values, even if the null hypothesis is true; that is, multiple t testing increases the Type I error rate so that the statement $P < \alpha$ may no longer be correct. To compare three or more means from one investigation, we require a method that takes into account the number of means being compared.

By far the most widely used statistical procedure for testing hypotheses about sets of three or more means is the analysis of variance, abbreviated ANOVA, developed primarily by R. A. Fisher in the 1920s (see Biography 12.1). The analysis of variance is a remarkably versatile technique, useful for data summary and hypothesis testing in a large class of statistical problems. Our survey, begun in Chapter 9, of fundamental methods in classical inferential statistics would not be complete without a chapter introducing some of the basic ideas of analysis of variance.

15.1 DRAWING INFERENCES ABOUT MEANS BY ANALYZING VARIATION

We begin with a brief explanation of the meaning of the term **analysis of variance (ANOVA)**. The explanation is necessary to ensure that you understand from the outset that in this chapter we are dealing with inferences about *means,* even though the procedure is called the analysis of variance.

To analyze something is to break it down into its parts or to identify its components. In the analysis of variance, we examine the total variation in a set of data and try to determine how much of it can be attributed to certain identifiable factors.

The variance to be analyzed is the sample variance:

$$s^2 = \frac{\Sigma\,(x - \bar{x})^2}{n - 1} \tag{15.1}$$

which measures the spread or dispersion of the observations x around the sample mean $\bar{x}$. In the context of analysis of variance, the numerator in Expression (15.1) is called the **total sum of squares**, SS_{total} (we have previously called it SS_x or just SS), and the denominator is the **total number of degrees of freedom**, $df_{total} = n - 1$, so we have

$$s^2 = \frac{SS_{total}}{df_{total}} \tag{15.2}$$

The sum of squares, or more fully the sum of squared deviations about the mean, is known as the **variation** (not variance) of the observations around the mean. The size of the difference $x_i - \bar{x}$ of an observation x_i from the sample mean $\bar{x}$ indicates the extent to which that observation deviates from a representative value for the sample. A deviation of +5, for example, indicates the same distance from the mean as a deviation of −5. Squaring each of the deviations equates the influence of the positive and negative deviations, so that when the squared deviations are summed, the sum of squares is an index of the total variation of the observations about the mean.

EXAMPLE 15.2 How often do students living in college residences go home during one semester? Data were obtained from a sample of 10 female students by counting the number of visits home for one semester, including the end of the semester, as follows:

3, 2, 8, 6, 7, 8, 6, 4, 8, 5

Representing each observation by an X above the appropriate point on the real-number line, we obtain a graphic display of the dispersion of the observations:

```
                    X
                X   X
    X X X X X X X
0 1 2 3 4 5 6 7 8 9
```

The mean of these observations is $\bar{x} = 5.7$ visits, and the display illustrates the variation around the mean. Given no further information, we attribute the variation in number of visits home to individual differences among the students.

However, if we know that the homes of some of these students are within an hour's drive of their college residence, others more than an hour but less than four hours, and still others more than a four-hour drive away, we obtain a different impression of the reasons for the dispersion. Replacing the X's in the graphic display by C's for students whose homes are close to the campus, F's for students whose homes are far, and M's for students from intermediate distances, we obtain the following display of number of visits home:

```
                    C
                M   C
    F F F M M C C
0 1 2 3 4 5 6 7 8 9
```

Now we see that some of the total variation in number of visits home is due to which of the three distance categories a student is in, although there are still some individual differences among students within each category.

Thus, we have analyzed, or broken down, the total variation in the data by identifying two **sources of variation** in number of visits home: (a) variation due to distance to the student's home and (b) residual variation, probably arising from many factors, but all of which we can lump together as individual differences among students. The variation (a), due to distance, is reflected in the variation of the three subsample means ($\bar{x}_C = 7.75$, $\bar{x}_M = 5.67$, and $\bar{x}_F = 3.00$) around the overall mean $\bar{x} = 5.70$. The residual variation (b) associated with individual differences is reflected in the variation of the four C's around 7.75, of the three M's around 5.67, and of the three F's around 3.00.

(Source: Student project by D. M. Mychaluk.)

If we can identify some **factor**, whether it be a treatment applied by the investigator or an attribute of the observational units, that distinguishes two or more subsamples within the total sample, we can partition each observation's deviation from the mean into two additive parts:

a. deviation of the subsample mean from the overall mean, and

b. deviation of the individual observation from the mean of its subsample.

Then, since variation is defined as the sum of squared deviations, squaring each observation's deviation and adding these squared deviations for the whole sample gives us two corresponding **sources of variation**:

a. variation due to the differences *between* the subsamples, and

b. variation due to differences among observational units *within* the same subsample.

The second of these, representing differences among observational units that are in the same category or that have been treated the same, is usually associated with random error. The analysis of variance helps us to decide if the variation due to differences between the subsamples is large enough, relative to the variation associated with random error, for us to conclude that the difference among the subsample means is statistically significant.

The term **factor** in analysis of variance refers to a dimension, other than the dependent variable, along which the observational units are categorized. Thus, any independent variable is a factor; the categories are the levels of the independent variable, or the levels of the factor. In addition, variation among observational units that are in the same subsample or that are treated the same is said to be due to the **replication factor**. **Replication** refers to the observation of more than one observational unit from the same subsample, and the associated variation is usually presumed to be random variation. In general, the term **factor** refers to any basis, within the design of an experiment, for distinguishing one observation from another.

In brief, the analysis of variance is used to break down the total variation in the values of the dependent variable into components due to identifiable sources of variation, and then to evaluate whether variation due to particular sources is large relative to random variation. Since variation due to a particular factor is a measure of the differences among the subsample means associated with the different levels of the factor, we are, in fact, using the analysis of variance to draw inferences about means.

15.2 ONE-WAY ANOVA WITH INDEPENDENT SAMPLES

Section 11.2 presented a t test for judging the significance of the difference between means of two independent samples. The **one-way analysis of variance (ANOVA)** introduced here is a generalization of the independent-samples t test in that it provides a test of the difference between means of k independent samples, where k is any integer 2 or more.

EXAMPLE 15.3 In Example 15.2, there were $k = 3$ means to be compared, corresponding to the three categories of distance to the student's home: close, intermediate, and far. The mean numbers of visits home by students in the three categories were $\bar{x}_C = 7.75$, $\bar{x}_M = 5.67$, and $\bar{x}_F = 3.00$, respectively. Are the three sample means significantly different from each other? That is, is there a significant difference in number of visits home in a semester by students in the three distance categories? The one-way analysis of variance addresses this question.

Hypotheses and Data

The null hypothesis to be tested in the one-way ANOVA is that k population means are all equal; in symbols,

$$H_0\text{: } \mu_1 = \mu_2 = \cdots = \mu_k \tag{15.3}$$

This null hypothesis includes as a special case $\mu_1 = \mu_2$, which is the null hypothesis for the independent-samples t test of the difference between two means. When $k = 2$, one-way ANOVA is equivalent to the t test.

The analysis of variance tests against the nondirectional, or two-sided, alternative that the means are not all equal:

$$H_A\text{: Not } H_0 \tag{15.4}$$

Notice that we do not obtain the alternative hypothesis by changing the equal signs in Expression (15.3) to not-equal signs. That notation would say that none of the k means is equal to any other, whereas the alternative hypothesis says only that they are not *all* equal to each other, leaving open the possibility that some subset of them may be equal to each other.

To test the null hypothesis in Expression (15.3), we obtain observations of k independent random samples, one from each of the k populations whose means are to be compared. The populations may be distinguished from each other by some characteristic of the observational units (such as distance to the student's home in Example 15.2) or by the treatment applied by the researcher. In either case, the different populations correspond to the different levels of the independent (treatment) variable, and the observations are the values of the dependent (response) variable. (You may wish to review the terminology introduced in Section 9.2, pages 328–29, at this point.)

Testing hypotheses in the analysis of variance involves certain assumptions. It is assumed that the underlying populations are Gaussian and that they all have the same variance σ^2. Fortunately, the analysis of variance may be used even when these assumptions are not met perfectly. The general situation is much the same as that described at the end of Section 11.2 for the t test, and the possibility of assumption violations should be kept in mind in the interpretation of results, as noted in Comment 4 at the end of Section 11.4.

In analysis of variance, it is conventional to use the symbol x to represent the dependent variable. This follows the practice introduced in Section 11.2 in connection with t tests on means, although it is contrary to the usage of y for the dependent variable and x for the independent variable in regression and correlation. In analysis of variance, we use x for the dependent or response variable and usually give verbal labels to independent variables. (In Example 15.2, for instance, the independent variable might be referred to as the distance factor, which may be abbreviated to Factor D for some purposes.)

Let us define the following notation:

k is the number of levels of the independent variable, that is, the number of categories, or treatments, being compared. We will use the general label Factor A to designate the independent variable whose levels distinguish the k samples from each other. To say that there are k samples being compared is to say that there are k levels of Factor A.

n_a is the number of observations in the ath sample, or at the ath level of Factor A.

$n = \sum n_a$ is the total number of observations.

x_{ai} represents the observation recorded for the ith observational unit in the ath sample. The first subscript identifies which sample the observation is in, and the second subscript identifies the observational unit within that sample.

$T_{a.} = \sum_{i=1}^{n_a} x_{ai}$ is the sum of the observations in the ath sample. The first subscript on the T (for "total") tells us that the total is for the ath level of the first subscript—the ath sample. The dot marks the place of the second subscript on T and indicates that the levels of the second subscript have been combined to arrive at this value. In other words, the dot indicates that this is a total of the observations at *all* the levels of the second subscript, not just for one specific level of the second subscript.

$\bar{x}_{a.} = \dfrac{T_{a.}}{n_a}$ is the mean of the observations in the ath sample. The first subscript on $\bar{x}$ tells us that this mean is for the ath sample, and the dot marking the place of the second subscript indicates that this value is not specific to a particular level of the second subscript but rather is a mean of the observations at all the levels of the second subscript.

$T = \sum T_{a.} = \sum \sum x_{ai}$ is the **grand total** of all the observations in the samples. (The double summation sign is explained in Example 14.21.)

$\bar{x} = \frac{T}{n}$ is the **grand mean** of all the observations in the samples. It is an ordinary arithmetic mean, since it is simply the sum of all the observations, divided by the number of observations.

EXAMPLE 15.4 Example 15.2 presented data from an investigation of the number of times that students living in college residences go home in one semester, including the end of the semester. Three populations of female students were sampled: those with homes close to the college, those with homes at an intermediate distance, and those whose homes were far from the college. (The three distances are defined in Example 15.2.) Does distance have a significant effect on the number of visits home?

Let us use this example to illustrate the notation that has been introduced. The independent variable is distance, and the dependent variable is number of visits home. There are $k = 3$ levels of the independent variable (close, intermediate, and far), and the null hypothesis is that the corresponding three populations of students do not differ in the mean number μ of visits home in a semester:

$$H_0\colon\ \mu_1 = \mu_2 = \mu_3$$

The alternative hypothesis is that distance has an effect on number of visits home:

$$H_A\colon \text{ Not } H_0$$

A total of $n = 10$ observations were made, with $n_1 = 4$ in the "close" sample, $n_2 = 3$ in the "intermediate" sample, and $n_3 = 3$ in the "far" sample. The notation for the observations, totals, and means is as follows.

Sample 1 (close)	**Sample 2 (intermediate)**	**Sample 3 (far)**	**Total sample**
$x_{11} = 8$	$x_{21} = 6$	$x_{31} = 3$	
$x_{12} = 7$	$x_{22} = 6$	$x_{32} = 2$	
$x_{13} = 8$	$x_{23} = 5$	$x_{33} = 4$	
$x_{14} = 8$			
			$k = 3$
$T_{1.} = 31$	$T_{2.} = 17$	$T_{3.} = 9$	$T = 57$
$n_1 = 4$	$n_2 = 3$	$n_3 = 3$	$n = 10$
$\bar{x}_{1.} = 7.75$	$\bar{x}_{2.} = 5.67$	$\bar{x}_{3.} = 3.00$	$\bar{x} = 5.7$

The Test Statistic

For a test of the null hypothesis in Expression (15.3), we require a test statistic whose sampling distribution is known when the null hypothesis is true. Such a test statistic may be obtained via the principle introduced in Section 15.1: If the deviation $x_{ai} - \bar{x}$ of each observation from the grand mean can be expressed as a sum of deviations associated with particular factors, then the total variation in the sample data can be partitioned into parts due to those factors.

In one-way ANOVA with independent samples, each total *deviation*

$x_{ai} - \bar{x}$ may be separated into two parts as

$$x_{ai} - \bar{x} = (\bar{x}_{a.} - \bar{x}) + (x_{ai} - \bar{x}_{a.}) \qquad \textbf{(15.5)}$$

By squaring both sides of Expression (15.5) and summing over all the observations in the sample, it can be shown that the total *variation*,

$$\text{SS}_{\text{total}} = \sum \sum (x_{ai} - \bar{x})^2 \qquad \textbf{(15.6)}$$

may likewise be separated into two parts, corresponding to the two terms on the right-hand side of Expression (15.5):

a. variation of the sample means $\bar{x}_{a.}$ around the grand mean $\bar{x}$, and

b. variation of the individual observations x_{ai} around their own sample mean $\bar{x}_{a.}$.

The following two paragraphs, (**a**) and (**b**), discuss these two parts of the total variation:

a. The variation of the sample means $\bar{x}_{a.}$ around the grand mean $\bar{x}$ is due to differences among the sample means $\bar{x}_{a.}$. Each sample mean is associated with a different level of Factor A. Hence, any of the following labels may be used to refer to this part of the total variation:

- the **variation between samples**, or
- the **variation due to Factor *A***, or
- the **sum of squares due to Factor *A***, which we will denote by

$$\text{SS}_A = \sum n_a(\bar{x}_{a.} - \bar{x})^2 \qquad \textbf{(15.7)}$$

(Remember that Factor A is simply a general label that we are using to refer to the independent variable. In any specific example, we would replace the general term with the name of the factor, such as "distance" in Example 15.4.) Since there are k sample means whose variation is being measured around the one grand mean, the sum of squares SS_A has $k - 1$ degrees of freedom.

b. The variation of the individual observations x_{ai} around their own sample mean $\bar{x}_{a.}$ is due to differences among the observational units within the same sample. Because it represents the random differences within samples, this part of the total variation is called

- the **variation within samples**, or
- the **error variation**, or
- the **sum of squares due to replications within levels of Factor *A***, which we will denote by

$$\text{SS}_{R(A)} = \sum \sum (x_{ai} - \bar{x}_{a.})^2 \qquad \textbf{(15.8)}$$

The subscript on $SS_{R(A)}$ stands for "replications within levels of Factor A." This sum of squares has $df_{R(A)} = n - k$ degrees of freedom, because it measures the variation of n observations (the x_{ai}) around the k sample means (the $\bar{x}_{a.}$).

Now we have identified two separate sums of squares [Expressions (15.7) and (15.8)], each with its own number of degrees of freedom. Recall that a variance estimate s^2, which is usually called a **mean square** (MS) in the context of analysis of variance, is obtained by dividing a sum of squares by its number of degrees of freedom (df):

$$MS = \frac{SS}{df} \tag{15.9}$$

The mean square for Factor A,

$$MS_A = \frac{SS_A}{df_A} = \frac{SS_A}{k - 1} \tag{15.10}$$

measures the dispersion of the sample means $\bar{x}_{a.}$ around the grand mean $\bar{x}$. If the differences among the sample means $\bar{x}_{a.}$ are due to random variation only (and not to differences in the underlying populations), then the mean square for Factor A, MS_A, is an estimate of the variance σ^2 of the population of observations. In other words, MS_A measures random variability if the null hypothesis in Expression (15.3) is true. If the null hypothesis is not true, MS_A measures more than random variability—the sample means will tend to differ by more than a random amount—and so MS_A will tend to be larger than a measure of just random variability. In any event, MS_A is a measure of the differences among the sample means. Now we need something against which to compare it, in order to determine whether those differences are larger than random variation can reasonably account for.

The within-sample mean square,

$$MS_{R(A)} = \frac{SS_{R(A)}}{df_{R(A)}} = \frac{SS_{R(A)}}{n - k} \tag{15.11}$$

measures the random variability of the individual observations around their own sample mean. Hence, it is sometimes called the **error mean square** or **error variance** in the one-way ANOVA. It is a measure of random variability that does not depend on the truth of the null hypothesis; $MS_{R(A)}$ is an unbiased estimate of the variance σ^2 of the population of observations whether or not the null hypothesis is true.

We now have two estimates of the same population variance σ^2, although the first one, in Expression (15.10), requires that the null hypothesis be true or else it will tend to be too large. Recalling from Section 12.4 that the ratio of two independent estimates of the same Gaussian-population variance follows the F distribution, we obtain the *test statistic for one-way ANOVA with independent samples:*

$$F_A = \frac{\text{MS}_A}{\text{MS}_{R(A)}} \quad \textbf{(15.12)}$$

which follows the F distribution with df_A and $\text{df}_{R(A)}$ degrees of freedom if the null hypothesis is true. (The equal-variance and normality assumptions were noted on page 513.)

Computational Formulas and the ANOVA Summary Table

The sum-of-squares formulas in Expressions (15.6), (15.7), and (15.8) are definitional formulas. They show what part of the total variation is being computed in each case—the variation of what around what—but they are generally not the most convenient formulas for computational purposes. The following are the *computational formulas* for the one-way analysis of variance with independent samples:

$$\text{SS}_A = \sum \frac{T_{a.}^2}{n_a} - \frac{T^2}{n} \quad \textbf{(15.13)}$$

$$\text{SS}_{R(A)} = \sum \sum x_{ai}^2 - \sum \frac{T_{a.}^2}{n_a} \quad \textbf{(15.14)}$$

$$\text{SS}_{\text{total}} = \sum \sum x_{ai}^2 - \frac{T^2}{n} \quad \textbf{(15.15)}$$

Only two of these formulas are needed, because the first two represent a partitioning of the third:

$$\text{SS}_A + \text{SS}_{R(A)} = \text{SS}_{\text{total}}$$

Their associated degrees of freedom have likewise been partitioned:

$$(k - 1) + (n - k) = n - 1$$

The results of an analysis of variance are often displayed in the form of an **analysis of variance summary table**. Such a table shows, for each source of variation in the data, the relevant sum of squares, number of degrees of freedom, mean square, and F ratio where applicable. The general form of the **ANOVA summary table for the one-way ANOVA with independent samples** is shown in Table 15.1. (Guidelines for how to round and how many digits to report are given in Appendix 1.)

We will consider the interpretation of the F ratio in the next section. In the rest of this section, we look at other information contained in the ANOVA summary table.

The ANOVA summary table provides information about the variability of the observations and about the precision of the sample means. The error variance $\text{MS}_{R(A)}$ is a pooled estimate (which we called s_{pooled}^2 in Section 11.2) of the population variance σ^2 of the observations. Its square root is the sample

TABLE 15.1 Analysis of variance table for one-way ANOVA with independent samples.

Source of variation	SS	df	MS	F
Factor A	SS_A	$k - 1$	MS_A	$F_A = \dfrac{MS_A}{MS_{R(A)}}$
Replications within A	$SS_{R(A)}$	$n - k$	$MS_{R(A)}$	—
Total	SS_{total}	$n - 1$		

standard deviation, say, s_{pooled}, of the observations within the samples:

$$s_{pooled} = \sqrt{\text{error mean square}} \tag{15.16}$$

$$= \sqrt{MS_{R(A)}} \quad \text{in one-way ANOVA with independent samples}$$

This is the standard or "average" amount by which the observations differ from their own sample mean.

The **estimated standard error of a sample mean** was given by the formula $s/\sqrt{n}$ in Expression (9.2) as an indication of the precision of a sample mean. In the present context, this formula may be written as

Estimated standard error of a sample mean

$$= \frac{s_{pooled}}{\sqrt{\text{number of observations on which the mean is based}}}$$

$$= \sqrt{\frac{\text{error mean square}}{\text{number of observations on which the mean is based}}} \tag{15.17}$$

Expression (15.17) gives us an estimate of the standard or "average" amount by which random sample means differ from their population mean.

The estimated standard error in Expression (15.17) summarizes important information about the precision of the sample mean. For example, we can use the estimated standard error to obtain confidence intervals for the corresponding population mean, as outlined in Section 10.3. The t-value required for the confidence interval is obtained from Appendix 4 with degrees of freedom appropriate to the error mean square, $df_{R(A)}$.

EXAMPLE 15.5 We will illustrate the computations with the data from Example 15.4, reprinted here for convenience of reference.

Sample 1	Sample 2	Sample 3	Total sample
8	6	3	
7	6	2	
8	5	4	
8			
			$k = 3$
$T_{1.} = 31$	$T_{2.} = 17$	$T_{3.} = 9$	$T = 57$
$n_1 = 4$	$n_2 = 3$	$n_3 = 3$	$n = 10$
$\bar{x}_{1.} = 7.75$	$\bar{x}_{2.} = 5.67$	$\bar{x}_{3.} = 3.00$	$\bar{x} = 5.7$

The three samples represent three categories of distance between a student's home and college residence. The independent variable, or Factor A, in this example therefore is distance. The variation due to distance is calculated from Expression (15.13) as

$$SS_A = \left(\frac{31^2}{4} + \frac{17^2}{3} + \frac{9^2}{3}\right) - \frac{57^2}{10} = 363.5833 - 324.9000 = 38.6833$$

The variation among students within the same distance category is calculated from Expression (15.14) as

$$SS_{R(A)} = (8^2 + 7^2 + 8^2 + 8^2 + 6^2 + 6^2 + 5^2 + 3^2 + 2^2 + 4^2) - 363.5833$$
$$= 367.0000 - 363.5833 = 3.4167$$

We obtain the corresponding mean squares by dividing each sum of squares by its number of degrees of freedom, as shown in Expressions (15.10) and (15.11):

$$MS_A = \frac{38.6833}{3 - 1} = 19.3417$$

$$MS_{R(A)} = \frac{3.4167}{10 - 3} = .4881$$

The F ratio for assessing whether MS_A is large relative to the measure of random variability $MS_{R(A)}$ is given by Expression (15.12):

$$F_A = \frac{19.3417}{.4881} = 39.63$$

The calculated values are presented in the following ANOVA summary table.

Source of variation	**SS**	**df**	**MS**	***F***
Distance	38.683	2	19.342	39.6
Students within distance categories	3.417	7	.488	—
Total	42.100	9		

We will see in a moment how to interpret the value of F_A. For now, we note that the error variance $MS_{R(A)}$ is interpretable as a pooled estimate s^2_{pooled} of the population variance σ^2 of the observations. Thus, $s_{\text{pooled}} = \sqrt{MS_{R(A)}} = \sqrt{.4881} = .70$ visits may be interpreted as the sample standard deviation of the observations within the samples. It is a measure of the inherent variability in the data, indicating a standard amount by which observations deviate from their own sample mean.

According to Expression (15.17), the estimated standard errors of the sample means are, respectively,

$$s_{\bar{x}_1} = \sqrt{\frac{.4881}{4}} = .3493$$

$$s_{\bar{x}_2} = \sqrt{\frac{.4881}{3}} = .4034$$

$$s_{\bar{x}_3} = \sqrt{\frac{.4881}{3}} = .4034$$

or approximately .35, .40, and .40. The estimated standard errors may be used to obtain confidence intervals for the true mean number of visits home by students in the three distance categories. For example, the 95% confidence limits for students living close to the college are given by Expression (10.25), using $t_{.025}(7\text{ df}) = 2.365$, as

$$7.75 \pm (2.365)(.3493)$$

indicating a true mean of between 6.92 and 8.58 visits home, at the 95% confidence level.

The Decision Rule

The null hypothesis is that the k population means are equal:

$$H_0\colon \mu_1 = \mu_2 = \cdots = \mu_k \tag{15.18}$$

If this null hypothesis is true, then the test statistic,

$$F_A = \frac{\text{MS}_A}{\text{MS}_{R(A)}} \tag{15.19}$$

may be regarded as a random selection from the $F(\text{df}_A, \text{df}_{R(A)})$ distribution. In this case, MS_A and $\text{MS}_{R(A)}$ both estimate the same parameter and therefore should not differ except by an amount attributable to chance. Their ratio, F_A, should not be too far from 1.00 if the null hypothesis is true.

On the other hand, if the null hypothesis is not true, then MS_A, which reflects any true differences among the population means in addition to random variation, will tend to be larger than $\text{MS}_{R(A)}$, which reflects random variation only. In this case, the F ratio in Expression (15.19) will tend to be greater than 1.00. Notice that even though the alternative hypothesis, given in Expression (15.4), is nondirectional, the F test in ANOVA is always one-tailed, because any differences among the means are squared in the computation of MS_A. As a result, large differences, whether positive or negative, are translated into F-values in the upper tail of the F distribution.

If the observed value of the test statistic F_A is equal to or greater than the critical value $F_\alpha(\text{df}_A, \text{df}_{R(A)})$, we reject the null hypothesis in Expression (15.18) at the α significance level. If the observed value of the test statistic is

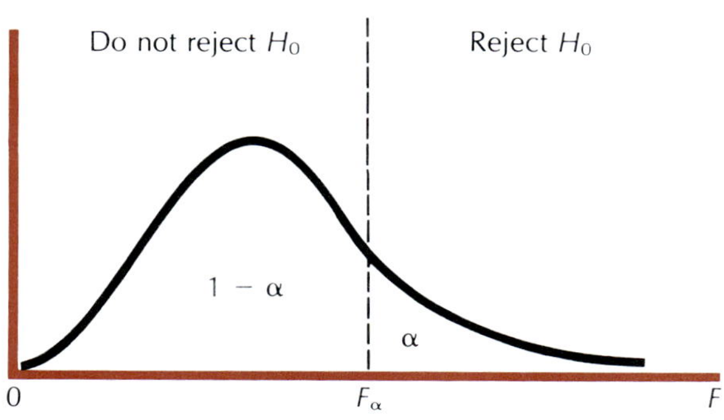

Figure 15.1 Sampling distribution of the F ratio in an analysis of variance if the null hypothesis is true. The critical value is shown for the decision about the null hypothesis against the general nondirectional (two-sided) alternative at the α significance level.

less than the critical value, we do not have enough evidence to reject the null hypothesis at the α significance level; the sample differences are within the realm of random variation from equal population means. The decision rule is illustrated in Figure 15.1.

EXAMPLE 15.6 Example 15.2 reported that the mean numbers of visits home per semester by female students living in college residences were 7.75, 5.67, and 3.00 for three samples of students whose homes were at progressively greater distances from the college. Are these means significantly different from each other?

The computations for this example were demonstrated in Example 15.5, where we found that the observed value of the test statistic was $F_A = 39.6$, based on 2 degrees of freedom for the numerator and 7 degrees of freedom for the denominator. For a .01-level test of the null hypothesis that the populations of students at the three distances do not differ with respect to mean number of visits home, the critical value is given in Appendix 6 as $F_{.01}(2, 7\text{ df}) = 9.55$ (see Figure 15.2). Since the observed value 39.6 exceeds the critical value 9.55, we reject the null hypothesis and conclude that the sample means are significantly different from each other ($P < .01$). The relative sizes of the sample means indicate that students from more distant homes visit home less frequently, and the F test indicates that this sample difference is statistically significant.

When there are just $k = 2$ independent-sample means to be compared, you may wonder whether to use the one-way ANOVA F test of this section or the independent-samples t test of Section 11.2. If the alternative hypothesis is one-sided, the t test must be used. But in the usual case where the alternative hypothesis is nondirectional, the two tests are exactly equivalent, since $F_A = t^2_{\text{obs}}$. [See Expression (12.26).]

Summary of the Test Procedure

The **one-way analysis of variance with independent samples** may be summarized as follows:

Assumptions.

a. Independent random samples.

b. Normality of the underlying populations.

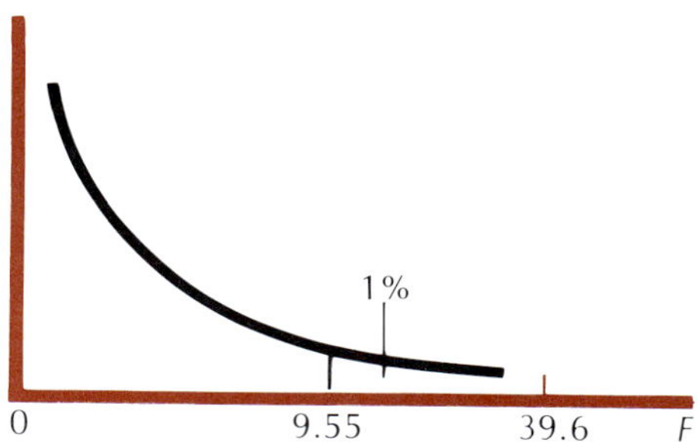

Figure 15.2

c. Equal population variances.

Null hypothesis. H_0: $\mu_1 = \mu_2 = \cdots = \mu_k$

Alternative hypothesis. H_A: The k population means are not all equal.

Step 1. Determine the values of k, n_a (for each of the k samples), and n.

Step 2. Calculate the k sample totals $T_{a.}$ and the grand total T.

Step 3. Calculate the k sample means $\bar{x}_{a.}$, in order to see what sample difference is being tested.

Step 4. Calculate the values required to complete the ANOVA summary table in the format of Table 15.1.

Step 5. Look up the critical value $F_\alpha(\text{df}_A, \text{df}_{R(A)})$.

Step 6. If the observed value F_A is greater than or equal to the critical value, conclude that the sample means are significantly different from each other ($P < \alpha$). In this case, examine the k sample means $\bar{x}_{a.}$ to discover the nature of the difference. If F_A is less than the critical value, conclude that the sample difference is not statistically significant ($P > \alpha$).

15.3 TWO-WAY ANOVA WITHOUT REPLICATION

Section 11.3 presented a t test for judging the significance of the difference between means of two paired samples. The **two-way analysis of variance without replication** is a generalization of the paired-samples t test in that pairs of two observations are replaced by **blocks** of k_A observations, and a test of the difference between two means is replaced by a test of the difference between k_A means, where k_A is any integer 2 or more.

To compare two levels of an independent variable using paired samples, we arrange to have one member of each pair in each of the two samples. Similarly, to compare k_A levels of Factor A using blocks, we arrange to have one member of each block in each of the k_A samples. To understand blocks, think of paired samples where a "pair" is not limited to two members. A **block** is a set of k_A observational units that are highly similar in ways related to the dependent variable, and where each of the units in a block is associated with a different one of the k_A levels of the independent variable, Factor A. A block of two observational units is a pair, in the sense of the paired-samples t test. More generally, **blocking** refers to arranging the observational units into relatively homogeneous sets called **blocks** (of k_A units each) and administering every one of the k_A treatments once in each block.

EXAMPLE 15.7 Sets of three litter-mates may serve as blocks (of size three) in animal research to compare three treatments. Within each litter, the three litter-mates are randomly assigned to the three treatment conditions. The similarity of litter-mates makes it more likely that differences in their responses will be due to the treatments rather than due to individual differences.

Sets of four adjacent plots of land may serve as blocks (of size four) to compare four treatments in agricultural research. The four treatments are applied in each block, a different treatment for each of the four plots. Adjacent plots of land are likely to be much more comparable than widely separated plots of land, so blocking makes it easier to distinguish the effects of the treatments from those due to differences in the land.

Subjects who perform under all five conditions in a perception task are serving as blocks of size five. By performing under all conditions, each subject serves as his own control, making it easier to detect any effects of the treatments.

Blocking may be done on the basis of any variable that is relevant to the dependent variable and that results in greater variability between blocks than within blocks.

Observational units within the same block are similar to start with, so any effects of the treatments (Factor A) may be more easily detected than they would be without blocking. Increased precision is possible because each block, in effect, serves as its own control. In general, **blocking** will increase the precision of the comparison of a set of k_A means if the similarity within the blocks is sufficiently greater than the similarity between blocks.

The term *two-way* in the title of this section indicates that the observational units are cross-classified according to two factors, say, Factor A and Factor B. Factor A is any independent variable or basis of classification of the observational units, just as in one-way ANOVA, and Factor B is another basis of classification that we have called **blocks** in the preceding paragraphs. In a two-way ANOVA, every one of the k_A levels of Factor A occurs in combination with every one of the k_B levels of Factor B, for a total of $k_A k_B$ **treatment combinations**.

The phrase *without replication* in the title of this section indicates that there is only one observational unit *in each treatment combination*. In one sense, there is still replication in a two-way design without replication. Indeed, if the blocks are used to obtain increased precision in testing Factor A (and not because of any interest in testing for differences between blocks), then the blocks are often referred to as **replications** or as the **replication factor**. They provide replications of the various levels of Factor A. It is the $k_A k_B$ treatment combinations that are "without replication" in the two-way ANOVA without replication: Each treatment combination is observed in one observational unit only.

Because the data consist of repeated observations of the same block under the various levels of Factor A, the two-way ANOVA without replication is often referred to as the **one-way ANOVA with repeated measurements**. The latter terminology is consistent with the fact that primary interest in such an analysis is usually in the difference between levels of Factor A and not in the differences between blocks, which play the role of a replication factor.

If the k_A levels of Factor A are randomly assigned to the k_A observational units in each block, the two-way ANOVA without replication is sometimes called a **randomized-blocks ANOVA**.

EXAMPLE 15.8 In a study comparing three supermarkets in the same city, prices were obtained on the same day for eight products sold at all three stores. The prices (in cents) were as follows for products of the same quantity and brand.

Product	Store 1	Store 2	Store 3
Lettuce	59	59	47
Mushrooms	71	79	94
Tomatoes	99	129	129
Eggs	112	110	105
Cottage cheese	195	179	193
Pork chops	205	195	199
Shrimp	389	349	375
Coffee	439	385	434

Is there a significant difference in the prices at the three stores?

The independent variable of interest here is store; let us label this variable as Factor A. The three different stores represent the $k_A = 3$ levels of Factor A to be compared.

The products are the blocks. Observing the same eight products at all three stores will make it easier to detect different price levels at the three stores than it would be if we had observed a different set of eight products at each store. Let us refer to product as Factor B. There are $k_B = 8$ levels of Factor B.

Thus, we have identified two factors, store and product, that might have an effect on the dependent variable, price.

(Source: Student project by G. Lazicki.)

Hypotheses and Data

We will use the following notation for the two-way ANOVA without replication:

k_A is the number of levels of Factor A. In the one-way case, we referred to this simply as k, since Factor A was the only factor there.

k_B is the number of levels of Factor B. If Factor B represents blocks, then k_B is the number of blocks.

$n = k_A k_B$ is the total number of observations.

x_{ab} represents the observation recorded for the treatment combination consisting of the ath level of Factor A and the bth level of Factor B. The first subscript identifies which level of Factor A is involved, and the second subscript identifies which level of Factor B.

$T_{a.} = \sum_{b=1}^{k_B} x_{ab}$ is the sum of the observations at the ath level of Factor A. The dot marks the place of the second subscript and indicates that this is a total of the observations at all k_B levels of Factor B.

$\bar{x}_{a.} = \dfrac{T_{a.}}{k_B}$ is the ordinary arithmetic mean of the observations at the ath level of Factor A.

$T_{.b} = \sum_{a=1}^{k_A} x_{ab}$ is the sum of the observations at the bth level of Factor B. The dot marks the place of the first subscript and indicates that this is a total of the observations at all k_A levels of Factor A.

$\bar{x}_{.b} = \dfrac{T_{.b}}{k_A}$ is the ordinary arithmetic mean of the observations at the bth level of Factor B.

$T = \sum\sum x_{ab} = \sum T_{a.} = \sum T_{.b}$ is the grand total of all the observations in the entire sample.

$\bar{x} = \dfrac{T}{n}$ is the grand mean of all the observations in the entire sample.

EXAMPLE 15.9 We will illustrate the preceding notation with the data from Example 15.8, concerning the prices of $k_B = 8$ products at $k_A = 3$ stores, as follows.

Product (Factor *B*)	Store (Factor *A*): Store 1	Store 2	Store 3		
Lettuce	$x_{11} = 59$	$x_{21} = 59$	$x_{31} = 47$	$T_{.1} = 165$	$\bar{x}_{.1} = 55.0$
Mushrooms	$x_{12} = 71$	$x_{22} = 79$	$x_{32} = 94$	$T_{.2} = 244$	$\bar{x}_{.2} = 81.3$
Tomatoes	$x_{13} = 99$	$x_{23} = 129$	$x_{33} = 129$	$T_{.3} = 357$	$\bar{x}_{.3} = 119.0$
Eggs	$x_{14} = 112$	$x_{24} = 110$	$x_{34} = 105$	$T_{.4} = 327$	$\bar{x}_{.4} = 109.0$
Cottage cheese	$x_{15} = 195$	$x_{25} = 179$	$x_{35} = 193$	$T_{.5} = 567$	$\bar{x}_{.5} = 189.0$
Pork chops	$x_{16} = 205$	$x_{26} = 195$	$x_{36} = 199$	$T_{.6} = 599$	$\bar{x}_{.6} = 199.7$
Shrimp	$x_{17} = 389$	$x_{27} = 349$	$x_{37} = 375$	$T_{.7} = 1113$	$\bar{x}_{.7} = 371.0$
Coffee	$x_{18} = 439$	$x_{28} = 385$	$x_{38} = 434$	$T_{.8} = 1258$	$\bar{x}_{.8} = 419.3$
$k_A = 3$	$T_{1.} = 1569$	$T_{2.} = 1485$	$T_{3.} = 1576$	$T = 4630$	
$k_B = 8$, $n = 24$	$\bar{x}_{1.} = 196.1$	$\bar{x}_{2.} = 185.6$	$\bar{x}_{3.} = 197.0$		$\bar{x} = 192.9$

The two-way ANOVA without replication provides answers to two separate questions. The questions correspond to two separate null hypotheses, each with its own significance test.

One question is whether the sample means $\bar{x}_{a.}$ at the different levels of Factor A differ significantly from each other. The Factor A null hypothesis may be written as

$$H_0(A)\colon\ \mu_{1.} = \mu_{2.} = \cdots = \mu_{k_A.} \tag{15.20}$$

where $\mu_{a.}$ is the mean of the population of observations at the ath level of Factor A. The F test in the analysis of variance tests against the nondirectional alternative that the means are not all equal:

$$H_A(A)\colon\ \text{Not } H_0(A) \tag{15.21}$$

The other question is whether the sample means $\bar{x}_{.b}$ at the different levels of Factor B differ significantly from each other. The Factor B null hypothesis may be written as

$$H_0(B)\colon\ \mu_{.1} = \mu_{.2} = \cdots = \mu_{.k_B} \tag{15.22}$$

where $\mu_{.b}$ is the mean of the population of observations at the bth level of

Factor B. Again the alternative is nondirectional:

$$H_A(B)\text{: Not } H_0(B) \tag{15.23}$$

If Factor B is a blocking factor, there is usually not much interest in testing for differences between blocks. Indeed, the blocks are deliberately chosen so that there will be greater differences between blocks than within blocks. In this case, the test of the Factor B null hypothesis is generally omitted.

EXAMPLE 15.10 In Example 15.8, the investigation is directed toward testing the Factor A null hypothesis that the true mean price per item is the same at the three stores:

$$H_0(A)\text{: } \mu_{1.} = \mu_{2.} = \mu_{3.}$$

The Factor B null hypothesis is that the mean prices of the eight products are all equal:

$$H_0(B)\text{: } \mu_{.1} = \mu_{.2} = \mu_{.3} = \mu_{.4} = \mu_{.5} = \mu_{.6} = \mu_{.7} = \mu_{.8}$$

However, product is a blocking factor in this investigation, and there is no interest in testing the Factor B null hypothesis. Different products were known in advance to be in different price ranges. The eight products were chosen not to compare prices of different *products* (Factor B null hypothesis), but rather to increase precision in comparing prices at different *stores* (Factor A null hypothesis).

The Test Statistics

Test statistics to test the null hypotheses in Expressions (15.20) and (15.22) are obtained via the same kind of derivation as in the one-way ANOVA: If the deviation $x_{ab} - \bar{x}$ of each observation from the grand mean can be expressed as a sum of deviations associated with particular sources of variation, then the total variation in the sample data can be partitioned into parts due to those sources. In two-way ANOVA without replication, each observation's total deviation $x_{ab} - \bar{x}$ may be separated into three parts as follows:

$$x_{ab} - \bar{x} = [\bar{x}_{a.} - \bar{x}] + [\bar{x}_{.b} - \bar{x}] + [x_{ab} - (\bar{x}_{a.} - \bar{x}) - (\bar{x}_{.b} - \bar{x}) - \bar{x}] \tag{15.24}$$

By squaring both sides of the equality and summing over all the observations in the sample, it can be shown that the total variation,

$$\text{SS}_{\text{total}} = \sum \sum (x_{ab} - \bar{x})^2 \tag{15.25}$$

is likewise separated into three corresponding sources of variation, which are identified in the following paragraphs.

First, the variation of the Factor A means $\bar{x}_{a.}$ around the grand mean $\bar{x}$ is due to differences among the means $\bar{x}_{a.}$ for the different levels of Factor A. This is exactly the same as the *variation due to Factor A* in the one-way analysis of variance, and it is again denoted by

$$\text{SS}_A = k_B \sum (\bar{x}_{a.} - \bar{x})^2 \tag{15.26}$$

Because this sum of squares measures the variation of the k_A Factor A means around the one grand mean, it has $\text{df}_A = k_A - 1$ degrees of freedom. If the differences among the Factor A means are due to random variation only—that is, if the null hypothesis in Expression (15.20) is true—then the mean square for Factor A,

$$\text{MS}_A = \frac{\text{SS}_A}{\text{df}_A} \tag{15.27}$$

measures random variability. If the null hypothesis in Expression (15.20) is not true, then MS_A reflects any true differences among the different levels of Factor A, in addition to random variability.

Second, the variation of the Factor B means $\bar{x}_{.b}$ around the grand mean $\bar{x}$ is due to differences among the means $\bar{x}_{.b}$ for the different levels of Factor B. The *variation due to Factor B* is measured by the sum of squares due to Factor B:

$$\text{SS}_B = k_A \sum (\bar{x}_{.b} - \bar{x})^2 \tag{15.28}$$

Because this sum of squares measures the variation of the k_B Factor B means around the one grand mean, it has $\text{df}_B = k_B - 1$ degrees of freedom. If the differences among the Factor B means are due to random variation only—that is, if the null hypothesis in Expression (15.22) is true—then the mean square for Factor B,

$$\text{MS}_B = \frac{\text{SS}_B}{\text{df}_B} \tag{15.29}$$

measures random variability. If the null hypothesis in Expression (15.22) is not true, then MS_B reflects any true differences among the different levels of Factor B, in addition to random variability.

To explain the third source of variation, let us refer to the levels of Factor A as *treatments* and the levels of Factor B as *blocks*. The third source of variation is associated with "differences of differences": It reflects the extent to which the differences between blocks are different for the different treatments. Equivalently, we can say that it reflects the extent to which the differences between treatments are different in the different blocks. Such differences in the treatment effects in different blocks are assumed to be due to random variability. In other words, this source of variation represents any unique contribution of a particular treatment combination's observation x_{ab} over and above the contribution $\bar{x}_{a.} - \bar{x}$ of the level of Factor A that is involved, and over and above the contribution $\bar{x}_{.b} - \bar{x}$ of the level of Factor B that is involved, and over and above the contribution $\bar{x}$ of the grand mean. Because this variation is due to particular *combinations* of treatments, it is called the variation due to the **interaction** of Factor A and Factor B, or the **variation due to the *A*-by-*B* interaction**, which is measured by the interaction sum of squares:

$$SS_{AB} = \sum \sum [x_{ab} - (\bar{x}_{a.} - \bar{x}) - (\bar{x}_{.b} - \bar{x}) - \bar{x}]^2 \tag{15.30}$$

EXAMPLE 15.11 To illustrate what "differences of differences" are measured by the interaction sum of squares, let us consider a small portion of the data from Example 15.8, as follows.

	Store (Factor *A*)		
Product (Factor *B*)	**Store 1**	**Store 2**	**Means**
Lettuce	59	59	59
Mushrooms	71	79	75
Means	65	69	67

To set the stage for "differences of differences," we first describe the simple "differences" that are *not* measured by the interaction sum of squares.

The Factor *A* sum of squares measures the variation of the Factor *A* means (such as 65 and 69, in the reduced example above) relative to the grand mean (67). The Factor *A* means can differ from the grand mean only if they differ from each other. Thus, the Factor *A* sum of squares reflects the extent to which the Factor *A* means (65 and 69) differ from each other, and thus it tells us about the difference between levels of Factor *A*.

Similarly, the Factor *B* sum of squares measures the variation of the Factor *B* means (59 and 75) relative to the grand mean (67). Thus, the Factor *B* sum of squares reflects the extent to which the Factor *B* means (59 and 75) differ from each other, and thus it tells us about the difference between levels of Factor *B*.

By way of contrast, the interaction sum of squares is not concerned with the differences in the margins of the preceding table. Rather, the interaction reflects the extent to which the difference between products is different in different stores: Mushrooms are 12 cents more expensive than lettuce in Store 1 but 20 cents more expensive in Store 2. The price difference between the products is different in the two stores (12-cent and 20-cent differences, respectively). This "difference in the differences" is reflected in the interaction sum of squares.

Equivalently, we can say that the price difference between the two stores is different for the two products: Store 2 charges 8 cents more than Store 1 for mushrooms but 0 cents more for lettuce. Again, the difference between levels of one factor is different at different levels of the other factor. Such "differences of differences" characterize an interaction.

Notice that we do not look at the marginal means to see if there is evidence of an interaction; the marginal means tell us about Factor *A* and about Factor *B*, but not about their interaction. An interaction is described in terms of cell values—particular combinations of levels of the factors. An interaction indicates that there is something about a particular combination of levels that does not show up in the marginal means. The levels "interact" so that the combination yields a value not predictable from the separate effects of the two factors. Consider, for example, the 71-cent mushrooms at Store 1. From the Factor *A* means, we would expect mushrooms (and everything else) to be 4 cents cheaper at Store 1 than at Store 2, but the actual difference in the price of mushrooms is 8 cents. From the Factor *B* means, we would expect mushrooms to be 16 cents more expensive than lettuce at Store 1 (and everywhere else), but the actual difference at Store 1 is only 12 cents. There is something about the combination of "mushrooms" and "Store 1" such that "mushrooms at Store 1" cost less than would be expected on the basis of the separate effects of the two factors. An interaction thus reflects the unique contribution of particular treatment *combinations*. If there is an interactive effect of particular combinations of levels, then the differences between levels of one factor are different at different levels of another factor.

The number of degrees of freedom df_{AB} associated with the interaction sum of squares SS_{AB} may be determined by noting, as shown in Expression (15.30), that SS_{AB} measures the variation of the $k_A k_B$ observations subject to $k_A - 1$ restrictions for the Factor A means, $k_B - 1$ restrictions for the Factor B means, and 1 restriction for the grand mean. This yields

$$\begin{aligned}\mathrm{df}_{AB} &= k_A k_B - (k_A - 1) - (k_B - 1) - 1 \\ &= k_A k_B - k_A - k_B + 1\end{aligned}$$

which is simply the product of the numbers of degrees of freedom for Factors A and B:

$$\mathrm{df}_{AB} = (k_A - 1)(k_B - 1) \tag{15.31}$$

We obtain a variance estimate from the variation due to interaction by dividing the sum of squares by its degrees of freedom. The resulting interaction mean square,

$$\mathrm{MS}_{AB} = \frac{\mathrm{SS}_{AB}}{\mathrm{df}_{AB}} \tag{15.32}$$

may be taken as a measure of random variability—the **error variance**—in the two-way ANOVA without replication.

We now have three estimates of the same underlying random variability σ^2. However, the first one, MS_A, will tend to be too large if the Factor A null hypothesis in Expression (15.20) is not true, and the second one, MS_B, will tend to be too large if the Factor B null hypothesis in Expression (15.22) is not true. The third one, MS_{AB}, may be taken as a measure of random variability only. Since the ratio of two independent estimates of the same Gaussian-population variance follows the F distribution, we obtain the *test statistics for the two-way ANOVA without replication*:

$$F_A = \frac{\mathrm{MS}_A}{\mathrm{MS}_{AB}} \tag{15.33}$$

follows the F distribution with df_A and df_{AB} degrees of freedom if the Factor A null hypothesis in Expression (15.20) is true, and

$$F_B = \frac{\mathrm{MS}_B}{\mathrm{MS}_{AB}} \tag{15.34}$$

follows the F distribution with df_B and df_{AB} degrees of freedom if the Factor B

null hypothesis in Expression (15.22) is true. (The equal-variance and normality assumptions were noted on page 513.)

As noted earlier in this section, the value of F_B is usually not reported if Factor B is a blocking factor. However, the investigator should still make an informal comparison of MS_B and MS_{AB} to see whether the blocking was successful. The purpose of blocking is to group observational units into relatively homogeneous blocks, so that precision is improved for comparisons within the blocks. Thus, in successful blocking, the variability within blocks, as measured by MS_{AB}, will be considerably less than that between blocks, as measured by MS_B.

Computational Formulas and the ANOVA Summary Table

The sum-of-squares formulas in Expressions (15.26), (15.28), and (15.30) are definitional formulas. You should examine them to see what part of the total variation is being computed in each case—the variation of what around what. However, they are generally not as convenient for computational purposes as are the following *computational formulas* for the two-way analysis of variance without replication:

$$SS_A = \frac{1}{k_B}\sum^{k_A} T_{a.}^2 - \frac{T^2}{n} \tag{15.35}$$

$$SS_B = \frac{1}{k_A}\sum^{k_B} T_{.b}^2 - \frac{T^2}{n} \tag{15.36}$$

$$SS_{AB} = \sum^{k_A}\sum^{k_B} x_{ab}^2 - SS_A - SS_B - \frac{T^2}{n} \tag{15.37}$$

$$SS_{\text{total}} = \sum^{k_A}\sum^{k_B} x_{ab}^2 - \frac{T^2}{n} \tag{15.38}$$

Only three of these formulas are needed, because the first three represent a partitioning of the fourth:

$$SS_A + SS_B + SS_{AB} = SS_{\text{total}} \tag{15.39}$$

Their associated degrees of freedom have likewise been partitioned:

$$\begin{aligned}(k_A - 1) + (k_B - 1) + (k_A - 1)(k_B - 1) &= k_A k_B - 1 \\ &= n - 1\end{aligned} \tag{15.40}$$

The general form of the **ANOVA summary table for the two-way ANOVA without replication** is shown in Table 15.2.

In addition to its role in the F tests, the error mean square MS_{AB} in the table provides information about the precision of the results. Substituting its value into Expression (15.17), we can obtain the estimated standard error of the sample means, from which we can calculate confidence intervals for the corresponding population means.

TABLE 15.2 Analysis of variance summary table for two-way ANOVA without replication.

Source of variation	SS	df	MS	F
Factor B	SS_B	$k_B - 1$	MS_B	$\left[F_B = \dfrac{MS_B}{MS_{AB}}\right]^a$
Factor A	SS_A	$k_A - 1$	MS_A	$F_A = \dfrac{MS_A}{MS_{AB}}$
A-by-B interaction	SS_{AB}	$(k_A - 1)(k_B - 1)$	MS_{AB}	—
Total	SS_{total}	$n - 1$		

[a] The F ratio for Factor B is often not reported if B is a blocking factor.

Significance tests follow the pattern described for one-way ANOVA although here two tests may be made. The observed value of the test statistic F_A is compared to the critical value $F_\alpha(\text{df}_A, \text{df}_{AB})$ for an α-level test of the Factor A null hypothesis in Expression (15.20). The observed value of the test statistic F_B may be compared to the critical value $F_\alpha(\text{df}_B, \text{df}_{AB})$ for an α-level test of the Factor B null hypothesis in Expression (15.22). These are separate tests of two different null hypotheses, and either test statistic may be found to be statistically significant regardless of the outcome of the other test. The decision rule in both cases is that illustrated in Figure 15.1.

EXAMPLE 15.12 Example 15.8 presented data for three competing supermarkets concerning the prices of eight products common to the three stores. The initial computations for the two-way ANOVA without replication were demonstrated in Example 15.9. We found that the mean price per item at Stores 1, 2, and 3 was 196.1 cents, 185.6 cents, and 197.0 cents, respectively. Is this observed difference in prices statistically significant, or is it the kind of sample difference that we could reasonably expect to obtain in a sample of this size if the true mean price were the same at the three stores?

The dependent variable in this example is price. The independent variable (Factor A) is store, with $k_A = 3$ levels, and product (Factor B) is a blocking factor with $k_B = 8$ levels. We can partition the total variation in price into three parts. Using Expression (15.35), we calculate the variation due to stores from the preliminary computations in Example 15.9:

$$SS_A = \frac{1}{8}(1{,}569^2 + 1{,}485^2 + 1{,}576^2) - \frac{4{,}630^2}{24} = 893{,}845.25 - 893{,}204.17 = 641.08$$

The variation due to products is calculated from Expression (15.36) as

$$\begin{aligned} SS_B &= \frac{1}{3}(165^2 + 244^2 + \cdots + 1{,}258^2) - 893{,}204.17 \\ &= 1{,}274{,}254.00 - 893{,}204.17 = 381{,}049.83 \end{aligned}$$

The variation due to the interaction of stores and products is calculated from Expression (15.37) as

$$\begin{aligned} SS_{AB} &= (59^2 + 71^2 + 99^2 + \cdots + 375^2 + 434^2) - SS_A - SS_B - 893{,}204.17 \\ &= 1{,}278{,}056.00 - 641.08 - 381{,}049.83 - 893{,}204.17 = 3{,}160.92 \end{aligned}$$

We obtain the corresponding mean squares by dividing each sum of squares by its number of degrees of freedom:

$$\text{MS}_A = \frac{641.08}{3 - 1} = 320.54$$

$$\text{MS}_B = \frac{381{,}049.83}{8 - 1} = 54{,}435.69$$

$$\text{MS}_{AB} = \frac{3{,}160.92}{(3 - 1)(8 - 1)} = 225.78$$

The error mean square in this case is $\text{MS}_{AB} = 225.78$, so the estimated standard error of each store's mean is $\sqrt{225.78/8} = 5.312$, or approximately 5.3 cents.

The F ratio to test the significance of the difference between the store means is given by Expression (15.33):

$$F_A = \frac{320.54}{225.78} = 1.42$$

Since products served as blocks in this investigation, there is no reason to test for differences between products (Factor B). An informal comparison reveals that MS_B is considerably larger than MS_{AB}, indicating that there was substantially more variability between blocks than within blocks. Thus the blocking was worthwhile, in that it increased the precision of the comparison between stores.

The results are summarized in the following ANOVA summary table.

Source of variation	**SS**	**df**	**MS**	***F***
Products	381,050	7	54,436	
Stores	641	2	321	1.42
P-by-*S* interaction	3,161	14	226	
Total	384,852	23		

The critical value for a .05-level test of the significance of the difference between stores is $F_{.05}(2, 14\text{ df}) = 3.74$ (see Figure 15.3). Since the observed value of the test statistic 1.42 is less than the critical value 3.74, we conclude that the observed

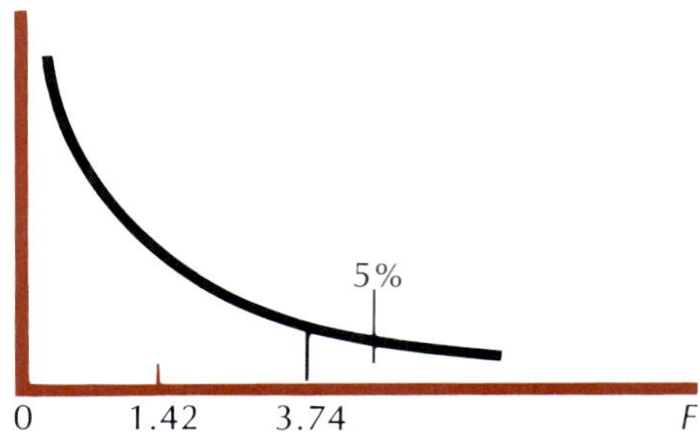

Figure 15.3

differences in mean price per item at the three stores were not statistically significant ($P > .05$). The sample did not provide sufficient evidence to conclude, at the 5% significance level, that the three stores differ with respect to true mean price per item.

A number of good computer programs are available to perform ANOVA calculations. A program in one of the major statistical packages, BMDP program 2V, produced the output shown in Figure 15.4 for this example. Some commentary is required: This program identifies the grand mean as a source of variation with one degree of freedom and accounts for a total of n degrees of freedom. In this book and in most applications, on the other hand, we partition the variation of the observations *around* the grand mean, and we omit the one-degree-of-freedom variation of the grand mean itself around zero, leaving a total of $n - 1$ degrees of freedom. Note that the sum of squares due to the mean on the printout equals T^2/n. The F test on the line labeled "mean" tests whether the grand mean is significantly different from zero. The first "error" source is blocks (products, in this example), and the second "error" source is the interaction of stores and products. The "tail probability" is the P-value. The last two columns on the right are refinements of the P-value that we will not consider here.

When there are just $k_A = 2$ Factor A means to be compared in blocks of size two, you may wonder whether to use the two-way ANOVA F test of this section or the paired-samples t test of Section 11.3. If the alternative hypothesis is one-sided, the t test must be used. However, in the usual case where the alternative hypothesis is nondirectional, the two tests are exactly equivalent, since $F_A = t^2_{\text{obs}}$. [See Expression (12.26).]

Summary of the Test Procedure

The **two-way analysis of variance without replication** may be summarized as follows:

Assumptions.

- **a.** Random sampling of blocks, or random assignment of the treatments or treatment combinations to the observational units.
- **b.** Additivity of the effects of the various factors: Treatment effects, block effects, and random error add together to yield particular observations.
- **c.** Normality of the populations underlying the various treatment combinations.
- **d.** Equal population variances.
- **e.** Equal population correlations for all possible pairs of treatments.

Hypotheses.

$$\begin{cases} H_0(A)\text{:} & \mu_{1.} = \mu_{2.} = \cdots = \mu_{k_A.} \\ H_A(A)\text{:} & \text{The } \mu_{a.} \text{ are not all equal.} \end{cases}$$

$$\begin{cases} H_0(B)\text{:} & \mu_{.1} = \mu_{.2} = \cdots = \mu_{.k_B} \\ H_A(B)\text{:} & \text{The } \mu_{.b} \text{ are not all equal.} \end{cases}$$

Step 1. Determine the values of k_A, k_B, and n.

```
ANALYSIS OF VARIANCE FOR   1-ST
DEPENDENT VARIABLE - PRICE     IN        CENTS

      SOURCE        SUM OF      DEGREES OF        MEAN           F      TAIL     GREENHOUSE    HUYNH
                    SQUARES      FREEDOM         SQUARE                PROB.     GEISSER      FELDT
                                                                                  PROB.       PROB.
      MEAN     893204.16667        1       893204.16667       16.41   0.0049
 1    ERROR    381049.83333        7        54435.69048

      STORE       641.08333        2          320.54167        1.42   0.2745      0.2763      0.2757
 2    ERROR      3160.91667       14          225.77976
```

Figure 15.4
BMDP2V printout for Example 15.12.

Step 2. Calculate the k_A Factor A totals $T_{a.}$, the k_B Factor B totals $T_{.b}$, and the grand total T.

Step 3. Calculate the k_A Factor A means, in order to see what sample difference is being tested. If Factor B is to be tested, do the same for the k_B Factor B means.

Step 4. Calculate the values required to complete the ANOVA summary table in the format of Table 15.2.

Step 5. If the observed value of the test statistic F_A is greater than or equal to the critical value $F_\alpha(\text{df}_A, \text{df}_{AB})$, conclude that there was a statistically significant effect due to Factor A ($P < \alpha$), and examine the k_A Factor A means to discover the nature of the effect. If the observed value F_A is less than the critical value, the effect of Factor A was not statistically significant ($P > \alpha$).

Step 6. If Factor B is a blocking factor designed to improve the precision of the test of Factor A, no formal test of Factor B is reported, although a large value of F_B indicates to the investigator that the blocking was successful. When a test of Factor B is required, it parallels that for Factor A: If F_B is greater than or equal to the critical value $F_\alpha(\text{df}_B, \text{df}_{AB})$, the effect of Factor B is significant ($P < \alpha$), and the Factor B means should be examined to discover the nature of the effect; otherwise, the effect was not significant ($P > \alpha$).

15.4 TWO-WAY ANOVA WITH INDEPENDENT SAMPLES OF EQUAL SIZE

We have been looking at the analysis of variance as a general procedure for testing hypotheses about sets of two or more means. These means may be arranged in a one-way classification suitable for analysis by the one-way ANOVA for independent samples. Or they may be arranged as the margins of a two-way classification, where significance tests may be made by the two-way ANOVA without replication. In this section, we consider situations where there is interest not only in testing the *marginal* means in a two-way classification, but also in testing for differences among the treatment combinations themselves, as measured by means in the *cells* of a two-way table. Such testing can readily be done if there is the same number of observations in every treatment combination. The method is called the **two-way analysis of variance with independent samples of equal size**.

Typically, three general questions are asked of the data in a two-way ANOVA with independent samples:

a. What is the effect of Factor A? Are the marginal means at the different levels of Factor A significantly different? This is called the **main effect** of Factor A. We have already met such effects in the previous two types of analysis of variance (Sections 15.2 and 15.3).

b. What is the effect of Factor B? Are the marginal means at the different levels of Factor B significantly different? This is the **main effect** of Factor B. In the two-way ANOVA with independent samples, both main effects (A and B) are normally tested, because neither factor is a blocking factor.

c. Is the effect of Factor A different for different levels of Factor B? Equivalently, is the effect of Factor B different for different levels of Factor A? This is called the A-by-B **interaction effect**. We talked about this interaction in the two-way ANOVA without replication (Section 15.3), but there was no test for its significance in that case because the treatment combinations were not replicated ("without replication"). The possibility of testing the significance of the interaction effect is the major new element in the two-way ANOVA with independent samples.

EXAMPLE 15.13 In the preparation of fiberglass skis for cross-country ski racing, a "grip" wax is applied to a region of the ski under the binding, for forward thrust, and a hard wax, known as "glider," is applied from the ends of the ski in to the grip-wax area, for speed. Two decisions that a racer must make are what type of glider wax to use and how far into the grip-wax area to extend the glider area.

An experiment was conducted to compare three brands of glider wax (Swix red, Falline red, and Toko white) that were prescribed for the prevailing conditions during testing (+4°C, moist snow). Each brand was tested with two different glider-wax areas that we will refer to as "short," in which glider wax was applied to all but a 24-inch grip-wax area of the ski, and "long," in which glider wax was applied to all but a 12-inch grip-wax area. (The grip-wax area was stripped bare of any wax for the test.) Four trials were conducted for each of the six treatment combinations of brand of wax and glider-wax area, in order to determine how far a skier would glide, without poling or pushing, along a level portion of a cross-country ski trail. For each trial, the same skier started from a standstill at the same place on a slight incline and passively rode the skis in a crouched position until they stopped. The distance traveled on each trial was measured from a stake at the base of the incline.

The dependent variable in this experiment was distance traveled. It was studied as a function of two independent variables. One independent variable, which we will call Factor A, was glider-wax area, with $k_A = 2$ levels, short and long. The other independent variable, which we will call Factor B, was brand of glider wax, with $k_B = 3$ levels. It is completely arbitrary which factor is labeled A and which one B.

The two-way analysis of variance with independent samples enables us to perform three significance tests and so to answer three different questions about what factors affect the distance traveled:

a. The main effect of glider-wax area (Factor A) refers to the difference between the mean distance traveled with a short glider-wax area and the mean distance traveled with a long glider-wax area, averaging over all three brands of wax. Did short and long glider-wax areas result in significantly different mean distances?

b. The main effect of brand of wax (Factor B) refers to the differences among mean distances traveled with the three brands, averaging over the two glider-wax areas. Did the three brands of wax result in significantly different mean distances?

c. The interaction effect of area and brand refers to any difference between brands in the effect of area or, equivalently, any difference between areas in the effect of brand. Was the *difference in distances for the short and long areas* significantly different for the different brands? Were the *differences in distances for the three brands* significantly different for the two areas?

Figure 15.5 shows two possible patterns of results in this experiment, in order to illustrate what is meant by **interaction**. Both panels show a main effect of area—distance traveled is greater with a long glider-wax area than with a short one—and a main effect of brand—distance traveled is greater with Brand 1 wax than with Brand 2 wax. The presence or absence of an interaction is reflected in whether or not the lines in the diagram are parallel.

In Part (a) of the figure, the lines are not parallel, indicating that there is an interaction between brand and area. (Whether it is a statistically significant interaction cannot be determined from inspection of the diagram.) The interaction in Part (a) may be described in various ways. The effect of brand—that is, the difference between Brand 1 and Brand 2—is greater with a long glider-wax area than with a short one. The effect of area—that is, the difference in distance traveled when one changes from a short glider-wax area to a long one—is greater for Brand 1 than for Brand 2. An interaction thus refers to "differences between differences": A difference between the levels of one factor is different at different levels of the other factor, resulting in nonparallel lines in diagrams such as Figure 15.5.

In Figure 15.5(b), the lines are parallel, indicating that there is no interaction between brand and area. The difference between Brand 1 and Brand 2 is the same whether the glider-wax area is short or long. Likewise, the increase in distance traveled that results from using a longer glider-wax area is the same for Brand 1 and Brand 2. The effect of each factor is the same regardless of which level of the other factor is considered, which is to say that there is no interaction.

Figure 15.5
Hypothetical results for two brands in the ski-wax experiment. The results in Part (a) involve an interaction between brand and area, while the results in Part (b) involve no interaction.

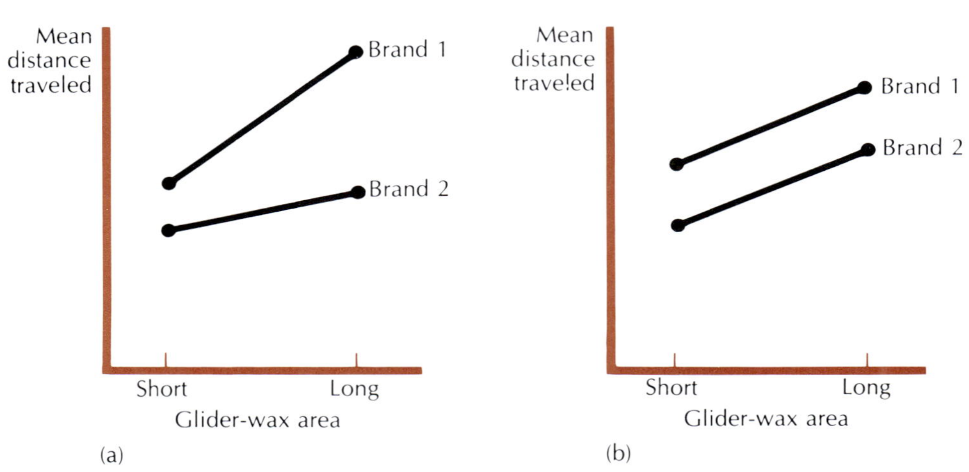

In many experiments, it is critically important to determine whether or not there is a significant interaction between factors. In this example, a cross-country ski racer would want to know not only about differences due to brands (Factor B) and differences due to glider-wax areas (Factor A), but also about their interaction. Should choice of brand depend on the extent of glider-wax area? Should the extent of the glider-wax area be different for different brands?

Figure 15.5 presented hypothetical data for illustrative purposes. The actual data from this experiment and their analysis will be presented after the introduction of the required notation and formulas.

(Source: Student project by G. Hall.)

Hypotheses and Data

We will use the following notation for the two-way ANOVA with independent samples of equal size:

k_A is the number of levels of Factor A.

k_B is the number of levels of Factor B. It is completely arbitrary which factor is labeled A and which one B.

n_R is the number of observations, or replications, of each treatment combination. The two-way ANOVA without replication is the special case in which there is just $n_R = 1$ observation in each treatment combination.

$n = k_A k_B n_R$ is the total number of observations.

x_{abi} represents the observation for the ith observational unit in the treatment combination consisting of the ath level of Factor A and the bth level of Factor B. The first subscript identifies which level of Factor A is involved, the second subscript identifies which level of Factor B is involved, and the third subscript identifies the observational unit within that treatment combination.

$T_{a..} = \sum_{b=1}^{k_B} \sum_{i=1}^{n_R} x_{abi}$ is the sum of the observations at the ath level of Factor A. The dots mark the places of the subscripts over which the summation is taken.

$\bar{x}_{a..} = \dfrac{T_{a..}}{k_B n_R}$ is the ordinary arithmetic mean of the observations at the ath level of Factor A.

$T_{.b.} = \sum_{a=1}^{k_A} \sum_{i=1}^{n_R} x_{abi}$ is the sum of the observations at the bth level of Factor B.

$\bar{x}_{.b.} = \dfrac{T_{.b.}}{k_A n_R}$ is the mean of the observations at the bth level of Factor B.

$T_{ab.} = \sum_{i=1}^{n_R} x_{abi}$ is the sum of the observations in the treatment combination consisting of the ath level of Factor A and the bth level of Factor B.

$\bar{x}_{ab.} = \dfrac{T_{ab.}}{n_R}$ is the mean of the observations in the treatment combination consisting of the ath level of Factor A and the bth level of Factor B.

$T = \sum \sum \sum x_{abi}$ is the grand total of all the observations.

$\bar{x} = \dfrac{T}{n}$ is the grand mean of all the observations.

The data for a two-way ANOVA with independent samples of size n_R consist of n_R observations for each of $k_A k_B$ independent samples, one sample for each combination of the k_A levels of Factor A with the k_B levels of Factor B.

EXAMPLE 15.14 Let us illustrate the notation using the data from the ski-wax experiment described in Example 15.13. There were $k_A = 2$ levels of glider-wax area, $k_B = 3$ levels of brand of glider wax, and $n_R = 4$ observations in each of the 6 treatment combinations, for a total of $n = 24$ observations. The dependent variable was distance traveled, measured (in meters) from a stake at the bottom of the starting slope, as shown in the following table.

Glider-wax area (Factor A)	*Brand of glider wax (Factor B)* Swix red	Falline red	Toko white	
Short	$x_{111} = 3.6$	$x_{121} = 4.0$	$x_{131} = 4.3$	
	$x_{112} = 3.8$	$x_{122} = 4.1$	$x_{132} = 4.5$	
	$x_{113} = 3.6$	$x_{123} = 3.9$	$x_{133} = 4.8$	
	$x_{114} = 3.8$	$x_{124} = 4.0$	$x_{134} = 4.5$	
	$T_{11.} = 14.8$	$T_{12.} = 16.0$	$T_{13.} = 18.1$	$T_{1..} = 48.9$
	$\bar{x}_{11.} = 3.70$	$\bar{x}_{12.} = 4.00$	$\bar{x}_{13.} = 4.52$	$\bar{x}_{1..} = 4.08$
Long	$x_{211} = 3.9$	$x_{221} = 4.3$	$x_{231} = 4.8$	
	$x_{212} = 4.0$	$x_{222} = 4.3$	$x_{232} = 5.2$	
	$x_{213} = 4.1$	$x_{223} = 4.2$	$x_{233} = 5.0$	
	$x_{214} = 4.0$	$x_{224} = 4.4$	$x_{234} = 5.0$	
	$T_{21.} = 16.0$	$T_{22.} = 17.2$	$T_{23.} = 20.0$	$T_{2..} = 53.2$
	$\bar{x}_{21.} = 4.00$	$\bar{x}_{22.} = 4.30$	$\bar{x}_{23.} = 5.00$	$\bar{x}_{2..} = 4.43$
	$T_{.1.} = 30.8$	$T_{.2.} = 33.2$	$T_{.3.} = 38.1$	$T = 102.1$
	$\bar{x}_{.1.} = 3.85$	$\bar{x}_{.2.} = 4.15$	$\bar{x}_{.3.} = 4.76$	$\bar{x} = 4.25$

(Source: Student project by G. Hall.)

On pages 535–36, we identified three significance tests of interest in a two-way classification with replication: the main effect of Factor A, the main effect of Factor B, and the A-by-B interaction effect. The test of each of these effects is a test of a different null hypothesis. The alternative hypothesis in each case is the nondirectional negative of the null hypothesis. The Factor A null hypothesis is that the population means for the different levels of Factor A are all equal:

$$H_0(A)\colon\ \mu_{1.} = \mu_{2.} = \cdots = \mu_{k_A.} \tag{15.41}$$

where $\mu_{a.}$ is the mean of the population of observations at the ath level of

Factor A. The Factor B null hypothesis is that the population means for the different levels of Factor B are all equal:

$$H_0(B)\colon\ \mu_{.1} = \mu_{.2} = \cdots = \mu_{.k_B} \tag{15.42}$$

where $\mu_{.b}$ is the mean of the population of observations at the bth level of Factor B. The interaction null hypothesis is that the population cell mean for each treatment combination involves simple addition of a contribution from Factor A and a contribution from Factor B to the grand mean:

$$H_0(AB)\colon\ \mu_{ab} = \mu + (\mu_{a.} - \mu) + (\mu_{.b} - \mu) \qquad \text{for all } a,\ b \tag{15.43}$$

where μ_{ab} is the mean of the population of observations in the treatment combination consisting of the ath level of Factor A and the bth level of Factor B. The interaction null hypothesis implies that for any given treatment combination, the contribution from Factor A does not depend on the particular level of Factor B; nor does the contribution from Factor B depend on which level of Factor A is involved.

EXAMPLE 15.15 In the ski-wax experiment in Examples 15.13 and 15.14, the Factor A null hypothesis is that the true mean distance traveled with a short glider-wax area is equal to that with a long glider-wax area:

$$H_0(A)\colon\ \mu_{1.} = \mu_{2.}$$

The Factor B null hypothesis is that the true mean distance traveled is the same for any of the three brands of glider wax:

$$H_0(B)\colon\ \mu_{.1} = \mu_{.2} = \mu_{.3}$$

The interaction null hypothesis is that the true mean distance traveled for any given treatment combination depends only on the true mean $\mu_{.b}$ for the relevant brand and on the true mean $\mu_{a.}$ for the relevant glider-wax area, but not on any quantity unique to that treatment combination:

$$H_0(AB)\colon\ \mu_{ab} = \mu_{a.} + \mu_{.b} - \mu \qquad \text{for } a = 1, 2, \text{ and } b = 1, 2, 3$$

The two-way ANOVA with independent samples provides F tests of these three null hypotheses against the corresponding nondirectional alternatives.

Partition of the Total Sum of Squares

There are four identifiable sources of variation into which the total variation may be partitioned for the two-way analysis of variance with independent samples. These four sources arise from the four terms on the right-hand side of the following equality representing the deviation of each observation from the grand mean:

$$x_{abi} - \bar{x} = [\bar{x}_{a..} - \bar{x}] + [\bar{x}_{.b.} - \bar{x}] + [\bar{x}_{ab.} - (\bar{x}_{a..} - \bar{x}) - (\bar{x}_{.b.} - \bar{x}) - \bar{x}] + [x_{abi} - \bar{x}_{ab.}] \tag{15.44}$$

The corresponding four parts into which the total variation is partitioned by squaring and summing as before are, respectively, as follows:

a. The sum of squares due to Factor A, SS_A, measures the variation of the k_A Factor A means around the grand mean; it has $k_A - 1$ degrees of freedom.

b. The sum of squares due to Factor B, SS_B, measures the variation of the k_B Factor B means around the grand mean; it has $k_B - 1$ degrees of freedom.

c. The interaction sum of squares, SS_{AB}, measures the variation of the $k_A k_B$ treatment-combination means $\bar{x}_{ab.}$ around the grand mean adjusted by the relevant Factor A and Factor B effects; it has $(k_A - 1)(k_B - 1)$ degrees of freedom.

d. The sum of squares due to replications within treatment combinations, $SS_{R(AB)}$, measures the variation of the individual observations x_{abi} around the cell means $\bar{x}_{ab.}$; it has $k_A k_B(n_R - 1)$ degrees of freedom.

Computational Formulas and the ANOVA Summary Table

The *computational formulas* for the sums of squares in the two-way ANOVA with independent samples of equal size are as follows:

$$SS_A = \frac{1}{k_B n_R} \sum^{k_A} T^2_{a..} - \frac{T^2}{n} \tag{15.45}$$

$$SS_B = \frac{1}{k_A n_R} \sum^{k_B} T^2_{.b.} - \frac{T^2}{n} \tag{15.46}$$

$$SS_{AB} = \frac{1}{n_R} \sum^{k_A} \sum^{k_B} T^2_{ab.} - SS_A - SS_B - \frac{T^2}{n} \tag{15.47}$$

$$SS_{R(AB)} = \sum^{k_A} \sum^{k_B} \sum^{n_R} x^2_{abi} - \frac{1}{n_R} \sum^{k_A} \sum^{k_B} T^2_{ab.} \tag{15.48}$$

$$SS_{\text{total}} = \sum^{k_A} \sum^{k_B} \sum^{n_R} x^2_{abi} - \frac{T^2}{n} \tag{15.49}$$

Only four of these formulas are needed, because the first four sums of squares represent a partitioning of the fifth:

$$SS_A + SS_B + SS_{AB} + SS_{R(AB)} = SS_{\text{total}} \tag{15.50}$$

Their associated degrees of freedom have likewise been partitioned:

$$(k_A - 1) + (k_B - 1) + (k_A - 1)(k_B - 1) + k_A k_B(n_R - 1) = n - 1 \tag{15.51}$$

Mean squares are calculated according to the usual relationship, MS = SS/df. As estimates of random variability, the values of MS_A, MS_B, and MS_{AB} depend on the truth of the Factor A, Factor B, and interaction null hypotheses [in Expressions (15.41), (15.42), and (15.43)], respectively, while $MS_{R(AB)}$ is a measure of random variability (**error variance**) only, regardless of the truth of any null hypothesis. Since the ratio of two independent estimates of the same Gaussian-population variance follows the F distribution, we have the following three *test statistics for the two-way ANOVA with independent samples of equal size*. The first,

$$F_A = \frac{MS_A}{MS_{R(AB)}} \tag{15.52}$$

TABLE 15.3 Analysis of variance summary table for two-way ANOVA with independent samples of equal size.

Source of variation	SS	df	MS	F
Factor A	SS_A	$k_A - 1$	MS_A	$F_A = \dfrac{MS_A}{MS_{R(AB)}}$
Factor B	SS_B	$k_B - 1$	MS_B	$F_B = \dfrac{MS_B}{MS_{R(AB)}}$
A-by-B interaction	SS_{AB}	$(k_A - 1)(k_B - 1)$	MS_{AB}	$F_{AB} = \dfrac{MS_{AB}}{MS_{R(AB)}}$
Replications within A-by-B	$SS_{R(AB)}$	$k_A k_B(n_R - 1)$	$MS_{R(AB)}$	—
Total	SS_{total}	$n - 1$		

follows the F distribution with df_A and $df_{R(AB)}$ degrees of freedom if the Factor A null hypothesis in Expression (15.41) is true. The second,

$$F_B = \frac{MS_B}{MS_{R(AB)}} \tag{15.53}$$

follows the F distribution with df_B and $df_{R(AB)}$ degrees of freedom if the Factor B null hypothesis in Expression (15.42) is true. The third,

$$F_{AB} = \frac{MS_{AB}}{MS_{R(AB)}} \tag{15.54}$$

follows the F distribution with df_{AB} and $df_{R(AB)}$ degrees of freedom if the interaction null hypothesis in Expression (15.43) is true. (The equal-variance and normality assumptions were noted on page 513.)

The general form of the **ANOVA summary table for the two-way ANOVA with independent samples of equal size** is shown in Table 15.3.

In addition to its role in the F tests, the error mean square $MS_{R(AB)}$ provides information about the precision of the results. Substituting its value into Expression (15.17), we can obtain the estimated standard error of the cell means as well as the estimated standard errors of the marginal means.

EXAMPLE 15.16 Example 15.13 described an experiment comparing three brands of glider wax applied to a more or less extensive area of cross-country skis. Factor A was the length of the area to which the glider wax was applied, and Factor B was the brand of wax. The dependent variable was the distance traveled (in meters) in a free glide on the skis.

The data and initial computations were presented in Example 15.14. The following calculations and ANOVA summary table illustrate the use of the formulas given in the present section, as applied to the data of Example 15.14.

$$SS_A = \frac{1}{(3)(4)}(48.9^2 + 53.2^2) - \frac{102.1^2}{24} = 435.120833 - 434.350417 = .770417$$

$$SS_B = \frac{1}{(2)(4)}(30.8^2 + 33.2^2 + 38.1^2) - \frac{102.1^2}{24} = 437.811250 - 434.350417 = 3.460833$$

$$SS_{AB} = \frac{1}{4}(14.8^2 + 16.0^2 + 18.1^2 + 16.0^2 + 17.2^2 + 20.0^2) - SS_A - SS_B - \frac{102.1^2}{24}$$
$$= 438.622500 - .770417 - 3.460833 - 434.350417 = .040833$$
$$SS_{R(AB)} = (3.6^2 + 3.8^2 + 3.6^2 + \cdots + 5.0^2) - 438.622500$$
$$= 438.930000 - 438.622500 = .307500$$
$$MS_A = \frac{.770417}{2-1} = .770417$$
$$MS_B = \frac{3.460833}{3-1} = 1.730417$$
$$MS_{AB} = \frac{.040833}{(2-1)(3-1)} = .020417$$
$$MS_{R(AB)} = \frac{.307500}{(2)(3)(4-1)} = .017083$$
$$F_A = \frac{.770417}{.017083} = 45.10$$
$$F_B = \frac{1.730417}{.017083} = 101.29$$
$$F_{AB} = \frac{.020417}{.017083} = 1.20$$

Source of variation	**SS**	**df**	**MS**	***F***
Glider-wax area (A)	.7704	1	.7704	45.1
Brand of wax (B)	3.4608	2	1.7304	101
A-by-B interaction	.0408	2	.0204	1.20
Replications within A-by-B	.3075	18	.0171	
Total	4.5796	23		

The estimated standard error of each cell mean is $\sqrt{.017083/4} = .06535$, or approximately .07 meter. The cell means for the short glider-wax area are 3.70, 4.00, and 4.52 m, while those for the long area are 4.00, 4.30, and 5.00 m, each with a standard error of .07 m.

Means should always be reported with the results of an analysis of variance. A table of means in the following format is usually convenient for a two-way design.

Mean distances traveled (meters).

Glider-wax area	***Brand of glider wax***			**Marginal means**
	Swix red	**Falline red**	**Toko white**	
Short	3.70	4.00	4.52	4.08
Long	4.00	4.30	5.00	4.43
Marginal means	3.85	4.15	4.76	4.25

Computer output from BMDP program 2V for this example is shown in Figure 15.6. The printout includes cell means and standard deviations and the ANOVA

```
GROUP STRUCTURE

 AREA       BRAND      COUNT
 SHORT      SWIX         4.
 SHORT      FALLINE      4.
 SHORT      TOKO         4.
 LONG       SWIX         4.
 LONG       FALLINE      4.
 LONG       TOKO         4.

          CELL MEANS FOR  1-ST DEPENDENT VARIABLE

                                                                            MARGINAL
AREA    =  SHORT      SHORT      SHORT      LONG       LONG       LONG
BRAND   =  SWIX       FALLINE    TOKO       SWIX       FALLINE    TOKO

DISTANCE     3.70000    4.00000    4.52500    4.00000    4.30000    5.00000    4.25417

COUNT        4          .4         4          4          4          4          24

STANDARD DEVIATIONS  FOR  1-ST DEPENDENT VARIABLE

AREA    =  SHORT      SHORT      SHORT      LONG       LONG       LONG
BRAND   =  SWIX       FALLINE    TOKO       SWIX       FALLINE    TOKO

DISTANCE     0.11547    0.08165    0.20616    0.08165    0.08165    0.16330


ANALYSIS OF VARIANCE FOR  1-ST
DEPENDENT VARIABLE - DISTANCE

     SOURCE                    SUM OF     DEGREES OF     MEAN          F        TAIL
                               SQUARES     FREEDOM      SQUARE                  PROB.

     MEAN                     434.35034      1        434.35034    25425.39    0.0000
     AREA                       0.77042      1          0.77042       45.10    0.0000
     BRAND                      3.46083      2          1.73042      101.29    0.0000
     AB                         0.04083      2          0.02042        1.20    0.3256
 1   ERROR                      0.30750     18          0.01708
```

Figure 15.6
BMDP2V printout for Example 15.16.

summary table. The inclusion of the grand mean as a source of variation with one degree of freedom was explained in Example 15.12.

Interpretation of the Results

The two-way analysis of variance with independent samples provides significance tests of three independent effects. Any one of the three effects may be found to be statistically significant regardless of the outcomes of the other two tests. The critical values for the three tests are given in the following paragraphs.

If the observed value of the test statistic F_A is equal to or greater than the critical value $F_\alpha(\text{df}_A, \text{df}_{R(AB)})$, we reject the Factor A null hypothesis in Expression (15.41) at the α significance level and conclude that the Factor A marginal means are significantly different from each other ($P < \alpha$). If F_A is less than the critical value, there is not enough evidence to reject the Factor A null hypothesis ($P > \alpha$).

If the observed value of the test statistic F_B is equal to or greater than the critical value $F_\alpha(\text{df}_B, \text{df}_{R(AB)})$, we reject the Factor B null hypothesis in Expression (15.42) at the α significance level and conclude that the Factor B marginal means are significantly different from each other ($P < \alpha$). If F_B is less than the critical value, there is not enough evidence to reject the Factor B null hypothesis ($P > \alpha$).

If the observed value of the test statistic F_{AB} is equal to or greater than the critical value $F_\alpha(\text{df}_{AB}, \text{df}_{R(AB)})$, we reject the interaction null hypothesis in Expression (15.43) at the α significance level. We conclude that the effects of Factor A depend on which level of Factor B we are considering, and the effects of Factor B depend on which level of Factor A we are considering ($P < \alpha$). If F_{AB} is less than the critical value, there is not enough evidence to reject the interaction null hypothesis ($P > \alpha$): The data are consistent with the idea that Factors A and B operate independently of each other and do not "interact."

Interpretation of the results of a two-way ANOVA normally begins with examination of the interaction effect. A **significant interaction** means that the effect of one factor depends on the particular level of the other factor. In other words, the *differences due to Factor A* are different at different levels of Factor B, and the *differences due to Factor B* are different at different levels of Factor A. The main effects, which simply average over the levels of the other factor, cannot reveal this interdependency of the factors. When the interaction is significant, we must look at the *cell* means—the means for the treatment combinations—to see how the effect of one factor depends on the level of the other factor. A graph of cell means, such as that in Figure 15.5(a), is usually helpful for this purpose. The main effects, which are based on the *marginal* means, may be misleading when the interaction is significant, so interpretation focuses on the *cell* means. If the interaction is significant, any significant main effect is interpreted only if it holds true at every level of the other factor. In general, we interpret a significant interaction by stating the effect of one factor (either one—whichever one seems easier to talk about) separately for each level of the other factor, pointing out how the effect is different at the different levels of the other factor.

If the interaction is not significant, we should not try to interpret a graph of the *cell* means; instead, we direct our attention to the *marginal* means corresponding to any significant main effects. A **nonsignificant interaction** means that no significant information is lost by averaging over the levels of the factors and looking only at marginal means.

EXAMPLE 15.17 Example 15.16 gives the table of means for the ski-wax experiment, as well as the ANOVA summary table with its three F ratios for the significance tests. Using the table of the F distribution in Appendix 6, we find the following:

a. $F_A = 45.1$ exceeds $F_{.01}(1, 18 \text{ df}) = 8.29$, as shown in Figure 15.7(a);

b. $F_B = 101$ exceeds $F_{.01}(2, 18 \text{ df}) = 6.01$, as shown in Figure 15.7(b); and

c. $F_{AB} = 1.20$ is less than $F_{.10}(2, 18 \text{ df}) = 2.62$, as shown in Figure 15.7(c).

We conclude that the main effect of glider-wax area and the main effect of brand of wax were both significant ($P < .01$), and that the interaction of area and brand was not statistically significant ($P > .10$).

What do these findings indicate with respect to the sample means? The means are graphed in Figure 15.8 from the table of means in Example 15.16. The fact that the interaction was not statistically significant indicates that the slight deviations from parallelism of the lines in Figure 15.8(a) are not significant. Therefore, our interpretation should be neither in terms of the cell means (3.70, 4.00, 4.52, 4.00, 4.30, and 5.00) nor in terms of Figure 15.8(a). Rather, we look at the marginal means corresponding to significant main effects. In this case, both main effects, as portrayed in Parts (b) and (c) of Figure 15.8, were significant. The mean distance traveled with a long glider-wax area (namely, 4.43 m) was significantly greater than the mean distance with a short glider-wax area (4.08 m), and the mean distances traveled with the three brands of glider wax (3.85 m for Swix, 4.15 m for Falline, and 4.76 m for Toko) were significantly different from each other. In general, a nonsignificant interaction indicates that the cell means, as in Figure 15.8(a), do not add any reliable information beyond that conveyed by the marginal means, as in Parts (b) and (c) of Figure 15.8.

Figure 15.7

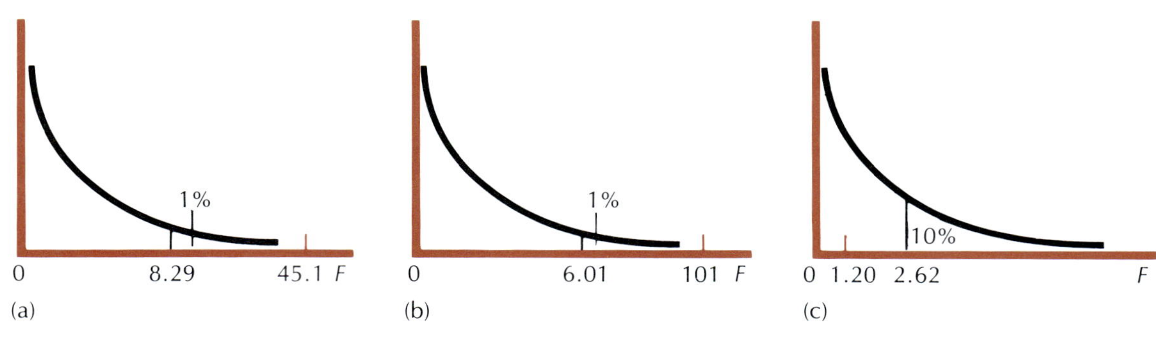

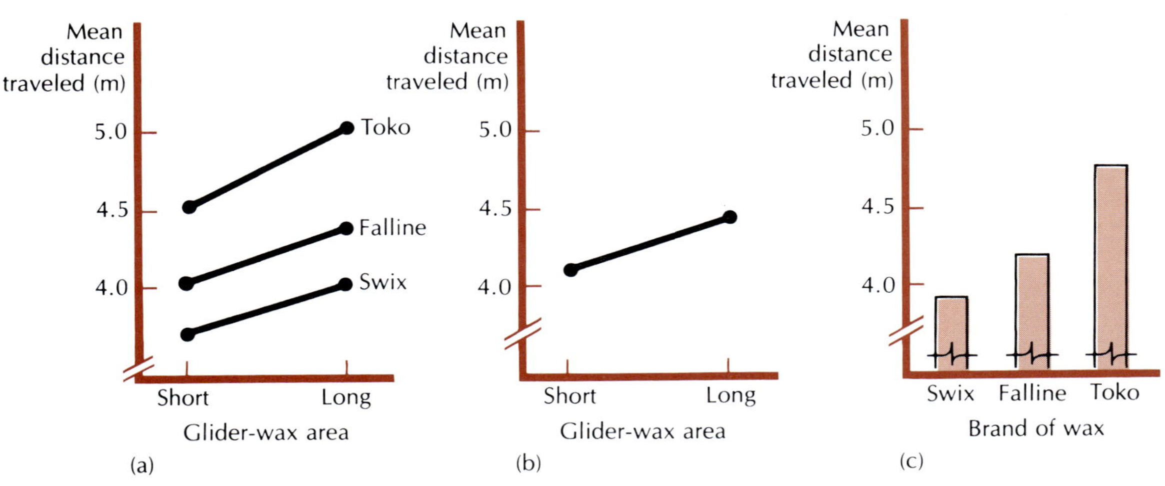

Figure 15.8
Graphs of means for the ski-wax experiment. The cell means are plotted in Part (a); the marginal means for glider-wax areas are plotted in (b); and the marginal means for the three brands of wax are plotted in (c).

EXAMPLE 15.18 For commercial sale to food stores, alfalfa sprouts are grown from seed in stainless steel trays. There are five stages in the entire process: soaking, bulking, spreading, lighting, and harvesting.

A firm in the alfalfa sprouting business conducted an experiment to determine whether adjustments to its standard procedure could be made in order to improve yield. One variable of interest was the amount of seed to grow on each tray, since the use of too little seed per tray is inefficient and too much inhibits proper growth due to crowding. A second variable examined was the length of time in the bulking stage, since it was felt that longer bulking time may improve yield even though the total time for all five stages was held constant at 93 hours. The experiment examined four levels of amount of seed (13.3, 14.0, 14.7, and 15.3 ounces per tray) and three bulking times ($25\frac{1}{2}$, $26\frac{1}{2}$, and $27\frac{1}{2}$ hours). There were 12 trays observed in each of the 12 treatment combinations, for a total of 144 observations. The dependent variable was the weight (in pounds) of sprouts harvested from each tray.

The data, in pounds harvested per tray, are summarized in the following table of means and in Figure 15.9.

	Amount of seed per tray				
Bulking time	**13.3 oz**	**14.0 oz**	**14.7 oz**	**15.3 oz**	**Marginal means**
$25\frac{1}{2}$ hr	7.71	9.75	9.75	9.25	9.11
$26\frac{1}{2}$ hr	8.83	9.08	10.12	11.92	9.99
$27\frac{1}{2}$ hr	9.12	10.42	9.79	8.50	9.46
Marginal means	8.56	9.75	9.89	9.89	9.52

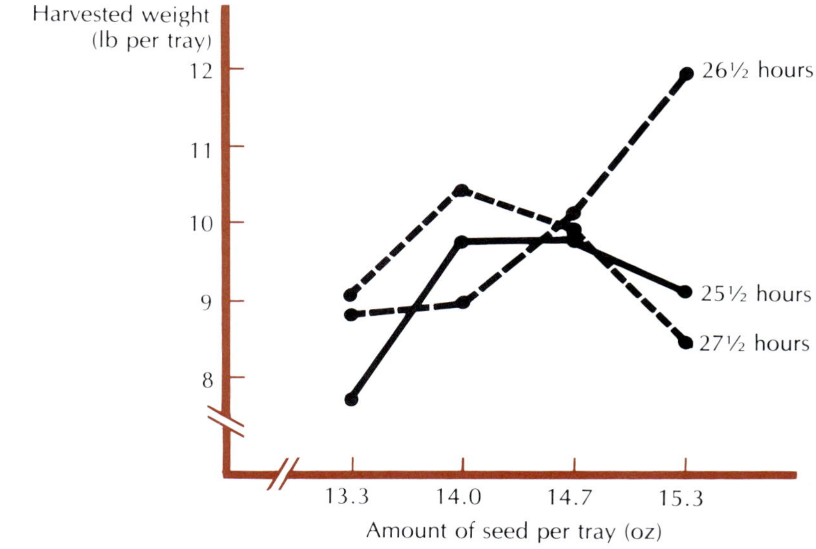

Figure 15.9 Cell means for the alfalfa sprout experiment, showing harvested weight as a function of amount of seed for three different bulking times.

A two-way analysis of variance for independent samples of equal size produced the following summary table.

Source of variation	SS	df	MS	F	
Amount of seed (A)	45.188	3	15.062	16.1	$P < .01$
Bulking time (B)	18.656	2	9.328	10.0	$P < .01$
A-by-B interaction	83.844	6	13.974	15.0	$P < .01$
Replications within A-by-B	123.250	132	.934		
Total	270.938	143			

Using Appendix 6, we find that all three F ratios are significant at the .01 level. The estimated standard error of each cell mean is $\sqrt{.934/12} = .28$ lb.

Interpretation begins with the significant interaction and the corresponding cell means as graphed in Figure 15.9. There we see that the optimal amount of seed depends on the bulking time used: 14.0 oz per tray is best at bulking times of $25\frac{1}{2}$ and $27\frac{1}{2}$ hours, but 15.3 oz is best if sprouts are in the bulking stage for $26\frac{1}{2}$ hours. Similarly, the optimal bulking time depends on how much seed is used: A bulking time of $25\frac{1}{2}$ hours is best with 13.3 or 14.0 oz of seed per tray, but $26\frac{1}{2}$ hours is best when there are 14.7 or 15.3 oz of seed per tray. The single treatment combination with the greatest yield in the experiment used 15.3 oz of seed and a bulking time of $26\frac{1}{2}$ hours.

Notice that when the interaction is significant, it can be misleading to interpret the marginal means, even though the main effects are significant. The marginal means for amount of seed indicate that 14.7 oz or 15.3 oz of seed per tray is best, but that is true only for a $26\frac{1}{2}$-hour bulking stage. The marginal means for bulking time indicate that $26\frac{1}{2}$

hours is best, but that is not true if 13.3 or 14.0 oz of seed per tray is used. In general, a significant interaction indicates that the marginal means overlook some important qualifying statements that must be made in stating the results.

(Source: Student project by N. Sinnamon.)

Summary of the Test Procedure

The **two-way analysis of variance with independent samples of equal size** may be summarized as follows:

Assumptions.

a. Independent random samples for the $k_A k_B$ treatment combinations.

b. Additivity: Main effects, interaction effect, and random error add together to yield particular observations.

c. Normality of the populations underlying the various treatment combinations.

d. Equal population variances.

Hypotheses.

$$\begin{cases} H_0(A)\text{:} & \mu_{1.} = \mu_{2.} = \cdots = \mu_{k_A} \\ H_A(A)\text{:} & \text{The } \mu_{a.} \text{ are not all equal.} \end{cases}$$

$$\begin{cases} H_0(B)\text{:} & \mu_{.1} = \mu_{.2} = \cdots = \mu_{.k_B} \\ H_A(B)\text{:} & \text{The } \mu_{.b} \text{ are not all equal.} \end{cases}$$

$$\begin{cases} H_0(AB)\text{:} & \mu_{ab} = \mu_{a.} + \mu_{.b} - \mu, \text{ for all } a, b \\ H_A(AB)\text{:} & \text{The population cell means are not all determined} \\ & \text{solely by the population marginal means.} \end{cases}$$

Step 1. Determine the values of k_A, k_B, n_R, and n.

Step 2. Calculate the $k_A k_B$ treatment-combination totals $T_{ab.}$, the k_A Factor A totals $T_{a..}$, the k_B Factor B totals $T_{.b.}$, and the grand total T.

Step 3. Calculate the k_A Factor A means, the k_B Factor B means, and the $k_A k_B$ treatment-combination means, in order to see what sample differences are being tested.

Step 4. Calculate the values required to complete the ANOVA summary table in the format of Table 15.3.

Step 5. Compare the observed values of the test statistics F_A, F_B, and F_{AB} with critical values from the F distribution with appropriate degrees of freedom.

a. If the interaction is significant, examine the treatment-combination means $\bar{x}_{ab.}$ to discover the nature of the effect, and interpret significant main effects only insofar as they are compatible with the treatment-combination means.

b. If the interaction is not significant but at least one main effect is, examine the relevant marginal means to discover the nature of the effect.

STUDY MATERIAL FOR CHAPTER 15

Terms and Concepts

Section 15.1

analysis of variance (ANOVA)
total sum of squares
total number of degrees of freedom
variation
source of variation
factor
replication

Section 15.2

one-way ANOVA with independent samples
grand total
grand mean
variation between samples
variation due to Factor *A*
sum of squares due to Factor *A*
variation within samples
error variation
sum of squares due to replications within levels of Factor *A*
mean square
error mean square in one-way ANOVA
error variance in one-way ANOVA
analysis of variance summary table
ANOVA summary table for one-way ANOVA with independent samples
estimated standard error of a sample mean
assumptions for one-way ANOVA with independent samples

Section 15.3

two-way ANOVA without replication
block
blocking
treatment combination
one-way ANOVA with repeated measurements
randomized-blocks ANOVA
interaction
variation due to the *A*-by-*B* interaction
error variance in two-way ANOVA without replication

ANOVA summary table for two-way ANOVA without replication
assumptions for two-way ANOVA without replication

Section 15.4

two-way ANOVA with independent samples of equal size
main effect
interaction effect
error variance in two-way ANOVA with independent samples
ANOVA summary table for two-way ANOVA with independent samples
assumptions for two-way ANOVA with independent samples

Key Formulas

Section 15.2

(15.9) $$\text{MS} = \frac{\text{SS}}{\text{df}}$$

(15.17) Estimated standard error of a sample mean

$$= \sqrt{\frac{\text{error mean square}}{\text{number of observations on which the mean is based}}}$$

Also see the formulas in Tables 15.1 and 15.4.

Section 15.3

See the formulas in Tables 15.2 and 15.4.

Section 15.4

See the formulas in Tables 15.3 and 15.4.

TABLE 15.4 Computational formulas for sums of squares.

Source of variation	One-way ANOVA with independent samples	Two-way ANOVA without replication	Two-way ANOVA with independent samples
Factor A	$\text{SS}_A = \sum \frac{T_{a.}^2}{n_a} - \frac{T^2}{n}$	$\text{SS}_A = \frac{1}{k_B} \sum T_{a.}^2 - \frac{T^2}{n}$	$\text{SS}_A = \frac{1}{k_B n_R} \sum T_{a..}^2 - \frac{T^2}{n}$
Factor B		$\text{SS}_B = \frac{1}{k_A} \sum T_{.b}^2 - \frac{T^2}{n}$	$\text{SS}_B = \frac{1}{k_A n_R} \sum T_{.b.}^2 - \frac{T^2}{n}$
A-by-B interaction		$\text{SS}_{AB} = \sum \sum x_{ab}^2 - \text{SS}_A - \text{SS}_B - \frac{T^2}{n}$	$\text{SS}_{AB} = \frac{1}{n_R} \sum \sum T_{ab.}^2 - \text{SS}_A - \text{SS}_B - \frac{T^2}{n}$
Replications within samples	$\text{SS}_{R(A)} = \sum \sum x_{ai}^2 - \sum \frac{T_{a.}^2}{n_a}$		$\text{SS}_{R(AB)} = \sum \sum \sum x_{abi}^2 - \frac{1}{n_R} \sum \sum T_{ab.}^2$

Exercises for Numerical Practice

Section 15.2

15/1. Twenty-seven observational units were randomly assigned to three levels of Factor A, with 9 units at each level. The data were analyzed by a one-way analysis of variance for independent samples. The sum of squares due to Factor A was 200.6, and the sum of squares due to replications within levels of Factor A was 953.0.

a. Complete the ANOVA summary table.

b. What is the critical value to test the effect of Factor A at the 5% significance level?

15/2. Six random samples were observed, each corresponding to one of six levels of Factor A. There were 11 observational units in each sample. The sum of squares due to Factor A was found to be 14.7, and the sum of squares due to replications within samples was 105.0.

a. Complete the ANOVA summary table.

b. What is the critical value to test the effect of Factor A at the 5% significance level?

15/3. The following observations were recorded in independent random samples at five levels of Factor A.

A_1	A_2	A_3	A_4	A_5
8	3	6	2	2
7	4	3	5	0
7	3	5	3	1

a. Complete the appropriate ANOVA summary table.

b. What is the critical value to test the effect of Factor A at the 5% significance level?

c. What does the result of the F test tell you about the data?

d. Calculate the estimated standard error of the sample mean at level A_1.

e. Calculate the 95% confidence limits for the population mean at level A_1.

15/4. The following observations were recorded in independent random samples at four levels of Factor A.

A_1	A_2	A_3	A_4
6	4	0	3
4	4	1	5
5	1	3	3
6	1	0	4
3	3	3	5
3	4	3	2
6	2	1	4

Answer Parts (a)–(e) in Exercise 15/3 for these data.

Section 15.3

15/5. A randomized-blocks experiment yielded the following ANOVA summary table.

Source of variation	SS	df	MS	F
Blocks	199.848	24		
Treatments	10.590	5		
Interaction	43.680			

a. Complete the ANOVA summary table.

b. How many blocks were observed in the experiment?

c. How many treatments were studied in the experiment?

d. What is the critical value for testing the effect of treatments at the 1% significance level?

e. Was the blocking successful in improving the precision of the experiment? How do you know?

15/6. A randomized-blocks experiment yielded the following ANOVA summary table.

Source of variation	SS	df	MS	F
Blocks	1682.20	10		
Treatments	119.76	4		
Interaction	218.00			

Answer Parts (a)–(e) in Exercise 15/5 for these data.

15/7. The following data were recorded for five treatments applied to three randomly selected blocks.

Block	A_1	A_2	A_3	A_4	A_5
B_1	9	3	5	6	2
B_2	9	4	6	7	3
B_3	6	1	4	4	0

a. Complete the appropriate ANOVA summary table.

b. What is the critical value to test the effect of Factor A at the 1% significance level?

c. What does the result of the F test for Factor A tell you about the data?

d. Calculate the estimated standard error of the sample mean at level A_3.

e. Calculate the 99% confidence limits for the population mean at level A_3.

f. Was the blocking successful in improving the precision of the experiment? How do you know?

15/8. The following data were recorded for three treatments applied to six randomly selected blocks.

Treatment	Block 1	Block 2	Block 3	Block 4	Block 5	Block 6
A_1	3	9	6	5	9	7
A_2	3	8	5	2	9	6
A_3	0	6	3	1	5	3

Answer Parts (a)–(f) in Exercise 15/7 for these data.

Section 15.4

15/9. An experiment with a two-way design yielded the following ANOVA summary table.

Source of variation	SS	df	MS	F
Factor *A*	26.72	4		
Factor *B*	8.62	1		
A-by-*B* interaction	123.48			
Replications within *A*-by-*B*	141.60	30		

a. Complete the ANOVA summary table.

b. Determine how many of each of the following there were in the experiment:

i. levels of Factor *A*.

ii. levels of Factor *B*.

iii. replications of each treatment combination.

c. At the 5% significance level, find the critical value for testing the effect of

i. Factor *A*.

ii. Factor *B*.

iii. the *A*-by-*B* interaction.

15/10. An experiment with a two-way design yielded the following ANOVA summary table.

Source of variation	SS	df	MS	F
Factor *A*	286	2		
Factor *B*	2901	3		
A-by-*B* interaction	3288			
Replications within *A*-by-*B*	7620	60		

Answer Parts (a)–(c) in Exercise 15/9 for these data.

15/11. Use the two-way ANOVA with independent samples to analyze the following data.

	A_1	A_2	A_3
B_1	8, 7, 9, 9	4, 6, 4, 5	1, 1, 2, 0
B_2	2, 0, 2, 0	4, 5, 5, 4	6, 8, 6, 7

a. Complete the appropriate ANOVA summary table.

b. At the 5% significance level, find the critical value for testing the effect of

i. Factor *A*.

ii. Factor *B*.

iii. the *A*-by-*B* interaction.

c. What do the results of the *F* tests tell you about the data?

d. Calculate the estimated standard error of the sample mean for treatment combination A_2B_2.

e. Calculate the 95% confidence limits for the population mean for treatment combination A_2B_2.

15/12. Use the two-way ANOVA with independent samples to analyze the following data.

	B_1	B_2	B_3	B_4
A_1	4, 4, 5	7, 5, 7	8, 6, 7	7, 8, 9
A_2	3, 3, 5	2, 4, 3	2, 1, 2	0, 0, 1

Answer Parts (a)–(e) in Exercise 15/11 for these data.

Problems for Applying the Concepts

15/13. Four samples of unmarried female university students aged 17 to 21 years took part in a study concerning family size. The four samples were distinguished by the number of children (2, 3, 4, or 5) in the family from which the student came. The students filled out a brief questionnaire in which they reported the number of children they would like to have in their future families, as follows.

Present family size	Desired number of children in future family
2 children	2, 4, 2, 4
3 children	3, 2, 2, 5, 5
4 children	3, 3, 4
5 children	4, 3, 0, 2, 3

Was the mean desired number of children significantly different for students coming from different sizes of families?

a. Identify the dependent and independent variables.

b. Complete the ANOVA summary table, indicating the outcome of the significance test.

c. State the values of the four means that were compared by the significance test, along with their estimated standard errors.

d. What does the significance test indicate about desired family size?

(Source: Student project by T. Kopperud.)

15/14. In a study of the relationship between athletic involvement and grade point average of university students, data were obtained from a sample of students on intercollegiate teams, a sample involved in intramural athletics only, and a sample with no athletic involvement. The following grade point averages were recorded (on a 9-point scale where 6 is average and 4 is the minimum passing grade) for six students in each of the three categories of participation in athletics:

Intercollegiate:	6.8	5.9	6.6	4.9	5.2	3.6
Intramural:	5.6	5.9	6.6	5.4	5.4	6.1
None:	6.5	6.7	6.2	4.1	5.2	6.7

Were the mean grades significantly different for the three categories of participation in athletics?

a. Identify the dependent and independent variables.

b. Complete the ANOVA summary table, indicating the outcome of the significance test.

c. State the values of the three means that were compared by the significance test, along with their estimated standard errors.

d. What does the significance test indicate about grade point averages?

(Source: Student project by B. Olanski.)

15/15. Scrabble is a word game in which players must form words from racks of seven letters drawn from a well-shuffled set of letters. Different letters have different point values. An experiment was conducted to study the relationship between the number of vowels (A, E, I, O, U) included among the seven letters and the number of points earned on the first play in a game. There were eight conditions in the experiment (0, 1, 2, 3, 4, 5, 6, and 7 vowels in the player's seven letters), with four replications of each condition, for a total of 32 observations, all in random order. It was as if 32 games were started, and the score on the first play of each game was recorded as follows.

Vowel/consonant ratio							
0/7	**1/6**	**2/5**	**3/4**	**4/3**	**5/2**	**6/1**	**7/0**
12	16	24	28	28	18	12	12
0	12	22	18	14	12	6	4
12	26	22	16	18	14	6	4
0	24	32	18	26	6	18	4

Were the mean numbers of points earned significantly different for the eight vowel/consonant ratios?

a. Identify the dependent and independent variables.

b. Complete the ANOVA summary table, indicating the outcome of the significance test.

c. State the values of the means that were compared by the significance test, along with their estimated standard errors.

d. What does the significance test indicate about Scrabble scores?

(Source: Student project by G. Hamilton.)

15/16. The psychological literature includes considerable discussion of the effect of birth order on individual development. A sample of 20 second-year college students was divided into five groups according to birth order in their family (first-, second-, third-, fourth-, and later-born). The following grade point averages from their previous year of studies was recorded (on a 9-point scale where 6 is average and 4 is the minimum passing grade):

First-born:	6.0	6.2	6.4	8.5	
Second-born:	3.5	5.6	6.0	6.1	
Third-born:	6.1	6.4	7.0	8.2	
Fourth-born:	7.3	7.4	8.4		
Later-born:	5.6	5.6	6.8	6.8	6.9

Were the mean grades significantly different for the different birth-order positions?

a. Identify the dependent and independent variables.

b. Complete the ANOVA summary table, indicating the outcome of the significance test.

c. State the values of the means that were compared by the significance test, along with their estimated standard errors.

d. What does the significance test indicate about grade point averages?

(Source: Student project by B. Weir.)

15/17. Use an analysis of variance to reanalyze the data of Problem 11/17, and verify that the observed value of the *F* test statistic is equal to the square of the *t* statistic calculated in the earlier problem. State the results of the *F* test of significance.

15/18. Use an analysis of variance to reanalyze the data of Problem 11/18, and verify that the observed value of the *F* test statistic is equal to the square of the *t* statistic calculated in the earlier problem. State the results of the *F* test of significance.

15/19. Nine female students in their second year of university studies agreed to keep track of the number of hours they spent watching television each day of the week. The following numbers of hours were recorded for Monday, Wednesday, and Friday by the nine students (who are labeled by the letters *A* to *I*).

	A	*B*	*C*	*D*	*E*	*F*	*G*	*H*	*I*
Monday	2	2	2	3	5	3	0	0	0
Wednesday	0	2	2	3	3	3	0	0	3
Friday	0	0	0	2	5	0	0	0	0

Was there a significant difference in the mean number of hours of television viewing on the three days of the week?

a. Identify the dependent and independent variables.

b. Complete the ANOVA summary table and significance test.

c. State the values of the means that were compared by the significance test, along with their estimated standard errors.

d. What does the significance test indicate about television viewing habits?

e. Was there an improvement in precision due to blocking? How do you know?

(Source: Student project by M. Diemert.)

15/20. A zoology experiment investigated the effect of ambient temperature on the rate of oxygen consumption by deer mice. Using a respirometer, the experimenter measured the rate of oxygen consumption in milliliters per hour per gram of body weight. Four mice were each observed at four different temperatures, giving the following 16 observations of oxygen consumption (in ml/h/g body weight).

Mouse	0°C	10°C	20°C	30°C
M_1	10.3	9.7	3.6	7.8
M_2	14.0	11.2	5.3	10.0
M_3	12.6	10.5	4.6	6.3
M_4	11.4	7.9	5.3	7.6

Did the rate of oxygen consumption differ significantly at the different temperatures?

a. Identify the dependent and independent variables.

b. Complete the ANOVA summary table and significance test.

c. State the values of the means that were compared by the significance test, along with their estimated standard errors.

d. What does the significance test indicate about rates of oxygen consumption?

e. Was there an improvement in precision due to blocking? How do you know?

(Source: Student project by I. Munkedal.)

15/21. An experiment was conducted to study the effect of a weight-lifter's hand position on his maximum lift in bench pressing. Eight male college students who lifted weights approximately once a week took part in the experiment. Four distances between the hands (0, 9, 18, and 27 inches) were tested in a different random order for each student, and the student's maximum lift (in pounds) at each distance was recorded as follows.

Student	0 in.	9 in.	18 in.	27 in.
S_1	140	180	205	200
S_2	115	165	190	200
S_3	105	140	180	165
S_4	130	180	210	205
S_5	90	120	155	160
S_6	100	140	165	165
S_7	85	110	155	150
S_8	115	155	170	160

Was there a significant difference in the mean maximum lift at the different distances?

a. Identify the dependent and independent variables.

b. Complete the ANOVA summary table and significance test.

c. State the values of the means that were compared by the significance test, along with their estimated standard errors.

d. What does the significance test indicate about a weight-lifter's maximum lift in bench pressing?

e. Was there an improvement in precision due to blocking? How do you know?

(Source: Student project by T. Kurtenbach.)

15/22. A physics experiment investigated the time that a falling weight took to reach the ground as a function of the distance that it fell. The weight was attached to a string wrapped around the circumference of a revolving wheel whose resistance slowed the weight's fall. There were three sets or blocks of observations, each block including seven distances, so that there were three replications of each distance. The following times were recorded in seconds.

Block	282 mm	564 mm	846 mm	1128 mm	1410 mm	1692 mm	1974 mm
B_1	9.8	13.0	15.1	18.0	20.1	22.2	24.5
B_2	9.9	12.6	15.2	17.9	20.9	22.1	25.0
B_3	9.6	12.8	15.4	18.1	20.8	22.6	24.8

a. Identify the dependent and independent variables.

b. Complete the ANOVA summary table and significance test.

c. State the values of the means that were compared by the significance test, along with their estimated standard errors.

d. What does the significance test indicate about the falling times?

e. Was there an improvement in precision due to blocking? How do you know?

(Source: Student project by C. Owre.)

15/23. Use an analysis of variance to reanalyze the data of Problem 11/21.

a. Verify that the observed value of the F statistic is equal to the square of the t statistic calculated in the earlier problem.

b. State the results of the F test of significance.

c. Was there an improvement in precision due to blocking? How do you know?

15/24. Use an analysis of variance to reanalyze the data of Problem 11/22.

a. Verify that the observed value of the F statistic is equal to the square of the t statistic calculated in the earlier problem.

b. State the results of the F test of significance.

c. Was there an improvement in precision due to blocking? How do you know?

15/25. In an investigation of sound transmission in an apartment building, 18 different people rated the loudness of a musical recording (the same recording for all subjects) played on a stereo in the apartment directly above the listener. The stereo was operated at three different volume-control settings and three different tone-control settings. Each subject rated the loudness in one of the nine conditions, using a scale from 1 (just barely audible) through 5 (almost completely audible) to 10 (completely and clearly audible).

Tone-control setting	Volume-control setting		
	Low	Medium	High
Strong treble	$1\frac{1}{2}$, 1	$2\frac{1}{2}$, 2	6, 5
Balanced	1, 1	2, 2	5, 6
Strong bass	$1\frac{1}{2}$, 2	3, $2\frac{1}{2}$	7, 6

Perform an analysis of variance to test the effects of tone and volume settings.

a. Identify the dependent and independent variables.

b. Complete the ANOVA summary table and significance tests.

c. What do the test results indicate about the loudness ratings?

(Source: Student project by R. Geyer.)

15/26. A tire manufacturer is experimenting with three processes for the production of radial tires. In one test, the tread life of random samples of tires produced by each process was measured either on pavement or on a gravel road surface. The results were as follows, in thousands of miles.

	Process 1	Process 2	Process 3
Pavement	28, 30, 29, 34	46, 43, 41, 44	50, 48, 49, 46
Gravel	29, 33, 32, 29	38, 33, 35, 36	49, 45, 48, 47

Perform an analysis of variance to test the effects of manufacturing process and road surface.

a. Identify the dependent and independent variables.

b. Complete the ANOVA summary table and significance tests.

c. What do the test results indicate about tread life?

15/27. A sample of 12 university students were interviewed in the second month of the academic year concerning their study habits during the previous month. Six of the students were in their first year of studies, and six were in their second or third year, and half of each group were male. The 12 students reported their average number of hours of study per week as follows.

	Female	Male
First-year students	21, 20, 14	15, 17, 12
Later-year students	30, 28, 30	23, 10, 29

Perform an analysis of variance to test the effects of year and gender.

a. Identify the dependent and independent variables.

b. Complete the ANOVA summary table and significance tests.

c. What do the results indicate about reported study times?

d. How does the ANOVA summary table in Part (b) compare with that for the same data in Problem 15/17? Explain.

(Source: Student project by I. Servold.)

15/28. Two common types of order at a local printing office are for 2″-by-3½″ business cards and for #10 windowless business envelopes. These may be ordered in either black or colored ink. Random samples of completed orders for 500 copies of cards or envelopes were selected from the printer's files. The printer had recorded, to the nearest five minutes, the time required to ink up, get a proper impression in the correct place on the stock, and run 500 copies for each of the 24 selected orders as follows.

	Black ink	Colored ink
Business cards	40, 35, 40, 30, 35, 35	45, 40, 35, 45, 35, 35
Envelopes	35, 35, 25, 30, 25, 30	35, 30, 40, 35, 30, 30

Perform an analysis of variance to test the effects of ink color and stock.

a. Identify the dependent and independent variables.

b. Complete the ANOVA summary table and significance tests.

c. What do the test results indicate about printing times?

(Source: Student project by D. Hutchinson.)

Reference

Kolata, G. B. "Clinical Trial of Psychotherapies Is Under Way." *Science* 212 (1981): 432–33.

Further Reading

Cox, D. R. *Planning of Experiments.* New York: Wiley, 1958. Chapter 6 includes an excellent discussion of the interpretation of interactions.

Dixon, W. J., et al., eds. *BMDP Statistical Software Manual, 1985 Reprinting.* Berkeley: University of California Press, 1985. Examples of BMDP output were shown in Figures 15.4 and 15.6.

Hill, M., ed. *BMDP User's Digest, 3rd edition.* Los Angeles: BMDP Statistical Software, Inc., 1984.

Moroney, M. J. *Facts from Figures.* Harmondsworth, Middlesex: Penguin, 1951. Chapter 19 is intended as a layman's introduction to the basic principles of analysis of variance.

*CHAPTER 16

NONPARAMETRIC METHODS

EXAMPLE 16.1 For many years, the treatment of alcoholism has been based on the assumption that, for individuals who have become physically addicted to alcohol, a successful outcome requires total abstinence from alcohol. Recently, however, studies have been reported claiming successful treatment of alcoholics by teaching them to control their drinking rather than to abstain altogether. These controversial findings have received considerable publicity because they have important practical implications for the treatment of one of society's most serious health problems.

Mary Pendery, Irving Maltzman, and Jolyon West reexamined the evidence in one of the most widely cited of the studies supporting controlled drinking. The original study involved 40 male alcoholic inpatients at Patton State Hospital in California. Twenty of the patients were assigned to an experimental program to teach controlled drinking, and the other 20 served as a control group, receiving traditional treatments directed toward total abstinence. The controlled-drinking subjects were reported to have had a significantly higher percentage of days without alcohol abuse following discharge than the abstinence subjects. In reviewing the hospital records and conducting follow-up interviews, Pendery, Maltzman, and West found several serious problems with this conclusion. For example, the dates of hospital admission (2 April 1970 to 26 February 1971) for the controlled-drinking (D) and abstinence (A) groups revealed the following order of admission of the 40 patients:

$$A \quad D \quad D \quad D \quad D \quad A \quad D \quad D \quad D \quad D \quad A \quad D \quad D \quad D \quad A \quad D \quad D \quad D \quad A \quad A$$

$$D \quad D \quad D \quad D \quad A \quad A \quad A \quad D \quad A \quad A \quad A \quad A \quad A \quad A \quad A \quad D \quad A \quad A \quad A \quad A$$

Using a significance test for rank-order data to be introduced in this chapter, Pendery, Maltzman, and West found that on the average, the controlled-drinking subjects were admitted significantly earlier than the abstinence subjects ($P < .01$). Thus, the treatment conditions were confounded with order of admission, so that the two groups may have differed even before the treatments were administered.

The follow-up investigation of the 20 controlled-drinking subjects confirmed the doubts of Pendery, Maltzman, and West about the original study: 9 continued to have problems due to excessive drinking (5 of the 9 were rehospitalized within a year of their treatment), 6 experienced further problems (including rehospitalization within a year for 4 of the 6) until they gave up controlled drinking in favor of total abstinence, and 4 died alcohol-related deaths. Only one of the 20 succeeded with a pattern of controlled

* This symbol identifies material that is not prerequisite for later topics in this book.

drinking, and he should not have been included in the study in the first place, because he had never been physically addicted to alcohol.

(Source: Pendery *et al.*, 1982. Copyright 1982 by the American Association for the Advancement of Science. Used with permission of the AAAS.)

16.1 PARAMETRIC VERSUS NONPARAMETRIC METHODS

Most of the statistical methods presented in this book fall under the general heading of *parametric* statistics. **Parametric methods** are those that revolve around the parameters, such as population means and variances, of a specific form of distribution, usually the Gaussian. For example, the F tests in the analysis of variance (Chapter 15) may be used to draw inferences about population means on the assumption that the observations were randomly sampled from Gaussian populations with equal variances. Similarly, correlation and regression analysis (Chapters 5 and 13) are usually considered to be parametric methods, even though the parametric assumptions of normality and homogeneity of variance are required only for inferential, but not for purely descriptive, purposes. Parametric methods were likewise presented in Chapters 10 through 12 for inferences about means and variances.

Nonparametric methods require less restrictive assumptions than do parametric methods. In particular, the population need not be describable in terms of the parameters of the Gaussian distribution. The term *nonparametric* should not be taken too literally, because nonparametric methods still involve the parameters of some, perhaps unspecified, distribution. Nonparametric methods do not all involve the same set of assumptions about the underlying distribution, but the assumption of independence is common for nonparametric methods as it is for parametric methods: The replications of any one treatment are sampled independently of each other. In addition, most of the nonparametric methods discussed in this chapter assume that the observations are sampled from an underlying distribution that is continuous (rather than discrete), but they do not stipulate the shape of the distribution (for example, Gaussian) as the parametric methods do. Some authors prefer the term **distribution-free methods** for nonparametric methods, to indicate that the methods do not depend on the shape of the underlying distribution.

Nonparametric methods also differ from parametric methods in terms of the type of information they use in the data. Parametric methods make use of the actual numerical magnitudes of the observations; for example, in the observed values 14, 13, and 18, the difference between 14 and 18 would be understood to be four times as large as the difference between 13 and 14. The units on the measurement scale are taken to be equal everywhere on the scale. (In some cases, data transformations can be used to make this assumption more appropriate.) On the other hand, nonparametric methods make use of no more than ordinal properties of the data, and they may be subdivided according to whether they are based on rank-order information, sign information, or categorical information. Methods using **rank-order information** would treat the observations 14, 13, and 18 in terms of their ranks 2, 1, and 3, which identify the smallest, second smallest, and third smallest but which ignore the sizes of the differences between observations. Methods using **sign information** consider only the direction of difference of each observation from a certain value. Thus, relative to the value 14, observations such as 13, 12, and 10 would be categorized as smaller and hence each would be replaced by a sign (−) indicating the direction of difference. Observations such as 18 or 180 would be categorized as larger, and each would be replaced by a sign (+) to indicate the direction of difference from the standard (14 in this example). Methods using **categorical information** treat each observation as falling in exactly one category and ignore any natural ordering that may exist among the categories; that is, observations are scored in terms of verbal labels rather than in terms of numerical magnitudes, order, or direction of difference from a reference value.

If the parametric assumptions are reasonably met, then nonparametric methods, whether of the rank-order or sign or categorical variety, will typically extract less information from numerically measured data than will the parametric methods. As a general rule, methods that extract the most information from the data are preferred. Nonparametric methods therefore are used chiefly for four purposes:

a. to analyze special types of data, such as those where the observational units are not so much measured as rank ordered or categorized-and-counted;

b. to test a null hypothesis different from that tested by a parametric test—the nonparametric null hypothesis is typically more general, with the parametric null hypothesis being one special case of it;

c. to sidestep severe violations of parametric assumptions of normality and homogeneity of variance; and

d. to provide computational convenience in cases where a quick, approximate answer is desired or where the effect is so large that the more powerful parametric methods are unnecessary.

The remaining sections of this chapter present nonparametric methods for several types of problems that we have previously examined with parametric methods. Specifically, we will look at methods for measuring correlation, for testing for a specified value of a measure of location, and for testing the significance of the difference between two or more treatments.

16.2 RANK CORRELATION

Section 5.2 introduced the Pearson product-moment correlation coefficient,

$$r = \frac{\sum xy - \frac{\sum x \sum y}{n}}{\sqrt{\sum x^2 - \frac{(\sum x)^2}{n}}\sqrt{\sum y^2 - \frac{(\sum y)^2}{n}}} \tag{16.1}$$

as a measure of linear relationship between two variables x and y. Pearson's r is sometimes called a *parametric measure of correlation,* in the sense that it makes use of the actual numerical magnitudes of the observed values x_i and y_i. In addition, testing the significance of the correlation coefficient (Section 13.1) requires the assumption that at least one of the variables is normally distributed. The statistical significance of r may then be tested simply by referring its absolute value to the critical values given in Appendix 8.

A correlation coefficient that makes use only of the rank-order information in the observations, and not their actual values, was proposed by the British psychologist Charles Spearman in 1904. **Spearman's rank correlation coefficient** r_S is an index of **monotonic relationship**. Two variables are monotonically related in the *positive* direction if increases in one variable are always accompanied by increases in the other variable, although the increases need not be in constant proportion to each other as is the case for a straight-line relationship. An *inverse* monotonic relationship is one in which increases in one variable are always accompanied by decreases in the other variable. A linear relationship is one type of monotonic relationship, but not the only type. The line depicting a monotonic relationship may be curved or irregular, but it never reverses direction (see Figure 16.1).

If we assign ranks $(1, 2, 3, \ldots, n)$ to the observational units with respect to the values of each variable separately and then compute the Pearson correlation of the ranks, the result is **Spearman's rank correlation coefficient**:

$$r_S = r_{\text{ranks}} \tag{16.2}$$

Spearman's r_S is a Pearson product-moment correlation coefficient calculated on ranks (r_{ranks}) instead of on raw scores (r_{xy}). It follows that Spearman's r_S is an index of the linear relationship of the *ranks* of the variables, that its sign indicates the direction of the relationship, and that its value is limited to the interval from -1.0 to $+1.0$, inclusive. Since it is computed on ranks rather than on the raw data x_i, y_i, it is sensitive only to whether the two variables are in the same rank order, that is, the extent to which the variables are monotonically related.

In assigning ranks to the n values of the two variables, it is immaterial whether the ranks $(1, 2, \ldots, n)$ are assigned from largest to smallest or from smallest to largest, as long as the direction of ranking is the same for both variables. However, it is conventional, and generally less conducive to confusion, to assign the rank of 1 to the smallest observation and successively larger rank numbers to successively larger observations. If there are tied observations, they should be assigned the mean of the ranks they would have received if they had not been exactly tied.

When there are no ties in either variable, Pearson's r formula in Expression (16.1) as applied to ranks reduces to a simpler but algebraically equivalent formula for Spearman's coefficient,

$$r_S = 1 - \frac{6 \sum d_i^2}{n(n^2 - 1)} \tag{16.3}$$

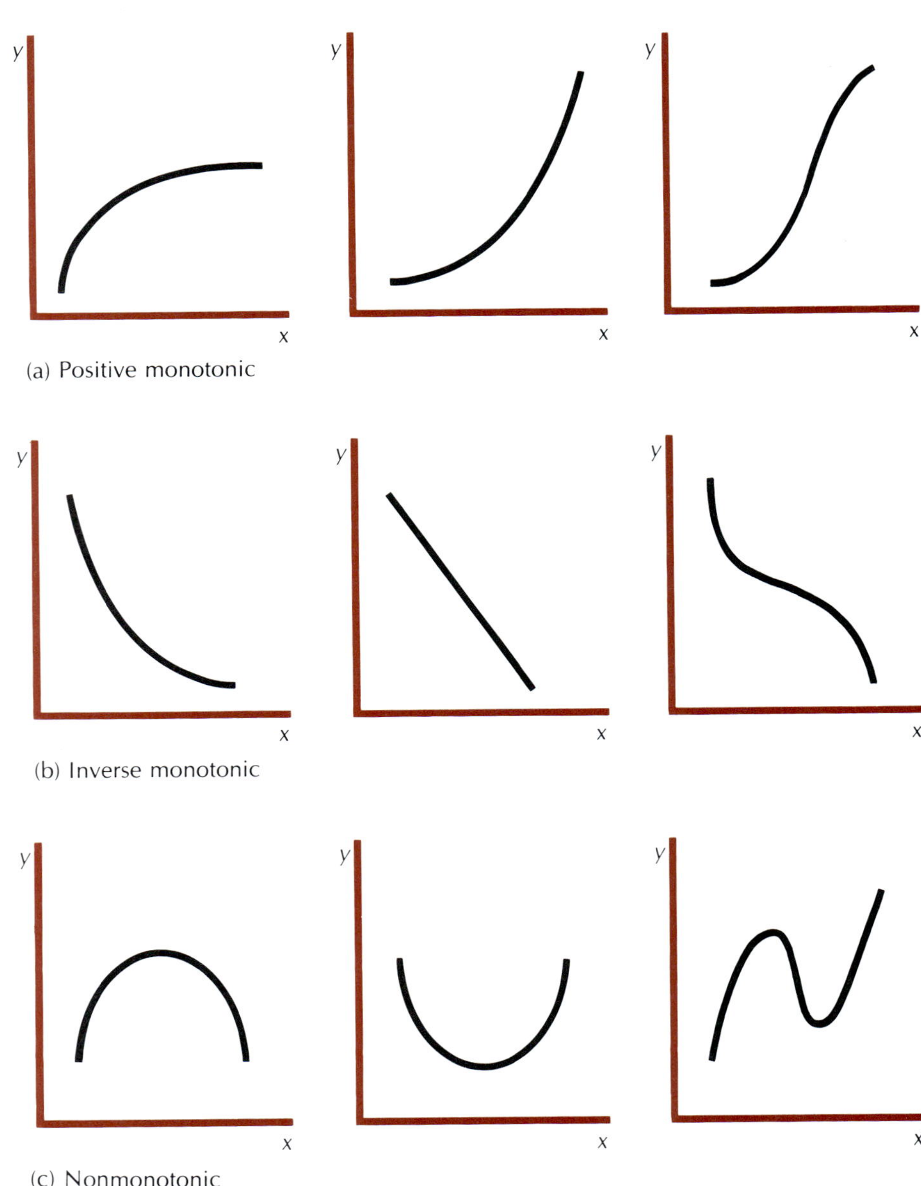

Figure 16.1 Schematic illustrations of (a) positive monotonic relationships, (b) inverse monotonic relationships, and (c) nonmonotonic relationships.

where d_i is the difference between the two ranks assigned to the two variables for the ith observational unit, and the 6 is a constant relating to the fact that we are dealing with two different orders of the first n integers. Even if there are some ties, the simpler formula in Expression (16.3) is sometimes used as an approximation, as long as the ties are not too numerous.

To test the null hypothesis that the true (population) relationship between variables x and y is not monotonic,

$$H_0\colon\ \rho_S = 0 \tag{16.4}$$

where ρ_S is the measure of monotonic relationship in the population, we compare the computed value of r_S with critical values of Spearman's r_S for $n - 2$ degrees of freedom as tabulated in Appendix 8. This is a distribution-free test, in that it requires no assumption about the specific form of the underlying distributions. Another aid to interpretation is r_S^2, which is the proportion of the variation in ranks that is attributable to the linear relationship between the ranks of the variables. This indicates the extent to which the ranks of one variable are linearly predictable from the ranks of the other variable, and thus reflects the extent of monotonic relationship between the variables.

EXAMPLE 16.2 Owners of show jumping competition horses generally insure their horses, in view of the unpredictability of injury and disease. To what extent does the amount of insurance depend on the number of competitions won by the horse in recent years?

Data were obtained for 10 show jumping horses that competed in recognized shows from 1980 through 1982. The dependent variable was the insured value of the horse (in thousands of dollars) in October 1982, and the independent variable was the number of winnings (placings in the top five) by the horse in recognized competitions in the previous two years. The horses are designated here by the letters A to J.

	A	*B*	*C*	*D*	*E*	*F*	*G*	*H*	*I*	*J*
Number of winnings 1980–82 (x)	19	33	35	20	4	16	15	12	20	29
Insured value 1982 (y)	7.5	10	15	25	3.5	5	5	4	10	10

One way of describing the relationship between insured value and number of winnings is by means of Pearson's correlation coefficient, which is found to be $r = .49$ according to Expression (16.1). This tells us that $r^2 = 24\%$ of the variation in insured values can be accounted for by the linear relationship with number of winnings. However, the scattergram in Figure 16.2 reveals that a linear description may not be most appropriate in this case. Perhaps the relationship is not linear at all, or perhaps the fourth observation, which is an outlier or atypical observation, is responsible for reducing the measure of relationship. It may be informative to measure the extent of monotonic relationship rather than of linear relationship.

Ranking the two variables separately from smallest to largest, we obtain the x-ranks and y-ranks shown in the following table. Notice, in the x-ranks, that the rank $6\frac{1}{2}$ replaces the tied ranks 6 and 7 and, in the y-ranks, that the rank $3\frac{1}{2}$ replaces the tied ranks 3 and 4, and the rank 7 replaces the three tied ranks 6, 7, and 8. After assigning ranks to ties, we must remember to skip the rank numbers that have been replaced; for example, ranks 3, 4, 6, and 8 do not appear in the y-ranks, because they have been

Horse	x-rank	y-rank	d	d^2
A	5	5	0	0
B	9	7	2	4
C	10	9	1	1
D	$6\frac{1}{2}$	10	$-3\frac{1}{2}$	$12\frac{1}{4}$
E	1	1	0	0
F	4	$3\frac{1}{2}$	$\frac{1}{2}$	$\frac{1}{4}$
G	3	$3\frac{1}{2}$	$-\frac{1}{2}$	$\frac{1}{4}$
H	2	2	0	0
I	$6\frac{1}{2}$	7	$-\frac{1}{2}$	$\frac{1}{4}$
J	8	7	1	1
Total	55	55	0	19

replaced by $3\frac{1}{2}$s and 7s. As a partial check on the assignment of ranks, we can verify that the highest rank assigned is n and that the sum of the x-ranks and the sum of the y-ranks both equal $n(n + 1)/2$, which is the sum of the first n integers.

The ranks may be substituted into Expression (16.1) to obtain Spearman's rank correlation coefficient:

$$r_S = r_{\text{ranks}} = .883$$

To illustrate the use of Expression (16.3), we use the last two columns in the preceding

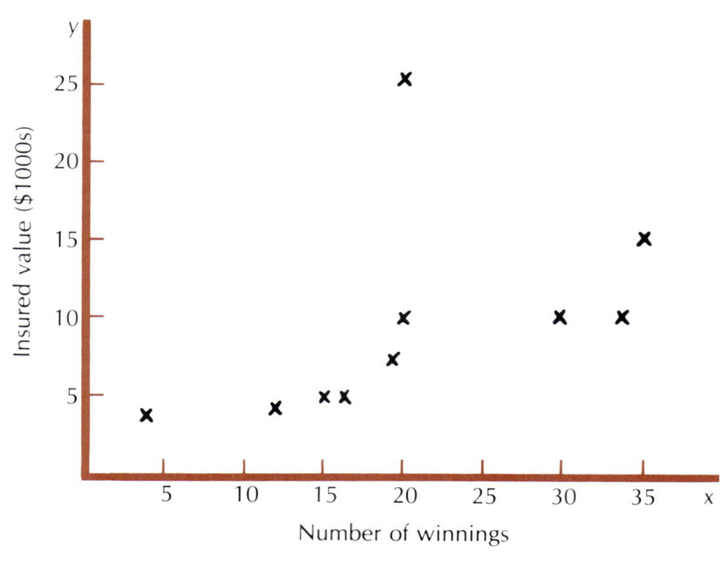

Figure 16.2 Scattergram of insured value as a function of number of winnings, for 10 show jumping competition horses.

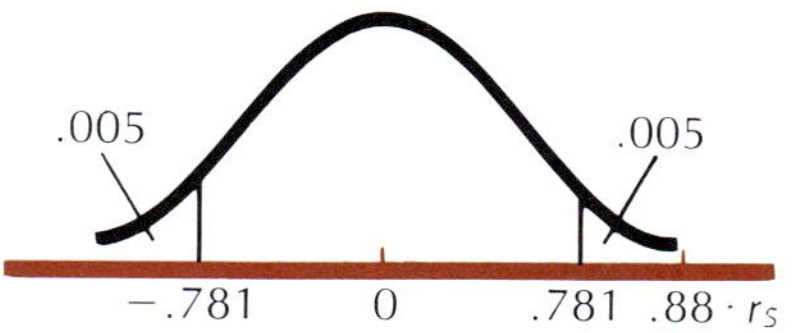

Figure 16.3

tabulation of ranks to calculate

$$r_S = 1 - \frac{(6)(19)}{(10)(99)} = .885$$

Even though there are several ties, the simpler formula gives a very good approximation to the value of r_{ranks}. The values would be identical if there were no ties.

Notice that Spearman's r_S (= .88) is substantially higher than Pearson's r (= .49) for these data, suggesting that there is a fairly strong monotonic, but not necessarily linear, relationship between the variables. Alternatively, we can say that Spearman's r_S is not as disturbed by the outlier (Horse D) as is Pearson's r. Checking Appendix 8 with $n - 2 = 8$ degrees of freedom, we find that the calculated value of r_S, which reflects the monotonic relationship between insured value and number of winnings, exceeds the 1%-level critical value .781 (see Figure 16.3), and therefore it is statistically significant ($P < .01$).

(Source: Student project by T. Arnold.)

16.3 TEST FOR A SPECIFIED VALUE OF A MEASURE OF LOCATION

Section 11.1 described the one-sample t test for a specified value of the mean. Such a test is used when we have data from a single random sample on the basis of which we wish to test whether the sample mean differs significantly from a specified value, say, μ_0, that had previously been conjectured to be the population mean. The null hypothesis is

$$H_0\colon \mu = \mu_0 \tag{16.5}$$

The t statistic given in Section 11.1 to test this null hypothesis is based on the assumption that the sampling distribution of the sample mean is Gaussian. This parametric assumption of normality will be satisfied if the sample size n is large enough or if the underlying population is Gaussian (Section 8.4).

The **sign test**, a form of which was used by the Scottish physician John Arbuthnot more than 250 years ago, is a nonparametric test based on sign information in the data. Used as an alternative to the one-sample t test, the sign test is referred to as the **one-sample sign test**. The null hypothesis is that the population median $\tilde{\mu}$ is equal to a specified value, say, $\tilde{\mu}_0$:

$$H_0\colon \tilde{\mu} = \tilde{\mu}_0 \tag{16.6}$$

If the population distribution is symmetrical, the null hypothesis in Expression (16.6) is the same as that in (16.5), but the sign test does not require symmetry. It assumes only that the population distribution is continuous.

The logic of the sign test is simple: If the null hypothesis in Expression (16.6) is true, then each observation has a probability of $\frac{1}{2}$ of being above the hypothesized median and a probability of $\frac{1}{2}$ of being below the hypothesized median. (Recall from Section 3.3 that the *median* is defined as the value that partitions a distribution into two equal halves.) The Gaussian approximation to the binomial distribution (for $p = q = \frac{1}{2}$) can be used to determine how unlikely the observed numbers above and below the hypothesized value would be under the null hypothesis. In practice, the procedure may be outlined as follows:

Step 1. Code each observation above the hypothesized value of the median with a plus sign and each observation below the hypothesized value with a minus sign. Values exactly equal to the hypothesized value are coded with a zero and are deleted from the analysis. (Such values represent a violation of the assumption of a continuous distribution, since the probability of any single exact value of a continuous distribution is zero.) Count the number of observations coded plus and the number coded minus, and call the difference between these two numbers D.

Step 2. Compute the value of the test statistic:

$$z_{\text{sign test}} = \frac{|D| - 1}{\sqrt{n}} \tag{16.7}$$

where n is the number of observations coded plus or minus, and not necessarily the original number of observations. This statistic may be compared with critical values from the standard normal distribution in Appendix 3; for example, $z_{\text{sign test}}$ must exceed 1.96 or 2.58 for (two-tailed) significance at the 5% or 1% level, respectively.

Note that Expression (16.7) is a continuity-corrected Gaussian approximation to the binomial distribution. According to the rule of thumb in Section 7.4, the approximation may be used when n is about 10 or larger. For smaller values of n or for more exact testing, binomial probabilities may be calculated directly by the methods of Section 7.2.

EXAMPLE 16.3 Consider an investigation to test the claim that students in college residences get an average of 7.5 hours of sleep each weeknight. Twenty-two male students living in college residences were asked about their average length of weeknight sleep. Their reported weeknight sleep durations had a mean of 7.05 hours and a standard deviation of .86 hour. Using the one-sample t test of Section 11.1, we find that the sample mean of 7.05 hours was significantly different from the hypothesized mean value of 7.50 hours ($P < .05$).

Let us use the sign test to test the null hypothesis that the population median is 7.5 hours. The 22 observations may be coded as being above or below the hypothesized

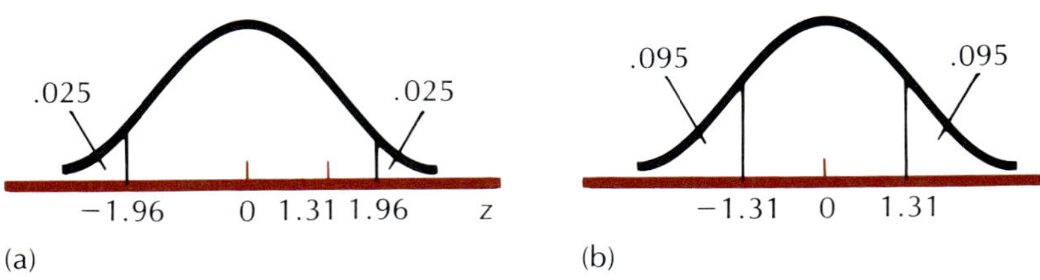

Figure 16.4

median as follows:

Reported sleep duration:	7	$6\frac{1}{2}$	6	$7\frac{1}{2}$	8	6	6	5	7	7	8	7	8	7	8	8	8	7	7	6	7	8
Sign:	−	−	−	0	+	−	−	−	−	−	+	−	+	−	+	+	+	−	−	−	−	+

There are 7 plus signs and 14 minus signs, for a total of $n = 21$ usable observations. The one observation exactly equal to the hypothesized median is dropped from the analysis. The test statistic,

$$z_{\text{sign test}} = \frac{|14 - 7| - 1}{\sqrt{21}} = 1.31$$

is less than the 5%-level critical value of 1.96, as shown in Figure 16.4(a). Specifically, from Appendix 3 we find that a z-value of 1.31 corresponds to a two-tailed P-value of $P = .19$, as shown in Figure 16.4(b). In any case, the sign test reveals no significant difference from the hypothesized median of 7.5 hours ($P > .05$). This result differs from that based on the t test because the sign test uses different information in the data. Indeed, for the sign test, the information used is that which would be obtained if each student were simply asked if he averaged more or less than 7.5 hours of sleep per night.

(Source: Student project by D. Stewart.)

Because it uses only sign information in the data, the sign test is often less powerful than the one-sample t test and thus less likely to detect real differences from the hypothesized value. However, it can be useful for situations where the parametric assumptions are particularly dubious and for quick inspection of data.

EXAMPLE 16.4 With small values of n, it is not difficult to compute the exact sign-test probability using the binomial distribution (Section 7.2). For example, if the null hypothesis in Expression (16.6) is true, the probability of obtaining a sample of size $n = 7$ with, say, 6 or more observations above the hypothesized median is

$$\binom{7}{6}\left(\frac{1}{2}\right)^6\left(\frac{1}{2}\right)^1 + \binom{7}{7}\left(\frac{1}{2}\right)^7\left(\frac{1}{2}\right)^0 = \frac{7!}{6!\,1!}\left(\frac{1}{2}\right)^6\left(\frac{1}{2}\right)^1 + \frac{7!}{7!\,0!}\left(\frac{1}{2}\right)^7\left(\frac{1}{2}\right)^0 = \frac{8}{128} = .0625$$

For a two-tailed test, the probability of obtaining 6 or more *or* 1 or fewer observations

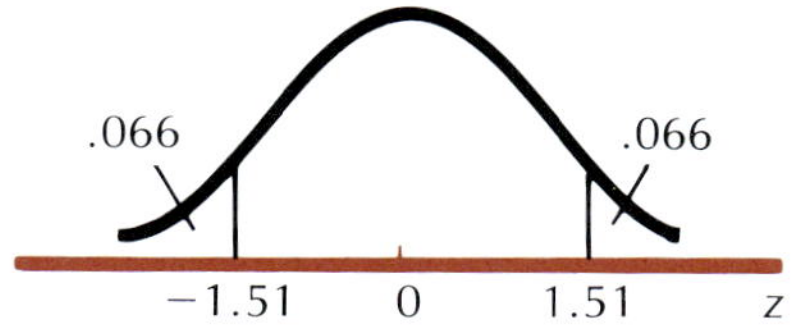

Figure 16.5

above the hypothesized median is twice .0625 (that is, .0625 for each tail of the distribution), or .125. We could say that the result is significant at the 12.5% level or simply that it is not significant at the conventional 5% level.

The Gaussian approximation in Expression (16.7) yields a z-value of

$$z_{\text{sign test}} = \frac{|6 - 1| - 1}{\sqrt{7}} = 1.51$$

for this example. You may use Appendix 3 to verify that even though n is less than 10 in this example, the Gaussian approximation still provides a fairly good approximation, in that a z-value of 1.51 cuts off an area of .066 in the upper tail of the Gaussian distribution (see Figure 16.5). This corresponds to a two-tail area of $P = .132$, or significance at the 13% level.

16.4 INDEPENDENT-SAMPLES TESTS

Section 11.2 presented a t test for judging the significance of the difference between means of two independent samples. The one-way ANOVA in Section 15.2 generalizes the independent-samples t test in that it provides a test of the difference between means of k independent samples, where k is any integer 2 or more. The independent-samples t test and the F test in ANOVA are based on the parametric assumptions of normality and homogeneity of variance, and they make use of the actual numerical values of a numerically measured dependent variable.

In this section we consider nonparametric methods for testing the significance of the difference between k independent samples. Two important nonparametric tests using rank-order information are the **Wilcoxon rank-sum test** as an alternative to the independent-samples t test for $k = 2$ treatments and the **Kruskal–Wallis rank-sum test** as an alternative to the one-way ANOVA F test for $k \geq 2$ treatments. The **Brown–Mood median test** is a nonparametric alternative using sign information for testing differences between two or more independent samples. **Pearson's chi-square test for two-way tables**, which was described in Section 14.5, uses categorical information about the dependent variable to test for differences between two or more independent samples. The chi-square test for two-way tables is especially useful when the dependent variable is recorded in terms of a few discrete categories, such as "healthy/diseased" or "for/against/undecided," so that each observational unit is not really measured but just counted in one category or another.

Wilcoxon Rank-Sum Test

The **Wilcoxon rank-sum test** was proposed by Frank Wilcoxon in 1945 and further developed by Henry B. Mann and D. Ransom Whitney in 1947. In a slightly different form, it is frequently referred to as the **Mann–Whitney test**. It is a nonparametric alternative to the independent-samples t test of the difference between two means.

The Wilcoxon rank-sum test is primarily a test of the difference in location (not necessarily means) of two independent random samples, but it is also somewhat sensitive to differences in shapes of the underlying populations. The null hypothesis is that the populations, which are assumed to be continuous (not necessarily Gaussian or even symmetric), are identical. If two random samples are drawn from identical populations and the resulting observations are arranged in ascending order, we should find that the observations from the two samples are randomly interspersed; that is, there should be no systematic tendency for observations from one sample to be concentrated at one end of the ascending order and those from the other sample to cluster together at the other end. The Wilcoxon rank-sum test evaluates the null hypothesis of identical underlying populations by considering whether any observed clustering of observations from one sample toward one end of the common ascending rank order is greater than would be expected by chance if the null hypothesis were true.

Consider the general situation in which we have two independent random samples of n_1 and n_2 observations, respectively, for a total of $n = n_1 + n_2$ observations. The Wilcoxon rank-sum test procedure for such data is outlined in the following steps:

Step 1. Assign the ranks $1, 2, \ldots, n$ to the n observations from smallest to largest (or largest to smallest). Because the procedure requires that the sum of the n ranks be equal to the sum of the first n integers, any tied observations must be assigned the mean of the ranks they would have received if they had not been exactly tied.

Step 2. Calculate the sum R_1 of the ranks in Sample 1. It is immaterial which sample is designated as Sample 1; the smaller sample is often chosen because it contains fewer ranks to be added. As an optional check on the ranking and summing, the total of all the ranks, $R_1 + R_2$, should equal $n(n + 1)/2$, which is the sum of the first n integers.

Step 3. Calculate the correction for ties, $\sum (t^3 - t)$, where t is the number of tied observations in a set of ties, and the summation extends over all sets of ties. (The presence of ties indicates a violation of the assumption of a continuous underlying distribution, because the probability of ties in a continuous distribution is zero. However, the test is fairly robust against such violations.)

Step 4. Compute the value of the test statistic:

$$z_{\text{rank-sum}} = \frac{\left| R_1 - \dfrac{n_1(n+1)}{2} \right| - \dfrac{1}{2}}{\sqrt{\dfrac{n_1 n_2}{12}\left(n + 1 - \dfrac{\Sigma\,(t^3 - t)}{n(n-1)}\right)}} \qquad \textbf{(16.8)}$$

This statistic may be compared with critical values from the standard normal distribution in Appendix 3, so that $z_{\text{rank-sum}}$ must exceed 1.96 or 2.58 for (two-tailed) significance at the 5% or 1% level, respectively.

A word of explanation concerning the rationale for Step 4 is in order: As the sample sizes n_1 and n_2 become large, the sampling distribution of R_1 under the null hypothesis approaches the Gaussian distribution with a mean of $n_1(n + 1)/2$ and a variance of $n_1 n_2(n + 1)/12$, if there are no ties. The number of standard deviation units by which the observed value R_1 differs from the mean of the sampling distribution thus could be compared with critical values (such as 1.96 and 2.58) from the standard normal distribution to evaluate its probability under the null hypothesis. This large-sample Gaussian approximation to the sampling distribution of R_1 can be modestly improved by two adjustments:

a. A **continuity correction** of one-half rank is made in the numerator to compensate for the fact that a continuous distribution (the Gaussian) is being used to approximate a discrete distribution (of rank-sums).

b. A **correction for ties** is made in the denominator to account for the reduction in variance produced by tied ranks.

The result is Expression (16.8).

Note that this is a large-sample approximation, but it turns out to be fairly accurate for samples even as small as $n_1 = n_2 = 4$ observations each. For smaller samples or for more exact testing, tabulations of the sampling distribution of R_1 under the null hypothesis are available (see the Further Reading at the end of this chapter).

EXAMPLE 16.5 Are the average lengths of popular songs different for country music and rock music? The mean lengths of songs on 10 country record albums and 11 rock albums were found to be the following numbers of minutes:

Country:	2.96	3.12	3.33	3.12	3.04	3.23	4.07	2.90	3.73	3.52	
Rock:	4.06	4.17	3.39	4.06	6.01	3.85	4.83	4.10	3.45	4.88	3.05

Let us apply the Wilcoxon rank-sum test to these data to test the difference between the two types of music. The ranks from 1 to 21 are assigned to the observations from smallest to largest as follows, with tied observations receiving the mean of the tied ranks:

Country:	2	$5\frac{1}{2}$	8	$5\frac{1}{2}$	3	7	16	1	12	11	
Rock:	$14\frac{1}{2}$	18	9	$14\frac{1}{2}$	21	13	19	17	10	20	4

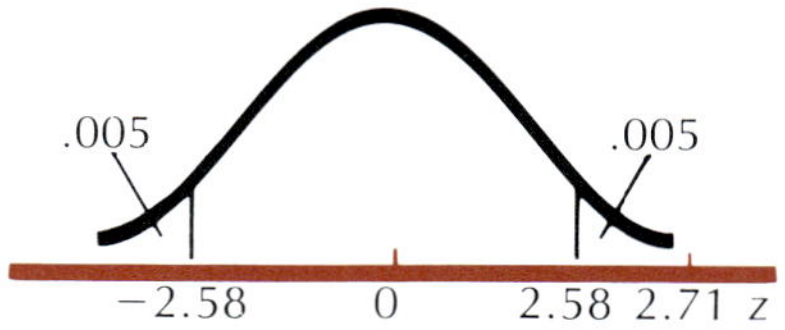

Figure 16.6

Designating the country sample as the first sample, we have

$$n_1 = 10, \qquad n_2 = 11, \qquad n = 21$$

$$R_1 = 71, \qquad R_2 = 160, \qquad R_1 + R_2 = 231 = \frac{n(n+1)}{2}$$

Since there are two sets of ties, with $t = 2$ observations tied in each set, the correction for ties is

$$\sum (t^3 - t) = (2^3 - 2) + (2^3 - 2) = 12$$

The test statistic is

$$z_{\text{rank-sum}} = \frac{\left|71 - \frac{(10)(22)}{2}\right| - \frac{1}{2}}{\sqrt{\frac{(10)(11)}{12}\left(22 - \frac{12}{(21)(20)}\right)}} = 2.71$$

which exceeds the 1%-level critical value of 2.58 (see Figure 16.6). We conclude that the rock songs were significantly longer than the country songs ($P < .01$).

(Source: Student project by K. E. George.)

Rank tests are also used when the dependent variable has been recorded not as a numerically measured variable, but only in terms of rank order. For example, the Wilcoxon rank-sum test may be applied to the order-of-admission data in Example 16.1 to test whether the two groups of alcoholics differed significantly with respect to date of admission (see Problem 16/27).

Kruskal–Wallis Rank-Sum Test

In 1952, William H. Kruskal and W. Allen Wallis published a generalization of the two-sample Wilcoxon rank-sum test. The resulting **Kruskal–Wallis rank-sum test** uses rank-order information in testing the significance of the difference between k (two or more) independent random samples; therefore, it is a nonparametric alternative to the F test in the one-way ANOVA for independent samples. Although the Kruskal–Wallis and Wilcoxon rank-sum tests are equivalent when $k = 2$, in practice the Wilcoxon rank-sum test is used when $k = 2$, and the Kruskal–Wallis version is used when $k \geq 3$.

The Kruskal–Wallis test is based on the same logic and assumptions as outlined for the Wilcoxon rank-sum test. The null hypothesis is that the k population distributions from which the observations have been randomly selected are identical, although the test actually is most sensitive to differences in location.

The following is an outline of the procedure for performing the Kruskal–Wallis rank-sum test with k independent samples of $n_1, n_2, \ldots, n_k$ observations, for a total of $n = n_1 + n_2 + \cdots + n_k$ observations:

Step 1. Assign the ranks $1, 2, \ldots, n$ to the n observations from smallest to largest (or largest to smallest). As usual, any ties must be assigned the mean of the ranks they would have received if they had not been tied, so that the sum of the assigned ranks remains equal to the sum of the first n integers.

Step 2. Calculate the sum R_j of the ranks in the jth sample, for $j = 1, \ldots, k$. As a check on the ranking and summing, verify that the total of the rank-sums, $\sum R_j$, is equal to $n(n + 1)/2$, which is the sum of the first n integers.

Step 3. Calculate the correction for ties, $\sum (t^3 - t)$, where t is the number of tied observations in a set of ties, and the summation extends over all sets of ties. (Again, the presence of ties indicates a violation of the assumption of a continuous underlying distribution, but the test is fairly robust against such violations.)

Step 4. Compute the value of the Kruskal–Wallis test statistic:

$$H = \frac{\dfrac{12}{n(n+1)}\left(\sum \dfrac{R_j^2}{n_j}\right) - 3(n+1)}{1 - \dfrac{\sum (t^3 - t)}{(n-1)(n)(n+1)}} \tag{16.9}$$

Notice that the denominator of Expression (16.9) is the tie correction, which equals 1.0 when there are no ties and hence may be ignored in such cases. If the statistic H exceeds the critical value $\chi^2_\alpha(k - 1 \text{ df})$ from the chi-square distribution with $k - 1$ degrees of freedom, we conclude that the difference among the k samples is significant at the α level.

The rationale for Step 4 may be summarized as follows: As the sample sizes n_j become large, the sampling distribution of H under the null hypothesis of identical populations approaches the chi-square distribution with $k - 1$ degrees of freedom. Step 4 therefore is based on a large-sample approximation, but in practice the chi-square approximation is acceptable even for samples as small as $n_j = 3$ observations each, when there are at least $k = 3$ samples. For smaller samples or for more exact testing, tabulations of the sampling distribution of H under the null hypothesis are available (see the Further Reading at the end of this chapter). For $k = 2$ samples, a continuity correction should be incorporated into the test statistic. To avoid the need to amend Expression (16.9) when $k = 2$, use the continuity-corrected Wilcoxon rank-sum test statistic in Expression (16.8), which is equivalent to the Kruskal–Wallis rank-sum statistic in the two-sample case.

EXAMPLE 16.6 A study was conducted concerning the number of times that students living in college residences go home in one semester. Three populations of female students were sampled: those with homes close to the college (within a one-hour drive), those with homes at an intermediate distance (between one and four hours), and those whose homes were far from the college (more than four hours). The numbers of visits home by the 10 participants in the study were as follows:

Close:	8	7	8	8
Intermediate:	6	6	5	
Far:	3	2	4	

Let us use the Kruskal–Wallis rank-sum test to see if the three samples differed significantly in number of visits home. We begin by ranking the 10 observations from smallest to largest:

Close:	9	7	9	9
Intermediate:	$5\frac{1}{2}$	$5\frac{1}{2}$	4	
Far:	2	1	3	

The sample sizes and rank-sums are as follows:

$$n_1 = 4, \qquad n_2 = 3, \qquad n_3 = 3, \qquad n = 10$$

$$R_1 = 34, \qquad R_2 = 15, \qquad R_3 = 6, \qquad \sum R_j = 55 = \frac{n(n+1)}{2}$$

There are $t = 3$ observations tied at 8 visits and $t = 2$ observations tied at 6 visits, so the correction for ties is

$$\sum (t^3 - t) = (3^3 - 3) + (2^3 - 2) = 30$$

The test statistic is

$$H = \frac{\dfrac{12}{(10)(11)}\left(\dfrac{34^2}{4} + \dfrac{15^2}{3} + \dfrac{6^2}{3}\right) - (3)(11)}{1 - \dfrac{30}{(9)(10)(11)}} = 8.27$$

and the critical value from the chi-square distribution with 2 degrees of freedom is 5.991 for a test at the 5% level (see Figure 16.7). Since H is greater than 5.991, we conclude that the three samples differed significantly in number of visits home ($P < .05$), with students making fewer visits home if they lived at greater distances from the college.

(Source: Student project by D. M. Mychaluk.)

Brown–Mood Median Test

The **Brown–Mood median test** was proposed by G. W. Brown and Alexander M. Mood in 1951 as a test of the difference between two or more independent

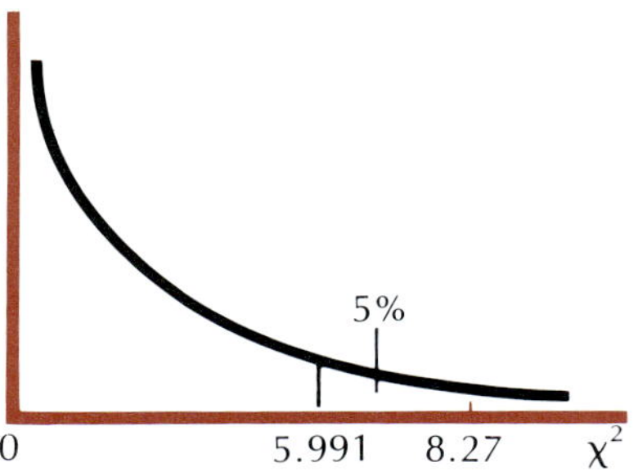

Figure 16.7

random samples, using only sign information in the data. Given the assumption of continuous underlying distributions, the median test evaluates the null hypothesis that the population medians are equal.

The logic of the test is straightforward: If the null hypothesis is true, then the overall median of the total sample should split each and every sample into approximately equal numbers of observations that are greater and smaller than the overall median. The Brown–Mood median test evaluates the null hypothesis of equal population medians by considering whether any observed excess of observations above or below the overall median in the separate samples is greater than would be expected by chance if the null hypothesis were true.

The Brown–Mood median test procedure for k independent samples involving a total of n observations is outlined in the following steps:

Step 1. Find the overall median of the n observations (see Section 3.3).

Step 2. Set up a k-by-2 table showing the numbers in each sample above and below the overall median. Any observations exactly equal to the overall median—even though inevitable if n is odd—may be considered as representing a violation of the assumption of a continuous underlying distribution, since the probability of any exact value of a continuous variable is taken to be zero. Probably the safest procedure for such observations is to delete them from the analysis; redefine the value of n accordingly.

Step 3. Perform a chi-square test for two-way tables, as described in Section 14.5, taking the critical value from the chi-square distribution with $k - 1$ degrees of freedom. A significant relationship between the independent variable and the direction of the observations from the median, so that some samples fall mostly above the median and others mostly below, suggests that the use of a common median is not appropriate for all levels of the independent variable and constitutes evidence against the null hypothesis of equal population medians.

Remember that Pearson's X^2 statistic is only approximated by the chi-square distribution. The approximation is good provided that none of the expected frequencies is too small, as specified in Expression (14.24). References cited under Further Reading at the end of this chapter provide methods for judging the significance of Pearson's X^2 statistic when there are too few observations—according to the criteria of Expression (14.24)—to use the chi-square approximation.

EXAMPLE 16.7 A hatchery received regular shipments of hatching eggs from two different suppliers for the same variety of broiler chicks. Bad eggs were culled from each shipment, and the rest were incubated for three weeks. The following numbers of live chicks hatched from each shipment were recorded for a sample consisting of 20 shipments from supplier C and 17 shipments from supplier E:

C:	3848	3883	2392	3838	2427	3767	2337	1679	3791	3773
	2344	3868	2420	2160	4184	3764	2080	3510	2012	2982
E:	2914	2299	2548	2445	1690	2584	2572	2104	2213	3044
	2769	3348	3162	3064	2886	3024	2269			

Was there a significant difference between the suppliers with respect to number of live chicks per shipment?

To test the difference using the Brown–Mood median test, we first find the overall median of the 37 observations. Since $37/2 = 18\frac{1}{2}$, we can find the median as the value of the observation such that 18 observations are smaller than the median and 18 observations are larger than the median, namely, 2769. The observation (2769) exactly equal to the median is deleted from the analysis, leaving a total of 36 observations. The observed frequencies are then recorded in the cells of the median-test contingency table, and the expected frequencies are computed as the product of the relevant marginals divided by 36 (as explained in Section 14.5) and recorded in parentheses.

	Supplier C	Supplier E	Totals
Above common median	11 (10)	7 (8)	18
Below common median	9 (10)	9 (8)	18
Totals	20	16	36

The calculation of the test statistic for the chi-square test for two-way tables may be summarized as follows.

Category	o_i	e_i	$o_i - e_i$	$(o_i - e_i)^2/e_i$
C/Above	11	10	1	.100
C/Below	9	10	−1	.100
E/Above	7	8	−1	.125
E/Below	9	8	1	.125
	$n = 36$	36		$X^2 = .45$

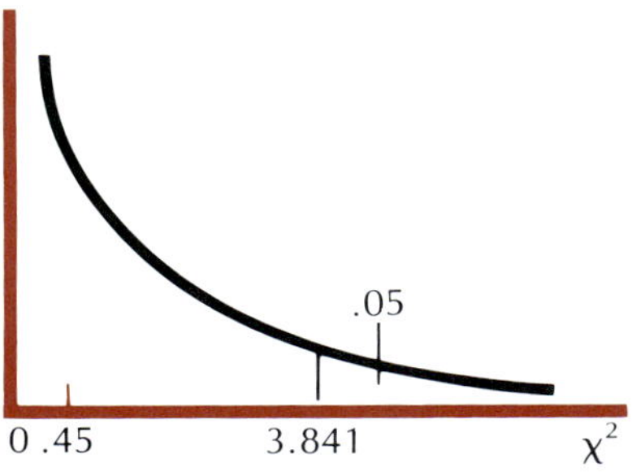

Figure 16.8

The test statistic $X^2 = .45$ is less than the critical value from the chi-square distribution with $k - 1 = 1$ degree of freedom, $\chi^2_{.05}(1 \text{ df}) = 3.841$, as shown in Figure 16.8. Thus, we conclude that the difference between suppliers with respect to number of live chicks per shipment was not significant according to the median test ($P > .05$).

Note that this sample was actually somewhat smaller than desirable for the chi-square test for two-way tables [see Expression (14.24)]. However, this would mean that the calculated value of X^2 is somewhat larger than it should be, so that in this case the result still would be not significant.

(Source: Student project by J. McLean.)

Even if the sample size is large enough to warrant use of the median test, the median test can lead to a different conclusion concerning significance than does the Wilcoxon or Kruskal–Wallis rank-sum test. This should not be surprising, since the two types of test (median and rank-sum) use different information in the data and are correspondingly sensitive to different sorts of differences between samples. The rank-sum tests take into account the actual rank of each observation, whereas the median test lumps together all observations above the median on the one hand and all observations below the median on the other hand. For most types of data, the fact that the median test uses only sign information implies that it has relatively low power to detect true differences in the populations when sample sizes are small. Other things being equal, if more than one test is appropriate for answering the question being asked of a given set of data, the more powerful test is preferred.

16.5 REPEATED-MEASURES TESTS

Section 11.3 presented a t test for judging the significance of the difference between means of two paired samples. The one-way ANOVA with repeated measurements of the same replicates in Section 15.3 generalizes the paired-samples t test in that it provides a test of the difference between means for k (two or more) treatments in cases where each replicate (or block) is observed repeatedly, once for each treatment. These tests are based on parametric assumptions, such as normality and homogeneity of variance, and they use the actual numerical values of a numerically measured dependent variable.

In this section we consider nonparameteric methods for testing the significance of the difference between k means in a one-way repeated-measures design. Two important nonparametric tests making use of rank-order information are the **Wilcoxon signed-rank test**, which is an alternative to the paired-samples t test for $k = 2$ treatments, and the **Friedman rank test**, which is an alternative to the one-way repeated-measures ANOVA F test for $k \geq 2$ treatments. The **sign test**, which we met in Section 16.3, provides a nonparametric method based on sign information or on categorical information for testing the difference between two paired samples.

Wilcoxon Signed-Rank Test

The **Wilcoxon signed-rank test** was introduced by Frank Wilcoxon in 1945 as a nonparametric alternative to the paired-samples t test of the difference between two means. Thus, the signed-rank test is intended for analyzing data from experiments in which each of n replicates (individuals or blocks) is observed under both of two treatments and the researcher wishes to evaluate the statistical significance of any observed differences between treatments. The null hypothesis is that the underlying populations, which are assumed to be continuous distributions, are identical for the two treatments, but the test is most sensitive to differences in location (for example, mean or median).

The signed-rank test involves rank ordering, from smallest to largest, the absolute values of the differences between treatments for the n replicates. If the null hypothesis is true, the sum of the ranks assigned to differences favoring one treatment should roughly equal the sum of the ranks assigned to the differences favoring the other treatment. The signed-rank test statistic is used to evaluate whether this is, in fact, the case. The details of the procedure are outlined in the following steps:

Step 1. For each replicate, obtain the difference d_i between the observations x_{1i} and x_{2i} for the two treatments:

$$d_i = x_{1i} - x_{2i} \qquad \text{for the } i\text{th replicate} \tag{16.10}$$

Step 2. Delete from the analysis any differences of zero, and redefine n as the number of nonzero differences. Assign the ranks $1, 2, \ldots, n$ to the absolute values of the nonzero differences, *from smallest to largest.* If there are any tied absolute values, assign each of them the mean rank of the tied set.

Step 3. Attach the sign of the difference d_i to the rank assigned to that difference, and calculate W, which is the sum of the signed ranks.

Step 4. Calculate the correction for ties, $\sum (t^3 - t)$, where t is the number of tied absolute values in a set of ties, and the summation extends over all sets of ties among the nonzero absolute differences.

Step 5. Compute the value of the test statistic:

$$z_{\text{signed-rank}} = \frac{|W| - \frac{1}{2}}{\sqrt{\frac{n(n+1)(2n+1)}{6} - \frac{\sum (t^3 - t)}{12}}} \qquad \textbf{(16.11)}$$

This statistic may be compared with critical values from the standard normal distribution, so that $z_{\text{signed-rank}}$ must exceed 1.96 or 2.58 for (two-tailed) significance at the 5% or 1% level, respectively.

The test statistic in Expression (16.11) is a large-sample approximation. As the number n of nonzero differences becomes large, the sampling distribution of W under the null hypothesis approaches the Gaussian distribution with a mean of zero and a variance of $n(n + 1)(2n + 1)/6$, if there are no ties. The number of standard deviation units by which the observed value W differs from zero thus could be compared with critical values (such as 1.96 and 2.58) from the standard normal distribution in Appendix 3 to evaluate its probability under the null hypothesis. This large-sample Gaussian approximation to the sampling distribution of W can be modestly improved by two adjustments:

a. A continuity correction of one-half rank is made in the numerator to compensate for the fact that a continuous distribution (the Gaussian) is being used to approximate a discrete distribution (of signed-rank sums).

b. A correction for ties is made in the denominator to account for the reduction in variance produced by tied ranks.

The result is Expression (16.11), a large-sample approximation that turns out to be fairly accurate even for n as small as 6. For smaller samples or for more exact testing, tabulations of the sampling distribution of W under the null hypothesis are available (see the Further Reading at the end of this chapter).

EXAMPLE 16.8 Twelve college students took part in an investigation of whether typing speed is affected by the nature of the material being typed. Each student typed two passages of equal length, one passage from a novel and the other from a textbook for a senior course in economics. Half of the students typed from the novel first and half from the text first. The times (in seconds) required to type the passages are recorded in the following table.

Let us use the Wilcoxon signed-rank test to see whether these data provide enough evidence to conclude that typing times for novel and text material are different. The preliminary calculations are shown in the three right-hand columns: the differences d_i between treatments, the ranks of the absolute values of the differences from smallest to largest, and the signed ranks. Notice that the data from two students (students 4 and 8) are deleted from the analysis because of differences of zero, leaving $n = 10$ observations in the analysis. There is only one set of tied absolute values, namely, the two observations with an absolute difference of 1 second. The correction for ties therefore is

$$\sum (t^3 - t) = 2^3 - 2 = 6$$

Student	Novel	Text	d_i	Rank of $\|d_i\|$	Signed rank
S_1	223	236	−13	6	−6
S_2	146	145	1	$1\frac{1}{2}$	$1\frac{1}{2}$
S_3	178	156	22	8	8
S_4	166	166	0		
S_5	165	164	1	$1\frac{1}{2}$	$1\frac{1}{2}$
S_6	115	120	−5	4	−4
S_7	246	312	−66	10	−10
S_8	168	168	0		
S_9	169	158	11	5	5
S_{10}	152	154	−2	3	−3
S_{11}	200	245	−45	9	−9
S_{12}	146	164	−18	7	−7
					$W = -23$ with $n = 10$

The test statistic is

$$z_{\text{signed-rank}} = \frac{23 - \frac{1}{2}}{\sqrt{\frac{(10)(11)(21)}{6} - \frac{6}{12}}} = 1.15$$

which is less than the critical value 1.96 for a test at the 5% significance level, as shown in Figure 16.9(a). Specifically, we find in Appendix 3 that a z-value of 1.15 corresponds to a two-tailed P-value of $P = .25$, as shown in Figure 16.9(b). The data do not provide enough evidence to conclude that the nature of the material, novel or text, makes a difference in typing speed in this population of students.

(Source: Student project by M. Saito.)

Friedman Rank Test

The **Friedman rank test** was introduced by Milton Friedman in 1937 as a nonparametric alternative to the F test in the one-way ANOVA with repeated measurements of the same replicates. In the general situation, there are n replicates (individuals or blocks) in the experiment, and every replicate is observed under all k treatments. The null hypothesis is that the underlying

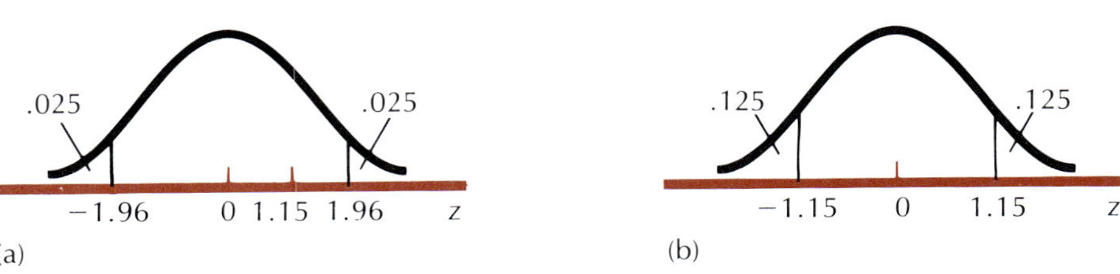

Figure 16.9

population distributions, which are assumed to be continuous, are identical for the k treatments, but the test is most sensitive to differences in location (for example, mean or median). In practice, the Friedman rank test is used only in cases with $k \geq 3$ treatments; when $k = 2$, the Friedman test is equivalent to the paired-samples sign test in the next subsection, which is used in its place.

The Friedman rank test involves rank ordering each replicate's k observations separately. If the null hypothesis of identical populations is true, the ranks will be randomly dispersed among the k treatments in each replicate, and the sums of the ranks for the k treatments will differ only by chance. The Friedman rank test evaluates whether or not this is the case. The details of the procedure are outlined in the following steps:

Step 1. For each of the n replicates, assign the ranks $1, 2, \ldots, k$ to the replicate's k observations from smallest to largest (or largest to smallest). Ties between *different* replicates require no special consideration because of the separate ranking for each replicate, but ties between treatments for the *same* replicate should be handled in the usual way, assigning the mean of the tied ranks.

Step 2. Calculate the sum R_j of the ranks for the jth treatment, for $j = 1, \ldots, k$. As a check on the ranking and summing, verify that the total of the rank-sums, $\sum R_j$, is equal to $nk(k + 1)/2$, which is n times the sum of the first k integers.

Step 3. Compute the value of the Friedman rank test statistic:

$$X^2_{\text{rank}} = \frac{12}{nk(k + 1)} \left(\sum R_j^2\right) - 3n(k + 1) \qquad \textbf{(16.12)}$$

Compare it with the critical value $\chi^2_\alpha(k - 1 \text{ df})$ from the chi-square distribution with $k - 1$ degrees of freedom for a test at the α level.

Under the null hypothesis of k identical populations, the sampling distribution of the test statistic X^2_{rank} approaches the chi-square distribution with $k - 1$ degrees of freedom as n becomes sufficiently large. This large-sample approximation is close enough for practical use even when n is as small as 5 if k is at least 3. [When $k = 2$, a continuity correction should be incorporated into the test statistic. To avoid the need to amend Expression (16.12) when $k = 2$, use the continuity-corrected sign test in the next subsection, which is equivalent to the Friedman rank test in the two-treatment case.]

As in most of the nonparametric tests we have met, the underlying distributions are assumed to be continuous for the Friedman test, and ties within a replicate represent a violation of the assumption. It is possible to incorporate a correction for ties into Expression (16.12), but the correction is rather complex and it has been omitted here. If there are ties, use of the uncorrected formula in Expression (16.12) will make the test slightly conservative: It will be slightly more difficult to attain significance than it should be, but not much more so if there are not too many ties.

EXAMPLE 16.9 In a study comparing three supermarkets in the same city, prices were obtained on the same day for eight products sold at all three stores. The prices (in cents) were as follows for products of the same quantity and brand.

Product	Store 1	Store 2	Store 3
Lettuce	59	59	47
Mushrooms	71	79	94
Tomatoes	99	129	129
Eggs	112	110	105
Cottage cheese	195	179	193
Pork chops	205	195	199
Shrimp	389	349	375
Coffee	439	385	434

The Friedman rank test can be used to test the null hypothesis that the corresponding three populations of prices are identical. The object is to compare $k = 3$ stores using data from $n = 8$ replicates (products). We begin by assigning ranks to the stores separately for each replicate. The following ranking is from lowest price to highest price, with ties handled in the usual way.

Product	Store 1	Store 2	Store 3
Lettuce	$2\frac{1}{2}$	$2\frac{1}{2}$	1
Mushrooms	1	2	3
Tomatoes	1	$2\frac{1}{2}$	$2\frac{1}{2}$
Eggs	3	2	1
Cottage cheese	3	1	2
Pork chops	3	1	2
Shrimp	3	1	2
Coffee	3	1	2
	$R_1 = 19\frac{1}{2}$	$R_2 = 13$	$R_3 = 15\frac{1}{2}$

As a check on the ranking, we verify that the total of the rank-sums,

$$R_1 + R_2 + R_3 = 19\tfrac{1}{2} + 13 + 15\tfrac{1}{2} = 48$$

is equal to n times the sum of the first k integers

$$\frac{nk(k+1)}{2} = \frac{(8)(3)(4)}{2} = 48$$

The Friedman rank test statistic is

$$X^2_{\text{rank}} = \frac{12}{(8)(3)(4)}(19.5^2 + 13^2 + 15.5^2) - (3)(8)(4) = 2.69$$

which is less than the critical value $\chi^2_{.05}(2\text{ df}) = 5.991$ (see Figure 16.10). There is not enough evidence to conclude that the populations of prices are different in the three stores ($P > .05$).

(Source: Student project by G. Lazicki.)

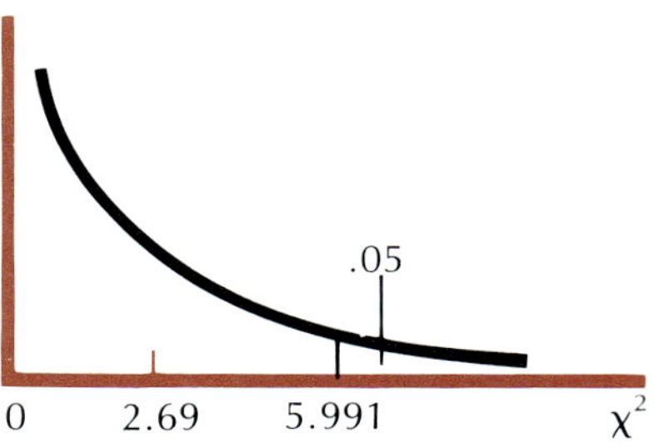

Figure 16.10

Paired-Samples Sign Test

In Section 16.3, we met the sign test as a nonparametric alternative to the one-sample t test for a specified value of the mean. Making use of either sign information or categorical information in the data, the sign test also provides an alternative to the paired-samples t test of the difference between two means. Used in this way, the sign test is referred to as the **paired-samples sign test**.

Data for the paired-samples sign test may be found in several different forms. First, if we have two numerical observations for each of n replicates, the only information we need to extract for the sign test is the number of replicates that changed in the positive direction between the two observations and the number that changed in the negative direction. Second, the sign test may be used even if the only information recorded is the direction of change—for example, if replicates are simply categorized as improved, deteriorated, or unchanged. Replicates showing no change do not enter the analysis. Finally, the paired-samples sign test may be used in situations where n replicates are categorized and counted in two categories (such as success and failure) on two occasions. The numbers of replicates changing categories from one occasion to the other (from success to failure and from failure to success) provide the relevant information for the sign test; replicates remaining in the same category on both occasions do not enter the analysis.

Assuming only that the population distribution is continuous, the paired-samples sign test evaluates the null hypothesis that the median difference between the treatments (in the population) is zero. If the null hypothesis is true, changes in the positive direction and changes in the negative direction should each occur with a probability of $\frac{1}{2}$. The Gaussian approximation to the binomial distribution for $p = q = \frac{1}{2}$ can be used to determine how unlikely the observed numbers of changes in each direction would be under the null hypothesis. In practice, the procedure may be outlined by the following steps:

Step 1. Count the number of replicates that changed in the positive direction and the number that changed in the negative direction, and call the difference between these two numbers D.

Step 2. Compute the value of the test statistic:

$$z_{\text{sign test}} = \frac{|D| - 1}{\sqrt{n}} \tag{16.13}$$

where n is the number of replicates that showed a change in one direction or the other, and not necessarily the total number of replicates observed. This statistic may be compared with critical values from the standard normal distribution; for example, $z_{\text{sign test}}$ must exceed 1.96 or 2.58 for (two-tailed) significance at the 5% or 1% level, respectively.

Note that Expression (16.13) is a continuity-corrected Gaussian approximation to the binomial distribution with $p = q = \frac{1}{2}$. According to the rule of thumb in Section 7.4, the approximation may be used when n is about 10 or larger. For smaller values of n or for more exact testing, binomial probabilities may be calculated directly by the methods of Section 7.2.

EXAMPLE 16.10 Example 16.8 described an investigation in which 12 college students each typed two passages of equal length, one passage from a novel and the other from an economics textbook. The Wilcoxon signed-rank test yielded a z-value of 1.15, for which the two-tailed P-value (from Appendix 3) is $P = .25$.

Let us reanalyze the data with the paired-samples sign test. There were 6 students who took longer to type the text material, 4 who took longer to type the material from the novel, and 2 who typed both passages at the same speed. Thus, we have $D = 6 - 4 = 2$, $n = 6 + 4 = 10$, and

$$z_{\text{sign test}} = \frac{|6 - 4| - 1}{\sqrt{10}} = .32$$

for which the two-tailed P-value (from Appendix 3) is $P = .75$ (see Figure 16.11). There is no significant difference between the two sorts of material in terms of typing speed, according to the paired-samples sign test.

Notice that the sign test provided even less evidence of a difference between the two conditions than the Wilcoxon signed-rank test did. Sign tests and rank tests use different information in the data.

Because it uses only sign information, the sign test is often less powerful than the Wilcoxon signed-rank test and thus is less likely to detect real differences in the populations. However, it is attractive both for its simplicity

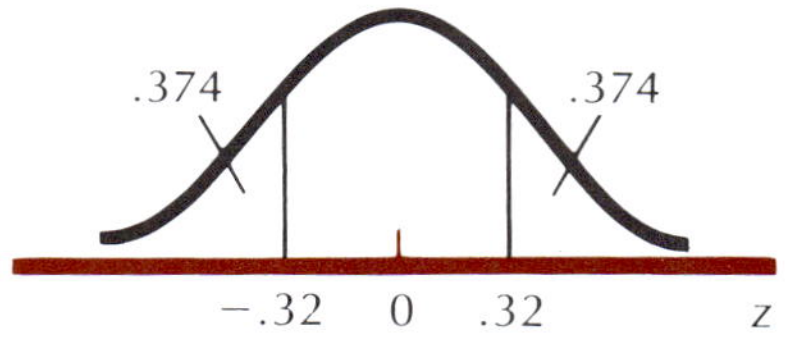

Figure 16.11

of calculation and for the fact that it may be used even where the data consist only of direction-of-change information (for example, improved or deteriorated) and not the amount of change from one condition to the other. Likewise, it may be used where the data consist only of change-of-category information rather than a measured amount of change on a numerical variable. In this latter form, it is sometimes known as the **McNemar test**.

EXAMPLE 16.11 As an example of the paired-samples sign test based on categorical information, consider a statistics class of 30 students who wrote two quizzes. The numbers passing and failing the two quizzes are shown in the following table.

An inadequate summary of the performance of a class of 30 students on two statistics quizzes. This summary of the data does not provide the information necessary for a correct analysis.

	Pass	Fail	Total
First quiz	23	7	30
Second quiz	15	15	30

Was there a significant difference in performance on the two quizzes? In other words, were the proportions of passing marks (23/30 on the first quiz and 15/30 on the second quiz) significantly different?

To answer the question, we must know how many students changed categories from the first to the second quiz. Such information is not available in the preceding table, which is in a format appropriate for an independent-samples design (as if there had been 60 different students) but not for a repeated-measures design with paired data (from 30 students, each observed twice). The required information is shown in the following table, which is in the format appropriate for a repeated-measures design (that is, an investigation where the same replicates are categorized twice).

An appropriate summary of the performance of a class of 30 students on two statistics quizzes. The summaries from the previous table of the two tests separately appear in the margins, and the cells show the categorization of the 30 students according to the four possible outcomes for any one student (*PP*, *PF*, *FP*, *FF*).

		First quiz		
		Pass	Fail	Totals
Second quiz	Pass	14	1	15
	Fail	9	6	15
	Totals	23	7	30

There were 9 students whose performance deteriorated, 1 student who improved, and $14 + 6 = 20$ students who remained in the same category. Thus, we have $D = 9 - 1 = 8$, $n = 9 + 1 = 10$, and

$$z_{\text{sign test}} = \frac{|9 - 1| - 1}{\sqrt{10}} = 2.21$$

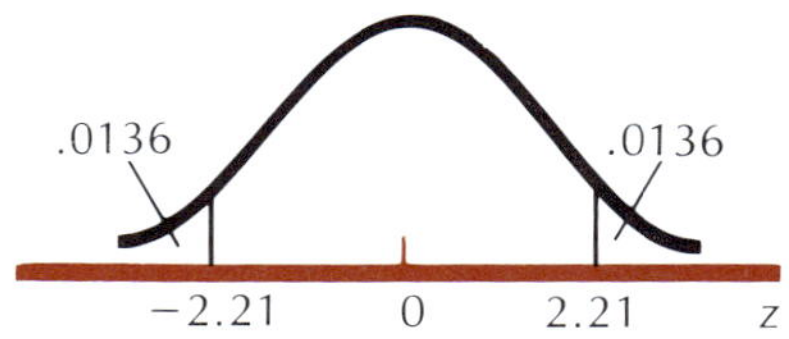

Figure 16.12

for which the two-tailed P-value (from Appendix 3) is $P = .027$ (see Figure 16.12). We conclude that performance was significantly worse on the second quiz ($P = .027$). The proportion (50%) who passed the second quiz was significantly less than the proportion (77%) who passed the first quiz.

16.6 ON DECIDING WHICH STATISTICAL METHOD TO USE

This chapter has considered four general types of statistical problem:

- measuring and testing the correlation between two variables,
- testing for a specified value of a measure of location in the one-sample case,
- testing the difference between k treatments in an independent-samples design, and
- testing the difference between k treatments in a repeated-measures design.

All are problems that we have examined in previous chapters by means of parametric methods. For each of these four problems, we have now presented at least one alternative statistical method, as summarized in Table 16.1. How do we decide which method to use in any particular application?

It is not possible to specify precisely how to decide which method to use in any and every situation that might arise—some degree of judgment based on experience is often required—but some general principles can be suggested as a starting point:

a. The most important rule is to begin by deciding which type of problem (which row of Table 16.1) we are dealing with, being especially careful to distinguish between independent-samples designs and repeated-measures designs.

b. For each type of problem, other things being equal, we prefer the method that makes the best use of the information in the data. Thus, for numerically measured dependent variables, the parametric methods, which use the actual numerical values of the observations, are generally preferred, unless the parametric assumptions are seriously violated.

c. Rather than being numerically measured, the dependent variable is sometimes recorded in a manner indicating only (1) the rank ordering of

TABLE 16.1 Summary of statistical methods that have been presented for four types of problems, categorized according to what information in the data is utilized by the method. (Parenthetical references identify the chapter and section where each method is discussed.)

	Information used in the data			
Type of problem	**Actual numerical magnitudes**	**Rank-order information**	**Sign information**	**Categorical information**
Correlation	Pearson's r (Sec. 5.2, 13.1)	Spearman's r_S (Sec. 16.2)		
One-sample tests	One-sample t test (Sec. 11.1)		One-sample sign test (Sec. 16.3)	Pearson's chi-square goodness-of-fit test (Sec. 14.4)
Tests on k independent samples	$k = 2$: Independent-samples t test (Sec. 11.2) $k \geq 2$: One-way ANOVA with independent samples (Sec. 15.2)	$k = 2$: Wilcoxon rank-sum test (Sec. 16.4) $k \geq 3$: Kruskal–Wallis rank-sum test (Sec. 16.4)	Brown–Mood median test (Sec. 16.4)	Pearson's chi-square test for two-way tables (Sec. 14.5)
Tests on k treatments in repeated-measures designs	$k = 2$: Paired-samples t test (Sec. 11.3) $k \geq 2$: One-way ANOVA with repeated measures (Sec. 15.3)	$k = 2$: Wilcoxon signed-rank test (Sec. 16.5) $k \geq 3$: Friedman rank test (Sec. 16.5)	$k = 2$: Paired-samples sign test (Sec. 16.5)	$k = 2$: Paired-samples sign test (Sec. 16.5)

observational units, or (2) the direction of change observed in each replicate, or (3) the number of replicates observed in each of several categories. For these special types of data, where the observational units are not so much measured as rank-ordered or categorized-and-counted, the appropriate nonparametric method should be chosen (from the last three columns of Table 16.1) according to the information available in the data.

d. Finally, it is sometimes informative to analyze the same data by more than one method, not to report the method that gives the answer one had hoped to find, but rather to obtain a more complete picture of the results. A more complete picture may be possible because the parametric and the various nonparametric methods for a given type of problem use different information in the data and test somewhat different null hypotheses. These differences in the methods provide important clues to understanding the data in cases where different methods yield different statements about the strength of the evidence, as illustrated in Example 16.2.

STUDY MATERIAL FOR CHAPTER 16

Terms and Concepts

Section 16.1

parametric methods
nonparametric methods

distribution-free methods
rank-order information
sign information
categorical information

Section 16.2

Spearman's rank correlation coefficient
monotonic relationship

Section 16.3

sign test
one-sample sign test

Section 16.4

Wilcoxon rank-sum test
Mann–Whitney test
continuity correction
correction for ties
Kruskal–Wallis rank-sum test
Brown–Mood median test

Section 16.5

Wilcoxon signed-rank test
Friedman rank test
paired-samples sign test
McNemar test

Key Formulas

Section 16.2

(16.3) $$r_S = 1 - \frac{6 \sum d^2}{n(n^2 - 1)}$$

Section 16.3

(16.7) $$z_{\text{sign test}} = \frac{|D| - 1}{\sqrt{n}}$$

Section 16.4

(16.8) $$z_{\text{rank-sum}} = \frac{\left| R_1 - \frac{n_1(n+1)}{2} \right| - \frac{1}{2}}{\sqrt{\frac{n_1 n_2}{12}\left(n + 1 - \frac{\sum (t^3 - t)}{n(n-1)}\right)}}$$

(16.9) $$H = \frac{\frac{12}{n(n+1)}\left(\frac{\sum R_j^2}{n_j}\right) - 3(n+1)}{1 - \frac{\sum (t^3 - t)}{(n-1)(n)(n+1)}}$$

Section 16.5

(16.11)
$$z_{\text{signed-rank}} = \frac{|W| - \frac{1}{2}}{\sqrt{\frac{n(n+1)(2n+1)}{6} - \frac{\Sigma(t^3 - t)}{12}}}$$

(16.12)
$$X^2_{\text{rank}} = \frac{12}{nk(k+1)}(\Sigma R_j^2) - 3n(k+1)$$

Exercises for Numerical Practice

Section 16.2

16/1. Seven bivariate observations were recorded for the variables x and y as follows:

x:	21	3	7	95	3	0	46
y:	15	18	15	12	27	19	15

a. Calculate Spearman's rank correlation coefficient, first using the method of Expression (16.2) and then using the method of Expression (16.3).

b. What is the critical value at the 1% significance level for a two-tailed test of the significance of the rank correlation coefficient in these data?

16/2. Eight bivariate observations were recorded for the variables x and y as follows:

x:	50	83	50	99	34	50	0	49
y:	76	24	190	3	9	68	88	88

a. Calculate Spearman's rank correlation coefficient, first using the method of Expression (16.2) and then using the method of Expression (16.3).

b. What is the critical value at the 5% significance level for a two-tailed test of the significance of the rank correlation coefficient in these data?

Section 16.3

16/3. A random sample of 15 observations was recorded as follows:

83 94 58 20 95 82 38 38 72 95 80 48 51 50 71

Calculate the test statistic and its P-value for a sign test of the hypothesis that the population median is 50.

16/4. A random sample of 19 observations was recorded as follows:

44 71 4 62 76 61 5 1 67 86 30 18 50 41 2 7 4 17 23

Calculate the test statistic and its P-value for a sign test of the hypothesis that the population median is 50.

Section 16.4

16/5. Calculate the test statistic and its P-value for a Wilcoxon rank-sum test of the null hypothesis of identical underlying populations, given the following two random samples:

Sample A:	6	36	65	60	38	23	3	13
Sample B:	94	92	45	73	38	84	0	

16/6. Calculate the test statistic and its P-value for a Wilcoxon rank-sum test of the null hypothesis of identical underlying populations, given the following two random samples:

Sample C:	62	81	80	32	93	45			
Sample D:	27	2	45	3	63	23	17	31	44

16/7. Calculate the test statistic for a Kruskal–Wallis rank-sum test of the null hypothesis of identical underlying populations, given the following three random samples:

Sample A:	47	20	48	36	37	
Sample B:	51	46	87	61		
Sample C:	22	15	29	36	32	46

What is the critical value for a test at the 5% level?

16/8. Calculate the test statistic for a Kruskal–Wallis rank-sum test of the null hypothesis of identical underlying populations, given the following four random samples:

Sample D:	87	98	82	67	
Sample E:	40	34	56		
Sample F:	26	56	10	14	39
Sample G:	69	56	82		

What is the critical value for a test at the 5% level?

16/9. Calculate the test statistic for a Brown–Mood median test of the null hypothesis of equal population medians, given the following four random samples:

Sample A:	21	37	36	38	13	24	23	29	29	69
Sample B:	39	44	30	69	36	30	37	34	62	68
Sample C:	82	58	86	75	81	72	64	70	60	71
Sample D:	82	96	93	99	97	79	92	99	86	80

What is the critical value for a test at the 5% level?

16/10. Calculate the test statistic for a Brown–Mood median test of the null hypothesis of equal population medians, given the following three random samples:

Sample E:	62	64	39	87	42	90	88	27	93	33	26	86
Sample F:	56	47	71	30	59	38	31	52	54	61	54	39
Sample G:	37	36	40	64	18	27	78	30	11	87	37	24

What is the critical value for a test at the 5% level?

Section 16.5

16/11. Calculate the test statistic and its P-value for a Wilcoxon signed-rank test of the significance of the difference between the following paired samples.

Replication	Treatment 1	Treatment 2
R_1	92	80
R_2	14	16
R_3	7	2
R_4	83	60
R_5	27	20
R_6	54	45
R_7	79	86
R_8	17	15
R_9	56	56

16/12. Calculate the test statistic and its P-value for a Wilcoxon signed-rank test of the significance of the difference between the following paired samples.

Replication	Treatment *A*	Treatment *B*
R_1	28	44
R_2	75	86
R_3	20	24
R_4	78	74
R_5	15	15
R_6	80	83
R_7	14	18
R_8	76	81

16/13. Three treatments were compared in a repeated-measures design with the following results.

Block	Treatment 1	Treatment 2	Treatment 3
B_1	21	27	24
B_2	38	98	99
B_3	2	61	13
B_4	52	51	75
B_5	74	80	78
B_6	23	16	14

Calculate the test statistic for a Friedman rank test of the difference between the three treatments. What is the critical value for a test at the 1% significance level?

16/14. Four treatments were compared in a repeated-measures design with the following results.

Block	Treatment 1	Treatment 2	Treatment 3	Treatment 4
B_1	26	21	17	16
B_2	77	78	75	74
B_3	52	40	43	41
B_4	69	67	59	64
B_5	16	14	15	9

Calculate the test statistic for a Friedman rank test of the difference between the four treatments. What is the critical value for a test at the 1% significance level?

16/15. Observations of the same variable were recorded on two different occasions for the same sample of 100 observational units. There was no change in 19 cases, a positive change in 46 cases, and a negative change in 35 cases. Calculate the test statistic and its P-value for a paired-samples sign test of the difference between the two occasions.

16/16. Observations of the same variable were recorded on two trials for a sample of 75 observational units. The value on the second trial was larger in 40 cases, smaller in 24 cases, and unchanged in the remaining 11 cases. Calculate the test statistic and its P-value for a paired-samples sign test of the difference between the two trials.

Problems for Applying the Concepts

16/17. Nine college students, who are identified here by the letters A through I, reported the following information about the total cost (in dollars) of their stereo equipment and the number of tape and record albums they owned.

	A	*B*	*C*	*D*	*E*	*F*	*G*	*H*	*I*
Cost	1800	710	750	400	1055	200	575	950	820
Number of albums	113	29	80	47	65	18	72	86	105

Pearson's correlation coefficient between cost and number of albums is $r = .746$.

a. Plot a scattergram of number of albums as a function of cost of stereo equipment.

b. Calculate Spearman's rank correlation coefficient, first using the method of Expression (16.2) and then using the method of Expression (16.3). In general, what is the relationship between the answers given by the two methods?

c. Is Spearman's rank correlation coefficient statistically significant in these data? What is the critical value for a two-tailed test at the 5% level?

(Source: Student project by K. Astner.)

16/18. The following data were recorded at a morning feeding for a sample of 9 healthy, egg-producing chickens, labeled A through I, of the same age. One variable was the hen's weight (in ounces), and the other variable was the time (in seconds) spent at the morning feeding, measured from the time the hen first ate at the feeding trough until it left the trough.

	A	*B*	*C*	*D*	*E*	*F*	*G*	*H*	*I*
Weight	40	47	35	39	41	41	45	33	39
Time	190	205	150	190	195	180	220	175	175

Pearson's correlation coefficient between weight and time is $r = .82$.

a. Plot a scattergram of time as a function of weight.

b. Calculate Spearman's rank correlation coefficient, first using the method of Expression (16.2) and then using the method of Expression (16.3). Why do the answers differ?

c. Is Spearman's rank correlation coefficient statistically significant in these data? What is the critical value for a two-tailed test at the 1% level?

(Source: Student project by P. Spitzer.)

16/19. A fifth-grade mathematics class of 12 students (identified here by the letters *A* through *L*) wrote two tests covering the same material. The test scores were as follows out of 100.

	A	*B*	*C*	*D*	*E*	*F*	*G*	*H*	*I*	*J*	*K*	*L*
1st test	52	14	66	74	64	71	77	78	65	66	76	30
2nd test	46	93	92	71	53	93	100	62	90	75	82	40

The product-moment correlation between scores on the two tests is $r = .22$.

a. Plot a scattergram of second-test marks as a function of first-test marks.

b. Calculate Spearman's rank correlation coefficient, first using the method of Expression (16.2) and then using the method of Expression (16.3). Why do the answers differ?

c. Is Spearman's rank correlation coefficient statistically significant in these data? What is the critical value for a two-tailed test at the 5% level?

(Source: Student project by M. Diemert.)

16/20. Ten pairs of college students were formed such that the individuals were matched with respect to their number of years of experience with the game of table tennis. Each pair then had a five-minute warm-up playing table tennis, after which they had five trials to see how many volleys they could keep the ball in play. The results for the 10 pairs, *A* through *J*, were as follows.

	A	*B*	*C*	*D*	*E*	*F*	*G*	*H*	*I*	*J*
Years of experience	0	1	3	5	6	7	9	7	5	3
Mean number of volleys	8.0	24.0	11.4	42.2	15.0	96.4	48.2	28.8	55.4	24.0

The product-moment correlation between years of experience and mean number of volleys is $r = .58$.

a. Plot a scattergram of mean number of volleys as a function of years of experience.

b. Calculate Spearman's rank correlation coefficient, first using the method of Expression (16.2) and then using the method of Expression (16.3). Why do the answers differ?

c. Is Spearman's rank correlation coefficient statistically significant in these data? What is the critical value for a two-tailed test at the 5% level?

(Source: Student project by C. Morris.)

16/21. A sample of 17 female college students reported the total amounts of their summer earnings (in dollars) from May through August 1980 as follows:

3700	1000	2500	1000	1400	1500	1000	3400	800
1800	1200	1200	400	2300	3500	400	300	

Use the sign test to test whether these observations may be viewed as a random sample from a population with a median of $2500.
(Source: Student project by C. Hay.)

16/22. A sample of 16 female college students reported their birth weights (in ounces) as follows:

112	118	118	118	101	93	128	118
140	108	118	123	112	127	112	134

Use the sign test to see if these data contradict the view that the median birth weight of female college students is 7 pounds (that is, 112 ounces).
(Source: Student project by C. Hochhausen.)

16/23. The statistician of a college hockey team kept track of how many goals each player scored as a percentage of the number of shots on the net by that player. The following percentages were recorded for 21 players after the first eight games of the regular season:

Right-handed players:	0.0	27.3	12.0	11.1	0.0	8.7	50.0	11.1	0.0		
Left-handed players:	50.0	25.0	18.8	0.0	0.0	0.0	20.0	52.6	31.2	10.7	12.1	33.3

Use the Wilcoxon rank-sum test to test whether the shooting accuracy of right-handed and left-handed players differed significantly.
(Source: Student project by L. McKinney.)

16/24. A large fast-food chain conducted an investigation of the extent to which its various outlets were adhering to company policies with respect to such matters as pricing and ordering of supplies. Random samples consisting of 10 outlets in the company's western district and 12 outlets in the eastern district were audited and given ratings that reflected the degree of adherence to company policies. The results of the study are tabulated as follows on a scale where a rating of 100 would mean perfect adherence to policy:

West:	60	58	34	51	63	73	68	59	57	61		
East:	89	71	58	49	72	69	70	66	73	65	77	67

Use the Wilcoxon rank-sum test to determine whether there is a significant difference in ratings between these two districts.

16/25. Four samples of unmarried female university students aged 17 to 21 years took part in a study concerning family size. The four samples were distinguished by the number of children (2, 3, 4, or 5) in the family from which the student came. The students filled out a brief questionnaire in which they reported the number of children they would like to have in their future families, as follows.

Present family size	Desired number of children in future family
2 children	2, 4, 2, 4
3 children	3, 2, 2, 5, 5
4 children	3, 3, 4
5 children	4, 3, 0, 2, 3

According to the Kruskal–Wallis rank-sum test, were the desired numbers of children significantly different for students coming from different sizes of family?

(Source: Student project by T. Kopperud.)

16/26. In a study of the relationship between athletic involvement and grade point average of university students, data were obtained from a sample of students on intercollegiate teams, a sample involved in intramural athletics only, and a sample with no athletic involvement. The following grade point averages were recorded (on a 9-point scale where 6 is average and 4 is the minimum passing grade) for six students in each of the three categories of participation in athletics.

Intercollegiate:	6.8	5.9	6.6	4.9	5.2	3.6
Intramural:	5.6	5.9	6.6	5.4	5.4	6.1
None:	6.5	6.7	6.2	4.1	5.2	6.7

Did the grades differ significantly for the three categories of participation in athletics, according to the Kruskal–Wallis rank-sum test?

(Source: Student project by B. Olanski.)

16/27. Example 16.1 includes information about the dates of hospital admission of 40 alcoholics involved in a study of controlled drinking. Test whether the time of admission differed significantly between patients assigned to the controlled-drinking condition and those assigned to the abstinence condition using

a. the Wilcoxon rank-sum test.

b. the Brown–Mood median test.

16/28. For commercial sale to food stores, alfalfa sprouts are grown in stainless steel trays. Since the use of too little seed per tray wastes space and too much inhibits proper growth due to crowding, an experiment was conducted to find the optimal amount of seed to grow on each tray. The following data are the weights (in pounds) of sprouts harvested from 48 trays, 12 trays for each of four amounts of seed per tray, with all other growing conditions held constant.

Ounces of seed per tray	Pounds of sprouts harvested per tray											
15.3	7.0	9.0	6.5	9.0	9.5	7.5	9.0	10.5	7.5	9.5	7.5	9.5
14.7	10.5	9.5	10.0	8.0	10.5	10.0	10.5	10.0	10.5	10.0	9.5	8.5
14.0	10.5	10.5	12.0	9.0	8.0	9.5	11.0	11.5	9.5	11.0	11.0	11.5
13.3	8.5	9.0	9.0	10.0	7.0	8.0	9.5	10.0	9.0	8.5	10.5	10.5

Use the Brown–Mood median test to see whether the different amounts of seed had a significant effect on the weight of sprouts harvested.

(Source: Student project by N. Sinnamon.)

16/29. In a study comparing two supermarkets in the same city, prices were obtained on the same day for 10 products sold at both stores (that is, products of the same brand and same size). The prices (in cents) were as follows for products *A* through *J*.

	A	B	C	D	E	F	G	H	I	J
Store *T*	155	249	109	169	69	47	252	165	117	163
Store *S*	145	237	99	155	73	43	247	167	109	167

Test whether the prices were significantly different at the two stores using

a. the Wilcoxon signed-rank test.

b. the paired-samples sign test.

(Source: Student project by C. F. Brown.)

16/30. Twelve college students, who are identified here by the letters *A* through *L*, took part in an investigation of the effect of distracting noise on reading speed. Each student read two passages, one with the radio on and one with the radio off. To control order effects, half of the students read first with the radio on, and half read first with the radio off, and to ensure that students would read for comprehension, the passages were selected from a book of jokes. The following reading speeds (in words per minute) were recorded.

	A	B	C	D	E	F	G	H	I	J	K	L
Radio off	233	255	255	265	278	271	300	265	278	258	290	303
Radio on	271	225	240	255	240	268	278	255	240	240	263	278

Test whether the radio had a significant effect on reading speed using

a. the Wilcoxon signed-rank test.

b. the paired-samples sign test.

(Source: Student project by A. MacNichol.)

16/31. An investigation was conducted among students successfully completing two years of university study, in order to determine whether there was any systematic change in the marks of such students as they completed high school and began university. The data reported here are average marks by each of four students (who are labeled *A, B, C,* and *D*) in their courses in the humanities (English, French, classics, music, drama, and so on) in three years of study.

	A	B	C	D
Grade 12	85	75	84	73
University 1	63	70	67	67
University 2	89	56	78	78

Use the Friedman rank test to see whether marks differed significantly over the three years.

(Source: Student project by D. Burzynski.)

16/32. Nine female students in their second year of university studies agreed to keep track of the number of hours they spent watching television each day of the week. The following numbers of hours were recorded for Monday, Wednesday, and Friday by the nine students (identified by the letters *A* through *I*).

	A	*B*	*C*	*D*	*E*	*F*	*G*	*H*	*I*
Monday	2	2	2	3	5	3	0	0	0
Wednesday	0	2	2	3	3	3	0	0	3
Friday	0	0	0	2	5	0	0	0	0

Use the Friedman rank test to see whether the television viewing times were significantly different on different days of the week.

(Source: Student project by M. Diemert.)

16/33. Twenty employees of a large manufacturing company were randomly selected to take part in a study of the effectiveness of a new assembly process intended to increase productivity. The workers were trained on the new process, used it on the job for a week of practice, and then were assessed for productivity. Thirteen of the workers exhibited higher productivity, six lower, and one no change. Test the significance of the difference in productivity. Report the test statistic and the *P*-value.

16/34. As part of a study of residential life on a college campus, 10 students were asked about their usual weekday bedtimes at home and in the college residence. Nine students reported later times at the college than at home, and one student reported no difference. Test the statistical significance of the difference between home and college bedtimes, and report the resulting (two-tailed) *P*-value according to

a. the Gaussian approximation.

b. the binomial probability function.

(Source: Student project by W. Vornbrock.)

16/35. Interviews were conducted with 19 male students who were living in college residences and who had early morning classes on some weekdays but not on others. Two of the 19 students always ate breakfast, whether or not they had an early morning class, and 11 never ate breakfast. The remaining 6 ate breakfast on days that they had an early class but not otherwise.

a. Test whether the proportion of students eating breakfast was significantly different depending on whether or not they had an early morning class. Report the (two-tailed) *P*-value according to both the Gaussian approximation and the binomial probability function.

b. Why would it be incorrect to analyze these data by applying a chi-square test for two-way tables to the 2-by-2 table showing breakfast/no breakfast crossed with early class/no early class?

(Source: Student project by I. Munkedal.)

16/36. A sample of 22 female students provided information for a study of social life on a college campus. Nine of the students reported having a boyfriend at the start of the academic year, and 13 did not. Seven of the 9 still reported having a boyfriend at the end of the first semester, and 6 of the 13 still reported having no boyfriend at the end of the first semester.

a. Test whether the proportion of female students who reported having a boyfriend was significantly different at the beginning and at the end of the semester. Report the (two-tailed) P-value according to both the Gaussian approximation and the binomial probability function.

b. Why would it be incorrect to analyze these data by applying a chi-square test for two-way tables to the 2-by-2 table showing boyfriend/no boyfriend crossed with September/December?

(Source: Student project by E. Nielsen.)

16/37. Ten houses with natural-gas heating were randomly selected from the same neighborhood. The total floor area of each house was measured (in square feet), and the amount of natural gas consumed in the previous month was recorded.

a. Which parametric method could be used to test the significance of the relationship between floor area and natural-gas consumption?

b. Which nonparametric method could be used?

16/38. Nineteen students wrote all four tests in a certain mathematics course. The student's mark on each test was expressed as a percentage.

a. Which parametric method could be used to determine whether the average marks differed significantly from one test to another?

b. Which nonparametric method could be used?

16/39. A horse owner kept track of the number of shoe replacements for each of 11 work horses in the past year. The owner wanted to know whether the average number of shoes required by these 11 horses differed significantly from 12 shoes per horse.

a. Which parametric method could be used?

b. Which nonparametric method could be used?

16/40. A random sample of students and of staff at a college were asked to specify which of two candidates for the U.S. presidency they preferred. Which statistical method could be used to test whether students and staff differed significantly in their preferences?

16/41. Two candidates for a major political office met in a televised debate. A random sample of viewers specified which of the two candidates they preferred at the start of the debate, and the same random sample specified their preference again at the end of the debate. Which statistical method could be used to test whether the debate had a significant effect on preferences?

16/42. A duck hunter sometimes went hunting by himself and sometimes with one or more friends. Each time, he recorded the size of the hunting party and the number of ducks brought back.

a. Which parametric method could be used to test the significance of the relationship between the size of the hunting party and the number of ducks brought back?

b. Which nonparametric method could be used?

16/43. A sample of 200 college students were cross-classified according to hair color and eye color. Which statistical method could be used to test the significance of the relationship between the two variables?

16/44. A sample of 16 individuals took the same physical fitness test both before and after taking a course in TaeKwon-Do (Korean art of self-defense).

- **a.** Which parametric method could be used to test whether there was a significant change in the average fitness score?
- **b.** Which nonparametric method could be used?

16/45. The numbers of students pictured with and without eyeglasses were counted separately for males and females in a high school yearbook. Which statistical method could be used to test whether males and females differed significantly with respect to wearing glasses for their yearbook picture?

16/46. A sample of students were cross-classified according to faculty (Arts, Science, or Business) and athletic participation (intercollegiate, intramural, or none), and each student's grade point average (GPA) was obtained. Which statistical method could be used to test the significance of the relationship between GPA and the other two variables?

16/47. A motorist kept careful track of the gas mileage obtained on each tank of gas in order to see whether gas mileage depended on which of four brands of gas she purchased. For each gas fill, she randomly selected one of the four gasoline brands.

- **a.** Which parametric method could be used to test whether gas mileage differed significantly among the four brands?
- **b.** Which nonparametric method could be used?

16/48. A sample of 15 students who classified themselves as smokers reported the number of cigarettes they each smoked in a typical day.

- **a.** Which parametric method could be used to determine whether the average number reported by this sample of students differed significantly from 20 cigarettes?
- **b.** Which nonparametric method could be used?

16/49. A sample of male college students reported their height to the nearest tenth of an inch for subsequent verification by actual measurement.

- **a.** Which parametric method could be used to determine whether the students had a significant bias to say that they were taller than they actually were?
- **b.** Which nonparametric method could be used?

16/50. As part of a study comparing Catholic and Protestant life-styles, samples of students claiming Catholic or Protestant religious backgrounds reported on a questionnaire as to whether they now "attend church regularly" or "do not attend church regularly." Which statistical method could be used to determine whether reported church attendance differed significantly between Catholics and Protestants?

16/51. As part of a study comparing Catholic and Protestant life-styles, samples of students claiming Catholic or Protestant religious backgrounds reported the size of the family (that is, the number of children) in which they grew up.

- **a.** Which parametric method could be used to test whether average family size is different in the relevant Catholic and Protestant populations?
- **b.** Which nonparametric method could be used?

16/52. A commercial popcorn popper was used to compare the popping time required by three different brands of popcorn. Five batches of each brand were popped, and the required time was recorded for each batch.

a. Which parametric method could be used to test whether the three brands differed significantly in popping time?

b. Which nonparametric method could be used?

16/53. Ten members of a jujitsu class were individually asked to rate three particular jujitsu exercises. They each did this by reporting which of the three they thought was the most useful exercise and which of the three they thought was the least useful. Which statistical method could be used to test whether the three exercises differed significantly in the ratings?

16/54. Twenty subjects were randomly assigned to each of two conditions in a memory experiment. The subjects in one condition were each given three minutes to study a photograph of a busy fairground scene, while those in the other condition each had three minutes to study a paragraph describing the same scene in detail. Subjects in both conditions were then allowed up to 20 minutes to write down as many items from the scene as they could remember.

a. Which parametric method could be used to test whether the average number of items recalled differed significantly between the two conditions?

b. Which nonparametric method could be used?

16/55. Many college students find that class tests in most of their courses tend to fall in the same week about halfway through the academic term. Are students who live in college residences more likely to go home for a visit on the weekend before or the weekend after the midterm exam week? Shortly after midterm exam week, a random sample of 50 students were asked whether they had visited home on the weekend before midterm exam week, on the weekend after, on both weekends, or on neither weekend. (Neither weekend was a long weekend.) Which statistical method could be used to test whether visits home were significantly more likely on one of the two weekends than the other?

Reference

Pendery, M. L.; Maltzman, I. M.; and West, L. J. "Controlled Drinking by Alcoholics? New Findings and a Reevaluation of a Major Affirmative Study." *Science* 217 (1982): 169–75.

Further Reading

Kendall, M. G. *Rank Correlation Methods*. 4th ed. London: Griffin, 1970. This authoritative reference work includes coverage of Spearman's rank correlation coefficient, along with tables for significance testing.

Mosteller, F., and Rourke, R. E. K. *Sturdy Statistics: Nonparametrics and Order Statistics*. Reading, MA: Addison-Wesley, 1973. This is a very readable text, intended for an introductory course in nonparametric methods. It provides more thorough coverage and tables for topics introduced in the present chapter.

Siegel, S. *Nonparametric Statistics for the Behavioral Sciences.* New York: McGraw-Hill, 1956. This is a convenient collection of nonparametric methods, including a worked example of each. Relevant tables for small sample sizes are included.

Snedecor, G. W., and Cochran, W. G. *Statistical Methods.* 7th ed. Ames, IA: Iowa State University Press, 1980. Chapters 7, 8, and 11 are relevant to the present chapter. They include practical suggestions and valuable comments arising out of the authors' considerable experience in data analysis.

PART IV

FURTHER TOPICS IN APPLIED STATISTICS

EXAMPLE 17.0 As a statistician, I receive calls from a variety of people who want to make sense of some data that are of interest to them. The calls illustrate the widespread interest in and use of statistics. Here are a few that come to mind:

a. A psychiatric hospital phoned to ask for a statistical consultant for an ongoing study concerning the life histories, treatments, and progress of its patients.

b. A newspaper office requested advice in planning and analyzing a readership survey that was expected to be of interest to potential advertisers.

c. An ecologist wanted to evaluate the effectiveness of a mosquito-control program.

d. A national church office wanted to obtain a profile of its church membership in terms of a large number of variables.

e. A cattle-breeding operation wanted to investigate some variables that might be good early predictors of selling price and meat quality.

f. A manufacturing firm wanted to assess the effects of changes in its procedure on the quality of its product.

g. A research psychologist requested advice in planning and analyzing a major experiment concerning memory processes.

h. A city's social services office had questions about a study it was conducting to evaluate and improve its programs for senior citizens.

i. An English professor wanted to determine the effectiveness of an English proficiency examination used to identify remedial needs of new students.

j. A radio station phoned for an interview related to gambling behavior.

Statistical principles and methods in the following chapters were among those needed to answer these questions.

In Chapter 1, we defined *statistics* as the science that deals with methods for answering questions through the proper collection and interpretation of empirical data. The field as a whole can be subdivided into descriptive statistics and inferential statistics, depending on whether one's purpose is to describe the observational units that were actually observed or, on the other hand, to generalize by drawing inferences about observational units that were not observed.

The first three parts of this book have provided an introduction to some of the major principles and methods of both descriptive and inferential statistics. In the area of descriptive statistics (Chapters 2–5), we saw how data may be described in terms of frequency distributions and by measures of location, dispersion, and relationship. After a brief excursion into mathematics to pick up the necessary tools of probability theory (Chapters 6–8), we examined (in Chapters 9–16) two general types of statistical inference—estimation and hypothesis testing—chiefly concerning means, variances, correlation and regression, and proportions.

The remainder of this book presents several further topics in both descriptive and inferential statistics. These topics will serve not only to fill in some gaps in the initial survey completed in the first 16 chapters, but also to increase your fluency in the use of many of the fundamental concepts previously introduced. The purpose of Part IV is to present a more complete picture of the field of applied statistics at the elementary level and, at the same time, to leave you with a more secure understanding of its basic principles.

*CHAPTER 17

TOPICS IN MATHEMATICAL EXPECTATION

EXAMPLE 17.1

Whenever you're called on to make up your mind,
and you're hampered by not having any,
the best way to solve the dilemma, you'll find,
is simply by spinning a penny.

No—not so that chance shall decide the affair
while you're passively standing there moping;
but the moment the penny is up in the air,
you suddenly know what you're hoping.

(Source: Hein, 1969. Used with permission of Paperjacks Ltd.)

The concept of mathematical expectation was introduced in Section 7.1. We saw that the expected value of a discrete random variable x is equal to the sum of its possible values, each weighted by its probability of occurrence:

$$E(x) = \sum x_i p_i \qquad (17.1)$$

This chapter begins by introducing an important application of mathematical expectation; then it outlines some definitions and principles that are useful in dealing with further statistical methods.

17.1 A CASE IN POINT: DECISION MAKING

EXAMPLE 17.2 A businessman is planning to open a drive-in restaurant in a small but rapidly growing city, but he must decide whether to locate his business on Road A, Road B, or Road C. The highway currently runs through the town on Road A, but there is a chance that it will soon be permanently rerouted along Road C. The businessman feels that the probability of rerouting is about 40%. The location on Road B would be accessible from both of the other two roads, but it would not have the visibility of an establishment right on the highway. Taking into account the cost of property and the

* This symbol identifies material that is not prerequisite for later topics in this book.

effect on business of the different locations, the businessman determines the following returns on his investment for the various possibilities.

Payoff matrix (in thousands of dollars)

		Action Locate restaurant on: Road *A*	Road *B*	Road *C*	Subjective probability
State of the world	Highway unchanged	70	25	−20	.60
	Highway rerouted	10	25	100	.40

Where should he locate the restaurant?

Many decision problems in everyday life have the same general format as that of Example 17.2. In each case, there are several alternative *actions* available, but the choice is complicated by our lack of knowledge about some relevant events, called the possible *states of the world* or *states of nature*. Our uncertainty about the possible states may be reflected in (subjective) *probabilities* of their occurrence. Associated with each state-action combination is a **payoff** that summarizes the consequences of that outcome. The decisions are not limited to business ventures as in Example 17.2. In another application, the actions may be identified as what to study for an examination (lecture notes, text, or a little of each), the states of the world as the possible examination questions, and the payoffs as the grades one would obtain for each combination of material studied and material on the exam. In yet other applications, the payoffs may be less tangible, such as personal satisfaction or happiness. In any case, **statistical decision theory** is concerned with methods for choosing the most appropriate action in light of the payoffs and the **state-of-the-world probabilities**.

Decision-Making Criteria

To begin, we consider three possible criteria that we might use for choosing which course of action to take in a decision problem with monetary payoffs. First, we might simply look over all of the possible payoffs and choose the action that includes the largest single payoff. This is called the **maximax criterion** because it favors the action whose maximum payoff is larger than for any other action—it maximizes the maximum payoff. Second, we might find the minimum payoff for each action and choose the action that guarantees the largest minimum. This is called the **maximin criterion** because it favors the action that maximizes the minimum possible payoff. Third, we might calculate the expected monetary value, or **EMV**, of each action (using the concept of mathematical expectation introduced in Section 7.1) and choose the action with the largest expected payoff. This is called the **EMV criterion**.

EXAMPLE 17.3 Example 17.2 presented a payoff matrix for a businessman considering three possible locations for a drive-in restaurant. In light of the information available, where should he locate the restaurant?

According to the maximax criterion, he should locate on Road C, because the payoff of \$100,000 on Road C, if the highway is rerouted, is larger than any other payoff. This decision would be a rather daring one, because he stands to lose \$20,000 on Road C if the highway is not rerouted.

According to the maximin criterion, he should locate on Road B, because the worst he could do on Road B is to gain \$25,000, whereas on Road A he might gain as little as \$10,000 and on Road C he might actually lose money. This decision is a rather cautious one: It guarantees a certain minimum payoff but provides no possibility of one of the larger payoffs.

According to the EMV criterion, he should locate on Road A. The expected payoffs associated with location on the three roads are

$$\text{EMV}(A) = (70)(.60) + (10)(.40) = 46$$
$$\text{EMV}(B) = (25)(.60) + (25)(.40) = 25$$
$$\text{EMV}(C) = (-20)(.60) + (100)(.40) = 28$$

respectively, in thousand-dollar units. The maximum expected payoff is \$46,000, corresponding to location on Road A. This decision would be a rational one, in that it makes use of all the available information.

No criterion for making decisions in the presence of uncertainty is completely foolproof. The maximax criterion is a **risk-taking strategy**, in that it ignores the losses that might result if the optimal state of the world does not occur. On the other hand, the maximin criterion is a **risk-avoiding strategy**. It considers only the worst possible outcome associated with each action. Both the maximax and the maximin criteria therefore can be faulted for not taking into account all of the payoffs and for not taking into account the probabilities of the various possible states of the world. The EMV criterion uses all the available information, but it can be faulted for assuming that monetary value is the only relevant aspect of the payoffs.

EXAMPLE 17.4 Suppose that in the payoff matrix of Example 17.2 there were only two possible actions, namely, B and C. It is not hard to imagine that a businessman may prefer action B over action C, even though the EMVs are \$25,000 for B and \$28,000 for C. It could be that he is not in a financial position where he can afford to take a chance—even if it were only a remote chance—of losing \$20,000. The monetary value may not accurately reflect the consequences of the outcome for this particular businessman.

The EMV criterion can be improved by replacing the monetary values in the payoff matrix with *utilities* for the particular decision maker. The **utility** of an outcome is its value to the decision maker. Utility is measured in arbitrary units called **utiles** that reflect the value of one outcome relative to the values of the other outcomes. For some decision problems and some decision makers, utilities may correspond directly to monetary values, but for other problems and/or other decision makers, utilities will reflect not only monetary values but also such factors as personal satisfaction accompanying various outcomes and the decision maker's attraction or aversion to gambling (that is, to taking a chance on a long shot).

Replacing monetary values by utilities yields a fourth criterion for decision making, according to which we calculate the **expected utility**, or **EU**, of each action and choose the action with the greatest expected utility. This procedure, called the *maximization of expected utility*, or the **EU criterion**, is the most widely applicable of the four decision-making criteria we have met. Like the EMV criterion, it takes into account all of the payoffs as well as the state-of-the-world probabilities, but unlike the EMV criterion, it is not limited to the consideration of the monetary values of the outcomes. (In fact, the EMV criterion is the special case of the EU criterion that occurs when utilities bear a linear relationship to monetary values.) In particular, the decision maker's utilities may reflect his/her risk-taking or risk-avoiding preferences, thus enabling the EU criterion to offer the best features of the maximax and maximin criteria, but with much greater flexibility. The EU criterion can be applied even when the payoffs are not monetary at all but correspond to different degrees of satisfaction or happiness, as long as the decision maker can assign to each outcome a number that reflects its value (for him/her) relative to the other outcomes.

EXAMPLE 17.5 Consider again the businessman of Example 17.2. Suppose that after taking into account his need for financial security and other nonmonetary aspects of the outcomes, he determines the following utilities of the six outcomes.

Payoff matrix (in utiles)

		Action Locate restaurant on: Road A	Road B	Road C	Subjective probability
State of the world	Highway unchanged	8	5	−30	.60
	Highway rerouted	3	5	10	.40

His expected utilities for location on Roads A, B, and C are, respectively,

$$\text{EU}(A) = (8)(.60) + (3)(.40) = 6$$
$$\text{EU}(B) = (5)(.60) + (5)(.40) = 5$$
$$\text{EU}(C) = (-30)(.60) + (10)(.40) = -14$$

According to the EU criterion, he should locate on Road A, because that action maximizes his expected utility.

The Value of Further Information

In many situations, it may be possible to obtain further information before making a decision. One may hire a consultant, or conduct some tests, or simply wait until more information is available, but in any case, there is typically some cost involved in obtaining the information. Both monetary and nonmonetary costs (for example, inconvenience or delay) may be included in the decision analysis if payoffs are measured in terms of utility. How much should one be willing to pay for further information?

To set an upper limit on the value of further information, consider the expected utility we would experience if we had perfect information about the true state of the world. With perfect information, we would choose, for each possible state of the world, the action with the greatest utility. The difference between the resulting **expected utility with perfect information** and the expected utility of the action that we would choose on the basis of our current information is the **expected value of perfect information**, or **EVPI**:

$$\text{EVPI} = \text{EU (perfect information)} - \text{EU (current information)} \quad (17.2)$$

The expected value of perfect information tells us how much we could expect to gain, relative to our current information, by obtaining perfect information about the state of the world.

EXAMPLE 17.6 Let us calculate the expected value of perfect information for the payoff matrix in Example 17.5. With perfect information, the businessman would locate on Road A if the highway were to remain unchanged (because the payoff of 8 utiles is the maximum for that state of the world), and he would locate on Road C if the highway were to be rerouted (because the payoff of 10 utiles is the maximum for that state of the world). The expected utility with perfect information is therefore

$$\text{EU (perfect information)} = (8)(.60) + (10)(.40) = 8.8 \text{ utiles}$$

We saw in Example 17.5 that the maximum expected utility based on current information is that associated with action A:

$$\text{EU (current information)} = \text{EU}(A) = 6.0 \text{ utiles}$$

The expected value of perfect information is therefore

$$\text{EVPI} = 8.8 - 6.0 = 2.8 \text{ utiles}$$

In other words, the businessman should be willing to "spend" up to 2.8 utiles (of time, effort, inconvenience, and/or money) to obtain perfect information about the future routing of the highway.

In practice, we typically must settle for imperfect, sample information rather than perfect information, and imperfect information is obviously worth less than perfect information: The more fallible the information, the less it is worth. Nonetheless, the expected value of perfect information (EVPI) provides a guideline by indicating the maximum possible amount that any information could be worth. Methods for determining the *expected value of sample information,* or *EVSI,* are available in references cited under Further Reading at the end of this chapter.

Limitations of Decision Theory

The results of statistical decision analysis are only as good as the information on which the analysis is based. The major **limitations of decision theory** relate to the specification of relevant actions, states, probabilities, and payoffs. Shortcomings in any of these areas may limit the usefulness of the conclusions derived from the analysis:

1. *Identification of possible actions.* Decision analysis can compare the alternative courses of action in an existing list of possible actions, but it is no substitute for creative thinking. There may be imaginative or novel courses of action that should also be included in the analysis.

2. *Identification of possible states of the world.* In some situations, it may be difficult to identify all the possible events that would be relevant to the amounts of the payoffs. (In recent years, for example, unforeseen changes in world oil prices have wreaked havoc with long-term planning by various governments and industries around the world.) Decision analysis is not robust against unforeseen developments or other omissions from the list of possible states of the world.

3. *Accuracy of probabilities.* Even if all relevant states of the world can be identified, it may be difficult to specify their probabilities with any confidence. The results of the analysis depend in part on the probabilities assigned to the various states.

4. *Accuracy of payoffs.* In most realistic problems, it is no trivial task to determine the payoffs corresponding to the various state-action combinations. Such is the case even when monetary values are the only relevant features of the payoffs; consider, for example, the practical difficulties the businessman in Example 17.2 would have in determining the dollar values for his payoff matrix. However, the difficulties are compounded when nonmonetary factors must be taken into account. Moreover, utilities are in arbitrary units that cannot be compared from one individual to another, with the result that there are additional complications when the costs and benefits associated with a particular course of action do not accrue to the same people. (Disputes between developers and conservationists, for example, often arise because positive payoffs for one group correspond to outcomes with negative payoffs for the other group. When governments intervene, they consider their own perceived political payoffs, both positive and negative.) Inaccurate specification of payoffs obviously can have serious effects on the decision analysis.

In spite of these limitations, statistical decision theory provides a rational framework that facilitates wise decision making in many important problems. Awareness of its limitations can go a long way toward preventing inappropriate applications.

17.2 DEFINITIONS OF BASIC CONCEPTS IN TERMS OF MATHEMATICAL EXPECTATION

We noted in Section 7.1 that the most general definitions of *mean* and *variance* are those expressed in terms of expected values, and that our usual computational formulas are derived as special cases of the expected-value definitions. In this section, we review this important concept by bringing together several definitions concerning expected values.

Throughout the rest of this chapter, x and y are discrete random variables; a, b, and k are constants; p_i is the probability that the variable x takes the value x_i; and p_{ij} is the joint probability that the variable x takes the value x_i and the variable y takes the value y_j.

The following are basic definitions of concepts relating to any random variables, but here the definitions are expanded only for discrete random variables x and y.

Definition 1

The **arithmetic mean** μ_x is defined as the expected value of the random variable x:

$$\begin{aligned}\mu_x &= E(x)\\ &= \sum x_i p_i \qquad \text{(17.3)}\end{aligned}$$

Definition 2

The **variance** σ_x^2 of the random variable x is defined as the expected value (or average) of the squared deviations from the mean:

$$\begin{aligned}\sigma_x^2 = \text{Var}\,(x) &= E(x - \mu_x)^2\\ &= \sum (x_i - \mu_x)^2 p_i \qquad \text{(17.4)}\end{aligned}$$

The notation Var (x) is read "the variance of x." By algebraic manipulation of Expression (17.4), it can be shown that the variance of x equals the mean of the squares minus the square of the mean; that is,

$$\sigma_x^2 = E(x^2) - \mu_x^2 \qquad \text{(17.5)}$$

Definition 3

The **standard deviation** σ_x of the random variable x is defined as the positive square root of the variance:

$$\sigma_x = \sqrt{\text{Var}\,(x)} \qquad \text{(17.6)}$$

Definition 4

The **covariance** Cov (x, y) of two random variables x and y is defined as the expected value of the product of their deviations from their means:

$$\begin{aligned}\text{Cov}\,(x, y) &= E(x - \mu_x)(y - \mu_y)\\ &= \sum \sum (x_i - \mu_x)(y_j - \mu_y)p_{ij} \qquad \text{(17.7)}\end{aligned}$$

which simplifies to

$$\text{Cov}(x, y) = E(xy) - E(x)E(y)$$
$$= (\sum \sum x_i y_j p_{ij}) - \mu_x \mu_y \quad \textbf{(17.8)}$$

The notation Cov (x, y) is read "the covariance of x and y." Like the correlation coefficient, the covariance measures how two variables vary together. From Expression (17.7), we see that if x tends to be above its mean when y is above its mean (and hence x tends to be below its mean when y is below its mean), then the product of deviations, and hence the covariance, will tend to be positive. On the other hand, if x tends to be below its mean when y is above its mean, and vice versa, then the product of the deviations, and hence the covariance, will tend to be negative. If x and y are independent, so that knowledge of whether x is above or below its mean tells us nothing about y, then the covariance is zero, and Expression (17.8) reduces to $E(xy) = E(x)E(y)$ in this special case.

In other words, the covariance has the same sign—positive, negative, or zero—as the correlation coefficient (see Section 5.2, page 164), the only difference being that the covariance is in the original units of measurement, not in standard units, and so is not bounded by -1 and $+1$. The fact that the covariance is not in standard units makes it very difficult to interpret directly, but it is an important building block in more easily interpreted measures of relationship, such as the correlation coefficient.

Definition 5

The **simple linear correlation** ρ_{xy} between the variables x and y is the covariance measured in standard deviation units:

$$\rho_{xy} = \frac{\text{Cov}(x, y)}{\sigma_x \sigma_y} \quad \textbf{(17.9)}$$

The Greek letter ρ (rho), which is used to designate the population correlation coefficient, corresponds to the Roman letter r, which is used for the sample correlation coefficient.

Notice that the terms of Expression (17.9) may be rearranged to show that covariance is an "unstandardized correlation," that is, the correlation coefficient multiplied by the standard deviations of the variables:

$$\text{Cov}(x, y) = \rho_{xy}\sigma_x\sigma_y \quad \text{for populations} \quad \textbf{(17.10)}$$

$$\text{cov}(x, y) = r_{xy}s_x s_y \quad \text{for samples} \quad \textbf{(17.11)}$$

Table 17.1 summarizes the preceding definitions and formulas and shows how they relate to material discussed in previous chapters.

TABLE 17.1 Summary of formulas for mean variance, covariance, and correlation. (Parenthetical references identify the chapter and section where the formulas have been discussed previously.)

Basic definition	Formulas for use with discrete probability distributions	Formulas for use with finite population data	Formulas for use with sample data
$\mu = E(x)$	$= \sum x_i p_i$ (Sec. 7.1)	$\mu = \frac{\sum x}{N}$ (Sec. 3.2)	$\bar{x} = \frac{\sum x}{n}$ (Sec. 3.2)
Definitional formulas $\sigma^2 = E(x - \mu)^2$ Computational formulas $\sigma^2 = E(x^2) - \mu^2$	$= \sum (x_i - \mu)^2 p_i$ (Sec. 7.1) $= (\sum x_i^2 p_i) - \mu^2$	$\sigma^2 = \frac{\sum (x - \mu)^2}{N}$ (Sec. 4.3) $\sigma^2 = \frac{\sum x^2}{N} - \mu^2$	$s^2 = \frac{\sum (x - \bar{x})^2}{n - 1}$ (Sec. 4.3) $s^2 = \frac{\sum x^2 - \frac{(\sum x)^2}{n}}{n - 1}$
Definitional formulas $\text{Cov}(x, y) = E(x - \mu_x)(y - \mu_y)$ Computational formulas $\text{Cov}(x, y) = E(xy) - E(x)E(y)$	$= \sum \sum (x_i - \mu_x)(y_j - \mu_y)p_{ij}$ $= (\sum \sum x_i y_j p_{ij}) - \mu_x \mu_y$	$\text{Cov}(x, y) = \frac{\sum (x - \mu_x)(y - \mu_y)}{N}$ $\text{Cov}(x, y) = \frac{\sum xy}{N} - \mu_x \mu_y$	$\text{cov}(x, y) = \frac{\sum (x - \bar{x})(y - \bar{y})}{n - 1}$ (Sec. 13.2) $\text{cov}(x, y) = \frac{\sum xy - \frac{\sum x \sum y}{n}}{n - 1}$
$\rho = \frac{\text{Cov}(x, y)}{\sigma_x \sigma_y}$	$= \frac{\text{Cov}(x, y)}{\sigma_x \sigma_y}$	$\rho = \frac{\text{Cov}(x, y)}{\sigma_x \sigma_y}$	$r = \frac{\text{cov}(x, y)}{s_x s_y}$ (Sec. 5.2)

EXAMPLE 17.7 Let us illustrate the preceding definitions using data from a group of college students described by the following table. The table shows the **joint probability distribution** of the number of children living at home (x) and the number of pets owned by the family (y) in a particular group of college students that we will take to be the population of interest.

		Number of children at home (x)				Marginal probabilities
		1	2	3	4	
Number of pets (y)	0	.1	.2	.1	0	.4
	1	.1	0	.2	.1	.4
	2	.1	0	0	.1	.2
Marginal probabilities		.3	.2	.3	.2	1.0

The mean number of children at home in the families of this population of college students is

$$\mu_x = (1)(.3) + (2)(.2) + (3)(.3) + (4)(.2) = 2.4 \text{ children}$$

and the mean number of pets is

$$\mu_y = (0)(.4) + (1)(.4) + (2)(.2) = .8 \text{ pets}$$

The variance of the number of children at home in this population may be obtained by Expression (17.4):

$$\sigma_x^2 = (1 - 2.4)^2(.3) + (2 - 2.4)^2(.2) + (3 - 2.4)^2(.3) + (4 - 2.4)^2(.2) = 1.24$$

or equivalently by Expression (17.5):

$$\sigma_x^2 = (1^2)(.3) + (2^2)(.2) + (3^2)(.3) + (4^2)(.2) - (2.4)^2 = 1.24$$

The variance of the number of pets in this population likewise may be obtained via either Expression (17.4) or (17.5) and is found to be $\sigma_y^2 = .56$.

The variance is not directly interpretable in any very useful way because it is in squared units—1.24 square-children and .56 square-pets! But using Expression (17.6), we find the standard deviation of the number of children at home to be $\sigma_x = 1.11$ children and the standard deviation of the number of pets to be $\sigma_y = .75$ pets. These are "standard" amounts by which the numbers deviate from their means.

The covariance of number of children and number of pets may be obtained from Expression (17.7)

$$\begin{aligned}\text{Cov}(x, y) &= (1 - 2.4)(0 - .8)(.1) + (1 - 2.4)(1 - .8)(.1)\\ &\quad + (1 - 2.4)(2 - .8)(.1) + (2 - 2.4)(0 - .8)(.2)\\ &\quad + (3 - 2.4)(0 - .8)(.1) + (3 - 2.4)(1 - .8)(.2)\\ &\quad + (4 - 2.4)(1 - .8)(.1) + (4 - 2.4)(2 - .8)(.1)\\ &= .18\end{aligned}$$

or equivalently from Expression (17.8):

$$\begin{aligned}\text{Cov}(x, y) &= (1)(0)(.1) + (1)(1)(.1) + (1)(2)(.1) + (2)(0)(.2) + (3)(0)(.1)\\ &\quad + (3)(1)(.2) + (4)(1)(.1) + (4)(2)(.1) - (2.4)(.8)\\ &= 2.10 - 1.92\\ &= .18\end{aligned}$$

The fact that the covariance is positive tells us that number of children and number of pets are positively correlated in this population. Except for its sign, the numerical value of the covariance is not directly interpretable in any very useful way because of its use of the original units of measurement: .18 "children-pets."

For interpretation, the covariance may be expressed in standard units as the correlation coefficient. The simple linear correlation between number of children at home and number of pets is given by Expression (17.9):

$$\rho_{xy} = \frac{.18}{\sqrt{1.24}\sqrt{.56}} = .22$$

indicating a very small, positive correlation between the variables in this population.

(Source: Student project by D. Pallant.)

17.3 USEFUL PRINCIPLES IN CALCULATING MEANS AND VARIANCES

Table 17.2 presents several important rules or principles that are useful in calculating means and variances of various types of functions. These principles can all be derived from the definitions in Section 17.2, but we will not take the space to write out the derivations here. Rather, we offer the following brief

TABLE 17.2 Means and variances of selected functions.

Function	Mean	Variance
Constant	$E(k) = k$	$\text{Var}(k) = 0$
Change of scale	$E(kx) = kE(x)$	$\text{Var}(kx) = k^2\,\text{Var}(x)$
Linear transformation	$E(bx + a) = bE(x) + a$	$\text{Var}(bx + a) = b^2\,\text{Var}(x)$
Standard score	$E\left(\frac{x - \mu}{\sigma}\right) = 0$	$\text{Var}\left(\frac{x - \mu}{\sigma}\right) = 1$
Sum of random variables	$E(x + y) = E(x) + E(y)$	$\text{Var}(x + y) = \text{Var}(x) + \text{Var}(y) + 2\,\text{Cov}(x, y)$
Difference of random variables	$E(x - y) = E(x) - E(y)$	$\text{Var}(x - y) = \text{Var}(x) + \text{Var}(y) - 2\,\text{Cov}(x, y)$

commentary:

a. The first line of Table 17.2 simply indicates that the expected value (or average value in the long run) of a *constant* is equal to the value of the constant, and that a constant does not vary (has zero variance).

b. The lines dealing with **changes of scale** (for example, a change from x to kx, or from feet to inches, or from pounds to kilograms) and **linear transformations** (for example, a change from x to $bx + a$, or from Fahrenheit to Celsius temperature) reflect the fact that *multiplying* every observation in a set by a constant has the effect of multiplying the mean and the standard deviation by the same constant. Hence, the variance (the square of the standard deviation) is multiplied by the square of the constant. The *addition* of a constant to every observation in a set affects the mean but not the variance or standard deviation. These results have the interesting and useful implication that standardized statistics, such as z, t, F, and r, are not changed by linear transformations of the observations, because the transformation constants all cancel in such statistics.

c. A **standard score**, $z = (x - \mu)/\sigma$, is the number of standard deviation units that an observation is away from the mean. The fourth line of Table 17.2 states that the mean and the standard deviation of standard scores are 0 and 1, respectively, as explained in Section 4.4.

d. The last two lines of the table deal with *sums and differences of random variables*. The expected value of a sum is equal to the sum of the expected values, and the expected value of a difference is equal to the difference of the expected values. However, the variance of a sum and the variance of a difference are both equal to the *sum* of the separate variances, with an appropriate adjustment by the covariance. To remember the formulas for variances of sums and differences, notice the analogy with the algebraic expansion of $(x + y)^2 = x^2 + 2xy + y^2$ and $(x - y)^2 = x^2 - 2xy + y^2$.

EXAMPLE 17.8 To illustrate the first line of Table 17.2, let us take an example where $k = 10$: There are 10 mm in 1 cm. What are the mean and variance of the number of millimeters in a centimeter? The mean, or average value in the long run, is $E(10) = 10$, and since a constant does not vary, the variance is $\text{Var}(10) = 0$. In other words, there are always 10 mm in 1 cm.

To illustrate the second line of Table 17.2, let $k = 10$ again, and let the variable x be height in centimeters. A group of students were found to have a mean height of $E(x) = 181.9$ cm with a standard deviation of 8.0 cm. What would the mean and standard deviation have been if the students had been measured in millimeters? Given that there are 10 mm in 1 cm, we hardly need Table 17.2 in order to realize that the mean and standard deviation would have been $(10)(181.9) = 1819$ mm and $(10)(8.0) = 80$ mm, respectively. The second line of Table 17.2 merely says that if we change the measurement units of the variable x, then the mean and standard deviation likewise are converted to the new measurement units. The table is presented in terms of the variance, which is the square of the standard deviation, so the variance in the present example is $(8.0\text{ cm})^2 = 64\text{ cm}^2$ or $(80\text{ mm})^2 = 6400\text{ mm}^2$. Notice that multiplication of the standard deviation by $k = 10$ is equivalent to multiplying the variance by $k^2 = 100$.

The third line (linear transformation) of the table adds an extra wrinkle to the second line. Now we not only multiply by a constant, but we add another constant to the result. To continue the height example, it is as if we wanted to know what the mean and standard deviation would have been if the students had been measured in millimeters ($b = 10$) while standing on a 200-mm footstool ($a = 200$). Clearly, the mean would have been 200 mm greater than without the footstool; that is, the mean would have been $(10)(181.9) + 200 = 2019$ mm. However, if the footstool were used for all the measurements, it would have no effect on the variability of the heights. The footstool changes the location, but not the dispersion, of the observations. It adds 200 mm to each observation, and hence to the mean, but leaves each observation's deviation from the mean unchanged, so the standard deviation remains at 80 mm (or 8.0 cm) and the variance remains at 6400 mm^2 (or 64 cm^2).

The fourth line (standard score) is just a special case of the third line. We may write a standard score as

$$z = \frac{x - \mu}{\sigma} = \frac{x}{\sigma} - \frac{\mu}{\sigma} = \left(\frac{1}{\sigma}\right)x + \left(-\frac{\mu}{\sigma}\right)$$

Thus, the fourth line of Table 17.2 is the special case of the third line with $b = 1/\sigma$ and $a = -\mu/\sigma$. Since $E(x) = \mu$ and $\text{Var}(x) = \sigma$, substitution into the third line tells us that

$$E(z) = \left(\frac{1}{\sigma}\right)\mu - \frac{\mu}{\sigma} = 0$$

and

$$\text{Var}(z) = \left(\frac{1}{\sigma^2}\right)\sigma^2 = 1$$

as spelled out in the fourth line. This result is no accident: Standard scores are defined by a formula that sets their mean at zero and makes their standard deviation to be the unit of measurement.

EXAMPLE 17.9 To illustrate the last two lines of Table 17.2, consider a statistics class that wrote two quizzes, each marked out of 50 points. The first quiz had a mean mark of 38 with a standard deviation of 5, the second quiz had a mean mark of 35 with a standard deviation of 9, and the correlation between marks on the two quizzes was .60. Each student was assigned a mark out of 100 by adding the two quiz marks. What were the mean and standard deviation of the combined marks out of 100?

The second last line of the table tells us that the mean of the combined marks is the sum of the separate means,

$$38 + 35 = 73$$

and the variance of the combined marks is the sum of the separate variances plus twice the covariance,

$$5^2 + 9^2 + (2)(.60)(5)(9) = 160$$

[The covariance is obtained from Expression (17.10) or (17.11).] Since $\sqrt{160} = 12.65$, we can report that the mean and standard deviation of the combined marks are 73% and 13 percentage points, respectively.

If the students in this example subtracted their mark on the second quiz from their mark on the first quiz, they would find out how much worse they did on the second quiz. What were the mean and standard deviation of these differences between the first and second quizzes?

The last line of Table 17.2 indicates that the mean of the differences is equal to the difference of the separate means,

$$38 - 35 = 3$$

and the variance of the differences is the sum of the separate variances minus twice the covariance,

$$5^2 + 9^2 - (2)(.60)(5)(9) = 52$$

Since $\sqrt{52} = 7.21$, we can report that the mean and standard deviation of the differences are 3 and 7, respectively. On the average, the marks dropped by 3 points out of 50, but the "standard" amount of deviation from the mean decrease of 3 points was 7 points.

STUDY MATERIAL FOR CHAPTER 17

Terms and Concepts

Section 17.1

payoff
statistical decision theory
state-of-the-world probability
maximax criterion
maximin criterion
EMV
EMV criterion
risk-taking strategy
risk-avoiding strategy
utility

utile
EU
EU criterion
expected utility with perfect information
expected value of perfect information, EVPI
limitations of decision theory

Section 17.2

arithmetic mean
variance
standard deviation
covariance
simple linear correlation
joint probability distribution

Section 17.3

change of scale
linear transformation
standard score

Key Formulas

(17.1) $E(x) = \sum x_i p_i$

Section 17.1

(17.2) EVPI = EU (perfect information) − EU (current information)

Section 17.2

See the formulas in Table 17.1.

Section 17.3

See the formulas in Table 17.2.

Exercises for Numerical Practice

Section 17.1

17/1. The following matrix shows monetary payoffs (in dollars) for each of three possible actions in three possible states of the world.

		Action 1	Action 2	Action 3	Subjective probability
	A	0	3	4	.2
State	*B*	10	2	−5	.3
	C	−5	1	6	.5

a. Calculate the EMV of each action.

b. Which action would be selected according to

  i. the maximax criterion?

  ii. the maximin criterion?

  iii. the EMV criterion?

  iv. the EU criterion, if this decision maker's utilities are obtained by squaring nonnegative payoffs and cubing negative payoffs?

c. Calculate the EVPI based on the utilities in Part (b).

17/2. The following matrix shows monetary payoffs (in dollars) for each of four available actions in two possible states of the world.

State		Action 1	Action 2	Action 3	Action 4	Subjective probability
	A	4.5	4	0	3	.9
	B	1	4	12	11	.1

a. Calculate the EMV of each action.

b. Which action would be selected according to

  i. the maximax criterion?

  ii. the maximin criterion?

  iii. the EMV criterion?

  iv. the EU criterion, if this decision maker's utilities are obtained by squaring the monetary payoffs?

c. Calculate the EVPI based on the utilities in Part (b).

Section 17.2

17/3. The following is the joint probability distribution of two discrete random variables x and y.

		Values of x 0	5	10	Marginal probabilities
Values of y	2	.2	.3	0	.5
	4	.4	0	.1	.5
Marginal probabilities		.6	.3	.1	1.0

Calculate the following:

a. the mean of x and the mean of y.

b. the variance of x and the variance of y.

c. the standard deviation of x and the standard deviation of y.

d. the covariance of x and y.

e. the correlation between x and y.

17/4. The following is the joint probability distribution of two discrete random variables x and y.

		Values of x			Marginal probabilities
		4	8	20	
Values of y	10	0	.20	.35	.55
	50	.25	.15	.05	.45
Marginal probabilities		.25	.35	.40	1.00

Calculate the following:

a. the mean of x and the mean of y.
b. the variance of x and the variance of y.
c. the standard deviation of x and the standard deviation of y.
d. the covariance of x and y.
e. the correlation between x and y.

Section 17.3

17/5. The mean and variance of a variable x are 50 and 64, respectively; the mean and variance of y are 20 and 25; and the correlation between x and y is .40. Calculate the following:

a. the mean, variance, and standard deviation of the constant 2.
b. the mean, variance, and standard deviation of $2x$.
c. the mean, variance, and standard deviation of $2x - 40$.
d. the mean, variance, and standard deviation of $3y$.
e. the mean, variance, and standard deviation of $3y + 12$.
f. the mean, variance, and standard deviation of $x + y$.
g. the mean, variance, and standard deviation of $x - y$.

17/6. The mean and variance of a variable x are 70 and 144, respectively; the mean and variance of y are 40 and 100; and the correlation between x and y is .90. Calculate the following:

a. the mean, variance, and standard deviation of the constant 5.
b. the mean, variance, and standard deviation of $5y$.
c. the mean, variance, and standard deviation of $5y + 47$.
d. the mean, variance, and standard deviation of $x/10$.
e. the mean, variance, and standard deviation of $20 - x/10$.
f. the mean, variance, and standard deviation of $x + y$.
g. the mean, variance, and standard deviation of $x - y$.

Problems for Applying the Concepts

17/7. A manufacturing company is about to sign a one-year contract with one of three suppliers of raw materials. A major source of uncertainty is the possibility of an airline strike that would have differential effects on the ability of each of the suppliers to deliver the material. The company estimates that the probability of a disruptive strike is .20. Taking into account the quality of the supplies and the transportation problems, the company determines the following payoff matrix (in thousands of dollars).

	Supplier *A*	Supplier *B*	Supplier *C*
Strike	−10	45	40
No strike	100	50	60

a. Calculate the EMV of each action.

b. Which supplier should the company choose according to

i. the maximax criterion?

ii. the maximin criterion?

iii. the EMV criterion?

iv. the EU criterion, if the company's utilities in the case of a strike are −10, 6, and 5 for *A*, *B*, and *C*, respectively, and 10, 7, and 8 in the no-strike case?

c. What is the maximum number of utiles that it might be worth to the company to obtain further information as to whether there will be a strike?

17/8. An investor plans to invest in one of three stock portfolios formulated with the assistance of a financial advisor. The net returns (dividends and capital gains) for an identical sum of money invested in each portfolio over a fixed time period are shown below (in thousands of dollars) for three possible states of the market during the investment period.

State of the market	Portfolio *A*	Portfolio *B*	Portfolio *C*	Subjective probability
Strong	20	30	80	.15
Average	18	10	20	.60
Weak	15	−20	−40	.25

a. Calculate the EMV of each portfolio.

b. Which portfolio should the investor choose according to

i. the maximax criterion?

ii. the maximin criterion?

iii. the EMV criterion?

iv. the EU criterion, if the investor's utilities in the case of a strong market are 20, 25, and 30 for *A*, *B*, and *C*, respectively; 15, 5, and 20 in the case of an average market; and 10, −50, and −100 in the case of a weak market?

c. What is the maximum number of utiles that it might be worth to the investor to obtain further information as to what the state of the stock market will be during the investment period?

17/9. Each member of a basketball team attempted 10 foul shots at the start and again at the end of a strenuous practice. The following table shows the joint probability distribution of number of successful foul shots when rested (x) and the number of successful foul shots when tired (y).

		Score when rested (x)			Marginal probabilities
		6	7	8	
Score when tired (y)	6	.2	.2	0	.4
	7	.1	.2	.1	.4
	8	0	.1	.1	.2
Marginal probabilities		.3	.5	.2	1.0

Treating these data as the description of a population, calculate the following:

a. the mean score when rested and when tired.

b. the variance of the scores when rested and when tired.

c. the standard deviation of the scores when rested and when tired.

d. the covariance and the correlation between the two variables.

(Source: Student project by J. Nasse.)

17/10. The following table shows the joint probability distribution of present family size (x) and desired number of children in one's future family (y) for a certain group of unmarried female university students.

		Desired number of children in future family (y)				Marginal probabilities
		2	3	4	5	
Present family size (x)	1	0	.1	0	0	.1
	2	.2	0	.2	0	.4
	3	.2	.1	0	.2	.5
Marginal probabilities		.4	.2	.2	.2	1.0

Treating these data as the description of a population, calculate the following:

a. the mean of each variable,

b. the variance of each variable.

c. the standard deviation of each variable.

d. the covariance and correlation between the two variables.

(Source: Student project by T. Kopperud.)

17/11. Measured in feet, the heights of the trees in an orchard have a mean of 15 and a variance of 4. Given that there are 12 inches in a foot:

a. What are the mean and standard deviation of the number of inches in a foot?

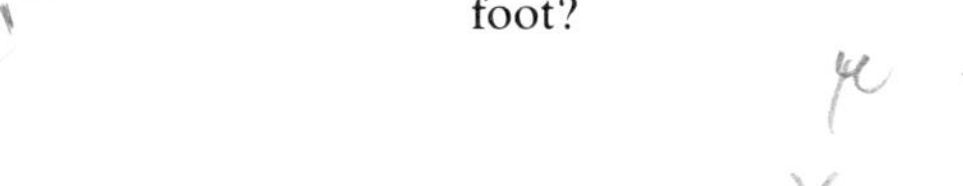

b. What would the mean and variance of the heights be if the trees were measured in inches?

17/12. A doctor found that the oral temperatures (in °F) of the patients she saw one morning had a mean of 101.0 and a standard deviation of 2.7. Given that Celsius temperature (C) may be calculated as $C = (5/9)(F - 32)$, what would the mean and variance of the temperatures have been if the temperatures had been measured on the Celsius scale?

17/13. The daily maximum temperatures (in °C) for a month at one weather station had a mean of 20.0 and a standard deviation of 4.0, while the daily minima had a mean of 5.0 and a standard deviation of 3.0. The correlation between daily maxima and minima was .55.

a. Given that Fahrenheit temperature (F) may be calculated as $F = 1.8C + 32$, what would the mean and variance of the maxima and minima have been if the temperatures had been measured on the Fahrenheit scale?

b. The difference between the maximum and minimum temperatures on a given day is called the *diurnal shift.* What were the mean and standard deviation of the diurnal shifts (in °C) recorded during the specified month at this weather station?

17/14. In a group of married couples, the weights (in pounds) of the husbands had a mean of 176 and a variance of 900, while the weights of the wives had a mean of 134 and a variance of 400. The correlation between the weights of husbands and wives was .38. Given that there are 2.2 pounds in a kilogram:

a. What are the mean and standard deviation of the number of pounds in a kilogram?

b. What would the mean and variance of the husbands' weights and of the wives' weights have been if the weights originally had been recorded in kilograms?

c. Consider the combined weight of each couple. Measured in pounds, what were the mean and standard deviation of the combined weights of the couples?

17/15. Example 11.4 illustrated the one-sample t test. The dependent variable was birth weight, measured in kilograms, and the test statistic was found to have a value of $t_{obs} = .97$. What would the value of the test statistic have been if the birth weights had been measured

a. in grams?

b. in ounces?

17/16. Example 15.16 used an analysis of variance to compare three brands of ski wax. The dependent variable was distance traveled, in meters, in a glide down a particular hill, and the test statistic was found to have a value of $F = 101$. What would the value of the test statistic have been if the distances had been measured

a. in centimeters?

b. in inches?

Reference

Hein, P. "A Psychological Tip." *Grooks I.* Markham, Ontario: Paperjacks, 1969, page 38. (Previously published by General Publishing Company, Toronto.)

Further Reading

Kotz, S., and Stroup, D. F. *Educated Guessing: How to Cope in an Uncertain World.* New York: Marcel Dekker, 1983. Assuming only high school mathematics, this book introduces mathematical expectation in Chapter 3 and decision making in Chapter 5.

Raiffa, H. *Decision Analysis: Introductory Lectures on Choices under Uncertainty.* Reading, MA: Addison-Wesley, 1968. Raiffa provides a thorough, yet largely nonmathematical, introduction to statistical decision theory.

Wilks, S. S. *Mathematical Statistics.* New York: Wiley, 1962. Further coverage of mathematical expectation generally involves more advanced mathematics and is provided in any text on mathematical statistics, such as in Wilks's Chapter 3.

CHAPTER 18

FURTHER TOPICS IN EXPERIMENTAL DESIGN AND ANOVA

EXAMPLE 18.1 During the early 1940s, a group of agricultural researchers in Virginia proposed a research project entitled "An Effort to Appraise the Effect of Fertilization of Land on the Growth and Physical Conditions of the Farm People." The researchers hoped to show that soil fertilization would ultimately lead to improvements in the health and well-being of the farm families. An isolated mountain valley in southwestern Virginia, with soil of very low fertility, was selected to be the site of the study.

"It was the plan of the agronomists of Virginia Polytechnic Institute and the Research Director of the Tennessee Valley Authority to evaluate the health of the people living in the valley by means of the latest and best medical procedures at the beginning of the project. This would be followed by fertilization of the farms yearly at Tennessee Valley Authority's expense and under the direction and guidance of the agronomists of the extension division of Virginia Polytechnic Institute. The farmers would be guided by the extension division in the most modern and best agricultural practices. At a fixed period one or more times during each year the people were to be examined and their health appraised. This study was to be conducted for a period of four years."

At this point in the planning, a statistician, Boyd Harshbarger, was asked "to review the project for its design and its adaptability for future statistical analysis of the data." Harshbarger identified several important problems that needed to be solved by the research committee:

a. What were the precise objectives of the experiment? For example, did the researchers want to know specifically about the effects of adequate fertilization or more generally about the use of "the most modern and best agricultural practices"? The research committee decided on the former, more specific, objective. If other changes in farming practice had been included along with fertilization, it may not have been possible to tell whether any improvements in the "growth and physical conditions of the farm people" were due to fertilization or to the other changes in farming practice.

b. How were the effects "on the growth and physical conditions of the farm people" to be measured? The research committee decided on a list of specific dependent variables, such as "changes in hemoglobin, plastic protein, ascorbic acid, serum calcium, food intake, . . . weights, color of skin, . . . [and] yield of crops . . . ."

c. Even if there were improvements in the growth and physical conditions of the farm people, how would it be possible to be sure that the improvements were due to the fertilization rather than to other changes (for example, the weather, the general state of the economy) during the four-year period of the experiment? To solve this problem, Harshbarger "insisted upon replication and randomization of the farms within each replication that were to furnish participants for the program. This led to the selection of two adjoining valleys which were isolated and similar in many respects. Each of the two valleys was divided in half. By a toss of a coin one-half of each valley was selected to receive complete fertilization. . . . Not all farms in each half of the replications supplied cooperators for the experiment. The farms were randomly drawn until the desired number of cooperators of the proper age and sex were found. . . . Drawing the cooperators from different replications and at random made it possible to compare the [data from the group of] cooperators receiving fertilizer with those of the non-fertilized group of cooperators."

At the end of the four-year project, "a summary of the data relating to the health of the participants of this project showed a marked improvement in all aspects for both the fertilized groups and the non-fertilized groups from the years 1944 to 1947. During this period there was a marked improvement in the economic condition throughout the valley. The twice-a-year physical examination possibly played a major role in the improvement of the health of the people. For most of the [variables] the improvement of the non-fertilized group exceeded by a slight margin the improvement of the fertilized group . . . .

"It was anticipated that this research would serve as a basis for recommendation for fertilizer practices as they may affect health in the programs of the Tennessee Valley Authority as well as the extension agents. Since there were very marked improvements in the health of the participants for all conditions for the cooperators in this research, it is obvious that, if there had been no checks and other conditions leading to valid conclusions, the claims for fertilizer practices could have been misleading. The fertilizer recommendations based on unfounded conclusions could have cost the farmer as well as the Tennessee Valley Authority many thousands of dollars as well as a loss of confidence in future recommendations."

(Source: Harshbarger, 1974. Used with permission of the American Statistical Association.)

As Example 18.1 illustrates, there is much more to statistics than cranking already gathered data through various computational formulas. Statistics is relevant to every step of an empirical investigation, from the formulation of the question to the presentation of the results. Indeed, the various stages of an investigation cannot be properly carried out independently of each other: The appropriate analysis and interpretation of the data depend on how the data were collected, and planning the collection of data requires clear formulation of the empirical question as well as an appreciation of what various types of statistical analysis can and cannot accomplish.

In this chapter, we bring together some of the issues in the planning of experiments and the analysis and interpretation of the resulting data. As in Chapter 15, the experiments considered here are ones in which two or more treatments are compared in terms of their effects on one or more dependent variables.

18.1 THE PLANNING OF EXPERIMENTS

Experimental design is the aspect of inferential statistics that is concerned with the planning of experiments so that an efficient analysis to achieve the experiment's purpose is possible. If inadequate attention is given to experimental design before the data are collected, it may turn out that no analysis is possible to answer the questions for which the experiment was run.

Each of the following subsections deals with a different one of the topics that must be considered in the proper planning of an experiment. The decisions made concerning these topics, whether deliberately or by default, determine what type of analysis will be appropriate and what kinds of conclusions will be permissible.

Dependent Variables

A **dependent variable** is a random variable that is observed in an experiment and whose observed values describe the observational units on a dimension of interest. It is a response variable that is observed for each observational unit.

The choice of specific dependent variables is often a nontrivial decision. We have already seen in Section 1.1 that some types of variables, such as retrospective report and verbal report, tend to be less trustworthy than direct measurement. In addition, it may not be obvious what particular sorts of observations would be relevant to answering the empirical question being asked. Interpretation of the experimental results must take into account any limitations in the choice of the dependent variables(s).

EXAMPLE 18.2 In Example 18.1, the project organizers began with the idea of assessing the long-term effects of soil fertilization on the "growth and physical conditions" of farm residents. However, the experiment could not be conducted until it was decided how to translate the all-encompassing phrase "growth and physical conditions" into specific dependent variables. Which specific aspects of growth and physical conditions did the researchers expect would be affected by the use of soil fertilizer, and what measurements should be made to reflect those aspects? As noted in Example 18.1, the researchers eventually drew up a list of dependent variables to be measured, including changes in hemoglobin, food intake, weight, and crop yield.

When more than one dependent variable is observed in an experiment, the researcher must decide between two basic approaches to analyzing the data. In **univariate analysis**, each dependent variable is analyzed separately, with the result that any correlations among the dependent variables are not explicitly taken into account. This approach may be appropriate if each dependent variable is of interest in its own right. However, multiple univariate analyses from the same experiment will tend to inflate the Type I error rate for the experiment and hence require caution in the interpretation of significance levels. (The problem is analogous to that in multiple t testing, as discussed on page 509.) Notice that in the context of designed experiments, the term *univariate analysis* refers to the analysis of a single *dependent* variable, even though there is also at least one independent variable involved. In this book, discussion is limited to univariate analysis, that is, analysis of only one dependent variable at a time.

The alternative approach is called **multivariate analysis**. In the context of designed experiments, multivariate analysis implies that two or more dependent variables are analyzed simultaneously, taking into account any correlations that may exist among the dependent variables. Thus, a multivariate analysis makes use of more information in the data than do separate univariate analyses when there are two or more dependent variables in an experiment. In addition, it provides better control of the Type I error rate in such experiments. However, both the computations and the interpretation of multivariate analyses are more complex than in the univariate case and are beyond the scope of this book.

Independent Variables

An **independent variable** is a dimension whose effects are under investigation; type of treatment and age of subject are examples of possible independent variables. As discussed in Section 9.2, the term *independent variable* is used in a general sense that includes both **classification variables** and **manipulated variables**. Variables such as age or sex of the subject are simply classifications of characteristics already possessed by the observational units before the study begins. If the values of the dependent variable are found to differ for different levels of a classification variable, we have demonstrated a correlation between the dependent variable and the classification variable, but we cannot determine what *caused* the differences. (For example, if we compare males and females in the performance of some task, we can describe how their performances differ, but we cannot say whether any differences are due to biological differences, or to social expectations, or to some other variable that distinguishes males and females.) On the other hand, manipulated variables, where the experimenter actually administers different treatments to different observational units (independent of the unit's own characteristics), allow the possibility of causal inferences. In the context of designed experiments, manipulated variables are sometimes thought of as the "true" independent variables, as distinguished from classification variables.

In experimental design, the notion of the independent variable is often subsumed under the more general term **factor**, which is a dimension (other than the dependent variable) used by the experimenter as a basis for distinguishing one observation from another. Observations may be distinguished according to which levels of the independent variables they were recorded under, and according to which observational unit they came from. Hence, each independent variable is a factor, and replication is another factor. Experiments are run to determine how the behaviour of a dependent variable may be related to the influences of two or more factors. Specifically, these influences are the main effects and interaction effects of the factors, together referred to as the **sources of variation** in the values of the dependent variable.

It is not always easy to translate the empirical question of interest into specific independent variables. This stage of the planning requires clarification of the precise objective of the investigation. (In Example 18.1, for example, the researchers had to decide whether they wanted to examine specifically the effects of adequate fertilization or more generally the effects of guidance "in the most modern and best agricultural practices.") Once again, interpretation of the experimental results must take into account any limitations in the choice of the independent variable(s).

EXAMPLE 18.3 A researcher was interested in determining whether people are more error-prone when they are under stress. To investigate this question, the researcher had to decide on a specific dependent variable (what kinds of errors were to be measured, and how) and a specific independent variable (what conditions of stress were to be observed).

For the dependent variable, the researcher chose to count the number of errors made in typing a given passage. The researcher manipulated stress by telling the subject either that she was to type the passage for practice to familiarize herself with the typewriter or that she was to type the passage while being timed with a stopwatch, as a test of her typing ability.

The interpretation of the results had to take into account the limitations in the dependent and independent variables. With respect to the dependent variable, the findings were limited to typing errors; the results may have been different if the effects of stress were examined on errors in other activities. With respect to the independent variable, perhaps the difference between the conditions was not so much in the amount of stress as in whether or not the typist attempted to type quickly. Even if the amount of stress actually differed between the two conditions, the findings are limited to the particular sort of stress induced by knowledge that one is being timed on a typing test. Obviously, no one experiment can lead to general conclusions about the effects of all forms of stress. The choice of the independent variable set the limits on the sort of stress to which the conclusions would apply.

(Source: Student project by L. Mills.)

Extraneous Variables

Besides the dependent and independent variables, there are many other variables, called **extraneous variables**, that might affect the outcome of an experiment. An extraneous variable is any variable that is neither a dependent variable nor a factor in the experiment. If the extraneous variables are not taken care of appropriately, the experiment is apt to be inconclusive or possibly misleading. In general, researchers would like these other variables to be **controlled variables**. An extraneous variable is said to be controlled when its value is held constant for all conditions in the experiment. On the other hand, researchers seek to avoid having any **confounding variables** among the extraneous variables. A confounding variable is an extraneous variable that takes on different values for different levels of an independent variable, so that the effects of the independent variable cannot be distinguished from those of the confounding variable.

In order to draw *causal* inferences about the effects of manipulation of an independent variable, two (or more) groups or conditions should be treated exactly the same except with respect to the independent variable to which any differences in the response variable are to be attributed. In the simplest situation, there are two groups, an **experimental group** and a **control group**, treated exactly the same except for the manipulation of the independent variable, but the principle is the same when there are more than two levels of the independent variable and hence more than two groups or conditions. Ideally, all extraneous variables should be held constant, but in practice this degree of control is unlikely to be achieved. Observational units are apt to differ with respect to their past histories and other extraneous variables that may be beyond the control of the experimenter. The solution to this problem is to perform **random assignment** of the observational units to the various treatments or conditions. In random assignment, an inanimate device, such as a table of random digits or a coin, is used to determine which condition will be applied to each observational unit.

It is difficult to overemphasize the importance of random assignment in establishing the effects of the manipulation of an independent variable. Without random assignment, the independent variable is inevitably vulnerable to the possibility of being confounded with some characteristic of the observational units. It then becomes impossible to know whether any observed effects are due to the manipulation of the independent variable or, on the other hand, to whatever characteristic made an observational unit more likely to receive one treatment rather than another. The crucial point is that random assignment ensures that no characteristic or property of an observational unit determines, or is correlated with, the treatment it receives. Thus, the benefit of random assignment is that any observed effects can be attributed to the manipulation of the independent variable rather than to some uncontrolled characteristic of the observational units.

EXAMPLE 18.4 The independent variable in Example 18.1 was amount of fertilizer provided: One group received no fertilizer, while the other group was provided with adequate soil fertilization. The dependent variables were changes in hemoglobin, food intake, crop yield, and other measures of health and well-being. In addition, there were numerous extraneous variables, which were not under investigation but which could affect the experimental results; for example, the weather, the general state of the economy, the height above sea level, the quality of the land, the particular location in the mountain valley, and so on.

In the original plan of the project, according to which *all* of the participating residents of the valley would have received fertilizer, there would have been numerous variables confounded with use of fertilizer. It would have been impossible to tell whether any improvements in health and well-being were due to the use of fertilizer or to such confounding variables as the weather or the general state of the economy.

Fortunately, the plan was revised before the project was conducted. The statistician Harshbarger insisted on the inclusion of a control group, whereby half of the cooperators did not receive fertilizer, so that there would be a standard against which to compare the experimental group, which received fertilizer. Potentially confounding variables such as the weather and the general state of the economy thus became controlled variables: They were the same for both the fertilized and the nonfertilized farms. Other extraneous variables were also controlled; for example, height above sea level and quality of the land were roughly the same for both groups. In addition, two valleys were used so that half of each valley could be randomly assigned (by a coin toss) to the two conditions (no fertilizer provided or adequate fertilizer provided). There were undoubtedly practical difficulties that led the researchers to rule out the more obvious and usual procedure, which would have involved random assignment of individual farms to the two conditions. In any case, the purpose of the random assignment was to ensure that the nature of the farm and of the farm family did not become confounding variables, whose effects would be indistinguishable from those of the independent variable.

As it turned out, the measures of health showed substantial improvement for both the experimental group and the control group. One possibility is that the provision of free medical examinations to all participants (so that the dependent variables could be measured) was responsible for this improvement. However, the experiment was not designed to evaluate this possibility, and the provision of medical examinations is completely confounded with such variables as the general state of the economy. Consequently, the experiment provides no basis for deciding whether the improvements in health should be attributed to the provision of free medical examinations, or to the general improvement in economic conditions, or to a change in some other variable.

Obviously, there are some independent variables where random assignment is impossible. We cannot randomly assign some observational units to be male and some to be female; nor can we randomly assign some to be young and some to be old. For ethical and practical reasons, we are unlikely to be able randomly to assign some subjects to be smokers and some to be nonsmokers, or some subjects to have four children and some to have no children. When random assignment is impossible, we have a classification variable rather than a manipulated variable. As previously noted, conclusions about classification variables are necessarily limited to being correlational rather than causal in nature. We can say only that changes in the dependent variable are or are not correlated with changes in the classification variable. The discussion of causality in connection with correlation in Section 5.2 is also relevant here, in connection with classification variables.

Levels of the Factors

The particular values of the independent variables are called their **levels**. If, for example, three different treatments are applied in an experiment, each treatment constitutes one level of the treatment factor. (In Example 18.1, the two levels of the independent variable are "no fertilizer provided" and "adequate fertilizer provided.")

Two sorts of factors are distinguished in terms of the reason for the selection of particular levels to be included in the experiment. In most of the experiments we will consider, the independent variables are **fixed factors**. A given factor, say, Factor *A*, is considered fixed if the levels of Factor *A* to be included in the experiment are chosen not randomly nor even haphazardly, but because of their intrinsic relevance to the experiment—precisely because they *are* those specific levels. In other words, the levels of a fixed factor are included because one of the purposes of the experiment is to see if those specific levels have different effects on the dependent variable.

The alternative to a fixed factor is a **random factor**. In most of the experiments we will consider, the only random factor is the replication factor. A given factor, say, Factor R, is considered random if the researcher wants to generalize results concerning other factors to levels of Factor R that are not included in the experiment. The specific levels of a random factor are of interest not in their own right, but only as representative of a larger population of levels.

EXAMPLE 18.5 In an experiment to compare three different treatments by applying each treatment to a different group of 10 subjects, treatment is a fixed factor, because the purpose of the experiment is to see if those three treatments have different effects on the dependent variable. The particular treatments to be compared in an experiment would not be chosen at random! If the researchers did not want their conclusions about the effects of the treatments to be limited to the particular 30 subjects actually observed, "subjects" would be considered a random factor. The resulting analysis would enable the researchers to draw conclusions about the effects of the treatments in whatever population the 30 subjects may be considered a random sample of.

Ideally, the particular levels of a random factor to be included in an experiment should be identified by **random selection** from the population of levels to which generalization is intended. (Random selection was defined and discussed in Section 6.2.) Thus, in the case of the replication factor, the replicates (for example, the subjects) should be randomly selected. Genuine random selection from the relevant population is often not feasible in practice. Nonetheless, the statistically supportable conclusions that may be drawn from experimental results are limited to the population from which the observational units may be considered to constitute a random sample. Generalizations beyond that population cannot be supported statistically, but may be justified by specialized knowledge in the area of application.

EXAMPLE 18.6 Much medical research concerning human health problems is performed with mice or other animals as subjects. There is no statistical basis for generalizing the results of such experiments to human populations, but experts in physiology or medicine may nevertheless correctly generalize certain results to humans on the basis of their specialized knowledge of the area of investigation.

Similarly, a disproportionate number of psychological experiments have been conducted with college students as subjects. Although there is no statistical basis for applying the results of such experiments to persons of other ages or abilities, knowledge of psychology may enable a person to make an informed judgment as to which experimental results apply to the general public and which are unique to college students.

Sometimes the distinction between the population from which the sample was selected and the population to which generalization is desired is more subtle than in the preceding two illustrations. One important, and too often overlooked, example is that involving subjects who volunteer to take part in an investigation. A population of volunteers (for anything) is different in many ways from a population of nonvolunteers and from a general, undifferentiated population. You are encouraged here to review Section 8.1, pages 286–289, which includes examples of problems due to nonrandom selection from the desired population.

Be careful not to confuse **random selection** of observational units with **random assignment** of observational units to treatments or conditions. In general, the purpose of random selection is to allow generalization of results to a particular population, while the purpose of random assignment is to avoid the introduction of confounding variables.

Combining the Levels of the Factors

The term **design** is used in two related senses:

a. the aspect of inferential statistics concerned with planning experiments, and

b. an arrangement of factors in an experiment.

Part of experimental planning involves deciding how to combine the levels of the factors into a particular design (in the latter sense). Even the simplest experiment has two factors—in general terms, a treatment factor and a replication factor—and more complex experiments may involve several factors. There are many possible ways of combining the levels of one factor with the levels of another factor, but two of these are basic and more frequently used than any others. We limit our discussion to designs based on these two; the Further Reading section at the end of this chapter directs you to presentations of more complex designs.

One method of combining the levels of two factors in an experiment is called **complete crossing**:

> Two factors are said to be **completely crossed** if every level of one factor occurs in combination with every level of the other factor. **(18.1)**

If Factor A has three levels A_1, A_2, and A_3, and Factor B has only two levels B_1 and B_2, Factors A and B are completely crossed in an experiment in which all six **treatment combinations** occur:

$$A_1B_1,\quad A_1B_2,\quad A_2B_1,\quad A_2B_2,\quad A_3B_1,\quad A_3B_2$$

The pattern of crossing is shown more clearly if the treatment combinations are displayed in a tabular cross-classification.

	A_1	A_2	A_3
B_1	A_1B_1	A_2B_1	A_3B_1
B_2	A_1B_2	A_2B_2	A_3B_2

The design of an experiment in which the factors are completely crossed is called a **factorial design**. The preceding design would be called a 3-by-2 (or 2-by-3) factorial design.

The other basic method of combining the levels of two factors in an experiment is called **nesting**:

> One factor is said to be **nested** under a second factor if each level of the **nesting factor** has a different subset of levels of the **nested factor** associated with it. **(18.2)**

If Factor A has three levels and Factor B has six levels, Factor B is nested under Factor A in an experiment in which each level of Factor B occurs in combination with only one level of Factor A, so that there are only six treatment combinations:

$$A_1B_1, \quad A_1B_2, \quad A_2B_3, \quad A_2B_4, \quad A_3B_5, \quad A_3B_6$$

A complete cross-classification does not exist in this case. Rather, we can think of B_1 and B_2 as being in the A_1 nest, B_3 and B_4 as being in the A_2 nest, and B_5 and B_6 as being in the A_3 nest.

A_1	A_2	A_3
A_1B_1	A_2B_3	A_3B_5
A_1B_2	A_2B_4	A_3B_6

In this design, Factor B is said to be nested under Factor A because each level of Factor A (each nest) is associated with (contains) different levels of Factor B. Nesting is, in fact, a particular kind of confounding that may sometimes be intentionally induced in the factors of an experiment. (Confounding with an *extraneous* variable is always to be avoided, but the confounding of two of the *factors* in an experiment is sometimes appropriate.) In the preceding design, the differences between A_1, A_2, and A_3 are confounded with differences between levels of B.

Chapter 15 presented the analysis of variance (ANOVA) for three different experimental designs. Let us review an example of each design in order to identify the factors and observe how the factors are combined in each case.

EXAMPLE 18.7 To illustrate the one-way ANOVA with independent samples, Example 15.4 presented data from an investigation of the number of times that students living in college residences go home in one semester. Three populations of female students were sampled: those with homes close to the college, those with homes at an intermediate distance, and those with homes far from the college. The dependent variable is number of visits home, and the independent variable is distance to the student's home, with three levels: close, intermediate, and far. There are two factors that distinguish among the observations: (1) the independent variable distance, which is a fixed factor, and (2) the replication factor, students, which is a random factor. The students factor is nested under the distance factor, because each student was observed at only one distance (in only one nest).

In general, a design consisting of one random factor nested under one fixed factor may be analyzed by the one-way ANOVA with independent samples.

(Source: Student project by D. M. Mychaluk.)

EXAMPLE 18.8 Example 15.8 illustrated the two-way ANOVA without replication. In order to compare average price levels at three different supermarkets, the researcher obtained prices on the same day for eight identical products sold at all three stores. The dependent variable is price, and the independent variable is stores. There are two factors that distinguish among the observations: (1) the independent variable, stores, which is a fixed factor, and (2) the replication factor, products, which is a random factor. The products factor is crossed with the stores factor, because every product was observed in all three of the stores.

In general, a design consisting of one random factor crossed with one fixed factor may be analyzed by the two-way ANOVA without replication (which is also called the one-way ANOVA with repeated measurements).

(Source: Student project by G. Lazicki.)

EXAMPLE 18.9 Example 15.14 presented data from a skiing experiment to illustrate the two-way ANOVA with independent samples. The experiment involved three brands of glider wax used in combination with each of two glider-wax areas, giving six treatment combinations in a 3-by-2 factorial design. There were four trials for each of the six treatment combinations, for a total of 24 observations. The dependent variable was distance traveled in a glide on the skis, and the independent variables were brand of wax and glider-wax area. There are three factors that distinguish among the observations: the two independent variables, brand and area, which are both fixed factors, and the replication factor, trials, which is a random factor. The brand factor and the area factor are completely crossed, because all six combinations of brand and area were observed, and the trials factor is nested under both brand and area, because each of the 24 trials was for only one of the six treatment combinations of brand and area. That is, there was a different set of four trials in each treatment combination (in each nest).

In general, a design consisting of one random factor, nested under two fixed factors that are crossed with each other, may be analyzed by the two-way ANOVA with independent samples.

(Source: Student project by G. Hall.)

The method of combining the levels of the different factors in an experiment determines not only which type of analysis is appropriate but also, to some extent, what sorts of conclusions may be drawn. In general, complete crossing is the preferred method of combining *two fixed factors*. Fixed factors are those whose effects an experiment is specifically designed to investigate, so it is generally preferable to avoid the confounding that would be introduced by nesting one fixed factor under another fixed factor. On the other hand, nesting is generally the preferred method of combining *two random factors*. The reason for this preference will be given in Expression (18.18). Finally, *a fixed factor and a random factor* may be combined either (a) by crossing or (b) by nesting the random factor under the fixed factor. The choice between these two alternatives distinguishes two major categories of experimental designs, described in the next subsection.

Replication

Replication is the repeated observation of a particular treatment or treatment combination, with a different observational unit being observed each time. Replication, or the replication factor, is like other factors in that the choice of replicates (observational units receiving a given treatment or treatment combination) is decided by the experimenter in designing the experiment, but it is not an independent variable. The replicates are chosen not to compare the performance of individual replicates, but rather to permit comparison of the levels of the independent variable(s).

Replication is necessary in order to obtain an estimate of random variation against which the effects of other factors can be evaluated, to see if they are significantly greater than random variation. Thus, replication is always a random factor, because its purpose is to permit generalization of the experimental results to a population of replicates, of which those actually observed may be considered a random sample. Without some form of replication, there is no basis for determining whether the effects of an independent variable are greater than would be expected by chance, and there is no statistical basis for generalizing the results beyond the particular observational units actually observed.

The manner in which levels of the replication factor are combined with levels of other factors is the basis for distinguishing two major categories of experimental designs. If the replication factor is *nested* under the levels of a fixed factor, the experiment has an **independent-samples design**, and the fixed factor is called an **independent-samples factor**. The nesting of replications under the different conditions of another factor implies that each subject or replicate is observed under only one level of the factor, so there is a separate sample of subjects under each level of the factor. Thus, the differences between levels of the independent variable are confounded with differences between samples or groups of subjects; independent-samples designs are also known as **between-subjects designs**. If subjects are randomly assigned to the various conditions, this particular type of confounding causes no problem: Because replication is a random factor, an analysis of variance for independent-samples designs can be used to determine whether the difference between samples is greater than can reasonably be attributed to random variation. Sections 15.2 and 15.4 presented the analysis of variance for two types of independent-samples designs, which were reviewed in Examples 18.7 and 18.9.

The second major category of experimental designs is that in which the replication factor is *crossed* with a fixed factor, in which case the replications are referred to as **blocks** and the experiment is said to have a **repeated-measures design**. (Blocks were introduced in Section 15.3.) The crossing of replications (blocks) with the different levels of another factor implies that each block (replicate) is observed under every level of the other factor. The block is observed repeatedly, once (or more) under each level of the fixed factor. The fixed factor with which the replication factor is crossed is called the

repeated-measures factor, because it is the dimension on which there are repeated measurements of the same block. A repeated-measures design in which the treatments are assigned randomly to the observational units within each block is called a **randomized-blocks design**. A repeated-measures design in which the blocks are the experimental subjects, so that each subject serves as its own control, is sometimes called a **within-subjects design**. Section 15.3 presented the analysis of variance for one type of repeated-measures design, which was reviewed in Example 18.8. Repeated-measures designs will be discussed more fully in Chapter 19, where we will see that there are some special problems associated with their use.

18.2 TESTING EXPERIMENTAL EFFECTS VIA ANOVA

Chapter 15 introduced analysis of variance (ANOVA) as a general technique for testing hypotheses about means. ANOVA is a highly versatile technique that can be used to test the significance of differences among means from virtually any experimental design encompassed within the framework of the decisions outlined in Section 18.1. This section explains some of the theory on which the methods in Chapters 15 and 18–20 are based.

The essence of ANOVA is the partitioning of the total variation (SS_{total} = sum of squared deviations around the grand mean) into variation (SS) due to two or more identifiable sources. A variance estimate, or mean square (MS = SS/df), is then obtained for each source of variation, and these can be compared in an F test to determine whether variation due to a certain source (say, due to an independent variable) is significantly greater than random variation.

Hypotheses Tested in ANOVA

There may be one or more F tests in ANOVA, depending on the design of the experiment. Each F test is a test of a different null hypothesis, but all fit under two general headings: main effects and interaction effects.

The null hypothesis in a test of the main effect of a factor says that the population means at the various levels of the factor are all equal; for Factor A with k_A levels,

$$H_0(A): \quad \mu_1 = \mu_2 = \cdots = \mu_{k_A} \tag{18.3}$$

The Factor A main-effect null hypothesis says that the k_A population means μ_a are all equal—that their variance is zero. Variance is a measure of dispersion, or of how spread out a set of values is. Let us use the symbol σ_A^2 to represent the true variance of the population means μ_a for the different levels of Factor A. This variance σ_A^2 is called the **variance component** due to the true effect of Factor A in the population. It measures how spread out the population means μ_a are for the different levels of Factor A. The null hypothesis in Expression (18.3) implies that σ_A^2, the variance component due to Factor A, is equal to zero. The alternative hypothesis is that the population means μ_a at the different levels of Factor A are not all equal—that there are some differences among the population means—that the variance component σ_A^2 is greater than zero.

The null hypothesis in a test of the interaction effect of two factors says that the population mean for any given treatment combination is determined completely by the marginal means of the factors. According to the interaction null hypothesis, the population mean for each treatment combination is simply the grand mean plus the main effects for the relevant levels of the factors; for Factor A with k_A levels and Factor B with k_B levels,

$$H_0(AB):\quad \mu_{ab} = \mu + (\mu_{a.} - \mu) + (\mu_{.b} - \mu)$$
$$\text{for } a = 1, \ldots, k_A, \text{ and } b = 1, \ldots, k_B \qquad \textbf{(18.4)}$$

Rearranging terms in Expression (18.4), we can say that the interaction null hypothesis says that the quantity $\mu_{ab} - \mu_{a.} - \mu_{.b} + \mu$ equals zero for all $k_A k_B$ treatment combinations. If the quantity $\mu_{ab} - \mu_{a.} - \mu_{.b} + \mu$ equals the same constant for all treatment combinations, it follows that the quantity has zero variance. Thus, the A-by-B null hypothesis proposes that the quantity $\mu_{ab} - \mu_{a.} - \mu_{.b} + \mu$ has zero variance. Let us use the symbol σ_{AB}^2 to represent the true variance of the quantity $\mu_{ab} - \mu_{a.} - \mu_{.b} + \mu$ for the different treatment combinations. This variance σ_{AB}^2 is called the **variance component** due to the A-by-B interaction, or due to the effects of particular combinations of levels of Factor A and of Factor B in the population. The null hypothesis in Expression (18.4) implies that σ_{AB}^2, the variance component due to the A-by-B interaction, is equal to zero. The alternative hypothesis is that the population means in the different treatment combinations cannot be described solely by the main effects—that particular combinations of levels "interact" to deviate from the main effects—that the variance component σ_{AB}^2 is greater than zero.

Null hypotheses in ANOVA are tested by F tests. The general form of the test statistic is

$$F_{\text{effect}} = \frac{\text{MS}_{\text{effect}}}{\text{MS}_{\text{error}}} \qquad \textbf{(18.5)}$$

where the "effect" might be a main effect or an interaction effect, and where MS_{error} is the **error term** for testing that particular effect. If the calculated value of the test statistic F_{effect} is greater than the critical value $F_\alpha(\text{df}_{\text{effect}}, \text{df}_{\text{error}})$, we say that the effect is statistically significant at the α level.

EXAMPLE 18.10 In the two-way ANOVA with independent samples, the A-by-B interaction is tested by

$$F_{AB} = \frac{\text{MS}_{AB}}{\text{MS}_{R(AB)}}$$

Here the A-by-B interaction is the effect being tested, and $\text{MS}_{R(AB)}$ is the appropriate error term to make that test.

Expression (18.5) indicates that to test an effect, the numerator of the F

ratio is the mean square for the effect to be tested. But what determines which mean square is the appropriate error term? The next section explains.

Determination of the Appropriate Error Term

The important point of theory on which ANOVA F tests are based was introduced in Section 12.4:

> The ratio of two independent estimates of the same Gaussian-population variance follows the F distribution. **(18.6)**

In the context of ANOVA, variance estimates are usually called *mean squares*. As we will see, Expression (18.6) provides the fundamental rationale for setting up ratios of mean squares in order to draw inferences about means.

The question of which is the appropriate error term for an F test depends on the expected values of the mean squares, often called the **expected mean squares**. (Expected values were discussed in Chapter 17.) The expected value of a statistic is the statistic's average value that would be obtained in the long run by repeated sampling and repeated calculation of the statistic. In other words, the expected value of a statistic is the value of the parameter that the statistic estimates without bias. The following paragraphs state, without proof, the expected values of the mean squares in the three designs discussed in Chapter 15.

In a one-way ANOVA with an independent sample of n_R observations at each level of Factor A, the expected mean squares for Factor A and for replications within A are

$$E(\mathrm{MS}_A) = \sigma^2 + n_R\sigma_A^2 \qquad \textbf{(18.7)}$$

$$E(\mathrm{MS}_{R(A)}) = \sigma^2 \qquad \textbf{(18.8)}$$

where σ^2 is the population variance of observations within each and every level of Factor A (the random variability of the observations), and σ_A^2 is the population variance component due to Factor A. As explained above, the variance component σ_A^2 due to Factor A is equal to zero if the Factor A null hypothesis is true and is positive otherwise. Thus, if the Factor A null hypothesis is true, MS_A and $\mathrm{MS}_{R(A)}$ both estimate the same population variance σ^2, but if the Factor A null hypothesis is not true, MS_A will tend to be larger than $\mathrm{MS}_{R(A)}$.

In view of Expression (18.6), the ratio $\mathrm{MS}_A/\mathrm{MS}_{R(A)}$ may be used for an F test of the Factor A null hypothesis, on the assumption that the underlying population is Gaussian. If the null hypothesis is true, the ratio $\mathrm{MS}_A/\mathrm{MS}_{R(A)}$ follows the F distribution; otherwise, the ratio will tend to be larger than is indicated by the F distribution. $\mathrm{MS}_{R(A)}$ is the appropriate error term for MS_A because, as shown in Expressions (18.7) and (18.8), the two mean squares estimate the same quantity except that $\mathrm{MS}_{R(A)}$ lacks the variance component due to the effect to be tested.

Consider a two-way ANOVA without replication, with k_A levels of a fixed Factor A and k_B levels of a random Factor B (blocks). The expected mean squares are as follows:

$$E(\text{MS}_A) = \sigma^2 + \sigma^2_{AB} + k_B\sigma^2_A \quad \textbf{(18.9)}$$

$$E(\text{MS}_B) = \sigma^2 + k_A\sigma^2_B \quad \textbf{(18.10)}$$

$$E(\text{MS}_{AB}) = \sigma^2 + \sigma^2_{AB} \quad \textbf{(18.11)}$$

The appropriate error term for testing the main effect of Factor A is the interaction mean square MS_{AB}, because MS_A and MS_{AB} both estimate the same population quantity when the Factor A null hypothesis is true (that is, when the variance component $\sigma^2_A = 0$, indicating that there is no variation among the population means for the various levels of Factor A). There is no exact error term for the main effect of Factor B, but MS_{AB} sometimes is used as an approximation, on the assumption that the variance component σ^2_{AB}, due to the interaction with the random Factor B, contributes little more than random variation. In any case, the test of Factor B is generally an informal check of the effectiveness of blocking rather than a formal significance test of the blocking factor.

There are four sources of variation and hence four mean squares in the usual ANOVA table for a two-way ANOVA with independent samples. If there are k_A levels of a fixed Factor A, k_B levels of a fixed Factor B, and n_R observations in each treatment combination, the expected mean squares are as follows:

$$E(\text{MS}_A) = \sigma^2 + n_R k_B \sigma^2_A \quad \textbf{(18.12)}$$

$$E(\text{MS}_B) = \sigma^2 + n_R k_A \sigma^2_B \quad \textbf{(18.13)}$$

$$E(\text{MS}_{AB}) = \sigma^2 + n_R \sigma^2_{AB} \quad \textbf{(18.14)}$$

$$E(\text{MS}_{R(AB)}) = \sigma^2 \quad \textbf{(18.15)}$$

If the Factor A null hypothesis is true (and hence the variance component $\sigma^2_A = 0$), MS_A and $\text{MS}_{R(AB)}$ both estimate the same population quantity σ^2. If the Factor B null hypothesis is true (and hence the variance component $\sigma^2_B = 0$), MS_B and $\text{MS}_{R(AB)}$ both estimate the same population variance σ^2. Likewise, if the interaction null hypothesis is true (and hence the variance component $\sigma^2_{AB} = 0$), MS_{AB} and $\text{MS}_{R(AB)}$ both estimate the same parameter σ^2. Thus, the within-cell mean square $\text{MS}_{R(AB)}$ is the appropriate error term for testing all three null hypotheses.

For our purposes in this book, you need not be concerned about the expressions for particular expected mean squares. They are given only as examples to illustrate the basis for determining appropriate error terms. From the theoretical perspective, the important point in the preceding discussion is summarized in the following **general principle defining the appropriate error term**:

> The appropriate error term for testing any effect V (main effect or interaction) is the mean square whose expected value is identical to $E(\text{MS}_V)$ except that it lacks the variance component due to V. **(18.16)**

In other words, if

$$E(\mathrm{MS}_V) = E(\mathrm{MS}_{\text{error}}) + m\sigma_V^2 \qquad \text{for some constant } m > 0 \tag{18.17}$$

then $\mathrm{MS}_{\text{error}}$ is the appropriate error term for testing the effect of V. This general principle applies for all ANOVA designs presented in this book. On the assumption of underlying Gaussian populations of observations, it implies that the ratio $F_V = \mathrm{MS}_V/\mathrm{MS}_{\text{error}}$ follows the F distribution if the effect's null hypothesis is true (and hence if the variance component $\sigma_V^2 = 0$), but not if it is false. Therefore, F_V can be compared to tabulated values of the F distribution for a test of the effect's null hypothesis.

EXAMPLE 18.11 What is the expected value of the appropriate error term for testing the effect of Factor C in a design where the expected mean square for C is

$$E(\mathrm{MS}_C) = \sigma^2 + n_R\sigma_{AC}^2 + n_R k_A \sigma_C^2$$

According to the principle in Expressions (18.16) and (18.17), the appropriate error term $\mathrm{MS}_{\text{error}}$ is the mean square whose expected value is given by

$$E(\mathrm{MS}_{\text{error}}) = \sigma^2 + n_R\sigma_{AC}^2$$

If the Factor C null hypothesis is true, MS_C and $\mathrm{MS}_{\text{error}}$ both estimate the same population variance, and their ratio follows the F distribution. If the Factor C null hypothesis is not true, the variance component σ_C^2 will tend to make MS_C larger than $\mathrm{MS}_{\text{error}}$, so that the ratio $\mathrm{MS}_C/\mathrm{MS}_{\text{error}}$ will tend to be larger than described by the F distribution.

The general principle in Expression (18.16) should help clarify why a particular mean square is appropriate as an error term for testing a given effect, but it does so only in terms of expected values. It does not indicate which mean square in an ANOVA summary table will have the required expected value. From a practical point of view, it is useful to be able to identify which mean square is the appropriate error term for testing a given effect. This can be done by finding the expected mean square for every effect in the design, but such an exercise is beyond the scope of this text. (The interested reader may pursue this approach in the references cited under Further Reading at the end of this chapter.) However, a pattern emerges from this exercise, and for our purposes it will be sufficient to know the following **guidelines for finding the appropriate error term**:

In a complete factorial design with or without replication, the appropriate error term for testing any effect V (main effect or interaction effect) is as follows: **(18.18)**

a. If V is crossed with no random factors, $\mathrm{MS}_{\text{within-cell}}$ is the appropriate error term for testing the effect of V.

b. If V is crossed with exactly one random factor W, then MS_{VW} is the appropriate error term for testing the effect of V.

c. If V is crossed with two or more (crossed) random factors, there is no exact error term for testing the effect of V. (More advanced texts describe approximate F tests in this situation, but it is generally preferable to avoid crossing random factors if the design includes a fixed factor of interest. See the Further Reading at the end of this chapter.)

d. There is no appropriate error term for testing some of the effects, such as the within-cell effect, whose mean squares serve as error terms for testing other effects.

The preceding guidelines are applicable for all ANOVA designs presented in this book, although the design in Section 19.3 will require a slightly more general wording of one guideline.

EXAMPLE 18.12 Consider a two-way factorial design with Factor A fixed and Factor B random, and with n_R observational units in each treatment combination. This design might arise, for example, in an investigation comparing the efficacy of three different medical therapies for a particular ailment. To ensure that the results are not limited to one particular therapist, the researchers randomly select five therapists trained in all three therapies. A different group of four patients is observed in each of the 15 therapy–therapist combinations, for a total of 60 patients. The fixed Factor A is type of therapy, the random Factor B is therapist, and the replications are the patients, nested under the therapy–therapist (A-by-B) combinations.

We can use the guidelines in Expression (18.18) to identify the appropriate error terms in this experiment. The appropriate error term for testing the main effect of A is MS_{AB}, because A is crossed with exactly one random factor, namely, B. (Notice that this error term is different from the error term given in Table 15.3, where it was assumed that Factors A and B were both fixed.) The appropriate error term for the main effect of B is $\text{MS}_{R(AB)}$, because B is crossed with no random factors. Similarly, the appropriate error term for testing the interaction is $\text{MS}_{R(AB)}$, because the interaction A-by-B is crossed with no (other) random factors. There is, as usual, no appropriate error term for testing the within-cell effect $R(AB)$.

18.3 THREE-WAY ANOVA WITH INDEPENDENT SAMPLES OF EQUAL SIZE

Chapter 15 introduced one-way and two-way ANOVA with independent samples. Situations sometimes arise where it is either necessary or desirable to examine three or more factors in a single experiment. There is no theoretical difficulty in generalizing the analysis of variance to handle completely crossed designs with any number of factors, but the computations become somewhat lengthier, and the interpretation of the results may be less obvious. Still, having studied the one-way and two-way cases, you should not find it difficult to come to some basic appreciation of what is involved in three-way and higher-order ANOVAs.

EXAMPLE 18.13 A farmer wanted to compare the yields from three varieties of barley (Bonanza, Klondike, and Fairfield) and to evaluate the effectiveness of his fertilization program. Data from his grain-production records over the years were used to examine the dependent variable, barley yield in bushels per acre, as a function of three independent variables: (1) variety of barley, (2) fertilization, and (3) weather. Fertilization was categorized into just two levels, according to whether or not any fertilizer was applied to a particular parcel of land, and weather was categorized into two levels designated as normal and dry, depending on whether or not there was at least average rainfall in a particular year. All three factors are fixed factors. The three levels of variety, two levels of fertilization, and two levels of weather were completely crossed, and data were available for two replications of each of the 12 treatment combinations. Thus, the design is a three-way factorial, 3-by-2-by-2, with independent samples of $n_R = 2$ observations per cell.

(Source: Student project by K. Astner.)

Computational Formulas and the ANOVA Summary Table

In general, a three-way factorial experiment with independent samples may be characterized as involving the following:

k_A levels of Factor A,
k_B levels of Factor B,
k_C levels of Factor C, for a total of
$k_A k_B k_C$ treatment combinations or **cells**, with an independent sample of
n_R observational units in each cell, for a total of
$n = k_A k_B k_C n_R$ observational units altogether.

The total variation in such an experiment can be partitioned into portions due to eight basic sources; namely, the three main effects (A, B, and C), the three two-way interactions (A-by-B, A-by-C, and B-by-C), the three-way interaction (A-by-B-by-C), and the random variation due to replications within cells, $R(ABC)$.

With terms defined by obvious extensions of the notation introduced for two-way ANOVA in Section 15.4, the computational formulas for the sums of squares in the **three-way ANOVA with independent samples of equal size** are as follows:

$$SS_A = \frac{1}{k_B k_C n_R} \sum T^2_{a...} - \frac{T^2}{n} \tag{18.19}$$

$$SS_B = \frac{1}{k_A k_C n_R} \sum T^2_{.b..} - \frac{T^2}{n} \tag{18.20}$$

$$SS_C = \frac{1}{k_A k_B n_R} \sum T^2_{..c.} - \frac{T^2}{n} \tag{18.21}$$

$$SS_{AB} = \frac{1}{k_C n_R} \sum \sum T^2_{ab..} - SS_A - SS_B - \frac{T^2}{n} \tag{18.22}$$

$$SS_{AC} = \frac{1}{k_B n_R} \sum \sum T^2_{a.c.} - SS_A - SS_C - \frac{T^2}{n} \tag{18.23}$$

$$SS_{BC} = \frac{1}{k_A n_R} \sum \sum T^2_{.bc.} - SS_B - SS_C - \frac{T^2}{n} \tag{18.24}$$

$$SS_{ABC} = \frac{1}{n_R} \sum \sum \sum T^2_{abc.} - SS_A - SS_B - SS_C - SS_{AB} - SS_{AC} - SS_{BC} - \frac{T^2}{n} \tag{18.25}$$

$$SS_{R(ABC)} = \sum \sum \sum \sum x^2_{abci} - \frac{1}{n_R} \sum \sum \sum T^2_{abc.} \tag{18.26}$$

$$SS_{\text{total}} = \sum \sum \sum \sum x^2_{abci} - \frac{T^2}{n} \tag{18.27}$$

The only "trick" to performing ANOVA calculations is to recognize that there is a pattern in the preceding computational formulas—indeed, the same pattern appears in the computational formulas for any ANOVA design. The following comments will help you to see the pattern:

a. There is a sum-of-squares formula for each source of variation, and each such formula may be thought of as having two parts: The first term identifies what we are calculating the variation *of,* and the remaining terms, which are subtracted from the first term, indicate what we are calculating the variation *around.* Each formula gives us the variation *of* something *around* something else. For example, SS_A is the variation *of* the Factor *A* means *around* the grand mean; SS_{AB} is the variation *of* the *A*-by-*B* treatment-combination means *around* the value that would be predicted on the basis of the Factor *A* means, the Factor *B* means, and the grand mean; $SS_{R(ABC)}$ is the variation *of* the individual observations *around* the cell means for the *A*-by-*B*-by-*C* treatment combinations.

b. The first term of each sum-of-squares formula directs us to square each of the totals for the separate levels (or treatment combinations) of the relevant source of variation, add those squared totals, and divide by the number of observations that entered into each total. For example, in the first term of SS_A, we square each Factor *A* total $T_{a...}$, add the squared totals, and divide by the number ($k_B k_C n_R$) of observations entering each Factor *A* total. The term T^2/n, which appears in several of the formulas, is the variation of the grand mean around zero. It follows the same format in

TABLE 18.1 Analysis-of-variance summary table for three-way ANOVA with independent samples of equal size.

Source of variation	SS	df	MS	F
Factor A	SS_A	$k_A - 1$	MS_A	F_A
Factor B	SS_B	$k_B - 1$	MS_B	F_B
Factor C	SS_C	$k_C - 1$	MS_C	F_C
A-by-B interaction	SS_{AB}	$(k_A - 1)(k_B - 1)$	MS_{AB}	F_{AB}
A-by-C interaction	SS_{AC}	$(k_A - 1)(k_C - 1)$	MS_{AC}	F_{AC}
B-by-C interaction	SS_{BC}	$(k_B - 1)(k_C - 1)$	MS_{BC}	F_{BC}
A-by-B-by-C interaction	SS_{ABC}	$(k_A - 1)(k_B - 1)(k_C - 1)$	MS_{ABC}	F_{ABC}
Replications within A-by-B-by-C	$SS_{R(ABC)}$	$k_A k_B k_C (n_R - 1)$	$MS_{R(ABC)}$	—
Total	SS_{total}	$n - 1$		

that we square the grand total (corresponding to the grand mean) and divide by the number (n) of observations entering that total.

You are encouraged to watch for the common pattern in the ANOVA computations for the examples and exercises that follow.

The general form of the **ANOVA summary table for the three-way ANOVA with independent samples** of equal size is shown in Table 18.1. The mean squares are again obtained as MS = SS/df. The appropriate error term for each F ratio may be determined according to the guidelines in Expression (18.18). In the usual case, where Factors A, B, and C are all fixed factors, the within-cell mean square $MS_{R(ABC)}$ is the appropriate error term for every effect. In addition to its role in the F tests, the error mean square provides information about the precision of the data. Substituting its value into Expression (15.17) in Section 15.2, we can obtain the estimated standard errors of the cell and marginal means.

EXAMPLE 18.14 An investigation to compare yields from three varieties of barley was described in Example 18.13. We should note at the outset that this is a **retrospective study** rather than a genuine designed experiment. Although we can use ANOVA to identify the major sources of variation in the data, the interpretation will be complicated by the fact that the treatments were not randomly assigned to the observational units (parcels of land in a particular year). It could be, for example, that the farmer chose to fertilize only when he thought it would do some good, thus exaggerating the effect of fertilization, or that he planted particular varieties of barley according to the amount of ground moisture in the spring, thus confounding the effect of variety with the condition of the land. Although ANOVA will identify the major sources of variation in the data, it does not tell us the reason for the observed differences. Because of the absence of random assignment of observational units to the various treatment combinations, it is always difficult and hazardous to draw causal conclusions from retrospective studies. P-values in such cases are really nothing more than descriptive summaries that may help us ask the right questions to uncover the reasons for a particular pattern of observations.

The following yields (in bushels per acre) were recorded for the 24 harvests.

Fertilization:	**Fertilizer**		**No fertilizer**	
Weather:	**Normal**	**Dry**	**Normal**	**Dry**
Variety:				
Bonanza	62	32	42	26
	58	37	40	24
Klondike	66	38	49	29
	69	31	39	20
Fairfield	70	36	54	27
	76	43	47	33

To help organize the computations, we record the following **tables of totals**, which will provide the totals required for the various sum-of-squares formulas.

Variety-by-fertilization-by-weather totals (and marginal totals)

	Fertilizer		**No fertilizer**		
	Normal	**Dry**	**Normal**	**Dry**	**Total**
Bonanza	120	69	82	50	321
Klondike	135	69	88	49	341
Fairfield	146	79	101	60	386
Total	401	217	271	159	1,048

Fertilization-by-weather totals (and marginal totals)

	Normal	**Dry**	**Total**
Fertilizer	401	217	618
No fertilizer	271	159	430
Total	672	376	1,048

Variety-by-fertilization totals (and marginal totals)

	Fertilizer	**No fertilizer**	**Total**
Bonanza	189	132	321
Klondike	204	137	341
Fairfield	225	161	386
Total	618	430	1,048

Variety-by-weather totals (and marginal totals)

	Normal	**Dry**	**Total**
Bonanza	202	119	321
Klondike	223	118	341
Fairfield	247	139	386
Total	672	376	1,048

The sums of squares are then calculated from Expressions (18.19) through (18.26):

$$SS_{\text{variety}} = \frac{321^2 + 341^2 + 386^2}{8} - \frac{1{,}048^2}{24} = 46{,}039.750 - 45{,}762.667 = 277.083$$

$$SS_{\text{fert.}} = \frac{618^2 + 430^2}{12} - 45{,}762.667 = 1{,}472.667$$

$$SS_{\text{weather}} = \frac{672^2 + 376^2}{12} - 45{,}762.667 = 3{,}650.667$$

$$SS_{VF} = \frac{189^2 + 204^2 + 225^2 + 132^2 + 137^2 + 161^2}{4}$$

$$- 277.083 - 1{,}472.667 - 45{,}762.667$$

$$= 6.583$$

$$\begin{aligned}
SS_{VW} &= \frac{202^2 + 223^2 + 247^2 + 119^2 + 118^2 + 139^2}{4} \\
&\quad - 277.083 - 3{,}650.667 - 45{,}762.667 \\
&= 46.583 \\
SS_{FW} &= \frac{401^2 + 271^2 + 217^2 + 159^2}{6} \\
&\quad - 1{,}472.667 - 3{,}650.667 - 45{,}762.667 = 216.000 \\
SS_{VFW} &= \frac{120^2 + 135^2 + 146^2 + 69^2 + 69^2 + 79^2 + 82^2 + 88^2 + 101^2 + 50^2 + 49^2 + 60^2}{2} \\
&\quad - 277.083 - 1{,}472.667 - 3{,}650.667 - 6.583 \\
&\quad - 46.583 - 216.000 - 45{,}762.667 \\
&= 51{,}437.000 - 51{,}432.250 = 4.750 \\
SS_{R(VFW)} &= (62^2 + 58^2 + 66^2 + \cdots + 33^2) - 51{,}437.000 \\
&= 51{,}666.000 - 51{,}437.000 = 229.000
\end{aligned}$$

The ANOVA summary table may then be completed as follows.

Source of variation	SS	df	MS	F	
Variety	277.1	2	138.5	7.26	$P < .01$
Fertilization	1,472.7	1	1,472.7	77.2	$P < .01$
Weather	3,650.7	1	3,650.7	191	$P < .01$
V-by-F interaction	6.6	2	3.3	.17	
V-by-W interaction	46.6	2	23.3	1.22	
F-by-W interaction	216.0	1	216.0	11.3	$P < .01$
V-by-F-by-W interaction	4.8	2	2.4	.12	
Replications within V-by-F-by-W	229.0	12	19.1		
Total	5,903.3	23			

The significant effects can be summarized by plotting the means for the fertilization-by-weather treatment combinations and the main-effect means for varieties as in Figure 18.1. The estimated standard errors of the means are obtained via Expression (15.17).

Fertilization-by-weather means in bu./acre (with estimated standard error = $\sqrt{19.1/6}$ = 1.8 bu./acre)

	Normal	Dry
Fertilizer	66.8	36.2
No fertilizer	45.2	26.5

Variety means in bu./acre (with estimated standard error = $\sqrt{19.1/8}$ = 1.5 bu./acre)

Bonanza	40.1
Klondike	42.6
Fairfield	48.2

The plots in Figure 18.1 show that yields were greater when fertilizer was applied (main effect of fertilization), and that the benefit of fertilization was greater if the weather was normal than if it was dry (fertilization-by-weather interaction). In other words, yields were greater in normal than in dry weather (main effect of weather), and

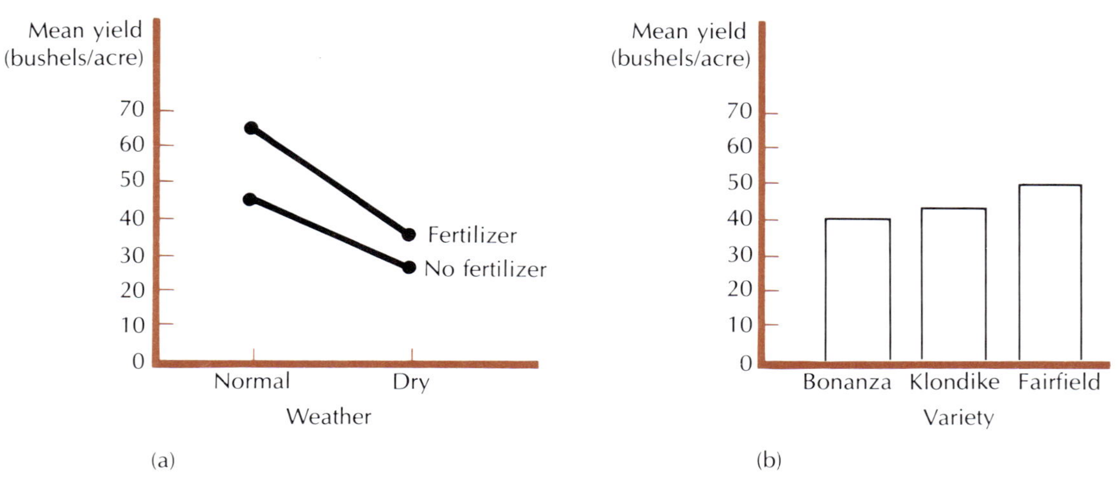

Figure 18.1
Plots summarizing the significant effects in the three-way ANOVA of barley yields.

the benefit of normal weather was greater if fertilizer was applied (fertilization-by weather interaction). Mean yields were greatest for Fairfield barley and least for Bonanza (main effect of variety).

(Source: Student project by K. Astner.)

Interpretation of Results

The meaning of a *main effect* is the same in three-way ANOVA as in the one-way and two-way cases. The main effect of Factor A, for example, refers to differences among the mean responses to the different levels of Factor A, averaging over all other factors.

The meaning of a two-way interaction is the same as in the case of two-way ANOVA with independent samples. The A-by-B interaction, for example, refers to the extent to which the effect of Factor A is different at different levels of Factor B (or, equivalently, the extent to which the effect of Factor B is different at different levels of Factor A).

The only new kind of effect in a three-factor design is the **three-way interaction**, which indicates the extent to which a two-way interaction differs at different levels of the third factor. The A-by-B-by-C interaction refers to the extent to which the A-by-B interaction is different at different levels of C or, equivalently, the extent to which the A-by-C interaction is different at different levels of B or, equivalently, the extent to which the B-by-C interaction is different at different levels of A. The three alternatives in the preceding sentence are three ways of describing the same phenomenon, and the experimenter should choose the way that provides the clearest answer to the question the experiment was designed to answer.

A significant three-way interaction means that two-way interactions and main effects should be interpreted with caution. It says, in effect, that the two-way interactions and main effects have averaged over the levels of a factor that should not be averaged over. Plotting the cell means is frequently necessary to help make sense of a three-way interaction. If we make separate plots of the cell means for the different levels of one factor, the difference in shapes of the two-way interactions in those separate plots reflects the three-way interaction.

EXAMPLE 18.15 Citing remarkable feats of memory by oral historians and poets in nonliterate societies, some people have suggested that literacy and schooling actually reduce one's ability to memorize. As part of a study of the relationship between memory ability and literacy, two groups of volunteers were obtained in the African nation of Botswana. The literate group consisted of 36 junior high school students, categorized into three equal-sized subgroups labeled as low, average, and high in "intelligence" according to their current year's grade average. The nonliterate group consisted of 36 adolescents and young adults recruited in various villages and likewise categorized into three subgroups of 12, the ratings of "intelligence" in this case being made by a group of (nonliterate) adults from the same villages. One at a time, the subjects listened to a lengthy tape-recorded story and, immediately afterwards, attempted to tell the experimenter as much of it as they could. For half of them the story was an African folk tale, and for the other half it was a European fairy tale. (The actual experiment was more complex than presented here. For simplicity, we are considering only what the subject heard first on the tape recorder.) The stories were much too long for verbatim recall, and subjects were scored for the number of episodes they recalled from the story.

As described above, there were three factors under investigation. One factor was the subject's literacy, with two levels: literate and nonliterate. A second factor was the subject's intelligence, with three levels: low, average, and high. The third factor was the type of story to be recalled: African or European. The portion of the investigation presented here has a three-way factorial design, specifically, 2-by-3-by-2 with independent samples of size $n_R = 6$.

The results of the experiment are portrayed schematically in Figure 18.2. In each part of the figure, the mean recall (dependent variable) is plotted for the appropriate treatments or treatment combinations. In practice, only those graphs representing statistically significant effects would be presented (as in Example 18.14), but all of the testable effects in the three-way ANOVA are presented here for illustration. The following paragraphs illustrate the kinds of statements that might be made *if the effect being described were found to be statistically significant.*

Parts (a) and (b) of Figure 18.2 portray the three-way interaction. The statement of a three-way interaction requires reference to all three factors. We could say that for literate subjects, African stories were recalled better than European stories, while for nonliterate subjects, the same was true for the high-intelligence group but not for the low-intelligence group. An alternative way of stating the three-way interaction is by noting that there is a greater interaction of intelligence and type of story for nonliterate subjects than for literate subjects (because the lines are more nearly parallel in the latter case).

Parts (c), (d), and (e) of Figure 18.2 show the three two-way interactions,

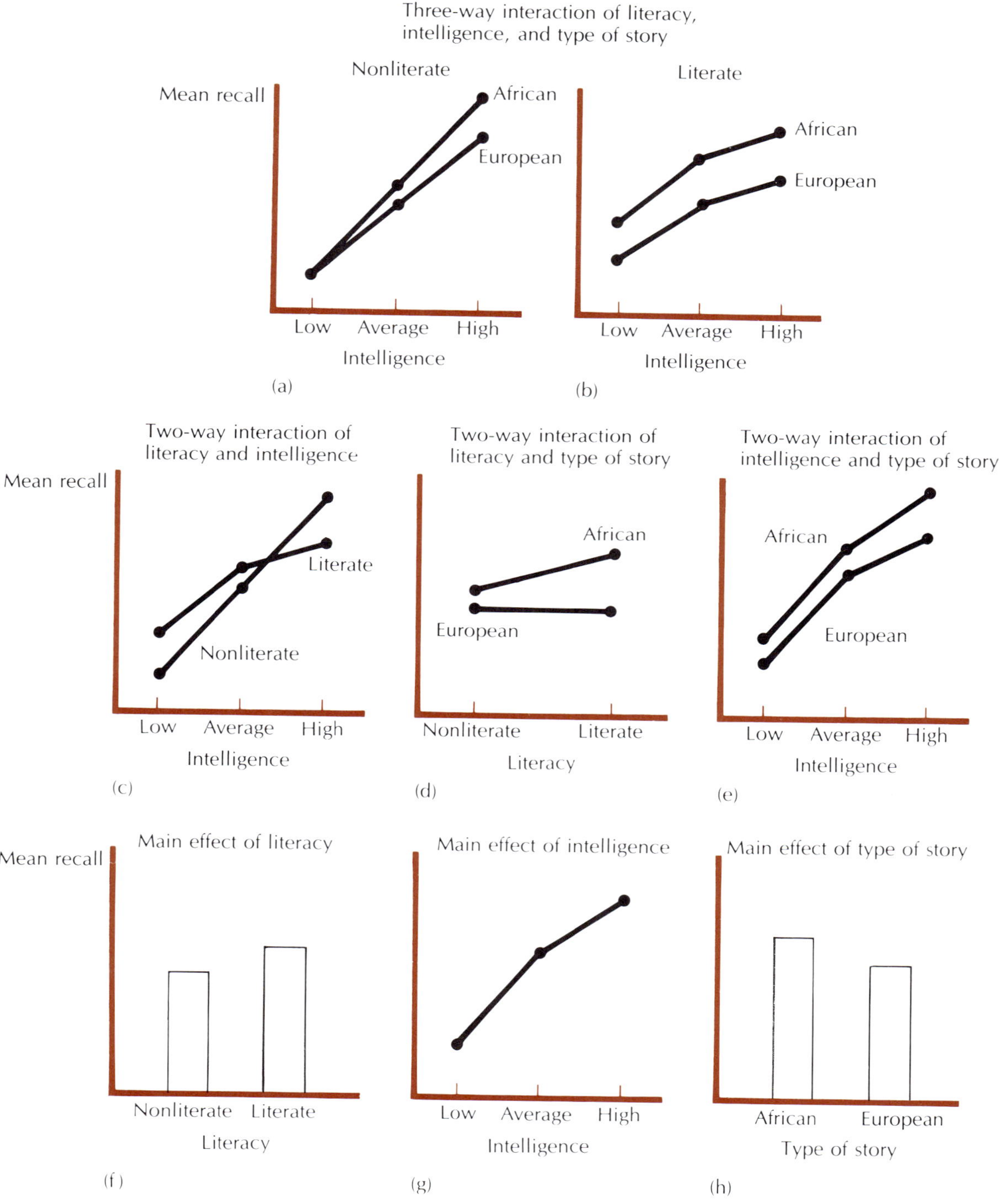

Figure 18.2
Graphical representation of main effects and interactions in a three-way factorial design.

described in turn by the following three statements: (c) The best mean recall scores were by the high-intelligence group of nonliterate subjects, but the literate subjects recalled more than the nonliterate subjects at the other two levels of intelligence. (d) African stories were recalled better than European stories, especially by the literate subjects. (e) African stories were recalled better than European stories, especially by subjects designated as more intelligent. Notice that the statement of each two-way interaction makes no reference to levels of the third factor. Statements of two-way interactions may have to be qualified or omitted if the three-way interaction is significant.

Parts (f), (g), and (h) of Figure 18.2 show the three main effects. The statement of each main effect makes no reference to the levels of the other factors: (f) Literate subjects recalled more than nonliterate subjects. (g) Mean recall scores increased with increasing intelligence. (h) African stories were recalled better than European stories. Such statements of main effects may have to be qualified or omitted if any interactions are significant.

(Source: Dube, 1982.)

Summary of the Test Procedure

The **three-way analysis of variance with independent samples of equal size** may be summarized as follows:

Assumptions: See Section 18.6.

Step 1. Determine the values of k_A, k_B, k_C, n_R, and n.

Step 2. Calculate the tables of totals (and their marginals) for A-by-B-by-C, A-by-B, A-by-C, and B-by-C.

Step 3. Calculate tables of means (and marginal means) corresponding to the tables of totals, in order to see what sample differences are being tested.

Step 4. Calculate the values required to complete the ANOVA summary table in the format of Table 18.1.

Step 5. Compare the observed values of the F test statistics with critical values from the F distribution with appropriate degrees of freedom.

a. If the three-way interaction is significant, examine the cell means $\bar{x}_{abc.}$ to discover the nature of the effect, and interpret significant main effects and two-way interactions only insofar as they are compatible with the cell means.

b. If the three-way interaction is not significant but at least one of the two-way interactions is, examine the relevant treatment-combination means for the significant interaction(s) to discover the nature of the effect(s), and interpret significant main effects only insofar as they are compatible with the treatment-combination means.

c. If none of the interactions is significant but at least one main effect is, examine the relevant marginal means to discover the nature of the effect.

Further statistical tests to aid in interpretation are considered in Chapter 20.

* 18.4 FACTORIAL DESIGNS WITH SAMPLES OF UNEQUAL SIZES

As long as the numbers of observations in the various cells of a factorial experiment are equal, the various sources of variation in the ANOVA table are **orthogonal**, that is, they do not overlap with each other. The opposite of orthogonal is **confounded**. If two of the sources in the table were confounded with each other, then the same variation would be measured twice (once as part of each of the two confounded sums of squares), and we would not know to which of the two sources the variation should really be attributed.

EXAMPLE 18.16 Consider a 2-by-2 factorial experiment to compare male and female performance on two tasks. Suppose there are three different subjects in each of the four cells as follows:

	Task 1	Task 2
Males	$n_R = 3$	$n_R = 3$
Females	$n_R = 3$	$n_R = 3$

We can compare the mean performance of the six males with the mean performance of the six females because the males and females have been observed under the same conditions, namely, half doing Task 1 and half doing Task 2. Similarly, we can compare the mean performance on Task 1 with the mean performance on Task 2 even if males and females differ in their performance, because the proportion of males assigned to Task 1 was the same as the proportion of males assigned to Task 2. Thus, the effects of the task factor and of the gender factor are not confounded here.

Suppose now that there were only two observations in one cell and that we obtained the following 11 observations:

	Task 1	Task 2	Means
Males	4, 6	2, 3, 4	3.8
Females	8, 9, 10	10, 11, 12	10.0
Means	7.4	7.0	

With unequal numbers in the cells, the two factors are confounded. Is the mean for Task 1 higher than that for Task 2 because Task 1 is easier? Or is it higher because a higher proportion of females worked on Task 1? The main effect of task (7.4 versus 7.0) is confounded with gender. Does the comparison of 3.8 versus 10.0 accurately reflect the difference in the mean performance of males and females? Or is the difference exaggerated because a higher proportion of males worked on Task 2 than on Task 1? The main effect of gender (3.8 versus 10.0) is confounded with task.

* This symbol identifies material that is not prerequisite for later topics in this book.

As illustrated in Example 18.16, unequal sample sizes in factorial experiments can cause the sources of variation in the ANOVA summary table to be confounded. (There is a particular pattern of unequal sample sizes—namely, that in which the *relative* numbers at the various levels of any one factor are the same for all levels of the other factors—where orthogonality remains. However, the orthogonal ANOVA for these so-called proportional sample sizes is appropriate only in certain special cases, and so it is omitted here. The method presented below may be used instead.) Several solutions to the problem of unequal sample sizes have been proposed; the references under Further Reading at the end of this chapter identify the major ones. One of the simplest and most easily interpreted of the approaches is the **unweighted-means analysis**.

All of the ANOVAs presented previously (in Chapter 15 and Section 18.3) have been based on the method of least squares. (The method of least squares was introduced on pages 153–155.) The general least-squares solution to the problem of unequal sample sizes is beyond the scope of this book, involving not only more advanced mathematics but also special problems in interpretation that sometimes limit its applicability. It should be noted, however, that the least-squares solution was given for the problem of unequal sample sizes in the one-way case in Section 15.2 because the solution has general applicability for one-way designs and does not entail any special problems in interpretation. The unweighted-means analysis is so named because the cell means are not weighted by the number of observations on which each is based, as they would be in a least-squares analysis. However, when samples sizes are equal, the unweighted-means analysis and the least-squares analysis yield identical ANOVAs.

The unweighted-means analysis takes each cell mean as the best estimate of the population mean for that particular treatment combination. The confounding due to unequal sample sizes is circumvented by analyzing the cell means as if they were the observations, thus creating equal sample sizes of one observation per cell. This analysis of cell means yields sums of squares for the main effects and interaction(s). The error variation is estimated in the usual way by pooling the variation within the various cells.

The following steps outline in more detail the procedure for performing an **unweighted-means ANOVA for a factorial design with unequal numbers of observations** in the various treatment combinations:

Step 1. Calculate the cell means, one for each treatment combination.

Step 2. Pretend the cell means are the only data you have. Treating the cell means as the observations in an experiment with $n_R = 1$ observation per cell, calculate the sums of squares for the main effects and interaction(s). Designate these sums of squares with a prime (SS'_A, and so on) to indicate that they were calculated while pretending.

Step 3. Stop pretending and calculate the sum of squares for replications

TABLE 18.2 Analysis-of-variance summary table for two-way unweighted-means ANOVA with independent samples of unequal sizes.

Source of variation	SS	df	MS	F
Factor A	$HM_{n_R}SS'_A$	$k_A - 1$	MS_A	F_A
Factor B	$HM_{n_R}SS'_B$	$k_B - 1$	MS_B	F_B
A-by-B interaction	$HM_{n_R}SS'_{AB}$	$(k_A - 1)(k_B - 1)$	MS_{AB}	F_{AB}
Replications within A-by-B	$SS_{R(AB)}$	$n - k_A k_B$	$MS_{R(AB)}$	—

within cells as

$$SS_{\text{repl. within cells}} = \sum x^2 - \sum \frac{(\text{cell total})^2}{\text{number of observations in the cell}} \tag{18.28}$$

This sum of squares has degrees of freedom equal to the difference between the total number n of observations and the number of treatment combinations.

Step 4. To put the "primed" sums of squares from Step 2 into the same units as the error variation from Step 3, multiply each value of SS′ by the harmonic mean (HM) of the sample sizes in the cells:

$$HM_{n_R} = \frac{\text{number of cells}}{\sum \dfrac{1}{\text{number of observations in a cell}}} \tag{18.29}$$

For example, $SS_A = HM_{n_R}SS'_A$. The sums of squares for main effects and interactions have their usual numbers of degrees of freedom.

Step 5. Use the equation MS = SS/df to obtain mean squares, and determine appropriate error terms for F ratios by the guidelines in Expression (18.18).

The preceding procedure applies to two-way, three-way, and higher-order factorial experiments with unequal sample sizes, as long as there is at least one observation in every cell. The general form of the **ANOVA summary table for the two-way unweighted-means ANOVA** with unequal sample sizes is shown in Table 18.2. The table for the three-way case follows the same pattern. Because of the lack of orthogonality in the original data, the sums of squares for the various sources of variation do not add up to SS_{total} in an unweighted-means analysis with unequal sample sizes.

EXAMPLE 18.17 To illustrate the computations in the unweighted-means analysis, let us analyze the hypothetical data from the 2-by-2 factorial experiment described in Example 18.16.

The cell means are calculated from the raw data in Example 18.16 as follows.

	Task 1	Task 2	Totals
Males	5.0	3.0	8.0
Females	9.0	11.0	20.0
Totals	14.0	14.0	28.0

Treating the cell means as observations in a 2-by-2 experiment with $n_R = 1$ observation per cell, we calculate the sums of squares for the main effects and interaction. Whether we use the formulas for the two-way ANOVA without replication (in Section 15.3) or for the two-way ANOVA with independent samples with $n_R = 1$ (in Section 15.4), we obtain the following:

$$SS'_{\text{gender}} = \frac{1}{2}(8^2 + 20^2) - \frac{28^2}{4} = 232 - 196 = 36$$

$$SS'_{\text{task}} = \frac{1}{2}(14^2 + 14^2) - \frac{28^2}{4} = 196 - 196 = 0$$

$$SS'_{GT} = (5^2 + 3^2 + 9^2 + 11^2) - 36 - 0 - 196 = 4$$

The harmonic mean of the actual sample sizes (2, 3, 3, and 3—see the data in Example 18.16) is

$$HM_{n_R} = \frac{4}{\frac{1}{2} + \frac{1}{3} + \frac{1}{3} + \frac{1}{3}} = 2.667$$

and we obtain the following sums of squares:

$$SS_G = (2.667)(36) = 96.00$$

$$SS_T = (2.667)(0) = 0$$

$$SS_{GT} = (2.667)(4) = 10.67$$

The within-cell variation is

$$SS_{R(GT)} = (4^2 + 6^2 + 2^2 + \cdots + 11^2 + 12^2) - \left(\frac{10^2}{2} + \frac{9^2}{3} + \frac{27^2}{3} + \frac{33^2}{3}\right)$$

$$= 691 - 683 = 8$$

The results are summarized in the following ANOVA summary table. All three F ratios use $MS_{R(GT)}$ as the error term, because gender and task are both fixed factors.

Source of variation	SS	df	MS	F	
Gender	96.00	1	96.00	84.00	$P < .05$
Task	0.00	1	0.00	0.00	Not sig.
G-by-*T* interaction	10.67	1	10.67	9.33	$P < .05$
Subjects within *G*-by-*T*	8.00	7	1.14		
Total		10			

Since the critical value $F_{.05}(1, 7\text{ df}) = 5.59$, we find that both the main effect of gender and the gender-by-task interaction are significant at the .05 level. Examining the cell means, we conclude that women score higher than men on both tasks, but especially on Task 2 ($P < .05$).

The unweighted-means analysis is also known as the **harmonic-mean method** of handling unequal sample sizes. The name comes from the use of the harmonic mean of the sample sizes, in place of the actual unequal sample sizes, in calculating the sums of squares for main effects and interactions.

The existence of a method for analyzing factorial experiments with unequal sample sizes should not lull you into thinking that such experiments can be handled routinely. You should always consider why the samples are of unequal size. In particular, if it is some characteristic or property of the observational units that has made some data unavailable, then any random assignment to conditions is negated by differential withdrawal rates from the different conditions. For example, the researcher may be left with more highly motivated subjects in some conditions than others, because the less motivated subjects would be more likely to withdraw from the more demanding conditions. Thus, unequal sample sizes may be a symptom of a confounding variable, in spite of what was initially random assignment. Subject withdrawal must be considered to be part of the assignment process, and it may result in a nonrandom assignment. The important point is that we must take into account the reason for the occurrence of unequal sample sizes.

18.5 MONOTONIC TRANSFORMATIONS OF DATA

Data are not always collected in the most convenient or appropriate form for purposes of analysis. For example, price data may be collected in dollars; what difference would it make if they were analyzed in cents? Or racing data may be collected in terms of elapsed time, perhaps in seconds; what difference would it make if they were analyzed in terms of speed, perhaps in feet per second? The process of converting each observation in a set of data to a new value by means of a well-defined mathematical function is known as **transformation of data**.

This section identifies several common transformations and their consequences in the analysis of variance and provides some comments as to when each might be appropriate. All of the transformations in this section are **monotonic transformations**; that is, either they leave the rank order of the observations unchanged (the largest observation remains largest, the second largest remains second largest, and so on), or they completely reverse the rank order (the largest observation becomes the smallest, the second largest becomes the second smallest, and so on).

Linear Transformation

A **linear transformation** consists of changing each observation x to a new value, say, y, by means of a function of the form $y = bx + a$, where b does not equal zero. Examples of linear transformations are addition of a (possibly negative) constant to every observation ($y = x + a$) and conversions from dollars to cents ($y = 100x$), from inches to feet ($y = x/12$), from pounds to kilograms ($y = x/2.2$), and from Celsius degrees to Fahrenheit degrees ($y = 1.8x + 32$).

The consequences of a linear transformation ($y = bx + a$) of the data for an analysis of variance are

a. to change the means by the same transformation, leaving their relative order unchanged, and

b. to multiply each sum of squares and mean square by b^2, but

c. to leave all ANOVA F ratios unchanged (see Section 17.3).

Thus, F ratios are said to be *invariant with respect to linear transformations*, and linear transformations may be applied whenever they would simplify the computations or be otherwise convenient.

Square-Root Transformation

A **square-root transformation** consists of changing each observation x (for $x \geq 0$) to its square root $\sqrt{x}$ or, if some values of x equal zero, perhaps to $\sqrt{x+1}$. It is commonly used for variables recorded as **counted data**, such as number of errors or number of successes. Such data often reflect an underlying Poisson distribution (Section 7.3), where the population variance for each treatment combination equals the corresponding population mean. Thus, treatment combinations that differ in location with counted data are apt to differ also in dispersion. For counted data from a Poisson distribution, the square-root transformation tends to equalize the variances, to improve the approximation of the underlying distribution to the Gaussian, and to permit simpler descriptions of any treatment effects.

Other consequences of this or any other nonlinear transformation of the data for an analysis of variance are

a. that the means must be calculated directly from the transformed data and may not even be in the same relative order as without the transformation and

b. that the sums of squares, mean squares, and F ratios must all be calculated directly from the transformed data and may not show the same effects as without the transformation.

In other words, when transformed data are analyzed, they are treated just as if they had been recorded on the transformed scale in the first place. Indeed, perhaps they should have been!

Logarithmic Transformation

A **logarithmic transformation** consists of changing each observation x (for $x > 0$) to $\log x$ or, if some values of x equal zero, to $\log(x + 1)$. For a given set of data, the logarithmic transformation is a somewhat stronger transformation than the square-root transformation. It is probably the most frequently used nonlinear transformation. It can be useful with counted data and in a wide variety of other applications involving positively skewed distributions. With experience, such data can sometimes be identified by visual inspection, perhaps aided by a box plot (Section 4.5) or other graphical methods. (If box plots for each treatment combination show outside values beyond the upper but not the lower adjacent values, there is evidence of positive skewness in the data.) In many cases with substantial positive skewness, the logarithmic transformation will tend to equalize the variances, to improve the approximation of the underlying distribution to the Gaussian, and to permit simpler descriptions of any treatment effects. The last paragraph in the preceding discussion of the square-root transformation identifies some other consequences of this nonlinear transformation.

The logarithmic transformation is, in fact, widely used, partly because positively skewed distributions are so common in practice. A few variables, such as sound intensity on the decibel scale, seismic disturbance on the Richter scale, and acidity and alkalinity on the pH scale, are routinely measured on a logarithmic scale, but many other variables can also be described more simply after a logarithmic transformation. A skewed distribution may be a sign that the data have been measured on an inappropriate scale, and the researcher should not hesitate to try replacing each observation by its logarithm to see if a better description of the data is possible on the logarithmic scale than in the original units of measurement.

Reciprocal Transformation

A **reciprocal transformation** consists of changing each observation x (for $x > 0$) to its reciprocal $1/x$. For a given set of data, the reciprocal transformation is a somewhat stronger transformation than the logarithmic. Data in which the dependent variable is time (for example, reaction time, time to perform a task, or time to move a certain distance) are often positively skewed. The reciprocal of time is speed (for example, speed of reaction, speed of performance, or speed of movement). The use of the reciprocal transformation often tends to reduce the skewness, to equalize variances, and to permit simpler descriptions of any treatment effects. The last paragraph in the preceding discussion of the square-root transformation identifies some other consequences of this nonlinear transformation.

There are also other variables besides time that have interpretable reciprocals. For example, the reciprocal of gas mileage (usually measured in mi/gal) is fuel consumption (usually measured in liters/100 km); the reciprocal of population density (people/square mile) is relative space (square miles/person); and so on. In each case, a judgment must be made as to which form of the data leads to more appropriate summary statistics for the data at hand. In general, the object is to summarize the data in a way that is informative and not misleading, and data transformations can be a useful tool in accomplishing this goal.

EXAMPLE 18.18 To illustrate the importance of choosing the appropriate transformation, let us consider the following hypothetical data. The original variable x represents the number of arithmetic questions answered correctly by fourth-grade students in one-hour tests.

	x (problems/hour)	$\sqrt{x}$	$\log x$	$1/x$ (hours/problem)
Student A:				
Test 1	100	10	2.00	.010
Test 2	52	7.2	1.72	.019
Test 3	4	2	.60	.250
Mean	52	6.4	1.44	.093
Student B:				
Test 1	64	8	1.81	.016
Test 2	50	7.1	1.70	.020
Test 3	36	6	1.56	.028
Mean	50	7.0	1.69	.021
Student C:				
Test 1	49	7	1.69	.020
Test 2	29	5.4	1.46	.034
Test 3	9	3	.95	.111
Mean	29	5.1	1.37	.055

Which student worked fastest on the average? Student A seems to be slightly faster than Student B if we look under x at the average number of problems per hour; but Student A is the slowest of all three students if we look under $1/x$ at the average number of hours per problem! Speed and time are two different ways of presenting the same data, and yet they seem to lead to different conclusions. More generally, arranging the students from fastest to slowest on the average, we have the following:

A, B, C, for the original data x;

B, A, C, for the square-root transformation $\sqrt{x}$;

B, A, C, for the logarithmic transformation $\log x$; and

B, C, A, for the reciprocal transformation $1/x$.

Which way should the data be presented? The problem is one of how best to represent a set of observations, the same problem that we met in Chapter 3 in comparing different types of averages or measures of central location. There is no easy answer to the question, except that we must try to determine which is the most informative and honest way to summarize the data. The next section includes a simple method that can provide some assistance.

The main point of this illustration is that failure to use the correct transformation can obscure the information that the data might otherwise convey to us.

18.6 ASSUMPTIONS UNDERLYING THE ANOVA F TESTS

The accuracy of the probability values assigned to statements of F-test results depends in large measure on how well certain assumptions of the underlying mathematical model are satisfied. There are four general **assumptions underlying the ANOVA *F* tests**: additivity, normality, homogeneity of variance, and independence.

Additivity

According to the **assumption of additivity**, the effects of the various factors combine with each other and with random error by adding. Such additivity does not always occur in real data, but violations of additivity can often be handled by means of nonlinear, monotonic transformations of the data, such as the square-root, logarithmic, and reciprocal transformations discussed in Section 18.5. For example, a multiplicative error structure can be converted to an additive structure by the use of the logarithmic transformation, whereby the analysis of variance is applied to the logs of the observations rather than to the observations themselves. (Linear transformations will not help here: *F* ratios are invariant with respect to linear transformations.)

Violations of the additivity assumption may often be detected in the form of violations of one or both of the next two assumptions.

Normality

According to the **assumption of normality**, the observations are sampled from a Gaussian population. In many practical applications, data come from positively skewed or otherwise non-Gaussian distributions, but if the researcher takes time to examine the data carefully before analysis, problems due to nonnormality can frequently be avoided by the judicious use of transformations. Moreover, ANOVA *F* tests are rather robust against nonnormality; that is, the results are not seriously distorted by moderate departures from normality.

Homogeneity of Variance

According to the **assumption of homogeneity of variance**, the populations corresponding to the various treatments or treatment combinations all have the same variance. Although ANOVA *F* tests are rather robust against violations of this assumption, heterogeneity of variance generally leads to somewhat too many significant results in the *F* test. Unequal sample sizes greatly exacerbate the problems due to heterogeneity if the smaller sample is more variable. In addition to the use of equal-sized samples, data transformations are again helpful to avoid serious problems.

One simple procedure to assist in choosing a transformation to improve homogeneity of variance involves identifying the largest and the smallest observation in each treatment combination. Perform various possible transformations, such as the square-root transformation, the logarithmic transformation, and the reciprocal transformation, on these two observations from each condition. Then calculate the range in each cell for the original data and for each pair of transformed values. The form of the data in which the ranges in the various cells are most uniform is apt to be the best form for stabilizing the variance across conditions. This method is known as the **range method for comparing transformations**. The results of this procedure should not be accepted unthinkingly, and small differences in the uniformity of the ranges for various forms of the data should be discounted. The procedure does not provide the last word in choosing a transformation, but rather one piece of helpful information. Section 18.5 offers further comments on the choice of data transformation.

EXAMPLE 18.19 The following are the selling prices, as published in the periodical *Arabian Horse World*, of 20 Arabian stallions shown on the Class A circuit, 10 sold in 1978 and 10 sold in 1981.

1978	1981
\$ 125,000	\$ 1,525,000
100,000	290,000
35,000	145,000
25,000	140,000
25,000	105,000
15,000	81,000
11,000	45,000
8,000	40,000
7,000	21,000
5,000	19,000

If we want to compare average prices in the two years, we note first that the dispersions of the prices are very different in the two samples. The extreme values were \$125,000 and \$5,000 in the 1978 sample and \$1,525,000 and \$19,000 in the 1981 sample. Would a data transformation improve homogeneity of variance in these data, and if so, which transformation is apt to be best?

In the range method for comparing transformations, we need transform only the extreme values in each sample to get an indication of what the dispersion would be under a given transformation. The difference between the transformed maximum and the transformed minimum is the range of the transformed data. Taking the difference after various transformations of the extreme values, we obtain the ranges in the two samples as follows.

	x	$\sqrt{x}$	$\log x$	$1/x$
1978 sample:	120,000	283	1.40	.000192
1981 sample:	1,506,000	1,097	1.90	.000052
Range ratio:	13	3.9	1.4	3.7

The ranges of the two samples differ by a factor of $1{,}506{,}000/120{,}000 \simeq 13$ on the original measurement scale, by a factor of $1{,}097/283 \simeq 3.9$ on the square-root scale, by a factor of $1.90/1.40 \simeq 1.4$ on the log scale, and by a factor of $.000192/.000052 \simeq 3.7$ on the reciprocal scale. The variability of the two samples evidently is made most similar by the logarithmic transformation. If there were more than two samples being compared, we would still calculate just one "range ratio" for each transformation—the largest range divided by the smallest range—to see which transformation would make the ranges most uniform.

(Source: Student project by H. Hoyland.)

Independence

According to the **assumption of independence**, the replications of a given treatment or treatment combination are all independent of each other. The score recorded for one observational unit does not provide any information about the scores recorded for other observational units receiving the same treatment combination. Violations of this assumption can result in serious distortion of the F test. ANOVA is *not* robust against violations of the independence assumption, and violations cannot be remedied by numerical manipulations such as data transformations.

Adherence to the independence assumption is largely a matter of good experimental design. The independence assumption is violated if treatment combinations are confounded with some other variable that induces similarity among the observational units in a given treatment combination. For example, if different treatments are compared in intact groups (such as school classes), with each treatment being administered to an entire group, the treatments are confounded with preexisting differences among the groups, and the individuals in the group have something in common other than the treatment. That is, the individuals are not independent replications. In this case, the intact group should perhaps be considered the observational unit, with the mean response of its members as the observation. In general, by careful planning and execution of the experiment, the experimenter can often ensure that the observations are independent of each other, or that the observational units can be redefined in such a way as to make the observations independent.

EXAMPLE 18.20 In an experiment concerning attitude change, six groups of four high school students each took part in discussions of a controversial issue, namely, busing to achieve racial balance in schools. Unknown to the other discussers, one of the members of each group was actually a confederate of the experimenter, with the task of using the discussion to reverse the attitudes that the other group members had previously recorded. In the three groups consisting of antibusing students, the confederate tried to increase probusing sentiment; in the three groups consisting of probusing students, the confederate tried to reduce probusing attitudes. After the discussion, the students were individually asked to recall the precise attitude toward busing that they had recorded on the prediscussion questionnaire.

Although there were 18 bona fide subjects in the two conditions, the six discussion groups (three in each condition) were treated as the observational units. The experimenters explained that "since subjects in groups were interacting, the post-discussion attitude scores of subjects within groups cannot be considered to be independent. Thus, the scores in each . . . group were averaged and the group means were used as the entry for all analyses."

(Source: Goethals & Reckman, 1973.)

STUDY MATERIAL FOR CHAPTER 18

Terms and Concepts

Section 18.1

experimental design
dependent variable
univariate analysis
multivariate analysis
independent variable
classification variable
manipulated variable
factor
source of variation
extraneous variable
controlled variable
confounding variable
experimental group
control group
random assignment
levels of a factor
fixed factor
random factor
random selection
design
complete crossing
treatment combination
factorial design
nesting
nesting factor
nested factor
replication
independent-samples design
independent-samples factor
between-subjects design
block
repeated-measures design
repeated-measures factor
randomized-blocks design
within-subjects design

Section 18.2

variance component
error term
expected mean square
general principle defining the appropriate error term
guidelines for finding the appropriate error term

Section 18.3

cell
three-way ANOVA with independent samples of equal size
ANOVA summary table for three-way ANOVA with independent samples
retrospective study
table of totals
three-way interaction

* Section 18.4

orthogonal
unweighted-means analysis
ANOVA summary table for unweighted-means analysis
harmonic-mean method

Section 18.5

transformation of data
monotonic transformation
linear transformation
square-root transformation
counted data
logarithmic transformation
reciprocal transformation

Section 18.6

assumptions underlying the ANOVA F tests
assumption of additivity
assumption of normality
assumption of homogeneity of variance
range method for comparing transformations
assumption of independence

Key Formulas

Section 18.2

(18.5) $$F_{\text{effect}} = \frac{\text{MS}_{\text{effect}}}{\text{MS}_{\text{error}}}$$

Section 18.3

See Expressions (18.19) through (18.27) and Table 18.1.

*Section 18.4

(18.28) $$\text{SS}_{\text{repl. within cells}} = \sum x^2 - \sum \frac{(\text{cell total})^2}{\text{number of observations in the cell}}$$

(18.29) $$\text{HM}_{n_R} = \frac{\text{number of cells}}{\sum \dfrac{1}{\text{number of observations in a cell}}}$$

Also see Table 18.2.

Exercises for Numerical Practice

Section 18.3

18/1. Use the three-way ANOVA with independent samples to analyze the following data.

	A_1		A_2	
	B_1	B_2	B_1	B_2
C_1	1, 2, 1	3, 0, 2	4, 5, 2	6, 3, 4
C_2	3, 2, 4	4, 5, 8	6, 4, 8	5, 8, 6
C_3	0, 1, 2	3, 1, 2	4, 6, 5	7, 5, 9

a. Complete the appropriate ANOVA summary table.

b. Give the critical value at the 5% significance level for each F ratio in the ANOVA table.

c. What do the results of the F tests tell you about the data?

d. Calculate the estimated standard error of the mean for each of the levels of:

i. Factor A.
ii. Factor B.
iii. Factor C.

18/2. Use the three-way ANOVA with independent samples to analyze the following data.

	A_1			A_2		
	B_1	B_2	B_3	B_1	B_2	B_3
C_1	2, 4	7, 6	7, 9	8, 6	4, 5	1, 3
C_2	5, 7	8, 6	4, 6	3, 1	2, 3	1, 0
C_3	9, 7	5, 6	3, 2	0, 2	3, 1	9, 8

Answer Parts (a) through (d) in Exercise 18/1 for these data.

* Section 18.4

18/3. Perform an unweighted-means ANOVA of the following data from a two-way design with independent samples.

	A_1	A_2	A_3
B_1	9, 7	7, 5, 3	2
B_2	0, 3, 0	2	1, 5

a. Complete the appropriate ANOVA summary table.

b. Give the critical value at the 5% significance level for each F ratio in the ANOVA table.

c. What do the results of the F tests tell you about the data?

18/4. Perform an unweighted-means ANOVA of the following data from a two-way design with independent samples.

	A_1	A_2	A_3
B_1	0, 2, 1	4, 5, 6, 5	9, 7, 8
B_2	6, 6	2, 4, 3	1, 2, 3

Answer Parts (a) through (c) in Exercise 18/3 for these data.

Section 18.5

18/5. Apply the square-root transformation to the following data, and then perform the one-way ANOVA with independent samples:

A_1: 9, 1, 16, 4

A_2: 16, 36, 9, 25

A_3: 64, 81, 36, 49

a. Complete the appropriate ANOVA summary table.

b. Give the critical value for the F test at the 5% significance level.

18/6. Apply the logarithmic transformation (common logarithms, base 10) to the following data, and then perform the one-way ANOVA with independent samples:

A_1: 20, 13, 16

A_2: 25, 40, 32

A_3: 63, 79, 50

a. Complete the appropriate ANOVA summary table.

b. Give the critical value for the F test at the 5% significance level.

Section 18.6

18/7. The following data were recorded in a one-way design with independent samples:

A_1: 10, 5, 15, 1, 10

A_2: 15, 35, 10, 25, 20

A_3: 65, 80, 35, 50, 40

Using the range method for comparing transformations, identify an appropriate transformation for improving homogeneity of variance in these data. You may limit your consideration to the following four forms of the data: original data, square-root transformation, logarithmic transformation, and reciprocal transformation.

18/8. The following data were recorded in a one-way design with independent samples:

A_1: 20, 10, 25, 15, 15

A_2: 25, 45, 20, 35, 30

A_3: 65, 95, 30, 50, 55

Using the range method for comparing transformations, identify an appropriate transformation for improving homogeneity of variance in these data. You may limit your consideration to the following four forms of the data: original data, square-root transformation, logarithmic transformation, and reciprocal transformation.

Problems for Applying the Concepts

18/9. "A mathematician compared two teaching methods by trying them on two different classes for a semester. Analyzing the resulting data as if students had been randomly assigned to the two classes, he concluded that method *A* was significantly better than method *B*." In the analysis performed by the mathematician:

a. What was the dependent variable?

b. List the factors in the experiment, and identify each as fixed or random.

c. Identify any crossing or nesting of factors.

d. Is this an independent-samples design or a repeated-measures design?

e. What is the major confounding variable in the experiment? What difference could it make in the results?

f. **i.** What did the mathematician treat as the observational unit in the analysis?

ii. Which of the ANOVA assumptions is violated in the stated analysis?

iii. What changes in the analysis could be made to avoid the violation, and what further difficulty would be encountered in that case?

(Source: Hooke, 1980.)

18/10. Twenty laboratory rats to be used in an experiment were housed in one large colony cage. When the time came to apply the experimental treatments, a lab technician removed the rats from the cage one at a time. The first 10 were put into a transport cage to be brought to receive Treatment 1, and the second 10 were put into an identical transport cage to receive Treatment 2. After receiving the appropriate treatment, each rat was observed on an activity wheel, and an activity score was recorded as a measure of the effect of the treatment.

a. What was the dependent variable?
b. List the factors in the experiment, and identify each as fixed or random.
c. Identify any crossing or nesting of factors.
d. Is this an independent-samples design or a repeated-measures design?
e. **i.** What is the major confounding variable in the experiment?
ii. What difference could it make in the results?
iii. How could it have been avoided?

(Source: Hebb, 1972, p. 148.)

18/11. An experiment was designed "to find out whether removing the frontal lobes of a monkey's brain affects his ability to learn a visual discrimination." Monkeys were randomly assigned to one of two groups, an experimental group and a control group, and the procedure then was "to test both groups, operate on one, test both groups again, and see whether the increases in score by the normals (due to practice effect) are significantly greater than the increases by the operates."

a. What was the dependent variable?
b. List the factors in the experiment, and identify each as fixed or random.
c. Identify any crossing or nesting of factors.
d. Is this an independent-samples design or a repeated-measures design?
e. **i.** What is the major confounding variable in the experiment?
ii. What difference could it make in the results?
iii. How could it have been avoided?

(Source: Hebb, 1972, pp. 147–48.)

18/12. An investigation was conducted to determine the effectiveness of a study skills program offered by a college's Department of Student Services. The study skills seminars and workshops were advertised throughout the campus and made available without restriction to any student wanting assistance in improving study skills. At the end of the year, two groups of students were identified: The experimental group consisted of all those who took part in the study skills program that year, and the control group contained an equal number of nonparticipating students identified by the registrar's office as being matched with the experimental group on the basis of previous grades and admissions test scores. Grades of the two groups after the study skills program were then compared, and the experimental group was found to have significantly higher grades than the control group.

a. What was the dependent variable?
b. List the factors in the experiment, and identify each as fixed or random.
c. Identify any crossing or nesting of factors.
d. Is this an independent-samples design or a repeated-measures design?
e. **i.** What is the major confounding variable in the experiment?
ii. What difference could it make in the results?
iii. How could it have been avoided?

18/13. Given the following expected mean squares, write the expected value of the appropriate error term to test the corresponding effect in each case:

a. $E(\text{MS}_A) = \sigma^2 + 30\sigma^2_{AC} + 60\sigma^2_A$

b. $E(\text{MS}_C) = \sigma^2 + 60\sigma^2_C$

18/14. Given the following expected mean squares, write the expected value of the appropriate error term to test the corresponding effect in each case:

a. $E(\text{MS}_{AB}) = \sigma^2 + 10\sigma^2_{ABC} + 20\sigma^2_{AB}$

b. $E(\text{MS}_{AC}) = \sigma^2 + 30\sigma^2_{AC}$

18/15. For each of the investigations described in

a. Problem 19/17,

b. Problem 19/19,

c. Problem 19/23

at the end of Chapter 19, answer the following questions:

i. List the factors and identify each as fixed or random.

ii. Identify any crossing and/or nesting of factors.

iii. List the sources of variation for the ANOVA summary table, give the number of degrees of freedom for each source, and identify the appropriate error term for each source.

18/16. For each of the investigations described in

a. Problem 19/16,

b. Problem 19/20,

c. Problem 19/24

at the end of Chapter 19, answer the following questions:

i. List the factors and identify each as fixed or random.

ii. Identify any crossing and/or nesting of factors.

iii. List the sources of variation for the ANOVA summary table, give the number of degrees of freedom for each source, and identify the appropriate error term for each source.

18/17. An investigator studied how color and detail in scenic photographs affect individuals' descriptions of the scenes. Each scene was printed in three levels of color, ranging from black-and-white to full-color. The investigator varied the amount of detail in the scene by displaying the photo at various distances from the subject; six different distances were investigated.

The investigator did not want her results to be limited to the particular scenes included in the study; rather, she wanted to make statements about the effects of color and detail in scenic photographs in general. Therefore, in choosing the five scenes to be displayed in the study, she made an effort to obtain scenic photographs that were representative of a population to which she wished her results to apply. A different group of four individuals was randomly assigned to each particular combination of scene, color, and detail (distance), and each individual viewed only one combination. The dependent variable was the number of words the subject used in his description of the scene.

a. List the factors in this experiment, and identify each as fixed or random.

b. Which factors are crossed and which nested?

c. List the sources of variation for the ANOVA summary table, give the number of degrees of freedom for each source, and identify the appropriate error term for each source.

18/18. An experiment was conducted to compare two different methods of teaching piano in individual lessons to beginning pupils of different ages (6, 7, 8, 9, and 10 years). Four piano teachers took part in the experiment, each one teaching six pupils at each age level, with half of the pupils at each age level being randomly assigned to receive each of the two methods.

The experimenter's primary interest was in comparing the two teaching methods and in determining whether the choice of method should be different for different age levels. Four teachers were included in the study, with the intention of generalizing the results concerning methods and ages to a population of comparable piano teachers. The dependent variable was a rating by an independent examiner of each pupil's progress after one year of lessons. The examiner did not know which method had been used with each pupil.

a. List the factors in this experiment, and identify each as fixed or random.

b. Which factors are crossed and which nested?

c. List the sources of variation for the ANOVA summary table, give the number of degrees of freedom for each source, and identify the appropriate error term for each source.

d. Which effect(s) would be tested to answer the experimenter's question about whether the choice of method should be different for different age levels?

18/19. Experiments in social psychology have found that a person's agreement with a statement of opinion may depend in part on the identity of the speaker or the source of the statement. In an investigation seeking to demonstrate this effect, a sample of 24 people were individually presented with a brief questionnaire containing the statement, "It is a good idea for a young person never to try booze or dope." On half of the questionnaires, the statement was said to be a quotation from the rock star Mick Jagger, and on the other half, a quotation from the evangelist Billy Graham. Half of the subjects were classified as young (college students) and half as old (over 40 years), and half of each age group was male. Subjects in each age–gender combination were randomly assigned to receive one of the two types of questionnaire. They responded to the quotation by circling a number from 0 (strongly disagree) to 10 (strongly agree) as follows.

	Young		Old	
Identity of speaker	**Male**	**Female**	**Male**	**Female**
Mick Jagger	6, 2, 1	1, 4, 9	3, 10, 10	10, 10, 9
Billy Graham	3, 4, 10	0, 10, 4	8, 7, 9	10, 9, 10

Perform an analysis of variance to test the effects of the factors.

a. Identify the replication factor, any other factors, and the dependent variable.

b. Complete the ANOVA summary table and significance tests.

c. Plot the appropriate means for portraying the significant effects, and state what the results indicate about agreement with the quotation.

(Source: Student project by T. Kopperud.)

18/20. The following data are weight losses (in kg) achieved in one year by 36 female volunteers randomly assigned to one of 12 combinations of diet, exercise program, and social context of the weight-loss program.

	Daily exercises		No prescribed exercises	
Diet	**Group support**	**No group support**	**Group support**	**No group support**
1	9, 6, 8	5, 8, 7	2, 3, 4	0, 2, 1
2	8, 8, 5	6, 9, 7	2, 5, 3	4, 0, 0
3	7, 6, 9	7, 8, 6	3, 1, 4	1, 1, 0

Perform an analysis of variance to test the effects of the factors.

a. Complete the ANOVA summary table and significance tests.

b. Plot the appropriate means for portraying the significant effects, and state what the results indicate about weight loss.

18/21. A car manufacturer conducted an experiment to compare fuel consumption (in liters/100 km) for the six types of vehicles defined by crossing type of fuel (gasoline or diesel) and type of transmission (three-speed automatic, $A3$; four-speed manual, $M4$; or five-speed manual, $M5$). Four vehicles of each type were chosen from the production line, and two of each type were randomly assigned to a city-driving test and the other two to a highway-driving test. All vehicles were subcompacts with a four-cylinder 1.6-liter engine. Fuel consumption was as follows for the 24 vehicles.

	Diesel		Gasoline	
Transmission	**City**	**Highway**	**City**	**Highway**
$A3$	7.6, 7.0	5.2, 5.0	10.0, 9.8	7.0, 6.8
$M4$	5.7, 5.3	3.6, 3.6	8.8, 9.0	5.1, 5.3
$M5$	6.6, 6.2	3.8, 4.0	9.1, 9.3	5.7, 5.5

Perform an analysis of variance to test the effects of the factors.

a. Complete the ANOVA summary table and significance tests.

b. Plot the appropriate means for portraying the significant effects, and state what the results indicate about fuel consumption.

18/22. What factors influence the relative humidity in a house? An experiment was conducted to examine the effects of showers and fireplace fires on the relative humidity in three levels of a split-level dwelling. Hygrometers were placed near the fireplace on the family-room level, in the living room on the middle level, and outside the bathroom on the top level. Readings were taken simultaneously on the three hygrometers, either after a fire had been burning for 30 minutes in the fireplace or after no fire, and either 15 minutes after a 15-minute running of the shower in the bathroom or after no shower. Each condition was replicated four times, yielding the following relative humidity readings (in %).

Level	No shower		Shower	
	No fire	Fire	No fire	Fire
Top	62, 63, 64, 72	55, 60, 63, 55	76, 71, 80, 72	80, 64, 76, 64
Middle	60, 68, 68, 72	64, 69, 64, 64	75, 72, 76, 68	72, 72, 72, 73
Bottom	71, 67, 72, 65	58, 56, 51, 65	71, 62, 62, 67	78, 72, 78, 69

Perform a three-way ANOVA for independent samples to test the effects of the factors.

a. Identify the replication factor, any other factors, and the dependent variable.

b. Complete the ANOVA summary table and significance tests.

c. Plot the appropriate means for portraying the significant effects, and state what the results indicate about relative humidity in this particular house.

(Source: Student project by H. Schnepf.)

*__18/23.__ Eighteen college students were classified according to their faculty (arts or commerce) and their level of athletic participation (none, intramural, or intercollegiate). The following grade point averages were recorded (on a 9-point scale where 6 is average and 4 is the minimum passing grade) for the 18 students.

Faculty	Participation in athletics		
	None	Intramural	Intercollegiate
Arts	5.8, 7.2	5.0, 6.1, 5.8, 7.3	4.3, 5.2, 7.5
Commerce	5.6, 6.6, 7.8	6.5, 7.2, 6.5, 7.3	7.2, 5.2

Perform an unweighted-means ANOVA to test the effects of the factors.

a. Complete the ANOVA summary table and significance tests.

b. Plot the appropriate means for portraying the significant effects, and state what the results indicate about grade point averages in the populations from which these observations represent random samples.

(Source: Student project by B. Kornfeld.)

*__18/24.__ A sample of 19 college students, cross-classified by age and sex, reported their hourly wage rates (in dollars) for summer employment in 1985 as follows:

	18-year-olds	19-year-olds	20-year-olds
Females	4.50	8.72	5.00
	4.50	5.00	5.85
	4.50	3.80	6.60
Males	8.50	4.75	9.10
	17.23	9.00	4.00
	7.00	11.69	
		6.25	
		7.50	

Perform an unweighted-means ANOVA to test the effects of the factors.

a. Complete the ANOVA summary table and significance tests.

b. Plot the appropriate means for portraying any significant effects, and state what the results indicate about summer wage rates for students.

(Source: Student project by D. Sweeney.)

***18/25.** Twenty-six individuals were interviewed in a study of attitudes toward equal treatment of men and women in the workplace. The participants were classified by age (under 30 years and over 30 years) and sex, and they were randomly assigned to receive either a message arguing in favor of equality or no message. After the message (if any), they rated their attitude toward equality for women as follows on a 7-point scale, where 7 indicates approval.

	Males		Females	
	Under 30	**Over 30**	**Under 30**	**Over 30**
Message	4, 6, 4, 5	4, 1	2, 7, 5, 6	7, 2
No message	3, 5, 4, 6, 3	1, 2	3, 4, 7, 7, 5	3, 2

Perform an unweighted-means ANOVA to test the effects of the factors.

a. Complete the ANOVA summary table and significance tests.

b. Plot the appropriate means for portraying the significant effects, and state what the results indicate about attitudes.

(Source: Student project by M. Seto.)

***18/26.** Opinions were collected from a sample of 20 students randomly selected from the two sections of a large introductory course. The instructor of one section taught in a lecture format, whereas the instructor of the other section adopted more of a discussion format. The purpose of the investigation was to determine whether a lecture format or a discussion format was preferred, and whether male and female students differed in their preferences. The following ratings were recorded on a scale from 0 (disapproval) to 9 (approval).

	Lecture format	Discussion format
Males	7, 4, 5, 1, 6, 4, 6	7, 7
Females	7, 6, 7, 7	4, 5, 6, 7, 7, 6, 8

Perform an unweighted-means ANOVA to test the effects of the factors.

a. Complete the ANOVA summary table and significance tests.

b. Plot the appropriate means for portraying the significant effects, and state what the results indicate about attitudes.

c. What are the major confounding variables in this investigation? What difference could they make in the results?

(Source: Student project by P. Kwok.)

18/27. In a study of hand-eye coordination, male and female subjects performed a star-tracing task, with a moderate time limit, while viewing the star either directly or in a mirror. Any deviation of more than 2 mm from the prescribed line was counted as an error. The following numbers of errors were recorded for 12 subjects in each condition.

	With mirror	No mirror
Females	20, 24, 19, 15, 22, 17, 17, 16, 19, 21, 20, 23	3, 3, 4, 2, 3, 1, 1, 1, 2, 2, 4, 2
Males	21, 22, 23, 20, 19, 17, 19, 19, 22, 18, 26, 20	1, 3, 2, 1, 2, 3, 4, 2, 4, 2, 2, 2

Using the range method for comparing transformations, identify an appropriate transformation for improving homogeneity of variance in these data. You may limit your consideration to the following four forms of the data: original data, square-root transformation, logarithmic transformation, and reciprocal transformation.

(Source: Student project by B. Schultz.)

18/28. The following data show the purchase weight and selling weight (in pounds) of four steers and four heifers randomly selected from recent sales by a cattle-feeding operation.

Steers		Heifers	
Purchase	Selling	Purchase	Selling
528	1250	485	1063
471	1371	503	997
494	1119	491	1025
495	1181	477	984

Using the range method for comparing transformations, identify an appropriate transformation for improving homogeneity of variance in these data. You may limit your consideration to the following four forms of the data: original data, square-root transformation, logarithmic transformation, and reciprocal transformation.

(Source: Student project by B. Dixon.)

References

Dube, E. F. "Literacy, Cultural Familiarity, and 'Intelligence' as Determinants of Story Recall." In *Memory Observed,* edited by U. Neisser. San Francisco: W. H. Freeman, 1982.

Goethals, G. R., and Reckman, R. F. "The Perception of Consistency in Attitudes." *Journal of Experimental Social Psychology* 9 (1973): 491–501.

Harshbarger, B. "An Example of How a Designed Experiment Saved Research Workers from False Recommendations." *The American Statistician* 28 (1974): 128–29.

Hebb, D. O. *Textbook of Psychology.* 3rd ed. Philadelphia: Saunders, 1972.

Hooke, R. "Getting People to Use Statistics Properly." *The American Statistician* 34 (1980): 39–42.

Further Reading

Cochran, W. G., and Cox, G. M. *Experimental Designs.* 2nd ed. New York: Wiley, 1957. The first three chapters of this standard reference work discuss general principles in designing and analyzing experiments, and Chapters 4 through 14 deal with a wide array of specific designs.

Cox, D. R. *Planning of Experiments.* New York: Wiley, 1958. This book provides an excellent discussion of the important concepts in experimental design, with emphasis on logic and understanding rather than on computations.

Hooke, R. *How to Tell the Liars from the Statisticians.* New York: Marcel Dekker, 1983. In a series of light, nonmathematical essays of one to three pages each, Hooke explains and illustrates a variety of statistical concepts. Essays 59–74 relate to experimental design.

Keppel, G. *Design and Analysis: A Researcher's Handbook.* 2nd ed. Englewood Cliffs, NJ: Prentice-Hall, 1982. Keppel provides more advanced and thorough coverage of the topics introduced in this chapter.

Kruskal, J. B. "Transformations of Data." *International Encyclopedia of Statistics.* Edited by W. H. Kruskal and J. M. Tanur. New York: The Free Press, 1978, pages 1044–56. This article is an excellent reference concerning data transformations.

Tatsuoka, M. M. *Multivariate Analysis.* New York: Wiley, 1971. This book is one of the more readable introductions to multivariate analysis.

Winer, B. J. *Statistical Principles in Experimental Design.* 2nd ed. New York: McGraw-Hill, 1971. Most of the more advanced topics referred to in this chapter can be found here.

CHAPTER 19

ANOVA FOR REPEATED-MEASURES DESIGNS

EXAMPLE 19.1 A number of researchers are studying the experience of pain with a view to finding better ways to alleviate pain in situations such as serious illness, accident, and surgery. As an example, let us consider an investigation of the effectiveness of fentanyl, a short-acting narcotic, in reducing dental pain.

Twenty dental patients who were scheduled to have a wisdom tooth extracted participated in the portion of the research described here. In a preliminary test, a series of electric stimuli were applied to the patient's tooth to determine the amount of current corresponding to his pain threshold and maximum pain tolerance levels. Then each patient was randomly assigned to one of two drug conditions involving the intravenous administration of either fentanyl (for the experimental group) or a saline solution as a **placebo** (for the control group). A **double-blind procedure** was used, whereby neither the patient nor the experimenter administering the treatment knew which condition the patient had been assigned to. In the experiment itself, each patient received seven levels of electrical stimulation of the tooth, ranging from pain threshold to maximum pain tolerance, as previously established. After each stimulus, the patient described the intensity of the pain on a specially constructed type of rating scale.

The experiment thus investigated the effects of two independent variables, or two fixed factors: drug condition and stimulus magnitude. The two levels of drug condition were crossed with the seven levels of stimulus magnitude, and all fourteen treatment combinations were observed. Drug condition was an independent-samples factor (different groups of ten patients were observed for the fentanyl and placebo conditions), and stimulus magnitude was a repeated-measures factor (each patient received all seven levels of stimulation). The design may be described as a 2-by-7 factorial (condition-by-stimulus) with repeated measurements on the stimulus-magnitude factor. The dependent variable was the rating of pain intensity.

The researchers reported the results as follows: "A two-way analysis of variance with repeated measures [on one factor] . . . showed that in comparison to the responses after placebo administration, sensory responses were significantly reduced after the administration of fentanyl ($P < .05$ . . .)."

(Source: Gracely *et al.*, 1979.)

Many experiments, such as that in Example 19.1, use repeated-measures designs. As defined in Section 18.1, a repeated-measures design is one in which the replication factor is crossed with a fixed factor. Each replicate, or block, is observed repeatedly, once (or more) under each level of the fixed factor, which is therefore called a *repeated-measures factor*.

The first three sections of this chapter look at three particular types of repeated-measures designs. During the course of the discussion, we will see how these repeated-measures designs relate to independent-samples designs and to each other. The fourth section of the chapter examines some special considerations involved in the design and analysis of repeated-measures experiments.

EXAMPLE 19.2 To facilitate a comparison among various experimental designs, this chapter includes several different analyses of the same set of numerical data. Hypothetical data are used for these comparative analyses, since only one of the analyses would be appropriate for any given set of real data. The data are the following 24 hypothetical observations:

$$\begin{matrix} 8 & 6 & 5 & 5 & 1 & 2 \\ 9 & 9 & 6 & 7 & 5 & 1 \\ 8 & 5 & 6 & 4 & 2 & 0 \\ 8 & 7 & 7 & 6 & 2 & 2 \end{matrix}$$

The appropriate analysis for these (or any other) data depends on the design of the experiment that produced the data. In this chapter, we will consider several designs that might have given rise to the preceding data. Given only a list of the $n = 24$ observations, we can calculate the grand mean as

$$\bar{x} = \frac{121}{24} = 5.04$$

The total variation of the observations around the grand mean is

$$\text{SS}_{\text{total}} = \sum x^2 - \frac{(\sum x)^2}{n} = 779 - 610.04 = 168.96$$

with

$$\text{df}_{\text{total}} = n - 1 = 23 \text{ degrees of freedom}$$

Recall from Section 15.1 that the sample variance is

$$s^2 = \frac{\text{SS}_{\text{total}}}{\text{df}_{\text{total}}}$$

This overall sample variance s^2 is the variance to be analyzed, or broken down into its constituent parts, in the analysis of variance. As we consider various designs in this chapter, we will compare the corresponding ways in which the total variation of 168.96 with 23 degrees of freedom in these data would be partitioned into the identifiable sources of variation in the design.

19.1 ONE-WAY ANOVA WITH REPEATED MEASUREMENTS OF THE SAME REPLICATES

The **one-way ANOVA with repeated measurements** was described in detail in Section 15.3. It is reviewed here in light of the concepts introduced in Chapter

18. The design is illustrated by a real-data application in Example 15.8, with calculations shown in Example 15.12.

In a one-way design with repeated measurements, two or more treatments—levels of a fixed factor—are compared at a number of levels of a random factor, called *blocks*. Equivalently, we can say that two or more treatments are compared within each of a number of blocks. The blocks are the replicates, and the treatments are crossed with the blocks, implying that each block or replicate is observed repeatedly, once for each treatment. If there are k_T treatments and k_R replications (blocks), then there are $n = k_T k_R$ observations altogether.

Computational Formulas and the ANOVA Summary Table

Three sources of variation may be identified in a one-way design with repeated measurements: variation due to differences between replications, variation due to differences between the treatments, and variation due to the treatment-by-replication interaction, which reflects the extent to which the differences between treatments are different in the different replications. The general form of the **ANOVA summary table for one-way ANOVA with repeated measurements** of the same replicates is shown in Table 19.1.

The sums of squares in Table 19.1 may be calculated from the formulas given in Section 15.3 for the two-way ANOVA without replication of treatment combinations, where treatments were labeled as Factor *A* and replications as Factor *B*. Equivalently, the formulas in Section 15.4 for the two-way ANOVA with independent samples may be used, since the one-way design with repeated measurements is formally identical to a two-way factorial arrangement with $n_R = 1$ observation per cell. With only $n_R = 1$ observation per cell, there is no variation within cells, and there are $n_R - 1 = 0$ degrees of freedom for such variation in each cell. Thus, the sources of variation in Table 19.1 do not include a line for within-cell variation.

Appropriate error terms for *F* ratios in Table 19.1 may be determined by the guidelines in Expression (18.18). Since the treatment factor is crossed with exactly one random factor (replications), the appropriate error term for treatments is MS_{TR}. In the usual situation, the treatments constitute a fixed factor, in which case there is no appropriate error term for testing replications. Likewise, there is no error term for the interaction. The fact that the effect of

TABLE 19.1 Analysis-of-variance summary table for one-way ANOVA with repeated measurements.

Source of variation	SS	df	MS	F
Replications (*R*)	SS_R	$k_R - 1$	MS_R	—
Treatments (*T*)	SS_T	$k_T - 1$	MS_T	$F_T = MS_T/MS_{TR}$
T-by-*R* interaction	SS_{TR}	$(k_T - 1)(k_R - 1)$	MS_{TR}	—
Total	SS_{total}	$n - 1$		

only one factor—the repeated-measures factor—can be tested explains the designation of this design as one-way with repeated measurements.

The steps in performing a one-way ANOVA with repeated measurements are summarized on pages 533–535.

Relationship to Independent-Samples ANOVA

The one-way ANOVA with repeated measurements and the one-way ANOVA with independent samples are two ways of testing the differences between means corresponding to the different levels of one independent variable, or fixed factor. To decide which analysis is appropriate, we must consider the design of the experiment. If each replicate (block) receives every level of the fixed factor, it is a repeated-measures design, and the total variation may be analyzed into the three sources shown in Table 19.1:

$$SS_{\text{total}} = SS_T + SS_R + SS_{TR} \tag{19.1}$$

with

$$df_{\text{total}} = df_T + df_R + df_{TR} \tag{19.2}$$

On the other hand, if each observation is independent of every other observation, so that each level of the fixed factor is observed in an independent group of replicates, it is an independent-samples design, and the total variation can be separated into only two sources:

$$SS_{\text{total}} = SS_T + SS_{R(T)} \tag{19.3}$$

with

$$df_{\text{total}} = df_T + df_{R(T)} \tag{19.4}$$

The repeated-measures analysis thus partitions the same total variation one step further than the independent-samples analysis. Specifically, Expressions (19.1) through (19.4) show that the repeated-measures analysis partitions the variation due to replications within treatments into a main effect of replications and a treatment-by-replication interaction:

$$SS_{R(T)} = SS_R + SS_{TR} \tag{19.5}$$

and

$$df_{R(T)} = df_R + df_{TR} \tag{19.6}$$

EXAMPLE 19.3 Let us suppose first that the data presented in Example 19.2 came from an experiment comparing $k_T = 6$ treatments with an independent set of $n_R = 4$ subjects receiving each treatment, for a total of $n = 24$ subjects. The data are as follows.

Treatment 1	Treatment 2	Treatment 3	Treatment 4	Treatment 5	Treatment 6
8	6	5	5	1	2
9	9	6	7	5	1
8	5	6	4	2	0
8	7	7	6	2	2

Using the formulas in Section 15.2, you should verify the ANOVA summary in the following table.

Analysis of the hypothetical data of Example 19.2 as if they came from a one-way design with independent samples.

Source of variation	SS	df	MS	F	
Treatments (T)	140.71	5	28.14	17.9	$P < .01$
Replications within T	28.25	18	1.57	—	
Total	168.96	23			

Now let us suppose that the data presented in Example 19.2 came from an experiment where the $k_T = 6$ treatments constituted a repeated-measures factor, such that each of the $k_R = 4$ subjects received all 6 of the treatments. In this case, the $n = 24$ observations all came from just 4 different subjects, each one observed repeatedly, once under each treatment. If the subjects (or, more generally, the replicates) are designated as R_1, R_2, R_3, and R_4, the data are as follows.

	Treatment 1	Treatment 2	Treatment 3	Treatment 4	Treatment 5	Treatment 6
R_1	8	6	5	5	1	2
R_2	9	9	6	7	5	1
R_3	8	5	6	4	2	0
R_4	8	7	7	6	2	2

Using the formulas in either Section 15.3 or 15.4, you should verify the ANOVA summary in the following table.

Analysis of the hypothetical data of Example 19.2 as if they came from a one-way design with repeated measurements.

Source of variation	SS	df	MS	F	
Replications (R)	14.46	3	4.82	—	
Treatments (T)	140.71	5	28.14	30.6	$P < .01$
T-by-R interaction	13.79	15	.92		
Total	168.96	23			

Notice that when replications are crossed with treatments, the variation due to replications within treatments (28.25 with 18 df from the independent-samples ANOVA table) can be partitioned into variation due to differences among the four replicates (14.46 with 3 df) and variation due to the treatment-by-replication interaction (13.79 with 15 df). The variation due to the main effect of replications is thus removed from the error term in the repeated-measures analysis.

We have applied two different analyses to the same set of numerical data. The numerical observations themselves do not determine which analysis is appropriate. To choose the correct method of analysis, we must know how the data were collected: from independent samples or, alternatively, with repeated measurements of the same replicates.

19.2 TWO-WAY ANOVA WITH TWO REPEATED-MEASURES FACTORS

In a two-way design with two repeated-measures factors, each replicate is observed in all of the treatment combinations of a two-way factorial design. The replication factor (blocks) is crossed with two other crossed factors, say, Factors A and B, so that each block constitutes a replication of the entire two-way array of A-by-B treatment combinations. Each block or replicate is observed repeatedly, once for each treatment combination. If there are k_A levels of Factor A, k_B levels of Factor B, and k_R replications (blocks), then there are $n = k_A k_B k_R$ observations altogether.

EXAMPLE 19.4 An experiment was conducted to investigate the action of yeast in homemade white bread. Ingredients for six loaves were simultaneously prepared according to a standard recipe, except for the amount of yeast: The batch of dough for two of the loaves contained the standard amount of yeast (2 T), another two-loaf batch contained half the yeast (1 T), and the third two-loaf batch contained double the standard amount of yeast (4 T). The amount of time allowed for the first rising of the dough was also manipulated: Each two-loaf batch was divided in half, with half being allowed to rise for two hours according to the recipe and the other half being allowed to rise for three hours. After the first rising, the standard recipe was followed for all loaves: The dough was punched down, put into regular-size bread pans, allowed to rise for one hour, and baked. The dependent variable was the height of the loaf after baking, with height defined as the mean of three height measurements taken at particular points near the two ends and the middle of the loaf. The entire procedure was repeated for four replications, yielding a total of 24 observations recorded as follows in millimeters.

Length of first rising:	**2 hours**			**3 hours**		
Amount of yeast per batch:	**1 T**	**2 T**	**4 T**	**1 T**	**2 T**	**4 T**
Replication:						
1	98	93	96	100	93	79
2	97	95	96	96	89	76
3	95	91	95	94	91	80
4	98	90	93	98	93	77

Length of first rising and amount of yeast are fixed factors, and the replications are the blocks, a random factor. All three factors are crossed, and hence the design is a 2-by-3 factorial (length-by-amount) with two repeated-measures factors.

(Source: Student project by D. Saltvold.)

Computational Formulas and the ANOVA Summary Table

A **two-way ANOVA with two repeated-measures factors** involves three completely crossed factors: the two repeated-measures factors (which are typically fixed factors, say, A and B) and the random replication factor (say, R). Because of the complete crossing, seven sources of variation may be identified in such a design: three main effects (A, B, and R), three two-way interactions (A-by-B, A-by-R, and B-by-R), and one three-way interaction (A-by-B-by-R).

The two-way ANOVA with two repeated-measures factors requires no new sum-of-squares formulas, because the design is formally identical to a three-way factorial arrangement with one observation per cell. The sums of squares may be obtained from Expressions (18.19) through (18.25) by treating the replication factor here as Factor C in the previous formulas and setting $n_R = 1$ observation per cell. These substitutions result in Expressions (19.7) through (19.13) below, as you can verify by comparing with Expressions (18.19) through (18.25). As in the one-way repeated-measures analysis, there is no variation within cells, since there is only one observation in each cell. The following computational formulas thus merely translate the three-way formulas from Section 18.3 into the notation of the **two-way ANOVA with two repeated-measures factors**:

$$\text{SS}_A = \frac{1}{k_B k_R} \sum T_{a..}^2 - \frac{T^2}{n} \tag{19.7}$$

$$\text{SS}_B = \frac{1}{k_A k_R} \sum T_{.b.}^2 - \frac{T^2}{n} \tag{19.8}$$

$$\text{SS}_R = \frac{1}{k_A k_B} \sum T_{..r}^2 - \frac{T^2}{n} \tag{19.9}$$

$$\text{SS}_{AB} = \frac{1}{k_R} \sum \sum T_{ab.}^2 - \text{SS}_A - \text{SS}_B - \frac{T^2}{n} \tag{19.10}$$

$$\text{SS}_{AR} = \frac{1}{k_B} \sum \sum T_{a.r}^2 - \text{SS}_A - \text{SS}_R - \frac{T^2}{n} \tag{19.11}$$

$$\text{SS}_{BR} = \frac{1}{k_A} \sum \sum T_{.br}^2 - \text{SS}_B - \text{SS}_R - \frac{T^2}{n} \tag{19.12}$$

$$\text{SS}_{ABR} = \sum \sum \sum x_{abr}^2 - \text{SS}_A - \text{SS}_B - \text{SS}_R - \text{SS}_{AB} - \text{SS}_{AR} - \text{SS}_{BR} - \frac{T^2}{n} \tag{19.13}$$

The general form of the **ANOVA summary table for the two-way ANOVA with two repeated-measures factors** is shown in Table 19.2. The table shows the appropriate error terms for the usual case, where A and B are both fixed factors and the replication factor is a random factor. In this case, the appropriate error terms for A, B, and the A-by-B interaction are the mean squares for their respective interactions with the replication factor. [Appropriate error terms may be determined by the guidelines in Expression (18.18).] Table 19.2 shows the sources of variation ordered so that error terms are listed immediately after the effects for which they are appropriate.

TABLE 19.2 Analysis-of-variance summary table for two-way ANOVA with two repeated-measures factors.

Source of variation	SS	df	MS	F
Replications (R)	SS_R	$k_R - 1$	MS_R	—
Factor A (fixed)	SS_A	$k_A - 1$	MS_A	$F_A = MS_A/MS_{AR}$
A-by-R interaction	SS_{AR}	$(k_A - 1)(k_R - 1)$	MS_{AR}	—
Factor B (fixed)	SS_B	$k_B - 1$	MS_B	$F_B = MS_B/MS_{BR}$
B-by-R interaction	SS_{BR}	$(k_B - 1)(k_R - 1)$	MS_{BR}	—
A-by-B interaction	SS_{AB}	$(k_A - 1)(k_B - 1)$	MS_{AB}	$F_{AB} = MS_{AB}/MS_{ABR}$
A-by-B-by-R interaction	SS_{ABR}	$(k_A - 1)(k_B - 1)(k_R - 1)$	MS_{ABR}	—
Total	SS_{total}	$n - 1$		

EXAMPLE 19.5 Example 19.4 presented heights of bread loaves for three different amounts of yeast and two different lengths of first rising, in four replications. The ANOVA calculations are facilitated by recording **tables of totals** that are required in the various sum-of-squares formulas.

Length-by-amount totals (and marginal totals)

Amount	Length 2 h	3 h	Marginal totals
1 T	388	388	776
2 T	369	366	735
4 T	380	312	692
Marginal totals	1,137	1,066	2,203

Length-by-replication totals (and marginal totals)

Replication	Length 2 h	3 h	Marginal totals
1	287	272	559
2	288	261	549
3	281	265	546
4	281	268	549
Marginal totals	1,137	1,066	2,203

Amount-by-replication totals (and marginal totals)

Replication	Amount 1 T	2 T	4 T	Marginal totals
1	198	186	175	559
2	193	184	172	549
3	189	182	175	546
4	196	183	170	549
Marginal totals	776	735	692	2,203

The sums of squares are then calculated from Expressions (19.7) through (19.13):

$$SS_{length} = \frac{1{,}137^2 + 1{,}066^2}{12} - \frac{2{,}203^2}{24} = 202{,}427.083 - 202{,}217.042 = 210.042$$

$$SS_{amount} = \frac{776^2 + 735^2 + 692^2}{8} - 202{,}217.042 = 441.083$$

$$SS_{repl.} = \frac{559^2 + 549^2 + 546^2 + 549^2}{6} - 202{,}217.042 = 16.125$$

$$\text{SS}_{LA} = \frac{388^2 + \cdots + 312^2}{4} - 210.042 - 441.083 - 202{,}217.042 = 369.083$$

$$\text{SS}_{LR} = \frac{287^2 + \cdots + 268^2}{3} - 210.042 - 16.125 - 202{,}217.042 = 19.792$$

$$\text{SS}_{AR} = \frac{198^2 + \cdots + 170^2}{2} - 441.083 - 16.125 - 202{,}217.042 = 20.250$$

$$\begin{aligned}\text{SS}_{LAR} &= (98^2 + \cdots + 77^2) - 210.042 - 441.083 - 16.125 \\ &\quad - 369.083 - 19.792 - 20.250 - 202{,}217.042 = 11.583\end{aligned}$$

The ANOVA summary table may now be completed as follows.

Source of variation	SS	df	MS	F	
Replications (R)	16.12	3	5.38	—	
Length of first rising (L)	210.04	1	210.04	31.8	$P < .025$
L-by-R	19.79	3	6.60	—	
Amount of yeast (A)	441.08	2	220.54	65.3	$P < .01$
A-by-R	20.25	6	3.38	—	
L-by-A	369.08	2	184.54	95.6	$P < .01$
L-by-A-by-R	11.58	6	1.93	—	
Total	1,087.96	23			

Computer output from BMDP program 2V for this example is shown in Figure 19.1. The inclusion of the grand mean as a source of variation with one degree of freedom was explained in Example 15.12. The printout identifies the replication factor and its interactions simply as four error terms. The "tail probability" is the P-value, and the last two columns on the right are refinements of the P-value that relate to the topic of the first subsection of Section 19.4.

Since the interaction is significant, interpretation begins with an examination of the cell means, rather than the marginal means, in a table of means.

Mean loaf heights (mm) (including marginal means)

Amount of yeast	Length of first rising 2 h	3 h	Marginal means
1 T	97.0	97.0	97.0
2 T	92.2	91.5	91.9
4 T	95.0	78.0	86.5
Marginal means	94.8	88.8	91.8

Figure 19.2 shows the length-by-amount interaction in a plot of the six cell means. The significant interaction indicates that the effect of varying the amount of yeast depends on the length of the first rising time: Amount of yeast (within the range studied here) has little effect if the standard 2-hour first-rising time is used, but increasing yeast yields flatter loaves after the second rising if the first-rising time is too long. Notice that neither excessive yeast nor excessive first-rising time is sufficient by itself to reduce final

ANALYSIS OF VARIANCE FOR 1-ST
DEPENDENT VARIABLE - HEIGHT OF THE LOAF AFTER BAKING

	SOURCE	SUM OF SQUARES	DEGREES OF FREEDOM	MEAN SQUARE	F	TAIL PROB.	GREENHOUSE GEISSER PROB.	HUYNH FELDT PROB.
	MEAN	202217.04167	1	202217.04167	37621.77	0.0000		
1	ERROR	16.12500	3	5.37500				
	LENGTH	210.04167	1	210.04167	31.84	0.0110		
2	ERROR	19.79167	3	6.59722				
	AMOUNT	441.08333	2	220.54167	65.35	0.0001	0.0029	0.0018
3	ERROR	20.25000	6	3.37500				
	LA	369.08333	2	184.54167	95.59	0.0000	0.0003	0.0000
4	ERROR	11.58333	6	1.93056				

Figure 19.1
BMDP2V printout for Example 19.5.

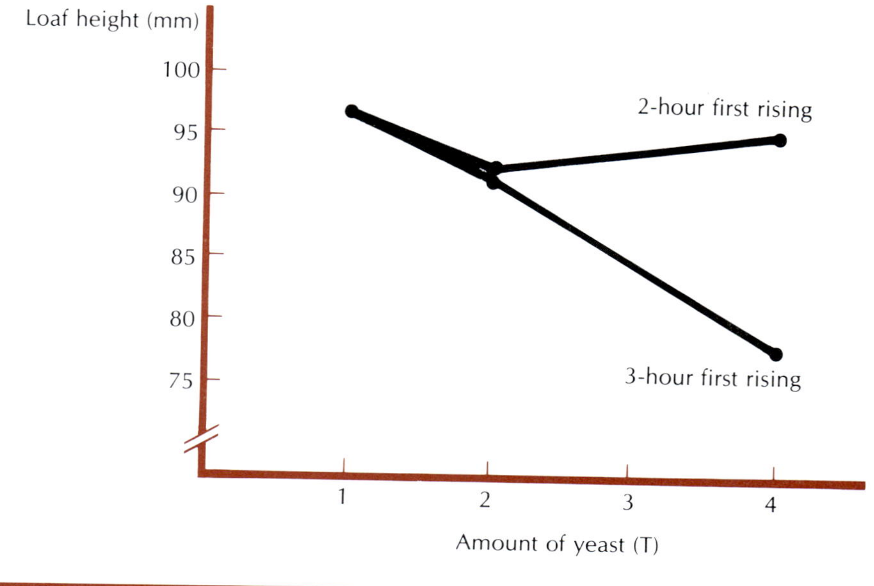

Figure 19.2 Mean loaf height as a function of amount of yeast for two different lengths of first rising. The standard recipe called for 2 T yeast and 2 hours first-rising time.

loaf height substantially; only their combination is—and thus we have an interaction effect.

Summary of the Test Procedure

The **two-way analysis of variance with two repeated-measures factors** may be summarized as follows:

Assumptions: See Sections 18.6 and 19.4.

Step 1. Determine the values of k_A, k_B, k_R, and n.

Step 2. Calculate the tables of totals (and their marginals) for A-by-B, A-by-R, and B-by-R.

Step 3. Calculate the table of means (and marginal means) corresponding to the A-by-B table of totals, in order to see what sample differences are being tested.

Step 4. Calculate the values required to complete the ANOVA summary table in the format of Table 19.2.

Step 5. Compare the observed values of the test statistics F_A, F_B, and F_{AB} with critical values from the F distribution with appropriate degrees of freedom.

a. If the interaction is significant, examine the treatment-combination means $\bar{x}_{ab.}$ to discover the nature of the effect, and interpret significant main effects only insofar as they are compatible with the treatment-combination means.

b. If the interaction is not significant but at least one main effect is, examine the relevant marginal means to discover the nature of the effect.

Further statistical tests to aid in interpretation are considered in Chapter 20.

Relationship to Independent-Samples ANOVA

The two-way ANOVA with two repeated-measures factors and the two-way ANOVA with independent samples are two ways of testing the main effects and interaction of two crossed factors. To decide which analysis is appropriate, we must consider the design of the experiment. If each replicate (block) receives every combination of the levels of the factors, it is a repeated-measures design, and the total variation may be analyzed into the seven sources shown in Table 19.2:

$$SS_{total} = SS_A + SS_B + SS_{AB} + SS_R + SS_{AR} + SS_{BR} + SS_{ABR} \tag{19.14}$$

with

$$df_{total} = df_A + df_B + df_{AB} + df_R + df_{AR} + df_{BR} + df_{ABR} \tag{19.15}$$

On the other hand, if each observation is independent of every other observation, so that each treatment combination is observed in an independent group of replicates, it is an independent-samples design, and the total variation can be separated into only four sources:

$$SS_{total} = SS_A + SS_B + SS_{AB} + SS_{R(AB)} \tag{19.16}$$

with

$$df_{total} = df_A + df_B + df_{AB} + df_{R(AB)} \tag{19.17}$$

The repeated-measures analysis thus breaks down the same total variation further than does the independent-samples analysis. Specifically, Expressions (19.14) through (19.17) show that the repeated-measures analysis partitions the variation due to replications within cells into a main effect of replications and three interactions with replications:

$$SS_{R(AB)} = SS_R + SS_{AR} + SS_{BR} + SS_{ABR} \tag{19.18}$$

and

$$df_{R(AB)} = df_R + df_{AR} + df_{BR} + df_{ABR} \tag{19.19}$$

EXAMPLE 19.6 Example 19.2 presented 24 hypothetical observations with a total variation of $SS_{total} = 168.96$ based on $df_{total} = 23$ degrees of freedom. In Example 19.3, we analyzed these data as if they represented four replications of each of six treatments, first breaking the variation down into two sources by assuming an independent-samples design and then into three sources by assuming a repeated-measures design.

Now let us suppose that those same data came from yet another type of design, namely, a two-way factorial with independent samples. If the six treatments constitute a factorial arrangement of $k_A = 2$ levels of Factor A and $k_B = 3$ levels of Factor B, with an independent set of $n_R = 4$ subjects receiving each treatment, the $n = 24$ observations are classified as follows.

	A_1			A_2	
B_1	B_2	B_3	B_1	B_2	B_3
8	6	5	5	1	2
9	9	6	7	5	1
8	5	6	4	2	0
8	7	7	6	2	2

Using the formulas in Section 15.4, you should verify the ANOVA summary in the following table (where Factors A and B are both assumed to be fixed factors).

Analysis of the hypothetical data of Example 19.2 as if they came from a two-way design with independent samples.

Source of variation	SS	df	MS	F	
Factor A	92.04	1	92.04	58.65	$P < .01$
Factor B	44.33	2	22.17	14.12	$P < .01$
A-by-B interaction	4.33	2	2.17	1.38	
Replications within A-by-B	28.25	18	1.57	—	
Total	168.96	23			

Comparing this table with the one-way, independent-samples analysis of the same data in the first ANOVA table in Example 19.3, we find that the variation due to Factor A (92.04 with 1 df), Factor B (44.33 with 2 df), and the A-by-B interaction (4.33 with 2 df) add up to equal the variation due to the same six treatments in a one-way design (140.71 with 5 df in the first ANOVA table in Example 19.3). Each analysis partitions the same total variation into as many orthogonal sources of variation as can be identified in the experimental design.

As a fourth possibility for the data presented in Example 19.2, let us suppose that they came from a repeated-measures design with $k_A = 2$ levels of Factor A, $k_B = 3$ levels of Factor B, and $k_R = 4$ subjects, each of whom received all six treatment combinations. In this case, the $n = 24$ observations all came from just four different subjects, each one observed repeatedly, once under each treatment combination, and we have a two-way design with two repeated-measures factors. If the subjects are labeled as R_1, R_2, R_3, and R_4, the data may be presented as follows.

		A_1			A_2	
	B_1	B_2	B_3	B_1	B_2	B_3
R_1	8	6	5	5	1	2
R_2	9	9	6	7	5	1
R_3	8	5	6	4	2	0
R_4	8	7	7	6	2	2

Using Expressions (19.7) through (19.13), you should verify the ANOVA summary in the following table (where Factors A and B are both assumed to be fixed factors).

Analysis of the hypothetical data of Example 19.2 as if they came from a two-way design with two repeated-measures factors.

Source of variation	SS	df	MS	F	
Factor A	92.04	1	92.04	602	$P < .01$
Factor B	44.33	2	22.17	15.3	$P < .01$
A-by-B interaction	4.33	2	2.17	2.79	
Replications (R)	14.46	3	4.82	—	
A-by-R interaction	.46	3	.15	—	
B-by-R interaction	8.67	6	1.44	—	
A-by-B-by-R interaction	4.67	6	.78	—	
Total	168.96	23			

The sources of variation in this table are listed in an order corresponding to the order of the sources in the independent-samples ANOVA table at the beginning of this example, to make it easier to compare these two analyses of the same data. Notice that the four sources of variation involving replications in the repeated-measures ANOVA table are together equal to the source "replications within A-by-B" in the independent-samples ANOVA table. The variation due to the main effect of replications (SS_R) thus is removed from the error terms (MS_{AR}, MS_{BR}, MS_{ABR}) in the repeated-measures analysis. Whether it is appropriate to do so depends on the design of the experiment. The same total variation can be analyzed or broken down in many different ways, the correct way(s) depending on the identifiable sources of variation in the design.

19.3 TWO-WAY ANOVA WITH ONE REPEATED-MEASURES FACTOR

A common type of factorial design in experimental research is that in which one or more repeated-measures factors are crossed with one or more independent-samples factors. Any number of factors may be involved, but the principles can be illustrated by the simplest case, which is the two-way design with one repeated-measures factor. (Example 19.1 describes an application of this design in pain research.) In this design, an independent group of replicates is observed under each level of the independent-samples factor, say, Factor A, but within each group every replicate receives all the levels of the repeated-measures factor, say, Factor B. The replication factor is nested under one factor (A) and crossed with the other factor (B). If there are k_A levels of Factor A, k_B levels of Factor B, and n_R replicates at each level of Factor A, then there are $n = k_A k_B n_R$ observations altogether.

Computational Formulas and the ANOVA Summary Table

A **two-way ANOVA with one repeated-measures factor** involves three factors: The independent-samples factor A is crossed with the repeated-measures

TABLE 19.3 Analysis-of-variance summary table for two-way ANOVA with one repeated-measures factor.

Source of variation	SS	df	MS	F
Factor A (fixed)	SS_A	$k_A - 1$	MS_A	$F_A = MS_A/MS_{R(A)}$
Replications within A	$SS_{R(A)}$	$k_A(n_R - 1)$	$MS_{R(A)}$	—
Factor B (fixed)	SS_B	$k_B - 1$	MS_B	$F_B = MS_B/MS_{BR(A)}$
A-by-B interaction	SS_{AB}	$(k_A - 1)(k_B - 1)$	MS_{AB}	$F_{AB} = MS_{AB}/MS_{BR(A)}$
B-by-R within A	$SS_{BR(A)}$	$k_A(k_B - 1)(n_R - 1)$	$MS_{BR(A)}$	—
Total	SS_{total}	$n - 1$		

factor B, and the replications factor R is nested under A and crossed with B. Five sources of variation can be identified in such a design, as shown in the **ANOVA summary table** in Table 19.3.

If Factor A is the independent-samples factor and Factor B is the repeated-measures factor, the computational formulas for the sums of squares in the **two-way ANOVA with one repeated-measures factor** are as follows:

$$SS_A = \frac{1}{k_B n_R} \sum T_{a..}^2 - \frac{T^2}{n} \tag{19.20}$$

$$SS_{R(A)} = \frac{1}{k_B} \sum \sum T_{a.r}^2 - \frac{1}{k_B n_R} \sum T_{a..}^2 \tag{19.21}$$

$$SS_B = \frac{1}{k_A n_R} \sum T_{.b.}^2 - \frac{T^2}{n} \tag{19.22}$$

$$SS_{AB} = \frac{1}{n_R} \sum \sum T_{ab.}^2 - SS_A - SS_B - \frac{T^2}{n} \tag{19.23}$$

$$SS_{BR(A)} = \sum \sum \sum x_{abr}^2 - SS_{AB} - SS_B - SS_{R(A)} - SS_A - \frac{T^2}{n} \tag{19.24}$$

$$SS_{\text{total}} = \sum \sum \sum x_{abr}^2 - \frac{T^2}{n} \tag{19.25}$$

Alternatively, the sums of squares can be calculated from the formulas in Section 18.3 for the three-way ANOVA (with one observation per cell) by making use of the principle introduced in Expression (19.5), to obtain

$$SS_{R(A)} = SS_R + SS_{AR} \tag{19.26}$$

and

$$SS_{BR(A)} = SS_{BR} + SS_{ABR} \tag{19.27}$$

The guidelines for finding the appropriate error term, as given in Expression (18.18), are not directly applicable in the two-way design with one repeated-measures factor. However, if we recognize the principle in Expression (19.26)—namely, that the A-by-R interaction source of variation is incorporated in the replications-within-A source—then the guidelines can be applied with only slight modification. Specifically, the appropriate error term for testing any effect V (main effect or interaction effect) in a factorial design involving independent-samples factors and/or repeated-measures factors may be determined by the following **modified guidelines for finding the appropriate error term**:

> **a.** The error term for V is the mean square incorporating the interaction of V with exactly one random factor.
>
> **b.** If there are two or more crossed random factors that are not part of the effect V, there is no exact error term for testing the effect of V. **(19.28)**

EXAMPLE 19.7 The essence of the modification in the guidelines may be illustrated by considering the appropriate error term for testing the A-by-B interaction in Table 19.3. Strictly speaking, the A-by-B effect is not crossed with the one random factor R in the design. However, A-by-B-by-R is incorporated in the B-by-$R(A)$ effect [see Expression (19.27)], so the appropriate error term for testing the A-by-B interaction is $MS_{BR(A)}$.

The error terms for the other two F ratios in Table 19.3 may be obtained from either the guidelines in Expression (18.18) or the modified guidelines in Expression (19.28). Factor A is crossed with no random factors, so the within-cell mean square $MS_{R(A)}$—which incorporates the A-by-R interaction—is the appropriate error term for testing A. Factor B is crossed with exactly one random factor $R(A)$, so $MS_{BR(A)}$—which incorporates the B-by-R interaction—is the appropriate error term for testing B.

EXAMPLE 19.8 The stall speeds of an aircraft are important items of information for the pilot of a small plane, as they determine the minimum speed to keep the plane in the air when cruising (with the wing flaps up) and when landing (with the flaps down). Stall speeds for the two flap positions were obtained from six single-engine and six twin-engine aircraft. It was known in advance that stall speeds would be higher with the flaps up than with the flaps down, and also that stall speeds would be higher for twin-engine than for single-engine aircraft. But do these factors combine to produce unexpected—and potentially very dangerous—effects that would not be predicted from the main effects of flap position and number of engines?

The question of interest is whether there is an interaction between flap position and number of engines: Is the effect of flap position different for single-engine and twin-engine aircraft?

The stall speeds (in miles per hour) were as follows.

Aircraft	Single-engine Flaps up	Single-engine Flaps down
1	62	54
2	61	57
3	48	43
4	48	47
5	49	47
6	58	51

Aircraft	Twin-engine Flaps up	Twin-engine Flaps down
7	75	66
8	79	72
9	82	74
10	70	60
11	57	55
12	80	72

Flap position and number of engines are fixed factors that are crossed with each other. The random replication factor, aircraft, is nested under number of engines and crossed with flap position. In other words, number of engines is an independent-samples factor, and flap position is a repeated-measures factor. The design is a 2-by-2 factorial (engines-by-flaps) with repeated measures on the flap-position factor. The dependent variable is stall speed.

The ANOVA calculations are facilitated by recording **tables of totals** that are required in the various sum-of-squares formulas.

Aircraft-within-engines totals

	Single	Twin
	116	141
	118	151
	91	156
	95	130
	96	112
	109	152
Marginal totals	625	842

Engines-by-flaps totals (and marginal totals)

	Single	Twin	Marginal totals
Flaps up	326	443	769
Flaps down	299	399	698
Marginal totals	625	842	1,467

The sums of squares are then calculated from Expressions (19.20) through (19.24):

$$SS_{\text{engines}} = \frac{625^2 + 842^2}{12} - \frac{1{,}467^2}{24} = 91{,}632.42 - 89{,}670.38 = 1{,}962.04$$

$$SS_{R(E)} = \frac{116^2 + \cdots + 152^2}{2} - 91{,}632.42 = 1{,}042.08$$

$$SS_{\text{flaps}} = \frac{769^2 + 698^2}{12} - 89{,}670.38 = 210.04$$

$$SS_{EF} = \frac{326^2 + 443^2 + 299^2 + 399^2}{6} - 1{,}962.04 - 210.04 - 89{,}670.38 = 12.04$$

$$SS_{FR(E)} = (62^2 + \cdots + 72^2) - 12.04 - 210.04 - 1{,}042.08 - 1{,}962.04 - 89{,}670.38 = 38.42$$

The ANOVA summary table is as follows.

Source of variation	SS	df	MS	F	
Engines (E)	1,962.04	1	1962.04	18.8	$P < .01$
Aircraft within E	1,042.08	10	104.21	—	
Flap position (F)	210.04	1	210.04	54.7	$P < .01$
E-by-F interaction	12.04	1	12.04	3.13	
F-by-aircraft within E	38.42	10	3.84	—	
Total	3,264.63	23			

As expected, there were significant main effects of number of engines (with mean stall speeds of 52 and 70 mph for single-engine and twin-engine aircraft) and flap position (with means of 64 and 58 mph for flaps up and flaps down, respectively), but the interaction was not significant at the .05 level. The investigation did not provide enough evidence to conclude that the effect of flap position is greater for one type of aircraft (single-engine or twin-engine) than for the other.

(Source: Student project by N. Daley.)

Summary of the Test Procedure

The **two-way analysis of variance with one repeated-measures factor** may be summarized as follows:

Assumptions: See Sections 18.6 and 19.4.

Step 1. Determine the number k_A of levels of the independent-samples factor, the number k_B of levels of the repeated-measures factor, the number n_R of replications of each treatment combination, and the total number n of observations.

Step 2. Calculate the tables of totals for R-within-A and A-by-B, along with the marginal totals for the latter table.

Step 3. Calculate the table of means (and marginal means) corresponding to the A-by-B table of totals, in order to see what sample differences are being tested.

Step 4. Calculate the values required to complete the ANOVA summary table in the format of Table 19.3.

Step 5. Compare the observed values of the test statistics F_A, F_B, and F_{AB} with critical values from the F distribution with appropriate degrees of freedom.

a. If the interaction is significant, examine the treatment-combination means $\bar{x}_{ab.}$ to discover the nature of the effect, and interpret significant main effects only insofar as they are compatible with the treatment-combination means.

b. If the interaction is not significant but at least one main effect is, examine the relevant marginal means to discover the nature of the effect.

Further statistical tests to aid in interpretation are considered in Chapter 20.

Relationship among Two-Way ANOVA Designs

We have now considered three different analyses for testing the main effects and interaction of two crossed factors:

a. In two-way ANOVA with independent samples, all differences between treatment combinations involve comparisons *between* independent groups of replicates; hence the term **between-subjects design**. The remaining variation $SS_{R(AB)}$, associated with replications within cells, cannot be further subdivided in this design. The distinguishing feature of an independent-samples (between-subjects) design is that replications are *nested* under one or more treatment factors.

b. In two-way ANOVA with repeated measurements of the same replicates on both factors, all differences between treatment combinations involve comparisons *within* the same replicates; hence the term **within-subjects design**. In this design, the variation associated with replications can be subdivided into four parts (SS_R, SS_{AR}, SS_{BR}, SS_{ABR}), as shown in Expressions (19.18) and (19.19). The distinguishing feature of a repeated-measures (within-subjects) design is that replications are *crossed* with one or more treatment factors.

c. In two-way ANOVA with repeated measurements of the same replicates on one factor, the differences between levels of the independent-samples factor (say, Factor A) involve comparisons *between* independent groups of replicates, while the differences between levels of the repeated-measures factor (say, Factor B) involve comparisons *within* the same replicates. Such designs are sometimes called **between-within designs**. The random variation associated with replications can be subdivided into two parts: one part $SS_{R(A)}$ for the independent-samples portion of the design, where replications are nested under the independent-samples factor A, and the other part $SS_{BR(A)}$ for the repeated-measures portion of the design, where replications are crossed with the repeated-measures factor B.

Which of these three analyses is appropriate depends on the design of the experiment, according to the principles discussed in Section 18.1.

EXAMPLE 19.9 To illustrate the relationships among the various designs mentioned in this chapter, let us continue our examination of the 24 hypothetical observations given in Example 19.2. The total variation in the data is $SS_{total} = 168.96$ based on $df_{total} = 23$ degrees of freedom. We used these data in Example 19.3 to compare two types of one-way analysis and in Example 19.6 to compare two types of two-way analysis.

For purposes of comparison, let us suppose that those same data originated in a two-way design with one repeated-measures factor. If the six treatments constitute a factorial arrangement of $k_A = 2$ levels of the independent-samples factor A and $k_B = 3$ levels of the repeated-measures factor B, with an independent set of $n_R = 4$ subjects at each level of Factor A, the $n = 24$ observations are classified as follows.

Subject	A_1 B_1	B_2	B_3
S_1	8	6	5
S_2	9	9	6
S_3	8	5	6
S_4	8	7	7

Subject	A_2 B_1	B_2	B_3
S_5	5	1	2
S_6	7	5	1
S_7	4	2	0
S_8	6	2	2

Using the formulas in Expressions (19.20) through (19.25), you should verify the ANOVA summary in the following table (where Factors A and B are both assumed to be fixed factors).

Analysis of the hypothetical data of Example 19.2 as if they came from a two-way design with one repeated-measures factor.

Source of variation	SS	df	MS	F	
Factor *A*	92.04	1	92.04	37.0	$P < .01$
Replications within *A*	14.92	6	2.49	—	
Factor *B*	44.33	2	22.17	20.0	$P < .01$
A-by-*B* interaction	4.33	2	2.17	1.95	
B-by-replications within *A*	13.33	12	1.11	—	
Total	168.96	23			

The preceding table documents the fifth analysis that we have performed on the data from Example 19.2. All five analyses are summarized in the following table to facilitate comparison. This table includes all of the ANOVA designs presented in this book, except the three-way design with independent samples in Section 18.3. The different analyses break down the total variation in the dependent variable according to whatever sources of variation can be identified in the design of the experiment.

Comparison of five analyses of the hypothetical data of Example 19.2.

One-way with independent samples			One-way with repeated measurements			Two-way with two repeated-measures factors			Two-way with one repeated-measures factor			Two-way with independent samples		
Source	SS	df	Source	SS	df	Source	SS	df	Source	SS	df	Source	SS	df
						A	92.04	1	*A*	92.04	1	*A*	92.04	1
T	140.71	5	*T*	140.71	5	*B*	44.33	2	*B*	44.33	2	*B*	44.33	2
						AB	4.33	2	*AB*	4.33	2	*AB*	4.33	2
			R	14.46	3	*R*	14.46	3	*R*(*A*)	14.92	6			
R(*T*)	28.25	18				*AR*	.46	3				*R*(*AB*)	28.25	18
						BR	8.67	6	*BR*(*A*)	13.33	12			
			TR	13.79	15	*ABR*	4.67	6						
Total	168.96	23	Total	168.96	23	Total	168.96	23	Total	168.96	23	Total	168.96	23

Figure 19.3 provides a schematic summary of all six ANOVA designs presented in this book. To determine which is the appropriate analysis for any given experiment, we identify the factors in the experimental design and classify each one as an independent-samples factor or a repeated-measures factor. This identification and classification of the factors is sufficient to specify which (if any) of the six ANOVA designs presented in this book is relevant. As illustrated in Figure 19.3, the distinction between independent-samples factors and repeated-measures factors resides in the nesting or crossing of the replication factor.

19.4 SPECIAL CONSIDERATIONS IN REPEATED-MEASURES DESIGNS

The preceding three sections have described the analysis of data from repeated-measures experiments. This section discusses some special considerations in the use and interpretation of such experiments.

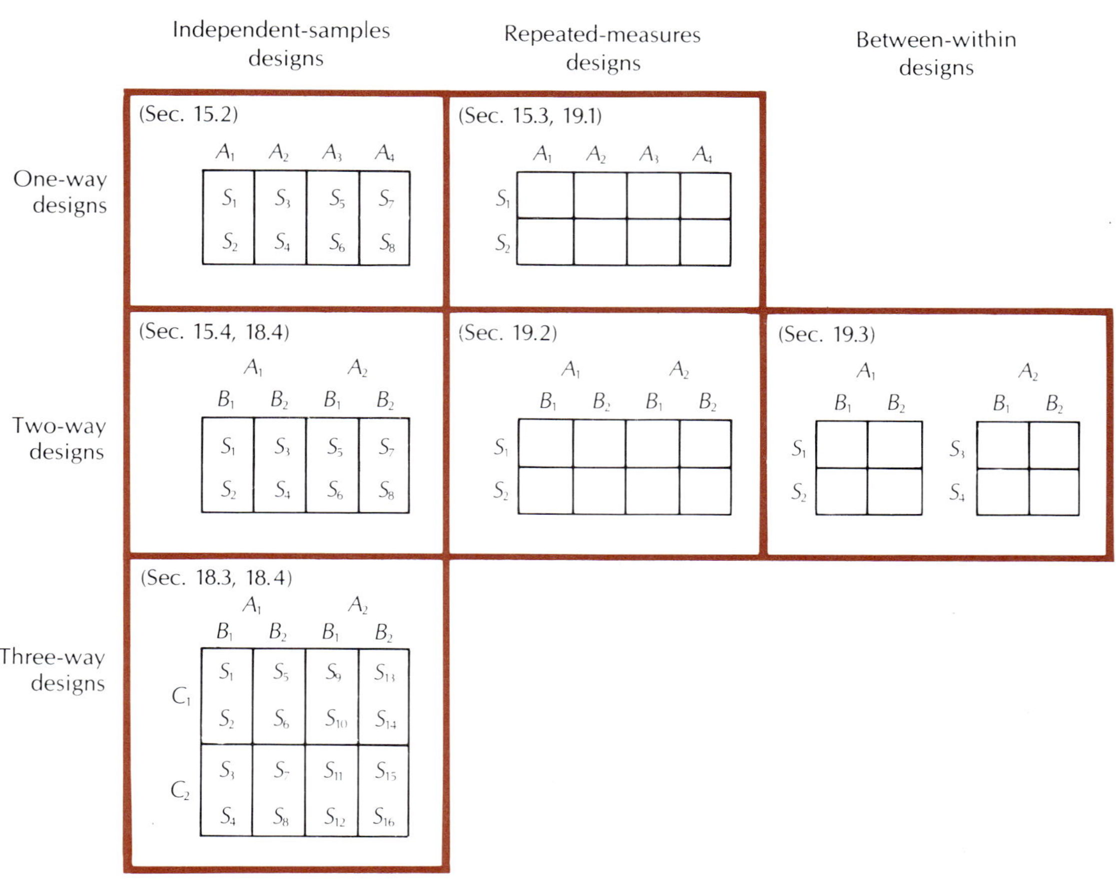

Figure 19.3
Schematic representation of ANOVA designs presented in this book. A, B, and C denote independent variables, and S (for subjects) represents the replication factor. The numbers of levels that are depicted for the factors were arbitrarily chosen for convenience and are in no way characteristic of any of the six designs. (Parenthetical references identify the chapter and section where each design is discussed.)

Assumption of Homogeneity of Covariance

The ANOVA F tests on repeated-measures factors require an additional assumption beyond those listed in Section 18.6. The **assumption of homogeneity of covariance** is concerned with the population covariance between observations in two different levels of a repeated-measures factor; in particular, it stipulates that such covariances should be the same for all possible pairs of levels of a repeated-measures factor. The assumption is more easily understood if we recall that covariance may be expressed in standard units as a correlation coefficient [see Expression (17.9).] Then, given the assumption of homogeneity of variance, the assumption of homogeneity of covariance is satisfied if all possible pairs of levels of a repeated-measures factor have equal correlations in the population. For example, if a repeated-measures factor has three levels labeled 1, 2, and 3, it is assumed that the population correlations of all pairs are equal: $\rho_{12} = \rho_{13} = \rho_{23}$.

The covariance-homogeneity assumption no doubt is frequently violated in real data, especially where the levels of the repeated-measures factor represent successive points in time. Examples of such factors would include trials in an experiment examining changes over time, or days of the week in an investigation of daily activity patterns. In such cases, levels (trials) that are adjacent to each other tend to be more highly correlated with each other than do levels that are separated by one or more intervening levels (trials).

The effect of heterogeneity of covariance is to introduce a positive bias into the F test, giving too many significant results. Unfortunately, it is difficult to determine whether the assumption has in fact been violated in any particular experiment. One case in which the assumption is never violated is that in which there are only two levels of the repeated-measures factor. With only two levels, there is only one correlation and therefore no possibility of unequal correlations. It can be shown that by reducing the degrees of freedom for the critical value of F to those appropriate for this special case, the researcher can compensate for the maximum possible bias that might be induced by covariance heterogeneity. The reduced degrees of freedom are called **conservative degrees of freedom**, because by compensating for the maximum possible bias, the reduced degrees of freedom tend, if anything, to introduce a negative bias, claiming too few significant results.

In light of the foregoing discussion, the following procedure is recommended as the general **procedure for *F* tests on repeated-measures factors**:

Step 1. Complete the ANOVA summary table (SS, df, MS, and F) as described in Sections 19.1 through 19.3. Notice in particular that the observed F ratio is calculated in the usual way and that the resulting value remains unchanged throughout the steps of this procedure.

Step 2. Compare each calculated F ratio to its tabulated critical value based on its full degrees of freedom (that is, the degrees of freedom as recorded in the ANOVA table). If the effect is not significant, this result should be reported, and Steps 3 and 4 are not required.

Step 3. If Step 2 yielded significance, then compare the calculated F ratio to the critical value tabulated for the conservative degrees of freedom. The conservative degrees of freedom are obtained by determining what the degrees of freedom would have been if there had been only two levels of each repeated-measures factor. If the effect is significant when tested with conservative degrees of freedom, this result should be reported, and Step 4 is not required.

Step 4. If the effect is significant according to the conventional F test in Step 2 but not significant according to the conservative F test in Step 3, the result is not clear-cut. In this case, there are two options available:

a. One option is to apply more advanced statistical methods. For example, it is possible to estimate a more accurate adjustment to the degrees of freedom. This has been done in the BMDP computer printouts in Figures 15.2 and 19.1, and the resulting adjusted P-value is shown in the column headed "Greenhouse Geisser Probability."

b. The other option is to make a significance judgment based on the information from the previous steps and on an inspection of the data for apparent violations of assumptions.

In connection with the latter alternative, remember that there is rarely any need for *exact* significance levels in practice—there is nothing magical about .05 or any other level. Rather, significance testing is always a matter of judgment—informed judgment, to be sure, but still a judgment—as to the probability of obtaining such discrepant sample means if the null hypothesis is true. When in doubt, it is probably safer to err in the conservative direction, by accepting the result using conservative degrees of freedom.

EXAMPLE 19.10 Data were obtained from a local bank to study the relationship between the day of the week and the type of business transacted at the bank by persons holding checking or savings accounts there. The data consisted of the number of transactions of each of two types, credits and debits, on each of three days of the week, Monday, Wednesday, and Friday. The numbers of transactions from three weeks, arbitrarily selected to provide a basis for generalizing the findings to other weeks, were analyzed.

The repeated-measures factors were type of transaction (T), with $k_T = 2$ levels, and day of the week (D), with $k_D = 3$ levels, and the replication factor was weeks (W), with $k_W = 3$ levels. The observed test statistics and the critical values with full degrees of freedom and with conservative degrees of freedom were as follows for F tests at the 5% level.

***F* test statistic**	**Full-df critical value**	**Conservative-df critical value**
$F_T = \text{MS}_T/\text{MS}_{TW} = 1.81$	$F_{.05}(1, 2) = 18.5$	$F_{.05}(1, 2) = 18.5$
$F_D = \text{MS}_D/\text{MS}_{DW} = 12.68$	$F_{.05}(2, 4) = 6.94$	$F_{.05}(1, 2) = 18.5$
$F_{TD} = \text{MS}_{TD}/\text{MS}_{TDW} = 13.21$	$F_{.05}(2, 4) = 6.94$	$F_{.05}(1, 2) = 18.5$

Using the procedure outlined above, we find first that the main effect of type of transaction is not significant at the .05 level. However, the results are ambiguous for the main effect of days and for the interaction, since the effects would be significant when tested with full degrees of freedom but not with conservative degrees of freedom. Since the levels of the repeated-measures factor (days) represented successive points in time, the assumption of homogeneity of covariance is immediately suspect, and calculation of sample correlation coefficients for pairs of days supports this suspicion.

Therefore, we conclude that although the data are suggestive, they are not sufficient to demonstrate any significant effects at the .05 level.

(Source: Student project by K. Ritchie.)

EXAMPLE 19.11 As a further illustration, consider an experiment similar to that in Example 19.1, but with three rather than two drug conditions. The modified experiment still has a two-way design with one repeated-measures factor. The repeated-measures factor is stimulus magnitude (S), with 7 levels; the independent-samples factor is drug condition (C), now with 3 levels; and there are 10 patients (P) in each drug condition. The critical values with full degrees of freedom and with conservative degrees of freedom would be as follows for F tests at the .05 significance level.

F test statistic	Full-df critical value	Conservative-df critical value
$F_C = \text{MS}_C/\text{MS}_{P(C)}$	$F_{.05}(2, 27) = 3.36$	$F_{.05}(2, 27) = 3.36$
$F_S = \text{MS}_S/\text{MS}_{SP(C)}$	$F_{.05}(6, 162) = 2.18$	$F_{.05}(1, 27) = 4.21$
$F_{SC} = \text{MS}_{SC}/\text{MS}_{SP(C)}$	$F_{.05}(12, 162) = 1.83$	$F_{.05}(2, 27) = 3.36$

Notice that the conservative degrees of freedom are the same as the full degrees of freedom for sources of variation (such as the main effect of C) that involve no repeated-measures factor.

Possibility of Carry-Over Effects

When a replicate in a repeated-measures design is an individual subject, the subjects' responses in such a design may depend in part on the other treatments they receive. Such dependencies are referred to as **carry-over effects**. Some carry-over effects, such as practice or fatigue, can sometimes be balanced out by presenting the conditions in different orders for different subjects. Careful planning may be necessary to avoid confounding due to such effects.

Other types of carry-over effects cannot be balanced out and remain as potential problems in repeated-measures designs. These effects include the following:

a. **asymmetrical transfer effects**, whereby Condition X has an effect on subsequent performance in Condition Y (perhaps by providing information or requiring physical exertion), but Condition Y has no effect on subsequent performance in Condition X; and

b. **range effects**, whereby subjects learn something about the range of levels that they receive of a factor and respond differently than they would if they did not know anything about the range.

Subjects in repeated-measures experiments are, in effect, "trained" in all the conditions included in the experiment, and it may be that the results can be safely generalized only to subjects who have been "trained" in the same way.

EXAMPLE 19.12 E. C. Poulton drew attention to a clear example of a range effect in an experiment reported by J. E. Kennedy and J. Landesman. Table height was a repeated-measures factor with six levels in an investigation to determine the optimal height of worktable for performing a simple manipulation task. Thus, the subjects worked on tables of six different heights during the course of the experiment. Different subjects were assigned to work at the various heights in different orders so as to balance out any order effects. When the table heights ranged from 25 cm below elbow height to 25 cm above elbow height, performance of the task was best at about elbow height. But when all the tables were made shorter (for a different group of subjects), with table heights ranging from 45 cm below elbow height to 5 cm above elbow height, performance of the task was best at about 15 cm below elbow height.

If we were not alert to the possibility of range effects, our conclusion about optimal table height would depend on the range of table heights that happened to be included in the experiment. Evidently, the subjects became most accustomed to working at a height that was about in the middle of the heights to which they were exposed in the experiment. To find the optimal table height for workers who will be working at only one height of table, subjects should be tested at only one height of table—that is, in an independent-samples design.

(Sources: Kennedy & Landesman, 1963; Poulton, 1973.)

Possibility of Increased Precision

In repeated-measures designs, the sum of squares due to replications (SS_R) is removed from the error sum of squares for at least some of the F tests. This removal leads to increased precision (that is, a smaller error term) if the reduction in the error sum of squares is large enough to offset the corresponding reduction in degrees of freedom for error. The ANOVA summary table provides evidence of increased precision if MS_R is considerably larger than the error term(s).

EXAMPLE 19.13 The statements in the preceding paragraph are illustrated in Example 19.3 for one-way designs and in Example 19.6 for two-way designs. In both the one-way and the two-way cases, the F ratios are larger in the repeated-measures analysis because their error terms are smaller than in the corresponding independent-samples analysis of the same data.

In the repeated-measures analyses themselves (Examples 19.3 and 19.6), the increased precision due to blocking is reflected in the fact that $MS_R = 4.82$ (which was part of the independent-samples error term) is considerably larger than the repeated-measures error terms (.92 in the one-way case, and .15, 1.44, and .78 in the two-way case). For the between-within design (Example 19.9), the increased precision due to blocking is manifest in the fact that the error term for the between-subjects part of the analysis ($MS_{R(A)} = 2.49$) is considerably larger than the error term for the within-subjects part of the analysis ($MS_{BR(A)} = 1.11$).

As previously noted in Sections 11.3 and 15.3, the possibility of increased precision exists in repeated-measures designs because, in effect, each replicate or block serves as its own control. Although the possibility of problems due to heterogeneity of covariance or carry-over effects must be carefully addressed in planning an investigation, blocking, which implies a repeated-measures analysis, remains an important tool of experimental design for obtaining precise answers to empirical questions.

STUDY MATERIAL FOR CHAPTER 19

Terms and Concepts

placebo
double-blind procedure

Section 19.1

one-way ANOVA with repeated measurements
ANOVA summary table for one-way ANOVA with repeated measurements

Section 19.2

two-way ANOVA with two repeated-measures factors
ANOVA summary table for two-way ANOVA with two repeated-measures factors
table of totals

Section 19.3

two-way ANOVA with one repeated-measures factor
ANOVA summary table for two-way ANOVA with one repeated-measures factor
modified guidelines for finding the appropriate error term
between-subjects design
within-subjects design
between-within design

Section 19.4

assumption of homogeneity of covariance
conservative degrees of freedom for testing repeated-measures factors
procedure for F tests on repeated-measures factors
carry-over effects
asymmetrical transfer effects
range effects

Key Formulas

Section 19.1

(19.5) $SS_{R(T)} = SS_R + SS_{TR}$

Section 19.2

(19.18) $SS_{R(AB)} = SS_R + SS_{AR} + SS_{BR} + SS_{ABR}$

Also see Expressions (19.7) through (19.13) and Table 19.2.

Section 19.3

(19.27) $SS_{BR(A)} = SS_{BR} + SS_{ABR}$

Also see Expressions (19.20) through (19.25) and Table 19.3.

Exercises for Numerical Practice

Section 19.1

19/1. Use the one-way ANOVA with repeated measurements to analyze the following data.

Replication	A_1	A_2	A_3	A_4	A_5	A_6	A_7	A_8
1	0	3	5	6	1	3	0	0
2	2	6	5	7	0	1	2	2
3	1	4	6	8	2	2	1	3

a. Complete the appropriate ANOVA summary table.
b. Give the critical F values at the 5% significance level
  i. with full degrees of freedom.
  ii. with conservative degrees of freedom.
c. What do the F test results tell you about these data?
d. Was there an improvement in precision due to blocking? How do you know?

19/2. Use the one-way ANOVA with repeated measurements to analyze the following data.

Replication	A_1	A_2	A_3	A_4	A_5	A_6	A_7	A_8	A_9	A_{10}	A_{11}	A_{12}
1	0	2	0	4	7	5	8	4	0	1	4	4
2	2	7	5	4	5	1	5	2	7	5	3	2

Answer Parts (a) through (d) in Exercise 19/1 for these data.

Section 19.2

19/3. Use the two-way ANOVA with two repeated-measures factors to analyze the following data.

	A_1				A_2			
Replication	B_1	B_2	B_3	B_4	B_1	B_2	B_3	B_4
1	0	3	5	6	1	3	0	0
2	2	6	5	7	0	1	2	2
3	1	4	6	8	2	2	1	3

Answer Parts (a) through (d) in Exercise 19/1 for these data.

19/4. Use the two-way ANOVA with two repeated-measures factors to analyze the following data.

Replication	A_1 B_1	A_1 B_2	A_1 B_3	A_2 B_1	A_2 B_2	A_2 B_3	A_3 B_1	A_3 B_2	A_3 B_3	A_4 B_1	A_4 B_2	A_4 B_3
1	0	2	0	4	7	5	8	4	0	1	4	4
2	2	7	5	4	5	1	5	2	7	5	3	2

Answer Parts (a) through (d) in Exercise 19/1 for these data.

Section 19.3

19/5. Given that A is an independent-samples factor, B is a repeated-measures factor, and R is the replication factor, use the two-way ANOVA with one repeated-measures factor to analyze the following data.

	A_1 B_1	A_1 B_2	A_1 B_3	A_1 B_4
R_1	0	3	5	6
R_2	2	6	5	7
R_3	1	4	6	8

	A_2 B_1	A_2 B_2	A_2 B_3	A_2 B_4
R_4	1	3	0	0
R_5	0	1	2	2
R_6	2	2	1	3

Answer Parts (a) through (d) in Exercise 19/1 for the these data.

19/6. Given that A is an independent-samples factor, B is a repeated-measures factor, and R is the replication factor, use the two-way ANOVA with one repeated-measures factor to analyze the following data.

	A_1 B_1	A_1 B_2	A_1 B_3
R_1	0	2	0
R_2	2	7	5

	A_2 B_1	A_2 B_2	A_2 B_3
R_3	4	7	5
R_4	4	5	1

	A_3 B_1	A_3 B_2	A_3 B_3
R_5	8	4	0
R_6	5	2	7

	A_4 B_1	A_4 B_2	A_4 B_3
R_7	1	4	4
R_8	5	3	2

Answer Parts (a) through (d) in Exercise 19/1 for these data.

Problems for Applying the Concepts

19/7. Which experimental design leads to an ANOVA table with the total sum of squares partitioned into the two sources A and $R(A)$?

19/8. Which experimental design leads to an ANOVA table with the total sum of squares partitioned into the eight sources A, B, C, AB, AC, BC, ABC, $R(ABC)$?

19/9. Which experimental design leads to an ANOVA table with the total sum of squares partitioned into the three sources A, R, AR?

19/10. Which experimental design leads to an ANOVA table with the total sum of squares partitioned into the seven sources A, B, R, AB, AR, BR, ABR?

19/11. Which experimental design leads to an ANOVA table with the total sum of squares partitioned into the five sources A, B, AB, $R(A)$, $BR(A)$?

19/12. Which experimental design leads to an ANOVA table with the total sum of squares partitioned into the four sources A, B, AB, $R(AB)$?

19/13. Which repeated-measures design has exactly the same format as a three-way factorial with one observation per cell?

19/14. Which repeated-measures design has exactly the same format as a two-way factorial with one observation per cell?

19/15. Four hockey players on an intercollegiate team took part in an investigation of the effects of hand preference and hand orientation on the accuracy of hockey shots. Each player shot a total of 80 pucks from the center dot toward the goal, 20 pucks in each of the four combinations of preferred/nonpreferred hand and forehand/backhand. The number of pucks that went into the net in each condition was recorded as follows.

Subject	Preferred hand		Nonpreferred hand	
	Forehand	Backhand	Forehand	Backhand
1	18	15	9	3
2	19	16	8	5
3	18	16	11	7
4	18	13	12	2

a. Identify the replication factor, any other factors, and the dependent variable.

b. Which experimental design is this?

c. Complete the appropriate ANOVA summary table, and give the critical value(s) for significance testing.

d. Plot the appropriate means for portraying any significant effects, and state what the results indicate about shooting accuracy.

e. Was there an improvement in precision due to blocking?

(Source: Student project by L. McKinney.)

19/16. A hand-grip dynamometer was used to measure the hand strength of 20 individuals sampled from four distinct populations, or types, defined by certain sociological characteristics. Each individual had two trials with the dynamometer, and the higher of the two scores was recorded to the nearest 5 pounds as the observation for that individual. The observations were as follows for the four types of individuals.

Type 1	Type 2	Type 3	Type 4
200	120	125	170
185	115	100	185
210	130	115	190
190	90	110	180
185	115	105	165

Perform an analysis of variance to test for differences among the four types of individuals.

a. Identify the replication factor, any other factors, and the dependent variable.

b. Which experimental design is this?

c. Complete the appropriate ANOVA summary table, and give the critical value(s) for significance testing.

d. Plot the appropriate means for portraying any significant effects, and state what the results indicate about hand strength.

e. Was there an improvement in precision due to blocking?

(Source: Student project by S. N. J. Watamaniuk.)

19/17. Three swimmers swam a 25-meter length three times, in order to compare three variations in swimming stroke. The variations, which consisted of using 0, 2, or 6 kicks per arm pull, were used in a different order by each swimmer. The following are the swimmers' times, rounded to the nearest second.

Swimmer	0 kicks	2 kicks	6 kicks
1	30	27	28
2	34	30	29
3	28	25	25

Perform an analysis of variance to test the effect of number of kicks per arm pull.

a. Identify the replication factor, any other factors, and the dependent variable.

b. Which experimental design is this?

c. Complete the appropriate ANOVA summary table, and give the critical value(s) for significance testing.

d. Plot the appropriate means for portraying any significant effects, and state what the results indicate about swimming speed.

e. Was there an improvement in precision due to blocking?

f. Why was it important that the various strokes be used in a different order by each swimmer?

(Source: Student project by W. Vornbrock.)

19/18. An investigation was undertaken concerning the number of advertisements for alcohol and tobacco products in two types of magazines, news and sports. The numbers of advertisements of each type were counted in one current issue of four magazines of each type, with the following results.

News magazines	Alcohol	Tobacco	Sports magazines	Alcohol	Tobacco
Time	8	2	*Sport*	4	6
Newsweek	8	7	*Inside Sports*	3	6
Maclean's	12	2	*Super Sports*	1	1
U.S. News & World Report	2	4	*Sports Illustrated*	4	6

a. Identify the replication factor, any other factors, and the dependent variable.

b. Which experimental design is this?

c. Use the range method for comparing transformations to identify an appropriate transformation for improving homogeneity of variance in these data. You may limit your consideration to the following four forms of the data: original data, square-root transformation, logarithmic transformation, and reciprocal transformation.

d. Perform an analysis of variance of the data in the form suggested in the answer to Part (c). Complete the appropriate ANOVA summary table, and give the critical value(s) for significance testing.

e. Plot the appropriate means for portraying any significant effects, and state what the results indicate about numbers of advertisements.

f. Was there an improvement in precision due to blocking?

(Source: Student project by D. Stewart.)

19/19. A male student conducted an investigation of factors involved in the formation of "first impressions" about a person. Using a college student directory, he selected the names of 16 female students, not including any whom he knew well. He then approached each subject individually and invited her to a movie on the following weekend. For half of the subjects he was dressed neatly (freshly laundered clothes, combed hair, and so on), and for the other half he was unkempt (soiled jeans, shirt untucked, shoelaces untied, and so on). His approach in asking the question was one of humility (as if asking a favor) for half of each dress group, and one of overconfidence (as if doing a favor) for the other half. The subjects' responses were rated on a 5-point scale as follows:

1 = a flat refusal
2 = a refusal with an excuse
3 = "some other time"
4 = a reluctant acceptance
5 = an enthusiastic acceptance

Attitude:	**Humble**		**Overconfident**	
Dress:	**Neat**	**Sloppy**	**Neat**	**Sloppy**
	5	4	3	1
	3	4	3	3
	4	3	2	2
	3	2	3	2

Perform an analysis of variance to test the effects of the factors.

a. Identify the replication factor, any other factors, and the dependent variable.

b. Which experimental design is this?

c. Complete the appropriate ANOVA summary table, and give the critical value(s) for significance testing.

d. Plot the appropriate means for portraying any significant effects, and state what the results indicate about responses to a movie invitation.

e. Was there an improvement in precision due to blocking?

(Source: Student project by R. Bishop.)

19/20. An investigation was conducted among students successfully completing two years of university study, in order to determine whether there was any systematic change in the marks of such students as they completed high school and began university. The data reported here are average marks by each of four students in their courses in the humanities (English, French, classics, music, drama, and so on) and in the sciences (mathematics, biology, chemistry, physics, and so on) in three years of study.

	Grade 12		University 1		University 2	
Student	Humanities	Sciences	Humanities	Sciences	Humanities	Sciences
1	85	89	63	94	89	78
2	75	73	70	83	56	89
3	84	89	67	83	78	89
4	73	74	67	83	78	89

Perform an analysis of variance to test the effects of the factors.

a. Identify the replication factor, any other factors, and the dependent variable.

b. Which experimental design is this?

c. Complete the appropriate ANOVA summary table, and give the critical value(s) for significance testing.

d. Plot the appropriate means for portraying any significant effects, and state what the results indicate about average marks in the population from which these students were sampled.

e. Was there an improvement in precision due to blocking?

(Source: Student project by D. Burzynski.)

19/21. Eighteen female students agreed to keep track of the number of hours they spent watching television each day of the week. The following numbers of hours were recorded for Monday, Wednesday, and Friday by the 18 students, 9 of whom were in their first year and 9 in their second year of university studies.

	First-year students		
Student	M	W	F
1	0	0	0
2	0	0	0
3	1	1	2
4	0	0	4
5	1	1	1
6	1	1	0
7	0	0	0
8	2	2	0
9	0	1	2

	Second-year students		
Student	M	W	F
10	2	0	0
11	2	2	0
12	2	2	0
13	3	3	2
14	5	3	5
15	3	3	0
16	0	0	0
17	0	0	0
18	0	3	0

Perform an analysis of variance to test the effects of the factors.

a. Identify the replication factor, any other factors, and the dependent variable.

b. Which experimental design is this?

c. Complete the appropriate ANOVA summary table, and give the critical value(s) for significance testing.

d. Plot the appropriate means for portraying any significant effects, and state what the results indicate about television-viewing habits.

e. Was there an improvement in precision due to blocking?

(Source: Student project by M. Diemert.)

19/22. Prompted by the claim that some types of music are "noisier" than others, an investigator randomly selected 10 rock music albums and 10 classical music albums to compare their power output as measured by a wattmeter in a stereo sound system. A randomly selected segment of each album was played three times, with the tone controls set differently each time: once at the extreme bass setting, once at the extreme treble setting, and once with bass and treble both set halfway through their ranges. At each setting, the power output was read on the wattmeter to the nearest tenth of a watt. The volume control was kept at a constant setting throughout the experiment. The readings (in watts) were as follows.

Album	Rock music			Album	Classical music		
	Bass	Treble	Neutral		Bass	Treble	Neutral
1	1.5	.6	.4	11	.4	.2	.2
2	1.0	.4	.4	12	.1	.4	.1
3	.6	.4	.2	13	.4	.2	.2
4	.8	.4	.4	14	1.0	.4	.4
5	1.3	.6	.4	15	.6	.2	.4
6	.8	.2	.2	16	.4	.2	.1
7	1.3	.6	.5	17	.2	.1	.1
8	.6	.4	.3	18	.3	.2	.2
9	1.8	.8	.5	19	.4	.2	.1
10	1.0	.4	.2	20	.5	.4	.2

Perform an analysis of variance to test the effects of the factors.

a. Identify the replication factor, any other factors, and the dependent variable.

b. Which experimental design is this?

c. Complete the appropriate ANOVA summary table, and give the critical value(s) for significance testing.

d. Plot the appropriate means for portraying any significant effects, and state what the results indicate about power output.

e. Was there an improvement in precision due to blocking?

(Source: Student project by P. Sartison.)

19/23. Four players were randomly selected from a college hockey team to take part in an experiment to study the accuracy of shooting as a function of distance from the net and curvature of the stick blade. Each player used both a straight stick and a stick with a half-inch curve, shooting from each of four distances: 10, 20, 30, and 40 feet. The data were obtained with the use of a shooting practice machine consisting of a goal net with four 6-inch-square targets attached in the four corners. Each target contained a light controlled by a computer to light up in random order with equal probability, and the player had 5 seconds to react to each lighted target. There were 20 shots for each of the eight treatment combinations, and the computer recorded the number of hits in each set of 20 as follows.

Player	Straight stick 10 ft	Straight stick 20 ft	Straight stick 30 ft	Straight stick 40 ft	Curved stick 10 ft	Curved stick 20 ft	Curved stick 30 ft	Curved stick 40 ft
1	15	10	7	3	16	13	9	5
2	16	11	5	2	17	12	8	6
3	9	6	2	0	10	7	3	1
4	13	9	6	2	14	12	10	4

Perform an analysis of variance to test the effects of the factors.

a. Identify the replication factor, any other factors, and the dependent variable.

b. Which experimental design is this?

c. Complete the appropriate ANOVA summary table, and give the critical value(s) for significance testing.

d. Plot the appropriate means for portraying any significant effects, and state what the results indicate about shooting accuracy.

e. Was there an improvement in precision due to blocking?

(Source: Student project by Y. de Champlain.)

19/24. A farmer was interested in comparing hay yields from two fields in which he made two cuttings per year. He knew in advance that first-cutting yields would be greater than second-cutting yields, but he wanted to know if the yield from one field was reliably greater than that from the other field, and whether the reduction in yield from first to second cutting was the same for the two fields. He recorded the following yields (in bales per acre) for three years.

Year	Field 1 Cutting 1	Field 1 Cutting 2	Field 2 Cutting 1	Field 2 Cutting 2
1977	190	68	120	56
1978	90	39	80	24
1979	175	60	108	42

a. Identify the replication factor, any other factors, and the dependent variable.

b. Which experimental design is this?

c. Use the range method for comparing transformations to identify an appropriate transformation for improving homogeneity of variance in these data. You may limit your consideration to the following four forms of the data: original data, square-root transformation, logarithmic transformation, and reciprocal transformation.

d. Perform an analysis of variance of the data in the form suggested in the answer to Part (c). Complete the appropriate ANOVA summary table, and give the critical value(s) for significance testing.

e. Plot the appropriate means for portraying any significant effects, and state what the results indicate about hay yields.

f. Was there an improvement in precision due to blocking?

g. What effect(s) would be tested to answer the farmer's question about whether the reduction in yield from first to second cutting was the same for the two fields?

(Source: Student project by S. Bailer.)

19/25. A random sample of 12 college students took part in an investigation of musical tastes. Half of the students were from a rural background, defined as a community of 500 or fewer, and the other half were classified as urban. Each student was asked to rate two popular musical performers, Billy Idol and Kenny Rogers, chosen to represent rock and country music, respectively. The ratings were as follows on a scale from 1 (negative) to 10 (positive).

	Rural	
Student	Idol	Rogers
1	7	9
2	6	5
3	9	5
4	9	7
5	4	8
6	6	8

	Urban	
Student	Idol	Rogers
7	10	2
8	8	6
9	8	4
10	9	6
11	7	3
12	8	4

Perform an analysis of variance to test the effects of the factors.

a. Identify the replication factor, any other factors, and the dependent variable.

b. Which experimental design is this?

c. Complete the appropriate ANOVA summary table, and give the critical value(s) for significance testing.

d. Plot the appropriate means for portraying any significant effects, and state what the results indicate about musical tastes.

e. Was there an improvement in precision due to blocking?

f. What effect(s) should be tested to determine whether rural and urban students differ in the type of music they prefer?

(Source: Student project by C. Holmes.)

19/26. In an investigation of college students' sleeping habits, 22 college students were asked to report (to the nearest half hour) the average number of hours they slept on weekday nights and on weekend nights (Friday and Saturday). The means for 11 females and 11 males were as follows.

	Weekdays	Weekends
Females	7.45	7.05
Males	6.55	6.82

a. Identify the replication factor, any other factors, and the dependent variable.

b. Which experimental design is this?

c. Calculate the sums of squares for the main effects and interaction of days and gender. (*Hint*: The table of totals used in the calculation of sums of squares can always be obtained from the table of means if the number of observations on which each mean is based is known. Note that the reports were given to the nearest half hour.)

d. When the data were analyzed by a three-way ANOVA for days (D) by gender (G) by students (S) in a 2-by-2-by-11 factorial arrangement, the ANOVA yielded the three sums of squares in the answer to Part (c), along with the following sums of squares:

$$SS_S = 66.01, \quad SS_{GS} = 25.26, \quad SS_{DS} = 35.01, \quad SS_{GDS} = 21.03$$

Complete the ANOVA summary table for the appropriate experimental design as specified in Part (b), and give the critical value(s) for significance testing.

e. Plot the appropriate means for portraying any significant effects, and state what the results indicate about sleeping habits.

f. Was there an improvement in precision due to blocking?

(Source: Student project by J. Sheets.)

19/27. Twelve university students took part in a study of reading speed. Each student was tested individually on two different readings, the same two readings for every student. They were timed as they read the first reading with a radio on in the same room, and then again as they read the second reading in relatively quiet surroundings, in order to assess the influence of distracting sounds on reading speed. Half of the students had grade point averages greater than 6.5 (on a 9-point scale where 6 is an average grade), and half had averages of 6.5 or less. The mean reading speeds (in words per minute) were as follows.

	Radio on	**Radio off**
GPA greater than 6.5	253	284
GPA of 6.5 or less	247	227

a. Identify the replication factor, any other factors, and the dependent variable.

b. Which experimental design is this?

c. Calculate the sums of squares for the main effects and interaction of grade point average and distraction. (*Hint*: The table of totals used in the calculation of sums of squares can always be obtained from the table of means if the number of observations on which each mean is based is known.)

d. When the data were analyzed by a three-way ANOVA for grade point average (G) by distraction (D) by students (S) in a 2-by-2-by-6 factorial arrangement, the ANOVA yielded the three sums of squares in the answer to Part (c), along with the following sums of squares:

$$SS_S = 121{,}550.3, \quad SS_{GS} = 54{,}218.5, \quad SS_{DS} = 4{,}663.0, \quad SS_{GDS} = 7{,}384.5$$

Complete the ANOVA summary table for the appropriate experimental design as specified in Part (b), and give the critical value(s) for significance testing.

e. Plot the appropriate means for portraying any significant effects, and state what the results indicate about reading speed.

f. Was there an improvement in precision due to blocking?

g. What kind of carry-over effect leads to confounding in this experiment? What other confounding variable is mentioned in the description of the experiment?

(Source: Student project by A. Warnke.)

References

Gracely, R. H.; Dubner, R.; and McGrath, P. A. "Narcotic Analgesia: Fentanyl Reduces the Intensity but Not the Unpleasantness of Painful Tooth Pulp Sensations." *Science* 203 (1979): 1261–63.

Kennedy, J. E., and Landesman, J. "Series Effects in Motor Performance Studies." *Journal of Applied Psychology* 47 (1963): 202–5.

Poulton, E. C. "Unwanted Range Effects from Using Within-Subject Experimental Designs." *Psychological Bulletin* 80 (1973): 113–21.

Further Reading

Dixon, W. J., *et al.*, eds. *BMDP Statistical Software Manual, 1985 Reprinting.* Berkeley: University of California Press, 1985. An example of BMDP output was shown in Figure 19.1.

Hill, M., ed. *BMDP User's Digest.* 3rd ed. Los Angeles: BMDP Statistical Software, Inc., 1984.

Keppel, G. *Design and Analysis: A Researcher's Handbook.* 2nd ed. Englewood Cliffs, NJ: Prentice-Hall, 1982. Part V provides more advanced and thorough coverage of the topics introduced in the present chapter.

Winer, B. J. *Statistical Principles in Experimental Design.* 2nd ed. New York: McGraw-Hill, 1971. Chapters 4 and 7 deal with repeated-measures designs.

*CHAPTER 20

SPECIALIZED COMPARISONS AMONG MEANS

EXAMPLE 20.1 Libel legislation is intended to protect public figures from false accusations that unjustly damage their reputation. However, sometimes a damaging statement may be qualified in such a way that, while it is not untrue, it still creates a negative impression by insinuation or innuendo.

Forty-eight university students in Texas took part in a study of the effect of innuendo in newspaper headlines. Each student received a booklet of newspaper headlines that were said to refer to candidates in former elections for the U.S. House of Representatives. Two of the headlines in each booklet served as filler to disguise the purpose of the study. The other four involved four different forms of headline, representing different levels of incrimination, with each subject receiving all four forms. The four forms, with an example of each, were as follows:

a. *Assertion*: "Bob Talbert Linked With Mafia"

b. *Question*: "Is Karen Downing Associated With Fraudulent Charity?"

c. *Denial*: "Andrew Winters Not Connected To Bank Embezzlement"

d. *Control*: "George Armstrong Arrives In City"

For half of the students, the headlines were said to be from the *New York Times* or the *Washington Post,* which were later rated by the students as being highly credible sources, and for the other half, from the *National Enquirer* or the *Midnight Globe,* rated as having low credibility. The basic design thus was a 2-by-4 factorial (source credibility by headline form) with repeated measures on the headline-form factor.

After examining the headlines, the students rated their impressions of each of the named individuals on a series of dimensions, such as good/bad and honest/dishonest. The ratings on the various dimensions were averaged to yield a single rating, on a scale from 1 (positive impression) to 7 (negative impression), that was used as the dependent variable. The mean ratings for the eight treatment combinations were as follows.

Credibility of source	Headline form			
	Assertion	Question	Denial	Control
High	4.65	4.28	3.54	2.84
Low	4.12	3.97	3.56	3.12

* This symbol identifies material that is not prerequisite for later topics in this book.

An analysis of variance indicated a significant interaction between source credibility and headline form. One way of interpreting this interaction is to say that the effect of credibility is different for the different forms of headline. But in what way is it different? Which specific forms of headlines showed a significant effect of source credibility and which did not? Using methods to be introduced in this chapter, the researchers found that high-credibility sources produced significantly more negative impressions than low-credibility sources if direct assertions were used (that is, 4.65 was significantly greater than 4.12), but that the differences between sources were not significant for any of the other three forms of headline.

A second way of interpreting the significant interaction is to say that the effect of headline form is different for the different levels of credibility. Again, methods that are used to determine the specific nature of such differences will be introduced in this chapter. The researchers found that the different headline forms led to significantly different ratings for both levels of source credibility. For the high-credibility sources, the effect of innuendo was demonstrated by the fact that students' impressions based on question headlines and denial headlines were significantly more negative than those based on the control headlines. For the low-credibility sources, ratings based on question headlines were significantly more negative than in the control condition, but responses to the denial headlines did not differ significantly from the control.

The results of the overall ANOVA were thus refined into more specific statements about the nature of the significant effects. One conclusion was that innuendo in the form of a questioning headline, whether in a high-credibility source or a low-credibility source, tends to create a negative impression about the person involved. Another conclusion was that the major difference between the levels of source credibility is in the greater impact of direct assertions of an incriminating nature when they are made by sources with high, rather than low, credibility.

(Source: Wegner *et al.*, 1981.)

In most experiments, such as that in Example 20.1, testing the statistical significance of the F ratios that appear in the standard ANOVA summary table does not constitute the end of the statistical analysis. Those **overall *F* tests** indicate whether there are significant main effects or interaction effects in the data, but if there are more than two means involved in the tested effect, a significant overall F test does not imply that every mean is significantly different from every other mean. It indicates only that at least one significant difference exists among the means in that effect.

In the case of main effects, a significant F value tells us that there are reliable differences among the k sample means being compared in the effect, but it does not tell us specifically which means differ significantly from which other ones. Is the significant main effect due to one particular treatment that differs from all the others, or is every treatment significantly different from every other treatment? Or is there some other pattern of treatment differences?

In the case of a two-way interaction, a significant F value tells us that the effect of one factor is reliably different at different levels of the other factor, but it does not tell us specifically which cell means are responsible for the interaction. Is there a single treatment combination that differs from all the rest? Or is Factor A effective at level B_1 but not at level B_2? Or is there some other pattern of cell means that produces the significant interaction?

To interpret experimental results, it is generally necessary to carry the statistical analysis beyond the overall F tests for main effects and interactions, in an attempt to find out specifically what is responsible for any significant effects that are found.

20.1 SIMPLE EFFECTS

Overall F tests are based on data from an entire experiment. One way of analyzing or breaking down a significant overall interaction is by testing **simple effects**, which are effects considered separately for each level of one of the factors involved in the overall interaction. A simple effect therefore is based on data from only one specific part of an experiment. Tests of simple effects are used to help interpret interactions that are found to be significant by an overall F test.

Simple Interaction Effects

Interpretation of an ANOVA summary table begins with the highest order of interaction that is significant—that is, it begins with the significant interaction involving the greatest number of factors. Suppose that the highest order of significant interaction is three-way, A-by-B-by-C. A significant three-way interaction implies that the two-way interaction of any two of those three factors is different for different levels of the third factor: We cannot average over the levels of any of the three factors without losing some statistically significant information.

To discover more precisely what information would be lost by averaging over one of the factors, we may plot one (or more) of the two-way interactions, say, A-by-B, at each level of the third factor, say, C. A two-way interaction examined at one specific level of a third factor is called a **simple interaction effect**. Thus, we may plot the simple interaction effect of A-by-B at level C_1, the simple interaction effect of A-by-B at C_2, and so on, for each of the k_C levels of Factor C. Likewise, we can test each of these simple interaction effects separately.

The tests of simple interaction effects in a three-way factorial consist of treating the data as if they came from several separate two-way factorial experiments, one for each level of the third factor. In fact, the sum of squares and degrees of freedom for a simple two-way interaction effect are obtained by applying the formulas for SS_{AB} and df_{AB} from the *two-way* ANOVA separately to the data for each level of the third factor. Using the two-way ANOVA formulas for SS_{AB}, then, we can compute $SS_{AB(C_1)}$, which is the sum of squares due to the A-by-B interaction at level 1 of Factor C. Applying the same formula repeatedly yields one sum of squares $SS_{AB(C_c)}$ for each level of Factor C, with $c = 1, \ldots, k_C$. Thus, the simple A-by-B interactions at each level of

Factor C provide one way of breaking down a significant three-way interaction into separate two-way interactions, which tend to be more easily interpreted.

A second way of breaking down the three-way interaction would be to examine the simple interaction effect of A-by-C at each level of Factor B, and yet another way would be to consider B-by-C at each level of Factor A. The relationship among these three ways of analyzing a significant three-way interaction may be clarified by noting the following equalities, based on the principle introduced in Expression (19.5):

$$SS_{AB(C_1)} + SS_{AB(C_2)} + \cdots + SS_{AB(C_{k_C})} = SS_{AB(C)} = SS_{AB} + SS_{ABC} \tag{20.1}$$

$$SS_{AC(B_1)} + SS_{AC(B_2)} + \cdots + SS_{AC(B_{k_B})} = SS_{AC(B)} = SS_{AC} + SS_{ABC} \tag{20.2}$$

$$SS_{BC(A_1)} + SS_{BC(A_2)} + \cdots + SS_{BC(A_{k_A})} = SS_{BC(A)} = SS_{BC} + SS_{ABC} \tag{20.3}$$

Thus, each of the three ways of analyzing a three-way interaction does so by attaching its variation to a different one of the two-way sums of squares. The question as to which one or more of these ways of breaking down the three-way interaction should be used cannot be given a general answer. The question must be answered in terms of what makes sense in a particular experiment: Which simple two-way interaction, examined separately at different levels of the third factor, provides the clearest answer to the questions that the experiment was designed to answer?

EXAMPLE 20.2 Figure 18.2 is a schematic portrayal of the results of a 2-by-3-by-2 factorial experiment concerning memory. The experiment involved two levels of literacy (L) of the subjects (nonliterate and literate), three levels of intelligence (I) of the subjects (low, average, and high), and two levels of type (T) of story (African and European). There were $n_R = 6$ replications of each treatment combination.

Parts (a) and (b) of Figure 18.2 depict the three-way interaction, L-by-I-by-T. Part (a) is the simple interaction effect of intelligence and type of story for the nonliterate subjects. Treating the 36 observations represented in this part as a separate experiment, we could calculate the sum of squares due to the I-by-T interaction in the nonliterate sample, say, $SS_{IT(L_1)}$. Part (b) is the simple interaction effect of intelligence and type of story for the literate subjects. Treating the 36 observations represented in this part as a separate experiment, we could calculate the sum of squares due to the I-by-T interaction in the literate sample, say, $SS_{IT(L_2)}$. These two simple-effect sums of squares together make up the intelligence-by-type-of-story interaction within levels of literacy, which is itself a composite of I-by-T and L-by-I-by-T from the original three-way ANOVA summary table [see Expressions (20.1) through (20.3)]:

$$SS_{IT(L_1)} + SS_{IT(L_2)} = SS_{IT(L)} = SS_{IT} + SS_{LIT}$$

If the three-way interaction L-by-I-by-T were significant, we could test the simple I-by-T interaction effects separately for nonliterates and literates as one way of analyzing the three-way interaction.

We will see how to test the significance of simple interaction effects after we have considered the other type of simple effects.

Simple Main Effects

When the highest order of interaction that is significant is a two-way interaction, it may be further analyzed in terms of the **simple main effects** of one factor at the separate levels of the other factor. (If the two-way interaction to be broken down is itself a simple interaction effect, we sometimes refer to the results of the analysis as **simple simple main effects**, but the procedure is the same.) A simple main effect is an effect of one factor examined at one specific level of a second factor.

If the two factors involved in a significant interaction are A and B, we could examine the simple main effects of A at the different levels of B, in which case

$$\mathrm{SS}_{A(B_1)} + \mathrm{SS}_{A(B_2)} + \cdots + \mathrm{SS}_{A(B_{k_B})} = \mathrm{SS}_{A(B)} = \mathrm{SS}_A + \mathrm{SS}_{AB} \qquad (20.4)$$

Or we could examine the simple main effects of B at the different levels of A, where

$$\mathrm{SS}_{B(A_1)} + \mathrm{SS}_{B(A_2)} + \cdots + \mathrm{SS}_{B(A_{k_A})} = \mathrm{SS}_{B(A)} = \mathrm{SS}_B + \mathrm{SS}_{AB} \qquad (20.5)$$

Again, either or both of these ways of breaking down the significant interaction may be used, depending on what makes the experimental results most clear. (Both ways were used in Example 20.1.) The sum of squares and degrees of freedom for a simple main effect are obtained by applying the formulas for SS_A and df_A from the *one-way* ANOVA separately to the data for each level of the second factor.

EXAMPLE 20.3 Example 19.5 analyzed data on the heights of homemade bread loaves as a function of length of first rising and amount of yeast. There were $k_L = 2$ levels of length of first-rising time (2 hours and 3 hours) crossed with $k_A = 3$ levels of amount of yeast (1 T, 2 T, and 4 T) in a 2-by-3 factorial design with two repeated-measures factors and $k_R = 4$ replications. The resulting analysis of variance, which was reported in full in Example 19.5, included sums of squares for three testable effects:

$\mathrm{SS}_{\text{length}} = 210.04$ with 1 degree of freedom

$\mathrm{SS}_{\text{amount}} = 441.08$ with 2 degrees of freedom

$\mathrm{SS}_{LA} = 369.08$ with 2 degrees of freedom

The length-by-amount interaction was significant at the 1% level.

To break down the significant two-way interaction, let us consider first the simple main effects of length of first-rising time for the different amounts of yeast. The length-by-amount table of totals (from Example 19.5) is as follows, where each of the six cells is a total over $k_R = 4$ replications.

Amount	Length 2 h	3 h	Marginal totals
1 T	388	388	776
2 T	369	366	735
4 T	380	312	692
Marginal totals	1,137	1,066	2,203

The sum of squares due to the simple main effect of length of first rising with 1 T yeast (Amount 1) is calculated by treating the 1-T data as if they constituted a separate experiment (that is, a one-way design testing the length effect, with $k_R = 4$ replications):

$$SS_{L(A_1)} = \frac{388^2 + 388^2}{4} - \frac{776^2}{8} = 0.000$$

The sum of squares due to the simple main effect of length of first rising with 2 T yeast (Amount 2) is calculated by treating the 2-T data separately:

$$SS_{L(A_2)} = \frac{369^2 + 366^2}{4} - \frac{735^2}{8} = 1.125$$

Similarly, we calculate the sum of squares due to the simple main effect of length of first rising with 4 T yeast (Amount 3) from the 4-T data alone:

$$SS_{L(A_3)} = \frac{380^2 + 312^2}{4} - \frac{692^2}{8} = 578.000$$

Each of these three sums of squares has $k_L - 1 = 1$ degree of freedom.

We will see in the next subsection how to use these sums of squares to test the significance of the simple main effects. For now, we limit ourselves to showing how they relate to the sums of squares in the ANOVA summary table, according to Expression (20.4). The total of the sums of squares for the three simple main effects of length with particular amounts of yeast is the sum of squares due to length within amount:

$$SS_{L(A)} = 0.000 + 1.125 + 578.000 = 579.125$$

with $1 + 1 + 1 = 3$ degrees of freedom. This equals the sum of the variation due to length of first rising (with 1 degree of freedom) and the variation due to the L-by-A interaction (with 2 degrees of freedom) from the calculations for the ANOVA table of Example 19.5:

$$SS_L + SS_{LA} = 210.042 + 369.083 = 579.125$$

with $1 + 2 = 3$ degrees of freedom. The simple main effects thus provide an alternative way of partitioning the variation associated with these same 3 degrees of freedom.

A second way of analyzing the significant two-way interaction is in terms of the simple main effects of amount of yeast, separately for the two lengths of first rising. The sum of squares due to the simple main effect of amount for the standard 2-hour rising time (Length 1) is calculated as if the 2-hour data were the only data collected (that is, for a one-way design testing the amount effect, with $k_R = 4$ replications):

$$SS_{A(L_1)} = \frac{388^2 + 369^2 + 380^2}{4} - \frac{1137^2}{12} = 45.500$$

with $k_A - 1 = 2$ degrees of freedom. Similarly, we calculate the sum of squares due to the simple main effect of amount for the 3-hour rising time (Length 2) from the 3-hour data alone:

$$SS_{A(L_2)} = \frac{388^2 + 366^2 + 312^2}{4} = \frac{1066^2}{12} = 764.667$$

with $k_A - 1 = 2$ degrees of freedom.

The total of the sums of squares for the simple main effects of amount within particular first-rising times is the sum of squares due to amount within length:

$$\text{SS}_{A(L)} = 45.500 + 764.667 = 810.167$$

with 2 + 2 = 4 degrees of freedom. This equals the sum of the variation due to amount (with 2 degrees of freedom) and the variation due to the L-by-A interaction (with 2 degrees of freedom) from the calculations for the ANOVA table of Example 19.5:

$$\text{SS}_{A} + \text{SS}_{LA} = 441.083 + 369.083 = 810.167$$

with 2 + 2 = 4 degrees of freedom. The simple main effects provide an alternative way of partitioning the variation associated with these same 4 degrees of freedom.

Tests of Simple Effects

Tests of simple effects take the same general form as overall F tests:

$$F_{\text{effect}} = \frac{\text{MS}_{\text{effect}}}{\text{MS}_{\text{error}}} \quad \textbf{(20.6)}$$

where $\text{MS}_{\text{effect}} = \text{SS}_{\text{effect}}/\text{df}_{\text{effect}}$, as usual. The only matter requiring discussion is the identification of the appropriate error term, MS_{error}.

The appropriate error term for testing any simple effect is determined by the ANOVA-table sources of which the simple effect is a part. Expressions (20.1) through (20.3) show that each simple two-way interaction effect is a part of the corresponding overall two-way interaction and of the three-way interaction. Similarly, Expressions (20.4) and (20.5) show that each simple main effect is a part of the corresponding overall main effect and the two-way interaction. The **rule for identifying the appropriate error term for testing a simple effect** is as follows:

a. If both (or all) of the overall sources of which the simple effect is a part have the same error term, that same error term is also appropriate for testing the simple effect.
b. If the overall sources of which the simple effect is a part have different error terms, a special error term must be calculated for testing the simple effect. **(20.7)**

Let us consider in turn the two parts of the rule in Expression (20.7). When do the sources of which a simple effect is a part have the same error term? This is the case in both the two-way and the three-way ANOVAs with independent samples, because all testable effects have the same error term $\text{MS}_{\text{replications-within-cells}}$ in independent-samples designs, provided the crossed factors are fixed factors. It is also the case for testing the simple main effect of the repeated-measures factor in the two-way ANOVA with one repeated-measures factor, because the repeated-measures factor (say, B) and its interaction with the independent-samples factor (A-by-B) both have the same error term $\text{MS}_{BR(A)}$. In these cases, the error term for testing simple effects is obtained directly from the ANOVA summary table.

Calculation of a special error term will be required for testing simple effects at specific levels of a repeated-measures factor. Among the designs presented in this book, this situation occurs in only two cases: simple main effects in the two-way design with two repeated-measures factors, and simple main effects of the independent-samples factor (at specific levels of the repeated-measures factor) in the two-way design with one repeated-measures factor. There are several ways of calculating the special error term in such cases. One conservative, or cautious, method is to compute a new error term based on only those data involved in the simple effect being tested. Thus, to test a simple main effect involving overall sources having different error terms, we compute the one-way ANOVA (main effect and error term) at the specified level of the second factor, as if that level constituted a separate experiment. This method is not always the best one for obtaining error terms for testing simple effects, but it is not likely to be seriously misleading and it has the advantages of requiring no new formulas and being relatively easy to understand. (See the references under Further Reading at the end of this chapter for other methods.)

EXAMPLE 20.4 In Example 18.14, we found a significant interaction of fertilization and weather with respect to barley yield (in bu/acre). To break down this interaction, we can examine the simple effects of fertilization at separate levels of weather and/or the simple effects of weather at separate levels of fertilization. In either case, we need the fertilization-by-weather table of totals, where each of the four cells is based on 6 observations.

	Normal (W_1)	Dry (W_2)	Marginal totals
Fertilizer	401	217	618
No fertilizer	271	159	430
Marginal totals	672	376	1,048

The sums of squares for the simple effects of fertilization in normal weather (W_1) and in dry weather (W_2) are

$$SS_{F(W_1)} = \frac{401^2 + 271^2}{6} - \frac{672^2}{12} = 39{,}040.333 - 37{,}632.000 = 1{,}408.333$$

$$SS_{F(W_2)} = \frac{217^2 + 159^2}{6} - \frac{376^2}{12} = 12{,}061.667 - 11{,}781.333 = 280.333$$

As a check on the computations, we verify that the total of these two sums of squares,

$$SS_{F(W_1)} + SS_{F(W_2)} = 1{,}408.333 + 280.333 = 1{,}688.667$$

equals the appropriate sum of squares from the ANOVA-table calculations of Example 18.14:

$$SS_{F(W)} = SS_F + SS_{FW} = 1{,}472.667 + 216.000 = 1{,}688.667$$

The simple effects $F(W_1)$ and $F(W_2)$ provide a repartitioning of the variation due to the effects F and F-by-W. Since F and F-by-W both have the same error term, namely, $MS_{R(VFW)}$, that error term is also appropriate for testing the simple effects $F(W_1)$ and

$F(W_2)$. The tests may be shown in an ANOVA summary table of simple effects, as follows.

Source of variation	SS	df	MS	F	
Fertilization (in normal weather)	1,408.3	1	1,408.3	73.8	$P < .01$
Fertilization (in dry weather)	280.3	1	280.3	14.7	$P < .01$
Replications within V-by-F-by-W	229.0	12	19.1		

Thus, fertilization was associated with significantly greater yields in both normal weather and in dry weather (as shown by the analysis of simple effects), but the improvement associated with fertilization was significantly greater in normal weather (as shown by the significant interaction in the overall analysis in Example 18.14). See Figure 18.1.

If we had wanted to focus on the effects of weather rather than on the effects of fertilization, we could have broken down the significant F-by-W interaction into the simple effects of weather at separate levels of fertilization, $W(F_1)$ and $W(F_2)$.

EXAMPLE 20.5 In Example 20.3, we calculated sums of squares due to simple effects in an experiment concerning the action of yeast in homemade bread. There were $k_L = 2$ lengths of first rising (2 hours and 3 hours) crossed with $k_A = 3$ amounts of yeast (1 T, 2 T, and 4 T), replicated $k_R = 4$ times in a 2-by-3 factorial design with repeated measures on both factors. An overall F test indicated that the length-by-amount interaction was significant at the 1% level.

Let us analyze the significant two-way interaction in terms of the simple main effects of length of first rising for the different amounts of yeast. These simple effects are each a part of the variation due to length within amounts, $SS_{L(A)}$, which is related to the ANOVA summary table as follows:

$$SS_{L(A)} = SS_L + SS_{LA}$$

according to Expression (20.4). The effects L and L-by-A have different error terms (namely, MS_{LR} and MS_{LAR}) in this design, and hence we must calculate a special error term for testing the simple main effects of length of first rising. It should be emphasized, however, that if L and L-by-A both had the same error term in the overall analysis, then that error term from the overall analysis would be the appropriate one for the simple effect tests, too.

In Example 20.3, we calculated the sum of squares due to the simple main effect of length of first rising with 1 T yeast (Amount 1) as

$$SS_{L(A_1)} = 0.000 \qquad \text{with 1 degree of freedom}$$

Although a sum of squares of zero can never be significant at any level, let us demonstrate the calculation of an error term for this type of simple effect. The error term may be obtained by treating the 1-T data as the results of a one-way experiment with repeated measures, where the appropriate error term for the effect of length is based on the length-by-replication interaction, which may be calculated from the individual observations given in Example 19.5:

$$SS_{LR(A_1)} = (98^2 + 97^2 + 95^2 + 98^2 + 100^2 + 96^2 + 94^2 + 98^2) - \frac{388^2 + 388^2}{4} - \frac{198^2 + 193^2 + 189^2 + 196^2}{2} + \frac{776^2}{8} = 3.000$$

with $(k_L - 1)(k_R - 1) = 3$ degrees of freedom. (You should verify this substitution of numerical values into the formula for two-way interaction, in order to learn why the final term is added rather than subtracted.) The analysis may be presented in an ANOVA summary table for the simple effect of length of first rising with 1 T yeast, as follows.

Source of variation	**SS**	**df**	**MS**	***F***
Length (with 1 T yeast)	0.00	1	0.00	0.00
L-by-*R* (with 1 T yeast)	3.00	3	1.00	—

There was no difference in final loaf height between 2-hour and 3-hour first-rising times when 1 T yeast was used.

Similarly, to test the simple main effect of length of first rising with 2 T yeast (Amount 2), we calculate the error term by treating the 2-T data as the results of a one-way experiment and obtaining the sum of squares for the length-by-replication interaction:

$$SS_{LR(A_2)} = (93^2 + 95^2 + 91^2 + 90^2 + 93^2 + 89^2 + 91^2 + 93^2) - \frac{369^2 + 366^2}{4} - \frac{186^2 + 184^2 + 182^2 + 183^2}{2} + \frac{735^2}{8} = 21.375$$

with $(k_L - 1)(k_R - 1) = 3$ degrees of freedom. The ANOVA summary table for the simple effect of length of first rising with 2 T yeast is as follows.

Source of variation	**SS**	**df**	**MS**	***F***	
Length (with 2 T yeast)	1.12	1	1.12	.16	$P > .10$
L-by-*R* (with 2 T yeast)	21.38	3	7.12	—	

(The sum of squares due to the simple effect of length was calculated in Example 20.3.) The difference in final loaf height between 2-hour and 3-hour first-rising times when 2 T yeast was used was not significant at the 10% level.

The same procedure yields the following ANOVA summary table for the simple effect of length of first rising with 4 T yeast.

Source of variation	**SS**	**df**	**MS**	***F***	
Length (with 4 T yeast)	578.00	1	578.00	248	$P < .01$
L-by-*R* (with 4 T yeast)	7.00	3	2.33	—	

Loaf height was significantly less with the 3-hour than with the 2-hour first-rising time when 4 T yeast was used ($P < .01$).

Without the tests of simple effects, we could make only rather general statements about the significant interaction in this experiment; for example, we could say that the effect of length of first rising was different with different amounts of yeast. The three tests of simple effects allow us to be more specific about the nature of the difference: The average final loaf height was significantly greater when a 2-hour first-rising time was used than when a 3-hour time was allowed, if 4 T yeast was used. But the difference in final heights between loaves with different first-rising times was not significant if 1 T or 2 T yeast was used. See Figure 19.2.

Another way of breaking down the significant two-way interaction is in terms of the simple main effects of amounts of yeast for 2-hour and 3-hour rising times separately. These simple effects could be reported either instead of or in addition to the simple effects of length of rising time for the different amounts separately that we have calculated in this example.

Tests of simple effects are used to break down significant interactions by specifying at which levels of one factor another factor has a significant effect. Trial and error may be required to find the most informative simple effects to report in describing a significant interaction; for example, effects of A at different levels of B, and/or effects of B at different levels of A.

It may be helpful to think of the examination of simple effects as a process for subdividing the whole experiment into a number of mini-experiments until all the conclusions can be stated in terms of main effects; for example, the main effect of A may be significant in one mini-experiment (say, level 2 of B), but not in another mini-experiment (say, treatment combination B_1C_3), and so on. Tests of simple effects thus can be used to reduce significant interactions to collections of statements about main effects restricted to particular treatments or treatment combinations.

20.2 SINGLE-DEGREE-OF-FREEDOM COMPARISONS

A main effect, whether simple or overall, refers to the differences among the k means corresponding to the k levels of a factor. Once the ANOVA results have been reduced to statements about main effects, with the aid of simple-effects tests if necessary, it remains only to determine which specific levels of the significant factor differ significantly from which other levels.

If the number k of levels of the factor is greater than 2, the F test of the main effect is called an **omnibus *F* test**, because it evaluates all the differences among the k means jointly to see if any significant variation exists among them. If the corresponding population means are designated as μ_j, the null hypothesis evaluated by the omnibus F test of a main effect has the following general form:

$$H_0\colon \mu_1 = \mu_2 = \cdots = \mu_k \qquad \textbf{(20.8)}$$

Rejection of the null hypothesis by the omnibus F test indicates a judgment or inference that at least one of the equal signs is incorrect, but it does not indicate which one(s). We evaluate which of the k population means may be inferred to be different from which others of them by testing specific comparisons among the means.

The Coefficients of a Comparison

Section 11.3 introduced a t test for comparing means of paired samples. It involves calculating a difference score for each pair in order to determine the mean difference between the samples,

$$\bar{d} = \bar{x}_1 - \bar{x}_2 \tag{20.9}$$

and then forming a t ratio, $t_{obs} = \bar{d}/s_{\bar{d}}$, to see if the observed mean difference is unusually large relative to its estimated standard error. The difference $\bar{d}$ in Expression (20.9) is a **comparison** of the two sample means.

Section 11.2 dealt with the t test for comparing means of two independent samples. In this case, the observations in the two samples cannot logically be paired to obtain difference scores, but the comparison of the two means again consists of evaluating the sample difference, $\bar{x}_1 - \bar{x}_2$, relative to its estimated standard error. The difference between means, $\bar{x}_1 - \bar{x}_2$, has the same form as in Expression (20.9), and we will now designate it too by the symbol $\bar{d}$ and call it a **comparison**.

When a factor has more than two levels, there are many different comparisons that could be made among the means for the various levels of the factor.

EXAMPLE 20.6 Example 20.1 described an experiment concerning the influence of innuendo in newspaper headlines. After examining a series of newspaper headlines, students rated their impressions of each of the individuals named in the headlines. Each student made ratings for four forms of headline representing four different levels of incrimination. The mean ratings for 24 students whose headlines were ascribed to a well-known and respected newspaper were as follows, where a higher number indicates a more negative impression of the incriminated person:

Assertion	Question	Denial	Control
4.65	4.28	3.54	2.84

If we wanted to see whether any form of incrimination (assertion, question, and/or denial) in headlines influenced impressions, we would examine the difference between the mean for the control condition and the mean for the other three conditions combined. Let us write this comparison as the mean difference

$$\begin{aligned}\bar{d}_1 &= \frac{4.65 + 4.28 + 3.54}{3} - 2.84 = \frac{1}{3}(4.65) + \frac{1}{3}(4.28) + \frac{1}{3}(3.54) - (1)(2.84)\\ &= 1.3167\end{aligned} \tag{20.10}$$

(The subscript on $\bar{d}$ is an arbitrary label serving only to distinguish one comparison from another.) To compare the effect of innuendo (question and denial) with the effect of direct assertion, we would examine the difference between the assertion mean and the mean for the two innuendo conditions. This comparison (say, $\bar{d}_2$) leaves out the fourth mean:

$$\bar{d}_2 = 4.65 - \frac{4.28 + 3.54}{2} = 1(4.65) - \frac{1}{2}(4.28) - \frac{1}{2}(3.54) + 0(2.84) = .74 \tag{20.11}$$

To see if the two forms of innuendo have different effects, we would compare the second and third means with each other in a comparison (say, $\bar{d}_3$) that leaves out the first and last means:

$$\bar{d}_3 = 4.28 - 3.54 = (0)(4.65) + (1)(4.28) - (1)(3.54) + (0)(2.84) = .74 \tag{20.12}$$

There are many other specific comparisons that could be made among four means, depending on the questions of interest in the experiment. As just one example, we could examine the difference between the mean of the first two means and the mean of the last two means, in order to compare headlines forms that directly suggest guilt with those that do not:

$$\begin{aligned}\bar{d}_4 &= \frac{4.65 + 4.28}{2} - \frac{3.54 + 2.84}{2} \\ &= \frac{1}{2}(4.65) + \frac{1}{2}(4.28) - \frac{1}{2}(3.54) - \frac{1}{2}(2.84) = 1.275\end{aligned} \tag{20.13}$$

In general, a **comparison** $\bar{d}$ among k means $\bar{x}_1, \bar{x}_2, \ldots, \bar{x}_k$ is defined as a weighted sum of the means, subject to the restriction that the weights must add up to zero:

$$\begin{aligned}\bar{d} &= c_1\bar{x}_1 + c_2\bar{x}_2 + \cdots + c_k\bar{x}_k \\ &= \sum c_j\bar{x}_j\end{aligned} \tag{20.14}$$

where

$$\sum c_j = 0 \tag{20.15}$$

and the c_j are not all equal to zero. The weights c_j are called the **coefficients of the comparison**, and the requirement that they add up to zero implies that some of them are positive and some negative and that the positive weights must balance out the negative weights. Expression (20.15) simply puts into mathematical notation a restriction that we impose on ourselves intuitively in calculating differences such as those in Example 20.6. The symbol $\bar{d}$ is used for a comparison, to emphasize that a comparison is basically a *difference* between two weighted means. One weighted mean is made up of the treatment means with positive coefficients, and it is being compared with the other weighted mean, which is made up of the treatment means with negative coefficients.

EXAMPLE 20.7 Let us identify the coefficients of the four comparisons in Example 20.6. The coefficient is the weight assigned to each mean in the weighted sum of means that constitutes the comparison. To determine the coefficient of any given mean, we ask what that mean was multiplied by in calculating the value of the comparison. The following lines show the coefficients of the comparisons in Expressions (20.10) through (20.13):

Coefficients of $\bar{d}_1$: $c_1 = \frac{1}{3}$, $c_2 = \frac{1}{3}$, $c_3 = \frac{1}{3}$, $c_4 = -1$,
Coefficients of $\bar{d}_2$: $c_1 = 1$, $c_2 = -\frac{1}{2}$, $c_3 = -\frac{1}{2}$, $c_4 = 0$.
Coefficients of $\bar{d}_3$: $c_1 = 0$, $c_2 = 1$, $c_3 = -1$, $c_4 = 0$.
Coefficients of $\bar{d}_4$: $c_1 = \frac{1}{2}$, $c_2 = \frac{1}{2}$, $c_3 = -\frac{1}{2}$, $c_4 = -\frac{1}{2}$.

In each case, you should verify that the sum of the coefficients is zero, as required in the definition of a comparison.

Notice that the multiplication of comparison coefficients by a constant yields another comparison. For example, if we multiply the coefficients of $\bar{d}_1$ by the constant 3 to get rid of the fractions, we obtain another comparison, say, $\bar{d}_5$, whose coefficients are as follows:

Coefficients of $\bar{d}_5$: $c_1 = 1, \quad c_2 = 1, \quad c_3 = 1, \quad c_4 = -3.$

The comparison $\bar{d}_5$ is a genuine comparison, since the sum of its coefficients is zero. Like the comparison $\bar{d}_1$ from which it was obtained, $\bar{d}_5$ compares the mean of the first three conditions (positive coefficients) with the mean for the last condition (negative coefficient). The value of the comparison $\bar{d}_5$, whose coefficients are three times those of $\bar{d}_1$, is

$$\bar{d}_5 = 4.65 + 4.28 + 3.54 - (3)(2.84) = 3.95 \tag{20.16}$$

which is three times the value of $\bar{d}_1$. Thus, multiplying the coefficients by a constant changes the scale of measurement but does not change what is being compared with what.

The Sum of Squares Due to a Comparison

We began Section 20.2 by noting that a main effect, whether simple or overall, refers to differences among the k means corresponding to the k levels of the factor. A main effect of Factor A, for example, refers to differences among the k_A Factor A means. These differences are measured in global fashion by SS_A, the variation due to Factor A, which has $k_A - 1$ degrees of freedom. A comparison on Factor A is a specific one of the differences among the k_A means. Associated with each comparison on Factor A is a portion of the overall variation due to Factor A (SS_A) and one of the $k_A - 1$ degrees of freedom. If we can determine how much of the overall variation among the k_A means is due to one particular comparison or difference, then we can test the significance of that specific difference.

The **sum of squares due to a comparison** $\bar{d}$ among k treatment means is called a **component of the treatment sum of squares** and is defined as

$$\text{SS}_{\bar{d}} = \frac{m\bar{d}^2}{\sum c_j^2} \tag{20.17}$$

where m is the number of observations from which each one of the k means has been computed. For example, in a one-way design with n_R replications of each treatment, m equals n_R, but in comparing marginal means for Factor A in a k_A-by-k_B factorial with n_R observations per cell, m equals $k_B n_R$ (which is the number of observations at each level of Factor A).

EXAMPLE 20.8 Example 20.6 presented four means, based on 24 observations each, of impressions formed about people on the basis of newspaper headlines that referred to them. Expression (20.10) is a comparison of the mean rating in three experimental conditions versus the mean rating in the control condition:

$$\bar{d}_1 = \frac{1}{3}(4.65) + \frac{1}{3}(4.28) + \frac{1}{3}(3.54) - 2.84 = 1.3167$$

The sum of squares due to this comparison is calculated from Expression (20.17) as

$$SS_{\bar{d}_1} = \frac{(24)(1.3167^2)}{(\frac{1}{3})^2 + (\frac{1}{3})^2 + (\frac{1}{3})^2 + (-1)^2} = 31.205$$

The value of m in Expression (20.17) is 24 in this case, because each of the four means is based on 24 observations.

If the coefficients of comparison $\bar{d}_1$ are multiplied by the constant 3, we obtain the comparison $\bar{d}_5$ in Expression (20.16). The sum of squares due to comparison $\bar{d}_5$ is

$$SS_{\bar{d}_5} = \frac{(24)(3.95^2)}{1^2 + 1^2 + 1^2 + (-3)^2} = 31.205$$

which is identical to the sum of squares due to comparison $\bar{d}_1$. This computation illustrates the general principle that although multiplication of a comparison by a constant changes the units in which the difference $\bar{d}$ is expressed, it does not change the value of $SS_{\bar{d}}$, the sum of squares due to the comparison.

Now let us consider more of the data from Example 20.1. The mean ratings for eight treatment combinations, along with the main-effect means for headline form, were as follows.

	Headline form			
Credibility of source	**Assertion**	**Question**	**Denial**	**Control**
High ($n_R = 24$)	4.65	4.28	3.54	2.84
Low ($n_R = 24$)	4.12	3.97	3.56	3.12
Main-effect means:	4.385	4.125	3.55	2.98

A comparison of the three experimental conditions versus the control condition in terms of the main-effect means (averaging over the two levels of source credibility) is given by

$$\bar{d} = 4.385 + 4.125 + 3.55 - (3)(2.98) = 3.12$$

The sum of squares due to this comparison is

$$SS_{\bar{d}} = \frac{(48)(3.12^2)}{1^2 + 1^2 + 1^2 + (-3)^2} = 38.94$$

Notice that the value of m [in Expression (20.17)] is 48 here, because each of the four means involved in the comparison is based on 48 observations.

The **number of degrees of freedom due to a comparison** is always one. Hence, the sum of squares due to a comparison has one degree of freedom, and this degree of freedom is one of the $k - 1$ degrees of freedom associated with the k levels of the factor represented by the k means. To understand why a comparison has one degree of freedom, recall that a comparison is basically a difference between two sets of means. One set receives positive signs in the comparison, and the other set receives negative signs. Regardless of the number of treatment means that are involved in the comparison, the comparison evaluates a difference between just two items: a weighted mean of the positive signed set versus a weighted mean of the negative signed set. Just as for any factor having only two levels, here too there is just one degree of freedom for the difference between the two weighted means.

Orthogonality

The concept of orthogonality was introduced in Section 18.4. Two sources of variation are said to be *orthogonal* if they are not confounded with each other. If two sources are orthogonal to each other, the sum of squares due to one source measures entirely different variation from that measured by the sum of squares for the other source. In this sense, orthogonal sources are ones that do not overlap with each other.

For a given set of means based on m observations each, two comparisons are **orthogonal comparisons** if and only if the sum of the products of their corresponding coefficients is zero. In symbols, comparison $\bar{d}_1$ having coefficients c_j and comparison $\bar{d}_2$ having coefficients c_j' are orthogonal if and only if

$$\sum c_j c_j' = 0 \tag{20.18}$$

When we say that $\bar{d}_1$ and $\bar{d}_2$ are orthogonal comparisons, we are saying that they reflect nonoverlapping sources of variation—sources that are not confounded with each other. The difference described by one such comparison contributes nothing to the difference described by the other comparison.

EXAMPLE 20.9 Comparison $\bar{d}_1$ in Example 20.6 represents the difference between three experimental conditions and one control condition, and comparison $\bar{d}_2$ represents the difference between the first experimental condition and the other two experimental conditions. The following lines give the coefficients of these two comparisons, with the conditions listed in the order given in Example 20.6 (assertion, question, denial, control):

Coefficients (c_j) of $\bar{d}_1$: $\frac{1}{3}$ $\frac{1}{3}$ $\frac{1}{3}$ -1
Coefficients (c_j') of $\bar{d}_2$: 1 $-\frac{1}{2}$ $-\frac{1}{2}$ 0

The products of corresponding coefficients are

$$c_j c_j': \quad \tfrac{1}{3} \quad -\tfrac{1}{6} \quad -\tfrac{1}{6} \quad 0$$

The sum of these products equals zero:

$$\sum c_j c_j' = \frac{1}{3} - \frac{1}{6} - \frac{1}{6} + 0 = 0$$

indicating that the comparisons $\bar{d}_1$ and $\bar{d}_2$ are orthogonal: They reflect nonoverlapping sources of variation. Examination of the coefficients of the comparisons suggests that this statement is intuitively reasonable: In comparison $\bar{d}_2$, the first mean is compared with the second and third means jointly, whereas in comparison $\bar{d}_1$ the first three means are not compared against each other at all but, as a unit, are compared against the fourth mean.

Comparison $\bar{d}_3$ in Example 20.6 represents the difference between the second and third means, and comparison $\bar{d}_4$ represents the difference between the mean of the first two means and the mean of the last two means. The following lines give the relevant coefficients:

Coefficients (c_j) of $\bar{d}_3$:	0	1	−1	0
Coefficients (c_j') of $\bar{d}_4$:	$\frac{1}{2}$	$\frac{1}{2}$	$-\frac{1}{2}$	$-\frac{1}{2}$

The products of corresponding coefficients are

$$c_j c_j': \quad 0 \quad \tfrac{1}{2} \quad \tfrac{1}{2} \quad 0$$

The sum of these products is not equal to zero:

$$\sum c_j c_j' = 0 + \frac{1}{2} + \frac{1}{2} + 0 = 1$$

indicating that the comparisons $\bar{d}_3$ and $\bar{d}_4$ are not orthogonal. In other words, they are at least partially redundant—or confounded—comparisons, as can be seen by noting that they both pit the second mean against the third mean.

Two components of a treatment sum of squares are **orthogonal components** if their comparisons are orthogonal; that is, $SS_{\bar{d}_1}$ and $SS_{\bar{d}_2}$ are orthogonal if the comparisons $\bar{d}_1$ and $\bar{d}_2$ are orthogonal. Orthogonal sums of squares reflect nonredundant sources of variation and hence can be added together to obtain the variation due to the comparisons jointly. For any treatment sum of squares, say, SS_A with $k_A - 1$ degrees of freedom, it is always possible to find a set of $k_A - 1$ mutually orthogonal comparisons. The corresponding $k_A - 1$ components of the treatment sum of squares, each component having one degree of freedom, will add up to SS_A, the value of the overall treatment sum of squares. Each orthogonal comparison accounts for a different one of the $k_A - 1$ degrees of freedom. It is never possible to find more than $k_A - 1$ mutually orthogonal comparisons among k_A means; the $k_A - 1$ mutually orthogonal comparisons use up all of the degrees of freedom and account for all of the variation among the k_A means.

Conversely, any sum of squares, say, SS_A, with $k_A - 1$ degrees of freedom, can be broken down into $k_A - 1$ orthogonal components, each with one degree of freedom. For values of k_A greater than two (that is, if there is more than one degree of freedom), this subdivision can be done in an infinite number of ways, each way resulting in one set of $k_A - 1$ mutually orthogonal comparisons. In practice, of course, we are not likely to be interested in more than one set of $k_A - 1$ orthogonal comparisons, since the different sets are not orthogonal to each other. Which of the infinite number of ways we choose to subdivide the treatment sum of squares depends on which comparisons make sense in a particular experiment. Indeed, we frequently are interested in only one or two specific comparisons, whether orthogonal or not, and not in an entire set of $k_A - 1$ orthogonal comparisons.

EXAMPLE 20.10 Let us return to the data in Example 20.6, portions of which are reprinted here for ease of reference. The means for the four levels of headline form (at the high-credibility level of the credibility factor) were based on 24 observations each:

	Assertion	Question	Denial	Control
Means:	4.65	4.28	3.54	2.84

The corresponding condition totals are readily obtained by multiplying each mean by the number of observations from which it was calculated, yielding

Totals:	111.60	102.72	84.96	68.16

with a grand total of 367.44. The omnibus sum of squares due to headline form may be calculated (from the one-way ANOVA formula) as

$$SS_{\text{form}} = \frac{1}{24}(111.60^2 + 102.72^2 + 84.96^2 + 68.16^2) - \frac{367.44^2}{96} = 46.5378$$

with $4 - 1 = 3$ degrees of freedom.

By computing sums of products of corresponding coefficients, you may verify that the following three comparisons are all mutually orthogonal:

Coefficients of $\bar{d}_5$:	1	1	1	−3
Coefficients of $\bar{d}_6$:	2	−1	−1	0
Coefficients of $\bar{d}_3$:	0	1	−1	0

These are basically the same as the first three comparisons in Example 20.6, except for multiplication by a constant to get rid of fractions and thus simplify the calculation of sums of squares. The numerical values of these comparisons are

$$\bar{d}_5 = 4.65 + 4.28 + 3.54 - (3)(2.84) = 3.95$$

$$\bar{d}_6 = (2)(4.65) - 4.28 - 3.54 + (0)(2.84) = 1.48$$

$$\bar{d}_3 = (0)(4.65) + 4.28 - 3.54 + (0)(2.84) = .74$$

The corresponding components of the omnibus sum of squares are

$$SS_{\bar{d}_5} = \frac{(24)(3.95^2)}{1^2 + 1^2 + 1^2 + (-3)^2} = 31.2050$$

$$SS_{\bar{d}_6} = \frac{(24)(1.48^2)}{2^2 + (-1)^2 + (-1)^2 + 0^2} = 8.7616$$

$$SS_{\bar{d}_3} = \frac{(24)(.74^2)}{0^2 + 1^2 + (-1)^2 + 0^2} = 6.5712$$

Because these three one-degree-of-freedom comparisons are mutually orthogonal, their components reflect three nonredundant sources of variation and hence add up to SS_{form}, the three-degree-of-freedom sum of squares representing differences among the means globally:

$$SS_{\bar{d}_5} + SS_{\bar{d}_6} + SS_{\bar{d}_3} = 31.2050 + 8.7616 + 6.5712 = 46.5378 = SS_{\text{form}}$$

For some purposes, it may be worth noting that $31.2050/46.5378 = 67\%$ of the variation due to headline form may be attributed to the comparison $\bar{d}_5$ of the three experimental conditions versus the control condition.

The subdivision of an omnibus sum of squares into single-degree-of-freedom components is sometimes directly useful for descriptive purposes, as a way of summarizing the data from an experiment. For instance, we can indicate what percentage of the variation among the treatment means is due to a specific one-degree-of-freedom comparison of interest, as illustrated in Example 20.10. This proportion is a potentially useful descriptive statistic.

In order to draw inferences that go beyond the sample data, however, we turn again from descriptive statistics to inferential statistics. In the next section, we consider how we might test whether the specific difference represented by a comparison is statistically significant.

20.3 SIGNIFICANCE TESTS FOR PLANNED COMPARISONS

Now that we know how to calculate the sum of squares [Expression (20.17)] due to a comparison $\bar{d}$, it is a straightforward matter to calculate the F ratio required to test the statistical significance of the comparison. The test statistic is

$$F_{\bar{d}} = \frac{\text{MS}_{\bar{d}}}{\text{MS}_{\text{error}}} \tag{20.19}$$

where $\text{MS}_{\bar{d}} = \text{SS}_{\bar{d}}/1 \text{ df} = \text{SS}_{\bar{d}}$, and MS_{error} is the error term (usually obtained from the ANOVA summary table) that would be used to test the omnibus effect of which the comparison $\bar{d}$ is a part. What is slightly less straightforward is the question as to what is the appropriate critical value with which to compare $F_{\bar{d}}$ to decide whether $F_{\bar{d}}$ reflects a statistically significant difference.

There are several approaches to testing the significance of comparisons. They all use $F_{\bar{d}}$ (or some function of $F_{\bar{d}}$) as the observed value of the test statistic, but they differ in what they use for the tabulated critical value against which $F_{\bar{d}}$ is tested.

The different approaches to significance testing for comparisons fall into two basic categories, and it is very important to understand the distinction between the two. The category known as **planned comparisons** is the topic of Section 20.3, and the category known as **postmortem comparisons** is defined and discussed in Section 20.4.

A **planned comparison** is one of a small number of particular comparisons that an experiment was especially designed to investigate. In this context, a small number typically means one or two. A planned comparison is formulated on the basis of the nature of the treatments or design, rather than on the basis of the observations or the observed means. Planned comparisons are sometimes called **a priori comparisons**, meaning that they are based on considerations that are logically prior to the data. In a planned comparison, the decision as to what to compare with what is determined by the design of the experiment and not by inspection of the experimental results.

Tests of planned comparisons among levels of an effect may be made regardless of whether the omnibus F for that effect is significant. For example, an experiment may be conducted to evaluate a new form of treatment relative to a control condition, and two traditional forms of treatment may be included in the study to obtain a more complete picture. In such an experiment, the difference between the new treatment and the control condition may be tested as a planned comparison, regardless of the outcome of the omnibus test of the treatment effect ($F_{\text{treatments}}$).

Critical Values for Planned-Comparison Tests

The test statistic for a planned comparison $\bar{d}$ is

$$F_{\bar{d}} = \frac{\text{MS}_{\bar{d}}}{\text{MS}_{\text{error}}}$$

with 1 and df_{error} degrees of freedom. To test $\bar{d}$ as a planned comparison at the α significance level, we simply compare $F_{\bar{d}}$ with the following:

$$\textbf{Planned-comparison critical } \boldsymbol{F}\textbf{-value} = F_\alpha(1, \text{df}_{\text{error}}) \qquad \textbf{(20.20)}$$

In other words, tests of planned comparisons are typical F tests in that their critical values are obtained from the F distribution with degrees of freedom corresponding to those for the numerator and denominator of the F ratio.

We noted in Section 11.2 that confidence intervals and significance tests provide alternative ways of looking at the same data: We use a confidence interval to try to specify what the relevant parameter is equal to, while we use a significance test to try to determine what it is not equal to. Thus, whether or not a particular comparison $\bar{d}$ is significant, a confidence interval for the corresponding population mean difference δ summarizes what the experiment tells us about the true mean difference δ. (The symbol δ is lower-case delta, the fourth letter of the Greek alphabet.) Based on principles in Table 17.2, it can be shown that the **estimated standard error of a comparison** $\bar{d}$ with coefficients c_j is

$$s_{\bar{d}} = \sqrt{\frac{(\sum c_j^2)\text{MS}_{\text{error}}}{m}} \qquad \textbf{(20.21)}$$

where m is the number of observations from which each mean that is involved in the comparison has been calculated, and MS_{error} is the pooled sample variance s^2_{pooled}. [Expression (11.34) is a special case of Expression (20.21).] The confidence limits for the population mean difference δ take the general form

(point estimate) ± (margin of error)

where the comparison $\bar{d}$ is the point estimate of the true mean difference δ, and the margin of error is a certain number, namely, $t_{\alpha/2}(\text{df}_{\text{error}})$, of estimated standard errors. Specifically, the **confidence interval for the population mean difference δ estimated by a planned comparison** $\bar{d}$ is given by

$$\bar{d} - t_{\alpha/2}s_{\bar{d}} < \delta < \bar{d} + t_{\alpha/2}s_{\bar{d}} \qquad \text{with confidence level } 1 - \alpha \qquad \textbf{(20.22)}$$

where $t_{\alpha/2}$ is obtained from the t distribution with df_{error} degrees of freedom. [Since F with one degree of freedom in the numerator is equal to t^2 (see

Section 12.4), the t-value in Expression (20.22) is equal to the square root of the planned-comparison critical F-value in Expression (20.20).]

EXAMPLE 20.11 Example 15.8 presented data from an investigation of average price levels at three different supermarkets. Prices were obtained on the same day for eight identical products sold at all three stores, yielding the following means:

	Store 1	Store 2	Store 3
Means:	196.125	185.625	197.000

The main effect of stores was found to be not significant [$F(2, 14) = 1.42$, $P > .05$] in Example 15.12. The error term in the omnibus F test was the mean square for the products-by-stores interaction, $MS_{PS} = 225.78$, with 14 degrees of freedom.

If the objective of the study was to evaluate the difference between independently owned supermarkets (Stores 1 and 3) and one owned by a large multinational corporation (Store 2), then we can test that difference as a planned comparison regardless of the outcome of the omnibus F test. Such a test would not be appropriate if the comparison had been suggested by inspection of the means rather than by the design. The relevant comparison is

$$\bar{d} = \frac{1}{2}(196.125 + 197.0) - 185.625 = \frac{1}{2}(196.125) - 185.625 + \frac{1}{2}(197.0)$$
$$= 10.9375 \text{ cents}$$

Notice that we have left the comparison coefficients as fractions in order to have cents as the units of measurement for the difference. Multiplying the comparison by a constant to get rid of fractional coefficients would not change what is being compared with what, but it would change the units of measurement, making the difference less easily interpreted.

The sum of squares due to the comparison $\bar{d}$ is

$$SS_{\bar{d}} = \frac{(8)(10.9375^2)}{(\frac{1}{2})^2 + (-1)^2 + (\frac{1}{2})^2} = 638.02$$

with 1 degree of freedom. Using the same error term as for the omnibus main effect of stores, we obtain the test statistic

$$F_{\bar{d}} = \frac{638.02}{225.78} = 2.83$$

with 1 and 14 degrees of freedom. The critical value for testing $\bar{d}$ as a planned comparison at the 5% level is $F_{.05}(1, 14) = 4.60$. Since the observed value $F_{\bar{d}}$ is less than the critical value, we conclude that the mean price difference of 10.9 cents per item between the multinational store (Store 2) and the independent stores (1 and 3) was not statistically significant ($P > .05$).

The estimated standard error of the difference $\bar{d}$ is calculated from Expression (20.21) as

$$s_{\bar{d}} = \sqrt{[(\tfrac{1}{2})^2 + (-1)^2 + (\tfrac{1}{2})^2]\frac{225.78}{8}} = 6.5064 \text{ cents}$$

Since $t_{.025}(14\ \text{df}) = 2.145$, the 95% confidence limits for the true mean difference are

$$10.9375 \pm (2.145)(6.5064) = 10.94 \pm 13.96 \text{ cents}$$

yielding a 95% confidence interval from −3.0 to 24.9 cents per item for the true mean amount by which prices in Stores 1 and 3 exceeded those in Store 2 on the day of the investigation. The fact that the interval includes a difference of zero reflects the fact that the comparison was not statistically significant. The width of the confidence interval indicates how precisely (or imprecisely) the investigation estimated the true mean difference. It should be noted, however, that the use of an error term involving a repeated-measures factor (stores) with more than two levels requires the assumption of homogeneity of covariance (see Section 19.4). If the assumption is suspect, as when the levels of the repeated-measures factor represent successive points in time, a conservative confidence interval may be obtained by using the conservative degrees of freedom; in this case, $t_{.025}(7\text{ df}) = 2.365$.

Trend Analysis

The primary reason for wanting to make planned comparisons is to see whether certain hypothesized differences appear in the data. **Trend analysis** is a special case of planned comparisons for examining the nature of the functional relationship between the dependent and the independent variables. Such an examination may be of interest if the values of the independent variable are specific numerical values of a continuous variable.

The aspect of trend analysis that distinguishes it from other analyses of planned comparisons lies in the choice of the coefficients for the various comparisons. When the levels of the independent variable are evenly spaced, the **coefficients for orthogonal trend comparisons** are those shown in Appendix 10. It is important to note that the coefficients in Appendix 10 apply to evenly spaced factor levels only. (For example, if dosage of a drug is the independent variable, then dosages of 5, 10, 15, and 20 mg represent even spacing of levels, whereas 5, 10, 50, and 100 mg are unevenly spaced levels.) Coefficients can be determined for the unevenly spaced case as well (see the references under Further Reading at the end of this chapter), but we will limit ourselves to the case of equal spacing in this book.

Orthogonal trend comparisons are useful for evaluating the shape of the curve relating the response variable to the independent variable. Each trend comparison may be tested to determine whether there are significant differences of a particular shape among the k means representing the k levels of the independent variable. Figure 20.1 shows various patterns of differences among means. If the k means were all identical, then all of the trend comparisons (which are differences exhibiting particular patterns) would be zero, as shown in Part (a) of Figure 20.1. The **linear comparison**, shown in Part (b) of the figure, reflects differences among the means such that the means tend to fall in a straight line with nonzero slope. (A linear equation is a polynomial of degree 1.) The **quadratic comparison** reflects differences among the means such that the means tend to lie along a smooth, one-bend curve (polynomial of degree 2), as shown in Part (c) of the figure. The **cubic comparison**, shown in Part (d), reflects differences among the means such that the means tend to lie along an S-shaped or two-bend curve (polynomial of degree 3). In general, the comparison for the polynomial of degree g reflects

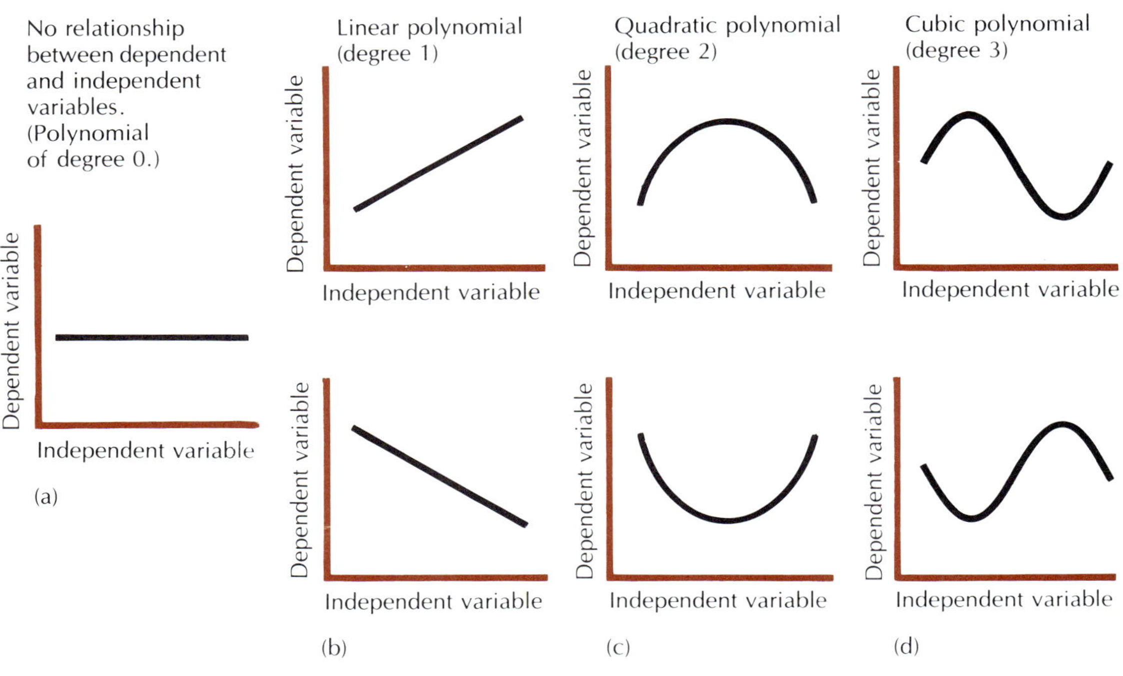

Figure 20.1
Schematic drawing showing various forms of relationship between the dependent and independent variables. The figure includes two examples of each degree of polynomial from 1 to 3.

differences such that the means follow a $(g - 1)$-bend curve, or **polynomial of degree g**, which may be written as follows:

$$\text{Dependent variable} = b_0 + b_1 I + b_2 I^2 + \cdots + b_g I^g \tag{20.23}$$

where I is the independent variable. Tests of orthogonal trend comparisons allow us to draw conclusions about any polynomial pattern that may exist in the dependent variable values as a function of the levels of the independent variable. [In addition, the actual coefficients b_j of the polynomial in Expression (20.23) may be estimated from the data; see the references under Further Reading at the end of this chapter.]

It may very well happen that two or more orthogonal trend comparisons are statistically significant. The degree of polynomial required to account for the significant variation in the data is the highest degree of trend that is significant. For example, if the linear and cubic comparisons were significant, a cubic polynomial would be required to describe the relationship between the dependent and independent variables.

The symbol k in Appendix 10 refers to the number of levels of the independent variable. As we have seen in Section 20.2, the omnibus variation among k means can be partitioned into $k - 1$ mutually orthogonal comparisons. For each value of k from 3 through 6, Appendix 10 gives the coefficients for the $k - 1$ mutually orthogonal comparisons representing polynomials of degree 1 through $k - 1$. For values of k from 7 through 10, Appendix 10 gives coefficients for only the first five mutually orthogonal polynomial comparisons: linear, quadratic, cubic, **quartic,** and **quintic**. In practice, the linear and quadratic—and perhaps the cubic—comparisons are most likely to be of interest; higher-order trends are usually lumped together as **deviations from linearity**, or **deviations from linear and quadratic trend**, or **higher-order trends**, as the case may be. In any event, the comparisons in Appendix 10 constitute one way of subdividing the sum of squares (for example, SS_A) due to differences among k means.

EXAMPLE 20.12 Four independent samples of used Honda Civics for sale were obtained from newspaper advertisements. The advertised prices (in dollars) of the five used cars in each of the four samples were as follows, the four samples being distinguished by the age of the car.

3 years old	4 years old	5 years old	6 years old
3,875	2,200	1,800	1,500
4,195	2,500	1,800	1,650
4,295	2,900	2,700	1,700
4,500	3,150	2,950	1,750
5,995	3,250	3,250	1,795

What is the nature of the relationship between the price of a used Honda Civic and its age?

An informal inspection of the data reveals that the dispersions in the four samples are rather unequal, suggesting that the data may violate the assumption of homogeneity of variance. The range varies over the four samples by a factor of 7.2, from \$295 for 6-year-old cars to \$2,120 for 3-year-old-cars. The discrepancy in the dispersions is reduced to a factor of 2.7 by the reciprocal transformation, a greater reduction than that provided by either the square-root or the logarithmic transformation, according to the range method of Section 18.6. Moreover, the reciprocal of car price can be expressed in easily interpretable units as the proportion of the car's price that is covered by, say, \$1,000, or the number of identical cars that, say, \$10,000 would buy. Thus, we will analyze the data in reciprocal form, as "purchasability" rather than price. Specifically, let us replace each price x by $1{,}000/x$, which is the "proportion of a car that \$1,000 will buy," as shown in the table on page 740.

The four sample means, which are plotted in Figure 20.2, show that the older the car in the sample, the larger was the proportion of its price that would be covered by \$1,000, from about 22% of the price for 3-year-old cars to 60% for 6-year-old cars. The significance of the difference due to age is tested in the following ANOVA summary

	3 years old	4 years old	5 years old	6 years old	Total
	.258	.455	.556	.667	
	.238	.400	.556	.606	
	.233	.345	.370	.588	
	.222	.317	.339	.571	
	.167	.308	.308	.557	
Total	1.118	1.825	2.128	2.989	8.060
Mean	.224	.365	.426	.598	

table for these reciprocal-transformed data in a one-way design with four independent samples.

Source of variation	SS	df	MS	F	
Age of car	.36056	3	.12019	22.5	$P < .01$
Cars within age levels	.08540	16	.00534	—	
Total	.44596	19			

The effect of age on the price (actually, purchasability) of used cars was significant at the 1% level.

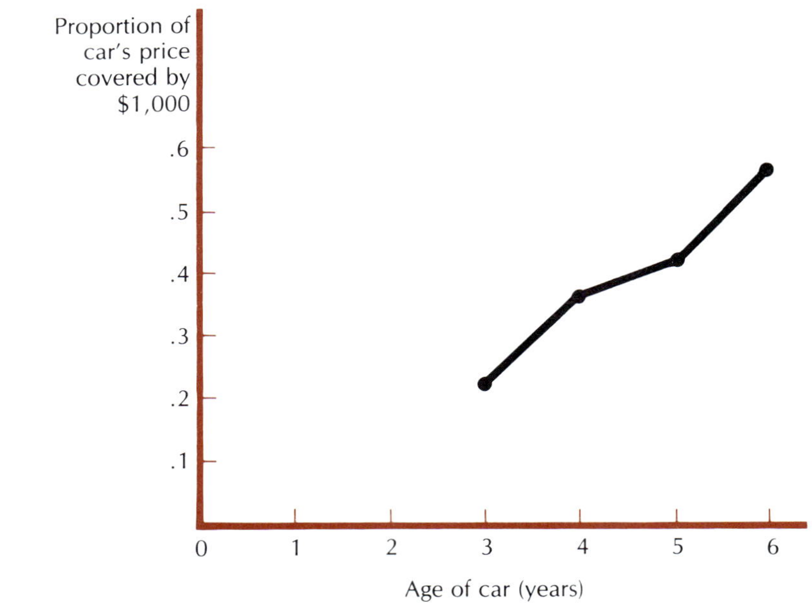

Figure 20.2 The proportion of a used car's price covered by $1,000, as a function of the age of the car, based on data from 20 Honda Civics.

We can use trend analysis to determine the nature of the relationship between purchasability and age. The sum of squares due to age, $SS_A = .36056$ with 3 degrees of freedom, can be partitioned into three components due to orthogonal polynomial comparisons:

$$SS_A = SS_{\text{linear}} + SS_{\text{quadratic}} + SS_{\text{cubic}}$$

where the three single-degree-of-freedom components are computed from Expression (20.17) using the coefficients given in Appendix 10 for $k = 4$ levels. The orthogonal trend comparisons (with coefficients from Appendix 10) are

$$\bar{d}_{\text{linear}} = (-3)(.224) + (-1)(.365) + (1)(.426) + (3)(.598) = 1.183$$
$$\bar{d}_{\text{quadratic}} = (1)(.224) + (-1)(.365) + (-1)(.426) + (1)(.598) = .031$$
$$\bar{d}_{\text{cubic}} = (-1)(.224) + (3)(.365) + (-3)(.426) + (1)(.598) = .192$$

The corresponding components of the sum of squares due to age are

$$SS_{\text{linear}} = \frac{(5)(1.183^2)}{(-3)^2 \quad (-1)^2 + 1^2 + 3^2} = .35013$$
$$SS_{\text{quadratic}} = \frac{(5)(.031^2)}{1^2 + (-1)^2 + (-1)^2 + 1^2} = .00120$$
$$SS_{\text{cubic}} = \frac{(5)(.192^2)}{(-1)^2 + 3^2 + (-3)^2 + 1^2} = .00922$$

Notice that the linear trend accounts for $.3501/.3606 = 97\%$ of the variation due to age of the car.

Each trend component may be tested individually as a planned comparison using the same error term as for the omnibus F test of the overall effect. The three F ratios are

$$F_{\text{linear}} = \frac{.35013}{.00534} = 65.6$$
$$F_{\text{quadratic}} = \frac{.00120}{.00534} = .23$$
$$F_{\text{cubic}} = \frac{.00922}{.00534} = 1.73$$

Critical values of the F distribution with 1 and 16 degrees of freedom indicate that the linear component is significant at the .01 level, and the other two components are not significant ($P > .10$). There is a significant linear relationship between the dependent and independent variables.

As an alternative approach, we may be interested in testing only whether the linear comparison is significant and whether there are any significant deviations from linearity. The latter test may be conducted by computing a sum of squares for deviations from linearity either as

$$SS_{\text{dev.lin.}} = SS_{\text{quadratic}} + SS_{\text{cubic}} = .0012 + .0092 = .0104$$

with $1 + 1 = 2$ degrees of freedom, or equivalently as

$$SS_{\text{dev.lin.}} = SS_{\text{age}} - SS_{\text{linear}} = .36056 - .35013 = .0104$$

with $(k_A - 1) - 1 = (4 - 1) - 1 = 2$ degrees of freedom. Under this approach, the ANOVA summary table would be as follows.

Source of variation	SS		df	MS		F		
Age of car	.36056		3	.12019		22.5		$P < .01$
Linear trend		.35013	1		.35013		65.6	$P < .01$
Higher-order trend		.01043	2		.00521		.98	Not significant
Cars within age	.08540		16	.00534				
Total	.44596		19					

The results indicate that on the average, for the type of car investigated, the proportion of a car's price that $1,000 would cover increases linearly with increasing age of the car, in the range from three to six years. There were no significant deviations from linearity within the age range investigated. The relationship in these data is shown in Figure 20.2.

(Source: Student project by E. Nielsen.)

In summary, **trend analysis** is a special case of planned comparisons where the coefficients for the comparisons are chosen so as to reflect treatment differences of a particular shape or pattern relative to the levels of the independent variable. It provides a way of subdividing a main-effect sum of squares to test hypotheses concerning the form of the polynomial that best represents the relationship of the dependent to the independent variable. Trend analysis is especially useful in testing a theory that predicts a particular form of relationship. We can test the significance of that form of representation, as well as the significance of higher-order deviations from a polynomial of the predicted degree.

One warning is in order: There is no statistical basis for extrapolating the observed polynomial relationship beyond the range of independent-variable values included in the investigation. Trend analysis is a form of regression analysis, and the warning concerning extrapolation in Section 5.1 applies here, too. The last paragraph in Example 20.12 illustrates how trend analysis results should be qualified to indicate the range of independent-variable values to which the results apply.

20.4 SIGNIFICANCE TESTS FOR POSTMORTEM COMPARISONS

Section 20.3 noted that there are two basic categories of significance tests for comparisons. They are called tests of **planned comparisons** and tests of **postmortem comparisons**. Both categories involve evaluation of the same test statistic:

$$F_{\bar{d}} = \frac{\text{MS}_{\bar{d}}}{\text{MS}_{\text{error}}}$$

However, they differ in the critical value that must be exceeded by $F_{\hat{a}}$ in order to claim significance. The difference in critical value reflects a fundamental and important difference with respect to why a particular comparison is being tested. A planned comparison is one of a small number of particular comparisons (typically one or two) that the experiment was especially designed to investigate. The following paragraphs show how a planned comparison differs from a postmortem comparison and how the significance testing procedure is affected by the difference.

Postmortem comparisons are used to help interpret a significant omnibus F test when the researcher has no planned comparisons to guide the search for which specific differences among the means are responsible for the omnibus effect. Postmortem comparisons sometimes consist of a complete set of comparisons of a particular type, such as in testing all $\binom{k}{2} = \frac{k(k-1)}{2}$ simple differences among k means taken two at a time, and sometimes they represent specific differences suggested by the data, such as in comparing the largest mean with the smallest mean. Because they are not limited to a small number of specific planned comparisons, procedures for postmortem comparisons are also known as procedures for **multiple comparisons**. In addition, postmortem comparisons may be referred to as **a posteriori comparisons**, meaning that they are not independent of the way the experiment came out but are logically posterior to the data.

Note that postmortem procedures are used for a detailed analysis of k means in cases where an omnibus test has already indicated that there are significant differences. If there are no global differences as indicated by an omnibus F test of the effect, then it is neither necessary nor appropriate to use postmortem comparisons to look for which means differ from which. The omnibus F test has already given the answer in that case.

The difference between planned and postmortem comparisons resides essentially in whether the comparison is to be evaluated unconditionally or conditionally. Test statistics (F-values) for planned comparisons are unconditional in the same sense that omnibus F ratios for main effects and interactions are unconditional: They are based on prior-to-the-data considerations and lead to P-values that apply only to single test results in isolation. With each test at the 5% level, we are in effect asking, What is the probability that in a single random sample, sight unseen, from a population with no treatment effect, we will obtain a sample that is among the 5% of the samples that look most as if there is a treatment effect in the population? (These 5% are the ones that lead to Type I errors for .05-level tests.) In such sight-unseen cases, the answer is 5%, and so we use a critical value cutting off 5% of the usual "unconditional" distribution to limit the Type I error rate to 5%.

Test statistics for postmortem comparisons are conditional on some aspect of the particular outcome of the experiment, such as the relative magnitudes of the means or the number of comparisons being tested. In this case, we might ask, If—rather than testing a single random sample, sight unseen—we look around a little bit for a likely looking sample (or part of a sample) to test, what is the probability that we will obtain a sample that is among the aforementioned 5% that look most as if there is a treatment effect in the population? Obviously, it is easier to find one of the samples showing large differences if we are allowed more than one try or if we are allowed to base our choice on the size of the differences rather than choosing sight unseen. Hence, a critical value from the usual "unconditional" distribution would lead to a Type I error rate greater than 5%. Consequently, the conditional nature of postmortem test statistics is taken into account in the critical value that must be attained to establish statistical significance: Critical values are higher for postmortem comparisons than for planned comparisons because postmortem statistics are conditional on some aspect of the data. They are not isolated selections, sight unseen.

There are several different postmortem procedures, distinguished by how much more conservative they are than planned-comparison tests. Here, the term **conservative** implies less likely to result in statistical significance. The more conservative a test is, the lower its Type I error rate for each single comparison and the lower its power to detect true differences in the population. The following paragraphs describe three postmortem procedures in order from most conservative to most liberal.

Scheffé's Test

Scheffé's test for postmortem comparisons ensures that no matter how many comparisons are made in a set of k means, the Type I error rate for all comparisons considered jointly does not exceed the chosen significance level α. Such a test is said to control the **experimentwise error rate**, which is the proportion of *experiments* yielding Type I errors in tests of comparisons. (By way of contrast, a planned-comparison test controls the **per-comparison error rate**, which is the proportion of *comparisons* yielding Type I errors. This type of control is appropriate if only one or two planned comparisons are being tested. If many comparisons in the same experiment were tested by the planned-comparison procedure, the experimentwise error rate would be much greater than α.) Scheffé's test is especially useful when the researcher wishes to make unplanned comparisons other than simple pairwise differences—for example, an unplanned comparison of three means jointly versus one other mean.

To test a comparison among k means at the α level by Scheffé's test, we calculate $F_{\hat{d}}$ as usual and then evaluate it against a critical value equal to $k - 1$ times the α-level critical value for testing the omnibus effect of which the comparison is a part; that is,

$$\textbf{Scheffé test critical } \boldsymbol{F}\textbf{-value} = (k - 1)F_\alpha(k - 1, \text{df}_{\text{error}}) \qquad \textbf{(20.24)}$$

Notice that k is the number of means in the total set of means; k remains constant for a given experiment whether or not some of the coefficients in the comparison are zero. Thus, for comparisons among levels of Factor A, k would be k_A, the number of levels of Factor A; for comparisons among the

$k_A k_B$ cell means in a k_A-by-k_B factorial design, k would be $k_A k_B$; and so on. It is worth noting that if the omnibus F test for a treatment effect is not significant, then it is impossible for any comparisons among those treatment means to reach significance using Scheffé's test.

Just as the critical value for a postmortem significance test is greater than that for a planned comparison, so also a confidence interval for the true mean difference δ estimated by a postmortem comparison $\bar{d}$ is wider than one based on a planned comparison. Specifically, the **confidence interval for the population mean difference δ estimated by a Scheffé comparison** $\bar{d}$ is given by

$$\bar{d} - \sqrt{(k-1)F_\alpha}\, s_{\bar{d}} < \delta < \bar{d} + \sqrt{(k-1)F_\alpha}\, s_{\bar{d}} \qquad \text{with confidence level } 1-\alpha \tag{20.25}$$

where F_α is obtained from the F distribution with $k - 1$ and df_{error} degrees of freedom. Notice that this interval has the same form as that based on planned comparisons in Expression (20.22), except that $t_{\alpha/2}$, which is the square root of the planned-comparison critical value,

$$t_{\alpha/2}(\text{df}_{\text{error}}) = \sqrt{F_\alpha(1, \text{df}_{\text{error}})} \tag{20.26}$$

is replaced here by the square root of the Scheffé test critical value.

EXAMPLE 20.13 Example 18.14 found a significant difference in the yields of three varieties of barley. The mean yields for the three varieties, each based on 8 observations, were as follows:

Bonanza: 321/8 = 40.125 bushels per acre

Klondike: 341/8 = 42.625 bushels per acre

Fairfield: 386/8 = 48.250 bushels per acre

(The means are left unrounded for purposes of calculation; see Appendix 1.)

The plot of the means for the three varieties in Figure 18.1(b) suggests that most of the difference among varieties may be due to the higher yield of Fairfield as compared to the other two varieties. Let us test this comparison, which is part of the significant main effect of variety. The comparison, say, $\bar{d}_1$, of Bonanza and Klondike jointly versus Fairfield is

$$\bar{d}_1 = \frac{40.125 + 42.625}{2} - 48.25 = \frac{1}{2}(40.125) + \frac{1}{2}(42.625) + (-1)(48.25) = -6.875$$

The corresponding component of the sum of squares due to variety is

$$\text{SS}_{\bar{d}_1} = \frac{(8)(-6.875)^2}{(\frac{1}{2})^2 + (\frac{1}{2})^2 + (-1)^2} = 252.083$$

with one degree of freedom. Because the comparison $\bar{d}_1$ is part of the main effect of variety, the appropriate error term is the error term for variety in the ANOVA summary table of Example 18.14, namely, $\text{MS}_{R(VFW)} = 19.1$ with 12 degrees of freedom. The test statistic therefore is

$$F_{\bar{d}_1} = \frac{252.083}{19.1} = 13.2$$

Since this was not a planned comparison but rather was suggested by an inspection of the data, it must be tested as a postmortem comparison. The critical value for Scheffé's test at the .05 level is given by Expression (20.24) as

$$2F_{.05}(2, 12) = (2)(3.89) = 7.78 \qquad \textbf{(20.27)}$$

The observed value 13.2 is greater than the critical value 7.78, and we conclude that the mean yield of Fairfield barley was significantly greater than the combined mean yield of Bonanza and Klondike ($P < .05$).

We could also use Scheffé's test to compare Fairfield with each of Bonanza and Klondike separately. In a comparison of Bonanza versus Fairfield, we have

$$\bar{d}_2 = 40.125 - 48.25 = -8.125$$

$$\text{SS}_{\bar{d}_2} = \frac{(8)(-8.125)^2}{1^2 + 0^2 + (-1)^2} = 264.0625$$

$$F_{\bar{d}_2} = \frac{264.0625}{19.1} = 13.8$$

In a comparison of Klondike versus Fairfield, we have

$$\bar{d}_3 = 42.625 - 48.25 = -5.625$$

$$\text{SS}_{\bar{d}_3} = \frac{(8)(-5.625)^2}{0^2 + 1^2 + (-1)^2} = 126.5625$$

$$F_{\bar{d}_3} = \frac{126.5625}{19.1} = 6.63$$

The critical value for Scheffé's test at the .05 level [from Expression (20.27)] is 7.78 for both comparisons $\bar{d}_2$ and $\bar{d}_3$, so we can conclude that the mean yield of Fairfield was significantly greater than that of Bonanza ($P < .05$) but not significantly greater than that of Klondike ($P > .05$).

The estimated standard error of each of the differences $\bar{d}_2$ and $\bar{d}_3$ is given by Expression (20.21) as

$$s_{\bar{d}} = \sqrt{\frac{[1^2 + (-1)^2]\,19.083}{8}} = 2.184 \text{ bushels per acre}$$

Under the conditions of this investigation, the 95% confidence limits for the true mean difference between Bonanza and Fairfield are

$$-8.125 \pm (\sqrt{7.78})(2.184) = -8.125 \pm 6.092$$

yielding a 95% confidence interval from −14.2 to −2.0 bushels per acre based on Scheffé's procedure. More clearly, we would say that the 95% confidence interval is from 2.0 to 14.2 bushels per acre in favor of Fairfield over Bonanza. The 95% confidence limits for the true mean difference between Klondike and Fairfield are

$$-5.625 \pm (\sqrt{7.78})(2.184) = -5.625 \pm 6.092$$

yielding a 95% confidence interval from −11.7 to .5 bushel per acre based on Scheffé's procedure. Thus, we can be 95% confident that the mean yield of Fairfield is between .5 bushel per acre less and 11.7 bushels per acre more than that of Klondike, under the conditions of this investigation.

Scheffé's test is recommended when an omnibus F test is significant and the researcher wants to explore the nature of the significant effect using unplanned comparisons that are not limited to simple pairwise differences of two means at a time. Scheffé's test controls the experimentwise error rate with no restrictions concerning the number, nature, or orthogonality of the comparisons to be tested. The price that the researcher pays for this control is a reduction in the power to detect true differences. The next subsection introduces a postmortem test with greater power for use when we are interested in a more restricted set of comparisons.

Tukey's Test

Tukey's test for postmortem comparisons is also known as **Tukey's honestly significant difference** method (see Biography 2.1). In its usual form—and the only form we discuss here—it is limited to making pairwise comparisons between two means at a time. That is, it is used to test the simple difference between one treatment mean and another treatment mean, selected from a set of k means. The maximum number of such differences in a set of k means $\binom{k}{2} = \dfrac{k(k-1)}{2}$. Tukey's test ensures that no matter how many of these pairwise comparisons are made, whether some of them or all of them, the Type I error rate for all such comparisons considered jointly does not exceed the chosen significance level α. In other words, Tukey's test controls the experimentwise error rate for the case where comparisons are limited to pairwise differences.

To perform an α-level test of a difference $\bar{d}$ between two means selected from a set of k means, Tukey's test evaluates the usual test statistic $F_{\bar{d}}$ against the critical value given by

$$\textbf{Tukey test critical } \boldsymbol{F}\textbf{-value} = \tfrac{1}{2}[Q_{\alpha}(k, \text{df}_{\text{error}})]^2$$

where the quantity that is squared is the value that cuts off the upper α-area tail of the studentized range (Q) distribution for k means estimated with df_{error} degrees of freedom for the error term. The studentized range distribution, which is tabulated in Appendix 11, is the sampling distribution of the **studentized range statistic**:

$$Q = \frac{\bar{x}_{\text{max}} - \bar{x}_{\text{min}}}{s_{\bar{x}}} \qquad \text{(2[illegible])}$$

under the null hypothesis of equal population means. (The term *range* reflects the fact that the statistic is based on the difference between the largest and the smallest means in a set of k means, and the term *studentized* indicates that the difference is evaluated with respect to the estimated standard error of the mean.) An alternative procedure for conducting Tukey's test is to calculate the value of the Q statistic directly for a comparison $\bar{d}$ of two means based on m

observations each by means of Expression (20.29):

$$Q_{\bar{d}} = |\bar{d}| \Big/ \sqrt{\frac{\text{MS}_{\text{error}}}{m}} = \sqrt{2F_{\bar{d}}} \qquad \textbf{(20.30)}$$

and then to test it against the critical Q-value $Q_\alpha(k, \text{df}_{\text{error}})$. Notice again that k is the number of means in the total set of means, as was discussed in connection with Scheffé's test.

When the number k of means is greater than two, confidence intervals based on Tukey's procedure will be wider than those based on planned comparisons but narrower than those based on Scheffé's procedure. Specifically, the **confidence interval for the population mean difference δ estimated by a Tukey comparison** $\bar{d}$ is given by

$$\bar{d} - \sqrt{\tfrac{1}{2}Q_\alpha^2}\, s_{\bar{d}} < \delta < \bar{d} + \sqrt{\tfrac{1}{2}Q_\alpha^2}\, s_{\bar{d}} \qquad \text{with confidence level } 1 - \alpha \qquad \textbf{(20.31)}$$

where Q_α is obtained from the studentized range distribution for k means and df_{error} degrees of freedom. Notice that all of the confidence limits in this chapter [see Expressions (20.22), (20.25), and (20.31)] have the same form, namely,

$$\bar{d} \pm (\text{square root of critical } F\text{-value})\,(\text{standard error of } \bar{d}) \qquad \textbf{(20.32)}$$

EXAMPLE 20.14 Example 20.13 used Scheffé's test to demonstrate postmortem comparisons among the $k = 3$ means involved in a significant main effect due to variety with respect to barley yield. There are

$$\binom{k}{2} = \frac{(3)(2)}{2} = 3 \text{ possible pairs of means}$$

whose differences can be tested. In Example 20.13, we calculated the test statistic $F_{\bar{d}_2} = 13.8$ for the Bonanza-versus-Fairfield difference and $F_{\bar{d}_3} = 6.63$ for the Klondike-versus-Fairfield difference. Using the same procedure, we obtain

$$\bar{d}_4 = 40.125 - 42.625 = -2.500 \text{ bushels per acre}$$

and $F_{\bar{d}_4} = 1.31$ for the Bonanza-versus-Klondike difference. All three of these pairwise comparisons can be tested by Tukey's procedure at the 5% level by evaluating the F test statistic against the critical F-value:

$$\frac{1}{2}[Q_{.05}(3, 12)]^2 = \frac{1}{2}(3.77)^2 = 7.11$$

based on the studentized range statistic for $k = 3$ means and $\text{df}_{\text{error}} = 12$. Checking which of the $F_{\bar{d}}$ values exceed the critical value 7.11, we conclude that the mean yield differed significantly ($P < .05$) between Bonanza and Fairfield, but not between Klondike and Bonanza or between Klondike and Fairfield.

The estimated standard error of each of the pairwise differences was calculated in Example 20.13 as $s_{\bar{d}} = 2.184$ bushels per acre. The 95% confidence limits for the true mean difference between Bonanza and Fairfield under the conditions of this study are

$$-8.125 \pm (\sqrt{7.11})(2.184) = -8.125 \pm 5.824$$

yielding a 95% confidence interval from −13.9 to −2.3 bushels per acre based on Tukey's procedure. The 95% confidence interval is from 2.3 to 13.9 bushels per acre in favor of Fairfield over Bonanza. The 95% confidence limits for the true mean difference between Klondike and Fairfield under the conditions of this study are

$$-5.625 \pm (\sqrt{7.11})(2.184) = -5.625 \pm 5.824$$

and the 95% confidence interval based on Tukey's procedure is from .2 bushel per acre in favor of Klondike to 11.4 bushels per acre in favor of Fairfield. Similarly, the 95% confidence limits for the true mean difference between Bonanza and Klondike are

$$-2.500 \pm 5.824$$

and the 95% confidence interval based on Tukey's procedure is from 3.3 bushels per acre in favor of Bonanza to 8.3 bushels per acre in favor of Klondike under the conditions of this investigation.

Notice that the margin of error (5.824) for Tukey's confidence intervals is smaller than the margin of error (6.092) for Scheffé's confidence intervals in Example 20.13, yielding narrower confidence intervals. This difference reflects the increased power that we gain with Tukey's method by examining only pairwise differences.

Tukey's test is recommended when an omnibus F test is significant and the researcher wants to determine which means differ significantly from which other means using unplanned comparisons of two means at a time selected from the set of k treatment means. Given this restriction on the type of comparison to be tested, Tukey's test controls the experimentwise error rate at the chosen significance level.

The Protected *t* Test

The **protected *t* test for postmortem comparisons** is also known as the **protected least significant difference** method. The test involves using the planned-comparison procedure, provided that the relevant omnibus F test was significant. The procedure is called a t test because F tests with one degree of freedom for the numerator are equivalent to t tests [see Expressions (12.26) and (20.26)]. The term *protected* indicates that the requirement of a significant omnibus F ratio provides some protection against inflation of the Type I error rate due to multiple comparisons.

To test a comparison $\bar{d}$ by the protected t test at the α level, we must first determine that the omnibus effect of which the comparison is a part is significant in the ANOVA. If the omnibus F test is significant, then we evaluate the usual test statistic $F_{\bar{d}}$ against the following:

$$\textbf{Protected } t \textbf{ test critical } F\textbf{-value} = F_{\alpha}(1, \text{df}_{\text{error}}) \qquad (20.33)$$

which is the same critical value as that used in tests of planned comparisons. **Confidence intervals for the population mean difference δ estimated by a protected *t* comparison** are identical to those for planned comparisons, as given in Expression (20.22).

EXAMPLE 20.15 Examples 20.13 and 20.14 demonstrated postmortem comparisons among $k = 3$ means involved in a main effect that was judged significant by an omnibus F test with 2 and 12 degrees of freedom. The same test statistic $F_{\hat{a}}$ is calculated for any given comparison, whether it is to be tested by Scheffé's test, Tukey's test, or the protected t test. Only the critical values differ, as shown here for comparative purposes for the data analyzed in Examples 20.13 and 20.14:

$$\text{Scheffé's test critical } F\text{-value} = 2F_{.05}(2, 12) = 7.78$$

$$\text{Tukey's test critical } F\text{-value} = \frac{1}{2}[Q_{.05}(3, 12)]^2 = 7.11$$

$$\text{Protected } t \text{ test critical } F\text{-value} = F_{.05}(1, 12) = 4.75$$

The critical values show that Scheffé's test is more conservative (less likely to yield significant results, whether rightly or wrongly) than Tukey's test, which is in turn more conservative than the protected t test.

The corresponding confidence intervals reflect the same pattern, as is illustrated by comparing the margins of error in an estimated pairwise difference as evaluated by the three procedures at the 95% confidence level:

Scheffé's procedure: $(\sqrt{7.78})(2.184) = 6.092$
Tukey's procedure: $(\sqrt{7.11})(2.184) = 5.824$
Protected t procedure: $(\sqrt{4.75})(2.184) = 4.760$

The last value, the margin of error claimed by the protected t test, is sometimes called the **least significant difference**. It is the smallest difference that would be judged a significant difference between two means according to the protected t test.) A smaller margin of error leads to a narrower confidence interval, which in turn indicates greater precision in our knowledge of the relevant parameter. However, some authorities question whether the confidence level is credible in the protected t procedure—whether the procedure claims more precise knowledge than is warranted.

The protected t test is somewhat controversial because it makes no adjustment for the number of comparisons being tested. Consequently, the Type I error rate does not exceed the chosen significance level α for each comparison considered singly, but if many comparisons are made, the error rate for all comparisons considered jointly may be very high. The critical values are set to control the per-comparison error rate, rather than the experimentwise error rate. There is some control of the experimentwise error rate, however, because of the requirement that the omnibus F test must be significant, but this does not provide as strict control as do Tukey's test and Scheffé's test (which also follow significant omnibus F tests).

Some statisticians maintain that the protected t test provides the best balance between adequate power to detect true mean differences and adequate protection against excessive Type I error rates. Others argue that its power advantage is purchased at the cost of too many Type I errors, and that the significance level claimed after the use of protected t tests is misleading. They recommend using Scheffé's test and Tukey's test so that reported significance levels are sure to be believable, and they further suggest that significance levels of 10% or 20% rather than 5% can be used if greater power is desired.

To some extent, a personal judgment is required in the choice of postmortem procedures. The three methods described in this section are three of the most widely used of the many postmortem procedures that have been proposed. If strict control of Type I error rates is desired in a particular application, Scheffé's test or Tukey's test should be used—Tukey's if comparisons are limited to pairwise differences and Scheffé's otherwise. On the other hand, if protected t tests can help to make sense of significant main effects, they may serve as important tools in the interpretation of experimental results, even if their experimentwise error rates are less rigorously defined. In any case, we should always specify which test procedure was used whenever postmortem comparisons are tested.

20.5 COMPARISONS AMONG MEANS BASED ON UNEQUAL SAMPLE SIZES

Up to this point, we have considered only the case where every one of the k means in the comparison is based on the same number (m) of observations. What happens if the sample sizes are unequal for the different means?

Unequal sample sizes require no change in the definition of a comparison; we still have

$$\bar{d} = \sum c_j \bar{x}_j \qquad \text{where } \sum c_j = 0$$

However, the formula for a component of the treatment sum of squares must be modified to allow for unequal sample sizes, so we have

$$\text{SS}_{\bar{d}} = \frac{\bar{d}^2}{\Sigma\,(c_j^2/m_j)} \tag{20.34}$$

where m_j is the number of observations on which the jth mean $\bar{x}_j$ is based. Similarly, the estimated standard error of a comparison $\bar{d}$ becomes

$$s_{\bar{d}} = \sqrt{\text{MS}_{\text{error}} \sum \frac{c_j^2}{m_j}} \tag{20.35}$$

With these modifications, the significance tests and confidence intervals described in this chapter may be computed with unequal sample sizes. When all the m_j are equal (say, to m), Expressions (20.34) and (20.35) reduce to the usual formulas for equal group sizes.

The major difficulty of unequal sample sizes, here as in the case of factorial experiments (Section 18.4), relates to the question of orthogonality. In general, two comparisons with coefficients c_j and c_j' are **orthogonal** if and only if

$$\sum \frac{c_j c_j'}{m_j} = 0 \tag{20.36}$$

This expression reduces to the usual criterion of orthogonality when all of the group sizes m_j are equal (say, to m). However, with unequal group sizes, it will generally be very difficult to set up meaningful comparisons that are orthogonal to each other. There is no particular problem if we are interested in specific comparisons regardless of orthogonality. Nonetheless, when comparisons are not orthogonal, their interpretation is more complicated because they reflect overlapping, or confounded, sources of variation. For example, in trend analysis it is desirable to have the linear component reflect only linear trend and for the quadratic component to be independent of any linear trend, yet this may be difficult to achieve with unequal group sizes.

The important point to note about the inequality of sample sizes is that it introduces extra complications in the analysis and interpretation of results, especially in matters relating to orthogonality. In addition, F tests are more sensitive to assumption violations when sample sizes are unequal. In most situations, the planning of an experiment should involve a serious effort to obtain samples of equal size.

STUDY MATERIAL FOR CHAPTER 20

Terms and Concepts

overall F test

Section 20.1

simple effect
simple interaction effect
simple main effect
simple simple main effect
rule for identifying the error term for testing a simple effect

Section 20.2

omnibus F test
comparison
coefficients of a comparison
sum of squares due to a comparison
component of the treatment sum of squares
number of degrees of freedom due to a comparison
orthogonal comparisons
orthogonal components

Section 20.3

planned comparison
a priori comparison
planned-comparison critical F-value
estimated standard error of a comparison
confidence interval for the population mean difference δ estimated by a planned comparison
trend analysis
coefficients for orthogonal trend comparisons
linear comparison
quadratic comparison
cubic comparison
polynomial of degree g
quartic comparison
quintic comparison
deviations from linearity
deviations from linear and quadratic trend
higher-order trends

Section 20.4

postmortem comparison
multiple comparisons
a posteriori comparison
conservative
Scheffé's test for postmortem comparisons
experimentwise error rate
per-comparison error rate
Scheffé test critical F-value
confidence interval for the population mean difference δ estimated by a Scheffé comparison
Tukey's test for postmortem comparisons
Tukey's honestly significant difference
Tukey test critical F-value
studentized range statistic
confidence interval for the population mean difference δ estimated by a Tukey comparison
protected t test for postmortem comparisons
protected least significant difference
protected t test critical F-value
confidence interval for the population mean difference δ estimated by a protected t comparison
least significant difference

Section 20.5

orthogonal comparisons with unequal sample sizes

Key Formulas

Section 20.1

(20.1) $SS_{AB(C_1)} + SS_{AB(C_2)} + \cdots + SS_{AB(C_{k_C})} = SS_{AB(C)} = SS_{AB} + SS_{ABC}$

(20.4) $SS_{A(B_1)} + SS_{A(B_2)} + \cdots + SS_{A(B_{k_B})} = SS_{A(B)} = SS_A + SS_{AB}$

(20.6) $F_{\text{effect}} = \dfrac{MS_{\text{effect}}}{MS_{\text{error}}}$

Section 20.2

(20.14) $\bar{d} = \sum c_j \bar{x}_j \qquad \text{where } \sum c_j = 0$

(20.17) $SS_{\bar{d}} = \dfrac{m\bar{d}^2}{\sum c_j^2}$

(20.18) $\sum c_j c_j' = 0 \qquad$ for orthogonal comparisons

Section 20.3

(20.19) $F_{\bar{d}} = \dfrac{MS_{\bar{d}}}{MS_{\text{error}}}$

(20.21) $s_{\bar{d}} = \sqrt{\dfrac{(\sum c_j^2)MS_{\text{error}}}{m}}$

(20.22) $\bar{d} - t_{\alpha/2}s_{\bar{d}} < \delta < \bar{d} + t_{\alpha/2}s_{\bar{d}} \qquad$ with confidence level $1 - \alpha$

Section 20.4

(20.25) $\bar{d} - \sqrt{(k-1)F_\alpha}\, s_{\bar{d}} < \delta < \bar{d} + \sqrt{(k-1)F_\alpha}\, s_{\bar{d}} \qquad$ with confidence level $1 - \alpha$

(20.31) $\bar{d} - \sqrt{\tfrac{1}{2}Q_\alpha^2}\, s_{\bar{d}} < \delta < \bar{d} + \sqrt{\tfrac{1}{2}Q_\alpha^2}\, s_{\bar{d}} \qquad$ with confidence level $1 - \alpha$

Section 20.5

(20.34) $SS_{\bar{d}} = \dfrac{\bar{d}^2}{\sum (c_j^2/m_j)}$

(20.35) $s_{\bar{d}} = \sqrt{MS_{\text{error}} \sum \dfrac{c_j^2}{m_j}}$

(20.36) $\sum \dfrac{c_i c_j'}{m_j} = 0 \qquad$ for orthogonal comparisons

Exercises for Numerical Practice

Section 20.1

20/1. Given the data in Exercise 18/1:

a. Calculate the sums of squares due to the simple A-by-B interaction separately for each level of Factor C. How many degrees of freedom does each of these sums of squares have?

b. When you calculate the sums of squares in Part (a), which sums of squares and degrees of freedom in the overall ANOVA summary table are being repartitioned?

c. Give the numerical values of (i) the appropriate error term and (ii) the critical F-value for testing the significance of each of the simple interaction effects in Part (a) at the 5% significance level.

20/2. Given the data in Exercise 18/2:

a. Calculate the sums of squares due to the simple B-by-C interaction separately for each level of Factor A. How many degrees of freedom does each of these sums of squares have?

b. When you calculate the sums of squares in Part (a), which sums of squares and degrees of freedom in the overall ANOVA summary table are being repartitioned?

c. Give the numerical values of (i) the appropriate error term and (ii) the critical F-value for testing the significance of each of the simple interaction effects in Part (a) at the 5% significance level.

20/3. Given the data in Exercise 15/11:

a. Calculate the sums of squares due to the simple main effect of B separately for each level of Factor A. How many degrees of freedom does each of these sums of squares have?

b. When you calculate the sums of squares in Part (a), which sums of squares and degrees of freedom in the overall ANOVA summary table are being repartitioned?

c. Give the numerical values of (i) the appropriate error term and (ii) the critical F-value for testing the significance of each of the simple main effects in Part (a) at the 5% significance level.

20/4. Given the data in Exercise 15/12, answer Parts (a) through (c) in Exercise 20/3.

20/5. Given the data in Exercise 19/5:

a. Calculate the sums of squares due to the simple main effect of B separately for each level of Factor A. How many degrees of freedom does each of these sums of squares have?

b. When you calculate the sums of squares in Part (a), which sums of squares and degrees of freedom in the overall ANOVA summary table are being repartitioned?

c. Give the numerical values of (i) the appropriate error term and (ii) the critical F-value for testing the significance of each of the simple main effects in Part (a) at the 5% significance level.

d. Calculate the sums of squares due to the simple main effect of A separately for each level of Factor B. How many degrees of freedom does each of these sums of squares have?

e. When you calculate the sums of squares in Part (d), which sums of squares and degrees of freedom in the overall ANOVA summary table are being repartitioned?

f. Give the numerical values of (i) the appropriate error term and (ii) the critical F-value for testing the significance of each of the simple main effects in Part (d) at the 5% significance level.

20/6. Given the data in Exercise 19/6, answer Parts (a) through (f) in Exercise 20/5.

20/7. Given the data in Exercise 19/3, answer Parts (a) through (c) in Exercise 20/5.

20/8. Given the data in Exercise 19/4 answer Parts (d) through (f) in Exercise 20/5.

Section 20.2

20/9. Given three means corresponding to the three levels of a factor:

a. Write the coefficients of a comparison of the first two means jointly against the third mean.

b. Write the coefficients of a comparison that is orthogonal to that in Part (a) and state in words what it compares.

20/10. Given three means corresponding to the three levels of a factor:

a. Write the coefficients of a comparison of the second mean against the third mean.

b. Write the coefficients of a comparison that is orthogonal to that in Part (a) and state in words what it compares.

20/11. Given four means corresponding to the four levels of a factor:

a. Write the coefficients of a comparison of the first and third means jointly against the second and fourth means jointly.

b. Write the coefficients of a comparison of the first mean against the third mean.

c. Write the coefficients of a comparison that is orthogonal to both those in Parts (a) and (b), and state in words what it compares.

20/12. Given four means corresponding to the four levels of a factor:

a. Write the coefficients of a comparison of the first three means jointly against the fourth mean.

b. Write the coefficients of a comparison of the first mean against the second and third means jointly.

c. Write the coefficients of a comparison that is orthogonal to both those in Parts (a) and (b) and state in words what it compares.

20/13. The following means were calculated in a one-way design with three independent samples of twelve observations each:

$$\bar{x}_1 = 9.0, \qquad \bar{x}_2 = 4.0, \qquad \bar{x}_3 = 3.0$$

Calculate the sum of squares due to the comparison of

a. the second mean against the third mean.

b. the first mean against the last two means jointly.

20/14. The following means were calculated in a one-way design with four independent samples of nine observations each:

$$\bar{x}_1 = 8.0, \qquad \bar{x}_2 = 1.0, \qquad \bar{x}_3 = 5.0, \qquad \bar{x}_4 = 6.0$$

Calculate the sum of squares due to the comparison of

a. the first mean against the last three means jointly.

b. the first two means jointly against the last two means jointly.

Section 20.3

20/15. The following means were calculated in a one-way design with four independent samples of seven observations each:

$$\bar{x}_1 = 7.0, \qquad \bar{x}_2 = 1.0, \qquad \bar{x}_3 = 4.0, \qquad \bar{x}_4 = 9.0$$

The error mean square in the one-way ANOVA was 3.00.

a. Test the difference between the fourth mean and the average of the first three means as a planned comparison. Calculate the F test statistic, and give the critical F-value for a test at the 5% level.

b. Calculate the 95% confidence limits for the population mean difference estimated by the planned comparison in Part (a).

20/16. The following means were calculated in a one-way design with five independent samples of six observations each:

$$\bar{x}_1 = 4.0, \qquad \bar{x}_2 = 5.0, \qquad \bar{x}_3 = 8.0, \qquad \bar{x}_4 = 3.0, \qquad \bar{x}_5 = 1.0$$

The error mean square in the one-way ANOVA was 2.00.

a. Test the difference between the fifth mean and the average of the first four means as a planned comparison. Calculate the F test statistic, and give the critical F-value for a test at the 5% level.

b. Calculate the 95% confidence limits for the population mean difference estimated by the planned comparison in Part (a).

20/17. The following means were calculated in a one-way design with five independent samples of six observations each at five equally spaced values (namely, 10, 20, 30, 40, 50) of a quantitative independent variable:

$$\bar{x}_1 = 4.0, \qquad \bar{x}_2 = 5.0, \qquad \bar{x}_3 = 8.0, \qquad \bar{x}_4 = 3.0, \qquad \bar{x}_5 = 1.0$$

The error mean square in the one-way ANOVA was 2.00.

a. Calculate the F test statistics for testing the linear, quadratic, cubic, and quartic trends, and give the critical F-value for tests at the 5% level.

b. Calculate the F test statistic for a single test of the deviations from linear trend, and give the critical F-value for a test at the 5% level.

20/18. The following means were calculated in a one-way design with four independent samples of seven observations each at four equally spaced values (namely, 25, 50, 75, 100) of a quantitative independent variable:

$$\bar{x}_1 = 7.0, \qquad \bar{x}_2 = 1.0, \qquad \bar{x}_3 = 4.0, \qquad \bar{x}_4 = 9.0$$

The error mean square in the one-way ANOVA was 3.00.

a. Calculate the F test statistics for testing the linear, quadratic, and cubic trends, and give the critical F-value for tests at the 5% level.

b. Calculate the F test statistic for a single test of the deviations from linear trend, and give the critical F-value for a test at the 5% level.

Section 20.4

20/19. Given the data in Exercise 20/15, where the main effect was significant at the 1% level, calculate the F test statistic for a postmortem comparison of the second mean against the fourth mean, give the critical F-value at the 5% level, and calculate the corresponding 95% confidence limits using each of the following:

- **a.** Scheffé's procedure.
- **b.** Tukey's procedure.
- **c.** the protected t procedure.

20/20. Given the data in Exercise 20/16, where the main effect was significant at the 1% level, calculate the F test statistic for a postmortem comparison of the third mean against the fifth mean, give the critical F-value at the 5% level, and calculate the corresponding 95% confidence limits using each of the following:

- **a.** Scheffé's procedure.
- **b.** Tukey's procedure.
- **c.** the protected t procedure.

Problems for Applying the Concepts

20/21. Given the data in Problem 18/21:

- **a.** Calculate the sums of squares due to the simple interaction of transmission and driving conditions for diesel and gasoline separately. How many degrees of freedom does each of these sums of squares have?
- **b.** When you calculate the sums of squares in Part (a), which sums of squares in the overall ANOVA summary table are being repartitioned?
- **c.** What condition would have to be met in order for it to be necessary and appropriate to test the significance of the simple effects in Part (a)?

20/22. Given the data in Problem 18/20:

- **a.** Calculate the sums of squares due to the simple interaction of exercise program and social context for the three diets separately. How many degrees of freedom does each of these sums of squares have?
- **b.** When you calculate the sums of squares in Part (a), which sums of squares in the overall ANOVA summary table are being repartitioned?
- **c.** What condition would have to be met in order for it to be necessary and appropriate to test the significance of the simple effects in Part (a)?

20/23. There was a significant interaction of fuel and driving condition in Problem 18/21.

- **a.** Test the significance of the simple main effect of fuel for city and highway driving separately. Which sums of squares in the overall ANOVA summary table measure the same variation as the sums of squares for these simple effects?
- **b.** Test the significance of the simple main effect of driving condition for gasoline and diesel fuel separately. Which sums of squares in the overall ANOVA summary table measure the same variation as the sums of squares for these simple effects?

c. In light of the results of Parts (a) and (b), how would you interpret the significant interaction of fuel and driving conditions?

20/24. There was a significant interaction of manufacturing process and road surface in Problem 15/26.

a. Test the significance of the simple main effect of road surface for the different manufacturing processes separately. Which sums of squares in the overall ANOVA summary table measure the same variation as the sums of squares for these simple effects?

b. Test the significance of the simple main effect of manufacturing process for the different road surfaces separately. Which sums of squares in the overall ANOVA summary table measure the same variation as the sums of squares for these simple effects?

c. In light of the results of Parts (a) and (b), how would you interpret the significant interaction of manufacturing process and road surface?

20/25. There was a significant interaction of background and performer in Problem 19/25.

a. Test the significance of the simple main effect of background for the different performers separately. Which sums of squares in the overall ANOVA summary table measure the same variation as the sums of squares for these simple effects?

b. Test the significance of the simple main effect of performer for the different backgrounds separately. Which sums of squares in the overall ANOVA summary table measure the same variation as the sums of squares for these simple effects?

c. In light of the results of Parts (a) and (b), how would you interpret the significant interaction of background and performer?

20/26. There was a significant interaction of musical genre and tone-control setting in Problem 19/22.

a. Test the significance of the simple main effect of musical genre for the different levels of tone-control setting separately. Which sums of squares in the overall ANOVA summary table measure the same variation as the sums of squares for these simple effects?

b. Test the significance of the simple main effect of tone-control setting for the different levels of genre separately. Which sums of squares in the overall ANOVA summary table measure the same variation as the sums of squares for these simple effects?

c. In light of the results of Parts (a) and (b), how would you interpret the significant interaction of musical genre and tone-control setting?

20/27. Problem 19/17 presented data from a study of three variations in swimming stroke.

a. Test the significance of the difference between the mean of the no-kicks condition and the mean of the two kicking conditions combined, as a planned comparison.

b. Construct a 95% confidence interval for the population difference that was estimated in Part (a), and state it in words.

20/28. Problem 19/20 presented data from an investigation of students' average marks in three consecutive years of study.

a. To provide information about difficulties in adjustment to university studies, test the significance of the planned comparison of mean first-year university marks versus mean marks from Grade 12 and second-year university jointly.

b. Construct a 95% confidence interval for the population difference that was estimated in Part (a), and state it in words.

20/29. A total of 42 students volunteered to take part in an experiment in which they would study a specified number of hours (no more, no less) per week for a particular course. The students were randomly assigned to six study-time groups, with seven students in each group. At the end of the experiment, the groups obtained the following mean test scores:

Number of hours per week:	1	2	3	4	5	6
Mean test score:	12	12	14	15	18	19

The sum of squares due to students within groups was 612.

a. Test the statistical significance of the linear trend, the quadratic trend, and the combined higher-order trends.

b. Outline two different ways of calculating the sum of squares due to the combined higher-order trends in Part (a).

c. What degree of polynomial is required to describe the relationship between study time and mean test score?

20/30. The following are mean numbers of errors on a task for five groups that were allowed differing amounts of time to practice the task. There were eight subjects in each group.

Practice time:	0 min	5 min	10 min	15 min	20 min
Mean number of errors:	9	6	5	3	2

The sum of squares due to subjects within groups was 1,050.

a. Test the statistical significance of the linear trend, the quadratic trend, and the combined higher-order trends.

b. Outline two different ways of calculating the sum of squares due to the combined higher-order trends in Part (a).

c. What degree of polynomial is required to describe the relationship between practice time and mean number of errors?

20/31. Eight subjects took part in a study of recall as a function of word length (L = 3, 4, 5, 6, or 7 letters) and modality of presentation (M = auditory or visual). The following mean recall scores were obtained by administering every treatment combination to each subject (S).

	Word length				
Modality	**3**	**4**	**5**	**6**	**7**
Auditory	10	8	5	5	5
Visual	16	12	9	7	5

Given $SS_S = 591.50$, $SS_{MS} = 218.47$, $SS_{LS} = 477.68$, $SS_{MLS} = 717.36$:

a. Complete the appropriate ANOVA summary table, and give the critical value(s) for significance testing.

b. Plot the appropriate means for portraying any significant effects.

c. Perform a trend analysis to determine what degree of polynomial is required to describe mean recall scores as a function of word length. Give the critical value(s) for significance testing.

20/32. Problem 15/20 described a zoology experiment concerning the effect of temperature on the rate of oxygen consumption by deer mice.

a. Plot the mean consumption rates at the four temperatures.

b. Perform a trend analysis to determine what degree of polynomial is required to describe mean rate of oxygen consumption as a function of temperature. Give the critical values for significance testing.

20/33. Problem 15/15 described an experiment concerning the relationship of a player's Scrabble score to the vowel/consonant ratio in the player's letters.

a. Plot the mean numbers of points earned for the eight vowel/consonant ratios.

b. Perform a trend analysis to determine what degree of polynomial is required to describe mean score as a function of number of vowels on the rack. Give the critical values for significance testing.

c. Calculate the margin of error at the 95% confidence level for a pairwise comparison of two means according to Tukey's procedure. Which pairs of means exhibited this "honestly significant difference"?

d. Calculate the margin of error at the 95% confidence level for a pairwise comparison of two means according to Scheffé's procedure. Which pairs of means differed significantly according to this criterion?

20/34. Problem 15/22 described a physics experiment concerning the relationship between time and distance traveled by a falling weight.

a. Plot the mean times for the seven distances.

b. Perform a trend analysis to determine what degree of polynomial is required to describe mean time as a function of distance. Give the critical values for significance testing.

c. Calculate the margin of error at the 95% confidence level for a pairwise comparison of two means according to Tukey's procedure. Which pairs of means exhibited this "honestly significant difference"?

d. Calculate the margin of error at the 95% confidence level for a pairwise comparison of two means according to Scheffé's procedure. Which pairs of means differed significantly according to this criterion?

20/35. There was a significant main effect of transmission in Problem 18/21.

a. Use Scheffé's test (i) to compare automatic transmissions against manual transmissions considered together and (ii) to compare 4-speed versus 5-speed manual transmissions. Construct 99% confidence intervals for these two comparisons.

b. Repeat Part (a) using the protected t test in place of Scheffé's test.

c. Which sum of squares in the overall ANOVA summary table is equal to the sum of the two sum-of-squares components in Part (a)?

d. Use Tukey's test to make all possible pairwise comparisons among the three types of transmission. Construct a 99% confidence interval for each comparison.

20/36. Problem 19/16 presented hand-strength data for four types of individuals.

a. Test the significance of the postmortem comparison of Types 1 and 4 combined versus Types 2 and 3 combined. Give the critical values for both Scheffé's test and the protected t test.

b. Construct 99% confidence intervals corresponding to the two postmortem tests in Part (a).

c. Use Tukey's test to make all possible pairwise comparisons between types. Which types differed significantly from each other?

Reference

Wegner, D. M.; Wenzlaff, R.; Kerker, R. M.; and Beattie, A. E. "Incrimination through Innuendo: Can Media Questions Become Public Answers?" *Journal of Personality and Social Psychology* 40 (1981): 822–32.

Further Reading

Keppel, G. *Design and Analysis: A Researcher's Handbook.* 2nd ed. Englewood Cliffs, NJ: Prentice-Hall, 1982. Keppel provides further coverage of the topics introduced in this chapter; for example, Appendix B deals with the problem of unequally spaced values of the independent variable in trend analysis.

Snedecor, G. W., and Cochran, W. G. *Statistical Methods.* 7th ed. Ames, IA: Iowa State University Press, 1980. Chapter 12 includes clear, concise discussions of planned comparisons and multiple comparisons.

Winer, B. J. *Statistical Principles in Experimental Design.* 2nd ed. New York: McGraw-Hill, 1971. Section 3.7 describes a method of calculating the regression coefficients of the best-fitting polynomial of degree g in trend analysis.

CHAPTER 21

FURTHER TOPICS IN REGRESSION ANALYSIS

EXAMPLE 21.1 Beginning in the 1950s and especially in 1964 with the release of the U.S. Surgeon General's report on smoking and health, there has been a growing public awareness of the health hazards due to cigarette smoke both for smokers and for nonsmokers who live or work with smokers. Is there any relationship between this growing public concern and the national level of cigarette consumption?

To answer this question, we must take into account the many factors that influence national cigarette consumption in any given year. One relevant factor, or independent variable, is the average price of cigarettes in that year. Another relevant factor is the immediately previous year's cigarette consumption, reflecting the influence of habit or persistence in smoking. These and other independent variables must all be in the equation describing the dependent variable, cigarette consumption.

Using methods of the type to be introduced in this chapter, Kenneth Warner of the University of Michigan analyzed adult per capita cigarette consumption in the United States from 1947 to 1978 as a function of 10 independent variables. The observational units were the 32 years from 1947 through 1978, each of which yielded one value of the dependent variable (adult per capita cigarette consumption for that year) and one value of each of the 10 independent variables. The 10 independent variables recorded for each year included cigarette price, the previous year's consumption, a measure reflecting the spread of smoking behavior (for example, among women in recent years), and seven variables reflecting various aspects of the antismoking movement. Warner explained the latter variables as follows: "I account for the effects of smoking-and-health publicity [such as that related to the Surgeon General's report in 1964] by inclusion in a demand regression equation of binary (dummy) variables with a value of 1 in years of adverse publicity and 0 in other years. In order to examine the decline in consumption since 1973, I include a measure of the effectiveness of the nonsmokers' rights movement, namely, the percentage in a given year of the adult population residing in states with laws that restrict smoking in public places." This percentage increased from 1.1% in 1964 to 72.3% in 1978. The resulting least-squares regression equation accounted for 97.3% of the variation in cigarette consumption over the years.

Warner's regression analysis led him to the following conclusions:

i. In the absence of the antismoking campaign, cigarette consumption would have been considerably higher than it is today.

ii. The major impact on consumption of publicizing smoking-and-health information has lagged behind knowledge and attitudinal changes. By 1971, . . . the impact on consumption of publicity alone was less than 10 per cent of actual consumption. By 1978, it was three or four times as great.

iii. The decreases in cigarette consumption since 1973 correlate highly with the legislative successes of the nonsmokers' rights movement. This correlation seems unlikely to reflect causation. Rather, both declining consumption and growth in legislation probably reflect a prevailing nonsmoking ethos and the conversion of modified knowledge and attitudes into behavioral change.

(Source: Warner, 1981. Copyright 1981 by the American Association for the Advancement of Science. Used with permission of the AAAS.)

Chapters 5 and 13 introduced methods of regression and correlation as ways of describing and analyzing relationships between two variables, namely, one dependent variable y as a straight-line function ($y = bx + a$) of one independent variable x. A mathematical description of a dependent variable in terms of one or more other variables is called a **model** of the dependent variable. For example, we could attempt to describe each year's per capita cigarette consumption y as a function of the average price x of cigarettes that year. The resulting equation would be a model of cigarette consumption.

In this chapter, we extend the methods of Chapters 5 and 13 beyond simple linear models to two other kinds of relationship that are frequently encountered. **Multiple linear regression** is used to construct an appropriate model in situations such as in Example 21.1, where several independent variables, rather than just one, must be taken into account to describe the dependent variable. **Nonlinear regression** is used for model building when the relationship between the variables is curvilinear, so that the straight-line description is inadequate.

21.1 MULTIPLE LINEAR REGRESSION

Example 21.1 presented an illustration where more than one independent variable had to be taken into account in any adequate model of the dependent variable. This section discusses regression techniques for applications with two or more independent variables.

In **multiple linear regression**, the dependent variable y is described as a function of two or more independent variables, or **predictors**, by an equation of the form

$$\hat{y} = b_0 + b_1x_1 + b_2x_2 + \cdots + b_gx_g \qquad \textbf{(21.1)}$$

where $\hat{y}$ is the fitted value of the dependent variable y as estimated on the basis of the predictors $x_1, x_2, \ldots, x_g$. The value b_0 is the **regression constant**, and the values $b_1, b_2, \ldots, b_g$ are the **regression coefficients**, calculated to yield the best linear combination of the predictors for describing the values of the dependent variable. The differences $y_i - \hat{y}_i$ between the observed values y_i and the corresponding fitted values $\hat{y}_i$ are the errors of estimation, or residuals.

Calculation of Regression Coefficients and Constant

According to the method of least squares (Section 5.1), the "best" values of $b_0, b_1, \ldots, b_g$ in Expression (21.1) are obtained by minimizing the least-squares criterion:

$$\sum (y - \hat{y})^2 \tag{21.2}$$

which is the sum of the squared errors of estimation. In the general case with g predictors, application of differential calculus leads to a set of $g + 1$ simultaneous equations, called the **normal equations**, in $g + 1$ unknowns $(b_0, b_1, \ldots, b_g)$:

$$\begin{aligned}
\sum y &= b_0 n + b_1 \sum x_1 + b_2 \sum x_2 + \cdots + b_g \sum x_g \\
\sum x_1 y &= b_0 \sum x_1 + b_1 \sum x_1^2 + b_2 \sum x_1 x_2 + \cdots + b_g \sum x_1 x_g \\
\sum x_2 y &= b_0 \sum x_2 + b_1 \sum x_1 x_2 + b_2 \sum x_2^2 + \cdots + b_g \sum x_2 x_g \\
&\vdots \\
\sum x_g y &= b_0 \sum x_g + b_1 \sum x_1 x_g + b_2 \sum x_2 x_g + \cdots + b_g \sum x_g^2
\end{aligned} \tag{21.3}$$

where n is the number of $(g + 1)$-variate observations in the sample.

A number of good packages of computer programs are available to solve the normal equations for multiple regression. However, the experience of working a few examples on a hand-held calculator greatly facilitates comprehension of the procedure. Fortunately, the principles of multiple regression can be illustrated by the case of $g = 2$ predictors. For higher values of g, the calculations become rather tedious (except in certain special cases), and computing facilities beyond those assumed in this book would usually be used.

Either the method of elimination or Cramer's rule may be used to solve the normal equations for $b_0, b_1, \ldots, b_g$. The use of Cramer's rule is demonstrated in the following example.

EXAMPLE 21.2 Example 13.8 described an investigation of the germination time of Lincoln–Homesteader peas as a function of the mean maximum daily temperature in the week beginning at the day of planting. The same investigator also recorded the date of planting, to see whether the amount of daylight might also be related to the speed of germination, since later plantings receive more daylight per day than do earlier plantings. The date of planting was recorded as the number of days after the arbitrary date of 15 April. In this example, the dependent variable is germination time (y), which is to be described as a function of two predictors: the mean maximum daily temperature (x_1) and the date of planting (x_2). The observations are recorded in the following table, along with some preliminary calculations for analyzing the relationship. Extra digits are retained for use in subsequent calculations.

Observation number	Germination time (days), y	Mean maximum temp. (°C), x_1	Planting date (days after 15 April), x_2
1	11	14.8	20
2	12	17.4	34
3	10	15.6	23
4	8	20.9	38
5	11	20.1	4
6	12	17.1	20
7	9	19.9	25
8	8	21.4	26
9	11	13.5	17
10	8	21.4	32
11	14	13.4	16
	$\bar{y}$ = 10.363636	$\bar{x}$: 17.772727	23.181818
	s_y = 1.963300	s_x: 3.123489	9.505979
	$\sum y$ = 114	$\sum x$: 195.5	255
	$\sum y^2$ = 1,220	$\sum x^2$: 3,572.13	6,815
	SS_y = 38.545455	SS_x: 97.561818	903.636364
		$\sum xy$: 1,978.7	2,546
		SP_{xy}: −47.390909	−96.727273
		r_{xy}: −.77280249	−.51828126
			$\sum x_1x_2$: 4,645.6
			SP_{12}: 113.554545
			$r_{x_1x_2}$: .38244389

To find the least-squares values of b_0, b_1, and b_2 in the regression equation

$$\hat{y} = b_0 + b_1x_1 + b_2x_2$$

we must solve the following normal equations [from Expression (21.3)]:

$$\begin{aligned} 114 &= 11\ b_0 + 195.5\ b_1 + 255\ b_2 \\ 1{,}978.7 &= 195.5b_0 + 3{,}572.13b_1 + 4{,}645.6b_2 \\ 2{,}546 &= 255\ b_0 + 4{,}645.6\ b_1 + 6{,}815\ b_2 \end{aligned}$$

Cramer's rule uses determinants to solve simultaneous equations such as these. We begin by calculating the determinant D of the matrix of coefficients of the unknowns on the right-hand side of the equations; that is,

$$D = \begin{vmatrix} 11 & 195.5 & 255 \\ 195.5 & 3{,}572.13 & 4{,}645.6 \\ 255 & 4{,}645.6 & 6{,}815 \end{vmatrix}$$

A convenient calculation procedure known as Sarrus' rule for 3-by-3 determinants involves adding the products of the top-left–to–bottom-right "extended" diagonals and subtracting the products of the bottom-left–to–top-right "extended" diagonals as follows:

$$\begin{aligned} D &= (11)(3{,}572.13)(6{,}815) + (195.5)(4{,}645.6)(255) + (255)(195.5)(4{,}645.6) \\ &\quad - (255)(3{,}572.13)(255) - (4{,}645.6)(4{,}645.6)(11) - (6{,}815)(195.5)(195.5) \\ &= 827{,}923.49 \end{aligned}$$

Next, according to Cramer's rule, we replace the coefficients of b_0 by the constants from the left-hand side of the normal equations, evaluate the resulting determinant, and divide by D to obtain the value of b_0:

$$b_0 = \frac{\begin{vmatrix} 114 & 195.5 & 255 \\ 1{,}978.7 & 3{,}572.13 & 4{,}645.6 \\ 2{,}546 & 4{,}645.6 & 6{,}815 \end{vmatrix}}{827{,}923.49} = \frac{15{,}839{,}218}{827{,}923.49} = 19.131258$$

Similarly, we replace the coefficients of b_1 (in the determinant D) by the constants from the left-hand side of the normal equations, evaluate the resulting determinant, and divide by D to obtain the value of b_1:

$$b_1 = \frac{\begin{vmatrix} 11 & 114 & 255 \\ 195.5 & 1{,}978.7 & 4{,}645.6 \\ 255 & 2{,}546 & 6{,}815 \end{vmatrix}}{827{,}923.49} = \frac{-350{,}243.6}{827{,}923.49} = -.423038651$$

Likewise, we replace the coefficients of b_2 (in the determinant D) by the normal-equation constants to obtain the value of b_2:

$$b_2 = \frac{\begin{vmatrix} 11 & 195.5 & 114 \\ 195.5 & 3{,}572.13 & 1{,}978.7 \\ 255 & 4{,}645.6 & 2{,}546 \end{vmatrix}}{827{,}923.49} = \frac{-44{,}609.79}{827{,}923.49} = -.05388154$$

The least-squares regression equation therefore is

$$\hat{y} = 19.1 - .423x_1 - .054x_2 \tag{21.4}$$

(Example 21.3 includes a printout to illustrate a computer solution for these data.)

For any given values of temperature (x_1) and planting date (x_2) within the observed range of the predictors, we may use the equation to estimate the mean number of days until germination ($\hat{y}$). Notice that the inclusion of a second predictor (x_2) in the equation changes the value of the regression coefficient for the first predictor, and the value of the regression constant, from their values in the simple linear regression on x_1 calculated in Example 13.8:

$$\hat{y} = 19.0 - .49x_1$$

One way to check that the values of b_0, b_1, and b_2 have been calculated correctly is to compute the residuals and verify that their sum is zero. For this purpose, it is wise to use the unrounded values of b_0, b_1, and b_2 in calculating the fitted values $\hat{y}$, to avoid undue rounding errors. The results of this procedure are recorded to four decimal places in the following table.

	Observed values, y	Fitted values, $\hat{y}$	Residuals, $y - \hat{y}$
	11	11.7927	−.7927
	12	9.9384	2.0616
	10	11.2926	−1.2926
	8	8.2423	−.2423
	11	10.4127	.5873
	12	10.8197	1.1803
	9	9.3658	−.3658
	8	8.6773	−.6773
	11	12.5043	−1.5043
	8	8.3540	−.3540
	14	12.6004	1.3996
Total	114	114.0000	0.0000

The sum of the residuals is zero, as it must be (barring rounding errors) if the values of b_0, b_1, and b_2 have been calculated correctly.

The values of the residuals should also be examined to see which observations are least well described by the equation, as evidenced by residuals with large absolute values. Were there any special circumstances (or errors) involved in those observations that might account for the discrepancies? Do observations with large residuals have anything in common? Observations whose residuals have large absolute values may give clues as to other factors that are important in describing the dependent variable.

We have yet to address the question as to how good a description of germination times is provided by the multiple linear regression equation in Expression (21.4). The next section will give us a way of answering the question.

(Source: Student project by G. Hamilton.)

The ANOVA Summary Table

Multiple linear regression analysis involves the partitioning of the total variation in the dependent variable y,

$$SS_y = \sum (y_i - \bar{y})^2 \tag{21.5}$$

into two orthogonal parts. One part,

$$SS_{\text{regr. on } x_1,\ldots,x_g} = \sum (\hat{y}_i - \bar{y})^2 \tag{21.6}$$

is the variation that may be attributed to the relationship of y to the best (in the least-squares sense) linear combination of the predictors $x_1, \ldots, x_g$. The other part,

$$SS_{\text{res.}|x_1,\ldots,x_g} = \sum (y_i - \hat{y}_i)^2 \tag{21.7}$$

is the residual variation due to the errors of estimation when y is described by the best linear combination of the predictors $x_1, \ldots, x_g$. [The vertical line in

TABLE 21.1 ANOVA summary table for multiple linear regression analysis.

Source of variation	SS	df	MS = SS/df	*F*
Regression on $x_1, \ldots, x_g$	$SS_{regr.}$	g	$MS_{regr.}$	$MS_{regr.}/MS_{res.}$
Residual given $x_1, \ldots, x_g$	$SS_{res.}$	$n - g - 1$	$MS_{res.}$	—
Total	SS_y	$n - 1$		

Expression (21.7) is read "given."] The partition of the total variation in y by multiple linear regression analysis may be summarized in an **ANOVA summary table** having the form of Table 21.1.

Table 13.2 for simple linear regression is the special case of Table 21.1 that occurs when there is $g = 1$ predictor. Moreover, the definitional formulas for sums of squares in Expressions (21.5) through (21.7) for multiple regression are identical to the definitional formulas in Expressions (13.21) through (13.23) for simple regression. Unfortunately, the more convenient computational formulas for $SS_{regr.}$ and $SS_{res.}$ that were presented for the case of $g = 1$ predictor do not apply with higher values of g. The sums of squares in Table 21.1 may be calculated from the definitional formulas, but care must be taken to avoid premature rounding in the calculations. An alternative method of obtaining the sums of squares that simplifies calculations in some cases will be presented in the next subsection, headed Partial Correlation, and methods that use matrix algebra are available in references cited under Further Reading at the end of this chapter. In practice, except for purposes of teaching the underlying principles, the calculations for multiple regression analysis are almost always performed with standard computer programs.

As in the case of simple linear regression, the ANOVA summary table facilitates a judgment as to how well the regression equation describes the relationship between the dependent and independent variables. For example, the **coefficient of multiple determination**,

$$R^2 = \frac{SS_{regr.}}{SS_{total}} \tag{21.8}$$

is the proportion of the total variation in the dependent variable that can be accounted for by its relationship with the best-fitting linear combination of the independent variables. The coefficient of determination r^2 is the special case of R^2 when there is just one independent variable. Like r^2, R^2 takes on values between 0 and 1 (0% and 100%), depending on how well the dependent variable is described by the regression equation in the sample at hand.

The positive square root of R^2 is called the **multiple correlation coefficient** R. Multiple correlation refers to the correlation of one variable, say, y, with the best-fitting (in the least-squares sense) linear combination of g other variables, say, $x_1, x_2, \ldots, x_g$. However, the linear combination of g variables that provides the best fit to the dependent variable y is the least-squares fitted value

$$\hat{y} = b_0 + b_1x_1 + b_2x_2 + \cdots + b_gx_g$$

as given in Expression (21.1). Thus, the multiple correlation of y with $x_1, \ldots, x_g$, written $R_{y \cdot x_1,\ldots,x_g}$, is equal to the simple correlation of y with $\hat{y}$:

$$R_{y \cdot x_1,\ldots,x_g} = r_{y\hat{y}} \qquad \textbf{(21.9)}$$

It follows that R, and hence R^2, may be calculated by applying the usual computational formula for r [Expression (5.19)] to the values of y and $\hat{y}$ (that is, with $\hat{y}$ substituted for x). However, if the ANOVA table is available, R^2 and R are more easily obtained via Expression (21.8).

Under the assumption that the errors of estimation are independent Gaussian deviates with constant variance throughout the range of the predictors, the F ratio in Table 21.1,

$$F_{\text{regr.}} = \frac{\text{MS}_{\text{regr.}}}{\text{MS}_{\text{res.}}} \qquad \textbf{(21.10)}$$

may be evaluated against the critical value F_α ($g, n - g - 1$ df). This is a test of whether the relationship described by the regression equation is statistically significant at the α level. The null hypothesis is that the population regression coefficients $\beta_1, \beta_2, \ldots, \beta_g$ considered jointly are all equal to zero. Equivalently, the test may be thought of as evaluating whether the multiple correlation coefficient R in Expression (21.9) is significantly greater than zero: Is the relationship between the observed values y and the fitted values $\hat{y}$ statistically significant?

EXAMPLE 21.3 In Example 21.2, we worked out the regression equation describing germination time (y) as a function of temperature (x_1) and planting date (x_2):

$$\hat{y} = 19.1 - .423x_1 - .054x_2$$

Example 21.2 also included a table showing the residuals $y - \hat{y}$ from this equation, along with the observed values y. Applying Expressions (21.5) (or its computational equivalent in Table 13.1) and (21.7) to values in the table in Example 21.2, we obtain

$$\text{SS}_y = 38.5454545$$

which we had previously calculated in Example 13.9, and

$$\text{SS}_{\text{res.}|x_1,x_2} = (-.7927)^2 + (2.0616)^2 + \cdots + (1.3996)^2 = 13.2854562$$

By subtraction, we have

$$\text{SS}_{\text{regr. on } x_1x_2} = 38.5454545 - 13.2854562 = 25.2599983$$

The ANOVA summary table for the multiple linear regression analysis of germination time on temperature and planting date is as follows.

Source of variation	SS	df	MS	F	
Regression on temperature and planting date (b_1, b_2)	25.2599983	2	12.63	7.61	$P < .05$
Residual given x_1, x_2	13.2854562	8	1.66	—	
Total	38.5454545	10			

Compare this table with the ANOVA table for the simple linear regression of germination time on temperature (x_1), as presented in Example 13.9. The inclusion of a second predictor in the equation removes from the previous residual source, one degree of freedom and a portion of the sum of squares, and these become part of the new regression source.

The coefficient of multiple determination is obtained from the ANOVA table, via Expression (21.8), as

$$R^2 = \frac{25.26}{38.55} = .6553$$

Approximately 66% of the variation in germination times can be accounted for by the best-fitting linear combination of temperature and planting date. Squaring the simple correlation coefficients given in Example 21.2, we find that the linear relationship with temperature alone could account for $r^2_{yx_1} = 60\%$ of the variation in germination times, and the linear relationship with planting date alone could account for $r^2_{yx_2} = 27\%$ of the variation in germination times. However, the two predictors are confounded (correlated with each other), each accounting for some of the same variation accounted for by the other. The total variation that the two predictors together can linearly account for is not $60\% + 27\% = 87\%$, as it would be if they were orthogonal (uncorrelated), but only $R^2 = 66\%$. The inclusion of planting date in the equation does not provide very much information not already provided by the temperature variable.

Comparing the ANOVA-table F ratio of 7.61 with critical values from the F distribution with 2 and 8 degrees of freedom, we find that the multiple regression is significant at the 5% level, but not at the 1% level. The multiple linear regression equation represents a statistically significant relationship. However, we may still want to know whether the inclusion of the second predictor produced a significant improvement in the description of germination times.

Figure 21.1 presents computer output for this example from the SAS/STAT™ software. Although we have not discussed all of the items in the SAS/STAT printout, you should be able to find the ANOVA summary table, the P-value for the F test of the regression, the standard error of estimation ("ROOT MSE"), the coefficient of multiple determination R^2, the regression coefficients and constant, the observed values, the fitted values, and the residuals.

Partial Correlation

The fact that the overall F ratio in multiple regression analysis is statistically significant does not imply that every one of the predictors included in the equation makes a significant contribution to the description of the dependent variable. It could be that a description based on some subset of the predictors would serve just as well. How can we determine whether one particular

```
                                SAS

                        ANALYSIS OF VARIANCE

                            SUM OF          MEAN
         SOURCE       DF   SQUARES         SQUARE       F VALUE      PROB>F

         MODEL         2  25.25999828   12.62999914       7.605       0.0141
         ERROR         8  13.28545627    1.66068203
         C TOTAL      10  38.54545455

            ROOT MSE       1.288675      R-SQUARE        0.6553
            DEP MEAN      10.36364       ADJ R-SQ        0.5692
            C.V.          12.43458

                        PARAMETER ESTIMATES

                     PARAMETER       STANDARD     T FOR HO:                     VARIABLE
VARIABLE     DF      ESTIMATE          ERROR      PARAMETER=0    PROB > !T!       LABEL

INTERCEP      1    17.13125814     2.35394688        8.127         0.0001     INTERCEPT
X1            1    -0.42303861     0.14120216       -2.996         0.0172     TEMPERATURE
X2            1    -0.05388154     0.04639642       -1.161         0.2790     PLANTING DATE

                                       PREDICT
                 OBS       ACTUAL       VALUE     RESIDUAL

                   1      11.0000     11.7927      -0.7927
                   2      12.0000      9.9384       2.0616
                   3      10.0000     11.2926      -1.2926
                   4       8.0000      8.2423      -0.2423
                   5      11.0000     10.4127       0.5873
                   6      12.0000     10.8197       1.1803
                   7       9.0000      9.3658      -0.3658
                   8       8.0000      8.6773      -0.6773
                   9      11.0000     12.5043      -1.5043
                  10       8.0000      8.3540      -0.3540
                  11      14.0000     12.6004       1.3996

                SUM OF RESIDUALS               6.75016E-14
                SUM OF SQUARED RESIDUALS          13.28546
```

Figure 21.1
SAS/STAT™ printout for Example 21.3.

predictor makes a significant contribution to the description, over and above the variation already accounted for by the other predictors in the equation?

The **partial correlation coefficient** $r_{yz|x_1,\ldots,x_c}$ measures the correlation between variables y and z, given that variables $x_1, \ldots, x_c$ have already been taken into account; that is, it is the correlation between those aspects of y and of z that cannot be accounted for by $x_1, \ldots, x_c$. The aspects of y that cannot be accounted for by the best-fitting linear combination of $x_1, \ldots, x_c$ are the residuals, or errors of estimation $y_i - \hat{y}_i$, in the regression of y on $x_1, \ldots, x_c$; and the aspects of z that cannot be accounted for by the best-fitting linear combination of $x_1, \ldots, x_c$ are the residuals, or errors of estimation $z_i - \hat{z}_i$, in the regression of z on $x_1, \ldots, x_c$. The partial correlation is the simple correlation between these two sets of residuals:

$$r_{yz|x_1,\ldots,x_c} = r_{y-\hat{y},z-\hat{z}} \tag{21.11}$$

Its statistical significance can be tested by reference to Appendix 8 with $n - c - 2$ degrees of freedom to see whether the regression equation based on $x_1, \ldots, x_c$ would be significantly improved by including the predictor z as well.

In terms of the ANOVA summary table, the **coefficient of partial determination** $r^2_{yz|x_1,\ldots,x_c}$ plays a role with respect to the *residual* sum of squares analogous to the role played by R^2 with respect to the *total* sum of squares. Recall that R^2 is the proportion of the *total* sum of squares that can be accounted for by the regression of y on $x_1, \ldots, x_c$. The residual sum of squares is the variation remaining after the effects of $x_1, \ldots, x_c$ have been removed, and hence $r^2_{yz|x_1,\ldots,x_c}$ is the proportion of the *residual* sum of squares (from the regression of y on $x_1, \ldots, x_c$) that would be accounted for by including z as an additional predictor in the regression equation describing y; that is,

$$\text{SS}_{\text{res.}|x_1,\ldots,x_c} - \text{SS}_{\text{res.}|x_1,\ldots,x_c,z} = r^2_{yz|x_1,\ldots,x_c}\text{SS}_{\text{res.}|x_1,\ldots,x_c} \tag{21.12}$$

In other words, the residual variation *with* z in the equation [the second term in Expression (21.12)] is smaller than the residual variation *without* z in the equation [the first term in Expression (21.12)], and the coefficient of partial determination tells us "how much smaller" [the right-hand side of Expression (21.12)]. The terms of Expression (21.12) may be rearranged to yield

$$\text{SS}_{\text{res.}|x_1,\ldots,x_c} = \frac{\text{SS}_{\text{res.}|x_1,\ldots,x_c,z}}{1 - r^2_{yz|x_1,\ldots,x_c}} \tag{21.13}$$

which we will later find convenient for some purposes.

Since the proportion of the residual sum of squares that would be accounted for by including z as an additional predictor would thereby become part of the *regression* sum of squares, the quantity in Expression (21.12) is called the **extra sum of squares due to the inclusion of a predictor** z in the equation (given that $x_1, \ldots, x_c$ are already in the equation) and is written as

$$\begin{aligned}\text{SS}_{\text{regr. on } z|x_1,\ldots,x_c} &= \text{SS}_{\text{regr. on } x_1,\ldots,x_c,z} - \text{SS}_{\text{regr. on } x_1,\ldots,x_c}\\ &= \text{SS}_{\text{res.}|x_1,\ldots,x_c} - \text{SS}_{\text{res.}|x_1,\ldots,x_c,z}\\ &= r^2_{yz|x_1,\ldots,x_c}\text{SS}_{\text{res.}|x_1,\ldots,x_c}\end{aligned} \tag{21.14}$$

Rearranging terms, we can write the coefficient of partial determination as

$$r^2_{yz|x_1,\ldots,x_c} = \frac{\mathrm{SS}_{\mathrm{regr.\ on\ } z|x_1,\ldots,x_c}}{\mathrm{SS}_{\mathrm{res.}|x_1,\ldots,x_c}}$$

$$= \frac{\text{extra sum of squares due to } z, \text{ given } x_1, \ldots, x_c}{\text{residual sum of squares given } x_1, \ldots, x_c \text{ (without } z\text{)}} \qquad \textbf{(21.15)}$$

The extra sum of squares due to the inclusion of one additional predictor has one degree of freedom, and its significance may be tested by the F ratio:

$$F_{\mathrm{regr.\ on\ } z|x_1,\ldots,x_c} = \frac{(\mathrm{SS}_{\mathrm{regr.\ on\ } z|x_1,\ldots,x_c})/1}{(\mathrm{SS}_{\mathrm{res.}\ |x_1,\ldots,x_c,z})/(n - c - 2)} \qquad \textbf{(21.16)}$$

with 1 and $n - c - 2$ degrees of freedom. This **test of the extra sum of squares** is equivalent to testing the significance of the partial correlation coefficient using Appendix 8.

Expression (21.11) is a definitional rather than computational formula for the partial correlation coefficient. In practice, the partial correlation $r_{yz|x}$ of variables y and z with x held constant is calculated from the simple correlations among the variables as follows:

$$r_{yz|x} = \frac{r_{yz} - r_{yx}r_{zx}}{\sqrt{1 - r^2_{yx}}\sqrt{1 - r^2_{zx}}} \qquad \textbf{(21.17)}$$

A correlation coefficient with the effects of *one* other variable removed is called a **first-order partial correlation coefficient. Second-order partial correlation coefficients**—those with the effects of *two* other variables removed—may be calculated from first-order coefficients, third-order from second-order, and so on. For example, the formula for the second-order partial correlation $r_{yz|xw}$ of variables y and z with x and w held constant is obtained by writing the variable w as a condition (that is, after a vertical line) in each of the correlations appearing in Expression (21.17):

$$r_{yz|xw} = \frac{r_{yz|w} - r_{yx|w}r_{zx|w}}{\sqrt{1 - r^2_{yx|w}}\sqrt{1 - r^2_{zx|w}}} \qquad \textbf{(21.18)}$$

The partial correlation coefficient, calculated from Expression (21.17), can be used to obtain the sums of squares for regression on two predictors, say, x_1 and x_2, from the sums of squares for regression on one predictor, say, x_1. Applying the principles summarized in Expression (21.14), we can say that the sum of squares due to regression on x_1 and x_2 is equal to the sum of squares due to regression on x_1 plus the proportion ($r^2_{yx_2|x_1}$) of the residual sum of squares given x_1 that can be accounted for by x_2 given that x_1 is already in the equation; that is,

$$\mathrm{SS}_{\mathrm{regr.\ on\ } x_1,x_2} = \mathrm{SS}_{\mathrm{regr.\ on\ } x_1} + r^2_{yx_2|x_1}\mathrm{SS}_{\mathrm{res.}|x_1} \qquad \textbf{(21.19)}$$

The accompanying residual sum of squares given x_1 and x_2 can be obtained by subtraction from the total sum of squares SS_y.

EXAMPLE 21.4 Example 13.9 presented the following ANOVA table for the regression of pea germination time y on temperature x_1.

Source of variation	SS	df	MS	F	
Regression on temperature (b_1)	23.0202584	1	23.02	13.3	$P < .01$
Residual given x_1	15.5251961	9	1.73	—	
Total	38.5454545	10			

Example 21.3 presented the following ANOVA table for the regression of germination time y on two predictors, temperature x_1 and planting date x_2.

Source of variation	SS	df	MS	F	
Regression on temperature and date (b_1, b_2)	25.2599983	2	12.63	7.61	$P < .05$
Residual given x_1, x_2	13.2854562	8	1.66	—	
Total	38.5454545	10			

The total sum of squares is the same in the two tables, but more of it is attributable to the regression when the second predictor is included in the equation. The extra sum of squares due to the inclusion of planting date in the equation may be obtained from either the first or the second line of Expression (21.14):

$$SS_{\text{regr. on } x_2|x_1} = 25.2599983 - 23.0202584 = 2.2397399$$

or equivalently

$$SS_{\text{regr. on } x_2|x_1} = 15.5251961 - 13.2854562 = 2.2397399$$

This extra sum of squares with its one degree of freedom is removed from the residual (unexplained) variation and added to the regression (explained) variation when the second predictor is included in the equation. The proportion of the residual sum of squares in the first analysis that can be accounted for by entering the second predictor into the equation is the coefficient of partial determination [Expression (21.15)]:

$$r^2_{yx_2|x_1} = \frac{2.2397399}{15.5251961} = .14426484$$

Approximately 14% of the residual variation from the regression on temperature can be accounted for by including planting date as a second predictor.

Expression (21.17) may be used to obtain the corresponding partial correlation coefficient from the simple linear correlations reported in Example 21.2. The correlation of germination time y and planting date x_2, holding temperature x_1 constant, is

$$r_{yx_2|x_1} = \frac{-.51828126 - (-.77280249)(.38244389)}{\sqrt{1 - (-.77280249)^2}\sqrt{1 - (.38244389)^2}} = -.37982211$$

The square of the partial correlation coefficient is the coefficient of partial determination:

$$r^2_{yx_2|x_1} = (-.37982211)^2 = .14426484$$

which is the same value as was calculated in the preceding paragraph as a ratio of sums of squares.

Calculating the coefficient of partial determination in the latter way, we could have used it to calculate the second ANOVA table from the first, without even solving the normal equations for the two-predictor regression. Given only the first ANOVA table and the coefficient of partial determination, we can calculate the extra sum of squares due to the inclusion of planting date x_2,

$$SS_{\text{regr. on } x_2|x_1} = (.14426484)(15.5251961) = 2.2397399$$

the sum of squares due to regression in the second ANOVA table,

$$SS_{\text{regr. on } x_1,x_2} = 23.0202584 + 2.2397399 = 25.2599983$$

and the residual sum of squares in the second ANOVA table,

$$SS_{\text{res.}|x_1,x_2} = 15.5251961 - 2.2397399 = 13.2854562$$

To test whether the inclusion of planting date accounts for significant additional variation, we calculate the F ratio in Expression (21.16):

$$F_{\text{regr. on } x_2|x_1} = \frac{2.2397399}{13.2854562/8} = 1.35$$

where the denominator is the residual mean square from the second ANOVA table. The critical value is $F_{.05}(1, 8 \text{ df}) = 5.32$, so we conclude that the inclusion of planting date in the regression equation does not account for significant variation beyond that already accounted for by the linear relationship with the temperature variable ($P > .05$).

Many of the standard computer programs for multiple regression show tests of extra sums of squares as t tests rather than as F tests. The tests are exactly equivalent since an F ratio with one degree of freedom for the numerator equals the square of a t statistic with the denominator's degrees of freedom, as shown in Expression (12.26). For example, the F ratio of 1.35 in the preceding paragraph is the square of $t = -1.161$, which is given in the SAS/STAT printout in Figure 21.1 as the test statistic for the additional contribution of planting date in the regression equation. The printout gives the corresponding P-value as .2790, indicating again that the extra contribution of planting date is not significant at the .05 level. Each of the t statistics in the printout provides an index of the extra contribution of one variable given that the other variables shown are included in the equation. As in Expression (13.32), each t statistic equals the regression coefficient b_j divided by its standard error, $t_{b_j} = b_j/s_{b_j}$. We use the F tests in this book because it is difficult to calculate the standard errors s_{b_j} without the use of matrix algebra when there is more than one predictor.

Types of Variables in Regression

Section 2.1 drew a distinction between quantitative (numerical) variables and qualitative (nominal) variables. For example, height and age are normally quantitative variables, because their values are usually stated in numerical form, whereas gender and eye color are qualitative variables, because their values—male/female and blue/brown/green/other—are designated by different verbal labels and are not intrinsically numerical. The analysis of variance (Chapters 15 and 18 through 20) and regression analysis (Chapters 5, 13, and 21) are both methods for analyzing the variation in a quantitative *dependent* variable, to see how that variation may be accounted for by one or more independent variables. They differ from each other in the type of *independent* variable(s) for which they are most suited.

Regression analysis was originally devised for *quantitative* independent variables [$x_1, x_2, \ldots, x_g$ in Expression (21.1)]. However, it is a very general procedure that can handle qualitative independent variables as well. To incorporate a qualitative independent variable having k levels or categories, into a regression equation of the form of Expression (21.1), we include $k - 1$ **dummy variables** (or **indicator variables**) that take on a different pattern of values to designate each of the k categories.

EXAMPLE 21.5 Consider gender as an example of a qualitative variable with $k = 2$ categories. To include gender as a predictor in a regression equation describing a dependent variable y as a function of a quantitative predictor, say, x_1, we define $k - 1 = 1$ dummy variable, say, x_2, with the following values:

$$x_2 = \begin{cases} 0 & \text{for females} \\ 1 & \text{for males} \end{cases}$$

Now when we obtain the multiple regression equation

$$\hat{y} = b_0 + b_1x_1 + b_2x_2$$

we can rewrite it as

$$\hat{y} = \begin{cases} b_0 + b_1x_1 + 0 = b_0 + b_1x_1 & \text{for females} \\ b_0 + b_1x_1 + b_2 = (b_0 + b_2) + b_1x_1 & \text{for males} \end{cases}$$

by substituting the values of x_2. Thus, b_2 is just a constant (the mean difference between males and females when the other predictors in the equation are held constant) that is added to the equation for males and omitted for females.

If we want to allow for the possibility that the lines relating y to the quantitative predictor x_1 for males and females are not parallel (that is, have different slopes), then in addition to the dummy variable x_2 we define a **cross-product variable**, say, x_3, as $x_3 = x_1x_2$. Then we compute the multiple regression equation:

$$\hat{y} = b_0 + b_1x_1 + b_2x_2 + b_3x_3$$

which reduces to

$$\hat{y} = \begin{cases} b_0 + b_1x_1 + 0 + 0 = b_0 + b_1x_1 & \text{for females} \\ b_0 + b_1x_1 + b_2 + b_3x_1 = (b_0 + b_2) + (b_1 + b_3)x_1 & \text{for males} \end{cases}$$

In other words, the dummy variable x_2 and the cross-product variable x_3 in this example allow for the possibility of separate lines for the male and female data, if the data warrant it.

EXAMPLE 21.6 Example 21.1 described an analysis of cigarette consumption y as a function of 10 independent variables. The least-squares regression equation was found to be

$$\hat{y} = -4621.2 - 16.36x_1 + .6896x_2 + 1824.7x_3 - 131.39x_4 - 305.41x_5 - 195.25x_6 - 110.05x_7 - 226.72x_8 - 79.3x_9 - 837.81x_{10}$$

Predictors x_1, x_2, x_3, and x_{10} are quantitative variables whose numerical values measure price, previous year's consumption, spread of smoking behavior in the population, and percentage of the adult population living in smoking-restricting states, respectively. With respect to variable x_2, for example, the equation tells us that on the average, *with the other* 9 *variables held constant,* 68.96% of the previous year's consumption is added in as one component of the current year's consumption, presumably reflecting effects of habit that are not included in the other 9 variables.

Predictors x_4, x_5, x_6, x_7, x_8, and x_9 are dummy variables included to capture the effects of antismoking publicity in particular years. For example, predictor x_6 is a dummy variable defined as

$$x_6 = \begin{cases} 1 & \text{for 1964} \\ 0 & \text{for all other years} \end{cases}$$

The equation tells us that on the average, with the other nine variables held constant (that is, taken into account statistically), "publicity surrounding release of the Surgeon General's report in 1964 was associated with a decrease in consumption of 195.25 cigarettes [per capita] in 1964." Thus, the coefficient $b_6 = -195.25$ of the dummy variable x_6 is just a constant that is added to the equation for 1964 and omitted for other years.

(Source: Warner, 1981. Copyright 1981 by the American Association for the Advancement of Science. Used with permission of the AAAS.)

In general, a dummy variable is a contrived numerical variable, devised in such a way as to add to the regression equation a constant appropriate to a particular level or category of a qualitative variable. It is not a measured variable but rather an indicator of the pertinent category of a qualitative variable. It is, in brief, a device for incorporating qualitative predictors into numerical equations.

EXAMPLE 21.7 As an example of a qualitative variable having $k = 3$ categories, consider an investigation involving three different treatments, say, *C, D,* and *E*. Treatments could be included as an independent variable in a regression equation by defining $k - 1 = 2$ dummy variables, say, x_1 and x_2, with the following patterns of values:

$$\begin{array}{lll} x_1 = 0 \text{ and } x_2 = 0 & & \text{for Treatment } C \\ x_1 = 1 \text{ and } x_2 = 0 & & \text{for Treatment } D \\ x_1 = 0 \text{ and } x_2 = 1 & & \text{for Treatment } E \end{array}$$

Dummy variable x_1 might be labeled as "Treatment *D* present/absent," and dummy variable x_2 as "Treatment *E* present/absent." What do these two variables contribute to the regression equation? When we compute the regression equation, these variables will appear as $b_1x_1 + b_2x_2$, which, upon substituting the values of x_1 and x_2, reduces to

$$\begin{array}{lll} b_1x_1 + b_2x_2 = 0 + 0 = 0 & & \text{for Treatment } C \\ b_1x_1 + b_2x_2 = b_1 + 0 = b_1 & & \text{for Treatment } D \\ b_1x_1 + b_2x_2 = 0 + b_2 = b_2 & & \text{for Treatment } E \end{array}$$

Thus, b_1 is just a constant (the mean contribution of Treatment D over and above that of Treatment C, with other variables in the equation held constant) that is added to the equation for Treatment D and omitted for the other two treatments, and b_2 is just a constant (the mean contribution of Treatment E over and above that of Treatment C, with other variables in the equation held constant) that is added to the equation for Treatment E and omitted for Treatments C and D.

The $k - 1$ dummy variables for a qualitative independent variable having k categories may be defined in any of a number of ways. The pattern introduced for $k = 3$ categories in Example 21.7 is often used. It is easily extended to higher values of k by adding one more dummy variable for each additional category and defining that variable to be equal to 1 for that category and 0 otherwise. Other ways of quantifying dummy variables are provided in the references cited under Further Reading at the end of this chapter.

The analysis of variance, which was discussed in Chapters 15 and 18 through 20, is the special case of multiple regression analysis that occurs when *all* of the independent variables are either qualitative variables or quantitative variables with values limited to k discrete categories for purposes of the investigation. The multiple regression methods discussed in this chapter can always be used in such cases. However, in an appropriately designed experiment, both the calculation and the interpretation tend to be simplified if the ANOVA methods of Chapters 15 and 18 through 20 are used when each independent variable or factor has a limited number of levels included in the investigation and each included level is observed two or more times. The more general methods of multiple regression analysis are called for if the investigation has not been appropriately designed for ANOVA. For example, multiple regression is needed if there is at least one independent variable that takes on a variety of numerical values, some of which may occur only once in the investigation. The four quantitative predictors in Example 21.6 are examples of independent variables that make multiple regression rather than ANOVA the appropriate method of analysis.

Stepwise Regression Analysis

One of the most difficult problems that arises in regression analysis is that of **variable selection**: Which predictor(s) should be included in the regression equation describing the dependent variable y? The answer to this question inevitably involves some degree of judgment based on experience and knowledge of the process being investigated. For one thing, the choice of which variables to record in the first place is always a matter of judgment. Once values have been recorded for the dependent variable and a set of predictors, there are several statistical procedures for variable selection that provide useful information and guidance, but even at this stage no one mechanical method is always best. In this section we discuss **stepwise regression analysis**, one of the most widely used of the variable-selection procedures. It is not a substitute for skill and experience, but it provides information that can enable the researcher to make a wise selection of predictors.

Stepwise regression analysis is a systematic procedure for entering, and sometimes removing, predictors one at a time until an equation is obtained in which

1. each of the included predictors accounts for significant variation that is not accounted for by the other included predictors, and
2. none of the excluded variables accounts for significant variation that is not already accounted for by the included variables.

As we proceed through the steps of stepwise regression analysis, we will compute several statistics repeatedly to assist in choosing the best set of predictors. Each of these statistics provides some useful information, but no one of them can be taken as providing the last word in variable selection. The following comments outline what to look for in five different statistics computed in stepwise regression analysis:

a. $F_{\text{regr.}}$ allows a test of whether the regression equation accounts for more variation in the dependent variable y than would be expected by chance if the true regression coefficients were all zero. (See Table 21.1.) The final regression equation should be statistically significant, to provide some assurance that the relationship observed in the sample would be unlikely to occur just by chance. Even so, a statistically significant regression equation may sometimes account for such a small proportion of the variation in the dependent variable that the relationship is of little practical importance.

b. R^2 is the proportion of variation in y that is accounted for by its relationship with the best-fitting linear combination of the predictors that are included in the equation. Entering an additional variable in the equation can never reduce the value of R^2, and it will increase R^2 substantially if the new variable makes a substantial *additional* contribution to the description of the dependent variable. Note that the increase in R^2 due to the entry of a given predictor depends on what other variables are already in the equation and whether or not they account for some of the same variation as the new predictor does. Just what constitutes a "substantial" increase in R^2 is a matter of judgment that depends on the problem at hand, the precision required in the final description, and the desirability of a simple model (that is, an equation with a small number of predictors). A given value of R^2 is always more impressive if it is based on an equation with a smaller number of predictors in it. As the number of degrees of freedom due to regression approaches the number of *different* observations in the data, R^2 will be large, not because the equation is a good description of the underlying process, but because there are too few observations for discrepancies to show up.

c. $\text{MS}_{\text{res.}}$ is a measure of the imprecision of the estimates $\hat{y}$ obtained from the equation. If the residual mean square becomes smaller when an additional predictor is entered into the equation—sufficiently smaller to make up for the loss of one degree of freedom from the error term—then the entry of the variable is said to result in improved precision. If the residual mean square becomes larger, as it will if the additional contribution of the new variable is relatively small, then the additional predictor should not be entered since it reduces the precision of the description. As the number of degrees of freedom due to regression approaches the number of *different* observations in the data, the residual mean square may become small, not because the equation is a good description of the underlying process, but because there are too few observations for discrepancies to show up. However, as long as the number of degrees of freedom for residual variation is reasonably large, the reduction in the residual mean square is a good indicator of what is gained by the entry of an additional predictor.

d. The coefficient of partial determination is calculated for each predictor not yet in the equation, holding constant those that are in the equation, to see what proportion of the residual variation would be accounted for by each predictor. The predictor that would contribute the *most* additional variation to the regression is entered next, the significance of the additional variation being tested by Expression (21.16). Used in this way to help decide if a predictor should be entered into the equation, the F ratio in Expression (21.16) is sometimes called the ***F* to enter**. Strictly speaking, this is not so much a significance test as simply a convenient criterion to assist in identifying the predictors making the greatest additional contribution to the description. The practice of selecting the best candidate for testing inflates the actual Type I error rate, but the practice is appropriate anyway, since finding the most useful predictors is more important here than knowing the actual Type I error rate at each stage of the procedure.

e. The coefficient of partial determination is calculated for each predictor already in the equation, holding constant the others in the equation, to see which variable accounts for the *least* variation not accounted for by the others. If the extra sum of squares due to the inclusion of this variable is not significant [Expression (21.16)], it is removed from the equation. This may happen even though the predictor contributed significantly when it was entered, because other predictors that were entered later may make it redundant. Used in this way to help decide if a predictor should be removed from the equation, the F ratio in Expression (21.16) is sometimes called the ***F* to remove**. Again, this is not so much a significance test as just a convenient criterion for variable selection. The practice of selecting the worst candidate for testing deflates the actual Type I error rate.

The procedure for stepwise regression analysis may be outlined as follows:

Step 0. Start with the equation

$$\hat{y} = \bar{y} \tag{21.20}$$

as the simplest model for describing the dependent variable y.

Step 1. Draw a set of scattergrams showing the dependent variable y as a function of each of the predictors, and check for systematic departures from linearity. Section 21.2 discusses how the set of available predictors can be revised to allow for nonlinear relationships, if necessary. If any such revisions are made, use the revised set of predictors throughout the stepwise analysis. Calculate the simple correlation r between y and each of the predictors. Select the predictor corresponding to the largest value of r^2. Since r^2 is the proportion of the variation in y that can be accounted for by the linear relationship with the independent variable, the selected independent variable is the single best predictor of y. Compute the simple linear regression equation, and test the significance of the regression. If the regression is not significant, reject the selected predictor and conclude that Expression (21.20) is the best available description of the dependent variable. If the regression is significant, the corresponding equation is a significantly better model than Expression (21.20) and we proceed to Step 2 to see if it can be further improved.

Step 2. Calculate the partial correlation between y and each of the predictors not currently in the equation, holding constant the predictor(s) currently in the equation. Select the predictor corresponding to the largest coefficient of partial determination. This is the predictor that makes the greatest additional contribution to the description. Enter this predictor into the equation by computing the multiple regression equation, and test the significance of the regression. If the regression is not significant, reject the selected predictor and go on to Step 3. If the regression is significant, leave the predictor in the equation and go on to Step 3 to see if it can be further improved.

Step 3. Calculate the partial correlation between y and each of the predictors currently in the equation, holding constant the other predictor(s) in the equation. Select the predictor corresponding to the smallest coefficient of partial determination. This is the predictor that contributes least to the description, given the other predictor(s) in the equation. Test the significance of the extra sum of squares due to the inclusion of this predictor. [If this predictor is not the one just entered, Expression (21.13) can be used with the second line of Expression (21.14) to calculate the extra sum of squares.] If the extra sum of squares is significant, leave the predictor in the equation and go on to Step 4 to see if the equation can be further improved. If it is not significant, remove the predictor from the equation and compute the regression equation without it. If the removed variable had just been entered as the most recent change in the equation, repeat Step 3 with the reduced equation to see if another predictor can be removed; if not, go on to Step 4.

Steps 4, 5, 6, ... Repeat Steps 2 and 3 until none of the predictors in the equation can be removed according to the criterion in Step 3 and none of the excluded predictors can be entered without being immediately removed. Compute the final regression equation along with its ANOVA summary table, coefficient of multiple determination, and standard error of estimation.

The set of predictors included in the final equation of a stepwise regression analysis should be critically examined in light of common sense, knowledge of the area of application, and careful study of the data. Scattergrams of the observations and plots of the residuals from the final regression can be especially helpful in revealing inadequacies in the selected set of variables. Stepwise regression analysis should be regarded as a valuable tool for variable selection, but one whose results can sometimes be further refined and may even be set aside altogether in favor of other methods of variable selection. For example, stepwise regression analysis may lead to the most appropriate selection from among the available predictors, but there may be other, more relevant predictors that were not observed, and the selected set may not generalize to other situations if the observational units were not randomly selected.

EXAMPLE 21.8 The following data were collected in the investigation described in Examples 13.8 and 21.2. The purpose of the investigation was to identify factors affecting the germination time of Lincoln–Homesteader pea seeds. The following list describes the variables that were recorded:

Dependent variable:

y = germination time = the number of days from planting until sprouting

Predictors:

x_1 = temperature = the mean maximum daily temperature (°C) over the seven days beginning with the day of planting
x_2 = planting date = the number of days after 15 April that planting occurred
x_3 = seeder = a dummy variable scored as $x_3 = 1$ for plantings using a mechanical seeder and $x_3 = 0$ for planting by hand
x_4 = cold spell = a dummy variable scored as $x_4 = 1$ if the weather turned cold (maximum less than 10°C on any one day) within the six days following planting and $x_4 = 0$ otherwise

The observations and some preliminary calculations are given on page 784.

The simple correlations r for all pairs of variables are conveniently shown in the following **correlation matrix**. Since the matrix is symmetric, correlations below the diagonal are omitted to reduce clutter. The left-most column, the bottom row, and the values of 1.0 on the diagonal could also be omitted with no loss of information.

		y	x_1	x_2	x_3	x_4
Germination time	y	1.00000000	−.77280249	−.51828126	.20817604	.66080079
Temperature	x_1		1.00000000	.38244389	−.34395147	−.34208922
Planting date	x_2			1.00000000	−.10236964	−.39047660
Seeder	x_3				1.00000000	−.03857584
Cold spell	x_4					1.00000000

Figure 21.2 shows scattergrams for the four predictors separately, and the stepwise regression analysis is outlined in the following paragraphs.

Step 0. The simplest model of germination times for these data, based on the mean of the y-values, is

$$\hat{y} = 10.4 \text{ days}$$

Observation number	Germination time, y	Temperature, x_1	Planting date, x_2	Seeder, x_3	Cold spell, x_4
1	11	14.8	20	0	1
2	12	17.4	34	0	1
3	10	15.6	23	0	0
4	8	20.9	38	0	0
5	11	20.1	4	0	1
6	12	17.1	20	0	0
7	9	19.9	25	0	0
8	8	21.4	26	0	0
9	11	13.5	17	1	0
10	8	21.4	32	1	0
11	14	13.4	16	1	1
	—	——	—	—	—
	$\bar{y} =$ 10.363636	$\bar{x}$: 17.772727	23.181818	.272727	.363636
	$s_y =$ 1.963300	s_x: 3.123489	9.505979	.467099	.504525
	$\sum y =$ 114	$\sum x$: 195.5	255	3	4
	$\sum y^2 =$ 1220	$\sum x^2$: 3572.13	6815	3	4
	$SS_y =$ 38.545455	SS_x: 97.561818	903.636364	2.181818	2.545455
		$\sum xy$: 1978.7	2546	33	48
		SP_{xy}: −47.390909	−96.727273	1.909091	6.545455
			$\sum x_1x_j$: 4645.6	48.3	65.7
			$SP_{:1j}$: 113.554545	−5.018182	−5.390909
				$\sum x_2x_j$: 65	74
				SP_{2j}: −4.545455	−18.727273
					$\sum x_3x_4 = 1$
					$SP_{34} = -.090909$

If none of the predictors yields significant improvement in the description of germination times, we will accept this simple model.

Step 1. None of the scattergrams reveals systematic departures from linearity, so we proceed with the set of predictors as defined above. From the correlation matrix, we see that the predictor having the strongest simple correlation with the dependent variable y is temperature x_1, whose linear relationship with y accounts for

$$r_{yx_1}^2 = (-.77280249)^2 = .59722368$$

of the total variation in y. The simple linear regression on x_1 was calculated in Example 13.8 as

$$\hat{y} = 19.0 - .49x_1 \tag{21.21}$$

The total variation in y is shown in the preliminary calculations to be

$$SS_y = 38.545455$$

so the sum of squares due to regression is

$$SS_{\text{regr. on } x_1} = (.59722368)(38.545455) = 23.0202584$$

and we have the following ANOVA summary table for Step 1.

Source of variation	SS	df	MS	F	
Regression on temperature (b_1)	23.0202584	1	23.02	13.3	$P < .01$
Residual given x_1	15.5251961	9	1.73	—	
Total	38.5454545	10			

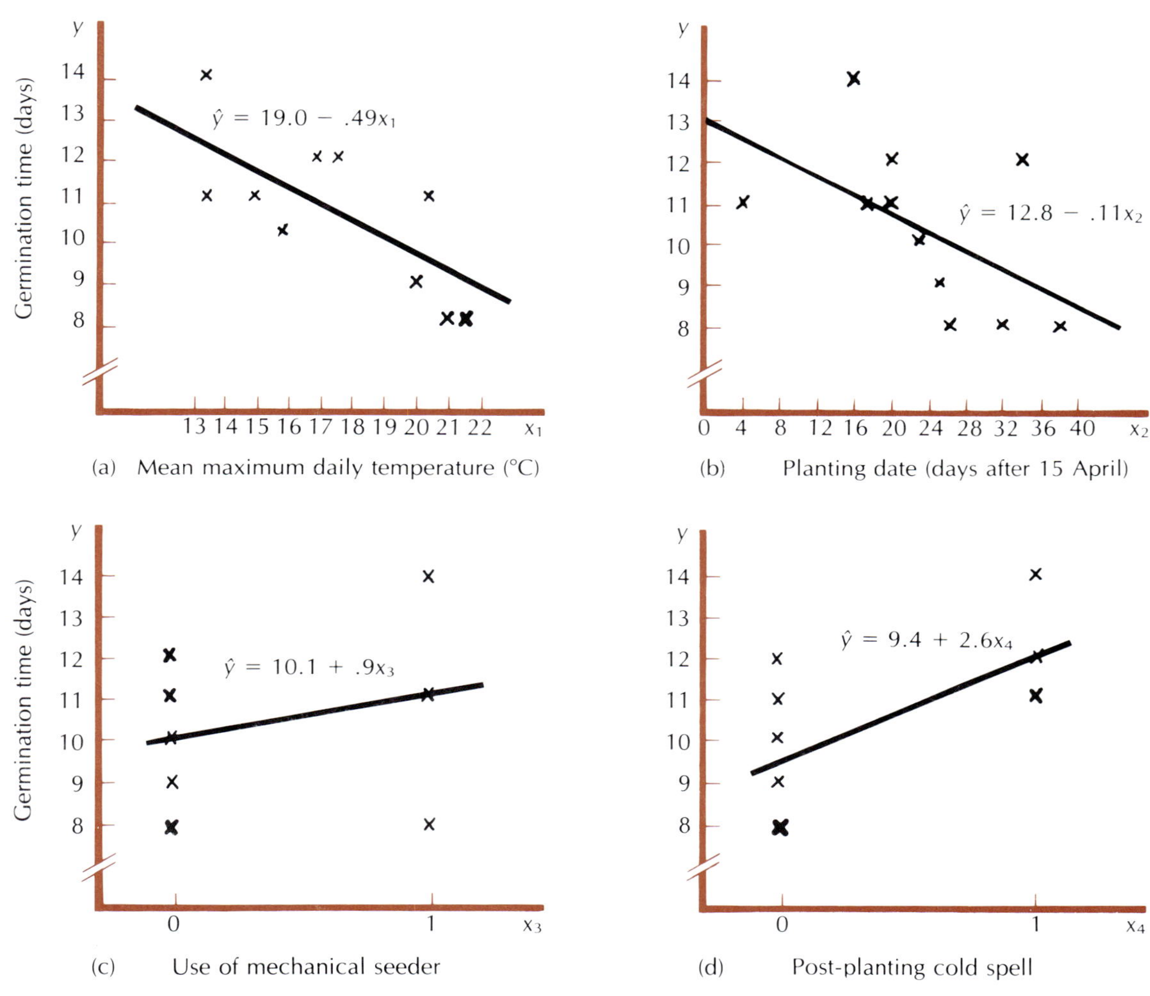

Figure 21.2
Scattergrams of germination time as a function of each of four predictors, for 11 plantings of peas. Simple linear regression lines are also shown.

The regression on x_1 [Expression (21.21)] accounts for

$$R^2 = \frac{23.0202584}{38.5454545} = 60\%$$

of the total variation in y. The F test in the ANOVA table shows that the regression is significant at the 1% level, so we go on to Step 2 to see if inclusion of an additional predictor will improve the equation.

Step 2. The partial correlations of y with the remaining predictors, given that x_1 is already in the equation, are calculated by substituting appropriate values from the correlation matrix into Expression (21.17):

$$r_{yx_2|x_1} = -.37982211$$
$$r_{yx_3|x_1} = -.09670763$$
$$r_{yx_4|x_1} = .66475875$$

The cold-spell variable x_4 makes the greatest additional contribution to the model, accounting for

$$r^2_{yx_4|x_1} = (.66475875)^2 = .44190420$$

or about 44% of the residual variation given x_1 (in the ANOVA table for Step 1). Entering x_4 into the equation, we must solve the following normal equations:

$$\begin{aligned} 114 &= 11\ b_0 + 195.5\ b_1 + 4\ b_4 \\ 1978.7 &= 195.5b_0 + 3572.13b_1 + 65.7b_4 \\ 48 &= 4\ b_0 + 65.7\ b_1 + 4\ b_4 \end{aligned}$$

The solution, according to Cramer's rule, is

$$b_0 = \frac{40150.22}{2412.05} = 16.645683$$

$$b_1 = \frac{-938.8}{2412.05} = -.38921250$$

$$b_4 = \frac{4214.17}{2412.05} = 1.747132$$

The multiple regression of germination time y on temperature x_1 and cold spell x_4 is

$$\hat{y} = 16.6 - .39x_1 + 1.7x_4 \qquad \textbf{(21.22)}$$

which reduces to

$$\hat{y} = 16.6 - .39x_1$$

if there is no post-planting cold spell ($x_4 = 0$), and to

$$\hat{y} = 18.4 - .39x_1$$

if there is a post-planting cold spell ($x_4 = 1$). The extra sum of squares due to x_4 given x_1 is

$$SS_{\text{regr. on } x_4|x_1} = (.44190420)(15.5251961) = 6.8606494$$

so the sum of squares due to regression on x_1 and x_4 is

$$SS_{\text{regr. on } x_1, x_4} = 23.0202584 + 6.8606494 = 29.8809078$$

and we have the following ANOVA summary table for Step 2.

Source of variation	SS	df	MS	F	
Regression on temperature and cold spell (b_1, b_4)	29.8809078	2	14.94	13.8	$P < .01$
Residual given x_1, x_4	8.6645468	8	1.08	—	
Total	38.5454545	10			

The regression on x_1 and x_4 [Expression (21.22)] accounts for

$$R^2 = \frac{29.8809078}{38.5454545} = 78\%$$

of the variation in germination times, which is substantially higher than the 60% based on a single predictor. Likewise, the residual mean square has been reduced from 1.73 to 1.08, indicating improved precision in the description. The F test in the ANOVA table shows that the regression is significant at the 1% level, so we go on to Step 3 to see if the equation contains any superfluous predictors.

Step 3. From Expression (21.17), the partial correlation of y with x_4 given x_1 is

$$r_{yx_4|x_1} = .66475875$$

and the partial correlation of y with x_1 given x_4 is

$$r_{yx_1|x_4} = -.77522566$$

from which we see that x_4 makes the smaller additional contribution. The extra sum of squares due to x_4 given x_1 was calculated in Step 2 as 6.8606494, so the F ratio to test its significance is, from Expression (21.16),

$$F_{\text{regr. on } x_4|x_1} = \frac{6.8606494}{1.083} = 6.33$$

This exceeds the critical value $F_{.05}(1, 8\text{ df}) = 5.32$, so the additional contribution of the cold-spell variable x_4, given x_1, is significant at the 5% level. Hence, we leave both predictors in the equation and go on to Step 4 to see if the inclusion of an additional predictor will improve the equation.

Step 4. The partial correlations of y with the remaining predictors, given that x_1 and x_4 are already in the equation, are calculated from Expression (21.18):

$$r_{yx_2|x_1x_4} = \frac{r_{yx_2|x_4} - r_{yx_1|x_4} r_{x_2x_1|x_4}}{\sqrt{1 - r^2_{yx_1|x_4}}\sqrt{1 - r^2_{x_2x_1|x_4}}} = -.25393780$$

$$r_{yx_3|x_1x_4} = \frac{r_{yx_3|x_4} - r_{yx_1|x_4} r_{x_3x_1|x_4}}{\sqrt{1 - r^2_{yx_1|x_4}}\sqrt{1 - r^2_{x_3x_1|x_4}}} = .02856601$$

The planting-date variable x_2 makes the greater additional contribution to the description, accounting for

$$r^2_{yx_2|x_1x_4} = (-.25393780)^2 = .06448441$$

or about 6% of the residual variation given x_1 and x_4 (in the ANOVA table for Step 2). The extra sum of squares due to regression on x_2 given x_1 and x_4 is

$$SS_{\text{regr. on } x_2|x_1,x_4} = (.06448441)(8.6645468) = .5587282$$

so the sum of squares due to regression on x_1, x_2, and x_4 is

$$SS_{\text{regr. on } x_1, x_2, x_4} = 29.8809078 + .5587282 = 30.4396359$$

and we have the following ANOVA summary table for Step 4.

Source of variation	SS	df	MS	F	
Regression on temperature, planting date, and cold spell (b_1, b_2, b_4)	30.4396359	3	10.15	8.76	$P < .01$
Residual given x_1, x_2, x_4	8.1058186	7	1.16	—	
Total	38.5454545	10			

The regression on x_1, x_2, and x_4 accounts for

$$R^2 = \frac{30.4396359}{38.5454545} = 79\%$$

of the variation in germination times, which is not much of an improvement over the 78% accounted for by just x_1 and x_4. Furthermore, the residual mean square has increased from 1.08 to 1.16, indicating less precision in the description. The F test in the ANOVA table shows that the regression on the three predictors is still significant at the 1% level, but in view of the changes in R^2 and the residual mean square, we must question whether the inclusion of x_2 is worthwhile. Step 5 answers the question.

Step 5. Using Expression (21.18), we check the additional contribution due to each of the predictors in the equation, given the other two:

$$r_{yx_1|x_2x_4} = -.75166126$$
$$r_{yx_2|x_1x_4} = -.25393780$$
$$r_{yx_4|x_1x_2} = .62439793$$

The smallest contribution is that due to x_2, whose extra sum of squares given x_1 and x_4 was given in Step 4 as .5587282. The F ratio to test its significance is, from Expression (21.16),

$$F_{\text{regr. on } x_2|x_1x_4} = \frac{.5587282}{1.158} = .48$$

which is not significant. (F ratios less than 1.0 are never significant at any conventional level of significance.) Thus, we reject x_2 from the equation, leaving x_1 and x_4 as predictors in the equation.

We have already seen in Step 3 that neither x_1 nor x_4 can be rejected from the equation, and in Steps 4 and 5 that the best remaining candidate for inclusion is immediately rejected. We conclude that the best available description of germination times is that given in Expression (21.22), namely,

$$\hat{y} = 16.6 - .39x_1 + 1.7x_4$$

The corresponding ANOVA summary table and related statistics are shown in Step 2.

Figure 21.3 presents computer output for this example from the SAS/STAT™ software. We will not be concerned with the quantities labeled "C(P)" and "Bounds on

```
                                         SAS

                  STEPWISE REGRESSION PROCEDURE FOR DEPENDENT VARIABLE Y

STEP 1     VARIABLE X1 ENTERED        R SQUARE = 0.59722368      C(P) =      4.49858836

                    DF          SUM OF SQUARES          MEAN SQUARE            F      PROB>F

REGRESSION           1             23.02025840          23.02025840        13.34      0.0053
ERROR                9             15.52519615           1.72502179
TOTAL               10             38.54545455

                     B VALUE         STD ERROR          TYPE II SS             F      PROB>F

INTERCEPT        18.99678526
X1               -0.48575262        0.13297113          23.02025840        13.34      0.0053

BOUNDS ON CONDITION NUMBER:             1,             1
--------------------------------------------------------------------------------------------

STEP 2     VARIABLE X4 ENTERED        R SQUARE = 0.77521223      C(P) =      1.41731387

                    DF          SUM OF SQUARES          MEAN SQUARE            F      PROB>F

REGRESSION           2             29.88090779          14.94045390        13.79      0.0026
ERROR                8              8.66454675           1.08306834
TOTAL               10             38.54545455

                     B VALUE         STD ERROR          TYPE II SS             F      PROB>F

INTERCEPT        16.64568313
X1               -0.38921250        0.11212793          13.04773896        12.05      0.0084
X4                1.74713211        0.69417835           6.86064939         6.33      0.0360

BOUNDS ON CONDITION NUMBER:      1.132535,       4.53014
--------------------------------------------------------------------------------------------

NO OTHER VARIABLES MET THE 0.0500 SIGNIFICANCE LEVEL FOR ENTRY INTO THE MODEL.

              SUMMARY OF STEPWISE REGRESSION PROCEDURE FOR DEPENDENT VARIABLE Y

          VARIABLE          NUMBER   PARTIAL    MODEL
STEP   ENTERED    REMOVED       IN      R**2     R**2      C(P)         F     PROB>F     LABEL

  1    X1                        1    0.5972   0.5972   4.49859   13.3449    0.0053     TEMPERATURE
  2    X4                        2    0.1780   0.7752   1.41731    6.3345    0.0360     COLD SPELL
```

Figure 21.3
SAS/STAT™ printout for Example 21.8.

condition number," but you should be able to identify all of the other values in the SAS/STAT printout. The "TYPE II SS" is the extra sum of squares due to the inclusion of a particular predictor, given that the other predictors shown for that step are also in the regression equation. Unlike the printout in Figure 21.1, this printout shows the F-value, rather than the t-value, as the test statistic for the extra sum of squares. As noted in Example 21.4, the F and t tests are equivalent in this case.

(Source: Student project by G. Hamilton.)

It should be emphasized that stepwise regression analysis is not always the best solution to the problem of variable selection, and that its results should therefore not be accepted without critical and thoughtful examination. Example 21.11 illustrates a case where stepwise regression analysis leads to a solution that is substantially inferior to a solution obtained by departing from the prescriptions of the stepwise procedure.

Stepwise regression analysis does, however, draw attention to the statistics that should be examined in any multiple regression analysis:

a. $F_{\text{regr.}}$, which tests the statistical significance of the regression as a whole;

b. R^2, which measures the proportion of variation accounted for by the regression model;

c. $\text{MS}_{\text{res.}}$, which is the square of the standard error of estimation, measuring the lack of precision of the fit; and

d. $F_{\text{extra SS}}$ (or t_{b_j}), which tests, for each regression coefficient b_j, the additional contribution of the corresponding predictor x_j, given that the other predictors are in the equation. (The last paragraph of Example 21.4 describes the alternative statistic, t_{b_j}, which provides an equivalent test of the extra sum of squares.)

The next subsection discusses these and other considerations in interpreting multiple regression equations.

Interpretation of the Multiple Regression Equation

Once we have obtained the multiple regression equation

$$\hat{y} = b_0 + b_1x_1 + b_2x_2 + \cdots + b_gx_g$$

what can we say about it? There are several important points to keep in mind when interpreting a multiple regression equation. You should refer to Example 21.8 for examples of these points.

Comment 1, Concerning the Significance of the Regression. Statistical significance should not be confused with practical importance. This point is discussed further under item (a) on page 780.

Comment 2, Concerning the Strength of the Relationship. The proportion R^2 of the variation in the dependent variable y accounted for by the regression is an important piece of information, but it must be interpreted in light of the number of degrees of freedom for regression (that is, the number of predictor terms in the equation) and the number of different observations in the data. The value of R^2 can be made arbitrarily close to 100% by increasing the number of predictor terms in the equation, but as the equation becomes more cumbersome, it is apt to be more descriptive of the idiosyncracies in a particular sample and less conducive to an understanding of the underlying process. Moreover, the value of R^2 may be an inflated value if there are many duplicate observations in the data, or if many of the observations are almost duplicates of each other. As an extreme example, suppose that the same two points (x, y) were recorded five times each. A straight line would fit the data perfectly (because any two points can be joined by a straight line), giving an R^2 value of 100% with 1 degree of freedom for regression and 8 degrees of freedom for residual. In spite of the value of R^2, the regression equation would not necessarily be a very good description of the underlying relationship between x and y, because there were too few *different* observations to allow any other pattern of relationship to be manifest. This point is also discussed under item (b) on page 780.

Comment 3, Concerning the Standard Error of Estimation. The positive square root of the residual mean square is the standard error of estimation s_e. It is an average or standard amount by which the errors of estimation $y_i - \hat{y}_i$ deviate from their mean, which is zero. Thus, it may be interpreted as a typical amount by which the fitted values $\hat{y}$ differ from the observed values y for the predictor values included in the sample. The standard error of estimation indicates, in the measurement units of the dependent variable, how close the fitted values tend to be to the observed values of the dependent variable. See also the remarks concerning the residual mean square under item (c) on page 780.

EXAMPLE 21.9 The residual mean square for the finally accepted model of germination times in Example 21.8 was given in the ANOVA table in Figure 21.3 as 1.08. The standard error of estimation was therefore $\sqrt{1.08} = 1.04$, indicating that the fitted values differed from the observed germination times by about 1.0 day on the average. This is confirmed by inspection of the following list of residuals $y - \hat{y}$ from the regression on x_1 and x_4.

Observation number	Residual, $y - \hat{y}$
1	−1.632470
2	.379482
3	−.573968
4	−.511142
5	.430356
6	2.009851
7	.099646
8	−.316536
9	−.391314
10	−.316536
11	.822632
	$\Sigma (y - \hat{y}) = 0.000000$
	$\Sigma (y - \hat{y})^2 = 8.664547$

Comment 4, Concerning Regression Coefficients. The regression coefficients in a multiple regression equation are often referred to as **partial regression coefficients**, because each one is the coefficient for one predictor, given that the other predictors are also in the equation. Multiple regression is for use when a researcher wants to know how well *a set of predictors considered jointly* describes a dependent variable. As a general rule, partial regression coefficients have no meaning except in the context of the particular set of predictors included in the equation. Indeed, the coefficient b_j of a given predictor x_j will change depending on which other predictors are included. The interpretation of a partial regression coefficient b_j must always include a qualification such as "given the other predictors included in the equation" or "holding constant the other predictors in the equation" or "with the other predictors' effects removed." (The exception to this rule is in the case of designed experiments where the levels of the independent variables have been chosen so as to yield orthogonal sources of variation—that is, uncorrelated predictors.) The coefficient b_j is the average increase in y accompanying an increase of one unit in x_j, when the effects of the other predictors in the equation have been statistically removed. (Example 21.6 includes correct interpretations of partial regression coefficients.) The answer to the question as to which is the single best predictor of the dependent variable y is provided not by the multiple regression equation, but rather by the simple correlations between y and the predictors. As noted under Step 1 on page 782, the predictor whose squared simple correlation with y is largest is the single best predictor of y.

Comment 5, Concerning Extrapolation. The multiple regression equation describes the dependent variable y within the observed range of each of the predictor variables, but it does not indicate whether it is safe to extrapolate to predictor values outside the observed range. Extrapolation is always risky and must be justified (if and when it can be) by knowledge of the process being described, not on statistical grounds. This same warning was discussed in connection with simple linear regression in Section 5.1, where it was illustrated by Example 5.7. (See also Example 13.12.)

Comment 6, Concerning Variables not Included in the Equation. In addition to the predictors included in the regression equation, there are many extraneous variables that might affect the value of the dependent variable. Extraneous variables include predictors that have been recorded but rejected for entry into the equation, as well as variables that have not been measured and recorded. Just as the applicability of the regression equation is limited to the observed ranges of the included predictors, so also is it limited to the extraneous variables' levels that were in effect when the observations were made. Under different conditions of observation, different relationships may exist and different predictors may be more useful. In the context of regression analysis, variables that are not included in the regression equation but that in fact have important effects on the dependent variable are called **lurking variables**. If a designed experiment is possible, random assignment of observational units to conditions and replication with a different sample can take care of problems due to some lurking variables. Otherwise, alertness to the possibility of lurking variables (for example, by looking for suspicious patterns in plots of the residuals as a function of possible lurking variables, such as the order in which the observations were made) and experience with the process being investigated are perhaps the best defenses against misinterpretations due to variables not included in the equation. This point is discussed further under Extraneous Variables in Section 18.1, with an example in Example 18.4.

Comment 7, Concerning Causality. The regression equation provides absolutely no information as to what causes what. The appropriateness of causal conclusions is never determined by the method of statistical analysis, whether multiple regression or ANOVA, but rather is determined by the design of the investigation and by knowledge of the process being investigated. The last paragraph of Example 21.1 includes an illustration.

21.2 NONLINEAR REGRESSION

So far in this chapter, we have considered only those situations in which a dependent variable y may be described by a linear, or straight-line, function of one or more predictors. This section introduces **nonlinear regression**, in which the dependent variable is described as a nonlinear function of one or more predictors.

There are a virtually unlimited number of different types of nonlinear functions that might be considered as candidates for describing particular nonlinear relationships. The choice of the most appropriate type of function to serve as the basis of the model, or description, of the underlying process being investigated is often difficult and usually requires expert knowledge in the area of application. In addition, many of the techniques of nonlinear regression analysis are beyond the level of this book. You should refer to the sources cited under Further Reading at the end of this chapter for the more advanced techniques.

In this section, we consider some simple methods by which the techniques introduced earlier in this chapter may be used to describe certain nonlinear relationships. The methods presented here all involve expressing the nonlinearly related variables in terms of new variables such that the relationship is linear in the new variables. Our discussion will be phrased in terms of a single predictor x, but the same principle may be applied when there is more than one predictor.

The first step is to identify whether the relationship is in fact nonlinear within the observed range of the predictor. We do so by examining any of the following three scattergrams:

a. the dependent variable y as a function of the predictor x, or

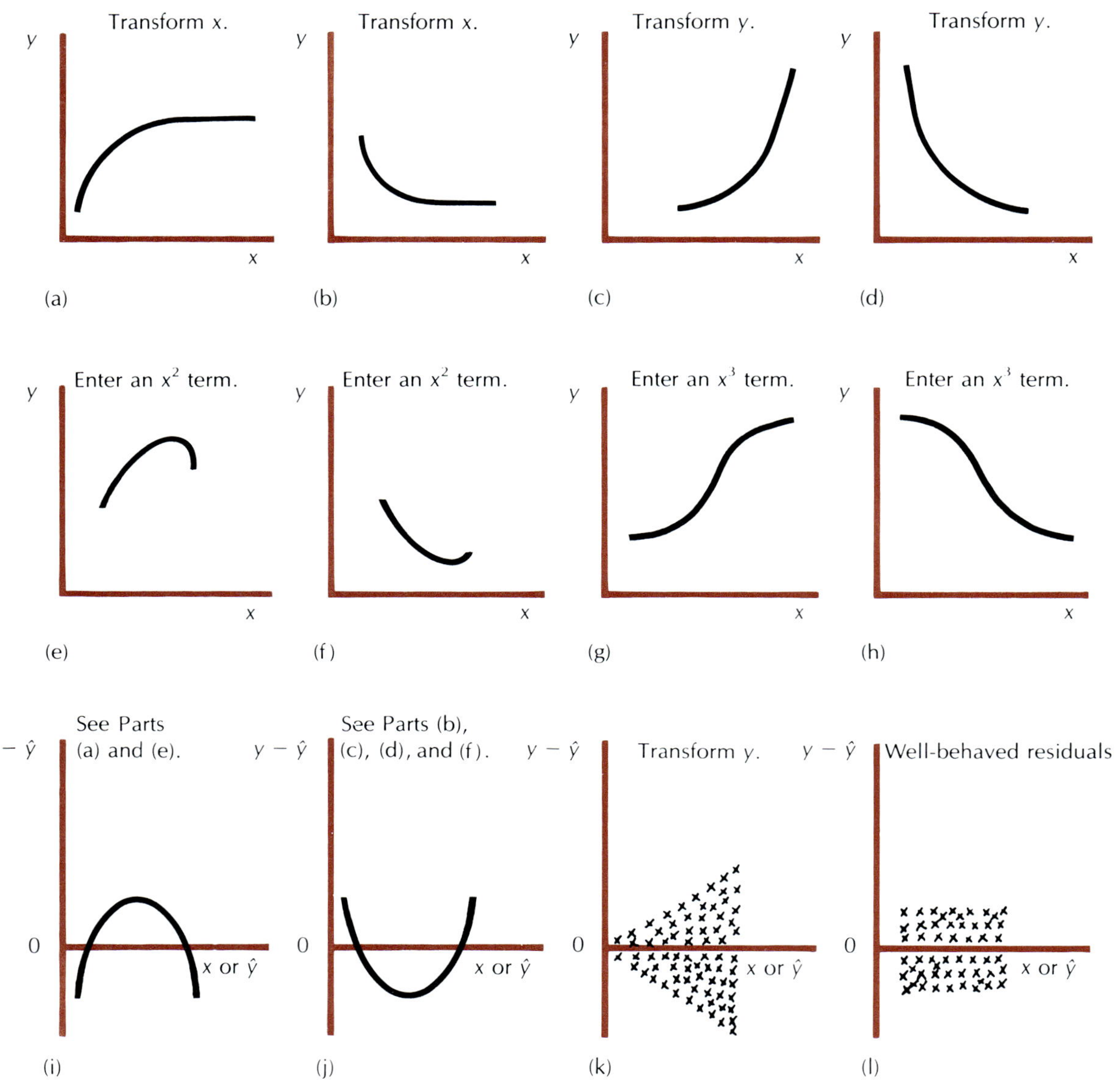

Figure 21.4
Schematic illustrations of some of the patterns that may be observed in scattergrams. The patterns in Parts (a) through (k) are indicative of nonlinear relationships, and the comments suggest how the relationship may be reexpressed in linear form. Part (l) illustrates the desired pattern of residuals resulting from a regression equation whose variables require no further transformation.

b. the linear regression residuals $y_i - \hat{y}_i$ as a function of the predictor x, or

c. the linear regression residuals $y_i - \hat{y}_i$ as a function of the fitted values $\hat{y}_i$.

A curvilinear pattern in any of these scattergrams indicates a nonlinear relationship between the original x- and y-values. Figure 21.4 illustrates some indications of nonlinearity that may appear in the scattergrams.

Data Transformations to Describe Nonlinear Relationships

If the scattergram of the dependent variable as a function of the predictor approximates one of the patterns in Parts (a) through (d) of Figure 21.4, it may be possible to obtain a better description of the relationship by **transforming** one of the variables. The square-root transformation is often helpful for variables recorded as **counted data** (such as number of errors or number of successes); the reciprocal transformation is often appropriate for time variables whose reciprocal is speed; and the logarithmic transformation—the most commonly used of these three—may be helpful with counted data or time data and in a wide variety of other applications. In each case, each of the original values of the variable that is to be transformed is replaced by its square root, or reciprocal, or logarithm (depending on which transformation is being used), and then ordinary linear regression methods are applied, just as if the data had been recorded in terms of the transformed variable in the first place. Two or more different transformations may be tried and the values of R^2 compared to see which one provides the best description of the relationship in the data at hand.

EXAMPLE 21.10 A study was conducted to investigate the relationship between the amount of grain lost in harvesting and the speed at which the combine is run. The testing was done by operating a combine at various speeds over a pan on the ground and counting the number of kernels of grain that fell into the pan at each speed. The data from 12 trials are recorded in the following table and plotted in Figure 21.5.

Speed (mph), x	Number of kernels lost, y
.5	0
1	0
1.5	2
2	3
2.5	3
3	2
3.5	4
4	2
4.5	5
5	6
5.5	9
6	13

The simple linear regression of number of kernels (y) on speed (x) is

$$\hat{y} = 1.83x - 1.85$$

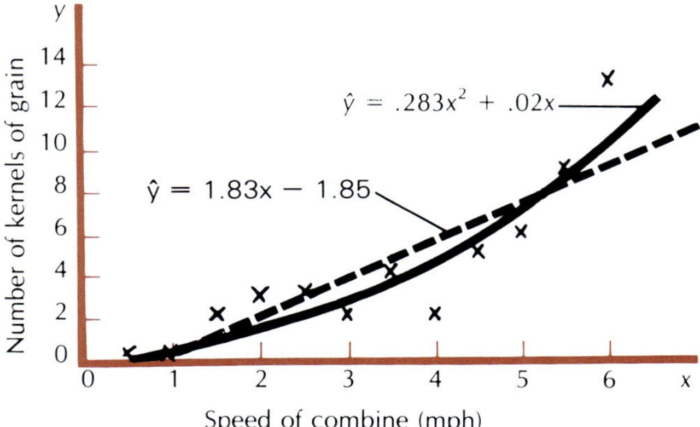

Figure 21.5 Scattergram of amount of grain lost as a function of combine speed.

accounting for $r^2 = 76\%$ of the variation in number of kernels. However, the scattergram shows a distinct curvilinear pattern resembling that in Part (c) of Figure 21.4. A transformation of the dependent variable is suggested.

Since we are dealing with counted data (number of kernels), let us try the square-root transformation. We replace each value of y with its square root, say, $z = \sqrt{y}$. The simple linear regression of z on x is then calculated via the usual formulas (Section 13.2), yielding $b = .531891$, $a = .021176$, and $r = .908362$. The resulting equation is

$$\hat{z} = .53x + .02$$

accounting for $r^2 = 83\%$ of the variation in the dependent variable. This is an improvement over the 76% accounted for in the untransformed data.

The ANOVA summary table for the regression of $\sqrt{y}$ on x is as follows.

Source of variation	SS	df	MS	F	
Regression of $\sqrt{y}$ on x	10.114	1	10.114	47.2	$P < .01$
Residual	2.144	10	.214		
Total for $\sqrt{y}$	12.258	11			$R^2 = 83\%$

The regression is significant at the .01 level. Squaring $\hat{z}$ to convert the equation back to the original units, we have

$$\hat{z}^2 = \hat{y} = (bx + a)^2 = .2829x^2 + .0225x + .0004$$

or approximately

$$\hat{y} = .283x^2 + .02x \qquad \textbf{(21.23)}$$

This equation is also plotted in Figure 21.5.

We could also try the logarithmic transformation of y to see whether it would give a better description than Expression (21.23). We cannot simply use $\log y$, however, because there are two observations equal to zero, the logarithm of which is undefined. The usual solution to this problem is to use the transformation as follows: Replace each value of y with $\log(y + 1)$. Taking common logarithms, let us call the transformed variable $w = \log(y + 1)$. The simple linear regression of w on x is then calculated by the usual formulas (Section 13.2), yielding $b = .174201$, $a = .025833$, and $r = .905064$. The resulting equation is

$$\hat{w} = .174x + .026$$

accounting for $r^2 = 82\%$ of the variation in the dependent variable, just slightly less than the 83% with the square-root transformation.

The ANOVA summary table for the regression of $\log(y + 1)$ on x is as follows.

Source of variation	SS	df	MS	F	
Regression of $\log(y + 1)$ on x	1.0849	1	1.0849	45.3	$P < .01$
Residual	.2395	10	.0240		
Total for $\log(y + 1)$	1.3244	11			$R^2 = 82\%$

Again, the regression is significant at the .01 level. Taking the antilog of $\hat{w}$ to convert the equation back to the original units, we have

$$10^{\hat{w}} = \hat{y} + 1 = 10^{bx+a} = 10^a 10^{bx} = 10^a(10^b)^x = 1.0613(1.49349^x)$$

or approximately

$$\hat{y} = 1.06(1.493^x) - 1$$

This equation is a form of **exponential curve** ($y = cd^x$). In this case, its predictions within the observed range of combine speeds are not very different from those of the quadratic polynomial in Expression (21.23). Collection of more data would be required to distinguish between these two models of grain loss.

(Source: Student project by D. Cameron.)

Polynomial-Fitting to Describe Nonlinear Relationships

If the scattergram of the dependent variable as a function of the predictor approximates one of the patterns in Parts (e) through (h) of Figure 21.4, it may be possible to obtain a better description of the relationship by fitting a quadratic (one-bend curve) or cubic (two-bend curve), rather than a linear (straight-line), polynomial to the data. This procedure is called **polynomial fitting**. To fit a polynomial, we do not *replace* the original predictor, say, x_1, but rather we define *additional* predictors to enter into the equation. If an x^2 term is needed, as in Parts (e) and (f) of Figure 21.4, we define a new variable, say, x_2, as $x_2 = x_1^2$ and then use ordinary multiple linear regression methods to compute the regression of y on x_1 and x_2:

$$\hat{y} = b_0 + b_1x_1 + b_2x_2 \tag{21.24}$$

Solved in the usual way, this equation may be rewritten as the quadratic polynomial

$$\hat{y} = b_0 + b_1x_1 + b_2x_1^2 \tag{21.25}$$

Similarly, if an x^3 term is needed, as in Parts (g) and (h) of Figure 21.4, we define a new variable, say, x_3, as $x_3 = x_1^3$ and then use ordinary multiple linear regression methods to compute the regression of y on x_1, x_2, and x_3:

$$\hat{y} = b_0 + b_1x_1 + b_2x_2 + b_3x_3 \tag{21.26}$$

Solved in the usual way, this equation may be rewritten as the cubic polynomial

$$\hat{y} = b_0 + b_1x_1 + b_2x_1^2 + b_3x_1^3 \tag{21.27}$$

Tests of the significance of the extra sum of squares due to additional x-terms may be used to help decide how many of the x-terms are required in the equation.

EXAMPLE 21.11 Ten different batches of dough were used in a study to investigate the effect of the amount of baking powder on the rising of baking powder biscuits. Each batch of dough received a different amount of baking powder, while the rest of the ingredients remained the same. The dough was rolled to a standard thickness of .75 inch before baking, and the height of the biscuit was measured after baking. Figure 21.6 is a scattergram of the data.

In view of the nonlinearity apparent in the scattergram, let us fit a quadratic polynomial to the data. The dependent variable y is the height of the biscuit in eighth-inch units, and the predictor x_1 is the amount of baking powder in half-teaspoon units. To fit a quadratic to the data, let us define another predictor $x_2 = x_1^2$. The data and preliminary calculations are as follows.

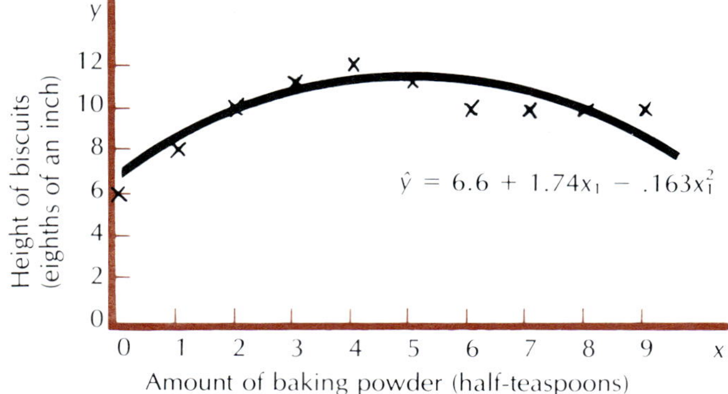

Figure 21.6 Scattergram of biscuit height as a function of amount of baking powder.

Observation number	Biscuit height, y	Amount of baking powder, x_1	$x_2 = x_1^2$
1	6	0	0
2	8	1	1
3	10	2	4
4	11	3	9
5	12	4	16
6	11	5	25
7	10	6	36
8	10	7	49
9	10	8	64
10	10	9	81
	—	—	—
	$\bar{y} =$ 9.8	$\bar{x}$: 4.5	28.5
	$s_y =$ 1.686548	s_x: 3.027650	28.304888
	$\sum y =$ 98	$\sum x$: 45	285
	$\sum y^2 =$ 986	$\sum x^2$: 285	15,333
	$SS_y =$ 25.6	SS_x: 82.5	7,210.5
		$\sum xy$: 464	2,914
		SP_{xy}: 23	121
		r_{xy}: .50047326	.28163245
			$\sum x_1x_2 =$ 2,025
			$SP_{12} =$ 742.5
			$r_{12} =$ .96269074

The simple linear regression of height y on x_1 is

$$\hat{y} = .28x_1 + 8.5$$

accounting for $r^2 = .25047349$ or approximately 25% of the variation in biscuit heights. The sum of squares due to the regression is

$$SS_{\text{regr. on } x_1} = (.25047349)(25.6) = 6.412$$

and thus the ANOVA summary table for the simple linear regression analysis is as follows.

Source of variation	SS	df	MS	F	
Regression on x_1	6.412	1	6.41	2.67	Not significant
Residual given x_1	19.188	8	2.40	—	
Total	25.600	9			

The normal equations for the regression of y on x_1 and x_2 are

$$98 = 10b_0 + 45b_1 + 285b_2$$
$$464 = 45b_0 + 285b_1 + 2{,}025b_2$$
$$2{,}914 = 285b_0 + 2{,}025b_1 + 15{,}333b_2$$

Cramer's rule gives the solution as

$$b_0 = \frac{2{,}871{,}000}{435{,}600} = 6.590909$$

$$b_1 = \frac{759{,}990}{435{,}600} = 1.744697$$

$$b_2 = \frac{-70{,}950}{435{,}600} = -.1628788$$

The multiple regression equation therefore is

$$\hat{y} = 6.6 + 1.74x_1 - .163x_2$$

which may be written in terms of the original predictor as the quadratic polynomial

$$\hat{y} = 6.6 + 1.74x_1 - .163x_1^2 \qquad \textbf{(21.28)}$$

which is shown in Figure 21.6

The partial correlation of y with x_2 given x_1 is [from Expression (21.17)]

$$r_{yx_2|x_1} = -.85441331$$

indicating that x_2 accounts for

$$r^2_{yx_2|x_1} = (-.85441331)^2 = .73002211$$

or about 73% of the residual variation given x_1 (in the preceding ANOVA table). The extra sum of squares due to x_2 given x_1 is therefore

$$SS_{\text{regr. on } x_2|x_1} = (.73002211)(19.188) = 14.008$$

and the sum of squares due to the regression of y on both x_1 and x_2 is

$$SS_{\text{regr. on } x_1,x_2} = 6.412 + 14.008 = 20.420$$

The ANOVA summary table for the regression on x_1 and x_2 may now be written as follows.

Source of variation	SS	df	MS	F	
Regression on x_1, x_2	20.420	2	10.21	13.8	$P < .01$
Residual given x_1, x_2	5.180	7	.74		
Total	25.600	9			

The regression on x_1 and x_2 [Expression (21.28)] accounts for

$$R^2 = \frac{20.420}{25.6} = 80\%$$

of the variation in biscuit heights. The F test in the second ANOVA table shows that the regression is significant at the 1% level, indicating that the equation in Expression (21.28) represents a statistically significant relationship.

This example illustrates the fact that stepwise regression analysis, if followed blindly, can lead to inappropriate conclusions. According to stepwise regression analysis, when the regression on the first predictor was not significant, we should have stopped and settled for the model $\hat{y} = 9.8$. However, the scattergram indicated that a curvilinear, rather than a linear, model was called for, so we went on to include the quadratic term $x_2 = x_1^2$. This led to a significant regression on two predictors, even though neither of the simple regressions was significant. Moreover, the multiple regression on x_1 and x_2 accounted for 80% of the variation in y, even though the simple regressions on x_1 and x_2 separately accounted for only 25% and 8%, respectively. The joint effect of two predictors can be greater than the sum of their separate effects. To avoid inappropriate conclusions, you should look for clues in the scattergrams, look for clues in plots of residuals (especially as a function of lurking variables), and bring to bear any expertise you have concerning the nature of the underlying process.

(Source: Student project by W. Sjogren.)

Examination of Residuals

Parts (i) through (l) of Figure 21.4 show some of the possible patterns that may appear in a plot of the residuals $y - \hat{y}$ as a function of either the predictor x or the fitted values $\hat{y}$. Any systematic departure from the pattern illustrated in Part (l) is an indication that the regression equation is in some way an inappropriate description of the relationship. Parts (i) and (j) show the pattern of residuals to be expected if a linear equation is fitted to one of the nonlinear relationships portrayed in Parts (a) through (f). Part (k) shows the pattern of residuals that often occurs if, contrary to the assumptions underlying the F tests, the variance is not constant throughout the observed range of the predictor. This heterogeneity of variance, which tends to give a positive bias to the F test, can sometimes be alleviated by applying an appropriate transformation to the dependent variable, as described previously, and reanalyzing the data in terms of the transformed variable.

In addition to the patterns in Figure 21.4, plots of residuals should be examined for any other systematic patterns. The residuals represent the unexplained variation in the data and hence provide information from which the observant researcher can identify what predictors may be missing from the regression model. Any systematic pattern in a plot of residuals indicates a systematic effect that has not been accounted for by the regression equation. The residuals should therefore be plotted not only against the fitted values $\hat{y}$ and against each of the predictors included in the equation, but also against the order in which the observations were made and against any other variables suspected of having an effect on the dependent variable. If there is a systematic departure from the pattern in Part (l) of Figure 21.4 in any plot of residuals, revision of the model to include an appropriate term should be considered.

STUDY MATERIAL FOR CHAPTER 21

Terms and Concepts

model

Section 21.1

multiple linear regression
predictor
regression constant
regression coefficient
normal equations
ANOVA summary table for multiple linear regression analysis
coefficient of multiple determination
multiple correlation coefficient
partial correlation coefficient
coefficient of partial determination
extra sum of squares due to the inclusion of a predictor
test of the extra sum of squares
first-order partial correlation coefficient
second-order partial correlation coefficient
dummy variable
indicator variable
cross-product variable
variable selection
stepwise regression analysis
F to enter
F to remove
correlation matrix
partial regression coefficient
lurking variable

Section 21.2

nonlinear regression
data transformations to describe nonlinear relationships
counted data
exponential curve
polynomial fitting to describe nonlinear relationships

Key Formulas

Section 21.1

(21.1) $\hat{y} = b_0 + b_1x_1 + b_2x_2 + \cdots + b_gx_g$

(21.3)
$$\left\{\begin{array}{lllll} \sum y = b_0 n & + b_1 \sum x_1 & + b_2 \sum x_2 & + \cdots & + b_g \sum x_g \\ \sum x_1 y = b_0 \sum x_1 & + b_1 \sum x_1^2 & + b_2 \sum x_1 x_2 & + \cdots & + b_g \sum x_1 x_g \\ \sum x_2 y = b_0 \sum x_2 & + b_1 \sum x_1 x_2 & + b_2 \sum x_2^2 & + \cdots & + b_g \sum x_2 x_g \\ \vdots & & & & \\ \sum x_g y = b_0 \sum x_g & + b_1 \sum x_1 x_g & + b_2 \sum x_2 x_g & + \cdots & + b_g \sum x_g^2 \end{array}\right\}$$

(21.5) $SS_y = \sum (y_i - \bar{y})^2$

(21.6) $SS_{\text{regr. on } x_1, \ldots, x_g} = \sum (\hat{y}_i - \bar{y})^2$

(21.7) $SS_{\text{res.}|x_1, \ldots, x_g} = \sum (y_i - \hat{y}_i)^2$

(21.8) $R^2 = \dfrac{SS_{\text{regr.}}}{SS_{\text{total}}}$

(21.9) $R_{y \cdot x_1, \ldots, x_g} = r_{y\hat{y}}$

(21.11) $r_{yz|x_1, \ldots, x_c} = r_{y-\hat{y},\, z-\hat{z}}$

(21.14)
$$\begin{aligned} SS_{\text{regr. on } z|x_1, \ldots, x_c} &= SS_{\text{regr. on } x_1, \ldots, x_c, z} - SS_{\text{regr. on } x_1, \ldots, x_c} \\ &= SS_{\text{res.}|x_1, \ldots, x_c} - SS_{\text{res.}|x_1, \ldots, x_c, z} \\ &= r^2_{yz|x_1, \ldots, x_c} SS_{\text{res.}|x_1, \ldots, x_c} \end{aligned}$$

(21.15) $r^2_{yz|x_1, \ldots, x_c} = \dfrac{SS_{\text{regr. on } z|x_1, \ldots, x_c}}{SS_{\text{res.}|x_1, \ldots, x_c}}$

(21.16) $F_{\text{regr. on } z|x_1, \ldots, x_c} = \dfrac{(SS_{\text{regr. on } z|x_1, \ldots, x_c})/1}{(SS_{\text{res.}|x_1, \ldots, x_c, z})/(n - c - 2)}$

(21.17) $r_{yz|x} = \dfrac{r_{yz} - r_{yx}r_{zx}}{\sqrt{1 - r^2_{yx}}\sqrt{1 - r^2_{zx}}}$

(21.19) $SS_{\text{regr. on } x_1, x_2} = SS_{\text{regr. on } x_1} + r^2_{yx_2|x_1} SS_{\text{res.}|x_1}$

Also see the formulas in Table 21.1.

Exercises for Numerical Practice

Section 21.1

21/1. Three variables were observed in a random sample of 6 observational units as follows.

x_1	x_2	y
9	8	3
6	3	2
5	7	6
3	4	8
7	2	1
2	6	9

Answer the following questions about the least-squares regression of y on x_1 and x_2:

a. Write the normal equations.

b. Write the least-squares regression equation.

c. Report the ANOVA summary table and the critical F-value for testing the significance of the regression at the 5% level.

d. Calculate the coefficient of multiple determination.

e. Calculate the standard error of estimation.

21/2. Three variables were observed in a random sample of 7 observational units as follows.

x_1	x_2	y
5	6	4
6	2	7
8	3	9
3	5	2
7	8	5
1	7	0
0	4	1

Answer Parts (a) through (e) in Exercise 21/1 for the least-squares regression of y on x_1 and x_2 in these data.

21/3. Given $r_{12} = .73$, $r_{13} = .91$, and $r_{23} = .58$, calculate the following:

a. $r_{12|3}$

b. $r_{13|2}$

c. $r_{23|1}$

21/4. Given $r_{12} = .84$, $r_{13} = .27$, and $r_{23} = .60$, calculate the following:

a. $r_{12|3}$

b. $r_{13|2}$

c. $r_{23|1}$

21/5. Given the data in Exercise 21/1:

a. Calculate the partial correlation coefficient $r_{yx_2|x_1}$, and give the critical value from Appendix 8 for testing its significance at the 5% level.

b. Calculate the extra sum of squares due to the inclusion of x_2 given that x_1 is already in the regression equation, and give the F test statistic and the critical F-value for testing the significance of the extra sum of squares at the 5% level. How does this test relate to that in Part (a)?

c. Calculate the partial correlation coefficient $r_{yx_1|x_2}$, and give the critical value from Appendix 8 for testing its significance at the 5% level.

d. Calculate the extra sum of squares due to the inclusion of x_1 given that x_2 is already in the regression equation, and give the F test statistic and the critical F-value for testing the significance of the extra sum of squares at the 5% level. How does this test relate to that in Part (c)?

21/6. Answer Parts (a) through (d) in Exercise 21/5 for the data in Exercise 21/2.

21/7. An investigation of a dependent variable y yielded the following ANOVA summary table and the partial correlation coefficient $r_{yx_2|x_1} = -.60$.

Source of variation	SS	df	MS	F
Regression on x_1	62.42	1	62.42	7.25
Residual	189.42	22	8.61	
Total	251.84	23		

a. How many observations were there in the investigation?

b. State the critical F-value for testing the significance of the regression on x_1 at the 5% level, and calculate the coefficient of determination and the standard error of estimation for the regression on x_1.

c. Record the ANOVA summary table for the regression of y on x_1 and x_2, state the critical F-value for testing the significance of the regression at the 5% level, and calculate the coefficient of multiple determination and the standard error of estimation.

d. Calculate the F test statistic for testing the significance of the extra sum of squares used in Part (c), and give the critical F-value for a test at the 5% level.

21/8. An investigation of a dependent variable y yielded the following ANOVA summary table and the partial correlation coefficient $r_{yx_2|x_1} = -.80$.

Source of variation	SS	df	MS	F
Regression on x_1	50.26	1	50.26	7.87
Residual	108.63	17	6.39	
Total	158.89	18		

Answer Parts (a) through (d) in Exercise 21/7 for these data.

Section 21.2

21/9. A random sample yielded the following 8 bivariate observations:

x:	8	3	1	6	2	7	4	5
y:	92	14	2	51	7	69	24	35

Calculate the regression of $\log y$ on x, and then express y as an exponential function of x. Plot the equation on a scattergram.

21/10. A random sample yielded the following 9 bivariate observations:

x:	5	8	4	1	9	0	6	7	3
y:	31	15	39	63	12	80	25	19	50

Calculate the regression of $\log y$ on x, and then express y as an exponential function of x. Plot the equation on a scattergram.

21/11. A random sample yielded the following 9 bivariate observations:

x:	3	1	9	4	7	6	2	5	8
y:	5	3	2	6	5	6	4	6	4

Use the method of polynomial fitting to calculate the regression equation describing y as a function of x and x^2. Plot the equation on a scattergram.

21/12. A random sample yielded the following 8 bivariate observations:

x:	6	2	9	8	1	5	7	4
y:	1	5	4	2	7	2	1	3

Use the method of polynomial fitting to calculate the regression equation describing y as a function of x and x^2. Plot the equation on a scattergram.

Problems for Applying the Concepts

21/13. The files of a medical center located in an industrial park were consulted to find the number of accidents reported in a recent month by each of seven companies served by the medical center. Each company was then contacted to find out how many years it had been in business and how long its work shifts were. The data were as follows.

Company	Number of accidents	Years in business	Hours/shift
1	5	3	12
2	1	4	8
3	4	5	8
4	3	3	10
5	0	7	8
6	7	2	10
7	2	6	8

Perform a multiple linear regression analysis of number of accidents as a function of years in business and shift length, and report the following information:

a. the regression equation. How many accidents does it predict for companies with 10-hour shifts that have been in business for four years? Is the use of a linear equation appropriate?

b. the ANOVA summary table. Is the regression significant?

c. the coefficient of multiple determination. What does it mean?

d. the standard error of estimation. What does it tell you?

(Source: Student project by S. N. J. Watamaniuk.)

21/14. Can the price of a used car in the classified ads be predicted from the age of the car and the number of options listed in the ad? One ad for each of 10 model years was randomly selected from a newspaper's classified ads, excluding those from dealers and those without prices shown. The number of options was determined by counting the number of items of information listed about the car, other than make/model and price. For each of the 10 cars in the sample, the age was recorded in years and the asking price in thousands of dollars.

Age of car	Number of options listed	Price of car
0	1	11.20
1	3	10.00
2	1	3.45
3	3	6.40
4	2	5.28
5	5	2.65
6	0	2.55
7	3	1.53
8	1	1.70
9	4	.78

Perform a multiple linear regression analysis of price as a function of age and number of listed options, and report the following information:

a. the regression equation. What price does it predict for eight-year-old cars with three options listed? Is the use of a linear equation appropriate?

b. the ANOVA summary table. Is the regression significant?

c. the coefficient of multiple determination. What does it mean?

d. the standard error of estimation. What does it tell you?

(Source: Student project by O. Dennis.)

21/15. The following table shows total personal savings and disposable personal income in the United States for the 15 years from 1935 through 1949. The table also shows the country's war status as a dummy variable with a value of $x_2 = 1$ when the country was at war and $x_2 = 0$ otherwise.

Year	Personal savings (billions of current dollars), y	Disposable personal income (billions of current dollars), x_1	War status, x_2
1935	2.1	58.5	0
1936	3.6	66.3	0
1937	3.8	71.2	0
1938	.7	65.5	0
1939	2.6	70.3	0
1940	3.8	75.7	0
1941	11.0	92.7	0
1942	27.6	116.9	1
1943	33.4	133.5	1
1944	37.3	146.3	1
1945	29.6	150.2	1
1946	15.2	160.0	0
1947	7.3	169.8	0
1948	13.4	189.1	0
1949	9.4	188.6	0

Perform a multiple linear regression analysis of personal savings as a function of disposable personal income and war status, and report the following information:

a. the regression equation. Plot the equation on a scattergram showing y as a function of x_1, using different symbols to distinguish between points for war years and for peace years.

b. the ANOVA summary table. Is the regression significant?

c. the coefficient of multiple determination. What does it mean?

d. the standard error of estimation. What does it tell you?

e. the F ratios for testing the additional contribution of each predictor, given that the other predictor is in the equation. Are the partial regression coefficients significant?

(Source: U.S. President, *Economic Report of the President,* February 1970, p. 197.)

21/16. The following table shows personal consumption expenditures and disposable personal income in the United States for the 15 years from 1935 through 1949. The table also shows the country's war status as a dummy variable with a value of $x_2 = 1$ when the country was at war and $x_2 = 0$ otherwise.

Year	Personal consumption expenditures (billions of current dollars), y	Disposable personal income (billions of current dollars), x_1	War status, x_2
1935	56	58.5	0
1936	62	66.3	0
1937	67	71.2	0
1938	64	65.5	0
1939	67	70.3	0
1940	71	75.7	0
1941	81	92.7	0
1942	89	116.9	1
1943	99	133.5	1
1944	108	146.3	1
1945	120	150.2	1
1946	144	160.0	0
1947	162	169.8	0
1948	175	189.1	0
1949	178	188.6	0

Perform a multiple linear regression analysis of personal consumption expenditures as a function of disposable personal income and war status, and report the information requested in Parts (a) through (e) in Problem 21/15.

(Source: U.S. President, *Economic Report of the President,* February 1970, pp. 177 and 184; February 1982, pp. 233 and 260.)

21/17. What features of a house are relevant to its market value? Information was obtained from a random sample of 10 houses listed for sale by a real estate office. Values of the following variables were recorded for each house:

y = asking price, in thousands of dollars
x_1 = floor area, in hundreds of square feet
x_2 = number of bedrooms
x_3 = number of bathrooms

The following sums of squares were calculated in simple linear regression analyses:

$$SS_{\text{regr. of } y \text{ on } x_1} = 516.6$$
$$SS_{\text{regr. of } y \text{ on } x_2} = 549.0$$
$$SS_{\text{regr. of } y \text{ on } x_3} = 2.4$$

The following is part of the ANOVA table from Step 2 of a stepwise regression analysis.

Source of variation	SS
Regression of y on x_1 and x_2	826.0
Residual	367.3
Total	1,193.3

a. What is the best single linear predictor of asking price? How do you know?

b. Calculate the coefficient of determination for the regression of price on its best single linear predictor.

c. Calculate the correlation between price and its best single linear predictor.

d. Compute the ANOVA table for the regression of price on its best single linear predictor, and test the significance of the regression.

e. Calculate the standard error of estimation for the regression in Part (d).

f. Calculate the extra sum of squares due to the inclusion of x_1 given that x_2 is already in the equation, and test its significance.

g. Test the significance of the multiple regression of y on x_1 and x_2.

h. Calculate the coefficient of determination for the regression of y on x_1 and x_2.

i. Calculate the standard error of estimation for the regression of y on x_1 and x_2.

j. If all tests were made at the 5% significance level, which predictor(s) would be in the final equation resulting from a stepwise regression analysis of these data? How do you know?

k. Calculate the partial correlation of y and x_1 with x_2 held constant.

(Source: Student project by W. Wong.)

21/18. What characteristics can be used to predict the price of a textbook for courses in the humanities and social sciences? Values of the following variables were obtained for a sample of 11 textbooks sold at a college bookstore:

y = price, in dollars
x_1 = type of cover, a dummy variable coded as paperback = 0, hardcover = 1
x_2 = number of pages
x_3 = number of diagrams, pictures, and other illustrations

The following sums of squares were calculated in simple linear regression analyses:

$$SS_{\text{regr. of } y \text{ on } x_1} = 864.3$$
$$SS_{\text{regr. of } y \text{ on } x_2} = 930.3$$
$$SS_{\text{regr. of } y \text{ on } x_3} = 655.1$$

The following is part of the ANOVA table from Step 2 of a stepwise regression analysis.

Source of variation	SS
Regression of y on x_2 and x_3	980.0
Residual	282.7
Total	1,262.7

a. What is the best single linear predictor of price? How do you know?

b. Calculate the coefficient of determination for the regression of price on its best single linear predictor.

c. Calculate the correlation between price and its best single linear predictor.

d. Compute the ANOVA table for the regression of price on its best single linear predictor, and test the significance of the regression.

e. Calculate the standard error of estimation for the regression in Part (d).

f. How can it be that x_2 and x_3 are included in Step 2 of a stepwise regression analysis, when x_3 accounts for less variation in y (665.1) than either x_1 or x_2?

g. Calculate the extra sum of squares due to the inclusion of x_3 given that x_2 is already in the equation, and test its significance.

h. Test the significance of the multiple regression of y on x_2 and x_3.

i. Calculate the coefficient of determination for the regression of y on x_2 and x_3.

j. Calculate the standard error of estimation for the regression of y on x_2 and x_3.

k. If all tests were made at the 5% significance level, which predictor(s) would be in the final equation resulting from a stepwise regression analysis of these data? How do you know?

l. Calculate the partial correlation of y and x_3 with x_2 held constant.

(Source: Student project by N. Sinnamon.)

21/19. An investigation was carried out concerning factors influencing the advertised price of used sports cars. Information was obtained from a current issue of the magazine *Auto Trader*. A random sample was selected of 12 sports car ads that included values of the following variables:

y = asking price, in thousands of dollars
x_1 = mileage on the car, in thousands of miles
x_2 = age of the car, in years
x_3 = origin of car, a dummy variable coded as domestic = 0, import = 1

The total sum of squares of y was 57.17, and the correlation matrix was as follows:

	x_1	x_2	x_3
y	−.73	−.91	.07
x_1		.82	−.08
x_2			.09

a. What is the best single linear predictor of price? How do you know?

b. Compute the ANOVA table for Step 1 of a stepwise regression analysis of these data, and test the significance of the regression.

c. Calculate the coefficient of determination for the regression in Part (b).

d. Calculate the standard error of estimation for the regression in Part (b).

e. Calculate the partial correlation of y and x_1, with x_2 held constant.

f. Calculate the partial correlation of y and x_3, with x_2 held constant.

g. Compute the ANOVA table for Step 2 of a stepwise regression analysis of these data, and test the significance of the regression.

h. Calculate the extra sum of squares due to the inclusion of the new variable [in Part (g)] in the equation, and test its significance.

i. Calculate the coefficient of determination for the regression in Part (g).

j. Calculate the standard error of estimation for the regression in Part (g).

k. If all tests were made at the 5% significance level, which predictor(s) would be in the final regression equation resulting from a stepwise regression analysis of these data? How do you know?

(Source: Student project by B. Reesor.)

21/20. Ten steers were randomly selected from the senior class at a county 4-H sale in 1982. The official judging at the sale resulted in a rank ordering of the steers in the senior class, and then the animals were sold by auction. Values of the following variables were obtained from the sale record:

y = sale price, in dollars per pound
x_1 = sale weight, in pounds
x_2 = rank in the judging, coded as first = 1, second = 2, and so on
x_3 = breed, a dummy variable coded as domestic = 0 (for Hereford, Angus) and exotic = 1 (for Charolais, Limousin, Simmental)

The total sum of squares of y was 2.954, and the correlation matrix was as follows:

	x_1	x_2	x_3
y	.83	−.75	.69
x_1		−.91	.75
x_2			−.82

a. What is the best single linear predictor of price? How do you know?
b. Compute the ANOVA table for Step 1 of a stepwise regression analysis of these data, and test the significance of the regression.
c. Calculate the coefficient of determination for the regression in Part (b).
d. Calculate the standard error of estimation for the regression in Part (b).
e. Calculate the partial correlation of y and x_2, with x_1 held constant.
f. Calculate the partial correlation of y and x_3, with x_1 held constant.
g. Compute the ANOVA table for Step 2 of a stepwise regression analysis of these data, and test the significance of the regression.
h. Calculate the extra sum of squares due to the inclusion of the new variable [in Part (g)] in the equation, and test its significance.
i. Calculate the coefficient of determination for the regression in Part (g).
j. Calculate the standard error of estimation for the regression in Part (g).
k. If all tests were made at the 5% significance level, which predictor(s) would be in the final regression equation resulting from a stepwise regression analysis of these data? How do you know?

(Source: Student project by L. Eleniak.)

21/21. Eight individuals took part in a study of the effect of skill-level setting on performance in a computer football game. In an attempt to take into account the differing abilities of the individuals, each individual first played one practice round at the lowest skill-level setting. Each individual was then randomly assigned to one of the four skill-level settings and played a game at his assigned setting, immediately after the practice round. The results follow.

Player	Practice score (x_1) at setting 1	Skill-level setting (x_2) for test game	Test game score (y) at assigned setting
1	42	1	59
2	56	1	70
3	63	2	48
4	56	2	48
5	59	3	38
6	56	3	28
7	62	4	21
8	35	4	15
	$\sum x$: 429	20	$\sum y =$ 327
	$\sum x^2$: 23,691	60	$\sum y^2 =$ 15,883
	$\sum xy$: 17,747	663	
	$\sum x_1x_2 =$ 1,069		

Perform a stepwise regression analysis of test game score as a function of practice score and skill-level setting, and interpret the results.
(Source: Student project by D. Hutchinson.)

21/22. Data from the April 1980 issue of *Consumer Reports* were analyzed in a study of highway gas mileage as a function of the car's weight and engine size. The following tabulation shows gasoline mileage in miles per gallon, weight in pounds, and engine size in cubic inches of displacement, for 10 cars randomly selected from those in the magazine report.

Car	Mileage (y)	Weight (x_1)	Engine size (x_2)
1	41.2	1,880	156
2	33.9	2,120	98
3	34.9	2,570	134
4	22.8	3,180	229
5	29.6	2,800	140
6	25.5	3,220	229
7	19.9	3,120	174
8	26.1	3,590	229
9	21.5	3,920	318
10	20.6	3,720	302
	$\sum y =$ 276.0	$\sum x$: 30,120	2,009
	$\sum y^2 =$ 8,074.74	$\sum x^2$: 94,781,800	451,423
		$\sum xy$: 793,210	52,128.4
			$\sum x_1x_2 =$ 6,438,010

Perform a stepwise regression analysis of gas mileage as a function of weight and engine size, and interpret the results.

(Source: Student project by C. Kooyman.)

21/23. The following data were recorded for 15 college students in a study of the relationship between height and shoe size. The dependent variable y was shoe size, recorded in terms of men's running shoe sizes for both men and women.

Observation number	Shoe size (y)	Height (x_1)	Gender (x_2)	x_1x_2 (x_3)
1	9	69	1	69
2	6.5	66	0	0
3	9	69	1	69
4	5.5	69	0	0
5	10	72.5	1	72.5
6	12	75	1	75
7	10	70	1	70
8	9.5	69.5	1	69.5
9	9	69	1	69
10	4	60.75	0	0
11	3	61	0	0
12	4	65	0	0
13	7	68	0	0
14	5.5	61	0	0
15	4	62	0	0
	$\sum y =$ 108.0	$\sum x$: 1,006.75	7	494.0
	$\sum y^2 =$ 886	$\sum x^2$: 67,837.0625	7	34,894.5
		$\sum xy$: 7,402.25	68.5	4,848.25
			$\sum x_1x_j$: 494	34,894.5
				$\sum x_2x_3 =$ 494

The first predictor x_1 was height in inches, and the second predictor x_2 was gender, coded as a dummy variable with $x_2 = 0$ for females and $x_2 = 1$ for males (to allow for different y-intercepts for males and females). The third predictor x_3 was defined as the product x_1x_2 of the first two predictors (to allow for different slopes for males and females).

The correlation matrix is as follows:

	x_1	x_2	x_3
y	.90254956	.89973694	.90888673
x_1		.76545720	.77923101
x_2			.99913483

Perform a stepwise regression analysis of shoe size as a function of the three predictors, and interpret the results. Plot a scattergram of shoe size as a function of height, using a different symbol for male and female data points, and plot the final regression equation.

(Source: Student project by J. Kruschel.)

21/24. The following table shows the total number of new cars, both domestic and

Year	New car sales (millions), y	Yearly increase in disposable personal income (billions of 1972 dollars), x_1	Yearly increase in number of family units (millions), x_2	Yearly increase in employment, including resident armed forces (millions), x_3	Yearly increase in unemployment (millions), x_4
1965	9.3	35.5	.5	1.7	−.4
1966	9.0	30.5	.7	2.0	−.5
1967	8.3	26.7	.9	1.6	.1
1968	9.7	27.8	.7	1.6	−.2
1969	9.6	21.2	.8	1.9	0
1970	8.4	29.1	.6	.7	1.3
1971	10.2	27.6	1.1	.5	.9
1972	10.9	31.1	1.1	2.7	−.1
1973	11.4	54.4	.7	2.8	−.5
1974	8.9	−7.2	.6	1.7	.8
1975	8.6	17.4	.5	−1.0	2.7
1976	10.1	31.9	.5	2.9	−.5
1977	11.1	36.1	.5	3.3	−.4
1978	11.3	45.9	.6	4.0	−.8
1979	10.7	26.9	1.8	2.7	−.1
1980	9.0	5.9	.7	.5	1.5
1981	8.5	33.1	.7	1.1	.7
1982	8.0	5.5	.4	−.9	2.4

(Sources: Motor Vehicle Manufacturers Association of the United States, *Motor Vehicle Facts and Figures, 1985,* p. 7; U.S. President, *Economic Report of the President,* February 1984, pp. 249–54.)

foreign, sold in the United States for each of the 18 years from 1965 through 1982. Data for four possible predictors of new car sales are also shown. Perform a stepwise regression analysis of new car sales as a function of the four predictors, and interpret the results. (The use of a computer is assumed for this problem.)

21/25. The following table shows the growth of the civilian labor force in the United States for the 14 years from 1929 through 1942.

Year	Years after 1929, x	Civilian labor force (millions of persons), y
1929	0	49.2
1930	1	49.8
1931	2	50.4
1932	3	51.0
1933	4	51.6
1934	5	52.2
1935	6	52.9
1936	7	53.4
1937	8	54.0
1938	9	54.6
1939	10	55.2
1940	11	55.6
1941	12	55.9
1942	13	56.4

a. Plot a scattergram of the data and check for evidence of nonlinearity. Calculate the regression equation describing the common (base 10) logarithm of y as a linear function of x.

b. Take antilogs of the equation in Part (a) in order to express y as an exponential function of x. Interpret this equation by giving the fitted value when $x = 0$ and the percentage increase in the labor force for each increment of one year in x.

c. Calculate the coefficient of determination.

d. Test the significance of the regression.

(Source: U.S. President, *Economic Report of the President*. February 1970, p. 202.)

21/26. The table on page 816 shows the growth of the civilian labor force in the United States for the 17 years from 1947 through 1963.

Answer Parts (a) through (d) in Problem 21/25.

(Source: U.S. President, *Economic Report of the President*, February 1982, p. 266.)

21/27. The continuing education office of the public school board in a large city advertised its fall program in a weekend edition of the local newspaper, after which there were 15 days on which registration could be made for the courses. The following data are the numbers (y) of people who registered for continuing

Year	Years after 1947, x	Civilian labor force (millions of persons), y
1947	0	59.4
1948	1	60.6
1949	2	61.3
1950	3	62.2
1951	4	62.0
1952	5	62.1
1953	6	63.0
1954	7	63.6
1955	8	65.0
1956	9	66.6
1957	10	66.9
1958	11	67.6
1959	12	68.4
1960	13	69.6
1961	14	70.5
1962	15	70.6
1963	16	71.8

education courses each day after the advertisement (x = number of days after advertisement):

x:	1	2	3	4	5	6	7	8	9	10	11	12	13	14	15
y:	712	621	421	318	263	382	217	96	53	18	39	42	18	33	21

a. Plot a scattergram of the data and check for evidence of nonlinearity. Which variable is the more likely candidate for transformation, to yield a linear relationship? Which transformation is suggested by the nature of the variable itself?

b. Calculate the coefficient of determination first for the original data and then for the data with the logarithmic transformation applied to the appropriate variable. Which form of the data results in the better description of the relationship?

c. Compute the regression equation for the data with the logarithmic transformation applied to the appropriate variable. Convert the equation to the original units of measurement, and plot it on the scattergram. Test the significance of the regression.
(Source: Student project by L. Youngberg.)

21/28. The following data were collected from 12 students living in college residences. The dependent variable was the number of times the student returned home during the first five weeks of the academic year, and the independent variable was the distance (in miles) from the college to the student's home.

a. Plot four scattergrams of the data and check for evidence of nonlinearity in each scattergram:

i. y as a function of x.
ii. square root of y as a function of x.
iii. y as a function of log x.
iv. square root of y as a function of log x.

Student	Distance (x)	Number of visits home (y)
1	1,200	0
2	180	1
3	60	5
4	60	4
5	150	2
6	85	3
7	50	4
8	60	2
9	75	3
10	35	10
11	210	2
12	20	5

b. Calculate the four coefficients of determination corresponding to the four scattergrams. Which form of the data results in the best description of the relationship?

c. Compute the regression equation for the best form of the data as determined in Part (b). Convert the equation to the original y-units, and plot it in the scattergram of the original data. Test the significance of the regression.

(Source: Student project by P. Van Bergen.)

21/29. Eleventh-grade boys were required to run around an outdoor track at the start of each physical education class period. The time was recorded for each student, and the average time for the total class was calculated. The following are the results for eight days differing in outdoor temperature.

Temperature (°C)	Average time (seconds)
0	385
5	351
10	335
15	325
20	287
25	341
30	368
35	389

a. Identify the dependent and independent variables, and plot a scattergram of the data to check for evidence of nonlinearity.

b. Compute the regression equation describing time as a quadratic polynomial function of temperature, and plot it on the scattergram. (*Hint*: Dividing one or more of the normal equations by a constant, such as 100, does not affect their solution, but it may make the numbers more manageable for hand-calculator analysis.)

c. Test the significance of the regression, and calculate the proportion of variation accounted for and the standard error of estimation.

(Source: Student project by V. Brandt.)

21/30. Ten male students took a typing test consisting of typing a 30-word passage. Their time in minutes was divided into 30 to give their speed in words per minute, and their number of errors was determined by counting the number of words (out of 30) with at least one mistake in it.

Student	Speed (x)	Number of errors (y)
1	13	4
2	35	6
3	15	9
4	47	10
5	28	5
6	11	18
7	8	5
8	6	7
9	7	15
10	9	0

a. Plot a scattergram showing number of errors as a function of speed, and check for evidence of nonlinearity.

b. Compute the regression equation describing number of errors as a quadratic polynomial function of speed, and plot it on the scattergram.

c. Test the significance of the regression, and calculate the proportion of variation accounted for and the standard error of estimation.

(Source: Student project by M. Tetreau.)

Reference

Warner, K. E. "Cigarette Smoking in the 1970's: The Impact of the Antismoking Campaign on Consumption." *Science* 211 (1981): 729–31.

Further Reading

Draper, N. R., and Smith, H. *Applied Regression Analysis.* 2nd ed. New York: Wiley, 1981. This is an excellent source of information and practical advice concerning regression problems. The discussion of the examination of residuals in Chapter 3 is more thorough than in most other references.

Ehrenberg, A. S. C. *Data Reduction: Analysing and Interpreting Statistical Data.* Revised reprint. Chichester: Wiley, 1978. Using interesting examples and very little mathematics, Chapters 5–10 provide an alternative approach to the problem of describing functional relationships among variables. Valuable practical advice concerning interpretation is included.

Joiner, B. L. "Lurking Variables: Some Examples." *The American Statistician* 35 (1981): 227–33. This article provides an interesting and enlightening discussion of lurking variables in the context of eight real-data examples.

Kleinbaum, D. G., and Kupper, L. L. *Applied Regression Analysis and Other Multivariable Methods.* North Scituate, MA: Duxbury, 1978. This text offers a thorough survey of regression methods without assuming that the reader has been exposed to calculus or matrix algebra.

SAS Institute Inc. *SAS® User's Guide: Statistics, Version 5 Edition*. Cary, NC: SAS Institute Inc., 1985. This guide describes the use of SAS/STAT computer programs, such as those whose output was shown in Figures 21.1 and 21.3.

Snedecor, G. W., and Cochran, W. G. *Statistical Methods*. 7th ed. Ames, IA: Iowa State University Press, 1980. Chapters 17–19 provide concise but thorough coverage extending the topics introduced in this chapter.

Tufte, E. R. *Data Analysis for Politics and Policy*. Englewood Cliffs, NJ: Prentice-Hall, 1974. A good portion of this highly readable book is devoted to the interpretation of multiple linear regression. The discussions of real-data examples are especially valuable.

APPENDIXES

Appendix 1 HOW AND WHEN TO ROUND

The question of how many digits to report in a numerical value arises repeatedly as we consider various computations throughout this book. To answer the question, we first distinguish four ways of describing the digits that make up a numerical value:

a. **Digit position** refers to the digit's position relative to the decimal point. When we identify digits as the hundreds digit, tens digit, units digit, tenths digit, and so on, we are identifying them in terms of their digit position.

b. **Decimal places** are digit positions to the right of the decimal point. The tenths digit occupies the first decimal place, the hundredths digit occupies the second decimal place, and so on.

c. **Significant figures** are the digits from the leftmost nonzero digit to the rightmost nonzero digit or zero that is considered to be exact.

d. **Effective digits** are digits that distinguish among the observations in a particular context. The *last effective digit* is the rightmost digit that is needed to distinguish a number from other numbers in the same context.

These terms are illustrated with examples in Figure A1.1.

A1.1 HOW TO ROUND

A few words about the proper **procedure for rounding** may help to prevent misunderstanding. Rounded to the nearest tenth, for example, a calculated value of 40.267 would become 40.3, because .267 is closer to .300 than to .200. If the calculated value were 40.24, it would become 40.2 rounded to the nearest tenth, because .24 is closer to .20 than to .30. We round to the closest value: Rounded to the nearest tenth, 40.250001 would become 40.3, and 40.249999 would become 40.2.

How should we round 40.250000 to the nearest tenth? There are two schools of thought as to how to round a value, such as 40.250000, that is exactly halfway between the two digits to which it might be rounded. Some statisticians always round *up* when the following digits are 5000...; others

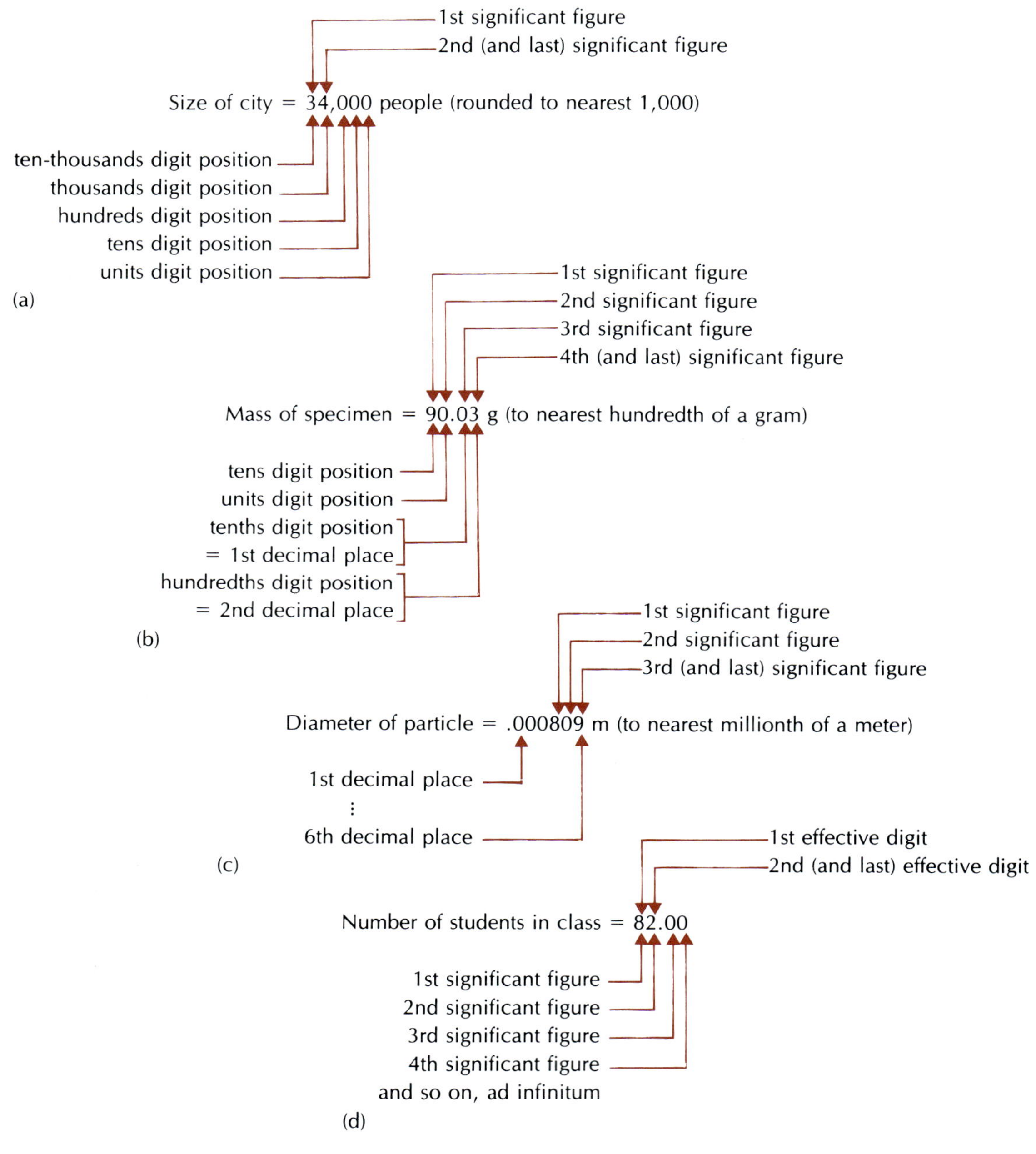

Figure A1.1
Illustrations of the terminology regarding digit positions, decimal places, significant figures, and effective digits. (a) Trailing zeros that result from rounding are not significant figures. (b) Intermediate zeros are significant figures. (c) Leading zeros are not significant figures. (d) Exact zeros are significant figures, but they are not effective digits except in digit positions where nonzero digits would be possible in the same context.

round *to the even digit* when the following digits are 5000..., so that the rounding errors will cancel out in the long run. According to the latter approach, 40.250000 to the nearest tenth would become 40.2 (because 2 is even), and 40.150000 to the nearest tenth would likewise become 40.2 (again rounding to the even digit). In this book, we follow the latter practice, rounding to the nearest digit wherever possible but preferring the even digit in those cases where the following digits are exactly, 5000....

A1.2 EFFECTS OF ROUNDING ON SUBSEQUENT CALCULATIONS

The result of any calculation will suffer from rounding error if the calculation involves even one quantity that has been excessively or prematurely rounded. How does rounding introduce error into a calculation? The following two statements identify the basic effects of rounding on subsequent calculations.

The first statement deals with the **effect of rounding in addition and subtraction**:

> If any of the quantities entering an addition or subtraction has been rounded, the sum or difference is reliable only to the digit position occupied by the last significant figure of the rounded quantity. **(A1.1)**

This statement implies that we specify the accuracy of a sum or difference in terms of digit positions (for example, thousands or tens or tenths) and not in terms of the number of significant figures. In particular, any digit positions that have been rounded off—that is, filled by a nonexact zero—in *any* of the quantities entering the addition or subtraction are not reliable in the sum or difference.

EXAMPLE A1.1 A student preparing an essay found a book that said dinosaurs became extinct 65,000,000 years ago. Since the book was published 13 years ago, the student concluded that dinosaurs became extinct 65,000,013 years ago.

The problem with the student's conclusion is the failure to take into account the principle in Expression (A1.1). The book's figure of 65,000,000 is rounded to the nearest million years, or perhaps even to the nearest 5 million years. Any sum or difference involving this figure will not have any reliable digits to the right of the millions digit position and therefore will be correct only to the nearest million years. The student should have concluded that dinosaurs became extinct 65,000,000 years ago.

The second statement of the basic effects of rounding on subsequent calculations deals with the **effect of rounding in multiplication and division**:

> If any of the quantities entering a multiplication or division has been rounded, the product or quotient is reliable only to the same number of significant figures as are in the rounded quantity. **(A1.2)**

This statement implies that we specify the accuracy of a product or quotient in terms of the number of significant figures and not in terms of digit positions. In particular, the number of reliable significant figures in a product or quotient is equal to the smallest number of significant figures in any of the quantities being multiplied or divided.

EXAMPLE A1.2 What is one third of the exact amount $8638? Given problems of this type, some people seek a solution by dividing by 3, while others multiply by .33. Are the answers the same?

Dividing by 3, we obtain a value of $2879.33 to the nearest cent. All six digits in this answer are correct, because neither of the quantities entering the division (8638/3) was rounded before calculating; both quantities have an infinite number of exact zeros to the right of the decimal point.

Multiplying $8638 by .33, we obtain a value of $2850.54. This answer is correct to only two significant figures according to Expression (A1.2) because the quantity .33 is rounded to two significant figures. The answer $2850.54 therefore should be reported as $2900, rounded to two significant figures. If $8638 is multiplied instead by .3333 (with four significant figures), the answer is $2879.05. Expression (A1.2) tells us that this answer is correct to four significant figures, so it could be reported as $2879. Comparison of the answers in this paragraph with that in the preceding paragraph confirms that the number of reliable digits in the answer depends on the rounding of the quantities entering the calculation.

The effects stated in Expressions (A1.1) and (A1.2) are relevant to the number of digits that should be retained during calculations as well as to the number of digits that should be retained in reporting computational results. The next two sections provide some guidelines to help avoid undue rounding errors in calculating and in reporting, respectively.

A1.3 WHEN NOT TO ROUND

To answer the question as to how many digits should be retained in a numerical quantity, we must distinguish between calculating and reporting. For **calculating**, we note that in any given application, each of the statistics defined by a formula in this book is, by definition, exactly equal to the value resulting from the use of that formula. For example, the mean of a set of numerical values is, mathematically, exactly equal to their sum divided by the number of values. This exact value, with all its decimal places, is what is referred to when the same statistic appears in a formula for calculating some further value.

It is possible to work out how many digits should be retained at each stage of a computation in order to ensure any desired degree of precision in the final answer, but it is not always easy. On the assumption of the availability of a pocket calculator, the following **general advice for calculating** is offered as a simple and safe expedient:

> When entering numerical values into a formula for purposes of computation, enter "exact" (not rounded) values—exact within the limits of your calculator, that is—and carry "exact" values through all stages of the computation. **(A1.3)**

In effect, the advice is to use rounding as little as possible in calculating, since premature rounding can lead to serious distortion of the final answer. This general advice applies to all calculations.

A1.4 WHEN TO ROUND

Although it is easiest and safest to carry extra digits during calculations, care should be exercised in how many digits are finally reported in presenting the results of an investigation. In **reporting** numerical values, our concern is no longer with the mathematical exactness of a definition but rather with what information the value conveys about the problem at hand. We must not pretend to know a numerical value more precisely than we do on the basis of our observations. Failure to round sufficiently not only paints a dishonest picture of the precision of the results but also tends to reduce their comprehensibility. On the other hand, overrounding can result in the loss of valuable information.

The purpose of this section is to present guidelines for the appropriate reporting of numerical values of various statistics introduced in this book. These guidelines are to be applied not legalistically but rather with a measure of judgment based on experience and common sense. Based on conventional practices in clear communication of statistical results, they are intended to be good advice but not dogma.

The guidelines are listed in the order in which they are needed in the text. A summary table at the end of Appendix 1 organizes the guidelines so that the pattern in the advice is evident.

Guideline for Reporting Index Numbers (Chapter 1)

Index numbers are conventionally reported to one decimal place (for example, 105.6).

Guideline for Reporting Percentages (Chapter 2)

Report percentages to two significant figures (for example, 13%, 1.3%, and .13%).

Interim Guideline for Reporting Means (Chapter 3)

For purposes of reporting, round the value of the mean to the digit position immediately to the right of the last effective digit of the least exact observation. (For example, if the observations are recorded to the nearest thousand, report the mean to the nearest hundred; if some of the observations are recorded to the nearest unit and others are recorded to the nearest thousand, report the mean to the nearest hundred; if all of the observations are recorded to the nearest unit, report the mean to the nearest tenth.) This interim guideline will be replaced by better guidelines in Chapters 4 and 8.

Guideline for Reporting Means and Standard Deviations (Chapter 4)

Report the standard deviation rounded to two significant figures, and report the mean rounded to the same digit position as the accompanying standard deviation is rounded (for example, $s = 2.3$ and $\bar{x} = 117.5$).

Guideline for Reporting *z*-Scores (Chapter 4)

Report z-scores rounded to three significant figures or two decimal places, whichever leaves fewer digits (for example, $z = 1.96$, $z = 10.3$, and $z = .02$).

Interim Guideline for Reporting Fitted Values in Regression (Chapter 5)

For purposes of reporting, round the fitted value $\hat{y}$ to the digit position immediately to the right of the last effective digit of the least exact y-observation. (For example, if the y-values are recorded to the nearest hundred, report the fitted values to the nearest ten; if some of the y-values are recorded to the nearest tenth and others are recorded to the nearest hundred, report the fitted values to the nearest ten; if all of the y-values are recorded to the nearest tenth, report the fitted values to the nearest hundredth.) This interim guideline will be replaced by a better guideline in Chapter 13.

Guideline for Reporting the *y*-Intercept in Regression (Chapter 5)

Report the y-intercept rounded to the same digit position as the fitted values are rounded. (For example, if the fitted values are to be reported to the nearest tenth, report the y-intercept to the nearest tenth.)

Guideline for Reporting Regression Coefficients (Chapter 5)

Report the regression coefficient b rounded so that bx_{max} is correct to the same digit position as the fitted values are rounded. It is not necessary to use the exact value of the largest x-value x_{max} in this guideline; it is sufficient to use the nearest power of 10 (for example, 1, 10, 100, 1000) in the vicinity of x_{max}. (For example, suppose that $x_{max} = 145$ and $b = .61749$. If we have determined that fitted values would be rounded to the nearest unit, then we need to round the regression coefficient b so that bx_{max} is correct to the nearest unit. Because the maximum observed value of x is in the vicinity of 100, as opposed to the vicinity of 10 or 1000, we report the slope as $b = .62$ to ensure that the product bx, approximately $.62 \times 100 = 62$, is correct to the nearest unit.)

Guideline for Reporting Coefficients of Correlation and Determination (Chapter 5)

Report coefficients of correlation and determination rounded to two decimal places, except as necessary to distinguish from ±1.00 (for example, $r = .25$ and $r^2 = .06$; $r = .998$ and $r^2 = .996$).

Guideline for Reporting Probabilities and Proportions (Chapter 6)

Report probabilities and proportions rounded to two significant figures, except as necessary to distinguish from 1.00 (for example, .37, .0037, and .9996).

Guideline for Reporting Sample Means and Their Standard Errors (Chapter 8)

Report standard errors rounded to two significant figures, and report the sample mean rounded to the same digit position as its standard error is rounded (for example, a sample mean of 102.3 with a standard error of 6.8).

Guideline for Reporting *t* Test Statistics (Chapter 11)

Report t test statistics rounded to three significant figures or to two decimal places, whichever leaves fewer digits (for example, $t = 2.43$ and $t = .61$).

Guideline for Reporting Chi-Square Test Statistics (Chapter 12)

Report chi-square test statistics rounded to three significant figures or to two decimal places, whichever leaves fewer digits (for example, $\chi^2 = 9.17$ and $\chi^2 = 20.9$). As an exception for values that are to be compared with a lower-tail critical value, simply report the chi-square value to three significant figures (for example, $\chi^2 = .00597$).

Guideline for Reporting *F* Test Statistics (Chapter 12)

Report F test statistics rounded to three significant figures or to two decimal places, whichever leaves fewer digits (for example, $F = 43.1$, $F = 3.46$, and $F = .87$).

Guideline for Reporting Standard Errors of Estimation (Chapter 13)

Report standard errors of estimation rounded to two significant figures (for example, $s_e = 1700$, $s_e = 2.3$, and $s_e = .0067$).

Guideline for Reporting Fitted Values in Regression (Chapter 13)

Report fitted values $\hat{y}$ rounded to the same digit position as the standard error of estimation is rounded. (For example, if $s_e = 2.3$, the fitted values would be reported to the nearest tenth.)

Guideline for Reporting Standard Errors of Proportions (Chapter 14)

Report standard errors of proportions rounded to two significant figures (for example, $\sigma_{\hat{p}} = .062$).

Guideline for Reporting ANOVA Summary tables (Chapter 15)

For purposes of reporting, round the error mean square to three significant figures and all other values in the SS and MS columns to the same digit position as the error mean square (see the examples in Chapter 15).

Summary of Guidelines for Rounding Reported Values

Table A1.1 provides a summary of the preceding guidelines for rounding reported values.

TABLE A1.1 Summary of guidelines for rounding.

A. REPORTED VALUES

Item	For purposes of reporting:
Quantities rounded to two significant figures	
Standard error	Round to 2 significant figures.
Standard deviation	Round to 2 significant figures.
Variance	Round to 2 significant figures.
Probability	Round to 2 significant figures (except as necessary to distinguish from 1.00).
Proportion	Round to 2 significant figures (except as necessary to distinguish from 1.00).
Quantities rounded to a certain digit position	
Sample mean	Round to *either* the digit position immediately to the right of the last effective digit of the least exact observation *or* (better) the same digit position as the standard error of the mean is rounded.
Population mean	Round to *either* the digit position immediately to the right of the last effective digit of the least exact observation *or* (better) the same digit position as the standard deviation is rounded.
Fitted value, $\hat{y}$	Round to *either* the digit position immediately to the right of the last effective digit of the least exact y-observation *or* (better) the same digit position as the standard error of estimation is rounded.
y-intercept	Round to the same digit position as the fitted value is rounded.
Regression coefficient, b	Round so that bx_{max} is correct to the same digit position as the fitted value is rounded.
Other quantities	
Test statistics (z, t, χ^2, F)	Round to 3 significant figures or 2 decimal places, whichever leaves fewer digits.
Correlation and determination coefficients	Round to 2 decimal places (except as necessary to distinguish from ±1.00).
ANOVA table	Round the error mean square to 3 significant figures and all other MS's and SS's to the same digit position as the error mean square is rounded.

B. VALUES TO BE USED IN CALCULATIONS

For purposes of calculation, the following advice is simple and generally safe:

When entering numerical values into a formula for computation, enter "exact" (not rounded) values—exact within the limits of your calculator—and carry "exact" values through all stages of the computation.

Otherwise, if calculations are based on rounded values, you may have to limit the reported values to fewer digits than specified in Part A of this table. The basic effects of rounding on subsequent calculations are as follows:

a. If any of the quantities entering an *addition or subtraction* has been rounded, the sum or difference is reliable only to the *digit position* occupied by the last significant figure of the rounded quantity.

b. If any of the quantities entering a *multiplication or division* has been rounded, the product or quotient is reliable only to the *number of significant figures* in the rounded quantity.

Appendix 2 RANDOM DIGITS

Row	00–04	05–09	10–14	15–19	20–24	25–29	30–34	35–39	40–44	45–49
00	66241	82072	36359	39074	87266	69303	41915	01524	67981	65300
01	89698	56630	50968	15842	03105	29841	66060	71815	80036	87387
02	41171	97682	14803	55187	71756	91954	34525	07612	41893	80773
03	40176	13043	64270	50830	25412	97076	31168	12737	26371	59680
04	40571	88714	03292	43736	83688	70206	62150	82523	85794	20849
05	20280	63877	62646	52520	01133	14998	92336	49006	20273	98940
06	18455	55505	59379	34457	17262	03321	99419	79223	93174	35294
07	61405	03948	64197	59215	24548	78490	09085	21420	35437	49093
08	54626	79858	28969	21101	52871	98072	60724	22094	66835	52009
09	81920	09365	98038	67302	54966	66641	22678	46146	48437	08524
10	62227	48454	20160	31572	92899	29415	47061	90457	85569	55041
11	50493	38619	32006	83696	40539	52844	14081	81046	33283	76354
12	49321	13575	62920	00724	47506	76257	86150	38377	12813	44205
13	51969	20117	81999	99650	48595	90711	75341	68328	50249	18249
14	36265	72249	99395	16146	47821	85596	07550	84611	52167	66654
15	68544	58538	51869	60137	65762	95560	13288	88262	23843	66834
16	98383	37635	36917	87140	89040	29255	97680	67444	66562	15537
17	41629	03786	72397	83593	69053	41031	56953	15564	30093	43957
18	94950	69846	61311	60965	25132	15485	99556	56381	37345	14746
19	76617	72242	44643	72978	21086	53483	53499	49884	20687	87299
20	50983	31993	51727	19832	18131	13060	92861	90454	80823	60732
21	67326	51593	78372	15690	16293	84459	09572	32434	52536	78997
22	84500	47405	93996	02934	09298	32418	93706	60181	96057	63129
23	69629	99616	86192	16918	82236	47813	43181	10789	50217	95972
24	45427	80621	64015	29877	59415	68889	43650	54068	45753	23874
25	04842	90686	62968	46564	77938	09747	42773	12460	44836	41988
26	60190	79290	32945	16872	49694	77467	62193	36652	71053	47972
27	92765	63512	46656	20048	63788	03192	97083	33615	04619	41765
28	47846	49625	57112	38574	32053	99708	65959	17295	62462	47261
29	81743	74308	29061	57561	26611	21209	46261	52710	76915	91415
30	39681	47404	92230	17205	61655	68488	27471	98319	61241	35487
31	68961	70218	13715	66055	40865	33533	20766	58662	05230	75106
32	67542	75202	99468	95321	547 5	32184	26215	84287	01726	51551
33	14374	09879	46825	84353	55861	13739	92348	14640	67591	24536
34	10460	90185	78344	15784	55746	55190	79116	16788	86504	49337
35	51282	87415	26064	05053	27906	09222	61320	98879	30825	50336
36	92403	68804	41198	69180	33027	42593	94438	80818	18903	72468
37	78291	15522	21300	40335	66128	63255	73629	69505	50344	41748
38	87215	72404	89599	15740	93907	15762	19878	37928	51665	67020
39	56470	50673	52912	66642	76092	38924	99427	73752	02765	34847
40	59469	37946	11912	32161	63864	41449	54738	01341	22549	38560
41	26484	66134	02903	12776	78687	00772	53585	87227	37004	41541
42	41903	75207	41897	34759	43284	85048	10112	13539	81625	06108
43	67053	89786	98020	76790	14015	58528	95742	15673	54205	63646
44	56835	79433	05468	80283	37499	29077	13072	78999	17146	34010
45	34299	20208	26358	12247	27316	71490	20933	05274	23962	64363
46	82451	23318	14230	71095	77512	71961	90958	23518	73872	69008
47	98044	52950	29316	62935	40153	34663	55002	21750	38889	99000
48	87059	14336	24469	44760	16470	28620	13192	35191	79835	16505
49	40267	19837	18161	40046	08280	51801	11155	17281	59255	71071

Row	50–54	55–59	60–64	65–69	70–74	75–79	80–84	85–89	90–94	95–99
00	84808	43031	48554	82458	87384	19590	06969	39413	00490	46371
01	75732	11167	81847	67665	62859	65585	36680	73201	53940	43287
02	84914	95208	39319	84116	87796	57141	94519	52247	11165	60096
03	72667	56841	62432	60656	43342	26220	92608	18328	32980	81617
04	18390	21474	84512	96440	04766	69049	65751	30281	78239	81369
05	13318	22031	13835	50508	26929	01665	45360	41295	08740	18617
06	09728	68854	89430	54618	63571	75724	73614	53595	95706	14704
07	54217	62829	07425	70136	97000	60189	60879	22515	38314	61529
08	58570	90187	91255	39523	48443	09426	38733	85223	47797	33018
09	71672	67977	36018	27377	55411	55640	38621	11145	30444	45297
10	64917	66341	80757	32247	25363	66505	23082	62579	94977	63445
11	12223	95956	79445	45645	73969	87528	03058	16936	68101	52699
12	68708	42538	38378	38118	20182	85867	84491	15485	60390	20316
13	73460	37140	72734	04144	82561	82752	93494	48507	16629	60373
14	99321	33832	12440	64249	04280	16799	70404	84303	26350	10273
15	75463	86244	38187	89003	31003	88562	91817	02069	19204	91970
16	66008	82094	02096	24799	66195	53896	44461	74904	74873	50640
17	29442	21353	62576	88604	72886	97588	75540	49080	78483	63536
18	12406	24728	27360	62888	49129	85255	13757	97367	26061	83547
19	48286	77396	81989	73584	42864	39132	37792	83695	40548	56832
20	53083	50542	03848	64942	43743	41942	33798	03064	39132	36488
21	96892	57009	63690	95371	19347	33715	50962	76426	92746	56077
22	43995	12197	80445	98603	92036	59567	86852	88842	61005	27222
23	67740	75694	32936	34038	33741	74345	56314	09761	02411	51896
24	47466	23182	23954	15369	93500	26580	70913	90378	02911	27425
25	66710	47453	17772	44528	01950	21711	41791	68131	02385	23176
26	18391	00912	94640	43488	47020	47584	01447	84400	13559	51632
27	52156	36861	31229	85868	83275	45152	75322	91962	44405	87033
28	64173	14597	38466	92943	71605	79843	46965	34527	34469	77054
29	83632	28935	44536	97770	62994	94353	72468	24548	52293	27024
30	85602	75788	84883	35231	88331	85273	23739	31370	69888	09007
31	56610	18964	42826	00447	30303	43632	09969	28540	66455	83999
32	33788	71359	41447	92560	18632	65375	09251	64684	64257	97168
33	79704	02936	79534	54777	87688	92474	91935	97536	50014	58777
34	82515	28962	53740	23583	68920	91591	78464	70903	96795	16822
35	80223	77642	29131	11656	29357	98885	72115	40082	96237	75506
36	72109	59902	96152	35299	77254	44999	44741	15814	97245	53367
37	76656	74256	93022	52309	07931	32243	52029	17912	58734	73931
38	65523	52849	82421	34211	81116	09902	76312	00428	21891	91809
39	51468	68349	43905	69053	33060	49048	86650	40404	78447	83966
40	73723	32825	63192	78713	69518	39899	82987	92956	71545	59099
41	68505	03166	68506	77924	61139	63491	34523	71448	66685	28487
42	40692	67295	17059	18489	01754	37909	21806	96590	37544	19776
43	28489	98954	88189	65791	46488	44483	65443	54561	55880	32310
44	29409	11662	72479	73065	71675	06015	12443	97592	66569	46773
45	15755	71956	22084	04784	30292	82839	92860	92070	37167	29411
46	17512	04000	22248	52834	06094	92698	76625	84248	79468	54038
47	96487	71990	14488	46343	05298	67548	89166	27843	83610	37814
48	83939	20296	35464	30465	03821	98114	34571	99957	15494	04518
49	98370	54035	97385	08583	15678	92224	44465	91994	15765	61535

Row \ Column:	00–04	05–09	10–14	15–19	20–24	25–29	30–34	35–39	40–44	45–49
50	99641	12631	88732	19870	96205	53512	41407	42416	92802	04033
51	61308	77241	01921	79629	44394	80695	62977	85337	73032	64308
52	87205	41236	71721	08285	09042	55618	19542	48260	93110	36038
53	06911	12617	94729	99272	69052	95765	64667	89094	73779	97986
54	13405	96644	93735	23248	93753	80180	89580	91182	39556	20365
55	45636	61732	47242	93127	99372	61336	47044	93350	71512	91277
56	73035	42168	21349	39020	60340	90158	53182	95230	65548	07303
57	08297	68536	55969	79322	57601	54258	87271	68671	13174	23287
58	26254	19690	37035	72678	97131	66561	03799	04276	43998	17070
59	68867	28745	36931	04192	79649	01732	92747	47509	26616	01251
60	24268	96244	38286	98741	27351	62198	90080	45578	03468	42995
61	52206	90947	44773	97944	92051	76161	88301	78783	37331	66909
62	68675	22517	52625	87163	29139	61193	77546	41111	80116	93077
63	46547	40520	47169	05378	33453	77568	58119	86531	63289	73287
64	72537	91079	75686	38388	12946	75624	05512	61744	74345	62173
65	56252	45528	00484	27082	66766	36427	40468	25755	82194	63818
66	65711	31454	47977	78863	99891	36474	93996	19649	32105	39637
67	76325	58388	48389	00655	86945	03935	18929	53385	14065	28934
68	32301	63874	88450	08966	18016	79968	83075	92704	90808	35485
69	83017	88254	55824	52962	16320	73178	29714	70569	67235	27057
70	46318	01330	13886	80044	49567	43885	57588	55760	33593	57002
71	67458	64229	48454	93535	71333	92908	66962	61620	86815	68708
72	21863	18057	60863	78935	58304	68313	28379	21250	98855	33644
73	43100	45950	71756	57866	76952	77855	31241	97238	55256	96796
74	54848	79562	74816	91407	20823	86480	44586	21702	87810	65903
75	84778	48723	24130	10402	02742	99232	86292	61863	53266	84796
76	24683	23733	23722	26571	80500	62377	27457	30711	37785	18135
77	42672	25838	02376	73951	11594	85636	56949	79720	10272	75359
78	50071	55670	45351	85848	67239	48381	47206	25461	56847	63347
79	41007	99628	29089	87080	68711	76148	71171	89335	21804	54064
80	23088	47607	57453	73795	06625	56375	30879	87792	70489	86525
81	46285	37426	29643	84872	74009	39526	24893	56387	72107	47316
82	33801	28058	53708	69838	19070	44239	59639	70558	65223	07143
83	36591	85097	79603	50138	40202	02530	88912	20779	86754	20299
84	75742	18166	00221	88682	80750	49621	07774	75575	28919	16130
85	60555	94102	39146	67795	05985	43280	97202	35613	25369	47959
86	58261	16861	39080	22820	46555	32213	38440	32662	48259	61197
87	98765	65802	44467	03358	38894	34290	31107	25519	26585	34852
88	39157	58231	30710	09394	04012	49122	26283	34946	23590	25663
89	08143	91252	23181	51183	52102	85298	52008	48688	86779	21722
90	66806	72352	64500	89120	13493	85813	93999	12558	24852	04575
91	08289	82806	36490	96421	81718	63075	54178	39209	03050	47089
92	12989	31280	71466	72234	26922	04753	61943	86149	26938	53736
93	44154	63471	30657	62298	56461	48879	54108	97126	43219	95349
94	63788	18000	10049	49041	28807	64190	39753	17397	48026	76947
95	39203	59841	91168	32021	82081	60164	37385	52925	91004	71887
96	39965	79079	97829	95836	26651	12495	68275	20281	73978	07258
97	17752	87692	07004	95860	89325	56997	70904	91993	13209	50274
98	04284	63927	07533	60557	41339	16728	96512	11116	92345	04612
99	03440	97786	37416	24541	36408	63936	36480	87028	05094	95318

Row	50–54	55–59	60–64	65–69	70–74	75–79	80–84	85–89	90–94	95–99
50	43880	79679	13863	94089	32831	95052	50386	00125	34839	93857
51	51879	97545	69161	63821	47985	48414	70709	22450	23654	56933
52	92220	98631	40834	28102	00386	56785	06654	81565	27022	31092
53	14391	51692	58939	50858	34172	96983	19661	44539	63956	20622
54	02642	79615	97584	82684	04227	12621	05561	79284	68421	91047
55	07466	12899	31434	06525	81175	38234	24468	30891	89620	50129
56	83343	72721	52695	36309	67961	73792	63300	89222	10618	24229
57	03745	48015	85373	77206	76214	85412	83510	73998	13500	65084
58	27975	70407	56983	07913	38682	89173	40739	40168	95705	46872
59	54284	28109	48080	80215	85753	64411	27938	56201	16005	49409
60	79521	93795	56291	03839	16098	44436	22678	37566	45822	26879
61	17817	48797	59971	28104	68171	05068	98190	33721	13991	73487
62	56213	82716	77356	91791	31267	19598	25159	28785	57736	72346
63	75194	03658	65212	50828	73031	12498	30153	80522	30866	05307
64	44549	28479	49939	43539	66337	61547	25104	27361	27060	17720
65	11543	45735	21121	46119	96548	48237	30815	01082	00715	18213
66	27327	47369	72686	74153	67849	91820	22255	91564	28009	19796
67	65332	83444	40231	84229	48713	46748	54693	63440	03439	97497
68	45214	30409	35466	73494	39421	86061	88928	55676	68453	66827
69	77929	36175	61017	71350	93393	32687	29040	74575	45306	22552
70	54366	88887	16301	19105	51147	31217	41907	42982	64904	63597
71	08535	65466	48869	58315	23905	24696	66332	22822	37808	78375
72	36947	67802	81864	59051	52076	34284	06530	51015	39540	61780
73	28323	33789	56413	16652	28571	53781	63579	42659	53203	29708
74	16748	41349	75175	66405	75745	33003	32043	01747	49361	61584
75	33178	69744	11252	49458	86585	85536	92257	24864	48761	31924
76	26466	93243	88962	31547	05650	29480	92795	39219	22342	60169
77	36535	14197	72029	40094	61100	17633	38541	08250	04353	13417
78	66835	93340	09121	97179	24446	47809	87930	83677	46036	07924
79	09357	02826	35480	92998	35244	39454	50956	36244	31511	40640
80	07296	75285	29833	78926	48012	97299	56635	57142	00203	77302
81	01106	48819	40679	96311	90666	91712	16907	65802	94408	76429
82	15742	99837	87999	36431	96530	84598	62879	82602	57911	18505
83	16523	51356	37907	65491	39889	49415	97503	09430	39471	12136
84	03536	42548	50478	54022	18614	03129	68513	08643	91870	93123
85	73445	35057	97928	83183	57729	35701	70757	28092	97686	90810
86	52017	99654	63051	87131	87755	29329	52001	24808	54075	48002
87	63724	57039	06679	46472	92762	75952	54470	88720	57702	61299
88	16675	01990	38803	84706	24066	41937	26551	58381	04810	35915
89	01377	36919	49327	24518	61098	25962	04427	33234	04480	02438
90	49752	61849	05823	84198	18174	74419	10322	95196	47893	77825
91	40734	81595	96763	68282	34155	29452	94005	23972	66115	40478
92	64213	91973	62604	00789	21825	25568	00981	89250	24446	86013
93	24505	41214	03031	34756	31600	84374	36871	83645	80482	22081
94	34248	31337	78109	49077	10187	84757	45754	51435	52726	24296
95	60229	06451	61294	53777	17640	85533	10178	23212	02002	08264
96	36712	16560	35055	99750	53169	58659	37377	53580	16829	10472
97	94150	42762	54989	58564	12434	81297	36197	84099	55629	03717
98	36402	94992	51794	59245	87178	84460	58370	34416	75064	07568
99	15853	95261	90876	66395	72788	66605	08718	96740	45414	81005

Source: The Rand Corporation. *A Million Random Digits with 100,000 Normal Deviates*. New York: The Free Press, 1955. Reprinted from pages 30–33. Copyright 1955 and 1983. Used with permission.

Appendix 3 STANDARD NORMAL DISTRIBUTION

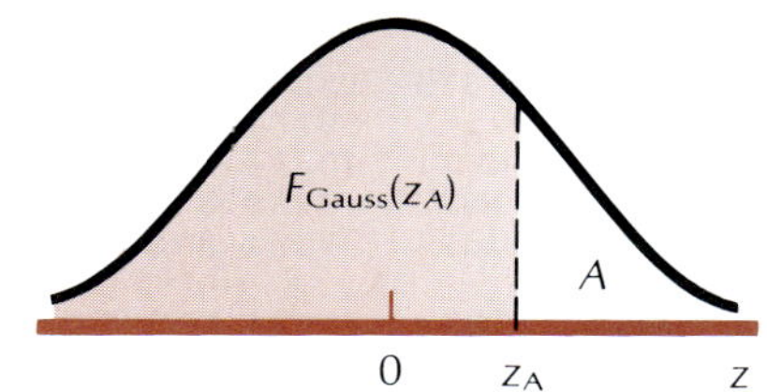

$F(z_A)$	z_A	A	$F(z_A)$	z_A	A	$F(z_A)$	z_A	A	$F(z_A)$	z_A	A
.500	.00	.500	.691	.50	.309	.841	1.00	.159	.933	1.50	.067
.504	.01	.496	.695	.51	.305	.844	1.01	.156	.934	1.51	.066
.508	.02	.492	.698	.52	.302	.846	1.02	.154	.936	1.52	.064
.512	.03	.488	.702	.53	.298	.848	1.03	.152	.937	1.53	.063
.516	.04	.484	.705	.54	.295	.851	1.04	.149	.938	1.54	.062
.520	.05	.480	.709	.55	.291	.853	1.05	.147	.939	1.55	.061
.524	.06	.476	.712	.56	.288	.855	1.06	.145	.941	1.56	.059
.528	.07	.472	.716	.57	.284	.858	1.07	.142	.942	1.57	.058
.532	.08	.468	.719	.58	.281	.860	1.08	.140	.943	1.58	.057
.536	.09	.464	.722	.59	.278	.862	1.09	.138	.944	1.59	.056
.540	.10	.460	.726	.60	.274	.864	1.10	.136	.9452	1.60	.0548
.544	.11	.456	.729	.61	.271	.867	1.11	.133	.9463	1.61	.0537
.548	.12	.452	.732	.62	.268	.869	1.12	.131	.9474	1.62	.0526
.552	.13	.448	.736	.63	.264	.871	1.13	.129	.9484	1.63	.0516
.556	.14	.444	.739	.64	.261	.873	1.14	.127	.9495	1.64	.0505
.560	.15	.440	.742	.65	.258	.875	1.15	.125	.9505	1.65	.0495
.564	.16	.436	.745	.66	.255	.877	1.16	.123	.9515	1.66	.0485
.567	.17	.433	.749	.67	.251	.879	1.17	.121	.9525	1.67	.0475
.571	.18	.429	.752	.68	.248	.881	1.18	.119	.9535	1.68	.0465
.575	.19	.425	.755	.69	.245	.883	1.19	.117	.9545	1.69	.0455
.579	.20	.421	.758	.70	.242	.885	1.20	.115	.9554	1.70	.0446
.583	.21	.417	.761	.71	.239	.887	1.21	.113	.9564	1.71	.0436
.587	.22	.413	.764	.72	.236	.889	1.22	.111	.9573	1.72	.0427
.591	.23	.409	.767	.73	.233	.891	1.23	.109	.9582	1.73	.0418
.595	.24	.405	.770	.74	.230	.893	1.24	.107	.9591	1.74	.0409
.599	.25	.401	.773	.75	.227	.894	1.25	.106	.9599	1.75	.0401
.603	.26	.397	.776	.76	.224	.896	1.26	.104	.9608	1.76	.0392
.606	.27	.394	.779	.77	.221	.898	1.27	.102	.9616	1.77	.0384
.610	.28	.390	.782	.78	.218	.900	1.28	.100	.9625	1.78	.0375
.614	.29	.386	.785	.79	.215	.901	1.29	.099	.9633	1.79	.0367
.618	.30	.382	.788	.80	.212	.903	1.30	.097	.9641	1.80	.0359
.622	.31	.378	.791	.81	.209	.905	1.31	.095	.9649	1.81	.0351
.626	.32	.374	.794	.82	.206	.907	1.32	.093	.9656	1.82	.0344
.629	.33	.371	.797	.83	.203	.908	1.33	.092	.9664	1.83	.0336
.633	.34	.367	.800	.84	.200	.910	1.34	.090	.9671	1.84	.0329
.637	.35	.363	.802	.85	.198	.911	1.35	.089	.9678	1.85	.0322
.641	.36	.359	.805	.86	.195	.913	1.36	.087	.9686	1.86	.0314
.644	.37	.356	.808	.87	.192	.915	1.37	.085	.9693	1.87	.0307
.648	.38	.352	.811	.88	.189	.916	1.38	.084	.9699	1.88	.0301
.652	.39	.348	.813	.89	.187	.918	1.39	.082	.9706	1.89	.0294
.655	.40	.345	.816	.90	.184	.919	1.40	.081	.9713	1.90	.0287
.659	.41	.341	.819	.91	.181	.921	1.41	.079	.9719	1.91	.0281
.663	.42	.337	.821	.92	.179	.922	1.42	.078	.9726	1.92	.0274
.666	.43	.334	.824	.93	.176	.924	1.43	.076	.9732	1.93	.0268
.670	.44	.330	.826	.94	.174	.925	1.44	.075	.9738	1.94	.0262
.674	.45	.326	.829	.95	.171	.926	1.45	.074	.9744	1.95	.0256
.677	.46	.323	.831	.96	.169	.928	1.46	.072	.9750	1.96	.0250
.681	.47	.319	.834	.97	.166	.929	1.47	.071	.9756	1.97	.0244
.684	.48	.316	.836	.98	.164	.931	1.48	.069	.9761	1.98	.0239
.688	.49	.312	.839	.99	.161	.932	1.49	.068	.9767	1.99	.0233

$F(z_A)$	z_A	A	$F(z_A)$	z_A	A	$F(z_A)$	z_A	A	$F(z_A)$	z_A	A
.9772	2.00	.0228	.9938	2.50	.0062	.99865	3.00	.00135	.999767	3.50	.000233
.9778	2.01	.0222	.9940	2.51	.0060	.99869	3.01	.00131	.999776	3.51	.000224
.9783	2.02	.0217	.9941	2.52	.0059	.99874	3.02	.00126	.999784	3.52	.000216
.9788	2.03	.0212	.9943	2.53	.0057	.99878	3.03	.00122	.999792	3.53	.000208
.9793	2.04	.0207	.9945	2.54	.0055	.99882	3.04	.00118	.999800	3.54	.000200
.9798	2.05	.0202	.9946	2.55	.0054	.99886	3.05	.00114	.999807	3.55	.000193
.9803	2.06	.0197	.9948	2.56	.0052	.99889	3.06	.00111	.999815	3.56	.000185
.9808	2.07	.0192	.9949	2.57	.0051	.99893	3.07	.00107	.999822	3.57	.000178
.9812	2.08	.0188	.9951	2.58	.0049	.99896	3.08	.00104	.999828	3.58	.000172
.9817	2.09	.0183	.9952	2.59	.0048	.99900	3.09	.00100	.999835	3.59	.000165
.9821	2.10	.0179	.9953	2.60	.0047	.99903	3.10	.00097	.999841	3.60	.000159
.9826	2.11	.0174	.9955	2.61	.0045	.99906	3.11	.00094	.999847	3.61	.000153
.9830	2.12	.0170	.9956	2.62	.0044	.99910	3.12	.00090	.999853	3.62	.000147
.9834	2.13	.0166	.9957	2.63	.0043	.99913	3.13	.00087	.999858	3.63	.000142
.9838	2.14	.0162	.9959	2.64	.0041	.99916	3.14	.00084	.999864	3.64	.000136
.9842	2.15	.0158	.9960	2.65	.0040	.99918	3.15	.00082	.999869	3.65	.000131
.9846	2.16	.0154	.9961	2.66	.0039	.99921	3.16	.00079	.999874	3.66	.000126
.9850	2.17	.0150	.9962	2.67	.0038	.99924	3.17	.00076	.999879	3.67	.000121
.9854	2.18	.0146	.9963	2.68	.0037	.99926	3.18	.00074	.999883	3.68	.000117
.9857	2.19	.0143	.9964	2.69	.0036	.99929	3.19	.00071	.999888	3.69	.000112
.9861	2.20	.0139	.9965	2.70	.0035	.99931	3.20	.00069	.999892	3.70	.000108
.9864	2.21	.0136	.9966	2.71	.0034	.99934	3.21	.00066	.999896	3.71	.000104
.9868	2.22	.0132	.9967	2.72	.0033	.99936	3.22	.00064	.999900	3.72	.000100
.9871	2.23	.0129	.9968	2.73	.0032	.99938	3.23	.00062	.999904	3.73	.000096
.9875	2.24	.0125	.9969	2.74	.0031	.99940	3.24	.00060	.999908	3.74	.000092
.9878	2.25	.0122	.9970	2.75	.0030	.99942	3.25	.00058	.999912	3.75	.000088
.9881	2.26	.0119	.9971	2.76	.0029	.99944	3.26	.00056	.999915	3.76	.000085
.9884	2.27	.0116	.9972	2.77	.0028	.99946	3.27	.00054	.999918	3.77	.000082
.9887	2.28	.0113	.99728	2.78	.00272	.99948	3.28	.00052	.999922	3.78	.000078
.9890	2.29	.0110	.99736	2.79	.00264	.99950	3.29	.00050	.999925	3.79	.000075
.9893	2.30	.0107	.99744	2.80	.00256	.99952	3.30	.00048	.999928	3.80	.000072
.9896	2.31	.0104	.99752	2.81	.00248	.99953	3.31	.00047	.999931	3.81	.000069
.9898	2.32	.0102	.99760	2.82	.00240	.99955	3.32	.00045	.999933	3.82	.000067
.9901	2.33	.0099	.99767	2.83	.00233	.99957	3.33	.00043	.999936	3.83	.000064
.9904	2.34	.0096	.99774	2.84	.00226	.99958	3.34	.00042	.999938	3.84	.000062
.9906	2.35	.0094	.99781	2.85	.00219	.99960	3.35	.00040	.999941	3.85	.000059
.9909	2.36	.0091	.99788	2.86	.00212	.99961	3.36	.00039	.999943	3.86	.000057
.9911	2.37	.0089	.99795	2.87	.00205	.99962	3.37	.00038	.999946	3.87	.000054
.9913	2.38	.0087	.99801	2.88	.00199	.99964	3.38	.00036	.999948	3.88	.000052
.9916	2.39	.0084	.99807	2.89	.00193	.99965	3.39	.00035	.999950	3.89	.000050
.9918	2.40	.0082	.99813	2.90	.00187	.99966	3.40	.00034	.999952	3.90	.000048
.9920	2.41	.0080	.99819	2.91	.00181	.99968	3.41	.00032	.999954	3.91	.000046
.9922	2.42	.0078	.99825	2.92	.00175	.99969	3.42	.00031	.999956	3.92	.000044
.9925	2.43	.0075	.99831	2.93	.00169	.99970	3.43	.00030	.999958	3.93	.000042
.9927	2.44	.0073	.99836	2.94	.00164	.99971	3.44	.00029	.999959	3.94	.000041
.9929	2.45	.0071	.99841	2.95	.00159	.99972	3.45	.00028	.999961	3.95	.000039
.9931	2.46	.0069	.99846	2.96	.00154	.99973	3.46	.00027	.999963	3.96	.000037
.9932	2.47	.0068	.99851	2.97	.00149	.99974	3.47	.00026	.999964	3.97	.000036
.9934	2.48	.0066	.99856	2.98	.00144	.99975	3.48	.00025	.999966	3.98	.000034
.9936	2.49	.0064	.99861	2.99	.00139	.99976	3.49	.00024	.999967	3.99	.000033

Source: Pearson, E. S., & Hartley, H. O. (Eds.). *Biometrika Tables for Statisticians*. Volume 1 (3rd ed., reprinted with corrections). London: Biometrika Trust, 1976. Table 1. With permission of the Biometrika Trustees.

Appendix 4 CRITICAL VALUES OF THE t DISTRIBUTION

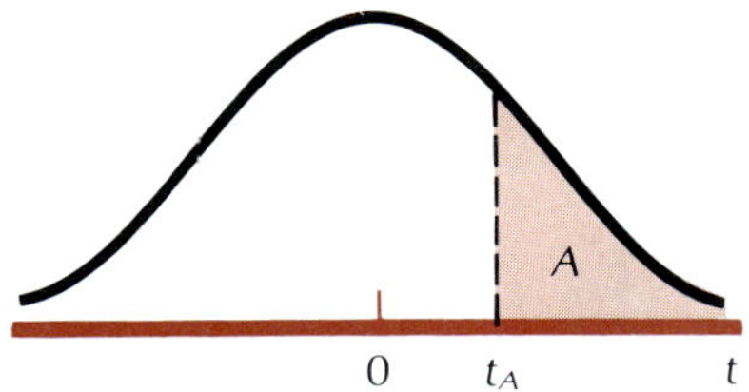

Degrees of freedom	$t_{.10}$	$t_{.05}$	$t_{.025}$	$t_{.01}$	$t_{.005}$	$t_{.0005}$
1	3·078	6·314	12·706	31·821	63·657	636·62
2	1·886	2·920	4·303	6·965	9·925	31·598
3	1·638	2·353	3·182	4·541	5·841	12·924
4	1·533	2·132	2·776	3·747	4·604	8·610
5	1·476	2·015	2·571	3·365	4·032	6·869
6	1·440	1·943	2·447	3·143	3·707	5·959
7	1·415	1·895	2·365	2·998	3·499	5·408
8	1·397	1·860	2·306	2·896	3·355	5·041
9	1·383	1·833	2·262	2·821	3·250	4·781
10	1·372	1·812	2·228	2·764	3·169	4·587
11	1·363	1·796	2·201	2·718	3·106	4·437
12	1·356	1·782	2·179	2·681	3·055	4·318
13	1·350	1·771	2·160	2·650	3·012	4·221
14	1·345	1·761	2·145	2·624	2·977	4·140
15	1·341	1·753	2·131	2·602	2·947	4·073
16	1·337	1·746	2·120	2·583	2·921	4·015
17	1·333	1·740	2·110	2·567	2·898	3·965
18	1·330	1·734	2·101	2·552	2·878	3·922
19	1·328	1·729	2·093	2·539	2·861	3·883
20	1·325	1·725	2·086	2·528	2·845	3·850
21	1·323	1·721	2·080	2·518	2·831	3·819
22	1·321	1·717	2·074	2·508	2·819	3·792
23	1·319	1·714	2·069	2·500	2·807	3·767
24	1·318	1·711	2·064	2·492	2·797	3·745
25	1·316	1·708	2·060	2·485	2·787	3·725
26	1·315	1·706	2·056	2·479	2·779	3·707
27	1·314	1·703	2·052	2·473	2·771	3·690
28	1·313	1·701	2·048	2·467	2·763	3·674
29	1·311	1·699	2·045	2·462	2·756	3·659
30	1·310	1·697	2·042	2·457	2·750	3·646
40	1·303	1·684	2·021	2·423	2·704	3·551
60	1·296	1·671	2·000	2·390	2·660	3·460
120	1·289	1·658	1·980	2·358	2·617	3·373
∞	1·282	1·645	1·960	2·326	2·576	3·291

Source: Pearson, E. S., & Hartley, H. O. (Eds.). *Biometrika Tables for Statisticians*. Volume 1 (3rd ed., reprinted with corrections). London: Biometrika Trust, 1976. Table 12. With permission of the Biometrika Trustees.

Appendix 5 CRITICAL VALUES OF THE CHI-SQUARE DISTRIBUTION

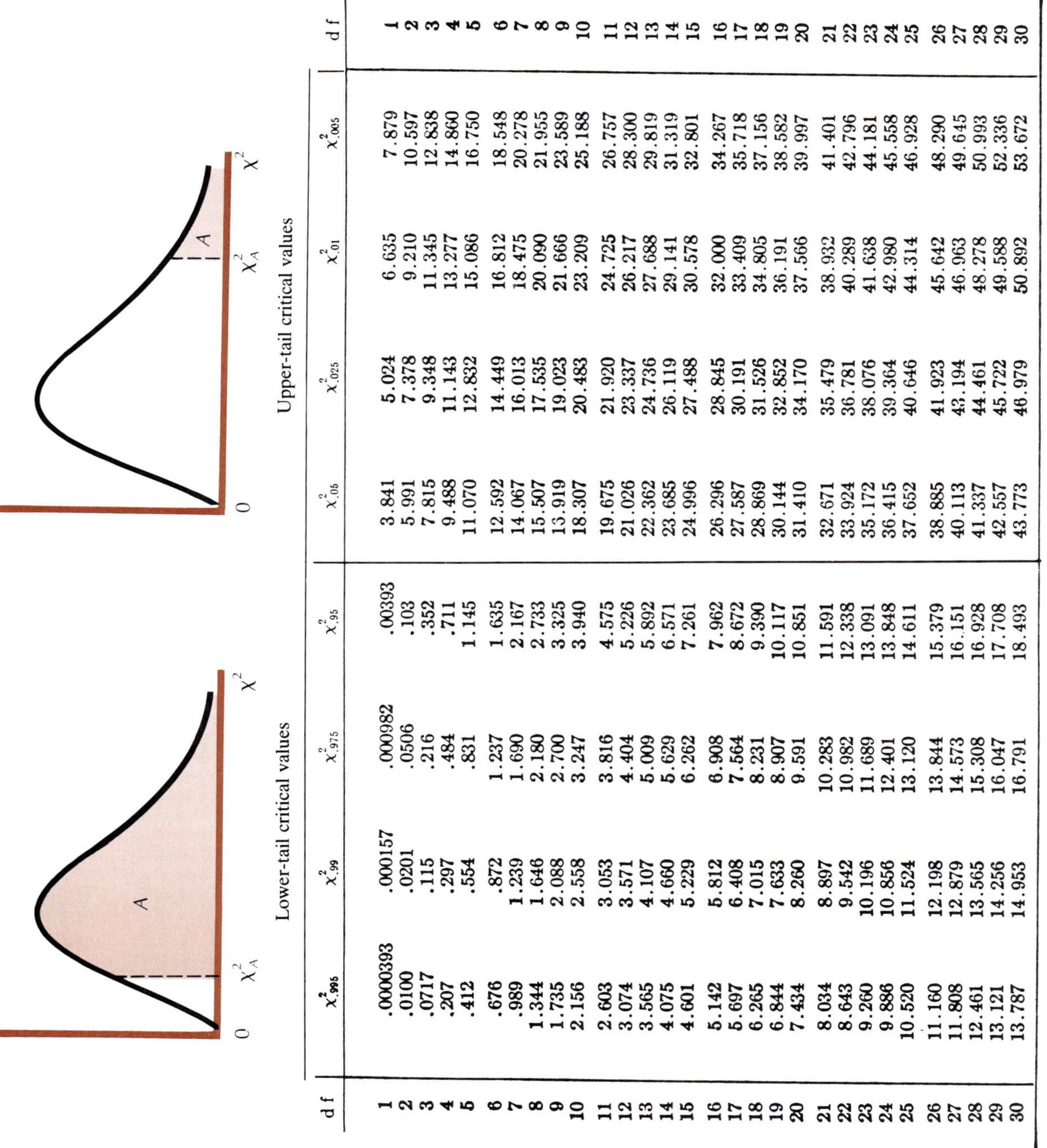

	Lower-tail critical values				Upper-tail critical values				
df	$\chi^2_{.995}$	$\chi^2_{.99}$	$\chi^2_{.975}$	$\chi^2_{.95}$	$\chi^2_{.05}$	$\chi^2_{.025}$	$\chi^2_{.01}$	$\chi^2_{.005}$	df
1	.0000393	.000157	.000982	.00393	3.841	5.024	6.635	7.879	1
2	.0100	.0201	.0506	.103	5.991	7.378	9.210	10.597	2
3	.0717	.115	.216	.352	7.815	9.348	11.345	12.838	3
4	.207	.297	.484	.711	9.488	11.143	13.277	14.860	4
5	.412	.554	.831	1.145	11.070	12.832	15.086	16.750	5
6	.676	.872	1.237	1.635	12.592	14.449	16.812	18.548	6
7	.989	1.239	1.690	2.167	14.067	16.013	18.475	20.278	7
8	1.344	1.646	2.180	2.733	15.507	17.535	20.090	21.955	8
9	1.735	2.088	2.700	3.325	13.919	19.023	21.666	23.589	9
10	2.156	2.558	3.247	3.940	18.307	20.483	23.209	25.188	10
11	2.603	3.053	3.816	4.575	19.675	21.920	24.725	26.757	11
12	3.074	3.571	4.404	5.226	21.026	23.337	26.217	28.300	12
13	3.565	4.107	5.009	5.892	22.362	24.736	27.688	29.819	13
14	4.075	4.660	5.629	6.571	23.685	26.119	29.141	31.319	14
15	4.601	5.229	6.262	7.261	24.996	27.488	30.578	32.801	15
16	5.142	5.812	6.908	7.962	26.296	28.845	32.000	34.267	16
17	5.697	6.408	7.564	8.672	27.587	30.191	33.409	35.718	17
18	6.265	7.015	8.231	9.390	28.869	31.526	34.805	37.156	18
19	6.844	7.633	8.907	10.117	30.144	32.852	36.191	38.582	19
20	7.434	8.260	9.591	10.851	31.410	34.170	37.566	39.997	20
21	8.034	8.897	10.283	11.591	32.671	35.479	38.932	41.401	21
22	8.643	9.542	10.982	12.338	33.924	36.781	40.289	42.796	22
23	9.260	10.196	11.689	13.091	35.172	38.076	41.638	44.181	23
24	9.886	10.856	12.401	13.848	36.415	39.364	42.980	45.558	24
25	10.520	11.524	13.120	14.611	37.652	40.646	44.314	46.928	25
26	11.160	12.198	13.844	15.379	38.885	41.923	45.642	48.290	26
27	11.808	12.879	14.573	16.151	40.113	43.194	46.963	49.645	27
28	12.461	13.565	15.308	16.928	41.337	44.461	48.278	50.993	28
29	13.121	14.256	16.047	17.708	42.557	45.722	49.588	52.336	29
30	13.787	14.953	16.791	18.493	43.773	46.979	50.892	53.672	30

When the number of degrees of freedom (df) is greater than 30, chi-square critical values may be approximated by

$$\chi^2_A(\text{df}) = \tfrac{1}{2}\left[z_A + \sqrt{2(\text{df}) - 1}\right]^2.$$

Source: Pearson, E. S., & Hartley, H. O. (Eds.). *Biometrika Tables for Statisticians.* Volume 1 (3rd ed., reprinted with corrections). London: Biometrika Trust, 1976. Table 8. With permission of the Biometrika Trustees.

Appendix 6 CRITICAL VALUES OF THE F DISTRIBUTION

Values of $F_{.10}$ — Degrees of freedom for numerator — $A = 10\%$

Rows: degrees of freedom for denominator (ν_2); columns: degrees of freedom for numerator (ν_1)

ν_2 \ ν_1	1	2	3	4	5	6	7	8	9	10	12	15	20	24	30	40	60	120	∞
1	39·86	49·50	53·59	55·83	57·24	58·20	58·91	59·44	59·86	60·19	60·71	61·22	61·74	62·00	62·26	62·53	62·79	63·06	63·33
2	8·53	9·00	9·16	9·24	9·29	9·33	9·35	9·37	9·38	9·39	9·41	9·42	9·44	9·45	9·46	9·47	9·47	9·48	9·49
3	5·54	5·46	5·39	5·34	5·31	5·28	5·27	5·25	5·24	5·23	5·22	5·20	5·18	5·18	5·17	5·16	5·15	5·14	5·13
4	4·54	4·32	4·19	4·11	4·05	4·01	3·98	3·95	3·94	3·92	3·90	3·87	3·84	3·83	3·82	3·80	3·79	3·78	3·76
5	4·06	3·78	3·62	3·52	3·45	3·40	3·37	3·34	3·32	3·30	3·27	3·24	3·21	3·19	3·17	3·16	3·14	3·12	3·10
6	3·78	3·46	3·29	3·18	3·11	3·05	3·01	2·98	2·96	2·94	2·90	2·87	2·84	2·82	2·80	2·78	2·76	2·74	2·72
7	3·59	3·26	3·07	2·96	2·88	2·83	2·78	2·75	2·72	2·70	2·67	2·63	2·59	2·58	2·56	2·54	2·51	2·49	2·47
8	3·46	3·11	2·92	2·81	2·73	2·67	2·62	2·59	2·56	2·54	2·50	2·46	2·42	2·40	2·38	2·36	2·34	2·32	2·29
9	3·36	3·01	2·81	2·69	2·61	2·55	2·51	2·47	2·44	2·42	2·38	2·34	2·30	2·28	2·25	2·23	2·21	2·18	2·16
10	3·29	2·92	2·73	2·61	2·52	2·46	2·41	2·38	2·35	2·32	2·28	2·24	2·20	2·18	2·16	2·13	2·11	2·08	2·06
11	3·23	2·86	2·66	2·54	2·45	2·39	2·34	2·30	2·27	2·25	2·21	2·17	2·12	2·10	2·08	2·05	2·03	2·00	1·97
12	3·18	2·81	2·61	2·48	2·39	2·33	2·28	2·24	2·21	2·19	2·15	2·10	2·06	2·04	2·01	1·99	1·96	1·93	1·90
13	3·14	2·76	2·56	2·43	2·35	2·28	2·23	2·20	2·16	2·14	2·10	2·05	2·01	1·98	1·96	1·93	1·90	1·88	1·85
14	3·10	2·73	2·52	2·39	2·31	2·24	2·19	2·15	2·12	2·10	2·05	2·01	1·96	1·94	1·91	1·89	1·86	1·83	1·80
15	3·07	2·70	2·49	2·36	2·27	2·21	2·16	2·12	2·09	2·06	2·02	1·97	1·92	1·90	1·87	1·85	1·82	1·79	1·76
16	3·05	2·67	2·46	2·33	2·24	2·18	2·13	2·09	2·06	2·03	1·99	1·94	1·89	1·87	1·84	1·81	1·78	1·75	1·72
17	3·03	2·64	2·44	2·31	2·22	2·15	2·10	2·06	2·03	2·00	1·96	1·91	1·86	1·84	1·81	1·78	1·75	1·72	1·69
18	3·01	2·62	2·42	2·29	2·20	2·13	2·08	2·04	2·00	1·98	1·93	1·89	1·84	1·81	1·78	1·75	1·72	1·69	1·66
19	2·99	2·61	2·40	2·27	2·18	2·11	2·06	2·02	1·98	1·96	1·91	1·86	1·81	1·79	1·76	1·73	1·70	1·67	1·63
20	2·97	2·59	2·38	2·25	2·16	2·09	2·04	2·00	1·96	1·94	1·89	1·84	1·79	1·77	1·74	1·71	1·68	1·64	1·61
21	2·96	2·57	2·36	2·23	2·14	2·08	2·02	1·98	1·95	1·92	1·87	1·83	1·78	1·75	1·72	1·69	1·66	1·62	1·59
22	2·95	2·56	2·35	2·22	2·13	2·06	2·01	1·97	1·93	1·90	1·86	1·81	1·76	1·73	1·70	1·67	1·64	1·60	1·57
23	2·94	2·55	2·34	2·21	2·11	2·05	1·99	1·95	1·92	1·89	1·84	1·80	1·74	1·72	1·69	1·66	1·62	1·59	1·55
24	2·93	2·54	2·33	2·19	2·10	2·04	1·98	1·94	1·91	1·88	1·83	1·78	1·73	1·70	1·67	1·64	1·61	1·57	1·53
25	2·92	2·53	2·32	2·18	2·09	2·02	1·97	1·93	1·89	1·87	1·82	1·77	1·72	1·69	1·66	1·63	1·59	1·56	1·52
26	2·91	2·52	2·31	2·17	2·08	2·01	1·96	1·92	1·88	1·86	1·81	1·76	1·71	1·68	1·65	1·61	1·58	1·54	1·50
27	2·90	2·51	2·30	2·17	2·07	2·00	1·95	1·91	1·87	1·85	1·80	1·75	1·70	1·67	1·64	1·60	1·57	1·53	1·49
28	2·89	2·50	2·29	2·16	2·06	2·00	1·94	1·90	1·87	1·84	1·79	1·74	1·69	1·66	1·63	1·59	1·56	1·52	1·48
29	2·89	2·50	2·28	2·15	2·06	1·99	1·93	1·89	1·86	1·83	1·78	1·73	1·68	1·65	1·62	1·58	1·55	1·51	1·47
30	2·88	2·49	2·28	2·14	2·05	1·98	1·93	1·88	1·85	1·82	1·77	1·72	1·67	1·64	1·61	1·57	1·54	1·50	1·46
40	2·84	2·44	2·23	2·09	2·00	1·93	1·87	1·83	1·79	1·76	1·71	1·66	1·61	1·57	1·54	1·51	1·47	1·42	1·38
60	2·79	2·39	2·18	2·04	1·95	1·87	1·82	1·77	1·74	1·71	1·66	1·60	1·54	1·51	1·48	1·44	1·40	1·35	1·29
120	2·75	2·35	2·13	1·99	1·90	1·82	1·77	1·72	1·68	1·65	1·60	1·55	1·48	1·45	1·41	1·37	1·32	1·26	1·19
∞	2·71	2·30	2·08	1·94	1·85	1·77	1·72	1·67	1·63	1·60	1·55	1·49	1·42	1·38	1·34	1·30	1·24	1·17	1·00

Degrees of freedom for denominator

Values of $F_{.05}$ Degrees of freedom for numerator $A = 5\%$

Degrees of freedom for denominator

ν_2 \ ν_1	1	2	3	4	5	6	7	8	9	10	12	15	20	24	30	40	60	120	∞
1	161·4	199·5	215·7	224·6	230·2	234·0	236·8	238·9	240·5	241·9	243·9	245·9	248·0	249·1	250·1	251·1	252·2	253·3	254·3
2	18·51	19·00	19·16	19·25	19·30	19·33	19·35	19·37	19·38	19·40	19·41	19·43	19·45	19·45	19·46	19·47	19·48	19·49	19·50
3	10·13	9·55	9·28	9·12	9·01	8·94	8·89	8·85	8·81	8·79	8·74	8·70	8·66	8·64	8·62	8·59	8·57	8·55	8·53
4	7·71	6·94	6·59	6·39	6·26	6·16	6·09	6·04	6·00	5·96	5·91	5·86	5·80	5·77	5·75	5·72	5·69	5·66	5·63
5	6·61	5·79	5·41	5·19	5·05	4·95	4·88	4·82	4·77	4·74	4·68	4·62	4·56	4·53	4·50	4·46	4·43	4·40	4·36
6	5·99	5·14	4·76	4·53	4·39	4·28	4·21	4·15	4·10	4·06	4·00	3·94	3·87	3·84	3·81	3·77	3·74	3·70	3·67
7	5·59	4·74	4·35	4·12	3·97	3·87	3·79	3·73	3·68	3·64	3·57	3·51	3·44	3·41	3·38	3·34	3·30	3·27	3·23
8	5·32	4·46	4·07	3·84	3·69	3·58	3·50	3·44	3·39	3·35	3·28	3·22	3·15	3·12	3·08	3·04	3·01	2·97	2·93
9	5·12	4·26	3·86	3·63	3·48	3·37	3·29	3·23	3·18	3·14	3·07	3·01	2·94	2·90	2·86	2·83	2·79	2·75	2·71
10	4·96	4·10	3·71	3·48	3·33	3·22	3·14	3·07	3·02	2·98	2·91	2·85	2·77	2·74	2·70	2·66	2·62	2·58	2·54
11	4·84	3·98	3·59	3·36	3·20	3·09	3·01	2·95	2·90	2·85	2·79	2·72	2·65	2·61	2·57	2·53	2·49	2·45	2·40
12	4·75	3·89	3·49	3·26	3·11	3·00	2·91	2·85	2·80	2·75	2·69	2·62	2·54	2·51	2·47	2·43	2·38	2·34	2·30
13	4·67	3·81	3·41	3·18	3·03	2·92	2·83	2·77	2·71	2·67	2·60	2·53	2·46	2·42	2·38	2·34	2·30	2·25	2·21
14	4·60	3·74	3·34	3·11	2·96	2·85	2·76	2·70	2·65	2·60	2·53	2·46	2·39	2·35	2·31	2·27	2·22	2·18	2·13
15	4·54	3·68	3·29	3·06	2·90	2·79	2·71	2·64	2·59	2·54	2·48	2·40	2·33	2·29	2·25	2·20	2·16	2·11	2·07
16	4·49	3·63	3·24	3·01	2·85	2·74	2·66	2·59	2·54	2·49	2·42	2·35	2·28	2·24	2·19	2·15	2·11	2·06	2·01
17	4·45	3·59	3·20	2·96	2·81	2·70	2·61	2·55	2·49	2·45	2·38	2·31	2·23	2·19	2·15	2·10	2·06	2·01	1·96
18	4·41	3·55	3·16	2·93	2·77	2·66	2·58	2·51	2·46	2·41	2·34	2·27	2·19	2·15	2·11	2·06	2·02	1·97	1·92
19	4·38	3·52	3·13	2·90	2·74	2·63	2·54	2·48	2·42	2·38	2·31	2·23	2·16	2·11	2·07	2·03	1·98	1·93	1·88
20	4·35	3·49	3·10	2·87	2·71	2·60	2·51	2·45	2·39	2·35	2·28	2·20	2·12	2·08	2·04	1·99	1·95	1·90	1·84
21	4·32	3·47	3·07	2·84	2·68	2·57	2·49	2·42	2·37	2·32	2·25	2·18	2·10	2·05	2·01	1·96	1·92	1·87	1·81
22	4·30	3·44	3·05	2·82	2·66	2·55	2·46	2·40	2·34	2·30	2·23	2·15	2·07	2·03	1·98	1·94	1·89	1·84	1·78
23	4·28	3·42	3·03	2·80	2·64	2·53	2·44	2·37	2·32	2·27	2·20	2·13	2·05	2·01	1·96	1·91	1·86	1·81	1·76
24	4·26	3·40	3·01	2·78	2·62	2·51	2·42	2·36	2·30	2·25	2·18	2·11	2·03	1·98	1·94	1·89	1·84	1·79	1·73
25	4·24	3·39	2·99	2·76	2·60	2·49	2·40	2·34	2·28	2·24	2·16	2·09	2·01	1·96	1·92	1·87	1·82	1·77	1·71
26	4·23	3·37	2·98	2·74	2·59	2·47	2·39	2·32	2·27	2·22	2·15	2·07	1·99	1·95	1·90	1·85	1·80	1·75	1·69
27	4·21	3·35	2·96	2·73	2·57	2·46	2·37	2·31	2·25	2·20	2·13	2·06	1·97	1·93	1·88	1·84	1·79	1·73	1·67
28	4·20	3·34	2·95	2·71	2·56	2·45	2·36	2·29	2·24	2·19	2·12	2·04	1·96	1·91	1·87	1·82	1·77	1·71	1·65
29	4·18	3·33	2·93	2·70	2·55	2·43	2·35	2·28	2·22	2·18	2·10	2·03	1·94	1·90	1·85	1·81	1·75	1·70	1·64
30	4·17	3·32	2·92	2·69	2·53	2·42	2·33	2·27	2·21	2·16	2·09	2·01	1·93	1·89	1·84	1·79	1·74	1·68	1·62
40	4·08	3·23	2·84	2·61	2·45	2·34	2·25	2·18	2·12	2·08	2·00	1·92	1·84	1·79	1·74	1·69	1·64	1·58	1·51
60	4·00	3·15	2·76	2·53	2·37	2·25	2·17	2·10	2·04	1·99	1·92	1·84	1·75	1·70	1·65	1·59	1·53	1·47	1·39
120	3·92	3·07	2·68	2·45	2·29	2·17	2·09	2·02	1·96	1·91	1·83	1·75	1·66	1·61	1·55	1·50	1·43	1·35	1·25
∞	3·84	3·00	2·60	2·37	2·21	2·10	2·01	1·94	1·88	1·83	1·75	1·67	1·57	1·52	1·46	1·39	1·32	1·22	1·00

Values of $F_{.025}$

Degrees of freedom for numerator

$A = 2\frac{1}{2}\%$

Degrees of freedom for denominator

ν_2 \ ν_1	1	2	3	4	5	6	7	8	9	10	12	15	20	24	30	40	60	120	∞
1	647·8	799·5	864·2	899·6	921·8	937·1	948·2	956·7	963·3	968·6	976·7	984·9	993·1	997·2	1001	1006	1010	1014	1018
2	38·51	39·00	39·17	39·25	39·30	39·33	39·36	39·37	39·39	39·40	39·41	39·43	39·45	39·46	39·46	39·47	39·48	39·49	39·50
3	17·44	16·04	15·44	15·10	14·88	14·73	14·62	14·54	14·47	14·42	14·34	14·25	14·17	14·12	14·08	14·04	13·99	13·95	13·90
4	12·22	10·65	9·98	9·60	9·36	9·20	9·07	8·98	8·90	8·84	8·75	8·66	8·56	8·51	8·46	8·41	8·36	8·31	8·26
5	10·01	8·43	7·76	7·39	7·15	6·98	6·85	6·76	6·68	6·62	6·52	6·43	6·33	6·28	6·23	6·18	6·12	6·07	6·02
6	8·81	7·26	6·60	6·23	5·99	5·82	5·70	5·60	5·52	5·46	5·37	5·27	5·17	5·12	5·07	5·01	4·96	4·90	4·85
7	8·07	6·54	5·89	5·52	5·29	5·12	4·99	4·90	4·82	4·76	4·67	4·57	4·47	4·42	4·36	4·31	4·25	4·20	4·14
8	7·57	6·06	5·42	5·05	4·82	4·65	4·53	4·43	4·36	4·30	4·20	4·10	4·00	3·95	3·89	3·84	3·78	3·73	3·67
9	7·21	5·71	5·08	4·72	4·48	4·32	4·20	4·10	4·03	3·96	3·87	3·77	3·67	3·61	3·56	3·51	3·45	3·39	3·33
10	6·94	5·46	4·83	4·47	4·24	4·07	3·95	3·85	3·78	3·72	3·62	3·52	3·42	3·37	3·31	3·26	3·20	3·14	3·08
11	6·72	5·26	4·63	4·28	4·04	3·88	3·76	3·66	3·59	3·53	3·43	3·33	3·23	3·17	3·12	3·06	3·00	2·94	2·88
12	6·55	5·10	4·47	4·12	3·89	3·73	3·61	3·51	3·44	3·37	3·28	3·18	3·07	3·02	2·96	2·91	2·85	2·79	2·72
13	6·41	4·97	4·35	4·00	3·77	3·60	3·48	3·39	3·31	3·25	3·15	3·05	2·95	2·89	2·84	2·78	2·72	2·66	2·60
14	6·30	4·86	4·24	3·89	3·66	3·50	3·38	3·29	3·21	3·15	3·05	2·95	2·84	2·79	2·73	2·67	2·61	2·55	2·49
15	6·20	4·77	4·15	3·80	3·58	3·41	3·29	3·20	3·12	3·06	2·96	2·86	2·76	2·70	2·64	2·59	2·52	2·46	2·40
16	6·12	4·69	4·08	3·73	3·50	3·34	3·22	3·12	3·05	2·99	2·89	2·79	2·68	2·63	2·57	2·51	2·45	2·38	2·32
17	6·04	4·62	4·01	3·66	3·44	3·28	3·16	3·06	2·98	2·92	2·82	2·72	2·62	2·56	2·50	2·44	2·38	2·32	2·25
18	5·98	4·56	3·95	3·61	3·38	3·22	3·10	3·01	2·93	2·87	2·77	2·67	2·56	2·50	2·44	2·38	2·32	2·26	2·19
19	5·92	4·51	3·90	3·56	3·33	3·17	3·05	2·96	2·88	2·82	2·72	2·62	2·51	2·45	2·39	2·33	2·27	2·20	2·13
20	5·87	4·46	3·86	3·51	3·29	3·13	3·01	2·91	2·84	2·77	2·68	2·57	2·46	2·41	2·35	2·29	2·22	2·16	2·09
21	5·83	4·42	3·82	3·48	3·25	3·09	2·97	2·87	2·80	2·73	2·64	2·53	2·42	2·37	2·31	2·25	2·18	2·11	2·04
22	5·79	4·38	3·78	3·44	3·22	3·05	2·93	2·84	2·76	2·70	2·60	2·50	2·39	2·33	2·27	2·21	2·14	2·08	2·00
23	5·75	4·35	3·75	3·41	3·18	3·02	2·90	2·81	2·73	2·67	2·57	2·47	2·36	2·30	2·24	2·18	2·11	2·04	1·97
24	5·72	4·32	3·72	3·38	3·15	2·99	2·87	2·78	2·70	2·64	2·54	2·44	2·33	2·27	2·21	2·15	2·08	2·01	1·94
25	5·69	4·29	3·69	3·35	3·13	2·97	2·85	2·75	2·68	2·61	2·51	2·41	2·30	2·24	2·18	2·12	2·05	1·98	1·91
26	5·66	4·27	3·67	3·33	3·10	2·94	2·82	2·73	2·65	2·59	2·49	2·39	2·28	2·22	2·16	2·09	2·03	1·95	1·88
27	5·63	4·24	3·65	3·31	3·08	2·92	2·80	2·71	2·63	2·57	2·47	2·36	2·25	2·19	2·13	2·07	2·00	1·93	1·85
28	5·61	4·22	3·63	3·29	3·06	2·90	2·78	2·69	2·61	2·55	2·45	2·34	2·23	2·17	2·11	2·05	1·98	1·91	1·83
29	5·59	4·20	3·61	3·27	3·04	2·88	2·76	2·67	2·59	2·53	2·43	2·32	2·21	2·15	2·09	2·03	1·96	1·89	1·81
30	5·57	4·18	3·59	3·25	3·03	2·87	2·75	2·65	2·57	2·51	2·41	2·31	2·20	2·14	2·07	2·01	1·94	1·87	1·79
40	5·42	4·05	3·46	3·13	2·90	2·74	2·62	2·53	2·45	2·39	2·29	2·18	2·07	2·01	1·94	1·88	1·80	1·72	1·64
60	5·29	3·93	3·34	3·01	2·79	2·63	2·51	2·41	2·33	2·27	2·17	2·06	1·94	1·88	1·82	1·74	1·67	1·58	1·48
120	5·15	3·80	3·23	2·89	2·67	2·52	2·39	2·30	2·22	2·16	2·05	1·94	1·82	1·76	1·69	1·61	1·53	1·43	1·31
∞	5·02	3·69	3·12	2·79	2·57	2·41	2·29	2·19	2·11	2·05	1·94	1·83	1·71	1·64	1·57	1·48	1·39	1·27	1·00

Values of $F_{.01}$

Degrees of freedom for numerator

$A = 1\%$

ν_2 \ ν_1 (Degrees of freedom for denominator)	1	2	3	4	5	6	7	8	9	10	12	15	20	24	30	40	60	120	∞
1	4052	4999·5	5403	5625	5764	5859	5928	5981	6022	6056	6106	6157	6209	6235	6261	6287	6313	6339	6366
2	98·50	99·00	99·17	99·25	99·30	99·33	99·36	99·37	99·39	99·40	99·42	99·43	99·45	99·46	99·47	99·47	99·48	99·49	99·50
3	34·12	30·82	29·46	28·71	28·24	27·91	27·67	27·49	27·35	27·23	27·05	26·87	26·69	26·60	26·50	26·41	26·32	26·22	26·13
4	21·20	18·00	16·69	15·98	15·52	15·21	14·98	14·80	14·66	14·55	14·37	14·20	14·02	13·93	13·84	13·75	13·65	13·56	13·46
5	16·26	13·27	12·06	11·39	10·97	10·67	10·46	10·29	10·16	10·05	9·89	9·72	9·55	9·47	9·38	9·29	9·20	9·11	9·02
6	13·75	10·92	9·78	9·15	8·75	8·47	8·26	8·10	7·98	7·87	7·72	7·56	7·40	7·31	7·23	7·14	7·06	6·97	6·88
7	12·25	9·55	8·45	7·85	7·46	7·19	6·99	6·84	6·72	6·62	6·47	6·31	6·16	6·07	5·99	5·91	5·82	5·74	5·65
8	11·26	8·65	7·59	7·01	6·63	6·37	6·18	6·03	5·91	5·81	5·67	5·52	5·36	5·28	5·20	5·12	5·03	4·95	4·86
9	10·56	8·02	6·99	6·42	6·06	5·80	5·61	5·47	5·35	5·26	5·11	4·96	4·81	4·73	4·65	4·57	4·48	4·40	4·31
10	10·04	7·56	6·55	5·99	5·64	5·39	5·20	5·06	4·94	4·85	4·71	4·56	4·41	4·33	4·25	4·17	4·08	4·00	3·91
11	9·65	7·21	6·22	5·67	5·32	5·07	4·89	4·74	4·63	4·54	4·40	4·25	4·10	4·02	3·94	3·86	3·78	3·69	3·60
12	9·33	6·93	5·95	5·41	5·06	4·82	4·64	4·50	4·39	4·30	4·16	4·01	3·86	3·78	3·70	3·62	3·54	3·45	3·36
13	9·07	6·70	5·74	5·21	4·86	4·62	4·44	4·30	4·19	4·10	3·96	3·82	3·66	3·59	3·51	3·43	3·34	3·25	3·17
14	8·86	6·51	5·56	5·04	4·69	4·46	4·28	4·14	4·03	3·94	3·80	3·66	3·51	3·43	3·35	3·27	3·18	3·09	3·00
15	8·68	6·36	5·42	4·89	4·56	4·32	4·14	4·00	3·89	3·80	3·67	3·52	3·37	3·29	3·21	3·13	3·05	2·96	2·87
16	8·53	6·23	5·29	4·77	4·44	4·20	4·03	3·89	3·78	3·69	3·55	3·41	3·26	3·18	3·10	3·02	2·93	2·84	2·75
17	8·40	6·11	5·18	4·67	4·34	4·10	3·93	3·79	3·68	3·59	3·46	3·31	3·16	3·08	3·00	2·92	2·83	2·75	2·65
18	8·29	6·01	5·09	4·58	4·25	4·01	3·84	3·71	3·60	3·51	3·37	3·23	3·08	3·00	2·92	2·84	2·75	2·66	2·57
19	8·18	5·93	5·01	4·50	4·17	3·94	3·77	3·63	3·52	3·43	3·30	3·15	3·00	2·92	2·84	2·76	2·67	2·58	2·49
20	8·10	5·85	4·94	4·43	4·10	3·87	3·70	3·56	3·46	3·37	3·23	3·09	2·94	2·86	2·78	2·69	2·61	2·52	2·42
21	8·02	5·78	4·87	4·37	4·04	3·81	3·64	3·51	3·40	3·31	3·17	3·03	2·88	2·80	2·72	2·64	2·55	2·46	2·36
22	7·95	5·72	4·82	4·31	3·99	3·76	3·59	3·45	3·35	3·26	3·12	2·98	2·83	2·75	2·67	2·58	2·50	2·40	2·31
23	7·88	5·66	4·76	4·26	3·94	3·71	3·54	3·41	3·30	3·21	3·07	2·93	2·78	2·70	2·62	2·54	2·45	2·35	2·26
24	7·82	5·61	4·72	4·22	3·90	3·67	3·50	3·36	3·26	3·17	3·03	2·89	2·74	2·66	2·58	2·49	2·40	2·31	2·21
25	7·77	5·57	4·68	4·18	3·85	3·63	3·46	3·32	3·22	3·13	2·99	2·85	2·70	2·62	2·54	2·45	2·36	2·27	2·17
26	7·72	5·53	4·64	4·14	3·82	3·59	3·42	3·29	3·18	3·09	2·96	2·81	2·66	2·58	2·50	2·42	2·33	2·23	2·13
27	7·68	5·49	4·60	4·11	3·78	3·56	3·39	3·26	3·15	3·06	2·93	2·78	2·63	2·55	2·47	2·38	2·29	2·20	2·10
28	7·64	5·45	4·57	4·07	3·75	3·53	3·36	3·23	3·12	3·03	2·90	2·75	2·60	2·52	2·44	2·35	2·26	2·17	2·06
29	7·60	5·42	4·54	4·04	3·73	3·50	3·33	3·20	3·09	3·00	2·87	2·73	2·57	2·49	2·41	2·33	2·23	2·14	2·03
30	7·56	5·39	4·51	4·02	3·70	3·47	3·30	3·17	3·07	2·98	2·84	2·70	2·55	2·47	2·39	2·30	2·21	2·11	2·01
40	7·31	5·18	4·31	3·83	3·51	3·29	3·12	2·99	2·89	2·80	2·66	2·52	2·37	2·29	2·20	2·11	2·02	1·92	1·80
60	7·08	4·98	4·13	3·65	3·34	3·12	2·95	2·82	2·72	2·63	2·50	2·35	2·20	2·12	2·03	1·94	1·84	1·73	1·60
120	6·85	4·79	3·95	3·48	3·17	2·96	2·79	2·66	2·56	2·47	2·34	2·19	2·03	1·95	1·86	1·76	1·66	1·53	1·38
∞	6·63	4·61	3·78	3·32	3·02	2·80	2·64	2·51	2·41	2·32	2·18	2·04	1·88	1·79	1·70	1·59	1·47	1·32	1·00

Source: Pearson, E. S., & Hartley, H. O. (Eds.). *Biometrika Tables for Statisticians.* Volume 1 (3rd ed., reprinted with corrections). London: Biometrika Trust, 1976. Table 18. With permission of the Biometrika Trustees.

Appendix 7 FISHER'S $z(r)$ TRANSFORMATION

$$z(r) = \frac{1}{2} \ln \frac{1+r}{1-r} \qquad r = \frac{e^{2z(r)} - 1}{e^{2z(r)} + 1}$$

r	$z(r)$	r	$z(r)$	r	$z(r)$	r	$z(r)$	r	$z(r)$	r	$z(r)$
.000	.000	.500	.549	.750	.973	.850	1.256	.900	1.472	.950	1.832
.010	.010	.505	.556	.752	.978	.851	1.260	.901	1.478	.951	1.842
.020	.020	.510	.563	.754	.982	.852	1.263	.902	1.483	.952	1.853
.030	.030	.515	.570	.756	.987	.853	1.267	.903	1.488	.953	1.863
.040	.040	.520	.576	.758	.991	.854	1.271	.904	1.494	.954	1.874
.050	.050	.525	.583	.760	.996	.855	1.274	.905	1.499	.955	1.886
.060	.060	.530	.590	.762	1.001	.856	1.278	.906	1.505	.956	1.897
.070	.070	.535	.597	.764	1.006	.857	1.282	.907	1.510	.957	1.909
.080	.080	.540	.604	.766	1.011	.858	1.286	.908	1.516	.958	1.921
.090	.090	.545	.611	.768	1.015	.859	1.290	.909	1.522	.959	1.933
.100	.100	.550	.618	.770	1.020	.860	1.293	.910	1.528	.960	1.946
.110	.110	.555	.626	.772	1.025	.861	1.297	.911	1.533	.961	1.959
.120	.121	.560	.633	.774	1.030	.862	1.301	.912	1.539	.962	1.972
.130	.131	.565	.640	.776	1.035	.863	1.305	.913	1.545	.963	1.986
.140	.141	.570	.648	.778	1.040	.864	1.309	.914	1.551	.964	2.000
.150	.151	.575	.655	.780	1.045	.865	1.313	.915	1.557	.965	2.014
.160	.161	.580	.662	.782	1.050	.866	1.317	.916	1.564	.966	2.029
.170	.172	.585	.670	.784	1.056	.867	1.321	.917	1.570	.967	2.044
.180	.182	.590	.678	.786	1.061	.868	1.325	.918	1.576	.968	2.060
.190	.192	.595	.685	.788	1.066	.869	1.329	.919	1.583	.969	2.076
.200	.203	.600	.693	.790	1.071	.870	1.333	.920	1.589	.970	2.092
.210	.213	.605	.701	.792	1.077	.871	1.337	.921	1.596	.971	2.110
.220	.224	.610	.709	.794	1.082	.872	1.341	.922	1.602	.972	2.127
.230	.234	.615	.717	.796	1.088	.873	1.346	.923	1.609	.973	2.146
.240	.245	.620	.725	.798	1.093	.874	1.350	.924	1.616	.974	2.165
.250	.255	.625	.733	.800	1.099	.875	1.354	.925	1.623	.975	2.185
.260	.266	.630	.741	.802	1.104	.876	1.358	.926	1.630	.976	2.205
.270	.277	.635	.750	.804	1.110	.877	1.363	.927	1.637	.977	2.227
.280	.288	.640	.758	.806	1.116	.878	1.367	.928	1.644	.978	2.249
.290	.299	.645	.767	.808	1.121	.879	1.371	.929	1.651	.979	2.273
.300	.310	.650	.775	.810	1.127	.880	1.376	.930	1.658	.980	2.298
.310	.321	.655	.784	.812	1.133	.881	1.380	.931	1.666	.981	2.323
.320	.332	.660	.793	.814	1.139	.882	1.385	.932	1.673	.982	2.351
.330	.343	.665	.802	.816	1.145	.883	1.389	.933	1.681	.983	2.380
.340	.354	.670	.811	.818	1.151	.884	1.394	.934	1.689	.984	2.410
.350	.365	.675	.820	.820	1.157	.885	1.398	.935	1.697	.985	2.443
.360	.377	.680	.829	.822	1.163	.886	1.403	.936	1.705	.986	2.477
.370	.388	.685	.838	.824	1.169	.887	1.408	.937	1.713	.987	2.515
.380	.400	.690	.848	.826	1.175	.888	1.412	.938	1.721	.988	2.555
.390	.412	.695	.858	.828	1.182	.889	1.417	.939	1.730	.989	2.599
.400	.424	.700	.867	.830	1.188	.890	1.422	.940	1.738	.990	2.647
.410	.436	.705	.877	.832	1.195	.891	1.427	.941	1.747	.991	2.700
.420	.448	.710	.887	.834	1.201	.892	1.432	.942	1.756	.992	2.759
.430	.460	.715	.897	.836	1.208	.893	1.437	.943	1.764	.993	2.826
.440	.472	.720	.908	.838	1.214	.894	1.442	.944	1.774	.994	2.903
.450	.485	.725	.918	.840	1.221	.895	1.447	.945	1.783	.995	2.994
.460	.497	.730	.929	.842	1.228	.896	1.452	.946	1.792	.996	3.106
.470	.510	.735	.940	.844	1.235	.897	1.457	.947	1.802	.997	3.250
.480	.523	.740	.950	.846	1.242	.898	1.462	.948	1.812	.998	3.453
.490	.536	.745	.962	.848	1.249	.899	1.467	.949	1.822	.999	3.800

Appendix 8 CRITICAL VALUES OF THE DISTRIBUTIONS OF r AND r_S WHEN $\rho = 0$

Main entries are critical values of Pearson's product-moment correlation coefficient r, under the assumption that at least one of the variables is normally distributed (see Section 13.1). The main entries are taken from Table VII of Fisher & Yates, *Statistical Tables for Biological, Agricultural and Medical Research,* published by Longman Group Ltd., London (previously published by Oliver & Boyd Ltd., Edinburgh), by permission of the authors and publishers.

Values in parentheses are critical values of Spearman's rank correlation coefficient r_S for small samples (see Section 16.2). For df > 8, the critical values of r may be used to approximate those of r_S. The parenthetical entries are taken from Table A11(ii) of Snedecor & Cochran, *Statistical Methods,* published by Iowa State University Press, by permission of the publisher.

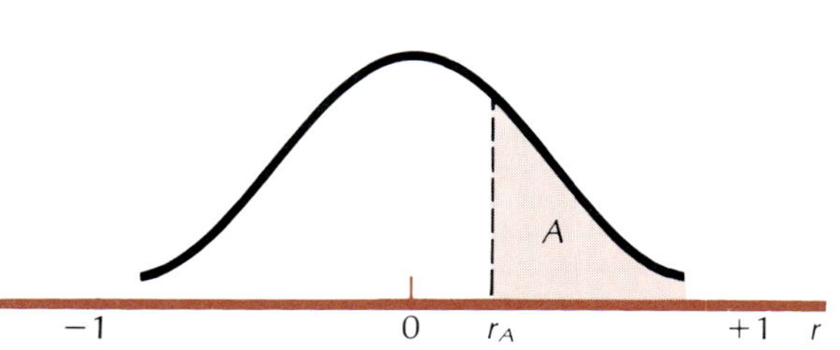

For a simple correlation coefficient,

$$\text{df} = n - 2.$$

For a partial correlation coefficient with c variables held constant,

$$\text{df} = n - c - 2.$$

The tabulated critical value of r for any given number of degrees of freedom (df) may be calculated from Student's t distribution (Appendix 4) according to the formula

$$r_A(\text{df}) = [t_A(\text{df})]/\sqrt{[t_A(\text{df})]^2 + \text{df}}.$$

df	$r_{.05}$	$r_{.025}$	$r_{.01}$	$r_{.005}$
1	.988(none)	.997(none)	.9995(none)	.9999(none)
2	.900(1.00)	.950(none)	.980(none)	.990(none)
3	.805(.900)	.878(1.00)	.934(1.00)	.959(none)
4	.729(.771)	.811(.886)	.882(.943)	.917(1.00)
5	.669(.714)	.754(.786)	.833(.892)	.874(.929)
6	.622(.643)	.707(.738)	.789(.810)	.834(.857)
7	.582(.600)	.666(.683)	.750(.783)	.798(.817)
8	.549(.564)	.632(.648)	.716(.733)	.765(.781)
9	.521	.602	.685	.735
10	.497	.576	.658	.708
11	.476	.553	.634	.684
12	.458	.532	.612	.661
13	.441	.514	.592	.641
14	.426	.497	.574	.623
15	.412	.482	.558	.606
16	.400	.468	.542	.590
17	.389	.456	.528	.575
18	.378	.444	.516	.561
19	.369	.433	.503	.549
20	.360	.423	.492	.537
21	.352	.413	.482	.526
22	.344	.404	.472	.515
23	.337	.396	.462	.505
24	.330	.388	.453	.496
25	.323	.381	.445	.487
26	.317	.374	.437	.479
27	.311	.367	.430	.471
28	.306	.361	.423	.463
29	.301	.355	.416	.456
30	.296	.349	.409	.449
35	.275	.325	.381	.418
40	.257	.304	.358	.393
45	.243	.288	.338	.372
50	.231	.273	.322	.354
60	.211	.250	.295	.325
70	.195	.232	.274	.303
80	.183	.217	.256	.283
90	.173	.205	.242	.267
100	.164	.195	.230	.254

Appendix 9 VALUES OF e^a

The base of the natural system of logarithms: $e = 2.7182818284...$

a	e^a	a	e^a	a	e^a	a	e^a
0	1.000	-2.5	.0821	-5.0	.00674	-7.5	.000553
-.1	.905	-2.6	.0743	-5.1	.00610	-7.6	.000500
-.2	.819	-2.7	.0672	-5.2	.00552	-7.7	.000453
-.3	.741	-2.8	.0608	-5.3	.00499	-7.8	.000410
-.4	.670	-2.9	.0550	-5.4	.00452	-7.9	.000371
-.5	.607	-3.0	.0498	-5.5	.00409	-8.0	.000335
-.6	.549	-3.1	.0450	-5.6	.00370	-8.1	.000304
-.7	.497	-3.2	.0408	-5.7	.00335	-8.2	.000275
-.8	.449	-3.3	.0369	-5.8	.00303	-8.3	.000249
-.9	.407	-3.4	.0334	-5.9	.00274	-8.4	.000225
-1.0	.368	-3.5	.0302	-6.0	.00248	-8.5	.000203
-1.1	.333	-3.6	.0273	-6.1	.00224	-8.6	.000184
-1.2	.301	-3.7	.0247	-6.2	.00203	-8.7	.000167
-1.3	.273	-3.8	.0224	-6.3	.00184	-8.8	.000151
-1.4	.247	-3.9	.0202	-6.4	.00166	-8.9	.000136
-1.5	.223	-4.0	.0183	-6.5	.00150	-9.0	.000123
-1.6	.202	-4.1	.0166	-6.6	.00136	-9.1	.000112
-1.7	.183	-4.2	.0150	-6.7	.00123	-9.2	.000101
-1.8	.165	-4.3	.0136	-6.8	.00111	-9.3	.000091
-1.9	.150	-4.4	.0123	-6.9	.00101	-9.4	.000083
-2.0	.135	-4.5	.0111	-7.0	.00091	-9.5	.000075
-2.1	.122	-4.6	.0101	-7.1	.00083	-9.6	.000068
-2.2	.111	-4.7	.0091	-7.2	.00075	-9.7	.000061
-2.3	.100	-4.8	.0082	-7.3	.00068	-9.8	.000055
-2.4	.091	-4.9	.0074	-7.4	.00061	-9.9	.000050

Appendix 10 COEFFICIENTS OF ORTHOGONAL POLYNOMIAL COMPARISONS

The entries in this table are taken from Table XXIII of Fisher & Yates, *Statistical Tables for Biological, Agricultural and Medical Research,* published by Longman Group Ltd., London (previously published by Oliver and Boyd Ltd., Edinburgh), by permission of the authors and publishers.

k	Polynomial	Coefficients										Σc_j^2
3	Linear	−1	0	1								2
	Quadratic	1	−2	1								6
4	Linear	−3	−1	1	3							20
	Quadratic	1	−1	−1	1							4
	Cubic	−1	3	−3	1							20
5	Linear	−2	−1	0	1	2						10
	Quadratic	2	−1	−2	−1	2						14
	Cubic	−1	2	0	−2	1						10
	Quartic	1	−4	6	−4	1						70
6	Linear	−5	−3	−1	1	3	5					70
	Quadratic	5	−1	−4	−4	−1	5					84
	Cubic	−5	7	4	−4	−7	5					180
	Quartic	1	−3	2	2	−3	1					28
	Quintic	−1	5	−10	10	−5	1					252
7	Linear	−3	−2	−1	0	1	2	3				28
	Quadratic	5	0	−3	−4	−3	0	5				84
	Cubic	−1	1	1	0	−1	−1	1				6
	Quartic	3	−7	1	6	1	−7	3				154
	Quintic	−1	4	−5	0	5	−4	1				84
8	Linear	−7	−5	−3	−1	1	3	5	7			168
	Quadratic	7	1	−3	−5	−5	−3	1	7			168
	Cubic	−7	5	7	3	−3	−7	−5	7			264
	Quartic	7	−13	−3	9	9	−3	−13	7			616
	Quintic	−7	23	−17	−15	15	17	−23	7			2184
9	Linear	−4	−3	−2	−1	0	1	2	3	4		60
	Quadratic	28	7	−8	−17	−20	−17	−8	7	28		2772
	Cubic	−14	7	13	9	0	−9	−13	−7	14		990
	Quartic	14	−21	−11	9	18	9	−11	−21	14		2002
	Quintic	−4	11	−4	−9	0	9	4	−11	4		468
10	Linear	−9	−7	−5	−3	−1	1	3	5	7	9	330
	Quadratic	6	2	−1	−3	−4	−4	−3	−1	2	6	132
	Cubic	−42	14	35	31	12	−12	−31	−35	−14	42	8580
	Quartic	18	−22	−17	3	18	18	3	−17	−22	18	2860
	Quintic	−6	14	−1	−11	−6	6	11	1	−14	6	780

Appendix 11 CRITICAL VALUES OF THE STUDENTIZED RANGE STATISTIC Q

df_{Error}	α	k (number of means in set) 2	3	4	5	6	7	8	9	10	11	12	13	14	15	16	17	18
1	.10	8.93	13.4	16.4	18.5	20.2	21.5	22.6	23.6	24.5	25.2	25.9	26.5	27.1	27.6	28.1	28.5	29.0
	.05	18.0	27.0	32.8	37.1	40.4	43.1	45.4	47.4	49.1	50.6	52.0	53.2	54.3	55.4	56.3	57.2	58.0
	.01	90.0	13.5	164	186	202	216	227	237	246	253	260	266	272	277	282	286	290
2	.10	4.13	5.73	6.78	7.54	8.14	8.63	9.05	9.41	9.73	10.0	10.3	10.5	10.7	10.9	11.1	11.2	11.4
	.05	6.09	8.3	9.8	10.9	11.7	12.4	13.0	13.5	14.0	14.4	14.7	15.1	15.4	15.7	15.9	16.1	16.4
	.01	14.0	19.0	22.3	24.7	26.6	28.2	29.5	30.7	31.7	32.6	33.4	34.1	34.8	35.4	36.0	36.5	37.0
3	.10	3.33	4.47	5.20	5.74	6.16	6.51	6.81	7.06	7.29	7.49	7.67	7.83	7.98	8.12	8.25	8.37	8.78
	.05	4.50	5.91	6.82	7.50	8.04	8.48	8.85	9.18	9.46	9.72	9.95	10.2	10.4	10.5	10.7	10.8	11.0
	.01	8.26	10.6	12.2	13.3	14.2	15.0	15.6	16.2	16.7	17.1	17.5	17.9	18.2	18.5	18.8	19.1	19.3
4	.10	3.01	3.98	4.59	5.04	5.39	5.69	5.93	6.14	6.33	6.50	6.65	6.78	6.91	7.03	7.13	7.23	7.33
	.05	3.93	5.04	5.76	6.29	6.71	7.05	7.35	7.60	7.83	8.03	8.21	8.37	8.52	8.66	8.79	8.91	9.03
	.01	6.51	8.12	9.17	9.96	10.6	11.1	11.5	11.9	12.3	12.6	12.8	13.1	13.3	13.5	13.7	13.9	14.1
5	.10	2.85	3.72	4.26	4.66	4.98	5.24	5.44	5.65	5.82	5.97	6.10	6.22	6.34	6.44	6.54	6.63	6.71
	.05	3.64	4.60	5.22	5.67	6.03	6.33	6.58	6.80	6.99	7.17	7.32	7.47	7.60	7.72	7.83	7.93	8.03
	.01	5.70	6.97	7.80	8.42	8.91	9.32	9.67	9.97	10.2	10.5	10.7	10.9	11.1	11.2	11.4	11.6	11.7
6	.10	2.75	3.56	4.07	4.44	4.73	4.97	5.17	5.34	5.50	5.64	5.76	5.88	5.98	6.08	6.16	6.25	6.33
	.05	3.46	4.34	4.90	5.31	5.63	5.89	6.12	6.32	6.49	6.65	6.79	6.92	7.03	7.14	7.24	7.34	7.43
	.01	5.24	6.33	7.03	7.56	7.97	8.32	8.61	8.87	9.10	9.30	9.49	9.65	9.81	9.95	10.1	10.2	10.3
7	.10	2.68	3.45	3.93	4.28	4.56	4.78	4.97	5.14	5.28	5.41	5.53	5.64	5.74	5.83	5.91	5.99	6.06
	.05	3.34	4.16	4.69	5.06	5.36	5.61	5.82	6.00	6.16	6.30	6.43	6.55	6.66	6.76	6.85	6.94	7.02
	.01	4.95	5.92	6.54	7.01	7.37	7.68	7.94	8.17	8.37	8.55	8.71	8.86	9.00	9.12	9.24	9.35	9.46
8	.10	2.63	3.37	3.83	4.17	4.43	4.65	4.83	4.99	5.13	5.25	5.36	5.46	5.56	5.64	5.74	5.83	5.87
	.05	3.26	4.04	4.53	4.89	5.17	5.40	5.60	5.77	5.92	6.05	6.18	6.29	6.39	6.48	6.57	6.65	6.73
	.01	4.74	5.63	6.20	6.63	6.96	7.24	7.47	7.68	7.78	8.03	8.18	8.31	8.44	8.55	8.66	8.76	8.85
9	.10	2.59	3.32	3.76	4.08	4.34	4.55	4.72	4.87	5.01	5.13	5.23	5.33	5.42	5.51	5.58	5.66	5.72
	.05	3.20	3.95	4.42	4.76	5.02	5.24	5.43	5.60	5.74	5.87	5.98	6.09	6.19	6.28	6.36	6.44	6.51
	.01	4.60	5.43	5.96	6.35	6.66	6.91	7.13	7.32	7.49	7.65	7.78	7.91	8.03	8.13	8.23	8.33	8.41
10	.10	2.56	3.28	3.70	4.02	4.26	4.47	4.64	4.78	4.91	5.03	5.13	5.23	5.32	5.40	5.47	5.54	5.61
	.05	3.15	3.88	4.33	4.65	4.91	5.12	5.30	5.46	5.60	5.72	5.83	5.93	6.03	6.11	6.19	6.27	6.34
	.01	4.48	5.27	5.77	6.14	6.43	6.67	6.87	7.05	7.21	7.36	7.48	7.60	7.71	7.81	7.91	8.00	8.08
11	.10	2.54	3.23	3.66	3.97	4.21	4.40	4.57	4.71	4.84	4.95	5.05	5.15	5.23	5.31	5.38	5.45	5.51
	.05	3.11	3.82	4.26	4.57	4.82	5.03	5.20	5.35	5.49	5.61	5.71	5.81	5.90	5.99	6.06	6.18	6.20
	.01	4.39	5.14	5.62	5.97	6.25	6.48	6.67	6.84	6.99	7.13	7.26	7.36	7.46	7.56	7.65	7.73	7.81

12	.10	2.52	3.20	3.62	3.92	4.16	4.35	4.51	4.65	4.78	4.89	4.99	5.08	5.16	5.24	5.31	5.37	5.44
	.05	3.08	3.77	4.20	4.51	4.75	4.95	5.12	5.27	5.40	5.51	5.62	5.71	5.80	5.88	5.95	6.02	6.09
	.01	4.32	5.04	5.50	5.84	6.10	6.32	6.51	6.67	6.81	6.94	7.06	7.17	7.26	7.36	7.44	7.52	7.50
13	.10	2.51	3.18	3.59	3.89	4.12	4.31	4.46	4.60	4.72	4.83	4.93	5.02	5.10	5.18	5.25	5.31	5.37
	.05	3.06	3.73	4.15	4.45	4.69	4.88	5.05	5.19	5.32	5.43	5.53	5.63	5.71	5.79	5.86	5.93	6.00
	.01	4.26	4.96	5.40	5.73	5.98	6.19	6.37	6.53	6.67	6.79	6.90	7.01	7.10	7.19	7.27	7.37	7.42
14	.10	2.49	3.16	3.56	3.83	4.08	4.27	4.42	4.56	4.68	4.79	4.88	4.97	5.05	5.12	5.19	5.26	5.32
	.05	3.03	3.70	4.11	4.41	4.64	4.83	4.99	5.13	5.25	5.36	5.46	5.55	5.64	5.72	5.79	5.85	5.92
	.01	4.21	4.89	5.32	5.63	5.88	6.08	6.26	6.41	6.54	6.66	6.77	6.87	6.96	7.05	7.13	7.20	7.27
16	.10	2.47	3.12	3.52	3.80	4.03	4.21	4.36	4.49	4.61	4.71	4.81	4.89	4.97	5.04	5.11	5.17	5.23
	.05	3.00	3.65	4.05	4.33	4.56	4.74	4.90	5.03	5.15	5.26	5.35	5.44	5.52	5.59	5.66	5.73	5.79
	.01	4.13	4.78	5.19	5.49	5.72	5.92	6.08	6.22	6.35	6.46	6.56	6.66	6.74	6.82	6.90	6.97	7.03
18	.10	2.45	3.10	3.49	3.77	3.98	4.16	4.31	4.44	4.55	4.66	4.75	4.83	4.91	4.98	5.04	5.10	5.16
	.05	2.97	3.61	4.00	4.28	4.49	4.67	4.82	4.96	5.07	5.17	5.27	5.35	5.43	5.50	5.57	5.63	5.69
	.01	4.07	4.70	5.09	5.38	5.60	5.79	5.94	6.08	6.20	6.31	6.41	6.50	6.58	6.65	6.73	6.79	6.85
20	.10	2.44	3.08	3.46	3.74	3.95	4.12	4.27	4.40	4.51	4.61	4.70	4.78	4.86	4.92	4.99	5.05	5.10
	.05	2.95	3.58	3.96	4.23	4.45	4.62	4.77	4.90	5.01	5.11	5.20	5.28	5.36	5.43	5.49	5.55	5.61
	.01	4.02	4.64	5.02	5.29	5.51	5.69	5.84	5.97	6.09	6.19	6.29	6.37	6.45	6.52	6.59	6.65	6.71
24	.10	2.42	3.05	3.42	3.69	3.90	4.07	4.21	4.34	4.45	4.54	4.63	4.71	4.78	4.85	4.91	4.97	5.02
	.05	2.92	3.53	3.90	4.17	4.37	4.54	4.68	4.81	4.92	5.01	5.10	5.18	5.25	5.32	5.38	5.44	5.49
	.01	3.96	4.54	4.91	5.17	5.37	5.54	5.69	5.81	5.92	6.02	6.11	6.19	6.26	6.33	6.39	6.45	6.51
30	.10	2.40	3.02	3.39	3.65	3.85	4.02	4.16	4.28	4.38	4.47	4.56	4.64	4.71	4.77	4.83	4.89	4.94
	.05	2.89	3.49	3.84	4.10	4.30	4.46	4.60	4.72	4.83	4.92	5.00	5.08	5.15	5.21	5.27	5.33	5.38
	.01	3.89	4.45	4.80	5.05	5.24	5.40	5.54	5.56	5.76	5.85	5.93	6.01	6.08	6.14	6.20	6.26	6.31
40	.10	2.38	2.99	3.35	3.61	3.80	3.96	4.10	4.22	4.32	4.41	4.49	4.56	4.63	4.70	4.75	4.81	4.86
	.05	2.86	3.44	3.79	4.04	4.23	4.39	4.52	4.63	4.74	4.82	4.91	4.98	5.05	5.11	5.16	5.22	5.27
	.01	3.82	4.37	4.70	4.93	5.11	5.27	5.39	5.50	5.60	5.69	5.77	5.84	5.90	5.96	6.02	6.07	6.11
60	.10	2.36	2.96	3.31	3.56	3.76	3.91	4.04	4.16	4.26	4.34	4.42	4.49	4.56	4.62	4.68	4.73	4.78
	.05	2.83	3.40	3.74	3.98	4.16	4.31	4.44	4.55	4.65	4.73	4.81	4.88	4.94	5.00	5.06	5.11	5.15
	.01	3.76	4.28	4.60	4.82	4.99	5.13	5.25	5.36	5.45	5.53	5.60	5.67	5.73	5.79	5.84	5.89	5.93
120	.10	2.34	2.93	3.28	3.52	3.71	3.86	3.99	4.10	4.19	4.28	4.35	4.42	4.49	4.59	4.60	4.65	4.69
	.05	2.80	3.36	3.69	3.92	4.10	4.24	4.36	4.48	4.56	4.64	4.72	4.78	4.84	4.90	4.95	5.00	5.04
	.01	3.70	4.20	4.50	4.71	4.87	5.01	5.12	5.21	5.30	5.38	5.44	5.51	5.56	5.61	5.66	5.71	5.75
∞	.10	2.33	2.90	3.24	3.48	3.66	3.81	3.93	4.04	4.13	4.21	4.29	4.35	4.41	4.47	4.52	4.57	4.61
	.05	2.77	3.31	3.63	3.86	4.03	4.17	4.29	4.39	4.47	4.55	4.62	4.68	4.74	4.80	4.85	4.89	4.93
	.01	3.64	4.12	4.40	4.60	4.76	4.88	4.99	5.08	5.16	5.23	5.29	5.35	5.40	5.45	5.49	5.54	5.57

Source: Pearson, E. S., & Hartley, H. O. (Eds.). *Biometrika Tables for Statisticians.* Volume 1 (3rd ed., reprinted with corrections). London: Biometrika Trust, 1976. Table 29. With permission of the Biometrika Trustees.

Selected bibliography: some useful references

American Statistician, The (published four times a year by the American Statistical Association).

Cleveland, W. S., & McGill, R. Graphical perception and graphical methods for analyzing scientific data. *Science,* 30 August 1985, *229*, 828–833.

Cochran, W. G. *Sampling Techniques* (3rd ed.). New York: Wiley, 1977.

Cochran, W. G., & Cox, G. M. *Experimental Designs* (2nd ed.). New York: Wiley, 1957.

Cox, D. R. *Planning of Experiments.* New York: Wiley, 1958.

Deming, W. E. *Sample Design in Business Research.* New York: Wiley, 1960.

Draper, N. R., & Smith, H. *Applied Regression Analysis* (2nd ed.). New York: Wiley, 1981.

Ehrenberg, A. S. C. The problem of numeracy. *American Statistician,* 1981, *35,* 67–71.

Feller, W. *An Introduction to Probability Theory and Its Applications* (Vol. 1, 3rd ed.). New York: Wiley, 1968.

Feller, W. *An Introduction to Probability Theory and Its Applications* (Vol. 2, 2nd ed.). New York: Wiley, 1971.

Goldstone, L. A. The pie chart: A piece of cake. *Teaching Statistics,* 1982, *4,* 76–79.

Hays, W. L. *Statistics* (3rd ed.). New York: Holt, Rinehart and Winston, 1981.

Hunter, W. G. Some ideas about teaching design of experiments, with 2^5 examples of experiments conducted by students. *The American Statistician,* 1977, *31,* 12–17.

Kendall, M. G. *Rank Correlation Methods* (4th ed.). London: Griffin, 1970.

Kendall, M. G., & Buckland, W. R. *A Dictionary of Statistical Terms* (4th ed.). London: Longman, 1982.

Kendall, M. G., & Stuart, A. *The Advanced Theory of Statistics.* Vol. 1. *Distribution Theory* (4th ed.). New York: Hafner, 1977.

Kendall, M. G., & Stuart, A. *The Advanced Theory of Statistics.* Vol. 2. *Inference and Relationship* (4th ed.). New York: Hafner, 1979.

Kendall, M. G., & Stuart, A. *The Advanced Theory of Statistics.* Vol. 3. *Design and Analysis, and Time-Series* (3rd ed.). New York: Hafner, 1976.

Keppel, G. *Design and Analysis: A Researcher's Handbook* (2nd ed.). Englewood Cliffs, N. J.: Prentice-Hall, 1982.

Kleinbaum, D. G., & Kupper, L. L. *Applied Regression Analysis and Other Multivariable Methods.* North Scituate, Mass.: Duxbury, 1978.

Kotz, S., & Johnson, N. L. (Eds.). *Encyclopedia of Statistical Sciences* (8 vols.). New York: Wiley, 1982–1985.

Kruskal, W. H., & Tanur, J. M. (Eds.). *International Encyclopedia of Statistics* (2 vols.). New York: Free Press, 1978.

Mosteller, F., & Rourke, R. E. K. *Sturdy Statistics: Nonparametrics and Order Statistics.* Reading, Mass.: Addison-Wesley, 1973.

Raiffa, H. *Decision Analysis: Introductory Lectures on Choices Under Uncertainty.* Reading, Mass.: Addison-Wesley, 1968.

Siegel, S. *Nonparametric Statistics for the Behavioral Sciences.* New York: McGraw-Hill, 1956.

Snedecor, G. W., & Cochran, W. G. *Statistical Methods* (7th ed.). Ames, Iowa: Iowa State University Press, 1980.

Teaching Statistics (published three times a year).

Tukey, J. W. *Exploratory Data Analysis.* Reading, Mass.: Addison-Wesley, 1977.

Winer, B. J. *Statistical Principles in Experimental Design* (2nd ed.). New York: McGraw-Hill, 1971.

Winkler, R. L. *An Introduction to Bayesian Inference and Decision.* New York: Holt, Rinehart and Winston, 1972.

Glossary

accuracy Lack of bias in an estimate; lack of systematic error; freedom from nonsampling errors. (Compare *precision.*)

additivity The property that is present when the effects of the various factors combine with each other and with random error by adding.

adjacent values The values of the most extreme observations within four quartile deviations below and above the midquartile.

alternative hypothesis A proposition that certain population values are different from each other, put forth as an alternative to the null hypothesis. (See also *directional alternative*; *nondirectional alternative.*)

analysis of variance (ANOVA) A general procedure for testing hypotheses about means; a partition of the total variation in a set of data into portions due to identifiable sources.

ANOVA summary table A table that shows, for each source of variation, the sum of squares, degrees of freedom, mean square, and F ratio where applicable.

a posteriori comparison See *postmortem comparison.*

applied statistics The aspect of statistics that focuses on the proper application of available statistical methods to answer questions in various substantive areas such as biology, business, and psychology; distinguished from *theoretical statistics.*

a priori comparison See *planned comparison.*

arithmetic mean The sum of the observations in a set, divided by the number of observations.

asymmetrical transfer effects Carry-over effects whereby Condition X has an effect on subsequent performance under Condition Y but Condition Y has little or no effect on subsequent performance under Condition X.

average Any single value that is, in some sense, most representative or typical of a set of observations. (See also *arithmetic mean*; *median*; *midquartile*; *midrange*; *mode*; *trimean.*)

bar chart A graphical representation of a nominal frequency distribution—namely, one in which frequencies are represented by bars of appropriate heights.

base period The reference point in time for an index number; distinguished from *given period.*

Bayesian inference Statistical inference in which parameters are treated as random variables, with probabilities assigned to various possible values of parameters; distinguished from *classical inference.*

Bayes' rule A formula for conditional probability that can be used to revise a prior probability in light of additional information to yield the posterior probability.

Bernoulli's theorem See *Law of Large Numbers.*

Bernoulli trial A trial having exactly two mutually exclusive outcomes with constant probabilities.

between-subjects design See *independent-samples design.*

between-within design A design having a repeated-measures factor and an independent-samples factor. (Compare *independent-samples design*; *repeated-measures design.*)

biased estimate of a parameter An estimate with a systematic tendency to be too large or too small; an estimate whose expected value is not equal to the parameter; distinguished from *unbiased estimate of a parameter.*

Bienaymé–Chebyshev inequality In any population or sample, no more than $1/z^2$ of the observations differ from the mean by z or more times the standard deviation, for $z > 1$.

bimodal distribution A distribution whose graph shows two peaks. (Compare *multimodal distribution.*)

binomial distribution The theoretical probability distribution of number of successes in a series of independent trials where the probability of success in each individual trial remains constant. (Compare *multinomial distribution*; *Poisson distribution.*)

binomial probability function A probability function that gives the probability of any specified number of successes in a given number of independent trials where the probability of success in each trial remains constant.

bivariate Pertaining to exactly two variables or, if independent variables and dependent variables are distinguished, exactly two dependent variables and any number of independent variables. (Compare *multivariate*; *univariate.*)

bivariate normal distribution A two-dimensional generalization of the univariate Gaussian distribution.

block A set of similar observational units such that each of the units in a block is associated with a different one of the levels of the independent variable; distinguished from *independent samples.* (See also *paired samples.*)

blocking The arrangement of observational units into relatively homogeneous sets, called blocks, for the purpose of increasing precision by administering every one of the treatments once in each block. (See also *block*; *paired samples.*)

boundary See *category boundary.*

box plot A graphical representation of a set of data, showing the quartiles, the adjacent values, and the outside values.

carry-over effects Any effects that appear in the results for one treatment but that are due to other treatments administered to the same subjects; e.g., practice effects, fatigue effects, asymmetrical transfer effects, range effects.

categorical data Records of the numbers of observational units observed in various categories; observations of categorized variables.

categorized variable A nominal variable or a quantitative variable whose values have been grouped into categories.

category boundary The hypothetical dividing point halfway between adjacent categories of a quantitative frequency distribution.

category interval The difference between successive category boundaries in a quantitative frequency distribution.

category interval of a distribution The common category interval in the case in which the category intervals are equal for all categories of a quantitative frequency distribution.

category limits, lower and upper The smallest and largest possible observations in each category of a quantitative frequency distribution.

category mark The midpoint of a category in a quantitative frequency distribution.

centiles Quantiles that divide a distribution into 100 segments, each containing 1% of the observations. (Compare *deciles*; *median*; *quartiles.*)

Central Limit Theorem For simple random samples that are sufficiently large, the sampling distribution of the mean is approximately Gaussian.

central location, measure of See *average.*

central tendency, measure of See *average.*

change of scale A transformation whereby each value of a variable is multiplied by the same constant. (Compare *linear transformation.*)

Chebyshev's rule See *Bienaymé–Chebyshev inequality.*

check count The number of leaves on a stem in a stem-and-leaf display, recorded to the right of the stem and used to check the total number of leaves.

chi-square distribution The theoretical sampling distribution of a Gaussian-population sum of squares divided by the population variance; the distribution of the sum of df independent squared standard normal deviates, where df is the number of degrees of freedom.

class A category in a frequency distribution.

classical inference Statistical inference based on the frequentistic interpretation of probability, whereby probabilities are attached to possible values of statistics but not to possible values of parameters; distinguished from *Bayesian inference.*

classical interpretation of probability The view that the probability of a given outcome is the number of equally likely outcomes of interest, divided by the total number of equally likely outcomes; distinguished from the *frequentistic interpretation of probability* and the *subjective interpretation of probability.*

classification variable An independent variable whose levels are distinguished in terms of characteristics already possessed by the observational units before the study begins; distinguished from *manipulated variable.*

coefficient See *comparison, coefficients of*; *correlation coefficient*; *determination, coefficient of*; *multiple correlation coefficient*; *multiple determination, coefficient of*; *partial correlation coefficient*; *partial determination, coefficient of*; *partial regression coefficient*; *regression coefficient.*

combinations of n items taken n_1 at a time, number of The number of ways of partitioning a set of n items into two sets, one having n_1 items and the other having the remaining items.

comparison A weighted sum of means, subject to the restriction that the weights must add up to zero; a difference between two weighted means; distinguished from *omnibus effect.* (See also *orthogonal comparisons*; *planned comparison*; *postmortem comparison.*)

comparison, coefficients of The weights applied to the means in a comparison.

complement of an event The remainder of the sample space, exclusive of the event.

complete crossing See *crossing.*

component of the treatment sum of squares Sum of squares due to a comparison. (See also *orthogonal components.*)

composite hypothesis An hypothesis that proposes an inequality; distinguished from *simple hypothesis.*

computational formula A formula that is algebraically equivalent to a definitional formula but is more convenient for computational purposes.

condition The occurrence of an event that defines the situation to which consideration is limited in conditional probability; the given event in conditional probability.

conditional probability The probability of an event given that consideration is limited to situations in which a certain other event—called the condition—also occurs.

confidence interval An interval estimate in classical inference.

confidence level The probability that a random sample will yield confidence limits that include the unknown parameter. (Compare *significance level.*)

confidence limits The endpoints of a confidence interval.

confounded Not independent; providing contaminated information about the items involved; distinguished from *orthogonal.*

confounding variable An extraneous variable that takes on different values for different levels of an independent variable; distinguished from *controlled variable.* (Compare *lurking variable.*)

conservative Having a reduced Type I error rate along with reduced power to detect true differences.

conservative degrees of freedom for testing repeated-measures factors The numbers of degrees of freedom that would have applied if there had been only two levels of each repeated-measures factor.

constant A quantity that does not vary in a particular problem.

contingency table A table showing the numbers of observational units classified in the various cells formed by cross-classification with respect to two or more variables.

continuity correction An adjustment that is often made to improve the approximation when a continuous distribution is used to approximate a discrete distribution.

continuous probability distribution A probability distribution in which the random variable is a continuous variable; distinguished from *discrete probability distribution.* (For examples, see *chi-square distribution*; *F distribution*; *Gaussian distribution*; *t distribution.*)

continuous variable A variable whose theoretically possible values include the infinite number of real numbers between two of its values; distinguished from *discrete variable.*

control group A group of observational units that are treated exactly the same as the experimental group, except without the experimental treatment of interest; distinguished from *experimental group.*

controlled variable An extraneous variable whose value is held constant for all conditions in the experiment; distinguished from *confounding variable.* (See also *double-blind procedure*; *placebo.* Compare *lurking variable.*)

convenience sampling Nonrandom sampling in which observational units that are more readily available are included in the sample and less convenient units have no chance of being included.

correlation coefficient See *Pearson product-moment correlation coefficient*; *Spearman's rank correlation coefficient.*

counted data Observations that consist of the number of items of a certain type for each observational unit.

covariance An average of the products of the deviations from the means of two variables.

critical value A value at the boundary between rejection and nonrejection regions in the sampling distribution of the test statistic under the null hypothesis.

crossing An arrangement of factors in which every level of one factor occurs in combination with every level of the other factor; distinguished from *nesting.* (See also *factorial design.*)

cross-product variable A predictor defined as the product of two other predictors and introduced to allow the inclusion of an interaction between the two predictors in the regression equation.

cubic Pertaining to a polynomial of degree 3, which is a 2-bend curve. (Compare *linear*; *quadratic*; *quartic*; *quintic.*)

cumulative distribution A distribution of a quantitative variable in which frequencies, percentages, or probabilities are cumulated either from low values of the variable to high or from high values to low. (See also *cumulative frequency distribution*; *cumulative percentage distribution*; *distribution function*; *"less than" distribution*; *"or more" distribution.*)

cumulative frequency distribution A quantitative frequency distribution in which the frequencies are cumulated either from low values of the variable to high or from high values to low.

cumulative percentage distribution A quantitative percentage distribution in which the percentages are cumulated either from low values of the variable to high or from high values to low.

data Facts; items of information.

deciles Quantiles that divide a distribution into ten segments, each containing 10% of the observations. (Compare *centiles*; *median*; *quartiles.*)

decimal places Digit positions to the right of the decimal point. (Compare *digit position*; *effective digits*; *significant figures.*)

decision theory, statistical A body of knowledge concerning methods for choosing the most appropriate action in light of the payoffs and the state-of-the-world probabilities.

definitional formula A formula that is a rather transparent symbolic statement of a definition but that is not necessarily convenient for computational purposes; distinguished from *computational formula.*

degree of confidence See *confidence level.*

degrees of freedom The number of values whose variation is being measured minus the number of independent sample-based restrictions on those values.

density The concentration of observations relative to an arbitrary reference interval on the relevant variable. (See also *frequency density*; *percentage frequency density*; *probability density function.*)

dependent variable An output variable, to be described as a function of one or more independent variables; a response variable.

descriptive statistics The aspect of statistics that deals with methods of characterizing or summarizing a given set of data; distinguished from *inferential statistics.*

design The aspect of inferential statistics that is concerned with the planning of experiments so that an efficient analysis to achieve the experiment's purpose is possible; also, the arrangement of factors in an experiment. (See also *between-within design*; *independent-samples design*; *repeated-measures design.*)

determination, coefficient of The proportion of the total variation in the dependent variable that can be accounted for by the linear relationship with the independent variable.

deviation from the mean The difference between an observation and the mean of the observations. (See also *mean deviation*; *quartile deviation*; *standard deviation.*)

difference score The difference between paired observations.

digit position The digit's position relative to the decimal point. (Compare *decimal places*; *effective digits*; *significant figures.*)

directional alternative An alternative hypothesis proposing a difference in a specified direction; distinguished from *nondirectional alternative.*

direct measurement Observation based on the actual experience of the data collector; distinguished from *verbal report.*

discrete probability distribution A probability distribution in which the random variable is a discrete variable; distinguished from *continuous probability distribution.* (For examples, see *binomial distribution*; *hypergeometric distribution*; *multinomial distribution*; *Poisson distribution.*)

discrete variable A variable such that each value of the variable on the real-number line is separated from the next larger or smaller value of the variable by real numbers that are not possible values of the variable; distinguished from *continuous variable.*

disjunctive event An event defined by the occurrence of one or both of two simple events; distinguished from *joint event* and *simple event.*

disjunctive probability The probability of a disjunctive event.

dispersion The spread of a set of observations along the real-number line. (See also *measure of dispersion.*)

distribution A description of the way in which a set of observations is distributed over the possible values of a variable. (For specific types of distributions, see *cumulative distribution*; *frequency distribution*; *percentage distribution*; *probability distribution.* For specific shapes of distributions, see *bimodal distribution*; *leptokurtic distribution*; *multimodal distribution*; *platykurtic distribution*; *reversed-J-shaped distribution*; *symmetric distribution*; *triangular distribution*; *uniform distribution*; *unimodal distribution*; *U-shaped distribution.*)

distribution-free methods See *nonparametric methods.*

distribution function A function giving the cumulative probabilities up to various values of the random variable; a cumulative "or less" probability function. (Compare *cumulative distribution.*)

double-blind procedure A control procedure whereby neither the subject nor the individual administering the treatment knows which treatment condition the subject is receiving.

dummy variable A contrived numerical variable devised to allow the inclusion of a qualitative variable as a predictor in a regression equation.

effective digits Digits that distinguish among the observations in a particular context. (Compare *decimal places*; *digit position*; *significant figures.*)

empirical Originating in observation or experience.

empirical probability distribution A probability distribution derived from actual observations; distinguished from *theoretical probability distribution.*

empirical rule See *rule of thumb for symmetric unimodal distributions.*

empirical sampling distribution A sampling distribution obtained by drawing repeated samples of the same size from the same population; distinguished from *theoretical sampling distribution.*

EMV criterion A criterion for decision making that favors the action having the highest expected monetary value.

error mean square See *error term.*

error of estimation See *residual.*

error rate See *experimentwise error rate*; *per-comparison error rate*; *Type I error*; *Type II error.*

error term The mean square that is the appropriate denominator of the F ratio in an analysis of variance.

error variance See *error term.*

error variation See *residual variation.*

estimate An approximate numerical description of a population, based on sample data and used when the exact value of the parameter is not known. (See also *biased estimate*; *interval estimate*; *point estimate*; *unbiased estimate.*)

estimation The aspect of inferential statistics that is concerned with specifying as precisely as possible the true value of an unknown parameter; distinguished from *significance testing.* (See also *interval estimate*; *point estimate.*)

EU criterion A criterion for decision making that favors the action having the highest expected utility.

EVPI See *expected value of perfect information.*

EVSI See *expected value of sample information.*

exclusive See *mutually exclusive.*

exhaustive Including all possibilities.

expectation See *expected value.*

expected frequency The mathematically expected number of observations in a given category under the null hypothesis; distinguished from *observed frequency.*

expected mean square The expected value of a mean square; the parameter estimated by a mean square.

expected utility with perfect information The expected utility that would accrue if the decision maker had perfect information as to when each state of the world would occur.

expected value The average value that would be obtained in the long run.

expected value of perfect information The difference between the expected utility with perfect information and the expected utility of the action that would be chosen on the basis of current information.

expected value of sample information The difference between the expected utility with additional imperfect information and the expected utility of the action that would be chosen on the basis of current information.

experimental design See *design.*

experimental group A group of observational units that receive an experimental treatment of interest; distinguished from *control group.*

experimentwise error rate The Type I error rate considered as the number of experiments exhibiting one or more Type I errors divided by the total number of experiments; distinguished from *per-comparison error rate.*

extraneous variable Any variable that is neither a dependent variable nor a factor in the experiment. (See also *confounding variable*; *controlled variable*; *lurking variable.*)

extra sum of squares The additional sum of squares that can be accounted for by the inclusion of a given predictor, over and above the sum of squares accounted for by the other predictors already in the regression equation.

extremes The minimum and the maximum.

factor Any basis, within the design of an experiment, for distinguishing one observation from another.

factorial design An arrangement of two or more factors that are completely crossed.

failure One of the two mutually exclusive and exhaustive categories of outcome in a Bernoulli trial; the other is *success.*

fair bet A bet whose payoffs are set so that the expected value of the payoff is zero.

***F* distribution** The sampling distribution of the ratio of two independent estimates of the same Gaussian-population variance; the distribution of the ratio of two independent chi-square variates, each divided by its degrees of freedom.

finite-population correction factor An adjustment that should be made in certain formulas for standard errors when the sampling fraction is about 10% or more.

Fisher's $z(r)$ transformation A transformation that converts correlation coefficients into values with an approximately Gaussian sampling distribution.

fitted value The described value of the dependent variable, as given by the regression equation.

fitting by eye A subjective procedure for obtaining a regression line by selecting the line that looks most appropriate when drawn on the scatter diagram.

fixed factor A factor whose levels to be included in the experiment are chosen because of their intrinsic relevance to the experiment; distinguished from *random factor.*

forecasting The use of a regression equation for genuine prediction, as distinguished from estimation or description of a population.

fractile See *quantile.*

***F* ratio** The test statistic in F tests, calculated as the ratio of two variance estimates.

frequency Number of observations. (See also *expected frequency*; *observed frequency.*)

frequency density The number of observations per reference interval. (Compare *probability density function.*)

frequency distribution A description of the frequencies with which various mutually exclusive and exhaustive categories of observation occur. (See also *cumulative frequency distribution*; *nominal frequency distribution*; *quantitative frequency distribution.* Compare *probability distribution.*)

frequency polygon A graphical representation of a quantitative frequency distribution—namely, one in which frequencies are represented by dots joined by straight lines.

frequency table Tabular representation of a frequency distribution.

frequentistic interpretation of probability The view that the probability of an event is the proportion of the time that the event will occur in the long run; distinguished from the *classical interpretation of probability* and the *subjective interpretation of probability.*

***F* to enter** A statistic calculated in stepwise regression analysis to reflect the additional contribution of a new predictor; F ratio for an extra sum of squares.

***F* to remove** A statistic calculated in stepwise regression analysis to reflect the unique contribution of a predictor already in the equation; F ratio for an extra sum of squares.

function A rule associating each possible item from one set—such as values of x—with exactly one item from another set—such as values of y. (See also *distribution function*; *probability density function*; *probability function.*)

gambler's fallacy The error of thinking that chance somehow makes up for past excesses.

Gaussian distribution A continuous theoretical probability distribution having a specific form of bell-shaped symmetry with infinite tails. (See also *bivariate normal distribution*; *normal probability density function*; *standard normal distribution.*)

given period The point in time being described by an index number; distinguished from *base period.*

global mode The value of the variable under the highest peak in the graph of a multimodal distribution; distinguished from *local modes.*

grand mean The arithmetic mean of all the observations in an investigation.

grand total The sum of all the observations in an investigation.

grouped data Observations on a quantitative variable that have been grouped into categories to be presented as categorical data.

harmonic mean The reciprocal of the arithmetic mean of the reciprocals of the observations.

heterogeneity of variance Inequality of population variances; violation of the assumption of homogeneity of variance.

histogram A graphical representation of a quantitative frequency distribution—namely, one in which frequencies are represented by a series of contiguous rectangles.

homogeneity of covariance Equality of population covariances for all possible pairs of levels of a repeated-measures factor.

homogeneity of variance Equality of population variances; distinguished from *heterogeneity of variance.*

honestly significant difference The margin of error of a comparison as evaluated by a studentized range statistic; distinguished from *least significant difference.* (See also *Tukey's test.*)

hypergeometric distribution The theoretical probability distribution of various possible patterns of outcomes in random sampling without replacement from a finite population. (Compare *multinomial distribution.*)

hypothesis A proposition put forward concerning unknown parameters in order to examine its consequences. (See also *alternative hypothesis*; *composite hypothesis*; *null hypothesis*; *simple hypothesis.*)

hypothesis testing See *significance testing.*

independent Pertaining to events for which the occurrence of one is irrelevant to whether or not the other occurs. (Compare *mutually exclusive.*)

independent samples Samples in which the selection of observational units for one sample has no bearing on which observational units are in any other sample; distinguished from *block* and *paired samples.*

independent-samples design A design in which the replication factor is nested under the levels of a fixed factor; distinguished from *repeated-measures design.* (Compare *between-within design.*)

independent-samples factor A fixed factor that has the replication factor nested under it; distinguished from *repeated-measures factor.*

independent variable An input variable, used as the basis for describing a dependent variable; a treatment variable; also, more strictly, a variable that is manipulated to see whether it has any effect on a dependent variable.

index number The value of a variable in a given time period, expressed as a percentage of its value in a base period, but with the percent sign dropped.

indicator variable See *dummy variable.*

inferential statistics The aspect of statistics that deals with methods for making appropriate inferences about populations on the basis of samples; distinguished from *descriptive statistics.* (See also *Bayesian inference*; *classical inference*; *estimation*; *significance testing.* Compare *probability theory.*)

interaction The extent to which the differences between the levels of one factor are different at different levels of another factor; the unique effect of the combination of two or more factors; an effect that cannot be accounted for by the sum of the effects of the separate factors. (Compare *main effect.*)

interaction, three-way The extent to which a two-way interaction differs at different levels of the third factor.

interquartile range A measure of dispersion calculated as the difference between the first and third quartiles.

intersection of two events The joint occurrence of the two events.

interval See *category interval*; *confidence interval*; *interval estimate.*

interval estimate A set of adjacent numerical values determined from sample data and accompanied by a statement of the probability that such an interval will include the unknown parameter. (Compare *point estimate.*)

joint event An event defined by the joint occurrence of two or more simple events; distinguished from *disjunctive event* and *simple event.*

joint probability The probability of a joint event.

judgment sampling Nonrandom sampling in which someone selects what appears to be a representative sample.

kurtosis The peakedness of a distribution relative to the Gaussian distribution as an arbitrary standard. (See also *leptokurtic distribution*; *platykurtic distribution*; *zero kurtosis.*)

Laspeyres price index An index number that compares the total cost of the base-period quantities at base-period prices with the total cost of the same quantities at given-period prices.

Law of Large Numbers If the number of trials is made sufficiently large, then the observed proportion of successes becomes arbitrarily close to the probability of success on each individual trial.

leaf The digit or digits following the starting part of an observation, used to represent the observation in a stem-and-leaf display.

least significant difference The margin of error of a comparison as evaluated by a t statistic; distinguished from *honestly significant difference.* (See also *protected t test.*)

least-squares criterion The sum of the squared errors of estimation.

least-squares line A regression line obtained by the method of least squares.

least-squares method A procedure for obtaining a regression line by minimizing the sum of the squared errors of estimation.

leptokurtic distribution A distribution with long tails, weak shoulders, and a peaked appearance near the middle; distinguished from *platykurtic distribution* and *zero kurtosis.*

"less than" distribution A cumulative distribution with cumulation from low to high; distinguished from *"or more" distribution.*

level of an independent variable A value of the independent variable; a treatment.

limits See *category limits*; *confidence limits*; *limits of summation.*

limits of summation, lower and upper The subscript values identifying the first and last observations to be included in a summation.

linear Pertaining to a polynomial of degree 1, which is a straight line. (Compare *cubic*; *quadratic*; *quartic*; *quintic.*)

linear equation in two unknowns An equation describing a dependent variable y as a straight-line function of an independent variable x.

linear transformation A monotonic transformation whereby each value of a variable x is replaced by $bx + a$ for specified constants b and a.

local modes The values of the variable under the lesser peaks in the graph of a multimodal distribution; distinguished from *global mode.*

location Position of a set of observations on the real-number line. (See also *measure of location.*)

logarithmic transformation A monotonic increasing transformation whereby each value of a positive-valued variable x is replaced by $\log x$.

lurking variable A variable that is not included in the regression equation but that in fact has an important effect on the dependent variable. (See also *extraneous variable.*)

main effect The discrepancy among the marginal means for the levels of a particular factor, averaging over the levels of all other factors. (Compare *interaction.*)

manipulated variable An independent variable whose levels are distinguished by the experimenter's administration of different treatments, independent of any characteristics of the observational units; distinguished from *classification variable.*

marginal probability The simple probability of an event that can occur jointly with each of the members of a set of mutually exclusive and exhaustive events.

margin of error of an estimate A certain number of standard errors or estimated standard errors, the number depending on the desired confidence level.

mark See *category mark.*

mathematical expectation See *expected value.*

mathematical statistics See *theoretical statistics.*

maximax criterion A criterion for decision making that favors the action that has a chance—even if remote—of resulting in the largest single payoff.

maximin criterion A criterion for decision making that favors the action that guarantees the largest minimum payoff.

maximization of expected utility See *EU criterion.*

maximum The largest in a set of observations.

mean See *arithmetic mean.* Compare *weighted mean.*

mean deviation A measure of dispersion calculated as the arithmetic mean of the absolute values of the deviations from the mean.

mean square A variance estimate; a sum of squares divided by its number of degrees of freedom. (See also *error mean square.*)

mean square deviation See *mean square.*

measure of central location See *average.*

measure of dispersion Any single number that describes a set of data by indicating how spread out the observations are along the real-number line. (See also *interquartile range*; *mean deviation*; *quartile deviation*; *range*; *standard deviation*; *variance.*)

measure of location Any single number that describes a set of data by indicating where the observations are located on the real-number line or among the possible values of the variable. (See also *average*; *extremes*; *quantile.*)

measure of relative location See *quantile.*

median A measure of central location such that half of the observations are greater than or equal to the median and half of the observations are less than or equal to the median.

midquartile A measure of central location calculated as the arithmetic mean of the first and third quartiles.

midrange A measure of central location calculated as the arithmetic mean of the extremes.

minimum The smallest in a set of observations.

mode A measure of central location equal to the observation value that occurs more frequently than any other value in a set of observations. (See also *global mode*; *local modes.*)

model A symbolic or analogical representation; for example, a mathematical description of a dependent variable in terms of other variables.

monotonic Pertaining to a relationship in which increases in one variable are always accompanied by increases in the other variable—monotonic increasing—or in which increases in one variable are always accompanied by decreases in the other variable—monotonic decreasing.

Monte Carlo methods The use of simulated data, randomly generated from known populations, to study certain phenomena or to solve problems.

multimodal distribution A distribution whose graph shows more than one peak; distinguished from *unimodal distribution.* (Compare *bimodal distribution.*)

multinomial distribution A generalization of the binomial distribution to accommodate two or more categories of outcome; the theoretical probability distribution of various possible patterns of outcomes in random sampling with replacement from a finite population. (Compare *hypergeometric distribution.*)

multiple comparisons See *postmortem comparison.*

multiple correlation coefficient The coefficient of correlation between a dependent variable and the best-fitting—in the least-squares sense—linear combination of a set of predictors. (Compare *partial correlation coefficient.*)

multiple determination, coefficient of The proportion of the total variation in the dependent variable that can be accounted for by the best linear combination of the predictors; sometimes abbreviated as coefficient of determination. (Compare *partial determination, coefficient of.*)

multiple linear regression Regression in which the dependent variable y is described as a linear function of two or more independent variables; distinguished from *nonlinear regression* and *simple linear regression.*

multivariate Pertaining to two or more variables or, if independent variables and dependent variables are distinguished, two or more dependent variables and any number of independent variables. (Compare *univariate*; *bivariate.*)

mutually exclusive Never occurring jointly; incompatible. (Compare *independent.*)

negative kurtosis See *platykurtic distribution.*

negatively skewed See *skewed left.*

nesting An arrangement of factors in which every level of the nesting factor has a different subset of levels of the nested factor associated with it; distinguished from *crossing.*

nominal frequency distribution A frequency distribution in which the categories are defined by nonnumerical verbal labels; distinguished from *quantitative frequency distribution.*

nominal variable A variable whose possible values are described by nonnumerical verbal labels; distinguished from *quantitative variable.*

nondirectional alternative An alternative hypothesis proposing inequality without specifying the direction of the difference; distinguished from *directional alternative.*

nonlinear regression Regression in which the dependent variable *y* is described as a curvilinear function of one or more independent variables; distinguished from *multiple linear regression* and *simple linear regression.*

nonparametric methods Methods that make use of no more than ordinal properties of the data and that are based on rather general assumptions such as independence and an underlying continuous distribution; distinguished from *parametric methods.*

nonprobability sampling See *nonrandom sampling.*

nonrandom sampling Any sampling procedure in which one or more observational units in the population have either an unknown probability or zero probability of being included in the sample; distinguished from *probability sampling.* (See also *convenience sampling*; *judgment sampling*; *quota sampling*; *systematic sampling.*)

nonrejection region The set of test-statistic values that would provide insufficient evidence to reject the null hypothesis; distinguished from *rejection region.*

nonsampling errors Discrepancies between a parameter and its estimate that arise from such sources as improper sampling procedure, inappropriate model, missing data, respondent errors, ambiguous wording, and interviewer errors; distinguished from *sampling error.* (Compare *accuracy.*)

normal distribution See *Gaussian distribution.*

normal equations A set of simultaneous equations that results from minimizing the least-squares criterion in regression analysis and whose solution consists of the least-squares regression coefficients and regression constant.

normal probability density function The mathematical equation that traces a Gaussian bell-shaped curve, which is a specific probability density function.

not statistically significant Judged to be within the realm of random error under the null hypothesis; providing insufficient evidence to reject the null hypothesis. (Compare *significant.*)

null hypothesis In significance testing, the working assumption of no difference in the relevant population values; the proposition against which evidence is sought in the sample data; distinguished from *alternative hypothesis.*

numerical frequency distribution See *quantitative frequency distribution.*

numerical variable See *quantitative variable.*

observation The piece of data from an individual item in a sample or population—from one observational unit.

observational unit An item or element that, when observed in an investigation, yields one univariate or multivariate observation.

observed frequency The number of observational units classified into a given category; distinguished from *expected frequency.*

odds against an event The ratio of the probability that the event will not occur to the probability that it will occur.

odds in favor of an event The ratio of the probability that the event will occur to the probability that it will not occur.

ogive Graphical representation of a cumulative distribution.

omnibus effect Any effect having more than one degree of freedom; distinguished from *comparison.*

one-sided alternative See *directional alternative.*

one-tailed test A significance test with a rejection region in only one tail of the relevant sampling distribution; distinguished from *two-tailed test.*

open category A category with either no lower limit or no upper limit.

"or more" distribution A cumulative distribution with cumulation from high to low; distinguished from *"less than" distribution.*

orthogonal Independent; providing uncontaminated information about the items involved; distinguished from *confounded.*

orthogonal comparisons Two comparisons that are independent of each other, as reflected in the sum of the products of their corresponding coefficients.

orthogonal components Sums of squares due to orthogonal comparisons.

outlier An observation that is so different from most of the observations in a set that one may wonder whether it came from a different population.

outside values Any observations beyond the adjacent values in a box plot.

overall effects Main effects and interactions based on data from an entire experiment; distinguished from *simple effects*.

paired samples Two samples in which the observational units come in pairs and in which one member of each pair is observed in one sample and the other member in the other sample; distinguished from *independent samples*. (See also *block*.)

parameter A numerical description that summarizes information about a population; distinguished from a *statistic*.

parametric methods Methods that make use of the actual numerical magnitudes of the observations and that are based on relatively restrictive assumptions such as normality and homogeneity of variance; distinguished from *nonparametric methods*.

partial correlation coefficient The coefficient of correlation between two variables, with one or more other variables held constant. (Compare *multiple correlation coefficient*.)

partial correlation coefficient, first-order The coefficient of correlation between two variables with one other variable held constant.

partial correlation coefficient, second-order The coefficient of correlation between two variables with two other variables held constant.

partial determination, coefficient of The proportion of the residual variation that can be accounted for by the inclusion of a certain additional predictor in the regression equation. (Compare *extra sum of squares*; *multiple determination, coefficient of*.)

partial regression coefficient Any one of the regression coefficients in multiple linear regression.

payoff The consequences of a particular action in a particular state of the world.

Pearson product-moment correlation coefficient An index of the linear relationship between two variables. (See also *perfect negative correlation*; *perfect positive correlation*.)

percentage A proportion expressed in units of one one-hundredth. (Compare *percentage point*.)

percentage distribution A description of the percentages of observations in various mutually exclusive and exhaustive categories. (See also *cumulative percentage distribution*. Compare *probability distribution*.)

percentage frequency density The percentage of observations per reference interval. (Compare *probability density function*.)

percentage point A difference of one unit on a percentage scale. (Compare *percentage*.)

percentiles See *centiles*.

per-comparison error rate The Type I error rate considered as the number of comparisons exhibiting Type I errors divided by the total number of comparisons; distinguished from *experimentwise error rate*.

perfect negative correlation A relationship represented by a Pearson product-moment correlation coefficient of −1.

perfect positive correlation A relationship represented by a Pearson product-moment correlation coefficient of +1.

pie chart A graphical representation of a nominal frequency distribution, in which frequencies are represented by radial sectors of a circle.

placebo A control treatment, lacking only the ingredient whose effectiveness is under investigation.

planned comparison One of a small number of particular comparisons that an experiment was especially designed to investigate; distinguished from *postmortem comparison*. (See also *trend analysis*.)

platykurtic distribution A distribution with short tails, heavy shoulders, and a flat appearance near the middle; distinguished from *leptokurtic distribution* and *zero kurtosis*.

point estimate A single numerical value calculated from sample data and taken to be indicative of the value of the unknown parameter. (Compare *interval estimate.*)

Poisson distribution A theoretical probability distribution used to describe the occurrence of independent random events occurring with a small, constant probability; the limiting form of the binomial distribution as the number of trials becomes large and the probability of success on each trial becomes small.

polynomial-fitting A procedure for describing a nonlinear relationship by means of a polynomial equation.

polynomial of degree *g* An equation that involves a number of terms added together, each term consisting of a coefficient multiplied by the independent variable raised to a nonnegative integer power, where the highest power is *g*. (See also *cubic*; *linear*; *quadratic*; *quartic*; *quintic.*)

population The set of all observations, or of all observational units, relevant to the question being asked; distinguished from *sample.*

positive kurtosis See *leptokurtic distribution.*

positively skewed See *skewed right.*

posterior probability In Bayes' rule, the revised probability, taking into account the additional evidence; distinguished from *prior probability.*

postmortem comparison A comparison used to help interpret a significant omnibus *F* test in the absence of planned comparisons; distinguished from *planned comparison.* (See also *protected t test*; *Scheffé's test*; *Tukey's test.*)

power of a test The probability that the significance test will detect a population difference of a given magnitude. (Compare *Type II error.*)

precision Lack of variability in an estimate as reflected in a small standard error; lack of random error; freedom from sampling error. (Compare *accuracy.*)

predictor An independent variable in regression analysis.

price relative The simplest form of price index, based on just one item.

prior probability In Bayes' rule, the probability that is to be revised in light of additional evidence; distinguished from *posterior probability.*

probability A number that is assigned to an event to indicate the likelihood that the event will occur. (For specific approaches to probability, see *classical interpretation of probability*; *frequentistic interpretation of probability*; *subjective interpretation of probability.* For specific types of probability, see *conditional probability*; *disjunctive probability*; *joint probability*; *marginal probability*; *posterior probability*; *prior probability*; *simple probability.*)

probability density function A representation of a continuous probability distribution, in which probability is portrayed by areas under the density curve. (For an example, see *normal probability density function.* Compare *frequency density.*)

probability distribution A description of the probabilities with which various values of a variable or categories of observation occur. (See also *continuous probability distribution*; *discrete probability distribution*; *empirical probability distribution*; *sampling distribution*; *theoretical probability distribution.* Compare *frequency distribution.*)

probability function A mathematical function giving the probability of occurrence of each value of a discrete random variable. (For an example, see *binomial probability function.*)

probability sampling Any sampling procedure in which every observational unit in the entire population has a known, nonzero probability of being included in the sample; distinguished from *nonrandom sampling.* (See also *simple random sampling*; *stratified random sampling.*)

probability theory A branch of mathematics that deals with the nature and likelihood of the possible samples that could be drawn from a known population. (Compare *inferential statistics.*)

proportion The number of outcomes of a given type divided by the number of trials.

protected least significant difference See *protected t test.*

protected *t* test A postmortem comparison procedure that uses the planned comparison procedure, provided that the relevant omnibus F test was significant. (Compare *Scheffé's test*; *Tukey's test.*)

purchasing power See *real income.*

***P*-value** The probability of obtaining a sample difference as large as the one actually observed, if the null hypothesis is true.

***P*-value inequality** A statement that the P-value is less than a specified significance level, meaning that the result is statistically significant, or that the P-value is greater than a specified significance level, meaning that the result is not statistically significant.

quadratic Pertaining to a polynomial of degree 2, which is a 1-bend curve. (Compare *cubic*; *linear*; *quartic*; *quintic.*)

qualitative frequency distribution See *nominal frequency distribution.*

qualitative variable See *nominal variable.*

quantile A value that cuts off a certain lower fraction of a distribution. (See *centiles*; *deciles*; *extremes*; *median*; *quartiles.*)

quantitative frequency distribution A frequency distribution in which the categories are defined by numerical values of a variable; distinguished from *nominal frequency distribution.*

quantitative variable A variable whose possible values are numerical; distinguished from *nominal variable.*

quartic Pertaining to a polynomial of degree 4, which is a 3-bend curve. (Compare *cubic*; *linear*; *quadratic*; *quintic.*)

quartile deviation A measure of dispersion calculated as one-half of the difference between the first and third quartiles.

quartiles Quantiles that divide a distribution into four segments, each containing 25% of the observations. (Compare *centiles*; *deciles*; *median.*)

quintic Pertaining to a polynomial of degree 5, which is a 4-bend curve. (Compare *cubic*; *linear*; *quadratic*; *quartic.*)

quota sampling Nonrandom sampling in which hired technicians are required to collect data from a certain number of observational units in specified categories but are free to select any observational units that satisfy the guidelines.

random Operating according to particular probabilities.

random assignment Allocation of observational units to the various treatments according to a random procedure based on an appropriate inanimate device. (Compare *random selection.*)

random digits A set of independently generated digits such that in each position any of the ten digits is equally likely to appear.

random factor A factor whose specific levels to be included in the experiment are not of interest in their own right but only as representative of a larger population of levels; distinguished from *fixed factor.*

randomized-blocks design A repeated-measures design in which the treatments are assigned randomly to the observational units within each block.

random sample A sample obtained by random sampling.

random sampling Sampling in which every member of the population has a known, nonzero probability of being selected; sometimes used as an abbreviated label for simple random sampling. (See also *simple random sampling*; *stratified random sampling.*)

random selection A selection procedure whereby every item in the population has an equal chance of being selected; simple random sampling. (Compare *random assignment.*)

random variable A variable whose values may be described by a probability distribution; more strictly, a function that assigns a numerical value to each outcome in an experiment whose results depend to some extent on chance.

range A measure of dispersion calculated as maximum minus minimum.

range effects Carry-over effects whereby subjects learn something about the range of levels that they receive of a factor and respond differently than they would if they did not know anything about the range.

rank An integer that indicates the ordinal position of an observation when the observations are arranged in ascending or descending order.

rank correlation See *Spearman's rank correlation coefficient.*

rank-order information Information as to which observation is largest, which is second largest, and so on.

raw data Numerical descriptions of individual items in a sample or population; the original observations.

real income Income adjusted for changes in a price index.

reciprocal transformation A monotonic decreasing transformation whereby each value of a positive-valued variable x is replaced by $1/x$.

rectangular distribution See *uniform distribution.*

regression The functional relationship between two or more variables as determined from observed values of those variables.

regression analysis The study of the functional relationships among observed variables.

regression coefficient The coefficient b of an independent variable x in a regression equation.

regression constant The y-intercept in a regression equation.

regression fallacy The error of giving an empirical explanation for the statistical phenomenon whereby extreme values of one variable tend to be associated with somewhat less extreme values of another variable.

rejection region The set of test-statistic values that would lead to rejection of the null hypothesis; distinguished from *nonrejection region.*

repeated-measures design A design in which the replication factor is crossed with a fixed factor; distinguished from *independent-samples design.* (Compare *between-within design.*)

repeated-measures factor A fixed factor that is crossed wtih the replication factor; distinguished from *independent-samples factor.*

replication One trial or repetition of an experimental situation.

residual The difference between the observed value of the dependent variable and the fitted value; error of estimation.

residual mean square The sum of the squared residuals, divided by their degrees of freedom. (Compare *standard error of estimation.*)

residual variation Variation due to the errors of estimation; the sum of the squared residuals.

response variable See *dependent variable.*

retrospective report Observation based on memory of a past event.

retrospective study An investigation of data that happened to be available relevant to the question at hand, rather than data deliberately collected by using random assignment of observational units to the various treatments.

reversed-J-shaped distribution A distribution with extreme positive skewness, whose graph vaguely resembles a backwards J.

risk-avoiding strategy A strategy that to some extent favors a small but safe payoff over a larger but uncertain payoff.

risk-taking strategy A strategy that to some extent favors a large but uncertain payoff over a smaller but safer payoff.

robustness The property characterizing any procedure that does not become seriously inaccurate when there are violations of the assumptions on which it is based.

root mean square See *standard deviation.*

root mean square deviation See *standard deviation.*

rounding Suppression of right-hand digits, leaving the rightmost remaining digit unchanged or increasing it by 1, whichever is closer to the exact value; distinguished from *truncation.*

rule of thumb for symmetric unimodal distributions For distributions that are more or less symmetric and unimodal, approximately 68% of the observations fall within 1.0 standard deviation above and below the mean, approximately 95% fall within 2.0 standard deviations above and below the mean, and approximately 99–100% of the observations fall within 3.0 standard deviations above and below the mean.

sample A set of some, but not all, of the observations—or of the observational units—relevant to the question being asked; distinguished from *population.* (See also *independent* *samples*; *paired samples*; *random sample*; *simple random sample*.)

sample space The entire set of possible events or outcomes of a given situation.

sample survey A method of inferential statistics for ascertaining the characteristics of a well-defined population on the basis of information from the members of an appropriately chosen sample.

sampling The process of selecting a sample from a population. (See also *nonrandom sampling*; *probability sampling*.)

sampling distribution A probability distribution in which the random variable is itself a statistic. (See also *empirical sampling distribution*; *theoretical sampling distribution*.)

sampling error The extent to which the discrepancy between a parameter and its estimate is due to observing only a sample rather than the whole population; distinguished from *nonsampling errors*. (Compare *precision*.)

sampling fraction The proportion of the population included in the sample.

sampling without replacement A sampling procedure in which the pool of available items becomes progressively smaller as each selection is made; distinguished from *sampling with replacement*.

sampling with replacement A sampling procedure in which each item selected is put back into the available pool of items before the next selection is made; sampling from an infinite pool; distinguished from *sampling without replacement*.

scatter diagram A plot in which each bivariate observation is represented by a dot at the appropriate point in the graph; also called a scattergram.

Scheffé's test A postmortem comparison procedure that ensures that no matter how many comparisons are made in a given set of means, the experimentwise error rate does not exceed the chosen significance level. (Compare *protected t test*; *Tukey's test*.)

semi-interquartile range See *quartile deviation*.

significance level The probability of rejecting the null hypothesis if it is in fact true. (Compare *confidence level*; *Type I error*.)

significance, statistical A judgment that a sample difference is not due solely to chance.

significance test A procedure for evaluating whether the difference between sample values should be attributed simply to random error or whether the difference is large enough to conclude that the corresponding population values are different.

signficance testing The aspect of inferential statistics that is concerned with demonstrating population differences; distinguished from *estimation*.

significant figures The digits from the leftmost nonzero digit to the rightmost digit that is either a known zero or a nonzero digit. (Compare *decimal places*; *digit position*; *effective digits*.)

significant, statistically Judged to be too improbable to be attributed to random error under the null hypothesis; providing sufficient evidence to reject the null hypothesis. (Compare *not statistically significant*.)

sign information Information concerning the direction of difference of each observation from a certain value.

simple effects Effects considered separately for each level of one of the factors involved in an overall interaction; distinguished from *overall effects*. (See also *simple interaction effect*; *simple main effect*; *simple simple main effect*.)

simple event An elemental event, not defined in terms of simpler events; distinguished from *disjunctive event* and *joint event*.

simple hypothesis An hypothesis that proposes an exact equality; distinguished from *composite hypothesis*.

simple interaction effect A two-way interaction examined at a specific level of a third factor. (Compare *simple main effect*.)

simple linear correlation The degree of linear relationship between two variables. (See also *Pearson product-moment correlation coefficient*. Compare *simple linear regression*.)

simple linear regression Regression in which the dependent variable y is described as a straight-line function of a single independent variable x. (Compare *simple linear correlation*.)

simple main effect An effect of one factor examined at a specific level of a second factor. (Compare *simple interaction effect.*)

simple probability The probability of a simple event.

simple random sampling Sampling in which every member of the population has an equal probability of being selected.

simple simple main effect An effect of one factor examined at a specific combination of levels of two other factors. (Compare *simple main effect.*)

simple stem-and-leaf Stem-and-leaf in which there is just one stem for each starting part, yielding category intervals of . . . , .01, .1, 1, 10, . . . ; distinguished from *squeezed stem-and-leaf* and *stretched stem-and-leaf.*

simulation Representation of a phenomenon by means of a model or imitation that facilitates the study of certain features of the phenomenon.

skewed left Having a longer tail to the left than to the right; distinguished from *skewed right* and *symmetric distribution.*

skewed right Having a longer tail to the right than to the left; distinguished from *skewed left* and *symmetric distribution.*

skewness Departure from symmetry in a distribution. (See also *skewed left* and *skewed right.*)

slope of a straight line Change in the dependent variable y divided by the corresponding change in the independent variable x; rise over run. (See also *regression coefficient.*)

Spearman's rank correlation coefficient The Pearson product-moment correlation coefficient calculated on the ranks of two variables.

split-plot design See *between-within design.*

square-root transformation A monotonic increasing transformation whereby each value of a nonnegative variable x is replaced by the square root of x.

squeezed stem-and-leaf Stem-and-leaf in which there are five stems for each starting part, yielding category intervals of . . . , .02, .2, 2, 20, . . . ; distinguished from *simple stem-and-leaf* and *stretched stem-and-leaf.*

standard deviation A measure of dispersion calculated as the positive square root of the variance; a standard amount of deviation in a set of observations.

standard error The standard deviation of the sampling distribution of a statistic. (Compare *precision.*)

standard error of estimation The positive square root of the sum of the squared errors of estimation divided by their degrees of freedom. (Compare *residual mean square.*)

standard normal deviate An observation from a standard normal distribution.

standard normal distribution A Gaussian distribution expressed in standard units, so that it has a mean of 0 and a standard deviation of 1.

standard score The number of standard units that an observation is away from the mean, a negative value indicating that the observation is below the mean.

standard unit A unit of measurement equal to one standard deviation. (Compare *standard score.*)

stanine scale A discrete nine-point scale derived by partitioning the standard normal distribution symmetrically into nine regions with category boundaries at ±.25, ±.75, ±1.25, and ±1.75.

starting part The leftmost digit or digits of an observation, used to label a category in a stem-and-leaf display.

state-of-the-world probability A subjective probability indicating the decision maker's certainty that a possible state of the world is actually the case.

statistic A numerical description that summarizes information about a sample; distinguished from *parameter.* (See also *test statistic.*)

statistics The science of answering questions on the basis of empirical data; the science of making sense of data. (See also *applied statistics*; *descriptive statistics*; *inferential statistics*; *theoretical statistics.*)

stem A line, consisting of a starting part and its leaves, in a stem-and-leaf display.

stem-and-leaf A procedure for sorting numerical observations into categories by using the initial digits of the observations to label the categories and the subsequent digits to represent the observations. (See also *simple stem-and-leaf*; *squeezed stem-and-leaf*; *stretched stem-and-leaf*.)

stem-and-leaf display A stem-and-leaf representation of a quantitative frequency distribution.

stepwise regression analysis A systematic procedure for entering, and sometimes removing, predictors one at a time until a regression equation is obtained that allows no more predictors to be entered or removed according to certain specified criteria.

stratified random sampling A sampling procedure in which a sample is randomly selected from each of several strata separately.

stratified random sampling with optimum allocation Stratified random sampling in which the number selected from each stratum depends on both the size and the variance of the stratum.

stratified random sampling with proportional allocation Stratified random sampling in which the number selected from each stratum is proportional to the size of the stratum.

stratum A relatively homogeneous subpopulation in stratified random sampling.

stretched stem-and-leaf Stem-and-leaf in which there are two stems for each starting part, yielding category intervals of . . . , .05, .5, 5, 50, . . . ; distinguished from *simple stem-and-leaf*; *squeezed stem-and-leaf*.

studentized range statistic The difference between the largest and the smallest means in a set of means, divided by the estimated standard error of a mean.

Student's *t* distribution See *t distribution*.

subject An individual specimen observed in an investigation.

subjective interpretation of probability The view that probability is a measure of a person's degree of certainty about an event; distinguished from the *classical interpretation of probability* and the *frequentistic interpretation of probability*.

success One of the two mutually exclusive and exhaustive categories of outcome in a Bernoulli trial; the other is *failure*.

summation notation A concise way of writing the sum of the values of a variable, consisting basically of the upper-case Greek letter sigma (Σ) followed by a symbol for the variable.

sum of products The sum of the products of corresponding deviations from the means of two variables.

sum of squares Sum of squared deviations. (See also *variation*.)

symmetric distribution A distribution that, if folded at its middle, is a left-right mirror image of itself; distinguished from *skewed left* and *skewed right*.

systematic sampling Nonrandom sampling in which every mth observational unit is selected from a sequential array of the members of the population, for some number m.

tallying A procedure for counting items in several categories simultaneously by making a tally mark for each item beside the relevant category label.

***t* distribution** The theoretical sampling distribution of the difference between a point estimate having a Gaussian sampling distribution and the parameter, divided by the estimated standard error of the point estimate; the distribution of the ratio of a standard normal deviate to the square root of a chi-square variate divided by its degrees of freedom.

test statistic The statistic calculated for comparison with a critical value in significance testing.

theoretical probability distribution A probability distribution derived mathematically from the rules of probability; distinguished from *empirical probability distribution*. (See also *binomial distribution*; *chi-square distribution*; *F distribution*; *Gaussian distribution*; *hypergeometric distribution*; *multinomial distribution*; *Poisson distribution*; *t distribution*.)

theoretical sampling distribution A sampling distribution derived from theoretical principles; distinguished from *empirical sampling distribution*.

theoretical statistics The aspect of statistics that focuses on the mathematical theory underlying statistical methods and on the development of new statistical methods; distinguished from *applied statistics*.

transformation The process of obtaining a new variable as a mathematical function of a given variable. (See also *change of scale*; *Fisher's z(r) transformation*; *linear transformation*; *logarithmic transformation*; *reciprocal transformation*; *square-root transformation*.)

treatment An experimental condition being investigated; a level of an independent variable.

treatment combination An experimental condition defined by the crossing of two or more independent variables.

treatment variable See *independent variable*.

trend analysis The special case of planned comparisons for examining the nature of the functional relationship between the dependent and the independent variables.

triangular distribution A distribution in which, at any point within the range of the random variable x, a given increase in x changes the associated probability by a fixed amount.

trimean A measure of location calculated as the arithmetic mean of the median and the midquartile.

true Existing in the underlying population.

true effects of the treatments The treatment effects that would be found if the entire—possibly hypothetical—population were observed.

true proportion The population proportion.

truncation Suppression of right-hand nonzero digits without making any adjustment in the remaining digits; distinguished from *rounding*.

Tukey's test A postmortem comparison procedure that ensures that no matter how many pairwise comparisons are made in a given set of means, the experimentwise error rate does not exceed the chosen significance level. (Compare *protected t test*; *Scheffé's test*.)

two-sided alternative See *nondirectional alternative*.

two-tailed test A significance test with rejection regions in both tails of the relevant sampling distribution; distinguished from *one-tailed test*.

Type I error Rejection of the null hypothesis when it is in fact true; a false alarm.

Type II error Failure to reject a false null hypothesis; a miss. (Compare *power of a test*.)

unbiased estimate of a parameter An estimate with no systematic tendency to be too large or too small; an estimate whose expected value is equal to the parameter; distinguished from *biased estimate*.

uniform density A probability density function in which the density takes on the same value at every value of the random variable.

uniform distribution A probability distribution in which every possible value of the variable is equally likely.

unimodal distribution A distribution whose graph shows a single peak; distinguished from *bimodal distribution* and *multimodal distribution*.

union of two events The disjunctive event consisting of the occurrence of either or both of the events.

univariate Pertaining to exactly one variable or, if independent variables and dependent variables are distinguished, exactly one dependent variable and any number of independent variables; distinguished from *bivariate* and *multivariate*.

U-shaped distribution A platykurtic distribution whose central peak is so flattened as to be a valley.

utiles The units of measurement of utility for a given decision maker—the units are not comparable across different decision makers.

utility The value of an outcome to the decision maker.

value of a variable A specific point on a dimension along which items may differ from each other.

variable A characteristic that may distinguish one item from another; a dimension along which items may differ from each other. (See also *categorized variable*; *continuous variable*; *dependent variable*; *discrete variable*; *independent variable*; *nominal variable*; *quantitative variable*; *random variable*.)

variable selection The choice of predictors to include in the regression equation.

variance A measure of dispersion calculated as an average of the squared deviations of a set of observations. (See also *analysis of variance*; *homogeneity of variance*.)

variance component The variance of the population means for the different levels of a factor.

variance-ratio distribution See *F distribution.*

variation The extent to which the observations in a set differ from each other, as measured by the sum of squares. (See also *residual variation.*)

Venn diagram A graphical representation of sets of items and relationships between sets.

verbal report Observation based on what people say is the case; distinguished from *direct measurement.*

weighted mean The sum of the products of the observations and relevant weights, divided by the sum of the weights.

within-subjects design See *repeated-measures design.*

***y*-intercept of a line** The value of y where the line intersects the y-axis. (See also *regression constant.*)

zero kurtosis The peakedness of the Gaussian distribution, which is neither leptokurtic nor platykurtic.

***z*-score** See *standard score.*

Answers to odd-numbered exercises and problems

Chapter 1

1/1. **a.** 24 **b.** 12 **c.** 13 **d.** 39 **e.** 39 **f.** 26 **g.** 23 **h.** 19 **i.** 69 **j.** 312 **k.** 85 **l.** 225

1/3. **a.** 6 **b.** 20 **c.** 12 **d.** −7 **e.** −7 **f.** 100 **g.** 37 **h.** 87 **i.** 71 **j.** 540 **k.** 175 **l.** 729

1/5. **a.** 28 **b.** 9 **c.** 23 **d.** 53 **e.** 53 **f.** 46 **g.** 27 **h.** 23 **i.** 124 **j.** 644 **k.** 209 **l.** 625

1/7. 121.3

1/9. **a.** 6.8% decrease **b.** $32,617

1/11. **a.** Actual age at which the child's bladder training began, retrospective report of the child's age when bladder training began, actual age at which bottle-feeding ended, and retrospective report of the child's age when bottle-feeding ended. Alternatively, one could consider these to be just two variables: error in the retrospective report of the child's age when bladder training began, and error in the retrospective report of the child's age when bottle-feeding ended. **b.** The mothers and the fathers who received the retrospective interviews. **c.** The sample consisted of the mothers and the fathers actually interviewed. The population is not specified clearly in the brief report in Example 1.3, but in general we can say that the population is all mothers and fathers to whom the results are to be generalized. **d.** An example of a statistic would be the average value of any of the variables in Problem 1/11(a) for the particular mothers and/or fathers interviewed. An example of a parameter would be the average value of any of the variables in Problem 1/11(a) for all the mothers and/or fathers in the relevant population.

1/13. **a.** Length and width (and height?) of his schoolroom. **b.** Evidently, schoolrooms. **c.** Observation, raw data, values of the variables in Part (a). The education office presumably wanted a direct measurement but in fact accepted Birmingham's verbal report.

1/15. **a.** Price in dollars. **b.** Menu items. **c.** The total cost in dollars of a hamburger, an order of French fries, and a soft drink; 2.70. **d.** The total cost in dollars of 6 hamburgers, 6 orders of French fries, and 6 soft drinks; 16.20.

1/17. The total area covered by the three rugs is 18,000 cm^2.

1/19. **a.** $96.47, $101.84, $104.95, $101.45, $101.15, $93.95 **b.** If measured in current dollars, average weekly earnings rose throughout the period from 1963 to 1983. However, real earnings in constant dollars rose from 1963 until the early 1970s and then declined to a lower level in 1983 than in 1963.

Chapter 2

2/1. **a.**

Stem	Leaves	
0	393071011513733713	(18)
1	632618068401356	(15)
2	6077337	(7)
		(40)

b.

Stem	Leaves	
0	3301011133313	(13)
0	97577	(5)
1	32104013	(8)
1	6686856	(7)
2	033	(3)
2	6777	(4)
		(40)

c.

Stem	Leaves	
0o	0101111	(7)
t	333333	(6)
f	5	(1)
s	777	(3)
e	9	(1)
1o	1001	(4)
t	323	(3)
f	45	(2)
s	6666	(4)
e	88	(2)
2o	0	(1)
t	33	(2)
f		(0)
s	6777	(4)
		(40)

2/3. **a.**

Category	Frequency	Percentage frequency
0.0– 4.9	13	32%
5.0– 9.9	5	12%
10.0–14.9	8	20%
15.0–19.9	7	18%
20.0–24.9	3	8%
25.0–29.9	4	10%
Total	40	100%

b.

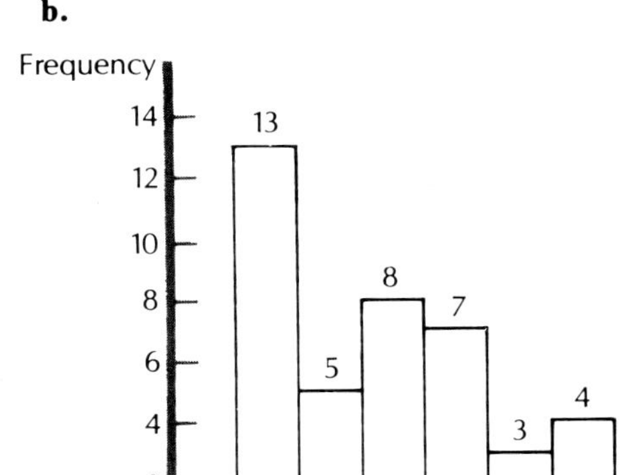

c.

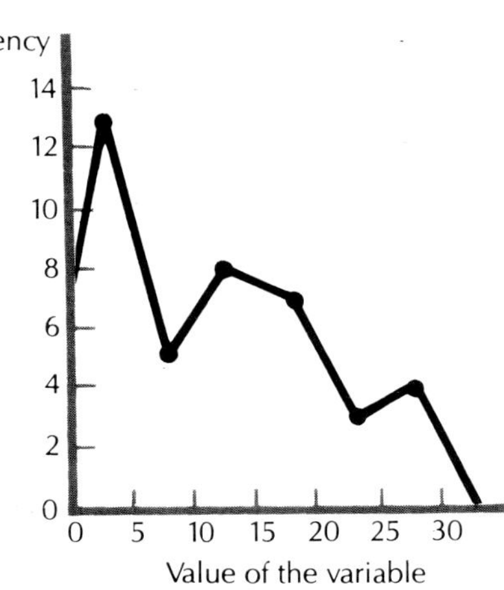

2/5.

Category	Frequency	Frequency density	Percentage frequency density
0– 9	36	36	18
10–19	98	98	49
20–49	52	17	8.7
50–99	14	2.8	1.4
Total	200		

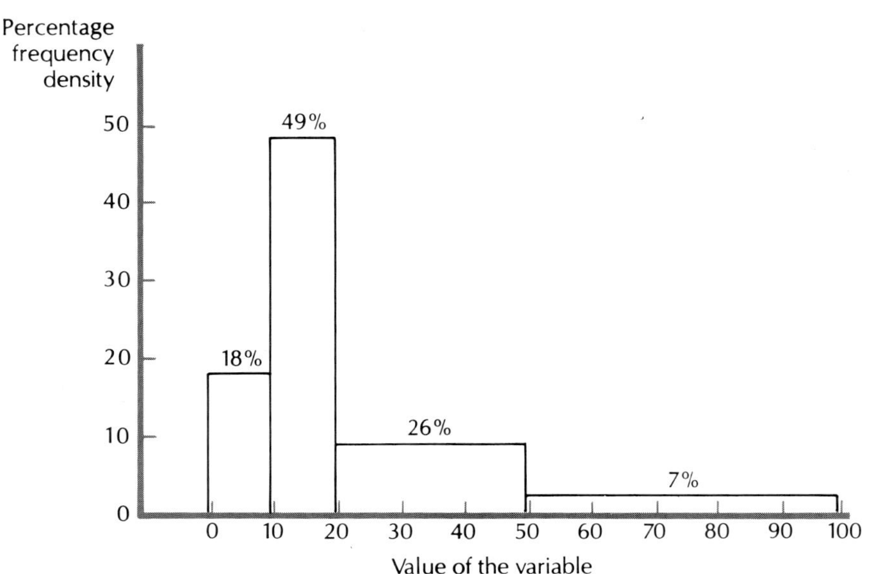

2/7.

Value of variable	Cumulative frequency
Less than 0.0	0
Less than 5.0	13
Less than 10.0	18
Less than 15.0	26
Less than 20.0	33
Less than 25.0	36
Less than 30.0	40

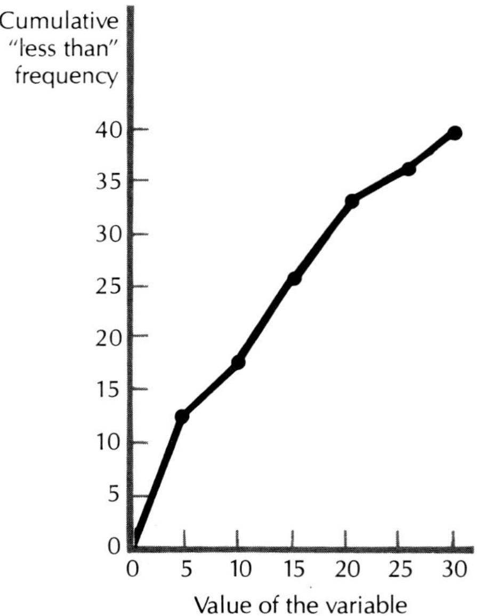

2/9. **a.** The categories are neither mutually exclusive nor exhaustive. **b.** Because the categories are not mutually exclusive, the frequencies cannot be added. Because the categories are not exhaustive, we cannot tell whether the frequencies are large or small relative to the total number of customers; for example, how many dissatisfied customers were there?

2/11. **a.** Pizza bills. **b.** Dollar amount of the pizza bill. **c.** The exact value cannot be determined from the table, but it was in the category \$25.00–\$29.99. **d.** 32 **e.** The exact number cannot be determined from the table, but there were at least 1 and no more than 32. **f.** 0.00, 4.99, 5.00, 9.99, 10.00, 14.99, 15.00, 19.99, 20.00, 24.99, 25.00, 29.99 **g.** −.005, 4.995, 9.995, 14.995, 19.995, 24.995, 29.995 **h.** 5.00 for all categories. **i.** 2.495, 7.495, 12.495, 17.495, 22.495, 27.495 **j.** Sample.

2/13. **a.** 15 **b.** 0–14, 15–29, 30–44, 45–59, 60–74 **c.** −.5, 14.5, 29.5, 44.5, 59.5, 74.5

2/15. a. Simple stem-and-leaf with category interval of 10:

0	265693405209026536791	(21)
1	096793907332305	(15)
2	854102242455	(12)
3	7555983	(7)
4	4	(1)
5	609	(3)
6	3	(1)
		(60)

b. Stretched stem-and-leaf with category interval of 5:

0	2340200231	(10)
0	65695965679	(11)
1	03033230	(8)
1	9679975	(7)
2	41022424	(8)
2	8555	(4)
3	3	(1)
3	755598	(6)
4	4	(1)
4		(0)
5	0	(1)
5	69	(2)
6	3	(1)
		(60)

c. Squeezed stem-and-leaf with category interval of 20:

0o	110001000100100011111000000101110001	(36)
t	2232233322322322232	(19)
f	5545	(4)
s	6	(1)
		(60)

d. Various features may be noted, such as the following: The display in Part (a) reveals a general trend for greater lengths to be less frequent, right from the first category. The display in Part (b) suggests that reigns have tended to concentrate in the vicinity of certain lengths; roughly 5–9, 20–24, 35–39, and 55–59 years, in order of decreasing frequency.

e.

Length of reign (in years)	Number of monarchs
0– 9	21
10–19	15
20–29	12
30–39	7
40–49	1
50–59	3
60–69	1
Total	60

The distribution is reversed-J-shaped (and skewed right).

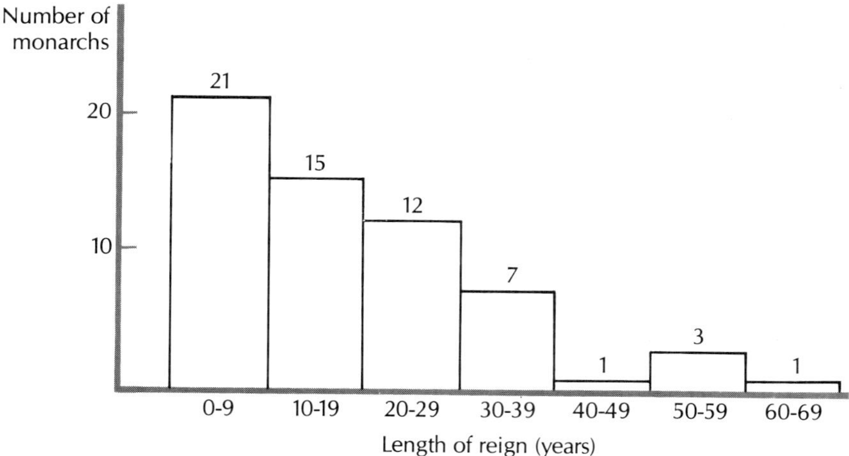

f. The frequency table is given in Example 2.2. The distribution is skewed right. Notice that the left-most leg of the frequency polygon shown in the figure points to −3, which is logically the next category mark to the left.

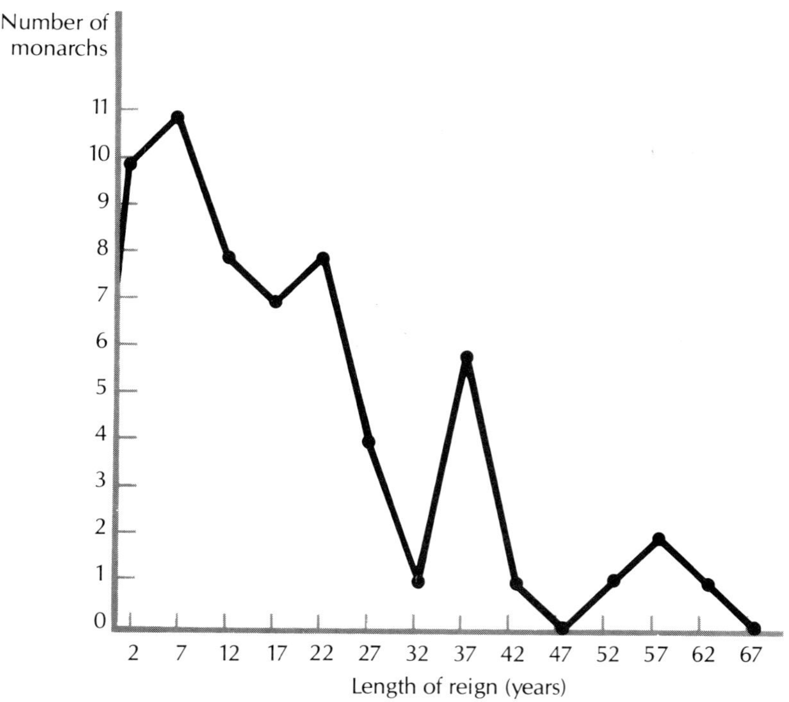

g.

Length of reign (in years)	Cumulative frequency
0 or more	60
10 or more	39
20 or more	24
30 or more	12
40 or more	5
50 or more	4
60 or more	1
70 or more	0

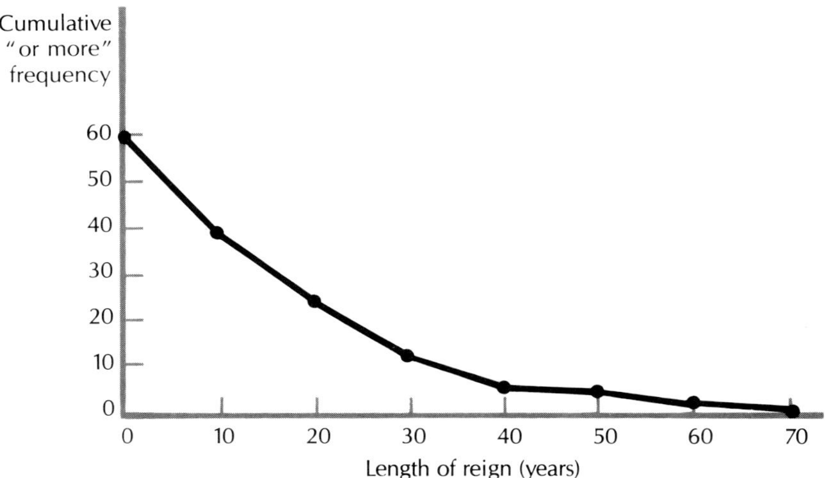

h.

Length of reign (in years)	Cumulative percentage
Less than 0	0%
Less than 5	17%
Less than 10	35%
Less than 15	48%
Less than 20	60%
Less than 25	73%
Less than 30	80%
Less than 35	82%
Less than 40	92%
Less than 45	93%
Less than 50	93%
Less than 55	95%
Less than 60	98%
Less than 65	100%

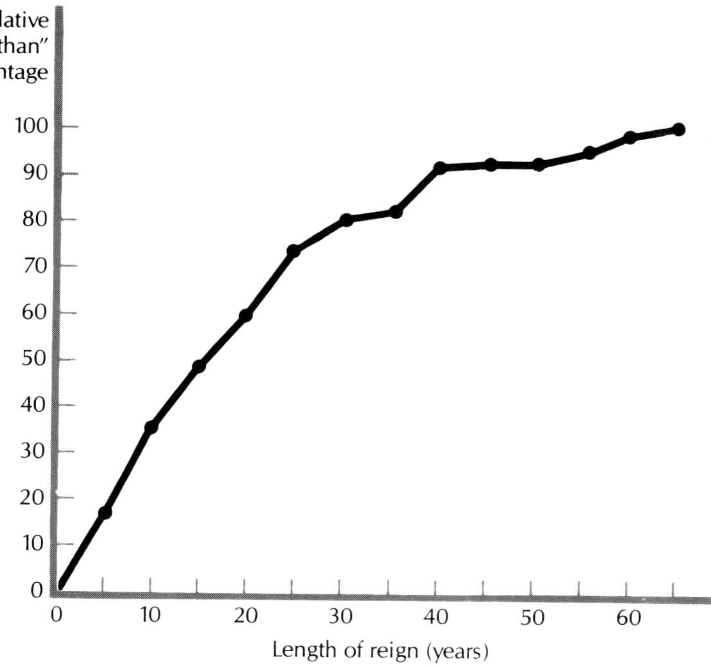

2/17. a. Textbooks. **b.** Price of new textbook. **c.** Simple stem-and-leaf:

Stem	Leaves	Count
0	681562518358747	(15)
1	469121	(6)
2	101	(3)
3	2	(1)
		(25)

d. Stretched stem-and-leaf:

Stem	Leaves	Count
0	12134	(5)
0	6856585877	(10)
1	4121	(4)
1	69	(2)
2	101	(3)
2		(0)
3	2	(1)
		(25)

e. Squeezed stem-and-leaf:

```
0o | 11      (2)
 t | 23      (2)
 f | 5554    (4)
 s | 6677    (4)
 e | 888     (3)
1o | 11      (2)
 t | 2       (1)
 f | 4       (1)
 s | 6       (1)
 e | 9       (1)
2o | 101     (3)

3o |         (0)
 t | 2       (1)
            ____
            (25)
```

f. Various features may be noted, such as the following: The simple stem-and-leaf gives a concise overall picture. The stretched version shows that most books are in the $5 to $10 range, and it draws attention to the one unusually expensive book. The squeezed version gives more detail, perhaps more than is necessary for only 25 observations.

g.

Price of book	Number of books
$ 0.00– 9.99	15
$10.00–19.99	6
$20.00–29.99	3
$30.00–39.99	1
Total	25

The distribution is reversed-J-shaped (and skewed right).

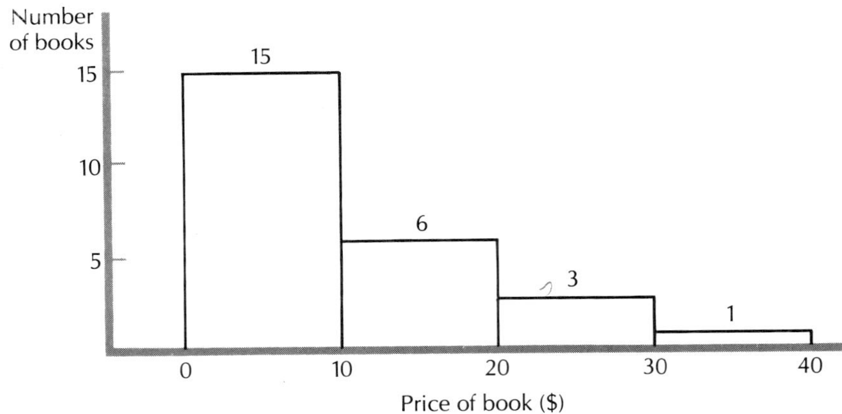

h.

Price of book	Number of books
$ 0.00– 4.99	5
$ 5.00– 9.99	10
$10.00–14.99	4
$15.00–19.99	2
$20.00–24.99	3
$25.00–29.99	0
$30.00–34.99	1
Total	25

The distribution is skewed right. Notice that the left-most leg of the frequency polygon points to −2.50, which is logically the next smaller category mark.

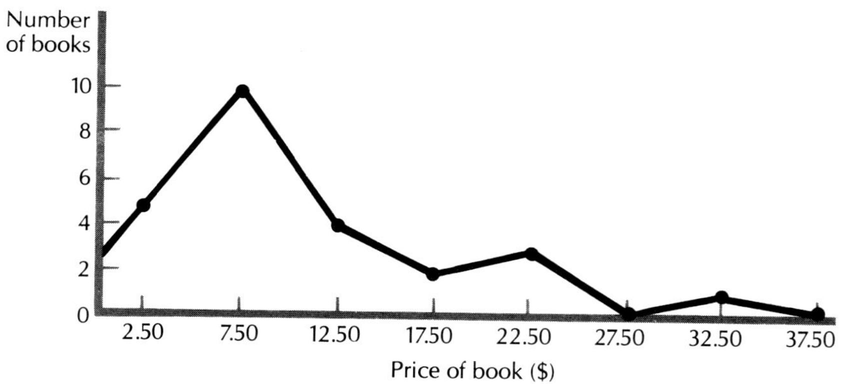

i.

Price of book	Cumulative percentage
Less than $ 5.00	20%
Less than $10.00	60%
Less than $15.00	76%
Less than $20.00	84%
Less than $25.00	96%
Less than $30.00	96%
Less than $35.00	100%

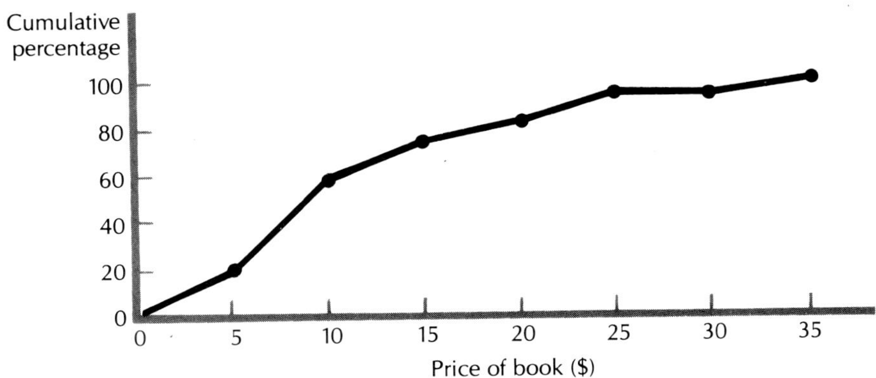

2/19. **a.** Type of season ticket. **b.** Regular season tickets. **c.** Student.

d.

Type of season ticket	Percentage of tickets sold
Adult	49%
Senior	32%
Family	17%
Student	2%
Total	100%

e.

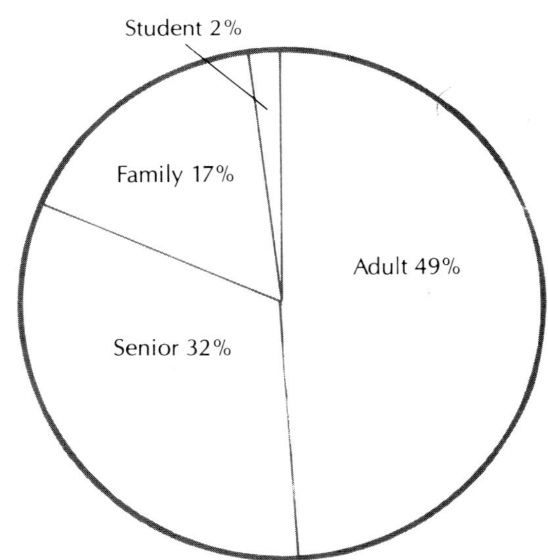

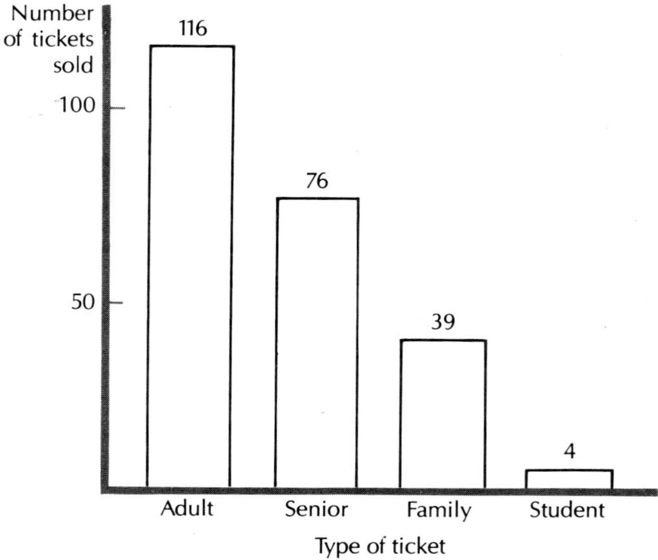

2/21. a. Frequency per $10,000: 3, 5, 9, 10, 23, 8, 5, 3, .33, .4.
b.

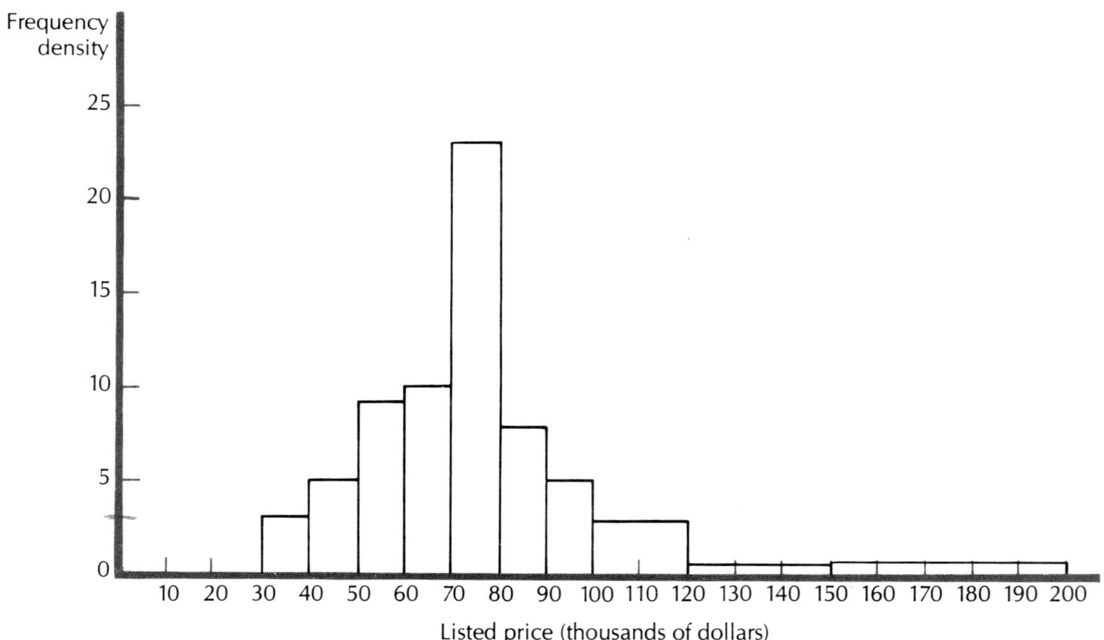

c.

Listed price	Cumulative percentage
Less than $ 30,000	0%
Less than $ 40,000	4%
Less than $ 50,000	11%
Less than $ 60,000	24%
Less than $ 70,000	38%
Less than $ 80,000	69%
Less than $ 90,000	81%
Less than $100,000	88%
Less than $120,000	96%
Less than $150,000	97%
Less than $200,000	100%

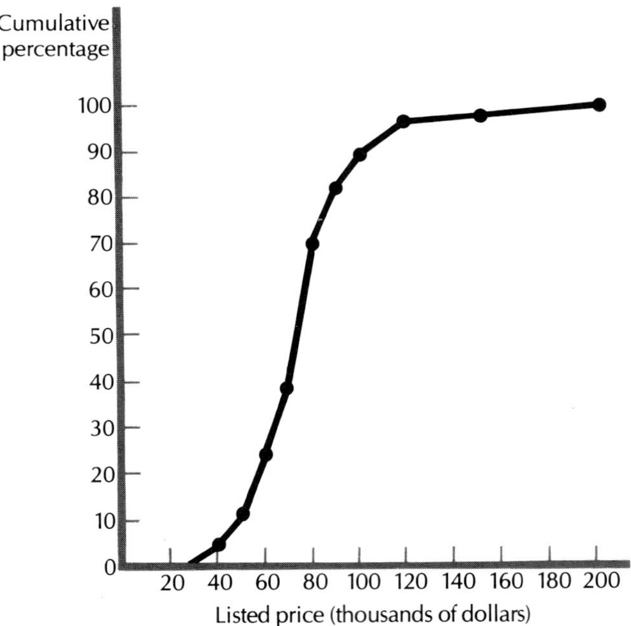

2/23. a.

Amount of adjusted gross income	Percentage frequency	Percentage frequency per $5,000
Less than $1,000 (incl. losses)	.01%	?
$ 1,000 but less than $ 5,000	7.7 %	9.6
$ 5,000 but less than $ 10,000	24.7 %	24.7
$ 10,000 but less than $ 15,000	21.7 %	21.7
$ 15,000 but less than $ 25,000	29.6 %	14.8
$ 25,000 but less than $ 50,000	14.1 %	2.8
$ 50,000 but less than $100,000	1.7 %	.2
$100,000 but less than $500,000	.4 %	.005
$500,000 or more	.01%	?
Total	100 %	

b. The densities make it clear that the greatest concentration of adjusted gross incomes is in the vicinity of $5,000–$10,000; the apparent concentration in the category $15,000–$25,000 is due to the larger interval of that category. **c.** Space may be saved by adding a verbal note about categories whose percentage frequency densities are too small to show up in the diagram.

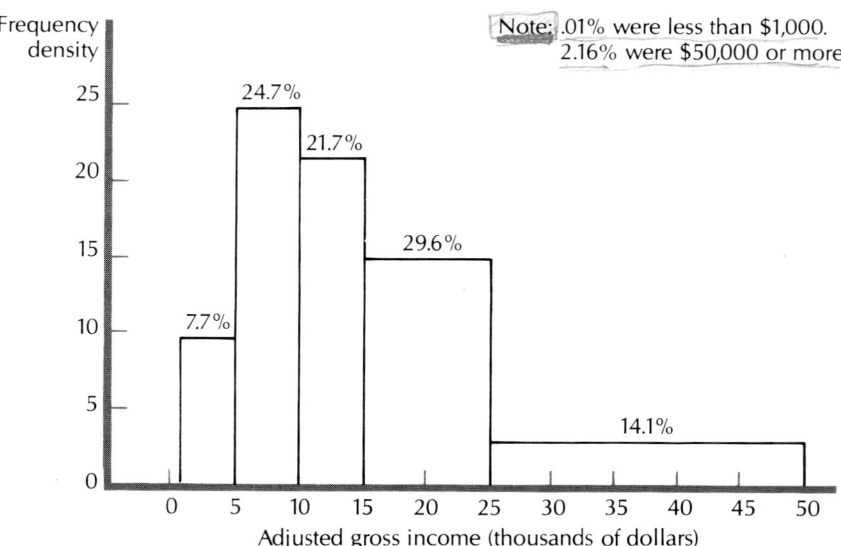

Chapter 3

3/1. **a.** 5.4 **b.** 10.1
3/3. **a.** 736 **b.** 179
3/5. 1.5
3/7. 4.3
3/9. **a.** Median = 7, mode = 9 **b.** Median = 11.5, modes = 7 and 12
3/11. **a.** $Q_1 = 53.5$, $Q_2 = 72$, $Q_3 = 81.5$ **b.** $C_{10} = 39$, $C_{80} = 82$ **c.** 51 **d.** 67.5 **e.** 69.75
3/13. **a.** $Q_1 = 1.9$, $Q_2 = 3.9$, $Q_3 = 6.1$ **b.** $C_{40} = 3.2$, $C_{95} = 10.3$

3/15. **a.** Mean. (The normal temperature is an arithmetic mean of temperatures recorded over a period typically of about 30 years.) **b.** Maximum. **c.** Weighted mean. **d.** Mode. (Reagan was the modal choice of voters.) **e.** Minimum. **f.** Median. (The median age was 29 years.)

3/17. **a.** Buyers of new Ford Mustangs. **b.** Age of the buyer. **c.** All buyers of new Ford Mustangs in the late 1970s. **d.** Mustang buyers whose ages were recorded for the investigation that led to the conclusion in statement (f). **e.** The median age, 29 years, of the Mustang buyers referred to in Part (d). **f.** The unknown median age of the Mustang buyers referred to in Part (c).

3/19. **a.** The 10 cows are the only ones being tested to obtain information about butterfat content in milk from a larger herd of which the 10 cows are a part. **b.** The 10 cows are being tested to evaluate the quality of those particular 10 cows.

3/21. **a.** Students. **b.** Verbal report of number of pages read. **c.** 175.9 **d.** 198.5 **e.** 308

3/23. **a.** Tables served. **b.** Amount of the tip. **c.** 200 cents. **d.** 0 cents. **e.** 72 cents. **f.** 50 cents. **g.** 50 cents.

3/25. **a.** Periods of business recovery. **b.** Duration of business recovery. **c.** 106 months. **d.** 12 months. **e.** 44.6 months. **f.** 38 months. **g.** There is no mode.

3/27. $31

3/29. $10.80

3/31. 17.2%

3/33. One-pizza orders, two-pizza orders, three-pizza orders, and so on.

3/35. **a.** $Q_1 = 130$, $Q_2 = 142$, $Q_3 = 177$ **b.** $D_4 = 136$, $D_5 = 142$, $D_6 = 149$ **c.** $C_{33} = 132$, $C_{50} = 142$, $C_{77} = 177$ **d.** C_{92}, C_{100}, C_{15}, respectively. **e.** 142 **f.** 161 **g.** 153.5 **h.** 147.75

3/37. **a.** $Q_1 = \$60{,}000$; $Q_2 = \$65{,}000$; $Q_3 = \$75{,}000$ **b.** $C_6 = \$44{,}500$; $C_{95} = \$91{,}000$ **c.** C_{18} or C_{19} would both be correct for Illinois; any value from C_{87} to C_{93} would be correct for California. **d.** $65,000 **e.** $67,500 **f.** $67,500 **g.** $66,250

3/39. **a.** $76,340 **b.** $Q_1 = \$60{,}950$; $Q_2 = \$73{,}860$; $Q_3 = \$84{,}950$ **c.** $C_2 = \$34{,}750$; $C_{98} = \$163{,}950$ **d.** $73,860 **e.** $72,950 **f.** $73,410

Chapter 4

4/1. 4

4/3. 1.25

4/5. **a.** 2.917 **b.** 2.917 **c.** 1.7

4/7. **a.** 3.5 **b.** 4.5

4/9. $\bar{x} = 5.7$, $s = 2.3$

4/11. **a.** 8 **b.** Approximately 2, or between 1.33 and 4. **c.** 3.6, which is consistent with the approximation.

4/13. **a.** No more than 25% of the observations. **b.** No more than 6.25% of the observations. **c.** At least 75% of the observations. **d.** At least 93.75% of the observations.

4/15. **a.** Approximately 68% of the observations. **b.** Approximately 95% of the observations. **c.** Approximately 99%–100% of the observations.

4/17. **a.** 2.17 **b.** −1.11 **c.** .02 **d.** −.86 **e.** 0.00

4/19. **a.** 250.0 **b.** 230.0 **c.** 296.8 **d.** 265.0 **e.** 246.6

4/21. **a.** $Q_1 = 40$, $Q_2 = 43$, $Q_3 = 50$ **b.** Interquartile range = 10, quartile deviation = 5 **c.** 33 and 65 **d.** 21 **e.** Box plot:

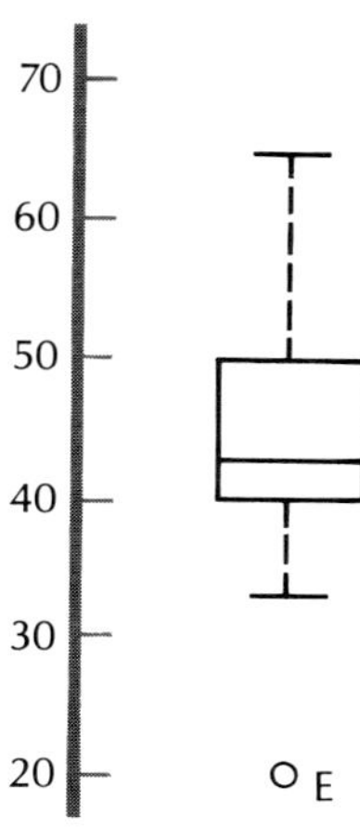

4/23. **a.** 11 **b.** Approximately 2.8, or between 1.8 and 5.5. **c.** 8.71 **d.** 3.0, which is consistent with the approximation. **e.** The students' opinions of an appropriate age for marriage had a mean of 24.4 years and a standard deviation of 3.0 years.

4/25. **a.** 22 **b.** Approximately 5.5, or between 3.7 and 11. **c.** 40.9 **d.** 6.4, which is consistent with the approximation. **e.** 6.4. The values are identical because the standard deviation is a measure of how spread out the observations are and is not affected by their location. Adding or subtracting a constant, which may simplify the calculations, does not affect the dispersion of the observations; it affects only their location on the real-number line. **f.** The resting heart rates in this sample had a mean of 71.1 beats per minute and a standard deviation of 6.4 beats per minute.

4/27. **a.** \$3.91 **b.** 391 cents **c.** \$3.91 equals 391 cents. The standard deviation is in the same units as the observations from which it is calculated. Multiplying or dividing by a constant changes the scale of measurement of the observations and hence changes the standard deviation by the same factor. **d.** The hourly wage rates of the students in this sample had a mean of \$8.99 and a standard deviation of \$3.91.

4/29. $\bar{x} = \$76{,}340$; $s = \$25{,}870$

4/31. **a.** −1.08, −.13, 2.11, .003, −.51, .69, −.70, −.38 **b.** $\bar{z} = 0.00$, $s_z = 1.00$

4/33. **a.** $z = .4$. Her mark is .4 standard deviation above the mean; this is above the class average, but not exceptionally so. **b.** $z = -.1$. His mark is below the class average, but only very slightly so, relative to the other marks.

4/35. Altex yield = 1.1 standard units; Candle yield = 1.0 standard units. Altex was slightly higher above its average than Candle was above its average.

4/37. **a.** At least 52% of the farmers had such yields. **b.** No more than 28% of the farmers had such yields. (It is not appropriate to divide this tail area in half, because the inequality does not assume symmetry.) **c.** No more than 25% of the farmers had such yields. (It is not appropriate to divide this tail area in half, because the inequality does not assume symmetry.)

4/39. **a.** Approximately 99%–100% of the farmers. **b.** Approximately 16% of the farmers. (The tail area is divided in half, because the empirical rule assumes symmetry.) **c.** Approximately 2%–3% of the farmers. (The tail area is divided in half, because the empirical rule assumes symmetry.)

4/41. **a.** Interquartile range = \$15,000, quartile deviation = \$7,500 **b.** \$40,000 and \$91,000 **c.** \$35,000 (Arkansas and Maine), \$100,000 (New York)
d. Box plot:

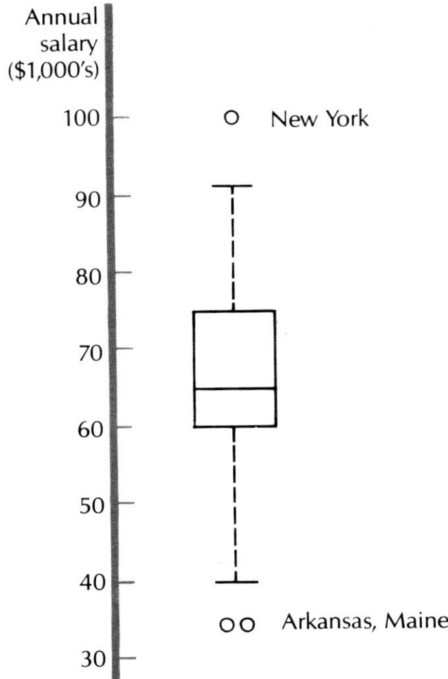

Chapter 5

5/1. **a.** $b = -1.1$ **b.** $a = 9.5$ **c.** $\hat{y} = -1.1x + 9.5$, or $\hat{y} = 9.5 - 1.1x$
5/3. $\hat{y} = .35x + 3.8$, or $\hat{y} = 3.8 + .35x$
5/5. $r = -.83$; $r^2 = .69 = 69\%$
5/7. $r = .29$; $r^2 = 9\%$
5/9. $r = -.96$
5/11. $\hat{y} = 14.2x - 4$
5/13. **a.** Incidence of tooth decay. **b.** Price. **c.** Number of residential construction starts. **d.** Yearly expenditures on consumer durables. **e.** Fair market value. **f.** University statistics grade. **g.** Number of accidents reported. **h.** Barley yield. **i.** Number of points scored. **j.** Fuel consumption.
5/15. **a.** DV = number of mistakes, with values of 13, 4, . . . , 3; IV = number of previous rehearsals, with values of 0, 1, . . . , 12 **b.** $\hat{y} = 8.6 - .47x$ **c.** $r = -.62$ **d.** The scattergram, which is shown in Figure 5.12(g), provides no strong evidence that the linear description is inappropriate for up to 12 prior rehearsals, but it is clear that the relationship cannot remain linear for increasing numbers of rehearsals, since 0 is the minimum number of mistakes possible. **e.** The regression equation suggests that this flute player made an average of about 9 mistakes upon first sight-reading and that every two rehearsals reduced the average number of mistakes by 1 (within the range of x-values observed). Only $r^2 = 38\%$ of the variation in number of mistakes can be accounted for by

the linear relationship with number of previous rehearsals, indicating a linear relationship of only moderate strength.

5/17. **a.** DV = height of biscuit, with values of 6, 8, . . . , 10; IV = amount of baking powder, with values of 0, 1, . . . , 9 **b.** $\hat{y} = 8.5 + .28x$ **c.** $r = .50$ **d.** $r = .50$ **e.** The scattergram suggests that a linear description is not appropriate for these data because the relationship between the variables appears distinctly curvilinear with a maximum near 4 half-teaspoons (2 teaspoons). **f.** Only $r^2 = 25\%$ of the variation in height can be accounted for by the linear relationship with amount of baking powder, indicating only a weak linear relationship. In view of the answer to Part (e), combined with the low coefficient of determination, the calculated slope and y-intercept are misleading descriptions of the relationship.

5/19. **a.** DV = personal consumption expenditures, with values of 430, 490, . . . , 1858, in billions of dollars; IV = disposable personal income, with values of 476, 548, . . . , 2015, in billions of dollars **b.** $\hat{y} = .932x - 25$ **c.** $r = .9996$ **d.** The scattergram provides no evidence that the linear description is inappropriate within the observed range of the variables. **e.** $r^2 = 99.93\%$ of the variation in personal consumption expenditures can be accounted for by the linear relationship with disposable personal income, indicating a very strong linear relationship within the observed range of incomes. The slope $b = .932$ indicates that for every dollar increase in disposable income, 93.2 cents is spent and 6.8 cents is saved, on the average. (In this type of application, economists refer to these values as the marginal propensity to consume and the marginal propensity to save, respectively.) The impossible value of −25 billion dollars of expenditures when income is zero suggests that the actual relationship is not linear over a wider range of income.

5/21. **a.** DV = pulse rate; IV = exercise duration **b.** 138 beats per minute. **c.** The regression equation suggests that the athlete's average pulse rate when standing alert, but not exercising, was 79.6 beats per minute and that each minute of performing step-ups added an average of 14.6 beats per minute to his pulse rate. **d.** $r = .99$ **e.** $r^2 = 98\%$ of the variation in pulse rate can be accounted for by the linear relationship with exercise duration. There is a strong linear relationship, within the observed range of the variables.

5/23. **a.** DV = average number of points scored; IV = distance **b.** $\hat{y} = 7.2 - .23x$ **c.** The regression equation seems to suggest that this dart player would average 7.2 points per dart when standing right at the board, and that each foot additional distance from the board would reduce his score by an average of .23 point. However, since the bull's-eye is worth 10 points, it seems unreasonable to predict only 7.2 points for a distance of 0 feet. This discrepancy suggests that the relationship is curvilinear, with scores rising more rapidly as the distance becomes very small, and that extrapolation outside the range of distances observed is not appropriate. **d.** $r^2 = 96\%$ of the variation in average points scored can be accounted for by the linear relationship with distance. There is a strong linear relationship within the observed range of the independent variable (8 to 28 feet).

5/25. Samples of $n = 2$ observations always produce perfect correlations, unless the two points have zero slope or infinite slope, in which case the correlation coefficient is not defined (because of division by zero). The correlation coefficient is an index of the extent to which the data points tend to fall in a perfect straight line. Any two points can be joined by a straight line and so represent perfect correlation.

5/27. Correlation itself never justifies a causal conclusion. It could be, for example, that a third variable, such as some aspect of personality, makes a person inclined to attend church frequently as well as less prone to high blood pressure, heart disease, and so on.

5/29. Although instruction and overconfidence may be appropriate explanations in some individual cases, these explanations are probably examples of the regression fallacy. Students who have an unusually bad day at midterm are likely to have a less bad day at the final, and students who have an unusually good day at midterm are likely to have a less good day at the final, simply because there is more room on the less extreme side of the distribution.

Chapter 6

6/1. −.13, 98, 300%, 4/3, 2.67
6/3. **a.** 7 to 3 in favor of A **b.** 11 to 9 in favor of B **c.** 9 to 1 against C **d.** 17 to 3 against D
6/5. **a.** 2/3 **b.** 1/3 **c.** 13/25 **d.** 12/25
6/7. **a.** 11/46 **b.** 11/62
6/9. **a.** .075 **b.** .125
6/11. **a.** .2 **b.** .4 **c.** .08 **d.** .52
6/13. **a.** 0 **b.** 0 **c.** 0 **d.** .70
6/15. **a.** .57 **b.** .63 **c.** .43 **d.** .37 **e.** .99 **f.** .58 **g.** .64 **h.** .79 **i.** .21 **j.** .36 **k.** .42 **l.** .01 **m.** 1/3 **n.** 36/37 **o.** 2/3 **p.** 1/37
6/17. .18
6/19. **a.** .66 **b.** .31 **c.** .031
6/21. **a.** .01 **b.** .01 **c.** .04
6/23. 16/31, or .52
6/25. C
6/27. 5 to 1 in favor of the discovery.
6/29. **a.** 4 to 1 against being in business and commerce. **b.** 9 to 1 against being in business and commerce.
6/31. **a.** 6/25, or .24 **b.** 3/10, or .3
6/33. 3/7, or .43
6/35. **a.** 42/93, simple. **b.** 71/93, disjunctive. **c.** 29/46, conditional. **d.** 5/93, joint. **e.** 67/93, disjunctive. **f.** 33/93, simple. **g.** 9/42, conditional.
6/37. **a.** .8 **b.** .06 **c.** .73 **d.** .27
6/39. **a.** .76 **b.** .078
6/41. **a.** .76 **b.** .95
6/43. To answer these questions on the basis of the data given, it is necessary to assume that month of birth and cause of death are independent. This assumption seems sufficiently reasonable to provide adequate approximations to the required probabilities: **a.** .016 **b.** .21 **c.** .075
6/45. **a.** .28 **b.** .82 **c.** .56
6/47. **a.** .15 **b.** .50 **c.** .75 **d.** .40 **e.** .25 **f.** No, because $P(H \text{ and } M) \neq 0$. **g.** No, because $P(H \text{ and } M) \neq P(H)P(M)$.
6/49. **a.** .97 **b.** .03
6/51. .217
6/53. .79
6/55. .77
6/57. .50
6/59. **a.** Success in one lottery has no bearing on one's chances of winning another lottery. **b.** It is true that any one ticket is just as likely to win as any other single ticket, but a person with several tickets is more likely to win than a person with one ticket. A lottery may be considered to have equally likely outcomes on a per-ticket basis but not on a per-person basis.
6/61. Several points could be mentioned, such as the following: (1) Occasions when a book was missing are apt to be better remembered (and hence to seem relatively more frequent) than occasions when a book is found. (2) Some books are taken out of a library much more often than others. A book that is sought after by one person may well be a book that is sought after by many and therefore one that is rarely on the shelf.

Chapter 7

7/1. **a.** No. The values of a probability function must total 1.0. **b.** No. The values of a probability function must total 1.0. **c.** Yes. **d.** No. The values of a probability function must be probabilities and hence between 0 and 1, inclusive **e.** No. The values of a probability function must be probabilities and hence nonnegative.

7/3. $F(1) = .29$, $F(2) = .82$, $F(3) = .84$, $F(4) = 1.00$
7/5. 2.05
7/7. **a.** 2.05 **b.** .9475 **c.** .97
7/9. **a.** 3.5 **b.** .45 **c.** .67
7/11. **a.** 16/81 **b.** 32/81 **c.** 24/81 **d.** 8/81 **e.** 1/81 **f.** 80/81
7/13. **a.** .45 **b.** .27
7/15. **a.** 720 **b.** 12
7/17. **a.** 32 **b.** 5.6
7/19. **a.** .10 **b.** .027
7/21. **a.** .48 **b.** .20
7/23. **a.** .18 **b.** .20
7/25. **a.** .9864 **b.** .440 **c.** .132 **d.** .374 **e.** .9070 **f.** .084 **g.** .552 **h.** .022 **i.** .0156 **j.** .99914
7/27. The Bienaymé–Chebyshev inequality would indicate "at least 64.57%," rather than "exactly 90.70%," because it does not make use of the information about the shape of the distribution, which in this case is Gaussian.
7/29. **a.** 1.34 **b.** 3.72 **c.** 2.58 **d.** 0.00 **e.** .84 **f.** −2.75 **g.** −2.33 **h.** −3.24
7/31. **a.** 1.96 **b.** −.84 **c.** 1.64 **d.** −.67
7/33. **a.** .374 **b.** .99869 **c.** .215 **d.** .924 **e.** .721
7/35. Amount of tips (**b**) and number of letters (**d**) are discrete; the rest are continuous.
7/37. .45
7/39. **a.** Grade, discrete. **b.** Probability function. **c.** 5.91 **d.** $\mu = 5.91$, $\sigma^2 = 2.80$, $\sigma = 1.67$ **e.** The probability of failing (obtaining a grade of 1, 2, or 3) is 8%. **f.** The probability of obtaining an excellent grade (8 or 9) is 17%.
7/41. $\mu = 3.50$, $\sigma = 1.71$. The standard deviation is 34% of the range, which is consistent with the rule of thumb.
7/43. **a.** .729 **b.** .243 **c.** .027 **d.** .001 **e.** .999
7/45. **a.** .19 **b.** .28 **c.** $\mu = 3.36$, $\sigma = 1.40$
7/47. **a.** .025 **b.** $\mu = .025$, $\sigma = .158$
7/49. .0024
7/51. 1,320/2,598,960 = .00051
7/53. **a.** .400 **b.** .395
7/55. **a.** .45 **b.** .36 **c.** 2 bags **d.** .0474
7/57. **a.** .51 **b.** .34 **c.** .12 **d.** .032
7/59. **a.** .00043 **b.** .00135 **c.** .9901 **d.** .749
7/61. **a.** .0068 **b.** .104
7/63. .0287
7/65. **a.** .0066 **b.** .797 **c.** Yes, because $np = 12$ and $nq = 38$ are both greater than 5.

Chapter 8

8/1. **a.** $\mu_{\bar{x}} = 56.9$, $\sigma_{\bar{x}} = 5.8$ **b.** 5.8 **c.** $z_{\bar{x}} = .71$ **d.** .239
8/3. **a.** 12.3 **b.** .582
8/5. **a.** $\mu_{\bar{x}} = 156$, $\sigma_{\bar{x}} = 11$ **b.** 11 **c.** $z_{\bar{x}} = -1.09$ **d.** .138
8/7. **a.** .43 **b.** .0204
8/9. **a.** Accuracy would be reduced due to systematic undercounting. Precision would suffer as well, since different books would be undercounted by differing amounts. **b.** Precision would be reduced. There may also be some loss of accuracy, since those in the "indifferent" (4) category on the

seven-point scale may have a bias in favor of either pro or con. **c.** Accuracy would be reduced, in that there is a systematic bias to understate the age by half a year on the average. Precision would not be affected.

8/11. **a. i.** $\mu_{\bar{x}} = 100.0$, $\sigma_{\bar{x}} = 5.0$ **ii.** $\mu_{\bar{x}} = 100.0$, $\sigma_{\bar{x}} = 2.5$ **b. i.** 225 **ii.** 57 **c. i.** .0038 **ii.** .092 **d. i.** .046 **ii.** .11

8/13. **a. i.** $\mu_{\bar{x}} = 4{,}500{,}000$, $\sigma_{\bar{x}} = 25{,}000$ (rounded from 24,749) **ii.** $\mu_{\bar{x}} = 4{,}500{,}000$, $\sigma_{\bar{x}} = 49{,}000$ (rounded from 49,497) **b. i.** 77 **ii.** 9 **c. i.** .18 **ii.** .035 **d. i.** .46 **ii.** .986

8/15. **a.** .37 **b.** .14

8/17. **a.** $\mu = 4.50$, $\tilde{\mu} = 4.50$, $\sigma = 2.87$ **b.** $\mu_{\bar{x}} = 4.50$, $\sigma_{\bar{x}} = 1.28$ **c.** Record 125 successive digits from a table of random digits, beginning at a blindly selected starting point and treating successive sets of 5 digits as samples of 5 observations. **e.** The mean and the standard deviation of the empirical sampling distribution should roughly approximate those of the theoretical sampling distribution, although the use of only 25 samples will limit the precision of the approximation. **g.** The mean of the sampling distribution of the median should roughly approximate the population median. You should find that the sample mean has a smaller standard error and hence greater precision than the sample median.

Chapter 9

9/1. 1.31

9/3. .87

9/5. **a.** 1018 to 1146 hours **b.** 1018 hours and 1146 hours **c.** 99%

9/7. **a.** μ_1 = the mean of the hypothetical population of all first-cutting yields that might possibly be obtained in the fields studied.
σ_1 = the standard deviation of the preceding population.
μ_2 = the mean of the hypothetical population of all second-cutting yields that might possibly be obtained in the fields studied.
σ_2 = the standard deviation of the preceding population.
b. $\bar{x}_1 = 119$ and $s_1 = 38$ are the mean and the standard deviation of the observed sample of 9 first-cutting yields.
$\bar{x}_2 = 47$ and $s_2 = 15$ are the mean and the standard deviation of the observed sample of 7 second-cutting yields.
c. First cutting: 119 ± 13 bales/acre. Second cutting: 47 ± 6 bales/acre. (Mean ± standard error.) **d.** DV = hay yield; IV = time of cutting, with levels first cutting and second cutting. **e.** Population standard deviations. **f.** The population variability of first-cutting yields is the same as the population variability of second-cutting yields. (Or the standard deviation of the hypothetical population of all possible first-cutting yields equals the standard deviation of the hypothetical population of all possible second-cutting yields.) **g.** Yes. The results are summarized in the phrase "significantly more variable." **h.** A Type I error might have been made, and a Type II error could not possibly have been made, given the reported results.

9/9. **a.** The proportion of all riverbank trees in the Fort McMurray area that are willows, and the proportion of all riverbank trees in the Camrose area that are willows. **b.** The sample proportions: 73% of the Fort McMurray sample of trees were willows, and 69% of the Camrose sample of trees were willows. **c.** DV = species of tree; IV = region, with levels Fort McMurray area and Camrose area. **d.** Population proportions. **e.** The proportion of all riverbank trees in the Fort McMurray area that are willows equals the proportion of all riverbank trees in the Camrose area that are willows. **f.** No. "The difference was not statistically significant." **g.** A Type II error might have been made, and a Type I error could not possibly have been made, given the reported results.

Chapter 10

10/1. **a.** .67 **b.** 1.96 **c.** 2.33 **d.** −1.28 **e.** −2.05
10/3. 2.7
10/5. 32.95 and 34.67
10/7. 41
10/9. **a.** 2.571 **b.** −2.947 **c.** 2.68 **d.** 1.960 **e.** −2.896 **f.** 1.69 **g.** 3.37 **h.** 2.326
10/11. 4.3
10/13. 331 and 461
10/15. **a.** We can be 95% confident that the sample mean differs from the population mean by less than .029 mm. **b.** 9.984 and 10.062 mm **c.** We can be 99% confident that the sample mean differs from the population mean by less than .049 mm. **d.** 9.934 and 10.008 mm **e.** 390 **f.** 959
10/17. **a.** We can be 95% confident that the sample mean differs from the mean of the relevant population of hens by less than 95 grams. **b.** 1039 and 1229 grams **c.** 995 to 1273 grams
10/19. 38.8 to 43.2 strokes
10/21. $16.29 to $68.96

Chapter 11

11/1. **a.** The sample mean was significantly less than the hypothesized value of 100.0 [t_{obs}(20 df) = −3.21, $P < .05$]. **b.** 92.3 to 98.3
11/3. The sample provides insufficient evidence to conclude that the population mean is less than 5.0 [t_{obs}(5 df) = −1.03, $P > .01$].
11/5. **a.** The sample means did not differ significantly [t_{obs}(22 df) = .75, $P > .10$]. **b.** We can be 90% confident that the population mean difference $\mu_1 - \mu_2$ is between −8 and +20.
11/7. **a.** The Sample 2 mean was significantly larger than the Sample 1 mean [t_{obs}(9 df) = −7.57, $P < .001$]. **b.** We can be 99.9% confident that the mean of Population 2 is between 2.50 and 11.07 larger than the mean of Population 1.
11/9. **a.** The Sample 1 mean was significantly larger than the Sample 2 mean [t_{obs}(3 df) = 4.38, $P < .05$]. **b.** We can be 95% confident that the mean of Population 1 is between 1.1 and 6.9 larger than the mean of Population 2.
11/11. **a.** H_0: $\mu = 2500$. According to the null hypothesis, the mean of the relevant population of female college students' summer earnings was $2500.
H_A: $\mu \neq 2500$. According to the alternative hypothesis, the population mean was not equal to $2500.
b. The sample mean of $1612 was significantly less than $2500 [$t_{obs}$(16 df) = −3.34, $P < .01$], thus favoring the alternative hypothesis. **c.** It is either a Type I error or a correct rejection. **d.** We can be 99% confident that the mean of the relevant population of female college students' summer earnings is between $835 and $2389.
11/13. **a.** H_0: $\mu = 37$. According to the null hypothesis, the mean pulse rate in the relevant population of mature light horses is 37 beats per minute.
H_A: $\mu \neq 37$. According to the alternative hypothesis, the mean pulse rate of such horses is not equal to 37 beats per minute.
b. The sample mean of 34.2 beats per minute did not differ significantly from 37 beats per minute [t_{obs}(9 df) = −1.28, $P > .20$], consistent with the null hypothesis. **c.** It is either a Type II error or a correct nonrejection. **d.** We can be 95% confident that the mean pulse rate of such horses is between 29.3 and 39.1 beats per minute.
11/15. **a.** DV = dressed weight; IV = species of deer.
b. H_0: $\mu_1 = \mu_2$. According to the null hypothesis, the mean dressed weight of the relevant population of mule deer bucks is equal to that of whitetail bucks.

H_A: $\mu_1 \neq \mu_2$. According to the alternative hypothesis, the mean dressed weights of the relevant populations of bucks of the two species are not equal.

c. The mean difference of 9 pounds between the two species in the sample is not statistically significant [t_{obs}(8 df) = .83, $P > .20$], consistent with the null hypothesis. **d.** We can be 95% confident that the mean dressed weight of the population of mule deer bucks is between 16 pounds lighter and 34 pounds heavier than that of whitetail bucks.

11/17. **a.** DV = reported average study time per week; IV = year of studies.

b. H_0: $\mu_1 = \mu_2$. According to the null hypothesis, the mean reported study times of the relevant first-year and later-year populations of students are equal.

H_A: $\mu_1 \neq \mu_2$. According to the alternative hypothesis, the mean reported study times of the relevant first-year and later-year populations of students are not equal.

c. The mean difference of 8.5 hours per week in the sample, with first-year students reporting less study time than later-year students, was statistically significant [t_{obs}(10 df) = -2.44, $P < .05$], thus supporting the alternative hypothesis. It is a separate question as to whether reported study times accurately reflect actual study times. **d.** We can be 95% confident that the mean weekly study times reported by first-year students would be between .7 and 16.3 hours less than those reported by later-year students in the populations from which these two samples were drawn.

11/19. **a.** DV = width of splenium; IV = sex of subject.

b. H_0: $\mu_m = \mu_f$. According to the null hypothesis, the mean splenial widths of the relevant male and female populations are equal.

H_A: $\mu_m \neq \mu_f$. According to the alternative hypothesis, men and women differ in mean splenial width.

c. The sample mean difference of .50 cm wider splenium in female brains was statistically significant [t_{obs}(12 df) = -5.07, $P < .001$]. **d.** We can be 99.9% confident that in the populations from which the samples were drawn, the mean splenial width is between .07 and .93 cm greater for females than for males.

11/21. **a.** DV = resting pulse rate; IV = consumption of tea.

b. H_0: $\mu_d = 0$. According to the null hypothesis, there is no difference in mean resting pulse rate before and after consumption of tea.

H_A: $\mu_d \neq 0$. According to the alternative hypothesis, there is a difference in mean resting pulse rate before and after consumption of tea.

c. The sample mean increase of 3.1 beats per minute after consumption of tea was statistically significant [t_{obs}(7 df) = 3.10, $P < .02$], thus favoring the alternative hypothesis. **d.** We can be 95% confident that the true mean increase in resting pulse rate is between .7 and 5.5 beats per minute under the conditions of this experiment.

11/23. **a.** DV = appraised value; IV = appraiser.

b. H_0: $\mu_A = \mu_B$. According to the null hypothesis, the two appraisers yield the same mean appraised values.

H_A: $\mu_A \neq \mu_B$. According to the alternative hypothesis, one appraiser tends to give higher appraised values on the average.

c. The sample mean difference of $5200 higher appraisals by Appraiser A was statistically significant [t_{obs}(8 df) = 2.41, $P < .05$]. **d.** We can be 95% confident that the true mean difference between Appraisers A and B is between $200 and $10,200 higher appraisals by Appraiser A.

11/25. **a.** DV = reported weight; IV = age (or generation).

b. H_0: $\mu_d = 10$. According to the null hypothesis, the mean difference between mother's weight and daughter's weight in this population of female college students is 10 pounds.

H_A: $\mu_d > 10$. According to the alternative hypothesis, mothers weigh more than 10 pounds more than their college-student daughters in this population.

c. The mean difference of 14.1 pounds observed in this sample was not significantly greater than 10 pounds [t_{obs}(14 df) = -1.17, $P > .10$, one-tailed], consistent with the null hypothesis. **d.** We can be 95% confident that, in the relevant population, female college students' mean weight is between 6.5 and 21.7 pounds less than that of their mothers.

11/27. Classical hypothesis testing does not involve probabilities that hypotheses are true; rather, it involves conditional probabilities of obtaining various kinds of sample data, on the condition that a certain hypothesis is true. Thus, a significance level of 2.8% does not mean that

"the probability that the increase resulted from a chance distribution of particularly rainy days was 2.8 percent,"

but rather that

the probability that such a large increase would occur if the cloud seeding were actually having no effect was 2.8 percent.

In symbols, a significance level of 2.8% does not mean

$P(H_0 \mid \text{the observed increase}) = .028$

as implied by the quoted statement, but rather

$P(\text{such a large increase} \mid H_0) = .028$

Chapter 12

12/1. **a.** 36.415 **b.** 20.090 **c.** 34.267 **d.** 40.646 **e.** 5.009 **f.** 7.434 **g.** 18.493 **h.** .0201

12/3. 16.4 to 31.5

12/5. 22.6 to 49.4

12/7. **a.** The sample value of 5.6 is significantly less than the hypothesized value of 10.0 [$\chi^2_{obs}(12\text{ df}) = 3.76, P < .05$]. **b.** The sample value of 13.0 is not significantly different from the hypothesized value of 10.0 [$\chi^2_{obs}(18\text{ df}) = 30.4, P > .05$].

12/9. **a.** 3.26 **b.** 3.29 **c.** 4.10 **d.** 2.88

12/11. The sample standard deviations are not significantly different [$F_{obs}(4, 9\text{ df}) = 4.12, P > .05$].

12/13. H_0: $\sigma = .60$. According to the null hypothesis, the distribution of recalled birth weights in the relevant population of female college students has a standard deviation of .60 kg.

H_A: $\sigma \neq .60$. According to the alternative hypothesis, the standard deviation of recalled birth weights in the relevant population is not .60 kg.

b. The sample standard deviation of .333 kg was significantly less than the hypothesized value of .60 kg [$\chi^2_{obs}(15\text{ df}) = 4.62, P < .02$], indicating that the recalled birth weights of female college students are less variable than actual birth weights in the general population. **c.** We can be 98% confident that the true standard deviation of the relevant population of recalled birth weights is between .23 and .56 kg.

12/15. **a.** H_0: $\sigma^2 = 30.0$. According to the null hypothesis, the variance of the relevant population of resting pulse rates is 30.0.

H_A: $\sigma^2 \neq 30.0$. According to the alternative hypothesis, the population variance is not equal to 30.0.

b. The sample variance of 47.5 did not differ significantly from the hypothesized value of 30.0 [$\chi^2_{obs}(9\text{ df}) = 14.3, P > .10$], consistent with the null hypothesis. **c.** We can be 95% confident that the true variance of resting pulse rates for such horses is between 22.5 and 158.

12/17. **a.** DV = price per page; IV = type of book (hardcover or paperback).

b. H_0: $\sigma_{hard} = \sigma_{paper}$. According to the null hypothesis, the true standard deviation of price per page is the same for hardcover and for paperback textbooks.

H_A: $\sigma_{hard} \neq \sigma_{paper}$. According to the alternative hypothesis, the population standard deviations are not the same.

c. The sample standard deviations of 1.41 cents per page for hardcovers and 1.62 cents per page for paperbacks were not significantly different [$F_{obs}(13, 10\text{ df}) = 1.32, P > .20$], consistent with the null hypothesis. **d.** It is either a Type II error or a correct nonrejection.

12/19. **a.** DV = average song length; IV = type of music.

b. H_0: $\sigma^2_{\text{country}} = \sigma^2_{\text{rock}}$. According to the null hypothesis, the true variance of average song lengths on country albums is equal to that on rock albums.
H_A: $\sigma^2_{\text{country}} \neq \sigma^2_{\text{rock}}$. According to the alternative hypothesis, the true variance of average song lengths on country albums is not equal to that on rock albums.
c. The sample standard deviations of .37 minute for the country albums and .83 minute for the rock albums were significantly dilfferent at the .05 significance level [$F_{\text{obs}}(10, 9\text{ df}) = 4.96$, $P < .05$], indicating greater variability among rock albums. **d.** It is either a Type I error or a correct rejection.

12/21. The thickness of supplier B's components had significantly greater variability than that of supplier A [$F_{\text{obs}}(5, 7\text{ df}) = 5.90$, $P < .05$].

Chapter 13

13/1. **a.** .377 **b.** −1.333 **c.** 2.994 **d.** −1.516 **e.** −.678 **f.** 1.099
13/3. **a.** −.53 and −.03 **b.** −.59 and .06
13/5. **a.** 50 **b.** Yes **c.** No
13/7. **a.** $SS_x = 63.3333$, $SS_y = 78.8333$, $SP_{xy} = -60.6667$; $s_x = 3.55903$, $s_y = 3.97073$, $\text{cov}(x, y) = -12.1333$ **b.** $\hat{y} = 19.0 - .96x$ **c.** $r = -.86$
d.

Source of variation	SS	df	MS	F	
Regression on x	58.11	1	58.11	11.2	$P < .05$
Residual	20.72	4	5.18	—	
Total	78.83	5			

e. $r^2 = 74\%$ **f.** $s_e = 2.3$ **g.** $s_b = .29$. The 95% confidence limits are −1.75 and −.16. **h.** $s_a = 1.5$. The 95% confidence limits are 14.7 and 23.3. **i.** $s_{\hat{y}|x=4} = .93$. The 95% confidence limits are 12.6 and 17.7. **j.** $s_f = 1.4$. The 95% confidence limits are 8.4 and 16.2. **k.** $s_f = 2.6$. The 95% confidence limits are 5.1 and 19.4.

13/9. **a.**

Source of variation	SS	df	MS	F	
Regression on x	571	1	571	56.3	$P < .01$
Residual	284	28	10.1	—	
Total	855	29			

b. $r^2 = 67\%$ **c.** $r = \pm.82$ **d.** $s_e = 3.2$

13/11. **a.**

Source of variation	SS	df	MS	F	
Regression on x	2,008	1	2,008	1.13	$P > .10$
Residual	44,536	25	1,781	—	
Total	46,544	26			

b. $s_e = 42$

13/13. **a.** 11
b. H_0: $\rho = 0$. According to the null hypothesis, the true correlation between reported grade point average and number of evenings per month spent at the bar is 0.
H_A: $\rho \neq 0$. According to the alternative hypothesis, the true correlation between the variables is not 0.
The observed correlation, $r = -.67$, is statistically significant ($P < .02$), indicating that higher

grades are associated with fewer evenings at the bar. **c.** We can be 98% confident that the true correlation is between −.08 and −.91.

13/15. **a.** 12 **b.** The sample correlation, $r = .85$, is statistically significant ($P < .01$). **c.** We can be 95% confident that the true correlation is between .58 and .95, and hence that between 34% and 90% of the true variation in placement can be accounted for by the linear relationship with amount of practice.

13/17. **a.** $\hat{y} = 13.8x - 30$, where x is fitness score. The slope indicates that each additional point on the fitness test is accompanied by an average increase of 13.8 in the year-end point total. The y-intercept of −30 is not directly interpretable because it involves unwarranted extrapolation to a fitness score of 0. It could be, for example, that fitness scores of 0 are not represented in the population of players from which the sample was drawn. **b.** $r = .87$

c.

Source of variation	SS	df	MS	F	
Regression (b)	5727	1	5727	16.1	$P < .025$
Residual	1782	5	356	—	
Total	7509	6			

d. $r^2 = 76\%$ of the variation in year-end point totals (in this sample) can be accounted for by the linear relationship with fitness scores. **e.** $s_e = 19$. The standard amount of discrepancy between the observed year-end point totals and those estimated by the regression equation is 19 points. **f.** $s_b = 3.4$. The 95% confidence limits are 5.0 and 22.7. **g.** $s_{\hat{y}|x=4.5} = 10$. The 95% confidence limits are 7 and 58. **h.** $s_f = 21$. The 95% confidence limits are 27 and 135.

13/19. **a.** $\hat{y} = 1120 + 188x$, where x is family income in $1000s. The slope indicates that each additional $1000 of annual family income is associated with an average increase of $188 in annual expenditures on durables. The y-intercept is not directly interpretable because it involves unwarranted extrapolation to $x = 0$ income. However, the fact that the equation indicates an expenditure of $1120 on durables when there is no income strongly suggests that the relationship between these variables is not linear in lower ranges of income. **b.** $r = .85$

c.

Source of variation	SS	df	MS	F	
Regression on income	20,764,000	1	20,764,000	20.2	$P < .01$
Residual	8,212,000	8	1,027,000	—	
Total	28,976,000	9			

d. $r^2 = 72\%$ of the variation in annual expenditures on consumer durables (in this sample) can be accounted for by the linear relationship with annual family income. **e.** $s_e = 1013$. The standard amount of discrepancy between the observed expenditures on durables and those estimated by the regression equation is $1013. **f.** $s_b = 42$. The 99% confidence limits are 48 and 327. **g.** $s_{\hat{y}|x=25} = 369$. The 99% confidence limits are 4570 and 7040. **h.** $s_f = 584$. The 99% confidence limits are 3840 and 7770.

Chapter 14

14/1. .80 and .032
14/3. .26 and .40
14/5. .029
14/7. .016
14/9. 752

14/11. The sample proportions of $37\frac{1}{2}\%$, 35%, 22%, and $5\frac{1}{2}\%$ did not differ significantly from the hypothesized proportions [$X^2(3 \text{ df}) = 6.43, P > .05$]. The critical value at the 5% level was 7.815.

14/13. **a.** 20 **b.** 15 **c.** 36 **d.** 14 **e.** 13 **f.** 30

14/15. The relationship between A and B in the sample is statistically significant [$X^2(3 \text{ df}) = 10.1, P < .05$]. The critical value at the 5% level was 7.815.

14/17. .027 and .133

14/19. **a.** 2.1 percentage points **b.** 6.7 percentage points **c.** Quadrupling sample sizes would cut the margins of error in half.

14/21. 385 vehicles

14/23. **a.** 78,369 persons **b.** 55,320 persons

14/25. **a.** H_0: $P(x) = .1$, for $x = 0, 1, 2, \ldots, 9$. According to the null hypothesis, every one of the 10 digits has a probability of 10% of occurring at any given position in the table.
H_A: Not H_0.
b. 8.4%, 9.4%, 8.6%, 10.4%, 8.6%, 11.8%, 9.8%, 10.8%, 11.0%, 11.2%, respectively.
c. $\chi^2_{.05}(9 \text{ df}) = 16.919$ **d.** The sample does not depart significantly from the hypothesized uniform distribution, $X^2(9 \text{ df}) = 6.68, P > .05$.

14/27. **a.** H_0: $p = .25$. According to the null hypothesis, the true proportion of vestigial-winged fruit flies resulting from this type of cross is .25.
H_A: $p \neq .25$
b. $\hat{p} = 23/98 = .23$ **c.** The sample proportion of vestigial-winged fruit flies (23%) is not significantly different from the hypothesized value of 25%, $X^2(1 \text{ df}) = .12, P > .05$. **d.** We can be 95% confident that the true proportion is between approximately .15 and .32.

14/29. **a.** H_0: The dice are fair; that is, each of the six faces is equally likely on each die.
H_A: The dice are not fair.
b.

Hypothesized:	.028	.056	.083	.111	.139	.167	.139	.111	.083	.056	.028
Estimates:	.011	.044	.067	.086	.117	.161	.150	.125	.108	.078	.053

c. The dice are not fair, being biased in favor of higher numbers, $X^2(10 \text{ df}) = 23.9, P < .01$.
d. We can be 95% confident that the true proportion of elevens for these dice is between approximately .050 and .105.

14/31. **a.** H_0: Usual time of arrival for mathematics lectures is independent of attitude toward mathematics.
H_A: Not H_0.
b. There is no significant relationship between time of arrival and attitude, $X^2(4 \text{ df}) = 1.22, P > .05$.

14/33. **a.**

	White	Yellow	Red or orange	Blue	Black	Gray	Green	Brown
Cranbrook:	.16	.08	.18	.22	.04	.09	.08	.14
Camrose:	.07	.04	.12	.18	.04	.12	.18	.26

b. The proportions of cars of various colors differed significantly between the two regions, $X^2(7 \text{ df}) = 16.2, P < .025$. (The major differences were that white, yellow, and red cars were relatively more common near Cranbrook, while brown and green cars were relatively more common near Camrose.)

14/35. **a.** H_0: $p_m = p_f$. According to the null hypothesis, male and female students are equally likely to own a vehicle.
H_A: $p_m \neq p_f$
b. $\hat{p}_m = .43, \hat{p}_f = .28$ **c.** Male students were more likely than female students to own a vehicle, $X^2(1 \text{ df}) = 13.4, P < .005$. **d.** Males: The 95% confidence limits are .37 and .49. Females: The 95% confidence limits are .22 and .33.

14/37. Chi-square test for two-way tables (test of homogeneity of populations).

14/39. Test of the significance of the correlation coefficient or of the regression.

14/41. Paired-samples t test of "before" and "after" scores, or one-sample t test of the differences.
14/43. One-sample t test.
14/45. Independent-samples F test.
14/47. Chi-square test for two-way tables.
14/49. Chi-square test for two-way tables (test of homogeneity of populations).
14/51. Independent-samples t test.

Chapter 15

15/1. **a.**

Source	SS	df	MS	F	
A	200.6	2	100.3	2.53	$P > .05$
$R(A)$	953.0	24	39.7		

b. 3.40

15/3. **a.**

Source	SS	df	MS	F	
A	64.27	4	16.07	12.7	$P < .01$
$R(A)$	12.67	10	1.27		

b. 3.48 **c.** The sample means for the five levels of Factor A differ by more than can be reasonably attributed to random error. **d.** .65 **e.** 5.89 and 8.78

15/5. **a.**

df	MS	F	
24	8.327		
5	2.118	5.82	$P < .01$
120	.364		

b. 25 **c.** 6 **d.** 3.17 **e.** Yes. The mean square between blocks (8.327) was much larger than the error variance (.364).

15/7. **a.**

Source	SS	df	MS	F	
Blocks	20.800	2	10.400		
A	75.600	4	18.900	126	$P < .01$
A-by-B	1.200	8	.150		

b. 7.01 **c.** The sample means for the five levels of Factor A differ by more than can reasonably be attributed to chance. **d.** .22 **e.** 4.25 and 5.75 **f.** Yes. The mean square between blocks (10.400) was much larger than the error variance (.150).

15/9. **a.**

df	MS	F	
4	6.68	1.42	$P > .10$
1	8.62	1.83	$P > .10$
4	30.87	6.54	$P < .01$
30	4.72		

b. **i.** 5 **ii.** 2 **iii.** 4 **c.** **i.** 2.69 **ii.** 4.17 **iii.** 2.69

15/11. **a.**

Source	SS	df	MS	F	
A	3.000	2	1.500	1.77	$P > .10$
B	2.042	1	2.042	2.41	$P > .10$
A-by-B	169.333	2	84.667	99.9	$P < .01$
$R(AB)$	15.250	18	.847		

b. **i.** 3.55 **ii.** 4.41 **iii.** 3.55 **c.** The effect of Factor A depends on which level of Factor B is involved, and vice versa. For example, the mean for B_1 is higher than for B_2 at level A_1, about the same as for B_2 at level A_2, and lower than for B_2 at level A_3. **d.** .46 **e.** 3.53 and 5.47

15/13. **a.** DV = desired number of children in future family; IV = present family size.

b.

Source of variation	SS	df	MS	F	
Present family size	2.93	3	.98	.55	$P > .10$
Students in same-sized families	23.07	13	1.77		

c. The four sample means ± standard error were 3.0 ± .7, 3.4 ± .6, 3.3 ± .8, and 2.4 ± .6 children, respectively. **d.** The mean desired family size was not significantly different for students coming from 2-, 3-, 4-, and 5-child families ($P > .10$).

15/15. **a.** DV = number of points earned; IV = vowel/consonant ratio.

b.

Source of variation	SS	df	MS	F	
Vowel/consonant ratio	1503.5	7	214.8	6.58	$P < .01$
Replications within conditions	784.0	24	32.7		

c. The eight sample means were 6.0, 19.5, 25.0, 20.0, 21.5, 12.5, 10.5, and 6.0 points, respectively, each with an estimated standard error of 2.9 points. **d.** First-turn Scrabble scores for this player dliffered significantly according to the vowel/consonant ratio of the letters on the rack ($P < .01$). The highest mean score in the sample occurred with 2 vowels, although mean scores with 1, 3, or 4 vowels were not much lower, and the lowest mean scores occurred with 0 or 7 vowels.

15/17. $F = 5.93$, whose square root is $t = \pm 2.44$. At the 5% significance level, the critical value for the F test is $F_{.05}(1, 10 \text{ df}) = 4.96$, whose square root is $\pm t_{.025}(10 \text{ df}) = \pm 2.23$, which is the critical value for the t test. The tests are exactly equivalent for comparing two means. The mean weekly study time of 16.5 hours reported by first-year students was significantly less than the mean of 25.0 hours reported by later-year students, $F(1, 10 \text{ df}) = 5.93$, $P < .05$.

15/19. **a.** DV = number of hours spent watching television per day; IV = day of the week.

b.

Source of variation	SS	df	MS	F	
Students	45.41	8	5.68		
Days	6.74	2	3.37	3.25	$.10 > P > .05$
S-by-D interaction	16.59	16	1.04		

c. The sample means were 1.89, 1.78, and .78 hours, respectively, each with an estimated standard error of .34 hour. **d.** The difference in mean number of hours of television viewing on Monday, Wednesday, and Friday was not statistically significant, $P > .05$. **e.** The mean square between blocks (5.68) was much larger than the error variance (1.04), indicating that the blocking succeeded in reducing the error variance and hence increasing precision.

15/21. a. DV = maximum lift; IV = distance between hands.
b.

Source of variation	SS	df	MS	F	
Students	12,849.2	7	1,835.6		
Distances	24,333.6	3	8,111.2	171	$P < .01$
S-by-D interaction	997.7	21	47.5		

c. The four sample means were 110.0, 148.8, 178.8, and 175.6 lb, respectively, each with an estimated standard error of 2.4 lb. **d.** The distance between the weight-lifter's hands had a significant effect on the maximum lift ($P < .01$). The best distances were 18 and 27 inches, and the worst was 0 inch. **e.** The mean square between blocks (1,835.6) was much larger than the error variance (47.5), indicating that the blocking succeeded in reducing the error variance and hence increasing precision.

15/23. a. $F = 9.62$, whose square root is $t = \pm 3.10$. At the 5% significance level, the critical value for the F test is $F_{.05}(1, 7 \text{ df}) = 5.59$, whose square root is $\pm t_{.025}(7 \text{ df}) = \pm 2.36$, which is the critical value for the t test. The tests are exactly equivalent for comparing two means. **b.** The sample mean increase of 3.1 beats per minute after consumption of tea was statistically significant, $F(1, 7 \text{ df}) = 9.62$, $P < .025$. **c.** The mean square between blocks (30.78) was much larger than the error variance (4.06), indicating that the blocking succeeded in reducing the error variance and hence increasing precision.

15/25. a. DV = loudness rating; IVs = tone-control setting and volume-control setting.
b.

Source of variation	SS	df	MS	F	
Volume setting	67.000	2	33.500	151	$P < .01$
Tone setting	2.333	2	1.167	5.25	$P < .05$
V-by-T interaction	.167	4	.042	.19	$P > .10$
Subjects within V-by-T	2.000	9	.222		

c. The mean loudness ratings of 1.3, 2.3, and 5.8 for the low, medium, and high volume settings, respectively, were significantly different ($P < .01$), as would be expected. The mean loudness ratings of 3.0, 2.8, and 3.7 for the three tone-control settings also differed significantly ($P < .05$), with loudness rated highest for the strong bass setting. The interaction of volume setting and tone setting was not statistically significant ($P > .10$).

15/27. a. DV = reported number of hours of study per week; IVs = year of studies and gender.
b.

Source of variation	SS	df	MS	F	
Year	216.8	1	216.8	7.45	$P < .05$
Gender	114.1	1	114.1	3.92	$.10 > P > .05$
Y-by-G interaction	18.8	1	18.8	.64	$P > .10$
Students within Y-by-G	232.7	8	29.1		

c. The mean weekly study time of 16.5 hours reported by first-year students was significantly less than the mean of 25.0 hours reported by later-year students ($P < .05$). The difference between mean study times reported by male and female students was not statistically significant, and there was no significant interaction of year and gender ($P > .05$).
d. The sum of squares for year is identical in the two cases (216.75), as is the total sum of squares (582.25), because the same data are being analyzed. The sum of squares for students-within-years

(365.50) in Problem 15/17 is partitioned into three parts in the present problem by taking into account the gender of the students. Thus, some of the original variation among students in the same year (365.50) can be attributed to gender (114.0833), and some to the interaction of gender and year (18.75); the remainder (232.6667) is associated with differences between students of the same year and gender. These three parts add up to the original 365.50. The present problem partitions the same total variation more finely than in Problem 15/17 by identifying additional sources of variation that were not considered in Problem 15/17.

Chapter 16

16/1. **a.** $r_S = -.90$, $r_S = -.81$ **b.** .929
16/3. $z_{\text{sign test}} = 1.34$, $P = .18$
16/5. $z_{\text{rank-sum}} = 1.74$, $P = .08$
16/7. $H = 7.79$, $\chi^2_{.05}(2\text{ df}) = 5.991$
16/9. $X^2 = 21.6$, $\chi^2_{.05}(3\text{ df}) = 7.815$
16/11. $z_{\text{signed-rank}} = 1.65$, $P = .10$
16/13. $X^2_{\text{rank}} = 2.33$, $\chi^2_{.01}(2\text{ df}) = 9.210$
16/15. $z_{\text{sign test}} = 1.11$, $P = .27$
16/17. **b.** $r_S = .75$ by both methods. The two methods yield identical answers as long as there are no ties. **c.** Significant at the 5% level. CV = .683.
16/19. **b.** $r_S = .281$, $r_S \simeq .283$. The second answer is approximate because there are ties in the data, but both could be reported as $r_S = .28$. **c.** Not significant at the 5% level. CV = .576.
16/21. The sample median is significantly less than \$2500 ($z_{\text{sign test}} = 2.25$, $P = .024$).
16/23. Handedness was not significantly related to accuracy ($z_{\text{rank-sum}} = 1.12$, $P = .26$).
16/25. The difference was not significant ($H = 1.11$, $P > .05$).
16/27. The controlled-drinking group had admission dates that were significantly earlier than those of the abstinence groups. **a.** $z_{\text{rank-sum}} = 3.18$, $P = .0015$ **b.** $X^2 = 6.40$, $P < .05$
16/29. **a.** The Wilcoxon signed-rank test found significant evidence of a difference between the stores ($z_{\text{signed rank}} = 2.07$, $P = .038$). **b.** The sign test failed to find significant evidence of a difference between the stores ($z_{\text{sign test}} = .95$, $P = .34$).
16/31. The marks did not differ significantly ($X^2_{\text{rank}} = 3.50$, $P > .05$).
16/33. There was no significant difference in productivity ($z_{\text{sign test}} = 1.38$, $P = .17$).
16/35. **a.** The difference was significant ($z_{\text{sign test}} = 2.04$, $P = .041$; $P = .031$ by an exact sign test). **b.** The chi-square test for two-way tables is for situations where each individual is categorized and counted only once (independent-samples design). In this case we have paired data, with each individual counted twice (repeated-measures design).
16/37. **a.** Pearson's *r*, or simple linear regression. **b.** Spearman's rank correlation.
16/39. **a.** One-sample *t* test. **b.** One-sample sign test.
16/41. Paired-samples sign test.
16/43. Chi-square test for two-way tables.
16/45. Chi-square test for two-way tables.
16/47. **a.** One-way ANOVA with independent samples. **b.** Kruskal–Wallis rank-sum test.
16/49. **a.** Paired-samples *t* test, which is equivalent to a one-way ANOVA with repeated measurements. **b.** Wilcoxon signed-rank test or the paired-samples sign test.
16/51. **a.** Independent-samples *t* test, which is equivalent to a one-way ANOVA with independent samples. **b.** Wilcoxon rank-sum test or the Brown–Mood median test.
16/53. Friedman rank test.
16/55. Paired-samples sign test.

Chapter 17

17/1. **a.** EMV(1) = .5, EMV(2) = 1.7, EMV(3) = 2.3 **b.** **i.** Action 1. **ii.** Action 2. **iii.** Action 3. **iv.** Action 2, because EU(2) = 3.5 is largest. **c.** 47.7

17/3. **a.** 2.5, 3.0 **b.** 11.25, 1.00 **c.** 3.35, 1.00 **d.** −.5 **e.** −.15
17/5. **a.** 2, 0, 0 **b.** 100, 256, 16 **c.** 60, 256, 16 **d.** 60, 225, 15 **e.** 72, 225, 15 **f.** 70, 121, 11 **g.** 30, 57, 7.5
17/7. **a.** EMV(A) = 78, EMV(B) = 49, EMV(C) = 56 **b. i.** A **ii.** B **iii.** A **iv.** C, because EU(C) = 7.4 is largest **c.** EVPI = 1.8 utiles
17/9. **a.** $\mu_x = 6.9$, $\mu_y = 6.8$ **b.** $\sigma_x^2 = .49$, $\sigma_y^2 = .56$ **c.** $\sigma_x = .7$, $\sigma_y = .75$ **d.** Cov $(x, y) = .28$, $\rho = .53$
17/11. **a.** 12 and 0 **b.** 180 and 576
17/13. **a.** 68.0 and 51.8 for the maxima, and 41.0 and 29.2 for the minima **b.** 15.0 and 3.4
17/15. **a.** .97 **b.** .97

Chapter 18

18/1. **a, b.**

Source	SS	df	MS	F	$F_{.05}$	
A	78.03	1	78.03	36.0	4.26	$P < .01$
B	12.25	1	12.25	5.65	4.26	$P < .05$
C	38.00	2	19.00	8.77	3.40	$P < .01$
A-by-B	.25	1	.25	.12	4.26	
A-by-C	11.55	2	5.78	2.67	3.40	
B-by-C	2.00	2	1.00	.46	3.40	
A-by-B-by-C	4.67	2	2.33	1.08	3.40	
$R(ABC)$	52.00	24	2.17			
Total		35				

c. The sample means corresponding to the two levels of Factor A differed significantly, the sample means corresponding to the two levels of Factor B differed significantly, and the sample means corresponding to the three levels of Factor C differed significantly. **d. i.** .35 **ii.** .35 **iii.** .42

18/3. **a, b.**

Source	SS	df	MS	F	$F_{.05}$	
A	6.55	2	3.27	.82	5.14	
B	22.09	1	22.09	5.52	5.99	$.05 < P < .10$
A-by-B	26.18	2	13.09	3.27	5.14	
$R(AB)$	24.00	6	4.00			
Total		11				

c. The differences in the sample were not large enough to conclude that either Factor A or Factor B has any effect on the dependent variable ($P > .05$).

18/5. **a.**

Source	SS	df	MS	F	
A	50.67	2	25.33	15.2	$P < .01$
$R(A)$	15.00	9	1.67		
Total		11			

b. 4.26

18/7. Square-root transformation.
18/9. **a.** It is not stated, but presumably student's grade. **b.** Teaching method was a fixed factor with levels A and B, and students were considered a random factor. **c.** Students were nested under teaching methods. **d.** Independent-samples design. **e.** Classes are confounded with teaching

methods. Students are not assigned randomly to classes, but choose particular classes because of personal preferences and/or conflicts with other classes. In the reference cited in the question, Hooke notes that "the 8:00 a.m. section of a course may be much better than the 10:00 a.m. section, for various reasons. (E.g., the early section may consist only of freshmen, while the sophomores, who may be mostly repeaters, tend to have a conflict at 8 and so take the later class.) The only logical conclusion from the cited experiment is that the class that received method A currently represents a population with a higher mean than that of the population represented by the other class. Unless we know that this was not true in the beginning, we are certainly in no position to claim that the observed difference is due to a difference in teaching methods." **f. i.** The student. **ii.** The assumption of independence. **iii.** Since students within a class are not independent, the class could have been considered the observational unit by using the class averages as the observations. However, in that case there was no replication (only one class received each method), and it would have been necessary to have at least four classes, two randomly assigned to each method, in order to test the significance of the difference between the methods.

18/11. **a.** Increase in a score reflecting learning of a visual discrimination. **b.** Condition of frontal lobes was a fixed factor with two levels (removed or intact), and monkeys constituted a random factor. **c.** Monkeys were nested under frontal-lobe conditions. **d.** Independent-samples design. **e. i.** Surgery is confounded with frontal-lobe conditions. **ii.** Any differences between the groups may have been because of the general effects of surgery, quite apart from any damage to the frontal lobes. **iii.** This confounding could have been avoided by having a control group that received a placebo operation, that is, surgery similar to that in the experimental group except that the frontal lobes would be left intact.

18/13. **a.** $E(\text{MS}_{\text{error}}) = \sigma^2 + 30\sigma^2_{AC}$ **b.** $E(\text{MS}_{\text{error}}) = \sigma^2$

18/15. **a. i.** Number of kicks (K) is a fixed factor, and swimmer (S) is a random factor. **ii.** K and S are crossed.

iii.

Source of variation	df	Error term
Swimmer	2	None
Kicks	2	MS_{KS}
K-by-S	4	None
Total	8	

b. i. Attitude (A) and dress (D) are fixed factors, and subject (S) is a random factor. **ii.** A and D are crossed. S is nested under A and D.

iii.

Source of variation	df	Error term
Attitude	1	$\text{MS}_{S(AD)}$
Dress	1	$\text{MS}_{S(AD)}$
A-by-D	1	$\text{MS}_{S(AD)}$
Subjects within A-by-D	12	None
Total	15	

c. i. Stick curvature (C) and distance (D) are fixed factors, and player (P) is a random factor. **ii.** C, D, and P are all crossed with each other.

iii.

Source of variation	df	Error term
Players	3	None
Curvature	1	MS_{CP}
C-by-*P*	3	None
Distance	3	MS_{DP}
D-by-*P*	9	None
C-by-*D*	3	MS_{CDP}
C-by-*D*-by-*P*	9	None
	31	

18/17. a. Color (C) and detail (D) are fixed factors, and scene (S) and subjects (I = individuals) are random factors. **b.** Color, detail, and scene are all crossed with each other. Subjects are nested under color, detail, and scene.

c.

Source of variation	df	Error term
Color	2	MS_{CS}
Detail	5	MS_{DS}
Scene	4	$MS_{I(CDS)}$
C-by-*D*	10	MS_{CDS}
C-by-*S*	8	$MS_{I(CDS)}$
D-by-*S*	20	$MS_{I(CDS)}$
C-by-*D*-by-*S*	40	$MS_{I(CDS)}$
Subjects within *C*-by-*D*-by-*S*	270	None
Total	359	

18/19. a. The subjects constitute the replication factor. The other factors are age of subject, gender of subject, and identity of speaker. The dependent variable is agreement rating.

b.

Source of variation	SS	df	MS	F	
Age	108.4	1	108.4	10.7	$P < .01$
Gender	7.0	1	7.0	.70	
Identity of speaker	3.4	1	3.4	.33	
A-by-*G*	3.4	1	3.4	.33	
A-by-*I*	2.0	1	2.0	.20	
G-by-*I*	3.4	1	3.4	.33	
A-by-*G*-by-*I*	2.0	1	2.0	.20	
Subjects within *A*-by-*G*-by-*I*	162.0	16	10.1		
Total		23			

c. The only significant effect is that due to age, with mean agreement ratings of 4.5 for young subjects and 8.8 for old subjects. Older subjects were significantly more likely to agree with the quotation. Neither the identity of the speaker nor the gender of the subject had a significant effect on agreement in this sample.

18/21. a.

Source of variation	SS	df	MS	F	
Fuel	32.2017	1	32.2017	773	$P < .01$
Transmission	9.4033	2	4.7017	113	$P < .01$
Driving condition	47.6017	1	47.6017	1140	$P < .01$
F-by-*T*	.1033	2	.0517	1.24	
F-by-*D*	2.2817	1	2.2817	54.8	$P < .01$
T-by-*D*	.2033	2	.1017	2.44	
F-by-*T*-by-*D*	.2633	2	.1317	3.16	$.05 < P < .10$
Cars within *F*-by-*T*-by-*D*	.5000	12	.0417		
Total		23			

b. The significant effects may be summarized by plots of just two effects: (1) the main effect of transmission (showing means of 7.3, 5.8, and 6.3 liters/100 km for *A*3, *M*4, and *M*5 transmissions, respectively), and (2) the interaction of fuel and type of driving (showing means of 6.4 in the city and 4.2 on the highway for diesel, and of 9.3 in the city and 5.9 on the highway for gasoline). The results indicate that (1) fuel consumption is less with a manual than with an automatic transmission, (2) fuel consumption is less on the highway than in the city, (3) fuel consumption is less with diesel than with gasoline, and (4) the difference between gasoline and diesel is greater in the city than on the highway (or, equivalently, the difference between city and highway consumption is greater with gasoline than with diesel).

18/23. a.

Source of variation	SS	df	MS	F
Faculty	1.07	1	1.07	.91
Participation	1.32	2	.66	.56
F-by-*P*	.30	2	.15	.13
Students within *F*-by-*P*	14.15	12	1.18	
Total		17		

b. The differences in the sample were not large enough to conclude that grade point averages are related either to faculty or to level of participation in athletics ($P > .10$).

18/25. a.

Source of variation	SS	df	MS	F	
Age	22.90	1	22.90	7.59	$P < .025$
Gender	6.23	1	6.23	2.06	
Message	3.87	1	3.87	1.28	
A-by-*G*	1.06	1	1.06	.35	
A-by-*M*	2.42	1	2.42	.80	
G-by-*M*	.02	1	.02	.01	
A-by-*G*-by-*M*	1.07	1	1.07	.35	
Individuals within *A*-by-*G*-by-*M*	54.30	18	3.02		
		25			

b. The marginal means computed from the cell means for age should be plotted: 4.79 for those under 30 and 2.75 for those over 30. The younger respondents were significantly more favorable toward the idea of equality than were the older respondents ($P < .025$).

18/27. Square-root transformation.

Chapter 19

19/1. **a, b.**

Source	SS	df	MS	F	CV (full df)	CV (cons. df)
R	5.58	2	2.79			
A	113.83	7	16.26	15.8	2.76	18.51
A-by-*R*	14.42	14	1.03			
Total	133.83	23				

c. The effect of Factor *A* is significant at the 5% level when tested with full degrees of freedom but not with conservative degrees of freedom. However, there is no particular evidence of assumption violations, and the observed *F* value is quite close to the conservative critical value; thus, the effect may be judged significant at the 5% level. Computer analysis, such as with BMDP, is advised if more exact testing is required. **d.** Yes. The mean square between replications (2.79) was considerably larger than the error term (1.03).

19/3. **a, b.**

Source	SS	df	MS	F	CV (full df)	CV (cons. df)
R	5.583	2	2.792			
A	54.000	1	54.000	61.7	18.51	18.51
R-by-*A*	1.750	2	.875			
B	34.833	3	11.611	20.4	4.78	18.51
R-by-*B*	3.417	6	.569			
A-by-*B*	25.000	3	8.333	5.41	4.78	18.51
R-by-*A*-by-*B*	9.250	6	1.542			
Total	133.833	23				

c. The *F* ratio for the *A*-by-*B* interaction just barely exceeds the critical value at the 5% level with full degrees of freedom but not with conservative degrees of freedom. In the absence of more exact testing, it may be wise to accept the conservative result and judge the interaction not significant at the 5% level. However, the main effects of Factors *A* and *B* are both significant at the 5% level.
d. Yes. The mean square between replications (2.792) was considerably larger than the error terms (.875, .569, and 1.542).

19/5. **a, b.**

Source	SS	df	MS	F	CV (full df)	CV (cons. df)
A	54.00	1	54.00	29.5	7.71	7.71
R(*A*)	7.33	4	1.83			
B	34.83	3	11.61	11.0	3.49	7.71
A-by-*B*	25.00	3	8.33	7.89	3.49	7.71
B-by-*R*(*A*)	12.67	12	1.06			
Total	133.83	23				

c. The main effects and interaction are all significant at the 5% level. **d.** No. The independent-samples error term (1.83) was not substantially larger than the repeated-measures error term (1.06).

19/7. One-way design with independent samples.

19/9. One-way design with repeated measurements.

19/11. Two-way design with one repeated-measures factor.

19/13. Two-way design with two repeated-measures factors.

19/15. **a.** The subjects constitute the replication factor. The other factors are hand preference and hand orientation. The dependent variable is number of pucks in the net in 20 tries. **b.** Two-way design with two repeated-measures factors.

c.

Source of variation	SS	df	MS	F		CV
Subject	8.25	3	2.75			
Preference	361.00	1	361.00	167	$P < .01$	34.12
P-by-*S*	6.50	3	2.17			
Orientation	81.00	1	81.00	18.0	$P < .025$	17.44
O-by-*S*	13.50	3	4.50			
P-by-*O*	6.25	1	6.25	5.77	$P < .10$	5.54
P-by-*O*-by-*S*	3.25	3	1.08			
Total		15				

d. Adopting the conventional 5% significance level, we conclude that the interaction is not significant, and plots of the main effects summarize the significant effects. The results show an advantage of the preferred hand over the nonpreferred hand (16.62 versus 7.12 successful shots out of 20, with an estimated standard error of .52) and an advantage of forehand over backhand (14.12 versus 9.62 successful shots out of 20, with an estimated standard error of .75). **e.** No. The mean square between blocks (2.75) was not substantially larger than the error terms (2.17, 4.50, 1.08).

19/17. a. The swimmers constitute the replication factor. The other factor is number of kicks per arm pull. The dependent variable is time to swim one length. **b.** One-way design with repeated measurements.

c.

Source	SS	df	MS	F		CV (full df)	CV (cons. df)
Swimmers	37.556	2	18.778				
Kicks	22.222	2	11.111	18.2	$P \simeq .05$	6.94	18.5
K-by-*S*	2.444	4	.611				
		8					

d. The effect of the kick factor is significant at the 5% level when tested with full degrees of freedom but not with conservative degrees of freedom. However, there is no particular evidence of assumption violations, and the observed *F* is very close to the conservative critical value; thus the effect is judged significant at the 5% level. The means to be plotted are 30.67, 27.33, and 27.33 seconds (with an estimated standard error of .45 second) for 0, 2, and 6 kicks per arm pull, respectively, indicating slower swimming without kicking. **e.** Yes. The mean square between blocks (18.778) was considerably larger than the error term (.611). **f.** Different orders were required to prevent the confounding of carry-over effects, such as warm-up effects or fatigue effects, with number of kicks.

19/19. a. The subjects constitute the replication factor. The other factors are attitude and dress. The dependent variable is rating of response. **b.** Two-way design with independent samples.

c.

Source	SS	df	MS	F		CV
Attitude	5.062	1	5.062	7.36	$P < .05$	4.75
Dress	1.562	1	1.562	2.27	$P > .10$	3.18
A-by-*D*	.062	1	.062	.09	$P > .10$	3.18
R(*AD*)	8.250	12	.688			
Total		15				

d. The significant effect of attitude may be summarized by a plot of the mean rating for the humble and overconfident attitudes (means of 3.5 and 2.4, with an estimated standard error of .3), indicating more positive responses in the former case. **e.** No. There was no blocking in this experiment.

19/21. a. The students constitute the replication factor. The other factors are year of studies and day of week. The dependent variable is number of hours of television viewing. **b.** Two-way design with one repeated-measures factor.

c.

Source of variation	SS	df	MS	F		CV (full df)	CV (cons. df)
Year	7.41	1	7.41	2.20	$P > .10$	3.05	3.05
Students within Y	53.93	16	3.37	—			
Day	1.33	2	.67	.66	$P > .10$	2.48	3.05
D-by-Y	6.37	2	3.19	3.16	$P > .05$	3.30	4.49
D-by-$S(Y)$	32.30	32	1.01	—			
Total		53					

d. The experiment demonstrated no statistically significant differences at the 5% level. **e.** Yes. The independent-samples error term (3.37) was much larger than the repeated-measures error term (1.01).

19/23. a. The players constitute the replication factor. The other factors are curvature of stick and distance from net. The dependent variable is number of hits in 20 shots. **b.** Two-way design with two repeated-measures factors.

c.

Source of variation	SS	df	MS	F		CV (full df)	CV (cons. df)
Players (P)	133.094	3	44.365	—			
Curvature (C)	30.031	1	30.031	34.7	$P < .01$	34.1	34.1
C-by-P	2.594	3	.865	—			
Distance (D)	529.594	3	176.531	104	$P < .01$	6.99	34.1
D-by-P	15.281	9	1.698	—			
C-by-D	2.594	3	.865	1.82	$P > .10$	2.81	5.54
C-by-D-by-P	4.281	9	.476	—			
Total		31					

d. Plots of the main effects summarize the significant effects. The main effect of curvature shows an advantage for the curved stick over the straight stick (means of 9.19 and 7.25 hits, with an estimated standard error of .23 hit) in a population of players commonly using curved sticks. The main effect of distance shows reduced accuracy as distance increases from 10 feet to 40 feet (means of 13.75, 10.00, 6.25, and 2.88 hits, with an estimated standard error of .46 hit). **e.** Yes. The mean square between blocks (44.365) was considerably larger than the error terms (.865, 1.698, .476).

19/25. a. The students constitute the replication factor. The other factors are background of student and musical performer to be rated. The dependent variable is rating of musical performer. **b.** Two-way design with one repeated-measures factor.

c.

Source of variation	SS	df	MS	F		CV
Background (B)	2.67	1	2.67	1.47	$P > .10$	3.29
Students within B	18.17	10	1.82	—		
Performer (P)	24.00	1	24.00	7.31	$P < .025$	6.94
P-by-B	28.17	1	28.17	8.58	$P < .025$	6.94
P-by-$S(B)$	32.83	10	3.28	—		
Total		23				

d. The cell means should be plotted with rural and urban values of 6.8 and 8.3 for Billy Idol, and 7.0 and 4.2 for Kenny Rogers. The significant interaction indicates that attitude ratings were significantly related to background: Rural students gave the two performers approximately equal mean ratings, whereas urban students gave a higher mean rating to Idol and a lower mean rating to Rogers. **e.** No. The repeated-measures error term (3.28) was larger than the independent-samples error term (1.82). **f.** The performer-by-background interaction.

19/27. **a.** The students constitute the replication factor. The other factors are grade point average and distraction condition. The dependent variable is reading speed. **b.** Two-way design with one repeated-measures factor.
c, d.

Source	SS	df	MS	F		CV
Grade	5,950	1	5,950	.34	$P > .10$	3.29
$S(G)$	175,770	10	17,580			
Distraction	180	1	180	.15	$P > .10$	3.29
D-by-G	3,900	1	3,900	3.24	$P > .10$	3.29
D-by-$S(G)$	12,050	10	1,200			
Total		23				

e. The experiment demonstrated no statistically significant differences. **f.** Yes. The independent-samples error term (17,580) was much larger than the repeated-measures error term (1,200).
g. Since the radio-on condition was always first, any warm-up or practice effect is confounded with the distraction factor. In addition, the readings themselves constitute a confounding variable: Since the same reading was always used for the radio-on condition and a second reading for the radio-off condition, any differences in the difficulty of the readings are confounded with the distraction factor.

Chapter 20

20/1. **a.** $SS_{AB(C_1)} = .083$, $SS_{AB(C_2)} = 4.083$, $SS_{AB(C_3)} = .75$. Each has 1 degree of freedom.
b. $SS_{AB(C_1)} + SS_{AB(C_2)} + SS_{AB(C_3)} = SS_{AB} + SS_{ABC} = 4.92$, with 3 degrees of freedom.
c. i. $MS_{R(ABC)} = 2.17$ (from the overall ANOVA table). **ii.** $F_{.05}(1, 24) = 4.26$

20/3. **a.** $SS_{B(A_1)} = 105.125$, $SS_{B(A_2)} = .125$, $SS_{B(A_3)} = 66.125$. Each has 1 degree of freedom.
b. $SS_{B(A_1)} + SS_{B(A_2)} + SS_{B(A_3)} = SS_B + SS_{AB} = 171.375$, with 3 degrees of freedom.
c. i. $MS_{R(AB)} = .847$ (from the overall ANOVA table). **ii.** $F_{.05}(1, 18) = 4.41$

20/5. **a.** $SS_{B(A_1)} = 57.58$, $SS_{B(A_2)} = 2.25$. Each has 3 degrees of freedom.
b. $SS_{B(A_1)} + SS_{B(A_2)} = SS_B + SS_{AB} = 59.83$, with 6 degrees of freedom.
c. i. $MS_{BR(A)} = 1.06$ (from the overall ANOVA table). **ii.** $F_{.05}(3, 12) = 3.49$ with full degrees of freedom, or $F_{.05}(1, 4) = 7.71$ with conservative degrees of freedom.
d. $SS_{A(B_1)} = 0$, $SS_{A(B_2)} = 8.17$, $SS_{A(B_3)} = 28.17$, $SS_{A(B_4)} = 42.67$. Each has 1 degree of freedom.
e. $SS_{A(B_1)} + SS_{A(B_2)} + SS_{A(B_3)} + SS_{A(B_4)} = SS_A + SS_{AB} = 79.00$, with 4 degrees of freedom.
f. i. $MS_{R(A(B_1))} = 1.00$, $MS_{R(A(B_2))} = 1.67$, $MS_{R(A(B_3))} = .67$, $MS_{R(A(B_4))} = 1.67$
ii. $F_{.05}(1, 4) = 7.71$

20/7. **a.** $SS_{B(A_1)} = 57.58$, $SS_{B(A_2)} = 2.25$. Each has 3 degrees of freedom.
b. $SS_{B(A_1)} + SS_{B(A_2)} = SS_B + SS_{AB} = 59.83$, with 6 degrees of freedom.
c. i. $MS_{RB(A_1)} = .694$, $MS_{RB(A_2)} = 1.42$ **ii.** $F_{.05}(3, 6) = 4.78$ with full degrees of freedom, or $F_{.05}(1, 2) = 18.51$ with conservative degrees of freedom.

20/9. **a.** $c_1 = 1$, $c_2 = 1$, $c_3 = -2$ (or any multiple thereof). **b.** $c_1 = 1$, $c_2 = -1$, $c_3 = 0$ (comparison of first versus second mean).

20/11. **a.** $c_1 = 1$, $c_2 = -1$, $c_3 = 1$, $c_4 = -1$ (or any multiple thereof). **b.** $c_1 = 1$, $c_2 = 0$, $c_3 = -1$, $c_4 = 0$ (or any multiple thereof). **c.** $c_1 = 0$, $c_2 = 1$, $c_3 = 0$, $c_4 = -1$ (comparison of second versus fourth mean).

20/13. **a.** $SS_{\bar{d}} = 6.00$ **b.** $SS_{\bar{d}} = 242.00$

20/15. **a.** $F_{\bar{d}} = 43.8$, $F_{.05}(1, 24) = 4.26$ **b.** 3.44 and 6.56

20/17. **a.** $F_{\text{linear}} = 19.2$, $F_{\text{quadratic}} = 42.0$, $F_{\text{cubic}} = .30$, $F_{\text{quartic}} = 18.9$, $F_{.05}(1, 25) = 4.24$ **b.** $F_{\text{dev. lin.}} = 20.4$, $F_{.05}(3, 25) = 2.99$

20/19. **a.** $F_{\bar{d}} = 74.7$, CV = 9.03; 5.22 and 10.78 **b.** $F_{\bar{d}} = 74.7$, CV = 7.605; 5.45 and 10.55
c. $F_{\bar{d}} = 74.7$, CV = 4.26; 6.09 and 9.91

20/21. **a.** $SS_{TD(F_1)} = .18$, $SS_{TD(F_2)} = .28667$. Each has 2 degrees of freedom. **b.** $SS_{TD(F_1)} + SS_{TD(F_2)} = SS_{TD} + SS_{FTD} = .467$ **c.** The three-way interaction would have to be significant.

20/23. **a.** City driving: $F = 620$, $P < .01$ (CV = 9.33). Highway driving: $F = 208$, $P < .01$ (CV = 9.33).
$SS_{F(D_1)} + SS_{F(D_2)} = SS_F + SS_{FD} = 34.483$
b. Diesel: $F = 348$, $P < .01$ (CV = 9.33). Gasoline: $F = 849$, $P < .01$ (CV = 9.33).
$SS_{D(F_1)} + SS_{D(F_2)} = SS_D + SS_{FD} = 49.883$
c. Fuel consumption was significantly less with diesel than with gasoline, whether in the city or on the highway, but the difference was significantly greater in city driving than on the highway. Fuel consumption was significantly less on the highway than in city driving for both gasoline and diesel, but the difference was significantly greater with gasoline than with diesel fuel.

20/25. **a.** Billy Idol: $F = 2.79$, $P > .10$ (CV = 3.29). Kenny Rogers: $F = 8.98$, $P < .025$ (CV = 6.94).
$SS_{B(P_1)} + SS_{B(P_2)} = SS_B + SS_{PB} = 30.83$
b. Rural students: $F = .03$, $P > .10$ (CV = 3.29). Urban students: $F = 15.9$, $P < .01$ (CV = 10.04). $SS_{P(B_1)} + SS_{P(B_2)} = SS_P + SS_{PB} = 52.17$
c. There was no significant difference between rural and urban students in their ratings of Billy Idol, but urban students rated Kenny Rogers significantly lower than did rural students. There was no significant difference between performers as rated by rural students, but urban students rated Billy Idol significantly higher than Kenny Rogers.

20/27. **a.** $F_{\bar{d}} = 36.4$, $P < .05$ (CV = 7.71 with full df or 18.5 with conservative df) **b.** We can be 95% confident that the true mean reduction in time associated with kicking (by the relevant population of swimmers, for a 25-m length) is between 1.8 and 4.9 seconds. (Making allowance for any possible heterogeneity of covariance, the interval widens to the limits 1.0 and 5.7 seconds.)

20/29. **a.**
$F_{\text{linear}} = 17.2$, $P < .01$ (CV = 7.41)
$F_{\text{quadratic}} = .40$, $P > .10$ (CV = 2.86)
$F_{\text{higher-order}} = .19$, $P > .10$ (CV = 2.25)
b. $SS_{\text{higher-order}} = SS_{\text{cubic}} + SS_{\text{quartic}} + SS_{\text{quintic}}$
$= SS_{\text{treatment}} - SS_{\text{linear}} - SS_{\text{quadratic}}$
c. A linear polynomial (degree 1), within the range of independent-variable values in this study.

20/31. **a.**

Source	SS	df	MS	F		CV (full df)	CV (cons. df)
S	591.5	7	84.5	—			
M	204.8	1	204.8	6.56	$P < .05$	5.59	5.59
M-by-S	218.5	7	31.2	—			
L	684.8	4	171.2	10.0	$P < .025$	3.29	8.07
L-by-S	477.7	28	17.1	—			
M-by-L	83.2	4	20.8	.81	$P > .10$	2.16	3.59
M-by-L-by-S	717.4	28	25.6	—			

b. The significant effects may be summarized by a plot of the main effect of modality (showing mean recall scores of 6.6 for auditory and 9.8 for visual presentations) and a plot of the main effect of word length (showing means of 13, 10, 7, 6, and 5, respectively).
c. $F_{\text{linear}} = 37.5$, $P < .01$ (CV = 7.64 will full df or 12.2 with conservative df)
$F_{\text{dev. lin.}} = .88$, $P > .10$ (CV = 2.29 with full df or 3.59 with conservative df)
Within the range of independent-variable values studied, a linear polynomial is required.

20/33. **a.** The plot should show mean scores (6.0, 19.5, 25.0, 20.0, 21.5, 12.5, 10.5, and 6.0 points, respectively) as a function of number of vowels on the rack.

$F_{\text{linear}} = 4.78, P < .05$ (CV = 4.26)
$F_{\text{quadratic}} = 30.9, P < .01$ (CV = 7.82)
$F_{\text{cubic}} = 7.60, P < .025$ (CV = 5.72)
$F_{\text{higher-order}} = .68, P > .10$ (CV = 2.19)

A cubic polynomial is required.
c. Tukey's margin of error = 13.4 points at the 95% confidence level. The 0- and 7-vowel conditions each differed significantly from each of the 1-, 2-, 3-, and 4-vowel conditions; and the 2-vowel condition differed significantly from the 6-vowel condition. **d.** Scheffé's margin of error = 16.6 points at the 95% confidence level. The 0- and 7-vowel conditions each differed significantly from the 2-vowel condition.

20/35. **a.** First comparison: $F = 204, P < .01$ (CV = 13.86). Confidence limits are .93 to 1.59 liters per 100 km greater consumption with automatic than with manual transmission.
Second comparison: $F = 21.7, P < .01$ (CV = 13.86). Confidence limits are .10 to .85 liter per 100 km greater consumption with 5-speed than with 4-speed transmission.
b. First comparison: $F = 204, P < .01$ (CV = 9.33). Confidence limits are .99 to 1.53 liters per 100 km greater consumption with automatic than with manual transmission.
Second comparison: $F = 21.7, P < .01$ (CV = 9.33). Confidence limits are .16 to .79 liter per 100 km greater consumption with 5-speed than with 4-speed transmission.
c. $SS_{\text{transmission}} = 9.403$
d. *A*3 versus *M*4: $F = 216, P < .01$ (CV = 12.70). Confidence limits are 1.14 to 1.86 liters per 100 km greater consumption with automatic than with 4-speed manual.
*A*3 versus *M*5: $F = 101, P < .01$ (CV = 12.70). Confidence limits are .66 to 1.39 liters per 100 km greater consumption with automatic than with 5-speed manual.
*M*4 versus *M*5: $F = 21.7, P < .01$ (CV = 12.70). Confidence limits are .11 to .84 liter per 100 km greater consumption with 5-speed than with 4-speed manual.

Chapter 21

21/1. **a.**
$$29 = 6b_0 + 32b_1 + 30b_2$$
$$118 = 32b_0 + 204b_1 + 163b_2$$
$$160 = 30b_0 + 163b_1 + 178b_2$$
b. $\hat{y} = 7.72 - 1.159x_1 + .660x_2$

c.

Source of variation	SS	df	MS	F	$F_{.05}$
Regression on x_1, x_2	52.410	2	26.205	32.4	9.55
Residual given x_1, x_2	2.423	3	.808		
Total	54.833	5			

d. $R^2 = 96\%$ **e.** $s_e = .90$

21/3. **a.** .60 **b.** .87 **c.** −.30

21/5. **a.** $r_{x_2y|x_1} = .91$, CV = ±.878 **b.** Extra SS = 12.077, $F_{\text{regr. on } x_2|x_1} = 15.0$, $F_{.05}(1, 3) = 10.13$. This test is equivalent to that in Part (a). **c.** $r_{x_1y|x_2} = -.97$, CV = ±.878 **d.** Extra SS = 44.374, $F_{\text{regr. on } x_1|x_2} = 54.9$, $F_{.05}(1, 3) = 10.13$. This test is equivalent to that in Part (c).

21/7. **a.** 24 **b.** CV = 4.30, $R^2 = 25\%$, $s_e = 2.9$

c.

Source of variation	SS	df	MS	F	$F_{.05}$
Regression on x_1, x_2	130.61	2	65.31	11.3	3.47
Residual given x_1, x_2	121.23	21	5.77		
Total	251.84	23			

$R^2 = 52\%$, $s_e = 2.4$
d. $F_{\text{regr. on } x_2|x_1} = 11.8$, CV = 4.32

21/9. $\hat{z} = .35 + .22x$, where $z = \log y$; $\hat{y} = 2.25(1.66^x)$

21/11. $\hat{y} = .76 + 2.184x - .2251x^2$

21/13. **a.** $\hat{y} = 4.2 - .86x_1 + .29x_2$, where x_1 is years in business and x_2 is shift length. For $x_1 = 4$ and $x_2 = 10$, $\hat{y} = 3.6$. The linear equation may be appropriate within the range of predictor values that were observed, but as the number of years in business continues to increase, the number of accidents cannot continue to decrease linearly indefinitely.
b.

Source of variation	SS	df	MS	F	
Regression (b_1, b_2)	21.76	2	10.88	3.32	$P > .10$
Residual given x_1, x_2	13.10	4	3.27	—	
Total	34.86	6			

c. $R^2 = 62\%$ of the variation in number of accidents can be accounted for by the relationship with the best-fitting linear combination of years in business and shift length. **d.** $s_e = 1.8$. The standard amount of discrepancy between the observed numbers of accidents and those estimated by the regression equation is 1.8 accidents.

21/15. **a.** $\hat{y} = -1.5 + .074x_1 + 23.4x_2$. This implies $\hat{y} = -1.5 + .074x_1$ for peace years and $\hat{y} = 21.8 + .074x_1$ for war years.
b.

Source of variation	SS	df	MS	F	
Regression on x_1, x_2	2,050.9	2	1,025.5	92.8	$P < .01$
Residual given x_1, x_2	132.6	12	11.0		
Total	2,183.5	14			

c. $R^2 = 94\%$ of the variation in personal savings can be accounted for by the linear relationship with disposable personal income, when a dummy variable is used to allow a different y-intercept for war years. **d.** $s_e = 3.3$. The standard amount of discrepancy between the observed values of personal savings and those estimated by the regression equation is \$3.3 billion.
e. $F_{\text{regr. on } x_2|x_1} = 135$. The extra contribution of x_2 given x_1 was significant at the 1% level.
$F_{\text{regr. on } x_1|x_2} = 15.0$. The extra contribution of x_1 given x_2 was significant at the 1% level.

21/17. **a.** Number of bedrooms, because it accounts for more variation (SS) in y than does either of the other two predictors. **b.** 46% **c.** ±.68 **d.** $F(1, 8) = 6.82$, $P < .05$ **e.** 9.0 (that is, \$9000) **f.** Extra SS = 277.0, $F(1, 7) = 5.28$, $P > .05$ **g.** $F(2, 7) = 7.87$, $P < .025$ **h.** 69% **i.** 7.2 (that is, \$7200) **j.** Number of bedrooms, because the regression in Part (d) is significant, but the extra sum of squares in Part (f) is not significant. **k.** ±.66

21/19. **a.** Age of the car, because .91 is larger than both .73 and .07. **b.** $F(1, 10) = 48$, $P < .01$ **c.** 83% **d.** .99 (that is, \$990) **e.** .07 **f.** .37 **g.** $F(2, 9) = 26$, $P < .01$ **h.** Extra SS = 1.3, $F(1, 9) = 1.4$, $P > .10$ **i.** 85% **j.** .97 (that is, \$970) **k.** Age of the car, because the regression in Part (b) is significant, but the extra sum of squares in Part (h) is not significant.

21/21. Let x_1 be practice score and x_2 be skill-level setting.

Simplest model, with no predictors: $\hat{y} = 41$, with $s_y = 19$.
First regression includes x_2: $F_{\text{regr.}} = 110$ ($P < .01$), $R^2 = 95\%$, $s_e = 4.7$.
Second regression includes x_1, x_2: $F_{\text{regr.}} = 64.7$ ($P < .01$), $R^2 = 96\%$, $s_e = 4.3$. Extra sum of squares due to x_1: $F = 1.94$ ($P > .10$).
Final equation: $\hat{y} = 79.5 - 15.45x_2$, as described for "first regression" above.
After one practice round, scores average about 64 points at the lowest skill-level setting for the relevant population, and they decrease by an average of about 15 points for each unit increase in the skill-level setting. Knowledge of practice-round scores was of no significant use in predicting scores in the next game.

21/23. Simplest model, with no predictors: $\hat{y} = 7.2$, with $s_y = 2.8$.
First equation includes x_3: $F_{\text{regr.}} = 61.7$ ($P < .01$), $R^2 = 83\%$, $s_e = 1.2$.
Second equation includes x_1, x_3: $F_{\text{regr.}} = 71.1$ ($P < .01$), $R^2 = 92\%$, $s_e = .84$. Extra sum of squares due to x_1: $F = 14.8$ ($P < .01$).
Third equation includes x_1, x_2, x_3: $F_{\text{regr.}} = 45.2$ ($P < .01$), $R^2 = 92\%$, $s_e = .86$. Extra sum of squares due to x_2: $F = .40$ (not significant).
Final equation: $\hat{y} = -15.26 + .3150x_1 + .0399x_3$ (see "second equation" above). This implies
$\hat{y} = -15.26 + .3150x_1$ for females and $\hat{y} = -15.26 + .3549x_1$ for males.
Each increase in height of 3 inches is accompanied, on the average, by an increase of about one unit in shoe size, although men of a given height wear larger shoes, on the average, than women of the same height. For these data, the difference between men and women in the height–shoe-size relationship appears in a different slope, but within the observed range of heights, the lines are so close to being parallel that this is virtually identical to what would be accomplished by the addition of a constant to the y-intercept instead. Thus, either x_2 or x_3 is needed in the model, but not both, to reflect male-female differences in the relationship. (Given x_1, the coefficient of partial determination is 57% for x_2 and 58% for x_3; either predictor would be about equally helpful in addition to x_1.)

21/25. **a.** $\hat{z} = 1.6937 + .00466x$, where $z = \log y$. **b.** $\hat{y} = (49.40)(1.01078^x)$. The fitted value for 1929 ($x = 0$) is 49.40 million persons, and the average rate of increase each year is 1.078%. **c.** $r^2 = 99.4\%$ **d.** The regression is significant ($P < .01$).

21/27. **a.** Number of registrations: Counted data generally respond well to either the square-root or the logarithmic transformation.
b. Original data: $r^2 = 82\%$.
Logarithm of y: $r^2 = 88\%$, preferred form of data.
c. $\hat{z} = 3.01 - .123x$, where $z = \log y$; $\hat{y} = (1024)(.753^x)$. The regression is significant at the 1% level, $r = -.94$, $F(1, 13) = 98.9$.

21/29. **a.** DV = time, IV = temperature. **b.** $\hat{y} = 387 - 8.52x + .250x^2$ **c.** $F(2, 5) = 12.5$ ($P < .025$), $R^2 = 83\%$, $s_e = 16$ seconds.

Index of names and data sources

Subject index